Lecture Notes in Computer Science 2337
Edited by G. Goos, J. Hartmanis, and J. van Leeuwen

Springer
Berlin
Heidelberg
New York
Barcelona
Hong Kong
London
Milan
Paris
Tokyo

William J. Cook   Andreas S. Schulz (Eds.)

Integer Programming and Combinatorial Optimization

9th International IPCO Conference
Cambridge, MA, USA, May 27-29, 2002
Proceedings

 Springer

Gerhard Goos, Karlsruhe University, Germany
Juris Hartmanis, Cornell University, NY, USA
Jan van Leeuwen, Utrecht University, The Netherlands

Volume Editors

William J. Cook
University of Princeton, Program in Applied and Computational Mathematics
Fine Hall, Washington Road, Princeton, NJ 08544-1000, USA

Andreas S. Schulz
Massachusetts Institute of Technology, Sloan School of Management
77 Massachusetts Avenue, Cambridge, MA 02139-4307, USA
E-mail: schulz@mit.edu

Cataloging-in-Publication Data applied for

Die Deutsche Bibliothek - CIP-Einheitsaufnahme

Integer programming and combinatorial optimization : proceedings / 9th
International IPCO Conference, Cambridge, MA, USA, May 27 - 29, 2002.
William J. Cook ; Andreas S. Schulz (ed.). - Berlin ; Heidelberg ; New York ;
Barcelona ; Hong Kong ; London ; Milan ; Paris ; Tokyo : Springer, 2002
   (Lecture notes in computer science ; Vol. 2337)
   ISBN 3-540-43676-6

CR Subject Classification (1998): G.1.6, G.2.1, F.2.2, I.3.5

ISSN 0302-9743
ISBN 3-540-43676-6 Springer-Verlag Berlin Heidelberg New York

Springer-Verlag Berlin Heidelberg New York
a member of BertelsmannSpringer Science+Business Media GmbH

http://www.springer.de

© Springer-Verlag Berlin Heidelberg 2002
Printed in Germany

Typesetting: Camera-ready by author, data conversion by Olgun Computergrafik
Printed on acid-free paper   SPIN 10869799   06/3142   5 4 3 2 1 0

Preface

This volume contains the papers selected for presentation at IPCO 2002, the Ninth International Conference on Integer Programming and Combinatorial Optimization, Cambridge, MA (USA), May 27–29, 2002. The IPCO series of conferences highlights recent developments in theory, computation, and application of integer programming and combinatorial optimization.

IPCO was established in 1988 when the first IPCO program committee was formed. IPCO is held every year in which no International Symposium on Mathematical Programming (ISMP) takes places. The ISMP is triennial, so IPCO conferences are held twice in every three-year period. The eight previous IPCO conferences were held in Waterloo (Canada) 1990, Pittsburgh (USA) 1992, Erice (Italy) 1993, Copenhagen (Denmark) 1995, Vancouver (Canada) 1996, Houston (USA) 1998, Graz (Austria) 1999, and Utrecht (The Netherlands) 2001.

In response to the call for papers for IPCO 2002, the program committee received 110 submissions, a record number for IPCO. The program committee met on January 7 and 8, 2002, in Aussois (France), and selected 33 papers for inclusion in the scientific program of IPCO 2002. The selection was based on originality and quality, and reflects many of the current directions in integer programming and combinatorial optimization research.

Starting with the Copenhagen 1995 meeting, IPCO proceedings have been published by Springer in the Lecture Notes in Computer Science Series. The volume numbers are LNCS 920 (Copenhagen 1995), LNCS 1084 (Vancouver 1996), LNCS 1412 (Houston 1998), LNCS 1610 (Graz 1999), and LNCS 2081 (Utrecht 2001). We are grateful for the help and support of Alfred Hofmann, Executive Editor at Springer, in publishing the current volume.

William J. Cook
Andreas S. Schulz

IPCO 2002 Organization

Program Committee

William J. Cook (Chair), Princeton University
Satoru Fujishige, Osaka University
Alexander Martin, Technische Universität Darmstadt
Gerhard Reinelt, Universität Heidelberg
Andreas S. Schulz, Massachusetts Institute of Technology
András Sebö, Laboratoire Leibniz, IMAG
Jens Vygen, Universität Bonn
David P. Williamson, IBM Almaden Research Center

Local Organizing Committee

Christine A. Liberty
Thomas L. Magnanti
Laura A. Rose
Andreas S. Schulz (Chair)
Nicolás E. Stier Moses

Conference Sponsors

Office of Naval Research
Sloan School of Management, Massachusetts Institute of Technology
School of Engineering, Massachusetts Institute of Technology
Bell Labs Innovations, Lucent Technologies
IBM Research
International Symposium on Mathematical Programming 2000
ILOG Direct Division – ILOG, Inc.
Akamai Technologies, Inc.
Sabre Inc.

External Referees

Dino Ahr
Susanne Albers
Kazutoshi Ando
Takao Asano
David Avis
Dimitris J. Bertsimas
Therese Biedl
Endre Boros
Ulrich Brenner
Chandra Chekuri
Joseph Cheriyan
José R. Correa
Lisa K. Fleischer
Armin Fügenschuh
Toshihiro Fujito
Michel X. Goemans
Stephan Held
Christoph Helmberg
Dorit S. Hochbaum
Yoshiko Ikebe
Hiro Ito
Satoru Iwata
Volker Kaibel
Kenji Kashiwabara
Naoki Katoh
Ekkehard Köhler
Sofia Kovaleva
Sven O. Krumke
Christian Liebchen
Thomas L. Magnanti
Kazuhisa Makino
Jens Maßberg
Shigeru Masuyama

Tomomi Matsui
Hiroyoshi Miwa
Markus Möller
Dirk Müller
Matthias Müller-Hannemann
Kazuo Murota
Hiroshi Nagamochi
Gianpaolo Oriolo
James B. Orlin
Marcus Oswald
Sven Peyer
Myriam Preissmann
André Rohe
Rainer Schrader
Rüdiger Schultz
Maiko Shigeno
Akiyoshi Shioura
David B. Shmoys
Martin Skutella
Frederik Stork
Benny Sudakov
Christian Szegedy
Akihisa Tamura
Shuji Tsukiyama
Marc Uetz
Takaaki Uno
Oliver Wegel
Jürgen Werber
Gerhard J. Woeginger
Laurence A. Wolsey
Mutsunori Yagiura
Yoshitsugu Yamamoto

Table of Contents

A Faster Scaling Algorithm
for Minimizing Submodular Functions*

Satoru Iwata

Department of Mathematical Informatics
University of Tokyo, Tokyo 113-0033, Japan
iwata@sr3.t.u-tokyo.ac.jp

Abstract. Combinatorial strongly polynomial algorithms for minimizing submodular functions have been developed by Iwata, Fleischer, and Fujishige (IFF) and by Schrijver. The IFF algorithm employs a scaling scheme for submodular functions, whereas Schrijver's algorithm exploits distance labeling. This paper combines these two techniques to yield a faster combinatorial algorithm for submodular function minimization. The resulting algorithm improves over the previously best known bound by an almost linear factor in the size of the underlying ground set.

1 Introduction

Let V be a finite nonempty set of cardinality n. A set function f on V is *submodular* if it satisfies

$$f(X) + f(Y) \geq f(X \cap Y) + f(X \cup Y), \qquad \forall X, Y \subseteq V .$$

Submodular functions are discrete analogues of convex functions [12]. Examples include cut capacity functions, matroid rank functions, and entropy functions.

The first polynomial-time algorithm for submodular function minimization is due to Grötschel–Lovász–Schrijver [7]. A strongly polynomial algorithm has also been described by Grötschel–Lovász–Schrijver [8]. These algorithms rely on the ellipsoid method, which is not efficient in practice.

Recently, combinatorial strongly polynomial algorithms have been developed by Iwata–Fleischer–Fujishige (IFF) [11] and by Schrijver [13]. Both of these algorithms build on the first combinatorial pseudopolynomial algorithm due to Cunningham [2]. The IFF algorithm employs a scaling scheme developed in capacity scaling algorithms for the submodular flow problem [5,9]. In contrast, Schrijver [13] achieves a strongly polynomial bound by distance labeling argument similar to [1]. In this paper, we combine these two techniques to yield a faster combinatorial algorithm.

Let γ denote the time required for computing the function value of f and M the maximum absolute value of f. The IFF scaling algorithm minimizes an

---

* This research is supported in part by a Grant-in-Aid for Scientific Research of the Ministry of Education, Science, Sports and Culture of Japan.

W.J. Cook and A.S. Schulz (Eds.): IPCO 2002, LNCS 2337, pp. 1–8, 2002.

integral submodular function in $O(n^5 \gamma \log M)$ time. The strongly polynomial version runs in $O(n^7 \gamma \log n)$ time, whereas an improved variant of Schrijver's algorithm runs in $O(n^7 \gamma + n^8)$ time [4].

The time complexity of our new scaling algorithm is $O((n^4 \gamma + n^5) \log M)$. Since the function evaluation oracle has to identify an arbitrary subset of V as its argument, it is quite natural to assume γ is at least linear in n. Thus the new algorithm is faster than the IFF algorithm by a factor of n. The strongly polynomial version of the new scaling algorithm runs in $O((n^6 \gamma + n^7) \log n)$ time. This is an improvement over the previous best bound by an almost linear factor in n.

These combinatorial algorithms perform multiplications and divisions, although the problem of submodular function minimization does not involve those operations. Schrijver [13] asks if one can minimize submodular functions in strongly polynomial time using only additions, subtractions, comparisons, and the oracle calls for function values. Such an algorithm is called 'fully combinatorial.' A very recent paper [10] settles this problem by developing a fully combinatorial variant of the IFF algorithm. Similarly, we can implement the strongly polynomial version of our scaling algorithm in a fully combinatorial manner. The resulting algorithm runs in $O(n^8 \gamma \log^2 n)$ time, improving the previous bound by a factor of n.

This paper is organized as follows. Section 2 provides preliminaries on submodular functions. In Sect. 3, we describe the new scaling algorithm. Section 4 is devoted to its complexity analysis. Finally, in Sect. 5, we discuss its extensions as well as a fully combinatorial implementation.

2 Preliminary

This section provides preliminaries on submodular functions. See [6,12] for more details and general background.

For a vector $x \in \mathbb{R}^V$ and a subset $Y \subseteq V$, we denote $x(Y) = \sum_{u \in Y} x(u)$. We also denote by x^- the vector in $\mathbb{R}^V$ with $x^-(u) = \min\{x(u), 0\}$. For each $u \in V$, let χ_u denote the vector in $\mathbb{R}^V$ with $\chi_u(u) = 1$ and $\chi_u(v) = 0$ for $v \in V \backslash \{u\}$.

For a submodular function $f : 2^V \to \mathbb{R}$ with $f(\emptyset) = 0$, we consider the *base polyhedron*

$$B(f) = \{x \mid x \in \mathbb{R}^V, x(V) = f(V), \forall Y \subseteq V : x(Y) \le f(Y)\} \ .$$

A vector in $B(f)$ is called a *base*. In particular, an extreme point of $B(f)$ is called an *extreme base*. An extreme base can be computed by the greedy algorithm of Edmonds [3] and Shapley [14] as follows.

Let $L = (v_1, \cdots, v_n)$ be a linear ordering of V. For any $v_j \in V$, we denote $L(v_j) = \{v_1, \cdots, v_j\}$. The greedy algorithm with respect to L generates an extreme base $y \in B(f)$ by

$$y(u) := f(L(u)) - f(L(u) \backslash \{u\}) \ .$$

Conversely, any extreme base can be obtained in this way with an appropriate linear ordering.

For any base $x \in \mathrm{B}(f)$ and any subset $Y \subseteq V$, we have $x^-(V) \leq x(Y) \leq f(Y)$. The following theorem shows that these inequalities are in fact tight for appropriately chosen x and Y. This theorem is immediate from the vector reduction theorem on polymatroids due to Edmonds [3]. It has motivated combinatorial algorithms for minimizing submodular functions.

Theorem 1. *For a submodular function $f : 2^V \to \mathbb{R}$, we have*

$$\max\{x^-(V) \mid x \in \mathrm{B}(f)\} = \min\{f(Y) \mid Y \subseteq V\} \ .$$

Moreover, if f is integer-valued, then the maximizer x can be chosen from among integral bases.   □

3   A Scaling Algorithm

This section presents a new scaling algorithm for minimizing an integral submodular function $f : 2^V \to \mathbb{Z}$.

The algorithm consists of scaling phases with a scale parameter $\delta \geq 0$. It keeps a set of linear orderings $\{L_i \mid i \in I\}$ of the vertices in V. We denote $v \preceq_i u$ if v precedes u in L_i or $v = u$. Each linear ordering L_i generates an extreme base $y_i \in \mathrm{B}(f)$ by the greedy algorithm. The algorithm also keeps a base $x \in \mathrm{B}(f)$ as a convex combination $x = \sum_{i \in I} \lambda_i y_i$ of the extreme bases. Initially, $I = \{0\}$ with an arbitrary linear ordering L_0 and $\lambda_0 = 1$.

Furthermore, the algorithm works with a flow in the complete directed graph on the vertex set V. The flow is represented as a skew-symmetric function $\varphi : V \times V \to \mathbb{R}$. Each arc capacity is equal to δ. Namely, $\varphi(u,v) + \varphi(v,u) = 0$ and $-\delta \leq \varphi(u,v) \leq \delta$ hold for any pair of vertices $u, v \in V$. The boundary $\partial\varphi$ is defined by $\partial\varphi(u) = \sum_{v \in V} \varphi(u,v)$ for $u \in V$. Initially, $\varphi(u,v) = 0$ for any $u, v \in V$.

Each scaling phase aims at increasing $z^-(V)$ for $z = x + \partial\varphi$. Given a flow φ, the procedure constructs an auxiliary directed graph $G_\varphi = (V, A_\varphi)$ with arc set $A_\varphi = \{(u,v) \mid u \neq v, \varphi(u,v) \leq 0\}$. Let $S = \{v \mid z(v) \leq -\delta\}$ and $T = \{v \mid z(v) \geq \delta\}$. A directed path in G_φ from S to T is called an *augmenting path*.

If there is an augmenting path P, the algorithm augments the flow φ along P by $\varphi(u,v) := \varphi(u,v) + \delta$ and $\varphi(v,u) := \varphi(v,u) - \delta$ for each arc (u,v) in P. This procedure is referred to as $\mathsf{Augment}(\varphi, P)$.

Each scaling phase also keeps a valid labeling d. A labeling $d : V \to \mathbb{Z}$ is *valid* if $d(u) = 0$ for $u \in S$ and $v \preceq_i u$ implies $d(v) \leq d(u) + 1$. A valid labeling $d(v)$ serves as a lower bound on the number of arcs from S to v in the directed graph $G_I = (V, A_I)$ with the arc set $A_I = \{(u,v) \mid \exists i \in I, v \preceq_i u\}$.

Let W be the set of vertices reachable from S in G_φ, and Z be the set of vertices that attains the minimum labeling in $V \backslash W$. A pair (u,v) of $u \in W$ and $v \in Z$ is called *active* for $i \in I$ if v is the first element of Z in L_i and u is the last element in L_i with $v \preceq_i u$ and $d(v) = d(u) + 1$. A triple (i, u, v) is also called *active* if (u, v) is active for $i \in I$. The procedure Multiple-Exchange(i, u, v) is applicable to an active triple (i, u, v).

For an active triple (i, u, v), the set of elements between v and u in L_i is called an *active interval*. Any element w in the active interval must satisfy $d(v) \leq d(w) \leq d(u)$. The active interval is divided into Q and R by $Q = \{w \mid w \in W, v \prec_i w \preceq_i u\}$ and $R = \{w \mid w \in V \backslash W, v \preceq_i w \prec_i u\}$.

The procedure Multiple-Exchange(i, u, v) moves the vertices in R to the place immediately after u in L_i, without changing the ordering in Q and in R. Then it computes an extreme base y_i generated by the new L_i. This results in $y_i(q) \geq y_i'(q)$ for $q \in Q$ and $y_i(r) \leq y_i'(r)$ for $r \in R$, where y_i' denotes the previous y_i.

Consider a complete bipartite graph with the vertex sets Q and R. The algorithm finds a flow $\xi : Q \times R \to \mathbb{R}_+$ such that $\sum_{r \in R} \xi(q, r) = y_i(q) - y_i'(q)$ for each $q \in Q$ and $\sum_{q \in Q} \xi(q, r) = y_i'(r) - y_i(r)$ for each $r \in R$. Such a flow can be obtained easily by the so-called northwest corner rule. Then the procedure computes $\eta = \max\{\xi(q, r) \mid q \in Q, r \in R\}$. If $\lambda_i \eta \leq \delta$, Multiple-Exchange$(i, u, v)$ is called *saturating*. Otherwise, it is called *nonsaturating*.

In the nonsaturating Multiple-Exchange(i, u, v), a new index k is added to I. The associated linear ordering L_k is the previous L_i. The coefficient λ_k is determined by $\lambda_k := \lambda_i - \delta/\eta$, and then λ_i is replaced by $\lambda_i := \delta/\eta$. Whether saturating or nonsaturating, the procedure adjusts the flow φ by $\varphi(q, r) := \varphi(q, r) - \lambda_i \xi(q, r)$ and $\varphi(r, q) := \varphi(r, q) + \lambda_i \xi(q, r)$ for every $(q, r) \in Q \times R$.

Let h denote the number of vertices in the active interval. The number of function evaluations required for computing the new extreme base y_i by the greedy algorithm is at most h. The northwest corner rule can be implemented to run in $O(h)$ time. Thus the total time complexity of Multiple-Exchange(i, u, v) is $O(h\gamma)$.

If there is no active triple, the algorithm applies Relabel to each $v \in Z$. The procedure Relabel(v) increments $d(v)$ by one. Then the labeling d remains valid.

The number of extreme bases in the expression of x increases as a consequence of nonsaturating Multiple-Exchange. In order to reduce the complexity, the algorithm occasionally applies a procedure Reduce(x, I) that computes an expression of x as a convex combination of affinely independent extreme bases chosen from the currently used ones. This computation takes $O(n^2|I|)$ time with the aid of Gaussian elimination.

We are now ready to describe the new scaling algorithm.

Step 0: Let L_0 be an arbitrary linear ordering. Compute an extreme base y_0 by the greedy algorithm with respect to L_0. Put $x := y_0$, $\lambda_0 := 1$, $I := \{0\}$, and $\delta := |x^-(V)|/n^2$.

Step 1: Put $d(v) := 0$ for $v \in V$, and $\varphi(u, v) := 0$ for $u, v \in V$.

Step 2: Put $S := \{v \mid z(v) < -\delta\}$ and $T := \{v \mid z(v) > \delta\}$. Let W be the set of vertices reachable from S in G_φ.

Step 3: If there is an augmenting path P, then do the following.

    (3-1) Apply Augment(φ, P).

    (3-2) Apply Reduce(x, I).

    (3-3) Go to Step 2.

Step 4: Find $\ell := \min\{d(v) \mid v \in V\backslash W\}$ and $Z := \{v \mid v \in V\backslash W, d(v) = \ell\}$. If $\ell < n$, then do the following.

(4-1) If there is an active triple (i, u, v), then apply Multiple-Exchange(i,u,v).

(4-2) Otherwise, apply Relabel(v) for each $v \in Z$.

(4-3) Go to Step 2.

Step 5: Determine the set X of vertices reachable from S in G_I. If $\delta \geq 1/n^2$, then apply Reduce(x, I), $\delta := \delta/2$, and go to Step 1.

We now intend to show that the scaling algorithm obtains a minimizer of f.

Lemma 1. *At the end of each scaling phase, $z^-(V) \geq f(X) - n(n+1)\delta/2$.*

Proof. At the end each scaling phase, $d(v) = n$ for every $v \in V\backslash W$. Since $d(v)$ is a lower bound on the number of arcs from S to v, this means there is no directed path from S to $V\backslash W$ in G_I. Thus we have $X \subseteq W \subseteq V\backslash T$, which implies $z(v) \leq \delta$ for $v \in X$. It follows from $S \subseteq X$ that $z(v) \geq -\delta$ for $v \in V\backslash X$. Since there is no arc in G_I emanating from X, we have $y_i(X) = f(X)$ for each $i \in I$, and hence $x(X) = \sum_{i\in I} \lambda_i y_i(X) = f(X)$. Therefore, we have $z^-(V) = z^-(X) + z^-(V\backslash X) \geq z(X) - \delta|X| - \delta|V\backslash X| = x(X) + \partial\varphi(X) - n\delta \geq f(X) - n(n+1)\delta/2$. □

Lemma 2. *At the end of each scaling phase, $x^-(V) \geq f(X) - n^2\delta$.*

Proof. Since $z = x + \partial\varphi$, we have $x^-(V) \geq z^-(V) - n(n-1)\delta/2$, which together with Lemma 1 implies $x^-(V) \geq f(X) - n^2\delta$. □

Theorem 2. *At the end of the last scaling phase, X is a minimizer of f.*

Proof. Since $\delta < 1/n^2$ in the last scaling phase, Lemma 2 implies $x^-(V) > f(X) - 1$. Then it follows from the integrality of f that $f(X) \leq f(Y)$ holds for any $Y \subseteq V$. □

4   Complexity

This section is devoted to complexity analysis of the new scaling algorithm.

Lemma 3. *Each scaling phase performs Augment $O(n^2)$ times.*

Proof. At the beginning of each scaling phase, the set X obtained in the previous scaling phase satisfies $z^-(V) \geq f(X) - 2n^2\delta$ by Lemma 2. For the first scaling phase, we have the same inequality by taking $X = \emptyset$. Note that $z^-(V) \leq z(Y) \leq f(Y) + n(n-1)\delta/2$ for any $Y \subseteq V$ throughout the procedure. Thus each scaling phase increases $z^-(V)$ by at most $3n^2\delta$. Since each augmentation increases $z^-(V)$ by δ, each scaling phase performs at most $3n^2$ augmentations. □

Lemma 4. *Each scaling phase performs Relabel $O(n^2)$ times.*

Proof. Each application of Relabel(v) increases $d(v)$ by one. Since Relabel(v) is applied only if $d(v) < n$, Relabel(v) is applied at most n times for each $v \in V$ in a scaling phase. Thus the total number of relabels in a scaling phase is at most n^2. □

Lemma 5. *The number of indices in I is at most $2n$.*

Proof. A new index is added as a result of nonsaturating Multiple-Exchange. In a nonsaturating Multiple-Exchange(i, u, v), at least one vertex in R becomes reachable from S in G_φ, which means the set W is enlarged. Thus there are at most n applications of nonsaturating Multiple-Exchange between augmentations. Hence the number of indices added between augmentations is at most n. After each augmentation, the number of indices is reduced to at most n. Thus $|I| \leq 2n$ holds.   □

In order to analyze the number of function evaluations in each scaling phase, we now introduce the notion of reordering phase. A *reordering phase* consists of consecutive applications of Multiple-Exchange between those of Relabel or Reduce. By Lemmas 3 and 4, each scaling phase performs $O(n^2)$ reordering phases.

Lemma 6. *There are $O(n^2)$ function evaluations in each reordering phase.*

Proof. The number of function evaluation in Multiple-Exchange(i, u, v) is at most the number of vertices in the active interval for (i, u, v). In order to bound the total number of function evaluations in a reordering phase, suppose the procedure Multiple-Exchange(i, u, v) marks each pair (i, w) for w in the active interval. We now intend to claim that any pair (i, w) of $i \in I$ and $w \in V$ is marked at most once in a reordering phase.

In a reordering phase, the algorithm does not change the labeling d nor delete a vertex from W. Hence the minimum value of d in $V \backslash W$ is nondecreasing. After execution of Multiple-Exchange(i, u, v), there will not be an active pair for i until the minimum value of d in $V \backslash W$ becomes larger. Let Multiple-Exchange(i, s, t) be the next application of Multiple-Exchange to the same index $i \in I$. Then we have $d(t) > d(v) = d(u) + 1$, which implies $v \prec_i u \prec_i t \prec_i s$ in the linear ordering L_i before Multiple-Exchange(i, u, v). Thus a pair (i, w) marked in Multiple-Exchange(i, u, v) will not be marked again in the reordering phase.

Since $|I| \leq 2n$ by Lemma 5, there are at most $2n^2$ possible marks without duplications. Therefore, the total number of function evaluations in a reordering phase is $O(n^2)$.   □

Theorem 3. *The algorithm performs $O(n^4 \log M)$ oracle calls for the function values and $O(n^5 \log M)$ arithmetic computations.*

Proof. Since $-2M \leq x^-(V)$ for $x \in B(f)$, the initial value of δ satisfies $\delta \leq 2M/n^2$. Each scaling phase cuts the value of δ in half, and the algorithm terminates when $\delta < 1/n^2$. Thus the algorithm consists of $O(\log M)$ scaling phases.

Since each scaling phase performs $O(n^2)$ reordering phases, Lemma 6 implies that the number of function evaluations in a scaling phase is $O(n^4)$. In addition, by Lemma 3, each scaling phase performs $O(n^2)$ calls of Reduce, which requires $O(n^3)$ arithmetic computations. Thus each scaling phase consists of $O(n^4)$ function evaluations and $O(n^5)$ arithmetic computations. Therefore, the total running time bound is $O((n^4 \gamma + n^5) \log M)$.   □

5   Discussions

A family $\mathcal{D} \subseteq 2^V$ is called a distributive lattice (or a ring family) if $X \cap Y \in \mathcal{D}$ and $X \cup Y \in \mathcal{D}$ for any pair of $X, Y \in \mathcal{D}$. A compact representation of $\mathcal{D}$ is given by a directed graph as follows. Let $D = (V, F)$ be a directed graph with the arc set F. A subset $Y \subseteq V$ is called an ideal of D if no arc enters Y in D. Then the set of ideals of D forms a distributive lattice. Conversely, any distributive lattice $\mathcal{D} \subseteq 2^V$ with $\emptyset, V \in \mathcal{D}$ can be represented in this way. Moreover, we may assume that the directed graph D is acyclic.

For minimizing a submodular function f on $\mathcal{D}$, we apply the scaling algorithm with a minor modification. The modified version uses the directed graph $G_\varphi = (V, A_\varphi \cup F)$ instead of $G_\varphi = (V, A_\varphi)$. The initial linear ordering L_0 must be consistent with D, i.e., $v \preceq_i u$ if $(u, v) \in F$. Then all the linear orderings that appear in the algorithm will be consistent with D. This ensures that the set X obtained at the end of each scaling phase belongs to $\mathcal{D}$. Thus the modification of our scaling algorithm finds a minimizer of f in $\mathcal{D}$.

Iwata–Fleischer–Fujishige [11] also describes a strongly polynomial algorithm that repeatedly applies their scaling algorithm with $O(\log n)$ scaling phases. The number of iterations is $O(n^2)$. Replacing the scaling algorithm by the new one, we obtain an improved strongly polynomial algorithm that runs in $O((n^6\gamma + n^7)\log n)$ time.

A very recent paper [10] has shown that the strongly polynomial IFF algorithm can be implemented by using only additions, subtractions, comparisons, and oracle calls for function values. Similarly, the new strongly polynomial scaling algorithm can be made fully combinatorial as follows.

The first step towards a fully combinatorial implementation is to neglect Reduce. This causes growth of the number of extreme bases for convex combination. However, the number is still bounded by a polynomial in n. Since the number of indices added between augmentations is at most n, each scaling phase yields $O(n^3)$ new extreme bases. Hence the number of extreme bases through the $O(\log n)$ scaling phases is $O(n^3 \log n)$.

The next step is to choose an appropriate step length in Multiple-Exchange, so that the coefficients should be rational numbers with a common denominator bounded by a polynomial in n. Let σ denote the value of δ in the first scaling phase. For each $i \in I$, we keep $\lambda_i = \mu_i \delta / \sigma$ with an integer μ_i. We then modify the definition of saturating Multiple-Exchange. Multiple-Exchange(i, u) is now called saturating if $\lambda_i \xi(q, r) \leq \varphi(q, r)$ for every $(q, r) \in Q \times R$. Otherwise, it is called nonsaturating. In nonsaturating Multiple-Exchange(i, u), let ν be the minimum integer such that $\nu \xi(q, r) > \varphi(q, r)\sigma/\delta$ for some $(q, r) \in Q \times R$. Then the new coefficients λ_k and λ_i are determined by $\mu_k := \mu_i - \nu$ and $\mu_i := \nu$. Thus the coefficients are rational numbers whose common denominator is σ/δ, which is bounded by a polynomial in n through the $O(\log n)$ scaling phases. Then it is easy to implement this algorithm using only additions, subtractions, comparisons, and oracle calls for the function values.

Finally, we discuss time complexity of the resulting fully combinatorial algorithm. The algorithm performs $O(n^2)$ iterations of $O(\log n)$ scaling phases. Since it keeps $O(n^3 \log n)$ extreme bases, each scaling phase requires $O(n^6 \log n)$ oracle calls for function evaluations and $O(n^6 \log n)$ fundamental operations. Therefore, the total running time is $O(n^8 \gamma \log^2 n)$. This improves the previous $O(n^9 \gamma \log^2 n)$ bound in [10] by a factor of n.

Acknowledgements

The author is grateful to Lisa Fleischer, Satoru Fujishige, Yasuko Matsui, and Kazuo Murota for stimulating conversations and very helpful comments on the manuscript. The idea of Multiple-Exchange was originally suggested by Satoru Fujishige as a heuristics to improve the practical performance of the IFF algorithm.

References

1. W. H. Cunningham: Testing membership in matroid polyhedra. J. Combinatorial Theory, Ser. B, **36** (1984) 161–188.
2. W. H. Cunningham: On submodular function minimization. Combinatorica **5** (1985) 185–192.
3. J. Edmonds: Submodular functions, matroids, and certain polyhedra. Combinatorial Structures and Their Applications, R. Guy, H. Hanani, N. Sauer, and J. Schönheim, eds., Gordon and Breach, pp. 69–87, 1970.
4. L. Fleischer and S. Iwata: A push-relabel framework for submodular function minimization and applications to parametric optimization. Discrete Appl. Math., to appear.
5. L. Fleischer, S. Iwata, and S. T. McCormick: A faster capacity scaling algorithm for submodular flow. Math. Programming, to appear.
6. S. Fujishige: Submodular Functions and Optimization, North-Holland, 1991.
7. M. Grötschel, L. Lovász, and A. Schrijver: The ellipsoid method and its consequences in combinatorial optimization. Combinatorica **1** (1981) 169–197.
8. M. Grötschel, L. Lovász, and A. Schrijver: Geometric Algorithms and Combinatorial Optimization, Springer-Verlag, 1988.
9. S. Iwata: A capacity scaling algorithm for convex cost submodular flows. Math. Programming **76** (1997) 299–308.
10. S. Iwata: A fully combinatorial algorithm for submodular function minimization. J. Combinatorial Theory, Ser. B, to appear.
11. S. Iwata, L. Fleischer, and S. Fujishige: A combinatorial strongly polynomial algorithm for minimizing submodular functions. J. ACM **48** (2001) 761–777.
12. L. Lovász: Submodular functions and convexity, Mathematical Programming – The State of the Art, A. Bachem, M. Grötschel, and B. Korte, eds., Springer-Verlag, 1983, pp. 235–257.
13. A. Schrijver: A combinatorial algorithm minimizing submodular functions in strongly polynomial time. J. Combinatorial Theory, Ser. B, **80** (2000) 346–355.
14. L. S. Shapley: Cores of convex games. Int. J. Game Theory **1** (1971) 11–26.

A Generalization of Edmonds' Matching and Matroid Intersection Algorithms

Bianca Spille[1] and Robert Weismantel[2],[*]

[1] EPFL-DMA, CH-1015 Lausanne, Switzerland
bianca.spille@epfl.ch
[2] Institute for Mathematical Optimization, University of Magdeburg,
Universitätsplatz 2, D-39106 Magdeburg, Germany
weismantel@imo.math.uni-magdeburg.de

Abstract. The independent path-matching problem is a common generalization of the matching problem and the matroid intersection problem. Cunningham and Geelen proved that this problem is solvable in polynomial time via the ellipsoid method. We present a polynomial-time combinatorial algorithm for its unweighted version that generalizes the known combinatorial algorithms for the cardinality matching problem and the matroid intersection problem.

1 Introduction

The matching problem as well as the matroid intersection problem are two of the most important combinatorial optimization problems for which polynomial-time combinatorial algorithms exist, both detected by Edmonds [3,4,5]. Cunningham and Geelen [2] proposed a common generalization of matching and matroid intersection as follows. Let $G = (V, E)$ be a graph, T_1, T_2 disjoint stable sets of G, and $R := V \setminus (T_1 \cup T_2)$. Let $\mathcal{M}_i$ be a matroid on T_i, for $i = 1, 2$.

Definition 1. *An* independent path-matching K *in* G *is a set of edges such that every component of $G(V, K)$ having at least one edge is a path from $T_1 \cup R$ to $T_2 \cup R$ all of whose internal nodes are in R, and such that the set of nodes of T_i in any of these paths is independent in $\mathcal{M}_i$, for $i = 1, 2$.*

In the special case where $\mathcal{M}_1$ and $\mathcal{M}_2$ are *free*, i.e., all subsets of T_i are independent, $i = 1, 2$, we refer to an independent path-matching as a *path-matching*. Figure 1 shows an example, the thick edges form a path-matching.

The problem of maximizing the cardinality of an independent path-matching is an NP-hard problem [2]. Instead, the independent path-matching problem is defined as follows.

Definition 2. *Let K be an independent path-matching.*
An edge e of K is a matching-edge *of K if e is an edge of a one-edge component of $G(V, K)$ having both ends in R, otherwise e is a* path-edge *of K.*

[*] Supported by the European DONET program TMR ERB FMRX-CT98-0202.

W.J. Cook and A.S. Schulz (Eds.): IPCO 2002, LNCS 2337, pp. 9–20, 2002.
© Springer-Verlag Berlin Heidelberg 2002

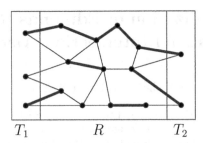

$$T_1 \qquad\qquad R \qquad\qquad T_2$$

Fig. 1. A path-matching

Define the corresponding independent path-matching vector $\psi^K \in \mathbb{R}^E$ *such that,* *for* $e \in E$,

$$\psi_e^K = \begin{cases} 2 & : & e \text{ matching-edge of } K; \\ 1 & : & e \text{ path-edge of } K; \\ 0 & : & e \text{ no edge in } K. \end{cases}$$

The size *of* K *is* $\psi^K(E) = \sum_{e \in E} \psi_e^K$ *which is the number of path-edges of* K *plus twice the number of matching-edges of* K.

The Independent Path-Matching Problem
Find an independent path-matching in G of maximum size.

In the special case when $R = V$, the maximum size of an independent path-matching is twice the maximum cardinality of a matching. Hence, we deal with the matching problem. In the special case when $R = \varnothing$ and G consists of a perfect matching joining copies T_1, T_2 of a set T, we are left with the matroid intersection problem: The independent path-matchings correspond to the common independent sets of $\mathcal{M}_1$ and $\mathcal{M}_2$.

Cunningham and Geelen [2] established a linear system of inequalities for the corresponding polyhedron $\text{conv}\{\psi^K : K \text{ independent path-matching}\}$ which is totally dual integral. Via the ellipsoid method one obtains a polynomial-time algorithm for the independent path-matching problem.

Frank and Szegö [6] algebraically proved the following min-max theorem.

Theorem 3. *(Frank and Szegö)*

$$\max\{\psi^K(E) : K \text{ independent path-matching}\}$$
$$= |R| + \min_{X \text{ cut}} (r_1(T_1 \cap X) + r_2(T_2 \cap X) + |R \cap X| - \text{odd}_G(X)),$$

where a cut *is a subset* $X \subseteq V$ *such that there is no path between* $T_1 \setminus X$ *and* $T_2 \setminus X$ *in* $G \setminus X$ *and* $\text{odd}_G(X)$ *denotes the number of connected components of* $G \setminus X$ *which are disjoint from* $T_1 \cup T_2$ *and have an odd number of nodes.*

We present the first polynomial-time combinatorial algorithm for the independent path-matching problem. It generalizes the known combinatorial algo-

Fig. 2. A matching-cycle

rithms for the cardinality matching problem and the matroid intersection problem and it provides an algorithmic proof of the above min-max theorem. It answers an open question in the design of combinatorial augmenting path algorithms and is a significant generalization of the augmenting path algorithms for matching and matroid intersection.

2   A Combinatorial Algorithm

Edmonds' matching algorithm [3] is based on the idea of augmenting along paths. More precisely, let M be a matching in a graph $G = (V, E)$. A path P in G is M-*augmenting* if its edges are alternately in and not in M and both end nodes are M-exposed. Berge [1] proved that a matching M in G is maximal if and only if there does not exist an M-augmenting path in G. The basic idea of the combinatorial algorithm for matching is to grow a tree of M-alternating paths rooted at an exposed node. If a leaf of the tree is also exposed, an M-augmenting path has been found. The main ingredient in constructing the tree is the method of shrinking (and expanding) odd cycles.

Since we mix up path-edges and matching-edges in the (independent) path-matching problem, we need a more sophisticated analogue to alternating odd cycles in the matching problem, so-called *matching-cycles*.

Definition 4. *A* matching-path *is an odd path in G on node set R of length $2k + 1$ that contains k matching-edges.*

A matching-cycle *is an odd cycle in G that consists of an even path in K (which could be empty) and a matching-path.*

Figure 2 shows a matching-cycle. It consists of an even path of length 4 between nodes v and w and a matching-path of length 7 with three matching-edges.

As we shrink alternating odd cycles for the matching problem, we will shrink matching-cycles for the independent path-matching problem.

For the matroid intersection problem, the method of augmenting along paths leads also to a combinatorial algorithm. Let $\mathcal{M}_1$ and $\mathcal{M}_2$ be two matroids on T and J a common independent set. Let $D_0(J)$ be the digraph with node set T and arcs

$$(w, v) \quad : \quad v \in J, w \notin J, \ J \cup \{w\} \notin \mathcal{M}_1, J \cup \{w\} \setminus \{v\} \in \mathcal{M}_1 \ ;$$
$$(v, w) \quad : \quad v \in J, w \notin J, \ J \cup \{w\} \notin \mathcal{M}_2, J \cup \{w\} \setminus \{v\} \in \mathcal{M}_2 \ .$$

A *J-augmenting path* is a dipath in $D_0(J)$ that starts in a node $w \notin J$ with $J \cup \{w\} \in \mathcal{M}_2$ and ends in a node $w' \notin J$ with $J \cup \{w'\} \in \mathcal{M}_1$. Any chordless such dipath P leads to an augmentation $J \triangle P$, [8]. Edmonds [4,5] and Lawler [7,8] showed that a common independent set J is maximal if and only if there does not exist a J-augmenting path.

The augmenting paths in the matroid intersection case are defined with the help of an augmentation digraph. We define an augmentation digraph for any independent path-matching in standard form.

Definition 5. *An independent path-matching K is in* standard form *if any path in K with more than one edge is a (T_1, T_2)-path.*

Lemma 6. *For any independent path-matching K of G there exists an independent path-matching in standard form of size larger than or equal to the size of K.*

Proof. Let K be an independent path-matching. For any path P in K with more than one edge which is not a (T_1, T_2)-path, let $v \in R$ be an end node of P. Along P in the direction of v to the other end node of P delete every second edge of P. The set of edges we are left with forms an independent path-matching in standard form of size larger than or equal to the size of K.   □

Let K be an independent path-matching in standard form. We define an *augmentation digraph* $D(K)$ with node set V, and an arc set that consists of

(i) *graph-arcs* corresponding to edges in G, and
(ii) *matroid-arcs* corresponding to the matroids $\mathcal{M}_1$ and $\mathcal{M}_2$.

We resort to this digraph in order to augment a non-maximal K or to verify that K is maximal. We direct the (T_1, T_2)-paths in K from T_1 to T_2, the (T_1, R)-one-edge paths from T_1 to R, and the (R, T_2)-one-edge paths from R to T_2. For $i = 1, 2$, let $J_i := K \cap T_i \in \mathcal{M}_i$.

Definition 7. *The* augmentation digraph $D(K)$ *has node set V and an arc set consisting of graph-arcs and matroid-arcs.*
Graph-arcs *are divided into $\ominus$-arcs $(v, w)^-$ and $\oplus$-arcs $(v, w)^+$ and are defined as follows:*

$$
\begin{array}{lll}
(v, w)^- & : & (v, w) \text{ path-arc of } K, \quad v \in T_1 \text{ or } w \in T_2; \\
(v, w)^-, (v, w)^+ & : & (v, w) \text{ path-arc of } K, \quad v, w \in R; \\
(v, w)^-, (w, v)^- & : & vw \text{ matching-edge of } K; \\
(v, w)^+, (w, v)^+ & : & vw \in E \text{ no edge of } K, \quad v, w \in R; \\
(w, v)^+ & : & vw \in E \text{ no edge of } K, \quad v \in T_1 \text{ or } w \in T_2.
\end{array}
$$

Matroid-arcs *are defined as follows:*

$$
\begin{array}{lll}
(w, v) & : & v \in J_1, w \in T_1 \setminus J_1, \ J_1 \cup \{w\} \notin \mathcal{M}_1, J_1 \cup \{w\} \setminus \{v\} \in \mathcal{M}_1 ; \\
(v, w) & : & v \in J_2, w \in T_2 \setminus J_2, \ J_2 \cup \{w\} \notin \mathcal{M}_2, J_2 \cup \{w\} \setminus \{v\} \in \mathcal{M}_2 .
\end{array}
$$

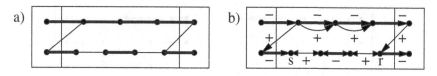

Fig. 3. A path-matching K and the augmentation digraph $D(K)$

In the case where both matroids are free, matroid-arcs do not occur.

In Fig. 3 we illustrate the construction of the augmentation digraph on a small example. The thick lines in a) form a path-matching K in standard form of size 8 (which is non-maximal). In b) we have the augmentation digraph $D(K)$, where a $\ominus$-arc $(v, w)^-$ is represented by $v \bullet\!\!\!\longrightarrow\!\!\bullet w$ and a $\oplus$-arc $(v, w)^+$ by $v \bullet\!\!\overset{+}{\longrightarrow}\!\!\bullet w$.

Definition 8. $w \in R$ is exposed if no edge of K is incident with w.
$w \in V$ is a source node if it satisfies one of the following three conditions:

  (i) $w \in T_2 \setminus J_2$ with $J_2 \cup \{w\} \in \mathcal{M}_2$,
  (ii) $w \in R$ is exposed,
  (iii) $w \in R$ and (w, v) is a one-edge-path in K for some $v \in T_2$.

$w \in V$ is a sink node if it satisfies one of the following three conditions:

  (i) $w \in T_1 \setminus J_1$ with $J_1 \cup \{w\} \in \mathcal{M}_1$,
  (ii) $w \in R$ is exposed,
  (iii) $w \in R$ and (v, w) is a one-edge-path in K for some $v \in T_1$.

In the matching case, the only source and sink nodes are the exposed nodes. For matroid intersection, only the source nodes and the sink nodes according to (i) exist. In Fig. 3 b), r is the only source node and s in the only sink node.

The source nodes are exactly the nodes in $R \cup T_2$ for which there is no $\ominus$-arc in $D(K)$ entering it, whereas the sink nodes are precisely the nodes in $T_1 \cup R$ for which there is no $\ominus$-arc in $D(K)$ leaving it.

Definition 9. An alternating dichain is a dichain in $D(K)$ whose graph-arcs are alternately $\oplus$-arcs and $\ominus$-arcs.
An alternating diwalk is an alternating dichain whose arcs are different.

The alternating dichain in Fig. 3 b) that starts in r with a $\oplus$-arc, uses then the $\ominus$-arc corresponding to the matching-edge, and ends in s with a $\oplus$-arc leads to an augmentation.

Let W be a K-alternating dichain in $D(K)$. For $vw \in E$, let ϕ_{vw}^W be the number of times W contains the $\oplus$-arc $(v, w)^+$ or the $\oplus$-arc $(w, v)^+$ (if existent) minus the number of times W contains the $\ominus$-arc $(v, w)^-$ or the $\ominus$-arc $(w, v)^-$ (if existent).

Definition 10. A K-alternating dichain W is feasible if $\psi^K + \phi^W \in \{0, 1, 2\}^E$.
The result of augmenting K along W is the edge set

$$K \triangle W := \{e \in E : (\psi^K + \phi^W)_e > 0\}.$$

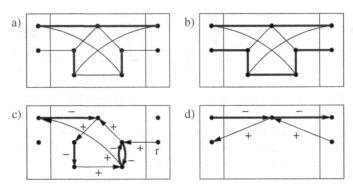

Fig. 4. Shrinking could be "bad"

Each alternating diwalk in $D(K)$ is feasible.

Lemma 11. *Let W be a feasible alternating dichain in $D(K)$ that starts with a $\oplus$-arc in a source node and ends with a $\oplus$-arc in a sink node. If $K \bigtriangleup W$ contains no cycle of G, then $K \bigtriangleup W$ is a path-matching in G of larger size than K. If W is in addition chordless w.r.t. matroid-arcs then $K \bigtriangleup W$ is an independent path-matching in G of larger size than K.*

Proof. Let W be a feasible alternating dichain in $D(K)$ that starts with a $\oplus$-arc in a source node and ends with a $\oplus$-arc in a sink node. Since W is an alternating dichain, $K \bigtriangleup W$ is the union of node-disjoint paths and cycles. All cycles are in R, all inner nodes of paths are in R, and all paths are directed from $T_1 \cup R$ to $R \cup T_2$ (because we have directed the path-edges for the augmentation digraph). Hence, if $K \bigtriangleup W$ contains no cycle of G then $K \bigtriangleup W$ is a path-matching in G. Since W contains more $\oplus$-arcs than $\ominus$-arcs, $K \bigtriangleup W$ is of larger size than K. If W is in addition chordless w.r.t. matroid-arcs then, using the same argumentation as for the augmenting paths in the matroid intersection case, the set of nodes of T_i in any of the paths in $K \bigtriangleup W$ is independent in $\mathcal{M}_i$, $i = 1, 2$. Consequently, $K \bigtriangleup W$ is an independent path-matching in G of larger size than K.   □

For the matching algorithm, we shrink certain odd cycles that are detected during the construction of the alternating tree. Similarly, for the independent path-matching algorithm, we shrink certain matching-cycles that are induced during the course of the algorithm. Roughly speaking, a K-alternating dichain W *induces a matching-cycle* C if it contains as subdichain the walk along the matching-path of C. The definition in Definition 12 is more strict. It covers all cases of induced matching-cycles that need to be shrink and at the same time it ensures that they can be detected in polynomial time.

We demonstrate on an example in Fig. 4 that by shrinking other matching-cycles something bad can happen. The thick lines in a) form a path-matching K in standard form, whereas the thick lines in b) form a path-matching K^* of larger size, i.e., K is non-maximal. The odd cycle C of length five is a matching-cycle. In c) we have a K-alternating diwalk W in $D(K)$. W contains as subdichain the

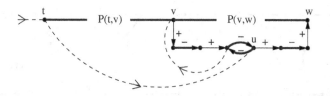

Fig. 5. An alternating diwalk (illustrated by dashed lines and ⊕- and ⊖-arcs) induces a matching-cycle $C(v, w, M)$ according to (ii)

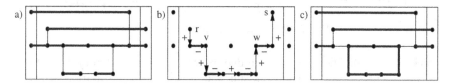

Fig. 6. Bad alternating diwalk

walk along the matching-path of C. If we shrink C we obtain the digraph in d) where K is a maximal path-matching.

Definition 12. *Let $C(v, w, M)$ be a matching-cycle in G that consists of an even path $P(v, w)$ in K from v to w (in that direction) and the matching-path M from v to w. We denote by M^r the reversed matching-path from w to v.*

An alternating diwalk W in $D(K)$ induces the matching-cycle $C(v, w, M)$ according to

(i) if W contains the matching-path M from v to w or W contains the ⊖-arc entering v and later the matching-path M^r from w to v.
Moreover, v is either a source node or the subdiwalk from the first node of W to the ⊖-arc entering v contains no arc of the matching-cycle.

(ii) if the following conditions are fulfilled (see Fig. 5):
  - *W contains the matching-path M from v to w,*
  - *the alternating diwalk W enters the matching-cycle the first time in a node u that is an end node of a ⊖-arc of M, and*
  - *there exists a node t in W such that the walk from t to u in W is a matching-path and there is an even path $P(t, v)$ in K from t to v.*

Consider the example in Fig. 6. The thick lines in a) form a path-matching K in standard form which is maximal. In b) we have a K-alternating diwalk starting in source node r and ending in sink node s that induces a matching-cycle according to (i). The augmentation along this diwalk would lead to the union of thick lines in c) which includes an odd cycle.

As we shrink odd cycles to solve the matching problem, we shrink induced matching-cycles to solve the independent path-matching problem. We consider an example in Fig. 7. The thick lines in a) form a path-matching K in standard form of size 9. There is only one path-matching K^* of larger size illustrated by

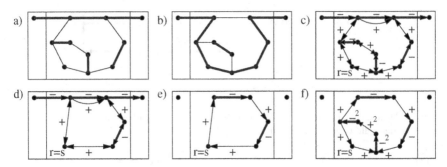

Fig. 7. Shrink and expand matching-cycle

thick lines in b). The augmentation digraph $D(K)$ is pictured in c). It contains only one source node r and only one sink node $s = r$. The augmentation structure that we are interested in is the one given in f) where we have to use certain arcs more than once ($-^2$ or $+^2$ means that the arc is used twice). If we try to find an alternating diwalk starting in r and ending in s, we obtain an alternating diwalk that induces a matching-cycle according to (i), namely, the odd cycle of length 5 located at node r. We shrink this matching-cycle and obtain the new digraph in d). It contains a "good" alternating diwalk: the one illustrated in e). The expansion of this diwalk leads to the augmentation structure in f).

Let $D' = (V, A')$ be a subdigraph of $D(K)$ and $C = C(v, w, M)$ a matching-cycle in D'. There is at most one $\ominus$-arc $(v', v)^-$ in D' entering C and at most one $\ominus$-arc $(w, w')^-$ leaving C. *Shrinking* C leads to the following digraph $D' \times C$: It has node-set $(V \setminus V(C)) \cup \{C\}$ and $\oplus$-arcs

$$(A')^+ \setminus \{(x, y)^+ \in A' : x, y \in V(C)\}$$
$$\cup \{(x, C)^+ : x \notin V(C), x \neq w', (x, y)^+ \in A' \text{ for some } y \in V(C)\}$$
$$\cup \{(C, y)^+ : y \notin V(C), y \neq v', (x, y)^+ \in A' \text{ for some } x \in V(C)\}.$$

Similarly we define the set of $\ominus$-arcs of $D' \times C$. There is at most one $\ominus$-arc in $D' \times C$ entering C and at most one leaving C. Shrinking C in G to obtain $G \times C$ is just the usual shrinking operation without parallel edges. $G \times C$ has node set $T_1 \cup R' \cup T_2$. Let W be an alternating dichain in $D(K) \times C$. The *expansion* of W to obtain an alternating dichain in $D(K)$ is a straightforward extension of the expansion of odd cycles in the matching case.

Definition 13. *Any digraph that we obtain from $D(K)$ by a sequence of shrinking operations of matching-cycles is a* derived digraph *of $D(K)$.*

A K-augmenting dichain is a feasible alternating dichain in $D(K)$ that is an expansion of an alternating diwalk W in some derived digraph D' of $D(K)$, where W starts with a $\oplus$-arc in a source node, ends with a $\oplus$-arc in a sink node, and does not induce a matching-cycle.

Lemma 14. *Let W be a K-augmenting dichain. Then there exists a K-augmenting dichain W^* such that $K \triangle W^*$ does not contain a cycle.*

The proof of this lemma requires quite involved proof techniques and is very lengthy. In order not to be beyond the scope of this paper, we refrain from presenting the proof here, we refer to [9]. The proof is constructive and this construction can be made in polynomial time.

We obtain

Lemma 15. *An independent path-matching K in standard form is not maximal if a K-augmenting dichain exists.*

Proof. Let K be an independent path-matching K in standard form and W a K-augmenting dichain. Lemma 14 implies the existence of a K-augmenting dichain W^* such that $K \triangle W^*$ does not contain a cycle. W.l.o.g. W^* is chordless w.r.t. matroid-arcs. Hence, using Lemma 11, $K \triangle W^*$ is an independent path-matching in G of larger size than K. Consequently, K is not maximal.   □

Lemma 15 suggests a possible approach to construct a maximal independent path-matching. Repeatedly find an augmenting dichain and obtain an independent path-matching of larger size using the dichain.

To find a K-augmenting dichain we use a construction similar to the construction of alternating trees rooted in exposed nodes for the matching algorithm. By a sequence of shrinking operations of matching-cycles of $D(K)$ we construct a digraph $\mathcal{D}$ (a subdigraph of a derived digraph of $D(K)$). For all arcs in $\mathcal{D}$ there is an alternating diwalk in $\mathcal{D}$ starting in a source node with a $\oplus$-arc that uses this arc and does not induce a matching-cycle. Certain matching-cycles are shrinked to obtain $\mathcal{D}$. As in the matching-case, we stop growing $\mathcal{D}$ if we detect a $\oplus$-arc that enters a sink node. Then $\mathcal{D}$ contains an alternating diwalk that starts with a $\oplus$-arc in a source node, ends with a $\oplus$-arc in the sink node, and does not induce a matching-cycle. If all possible $\oplus$-arcs are checked and no sink node can be reached in $\mathcal{D}$ via an entering $\oplus$-arc in $\mathcal{D}$, then we will prove that K is maximal.

We first consider the case where both matroids are free, i.e., we deal with the path-matching problem. A node v in $\mathcal{D}$ is $\ominus$-*reachable* if v is a source node or there is an alternating diwalk in $\mathcal{D}$ starting with a $\oplus$-arc in a source node and ending in v with a $\ominus$-arc. v is $\oplus$-*reachable* if there is an alternating diwalk in $\mathcal{D}$ starting with a $\oplus$-arc in a source node and ending in v with a $\oplus$-arc. v is *reachable* if it is $\ominus$-reachable or $\oplus$-reachable.

Algorithm 16. *(Path-Matching Algorithm)*

Input. K path-matching in standard form.
Output. K^* maximum path-matching.

1. *Let A^+ (A^-) be the set of $\oplus$-arcs ($\ominus$-arcs) in $D(K)$ and for $v \in V$ we denote by $s(v)$ the unique node (if it exists) such that $(v, s(v))^- \in A^-$.*
2. *Set $\mathcal{V} := V$, $\mathcal{A} := \varnothing$, $\mathcal{D} := (\mathcal{V}, \mathcal{A})$, $A^* := A^+$.*
3. *While there exists $(x, y)^+ \in A^*$ such that x is $\ominus$-reachable in $\mathcal{D}$ perform the following steps:*
   (a) If $(y, s(y))^- \in \mathcal{A}$, goto (e).

(b) If $(x,y)^+$ induces a matching-cycle (i.e., the alternating diwalk including $(x,y)^+$ that starts with a $\oplus$-arc in a source node induces a matching-cycle but the subdiwalk ending in x induces none), shrink the matching-cycle and update. Goto (e).

(c) If y is a sink node, $\mathcal{A} := \mathcal{A} \cup \{(x,y)^+\}$, $\mathcal{D} := (\mathcal{V}, \mathcal{A})$, goto 5.

(d) $\mathcal{A} := \mathcal{A} \cup \{(x,y)^+, (y, s(y))^-\}$, $\mathcal{D} := (\mathcal{V}, \mathcal{A})$.

(e) $A^* := A^* \setminus \{(x,y)^+\}$.

4. K is a maximum path-matching. Return $K^* := K$. Stop.

5. Find the alternating diwalk $\mathcal{W}$ in $\mathcal{D}$ that starts with a $\oplus$-arc in a source node, ends with a $\oplus$-arc in the sink node y, and does not induce a matching-cycle.

6. Expand $\mathcal{W}$ to a K-augmenting dichain W.

7. Transform W to a K-augmenting dichain W^* such that $K \triangle W^*$ does not contain a cycle.

8. Replace K by a standard form of $K \triangle W^*$ and goto Step 1.

With 3.(a) we obtain that each node has at most one $\oplus$-arc entering it. Since each node also has at most one $\ominus$-arc entering it, we obtain the uniqueness of alternating diwalks in $\mathcal{D}$. This turns the algorithm polynomial.

In 3.(b) $(x,y)^+$ induces a matching-cycle C. We shrink the matching-cycle. This leads to updated sets $\mathcal{V}$, $\mathcal{A}$, V, A^+, A^-, and A^*. We perform an additional *update* which is as follows:

(a) If there are already $\oplus$-arcs in $\mathcal{A}$ entering the pseudonode, delete all from $\mathcal{A}$. If (before shrinking C) $\mathcal{A}$ contains a $\ominus$-arc leaving C, then there is a $\oplus$-arc $(a,b)^+$ entering C on the unique alternating diwalk in $\mathcal{D}$ that starts in a source node with a $\oplus$-arc and contains this $\ominus$-arc. Add $(a, C)^+$ to $\mathcal{A}$. Then the $\ominus$-arc is again reachable. Add all $\oplus$-arcs in A^+ leaving or entering the pseudonode to A^*.

(b) If $(x,y)^+$ induces a matching-cycle C according to (ii), we have to make sure that we reach the pseudonode via an alternating diwalk in $\mathcal{D}$ that starts in a source node with a $\oplus$-arc and enters the pseudonode with a $\ominus$-arc:

Let W be the alternating diwalk that induces the matching-cycle $C = C(v, w, M)$ according to (ii). Let $P(t, v) = (t, v_1, \ldots, v_{2k}, v_{2k+1}, v)$ be the even path from t to v in Fig. 5. Then $(v_{2k+1}, v)^-$ is the $\ominus$-arc entering the matching-cycle and $(v_{2k+1}, C)^- \in \mathcal{A}$. For $z \in V$ we denote by $p(z)$ the unique node in $\mathcal{V}$ (if it exists) such that $(p(z), z)^+ \in \mathcal{A}$. Set

$$\mathcal{A} := \mathcal{A} \setminus \{(p(v_{2i+1}), v_{2i+1})^+ : i = 0, \ldots, k\}$$
$$\cup \left(\{(t, v_1)^+\} \cup \{(v_{2i-1}, v_{2i})^-, (v_{2i}, v_{2i+1})^+ : i = 1, \ldots, k\} \right).$$

Since t is $\ominus$-reachable, it follows that the pseudonode C is $\ominus$-reachable.

If a sink node y is reached, then $\mathcal{D}$ contains an alternating diwalk $\mathcal{W}$ that starts with a $\oplus$-arc in a source node, ends with a $\oplus$-arc $(x,y)^+$ in the sink node y, and does not induce a matching-cycle. We can find $\mathcal{W}$ by tracing back from y in $\mathcal{D}$ (starting with $(x,y)^+$) until we reach a $\oplus$-arc leaving a source node. We then expand $\mathcal{W}$ to a K-augmenting dichain W. Using the construction in the proof of Lemma 14, we next transform W in polynomial time to a K-augmenting dichain W^* such that $K \triangle W^*$ does not contain a cycle. Lemma 11 implies that $K \triangle W^*$ is a path-matching in G of larger size than K.

If all possible $\oplus$-arcs are checked and no sink node can be reached in $\mathcal{D}$ via an entering $\oplus$-arc, then K is maximal. To see this, let N be the set of nodes in $\mathcal{D}$ that are reachable. Define

$$X := \{v \in N \cup J_2 : v \text{ is not } \ominus\text{-reachable}\}.$$

Then X is a cut in G such that $\psi^K(E) = |R| + |X| - \mathrm{odd}_G(X)$, see [9]. By the min-max formula of Theorem 3, this proves the maximality of K.

In case of general matroids there are some modifications necessary. For a matroid $\mathcal{M}$, $J \in \mathcal{M}$, and $w \notin J$ with $J \cup \{w\} \notin \mathcal{M}$, let $\mathcal{C}(J, w)$ be the corresponding circuit in $J \cup \{w\}$ w.r.t $\mathcal{M}$, i.e.,

$$\mathcal{C}(J, w) = \{v \in J : J \cup \{w\} \setminus \{v\} \in \mathcal{M}\} \cup \{w\}.$$

A node v in $\mathcal{D}$ is called $\ominus$-*reachable* if it is $\ominus$-reachable or there is a node u in $\mathcal{D}$ that is $\ominus$-reachable such that (u, v) is a matroid-arc in $\mathcal{D}$. Similarly for $\oplus$-reachability.

In the Path-Matching Algorithm replace (a) by (a') and (d) by (d'):

(a') *If $(y, s(y))^- \in \mathcal{A}$, goto (e).*
  If $y \in T_1 \setminus J_1$ with $J_1 \cup \{y\} \notin \mathcal{M}_1$
    If for all $v \in \mathcal{C}_1(J_1, y)$ we have $(v, s(v))^- \in \mathcal{A}$, goto (e).
(d') *If $y \in J_1 \cup R$, $\mathcal{A} := \mathcal{A} \cup \{(x, y)^+, (y, s(y))^-\}$*
  If $s(y) \in T_2$ / add matroid-arcs leaving $s(y)$ */*

$$M := \{(s(y), w) : w \in T_2 \setminus J_2, J_2 \cup \{w\} \notin \mathcal{M}_2,$$
$$s(y) \in \mathcal{C}_2(J_2, w), (s(y), w) \notin \mathcal{A}\},$$
$$\mathcal{A} := \mathcal{A} \cup M,$$

  $\mathcal{D} := (\mathcal{V}, \mathcal{A})$, *goto (e).*
  / $y \in T_1 \setminus J_1$ with $J_1 \cup \{y\} \notin \mathcal{M}_1$,*
    *add matroid-arcs leaving y and the succeeding $\ominus$-arcs */*
  $N := \{v \in \mathcal{C}_1(J_1, y) : (v, s(v))^- \notin \mathcal{A}\}$
  $\mathcal{A} := \mathcal{A} \cup \{(x, y)^+, (y, v), (v, s(v))^- : v \in N\}.$
  While there is $v \in N$ such that $s(v) \in T_2$ and there is $w \in T_2 \setminus J_2$ such that $J_2 \cup \{w\} \notin \mathcal{M}_2$, $s(v) \in \mathcal{C}_2(J_2, w)$, and $(s(v), w) \notin \mathcal{A}$
  / add matroid-arcs leaving $s(v)$ */*

$$M := \{(s(v), w) : w \in T_2 \setminus J_2, J_2 \cup \{w\} \notin \mathcal{M}_2,$$
$$s(v) \in \mathcal{C}_2(J_2, w), (s(v), w) \notin \mathcal{A}\},$$
$$\mathcal{A} := \mathcal{A} \cup M.$$

$\mathcal{D} := (\mathcal{V}, \mathcal{A}).$

Again, if all possible $\oplus$-arcs are checked and no sink node can be reached in $\mathcal{D}$ via an entering $\oplus$-arc, then K is a maximum independent path-matching. To see this, let N be the set of nodes in $\mathcal{D}$ that are reachable. Define

$$X := \{v \in N \cup J_2 : v \text{ is not } \ominus\text{-reachable}\} \cup \{w \in T_2 \setminus J_2 : J_2 \cup \{w\} \notin \mathcal{M}_2\}.$$

Then X is a cut in G such that

$$\psi^K(E) = |R| + r_1(T_1 \cap X) + r_2(T_2 \cap X) + |R \cap X| - \mathrm{odd}_G(X),$$

see [9]. By the min-max formula of Theorem 3, this proves the maximality of K. Notice that this argument provides an algorithmic proof of the min-max theorem.

As a generalization of Berge's augmenting path theorem for matchings and Edmonds' and Lawler's augmenting path theorem for matroid intersection we have the following characterization of maximal independent path-matchings via augmenting dichains.

Theorem 17. *An independent path-matching K in standard form is maximal if and only if there does not exist a K-augmenting dichain.*

Proof. Let K be an independent path-matching in standard form. If a K-augmenting dichain exists, Lemma 15 implies that K is not maximal. If K is not maximal, the Independent Path-Matching Algorithm finds a K-augmenting dichain.   □

The Independent Path-Matching Algorithm is a combinatorial algorithm for the independent path-matching problem that runs in (oracle) polynomial time. It is a significant generalization of the known combinatorial algorithms for the matching problem and the matroid intersection problem.

References

1. C. Berge, *Two theorems in graph theory,* Proceedings of the National Academy of Sciences (U.S.A.) **43**, 842–844, 1957
2. W. H. Cunningham and J. F. Geelen, *The optimal path-matching problem,* Combinatorica **17**, no. 3, 315–337, 1997
3. J. Edmonds, *Paths, trees, and flowers,* Canadian Journal of Mathematics **11**, 449–467, 1965
4. J. Edmonds, *Matroid partition,* Math. Decision Sciences, Proceedings 5th Summer Seminary Stanford 1967, Part 1 (Lectures of Applied Mathematics 11), 335–345, 1968
5. J. Edmonds, *Matroid Intersection,* Annals of Discrete Mathematics **4**, 39–49, 1979
6. A. Frank and L. Szegö, *A note on the path-matching formula,* EGRES Technical Report No. 2001-03, 2001
7. E. L. Lawler, *Matroid intersection algorithms,* Mathematical Programming **9**, 31–56, 1975
8. E. L. Lawler, *Combinatorial Optimization: Networks and Matroids,* Holt, Rinehart and Winston, New York etc., 1976
9. B. Spille and R. Weismantel, *A Combinatorial Algorithm for the Independent Path-Matching Problem,* Manuscript, 2001

A Coordinatewise Domain Scaling Algorithm for M-convex Function Minimization*

Akihisa Tamura

Research Institute for Mathematical Sciences, Kyoto University,
Kyoto 606-8502, Japan
tamura@kurims.kyoto-u.ac.jp
http://www.kurims.kyoto-u.ac.jp/

Abstract. We present a polynomial time domain scaling algorithm for the minimization of an M-convex function. M-convex functions are non-linear discrete functions with (poly)matroid structures, which are being recognized to play a fundamental role in tractable cases of discrete optimization. The novel idea of the algorithm is to use an individual scaling factor for each coordinate.

1 Introduction

Linear optimization problems on $0-1$ vectors are NP-hard in general, much more are nonlinear optimization problems on integral lattice points. It is vital to identify subclasses of tractable, or polynomially solvable, problems by extracting "good" structures.

In the linear case, network flow problems, such as the maximum flow problem and the minimum cost flow problem, are typical of tractable problems with wide applications. These problems can be extended to the frameworks of matroids and submodular systems. The submodular flow problem [6], which is an extension of the minimum cost flow problem, is one of the general problems solvable in polynomial time. Many efficient algorithms for the problem are known (e.g., [1,8,10]). In the nonlinear case, submodular functions and separable discrete convex functions constitute classes of tractable nonlinear functions. The development of combinatorial strongly polynomial time algorithms [13,15,28] for submodular function minimization is a recent news. Separable discrete convex functions also have appeared in many tractable cases (e.g., [11,12]).

Recently, certain classes of discrete functions, which include submodular functions and separable discrete convex functions as special cases, were investigated by Murota [19,20]. The framework introduced by Murota, "discrete convex analysis," is being recognized as a unified framework for tractable discrete optimization problems with reference to existing studies on submodular functions [5,10,16], generalized polymatroids [7,8,10], valuated matroids [3,4,22] and convex analysis [27]. In discrete convex analysis, the concepts of L-/M-convex functions play a central role. Submodular functions are a subclass of L-convex functions. Structures of matroids and polymatroids are generalized to

---

* This work is supported by a Grant-in-Aid of the Ministry of Education, Culture, Sports, Science and Technology of Japan.

W.J. Cook and A.S. Schulz (Eds.): IPCO 2002, LNCS 2337, pp. 21–35, 2002.

M-convex functions. A separable discrete convex function is a function with both L-convexity and M-convexity (in their variants). Discrete convex analysis is not only a general framework but also a fruitful one with applications in the areas of mathematical economics and engineering [2,23,25,26]. With the concept of M-convex functions, the submodular flow problem is extended to the M-convex submodular flow problem [21], which is one of the most general problems solvable in polynomial time [14]. As an application of the M-convex submodular flow problem, an economic problem of finding a competitive equilibrium of an economy with indivisible commodities can be solved in polynomial time [26]. L-convex function minimization and M-convex function minimization are fundamental and important problems in discrete convex analysis. The former is an extension of submodular function minimization and the latter is an extension of separable discrete convex function minimization over an integral base polyhedron studied in [9,11,12]. Moreover, the polynomial time algorithm [14] for the M-convex submodular flow problem needs M-convex function minimization at each iteration.

This paper addresses the M-convex function minimization problem. A number of algorithms are already available for minimizing an M-convex function $f : \mathbb{Z}^n \to \mathbb{R}$. Since a locally minimal solution of f is globally minimal, a descent algorithm finds an optimal solution, but it does not terminate in polynomial time. Shioura [29] showed the "minimizer cut property" which guarantees that any nonoptimal solution in the effective domain dom f can be efficiently separated from a minimizer, and exploited this property to develop the domain reduction algorithm, the first polynomial time algorithm for M-convex function minimization. Time complexity of the domain reduction algorithm is $O(Fn^4(\log L)^2)$, where F is an upper bound of the time to evaluate f and $L = \max\{|x(u) - y(u)| \mid u \in \{1, \dots, n\},\ x, y \in \text{dom } f\}$. Moriguchi, Murota and Shioura [17,18] showed the "proximity property" which guarantees that a globally minimal solution of the original function f exists in a neighborhood of a locally minimal solution of "scaled" function

$$f_\alpha(x) = f(x^0 + \alpha x) \qquad (x \in \mathbb{Z}^n)$$

for an $x^0 \in \text{dom } f$ and a positive integer α, where f_α is not necessarily M-convex. They also gave an $O(Fn^3 \log(L/n))$-time algorithm for a subclass of M-convex functions f such that f_α is M-convex for any $x^0 \in \text{dom } f$ and any positive integer α. Although their algorithm, being a descent algorithm, works for any M-convex function, it does not terminate in polynomial time in general. This is because M-convexity is not inherited in the scaling operation.

We develop an $O(Fn^3 \log(L/n))$-time scaling algorithm applicable to any M-convex function by exploiting a new property which is a common generalization of the minimizer cut property and the proximity property. The novel idea of the algorithm is to use an individual scaling factor α_v for each $v \in \{1, \dots, n\}$.

2   M-convexity

Here we define M-convexity and review known results which are utilized in the next section.

Let V be a nonempty finite set of cardinality n, and $\mathbb{R}$, $\mathbb{Z}$ and $\mathbb{Z}_{++}$ be the sets of reals, integers and positive integers, respectively. Given a vector $z = (z(v) : v \in V) \in \mathbb{Z}^V$, its positive support and negative support are defined by

$$\text{supp}^+(z) = \{v \in V \mid z(v) > 0\} \quad \text{and} \quad \text{supp}^-(z) = \{v \in V \mid z(v) < 0\}.$$

For $v \in V$, we denote by χ_v the characteristic vector of v defined by $\chi_v(v) = 1$ and $\chi_v(u) = 0$ for $u \neq v$. For a function $f : \mathbb{Z}^V \to \mathbb{R} \cup \{+\infty\}$, we define the effective domain of f and the set of minimizers of f by

$$\text{dom}\, f = \{x \in \mathbb{Z}^V \mid f(x) < +\infty\},$$
$$\arg\min f = \{x \in \mathbb{Z}^V \mid f(x) \leq f(y) \quad (\forall y \in \mathbb{Z}^V)\}.$$

A function $f : \mathbb{Z}^V \to \mathbb{R} \cup \{+\infty\}$ with $\text{dom}\, f \neq \emptyset$ is called M-convex [19,20] if it satisfies

(M-EXC) $\forall x, y \in \text{dom}\, f$, $\forall u \in \text{supp}^+(x - y)$, $\exists v \in \text{supp}^-(x - y)$:

$$f(x) + f(y) \geq f(x - \chi_u + \chi_v) + f(y + \chi_u - \chi_v).$$

We note that (M-EXC) is also represented as: for $x, y \in \text{dom}\, f$,

$$f(x) + f(y) \geq \max_{u \in \text{supp}^+(x-y)} \min_{v \in \text{supp}^-(x-y)} [\, f(x - \chi_u + \chi_v) + f(y + \chi_u - \chi_v)\,],$$

where the maximum and the minimum over an empty set are $-\infty$ and $+\infty$, respectively. By (M-EXC), $\text{dom}\, f$ lies on a hyperplane whose normal vector is the vector of all ones, that is, for any $x, y \in \text{dom}\, f$, $x(V) = y(V)$ holds, where $x(V) = \sum_{v \in V} x(v)$.

Remark 1. An ordinary convex function $\psi : \mathbb{R}^V \to \mathbb{R}$ can be characterized by the following property: for any $x, y \in \mathbb{R}^V$ and $\lambda \in \mathbb{R}$ with $0 \leq \lambda \leq 1$, $\psi(x) + \psi(y) \geq \psi(\hat{x}) + \psi(\hat{y})$ holds, where $\hat{x} = x - \lambda(x - y)$ and $\hat{y} = y + \lambda(x - y)$. That is, two points $\hat{x}$ and $\hat{y}$, which are obtained by approaching the same distance from x and y to each other in a straight line, do not increase the sum of function values. Unfortunately, two points on an integral lattice $\mathbb{Z}^V$ may not approach to each other in a straight line on $\mathbb{Z}^V$. Property (M-EXC) extracts the essence of the above characterization of ordinary convex functions. Fig. 1 draw an example of (M-EXC) with $|V| = 3$, where $x' = x - \chi_u + \chi_v$ and $y' = y + \chi_u - \chi_v$.

It is known that $\arg\min f$ and $\text{dom}\, f$ are an M-convex set, where a nonempty set $B \subseteq \mathbb{Z}^V$ is said to be an M-convex set if it satisfies

(B-EXC) $\forall x, y \in B$, $\forall u \in \text{supp}^+(x - y)$, $\exists v \in \text{supp}^-(x - y)$:

$$x - \chi_u + \chi_v \in B, \quad y + \chi_u - \chi_v \in B.$$

An M-convex set is a synonym of the set of integer points of the base polyhedron of an integral submodular system (see [10] for submodular systems). Thus, the M-convexity is a quantitative generalization of the base polyhedron of an integral submodular system.

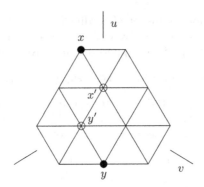

Fig. 1. An example of (M-EXC)

Lemma 1 ([19,20]). *For an M-convex function* f, *dom* f *is an M-convex set, and if* arg min $f \neq \emptyset$ *then it is also an M-convex set.*

Proof. The assertion for dom f immediately follows from (M-EXC). Let x and y be any points in arg min f. By (M-EXC), for any $u \in \text{supp}^+(x - y)$, there exists $v \in \text{supp}^-(x - y)$ such that

$$f(x) + f(y) \geq f(x - \chi_u + \chi_v) + f(y + \chi_u - \chi_v).$$

Since $x, y \in$ arg min f, we have $x - \chi_u + \chi_v, y + \chi_u - \chi_v \in$ arg min f. Hence, if arg min $f \neq \emptyset$, then it satisfies (B-EXC). □

Lemma 1 implies the following property.

Proposition 1. *Let* $f : \mathbb{Z}^V \to \mathbb{R} \cup \{+\infty\}$ *be an M-convex function. For* $u, v \in V$ *and* $\gamma_1, \gamma_2 \in \mathbb{Z}$, *suppose that there exist* $x_1, x_2 \in$ arg min f *with* $x_1(u) \leq \gamma_1$ *and* $x_2(v) \geq \gamma_2$, *where* $\gamma_2 \leq \gamma_1$ *if* $u = v$. *Then, there exists* $x_* \in$ arg min f *with* $x_*(u) \leq \gamma_1$ *and* $x_*(v) \geq \gamma_2$.

Proof. If $x_2(u) < x_1(u)$ then there is nothing to prove. Let x_* be a point in arg min f such that x_* minimizes $x_*(u)$ among all points $x \in$ arg min f with $x(v) \geq \gamma_2$ and $x(u) \geq x_1(u)$. We prove that $x_*(u) \leq \gamma_1$. Suppose to the contrary that $x_*(u) > \gamma_1$. Since $u \in \text{supp}^+(x_* - x_1)$ and arg min f is an M-convex set, there exists $w \in \text{supp}^-(x_* - x_1)$ such that $x' = x_* - \chi_u + \chi_w \in$ arg min f. Obviously, $x'(v) \geq \gamma_2$ and $x_1(u) \leq x'(u) < x_*(u)$ hold. However, this contradicts the definition of x_*. □

M-convex functions have various features of convex functions. For example, a locally minimal solution of an M-convex function is globally minimal. The next theorem shows how to verify optimality of a given solution with $O(n^2)$ function evaluations.

Theorem 1 ([19,20]). *For an M-convex function* f *and* $x \in$ dom f, $f(x) \leq f(y)$ *for any* $y \in \mathbb{Z}^V$ *if and only if* $f(x) \leq f(x - \chi_u + \chi_v)$ *for all* $u, v \in V$.

Proof. The "only if" part is trivial. We prove the "if" part. Suppose to the contrary that there is a point y with $f(y) < f(x)$, and in addition suppose that y minimizes $\| y - x \|$ among such points. Since $y \neq x$ and $y(V) = x(V)$, there exists an element $u \in \mathrm{supp}^+(x - y)$. By (M-EXC), for some $v \in \mathrm{supp}^-(x - y)$,

$$f(x) + f(y) \geq f(x - \chi_u + \chi_v) + f(y + \chi_u - \chi_v)$$

holds. Since $f(x) \leq f(x - \chi_u + \chi_v)$, we have $f(y + \chi_u - \chi_v) \leq f(y) < f(x)$. Obviously, $\| (y + \chi_u - \chi_v) - x \| < \| y - x \|$ holds. However, this contradicts the definition of y. Hence we have $f(x) \leq f(y)$ for any $y \in \mathbb{Z}^V$. $\qquad\square$

The following properties, Theorems 2 and 3, play important roles in our algorithm. The former is due to Shioura [29] and the latter is implicit in the proof of the proximity theorem for M-convex functions in [17,18]. The minimizer cut property, Theorem 2, suggests how to separate a given nonoptimal solution from some minimizer, and serves as a basis of the domain reduction algorithm in [29]. Both (a) and (b) of Theorem 3 together with Proposition 1 show that some minimizer of the original function f exists in a neighborhood of a locally minimal solution of a scaled function with respect to a fixed coordinate.

Theorem 2 (minimizer cut property, [29]). *Let $f : \mathbb{Z}^V \to \mathbb{R} \cup \{+\infty\}$ be an M-convex function with $\arg\min f \neq \emptyset$. For $x \in \mathrm{dom}\, f$ and $v \in V$, the following statements hold.*

(a) *If $f(x - \chi_v + \chi_u) = \min_{s \in V} f(x - \chi_v + \chi_s)$, then there exists $x_* \in \arg\min f$ such that $x(u) - \chi_v(u) + 1 \leq x_*(u)$.*
(b) *If $f(x + \chi_v - \chi_u) = \min_{s \in V} f(x + \chi_v - \chi_s)$, then there exists $x_* \in \arg\min f$ such that $x_*(u) \leq x(u) + \chi_v(u) - 1$.*

Theorem 3 (proximity property). *Let $f : \mathbb{Z}^V \to \mathbb{R} \cup \{+\infty\}$ be an M-convex function with $\arg\min f \neq \emptyset$. For $x \in \mathrm{dom}\, f$, $v \in V$ and $\alpha \in \mathbb{Z}_{++}$, the following statements hold.*

(a) *If $f(x) = \min_{s \in V} f(x - \alpha(\chi_v - \chi_s))$, then there exists $x_* \in \arg\min f$ such that $x(v) - (n-1)(\alpha - 1) \leq x_*(v)$.*
(b) *If $f(x) = \min_{s \in V} f(x + \alpha(\chi_v - \chi_s))$, then there exists $x_* \in \arg\min f$ such that $x_*(v) \leq x(v) + (n-1)(\alpha - 1)$.*

Here we show a new property, Theorem 4 below, as a common generalization of Theorems 2 and 3. A special case of Theorem 4 with $\alpha = 1$ is equivalent to Theorem 2. Another special case of Theorem 4 with $u = v$ is identical with Theorem 3.

Theorem 4 (minimizer cut property with scaling). *Let f be an M-convex function with $\arg\min f \neq \emptyset$. For $x \in \mathrm{dom}\, f$, $v \in V$ and $\alpha \in \mathbb{Z}_{++}$, the following statements hold.*

(a) *If $f(x - \alpha(\chi_v - \chi_u)) = \min_{s \in V} f(x - \alpha(\chi_v - \chi_s))$, then there exists $x_* \in \arg\min f$ such that $x(u) - \alpha(\chi_v(u) - 1) - (n-1)(\alpha - 1) \leq x_*(u)$.*

(b) If $f(x + \alpha(\chi_v - \chi_u)) = \min_{s \in V} f(x + \alpha(\chi_v - \chi_s))$, then there exists $x_* \in$ arg min f such that $x_*(u) \leq x(u) + \alpha(\chi_v(u) - 1) + (n-1)(\alpha - 1)$.

Proof. Here we prove the assertion (a) because we can similarly prove (b). It is sufficient to consider the case where there exists $x_* \in$ arg min f such that $x_*(u)$ is maximum. Let $\widehat{x} = x - \alpha(\chi_v - \chi_u)$. Assume that $\widehat{x}(u) > x_*(u)$ and $k = \widehat{x}(u) - x_*(u)$.

CLAIM A: There exist $w_1, w_2, \ldots, w_k \in V \setminus \{u\}$ and $y_0(= \widehat{x}), y_1, \ldots, y_k \in$ dom f such that $y_i = y_{i-1} - \chi_u + \chi_{w_i}$ and $f(y_i) < f(y_{i-1})$ for $i = 1, 2, \ldots, k$.

[Proof of Claim A] Let $y_{i-1} \in$ dom f. By (M-EXC), for y_{i-1}, x_* and $u \in$ supp$^+(y_{i-1} - x_*)$, there exists $w_i \in$ supp$^-(y_{i-1} - x_*) \subseteq V \setminus \{u\}$ such that

$$f(x_*) + f(y_{i-1}) \geq f(x_* + \chi_u - \chi_{w_i}) + f(y_{i-1} - \chi_u + \chi_{w_i}).$$

Since $f(x_*) < f(x_* + \chi_u - \chi_{w_i})$, we have $f(y_{i-1}) > f(y_{i-1} - \chi_u + \chi_{w_i}) = f(y_i)$.

CLAIM B: For any $w \in V \setminus \{u\}$ with $y_k(w) > \widehat{x}(w)$ and $\beta \in \mathbb{Z}$ with $0 \leq \beta \leq y_k(w) - \widehat{x}(w) - 1$, $f(\widehat{x} - (\beta + 1)(\chi_u - \chi_w)) < f(\widehat{x} - \beta(\chi_u - \chi_w))$ holds.

[Proof of Claim B] We prove the claim by induction on β. For β with $0 \leq \beta \leq y_k(w) - \widehat{x}(w) - 1$, put $x' = \widehat{x} - \beta(\chi_u - \chi_w)$ and assume $x' \in$ dom f. Let j_* ($1 \leq j_* \leq k$) be the maximum index with $w_{j_*} = w$. Since $y_{j_*}(w) = y_k(w) > x'(w)$ and supp$^-(y_{j_*} - x') = \{u\}$, we have $f(x') + f(y_{j_*}) \geq f(x' - \chi_u + \chi_w) + f(y_{j_*} + \chi_u - \chi_w)$ by (M-EXC). Claim A guarantees that $f(y_{j_*-1}) = f(y_{j_*} + \chi_u - \chi_w) > f(y_{j_*})$, and hence, $f(x') > f(x' - \chi_u + \chi_w)$.

The hypothesis of (a) and Claim B imply $\mu_w = y_k(w) - \widehat{x}(w) \leq \alpha - 1$ for any $w \in V \setminus \{u\}$, because

$$f(\widehat{x} - \mu_w(\chi_u - \chi_w)) < \cdots < f(\widehat{x} - (\chi_u - \chi_w)) < f(\widehat{x}) \leq f(\widehat{x} - \alpha(\chi_u - \chi_w))$$

holds for any w with $\mu_w > 0$. Thus, we have

$$\widehat{x}(u) - x_*(u) = \widehat{x}(u) - y_k(u) = \sum_{w \in V \setminus \{u\}} \{y_k(w) - \widehat{x}(w)\} \leq (n-1)(\alpha - 1),$$

where the second equality follows from $\widehat{x}(V) = y_k(V)$.   □

3   Proposed Algorithm

This section describes a scaling algorithm of time complexity $O(Fn^3 \log(L/n))$ for the M-convex function minimization. It is assumed that the effective domain of a given M-convex function f is bounded and that a vector $x^0 \in$ dom f is given.

We preliminarily show that $L = \max\{|x(u) - y(u)| \mid u \in V, x, y \in$ dom $f\}$ can be computed in $O(Fn^2 \log L)$ time. For $x \in$ dom f and $u, v \in V$, the exchange capacity associated with x, u and v is defined as

$$\tilde{c}_f(x, v, u) = \max\{\alpha \mid x + \alpha(\chi_v - \chi_u) \in \text{dom } f\},$$

which can be computed in $O(F \log L)$ time by the binary search because $0 \leq \tilde{c}_f(x, v, u) \leq L$. For each $w \in V$, define

$$l_f(w) = \min\{x(w) \mid x \in \mathrm{dom}\, f\}, \quad u_f(w) = \max\{x(w) \mid x \in \mathrm{dom}\, f\}.$$

The values $l_f(w)$ and $u_f(w)$ can be calculated by the following algorithm in $O(Fn \log L)$ time.

function CALCULATE_BOUND(f, x, w)
   input: f : M-convex function, $x \in \mathrm{dom}\, f$, $w \in V$;
   output: $(l_f(w), u_f(w))$;

$\ell 1$:   number $V \setminus \{w\}$ from v_2 to v_n ;
$\ell 2$: $y := z := x$;
$\ell 3$:   **for** $i := 2$ **to** n **do** {
$\ell 4$: $y := y + \tilde{c}(y, v_i, w)(\chi_{v_i} - \chi_w), \quad z := z + \tilde{c}(z, w, v_i)(\chi_w - \chi_{v_i})$
$\ell 5$:   } ;
$\ell 6$:   **return** $(y(w), z(w))$.

The correctness of the above algorithm can be verified from the fact that $\mathrm{dom}\, f$ satisfies (B-EXC) (see also [10,29]).

Lemma 2. *Values $l_f(w)$ and $u_f(w)$ for a fixed $w \in V$ can be computed in $O(Fn \log L)$ time by* CALCULATE_BOUND, *and L in $O(Fn^2 \log L)$ time.*

Proof. We first show the correctness of CALCULATE_BOUND. Let y_i denote the point y defined at iteration i of CALCULATE_BOUND. Suppose to the contrary that $l_f(w) < y_n(w)$, i.e., there exists $\hat{y} \in \mathrm{dom}\, f$ with $\hat{y}(w) < y_n(w)$. Since $\mathrm{dom}\, f$ is an M-convex set, there exists $v_j \in \mathrm{supp}^-(y_n - \hat{y})$ such that $y' = y_n + \chi_{v_j} - \chi_w \in \mathrm{dom}\, f$. By applying (B-EXC) for y', y_j and $v_i \in \mathrm{supp}^+(y' - y_j)$, there exists $v \in \mathrm{supp}^-(y' - y_j)$ with $y_j + \chi_{v_j} - \chi_v \in \mathrm{dom}\, f$. Moreover, v must be equal to w because $\mathrm{supp}^-(y' - y_j) \subseteq \mathrm{supp}^-(y_n - y_j) \cup \{w\} = \{w\}$ holds by the definition of y_i. However, this contradicts the definition of $\tilde{c}(y_{j-1}, v_j, w)$.

Time complexity immediately follows from the above discussion. □

For any two vectors $a, b \in \mathbb{Z}^V$, let $[a, b]$ denote the set $\{x \in \mathbb{Z}^V \mid a \leq x \leq b\}$ and f_a^b be defined by

$$f_a^b(x) = \begin{cases} f(x) & \text{if } x \in [a, b] \\ +\infty & \text{otherwise.} \end{cases}$$

Condition (M-EXC) directly yields the next property.

Proposition 2. *For an M-convex function f and $a, b \in \mathbb{Z}^V$, if $\mathrm{dom}\, f_a^b \neq \emptyset$ then f_a^b is also M-convex.*

We go on to the main topic of describing our algorithm for minimizing f. The novel idea of the algorithm is to use an individual scaling factor $\alpha_v \in \mathbb{Z}_{++}$ for each $v \in V$. Besides the factors α_v $(v \in V)$, the algorithm maintains a current vector $x \in \mathrm{dom}\, f$, two vectors $a, b \in \mathbb{Z}^V$ and a subset $V' \subseteq V$, and it preserves the following four conditions:

(c1) $x \in \operatorname{dom} f \cap [a, b] = \operatorname{dom} f_a^b$,
(c2) $b(v) - a(v) \leq 2n\alpha_v$ for $v \in V$,
(c3) $\arg\min f \cap [a, b] \neq \emptyset$, (i.e., $\arg\min f_a^b \subseteq \arg\min f$),
(c4) there exists $x_* \in \arg\min f$ such that $x_*(v) = x(v)$ for all $v \in V \setminus V'$.

These parameters are initially put as $x := x^0$, $V' := V$, $a(v) := l_f(v)$, $b(v) := u_f(v)$ and $\alpha_v := 2^{\lceil \log_2\{(u_f(v) - l_f(v))/n\}\rceil}/2$ for $v \in V$. Thus, conditions (c1) to (c4) are initially satisfied. At each iteration of the algorithm, interval $[a, b]$ is strictly reduced and V' is not increased. By (c4) and the fact that $y(V)$ is a constant for any $y \in \operatorname{dom} f$, the algorithm stops if $|V'| \leq 1$; then the current x is a minimizer of f. The algorithm terminates in $O(n^2 \log(L/n))$ iterations and requires $O(Fn)$ time at each iteration. Hence, the total time complexity of the algorithm is $O(Fn^3 \log(L/n))$.

Before entering into a precise description of the algorithm, we briefly explain its typical behaviour. First, take $v \in V'$ arbitrarily, and find $u \in V'$ minimizing $f_a^b(x - \chi_u + \chi_v)$. Here we explain what the algorithm does in the case where $u \neq v$. Then, there exists $x_1 \in \arg\min f_a^b$ with $x_1(u) \leq x(u) - 1$ by (b) of Theorem 2. This inequality suggests that an optimal solution can be found by decreasing $x(u)$. Next find $w \in V'$ minimizing $f_a^b(x - \alpha_u(\chi_u - \chi_w))$. When u can be chosen as w attaining the minimum value, (a) of Theorem 3 guarantees that there exists $x_2 \in \arg\min f_a^b$ with $x(u) - (n-1)(\alpha_u - 1) \leq x_2(u)$. By Proposition 1, there exists $x_* \in \arg\min f_a^b$ with $x(u) - (n-1)(\alpha_u - 1) \leq x_*(u) \leq x(u) - 1 < x(u)$. Thus, we can put $a(u) := \max[a(u), x(u) - (n-1)(\alpha_u - 1)]$ and $b(u) := x(u)$. Since $b(u) - a(u) \leq n\alpha_u$, the scaling factor α_u can be divided by 2 without violating (c2). In the other case with $w \neq u$, we update a, b and x as $b(u) := x(u) - 1$, $x := x - \alpha_u(\chi_u - \chi_w)$ and $a(w) := \max[a(w), x(w) - (n-1)(\alpha_u - 1)]$, where the update of a is justified by (a) of Theorem 4. This is a part of our algorithm described below (see CASE2). A complete description of our algorithm is now given in Fig. 2.

In the sequel of this section, we show the correctness and the time complexity of algorithm COORDINATEWISE_SCALING.

Lemma 3. COORDINATEWISE_SCALING *preserves conditions* (c1) *to* (c4).

Proof. As we mentioned above, conditions (c1) to (c4) are satisfied just after the execution of line $\ell03$. We note that Proposition 2 guarantees that f_a^b is M-convex. The while loop, which is lines $\ell04$ to $\ell12$, consists of three cases.

The first case at line $\ell08$ implies that there exists $x_* \in \arg\min f$ with $x_*(v) = x(v)$ by condition (c3), Proposition 1 and Theorem 2 for f_a^b. Trivially, conditions (c1) to (c4) are satisfied after the execution of line $\ell09$.

We next consider the second case at line $\ell10$. By (c3) and (b) of Theorem 2 for f_a^b, there exists $x_1 \in \arg\min f_a^b \subseteq \arg\min f$ with $x_1(u_1) \leq x(u_1) - 1$. Let us consider function CASE2, in which we have either $u \in W_u$ or $u \notin W_u$. Assume first that $u \in W_u$. By (c3) and (a) of Theorem 3 for f_a^b, there exists $x_2 \in \arg\min f$ with $a(u) \leq x_2(u)$, where a is the updated one. We note that the updated a and b have $a(u) \leq b(u)$. By Proposition 1, there exists $x_* \in \arg\min f$ with $a(u) \leq x_*(u) \leq b(u)$. Thus, (c1) to (c4) are satisfied. Assume next that $u \notin W_u$. By (c3)

algorithm $\textsc{Coordinatewise\_Scaling}(f, V, x^0)$
   input: f : M-convex function with bounded dom $f \subset \mathbb{Z}^V$, $x^0 \in \mathrm{dom}\, f$;
   output: a minimizer of f ;

$\ell01$: $n := |V|$, $V' := V$, $x := x^0$;
$\ell02$:   **for each** $v \in V$ **do** $(a(v), b(v)) := \textsc{Calculate\_Bound}(f, x, v)$;
$\ell03$:   **for each** $v \in V$ **do** $\alpha_v := 2^{\lceil \log_2\{(b(v)-a(v))/n\}\rceil}/2$;
$\ell04$:   **while** $|V'| \geq 2$ **do** {
$\ell05$:   take $v \in V'$;
$\ell06$:   find $u_1 \in V' : f_a^b(x + \chi_v - \chi_{u_1}) = \min_{s \in V' \setminus \{v\}} f_a^b(x + \chi_v - \chi_s)$;
$\ell07$:   find $u_2 \in V' : f_a^b(x - \chi_v + \chi_{u_2}) = \min_{s \in V' \setminus \{v\}} f_a^b(x - \chi_v + \chi_s)$;
$\ell08$:   **if** $f_a^b(x) \leq f_a^b(x + \chi_v - \chi_{u_1})$ **and** $f_a^b(x) \leq f_a^b(x - \chi_v + \chi_{u_2})$ **then**
$\ell09$:   { $a(v) := b(v) := x(v)$, $V' := V' \setminus \{v\}$ } ;
$\ell10$:   **else if** $f_a^b(x) > f_a^b(x + \chi_v - \chi_{u_1})$ **then** $\textsc{Case2}(u_1)$;
$\ell11$:   **else** $(f_a^b(x) > f_a^b(x - \chi_v + \chi_{u_2}))$ **then** $\textsc{Case3}(u_2)$;
$\ell12$:   } ;
$\ell13$:   **return** (x) ;

function $\textsc{Case2}(u)$ $(\exists x_1 \in \arg\min f$ with $x_1(u) \leq x(u) - 1)$
$\ell1$: $W_u := \arg\min_{s \in V'} f_a^b(x - \alpha_u(\chi_u - \chi_s))$;
$\ell2$:   **if** $u \in W_u$ **then**
$\ell3$:   { $a(u) := \max[a(u), x(u) - (n-1)(\alpha_u - 1)]$, $b(u) := x(u)$ } ;
$\ell4$:   **else** $(u \notin W_u)$ {
$\ell5$:   take $w \in W_u$, $b(u) := x(u) - 1$, $x := x - \alpha_u(\chi_u - \chi_w)$;
$\ell6$: $a(w) := \max[a(w), x(w) - (n-1)(\alpha_u - 1)]$;
$\ell7$: $\textsc{Update\_Factor}(w)$ } ;
$\ell8$: $\textsc{Update\_Factor}(u)$;
$\ell9$:   **return** ;

function $\textsc{Case3}(u)$ $(\exists x_2 \in \arg\min f$ with $x_2(u) \geq x(u) + 1)$
$\ell1$: $W_u := \arg\min_{s \in V'} f_a^b(x + \alpha_u(\chi_u - \chi_s))$;
$\ell2$:   **if** $u \in W_u$ **then**
$\ell3$:   { $a(u) := x(u)$, $b(u) := \min[b(u), x(u) + (n-1)(\alpha_u - 1)]$ } ;
$\ell4$:   **else** $(u \notin W_u)$ {
$\ell5$:   take $w \in W_u$, $a(u) := x(u) + 1$, $x := x + \alpha_u(\chi_u - \chi_w)$;
$\ell6$: $b(w) := \min[b(w), x(w) + (n-1)(\alpha_u - 1)]$;
$\ell7$: $\textsc{Update\_Factor}(w)$ } ;
$\ell8$: $\textsc{Update\_Factor}(u)$;
$\ell9$:   **return** ;

function $\textsc{Update\_Factor}(s)$
$\ell1$:   **while** $\alpha_s > 1$ **and** $b(s) - a(s) \leq n\alpha_s$ **do** $\alpha_s := \alpha_s/2$;
$\ell2$:   **if** $a(s) = b(s)$ **then** $V' := V' \setminus \{s\}$;
$\ell3$:   **return** .

Fig. 2. A complete description of our algorithm

and (a) of Theorem 4 for f_a^b, there exists $x_2 \in \arg\min f$ with $a(w) \le x_2(w)$, where a is the updated one. By Proposition 1, there exists $x_* \in \arg\min f$ with $a(w) \le x_*(w)$ and $x_*(u) \le b(u)$ for the updated a and b, and hence, (c3) holds. Obviously, the updated a, b and x violate no other conditions at the end of line $\ell 6$. At lines $\ell 7$ and $\ell 8$, UPDATE_FACTOR reduces α_w and α_u respectively, preserving (c1) to (c4).

Similarly, the third case at $\ell 11$ also preserves (c1) to (c4). $\square$

Lemma 4. *The while loop in lines $\ell 04$ to $\ell 12$ of* COORDINATEWISE_SCALING *terminates in* $O(n^2 \log(L/n))$ *iterations.*

Proof. We divide the iterations of the while loop into three cases.

CLAIM A: The case at line $\ell 08$ occurs at most $n - 1$ times.
[Proof of Claim A] Every time this case occurs, the set V' is reduced by one element. Thus, this case occurs at most $n - 1$ times.

CLAIM B: The case of $u \in W_u$ in CASE2 or CASE3 occurs $O(n \log(L/n))$ times.
[Proof of Claim B] In this case, α_u is greater than 1 from the hypothesis at line $\ell 10$ or $\ell 11$ of COORDINATEWISE_SCALING. Hence, α_u must be updated at line $\ell 8$ of CASE2 or CASE3 at least once because $b(u) - a(u) < n\alpha_u$ for the updated a and b. Thus, this case occurs at most $\lceil \log(L/n) \rceil$ times for a fixed $u \in V$.

CLAIM C: The case of $u \notin W_u$ in CASE2 or CASE3 occurs $O(n^2 \log(L/n))$ times.

By Claims A, B and C, the while loop terminates in $O(n^2 \log(L/n))$ iterations. In the following, we prove Claim C.

[Proof of Claim C] We assume that for each $s \in V$, the s-axis is marked by "$\triangle$" at α_s intervals so that $x(s)$ lies on a mark (see Fig. 3). To prove Claim C, we adopt two auxiliary vectors $a', b' \in \mathbb{Z}^V$ such that

$$b'(s) - x(s) \text{ and } x(s) - a'(s) \text{ are divisible by } \alpha_s, \tag{1}$$

$$b'(s) - a'(s) + 2\alpha_s \ge \alpha_s \lfloor (b(s) - a(s))/\alpha_s \rfloor, \tag{2}$$

$$a'(s) - 2\alpha_s < a(s) \le a'(s) \le x(s) \le b'(s) \le b(s) < b'(s) + 2\alpha_s \tag{3}$$

for any $s \in V$. For instance, the situation of Fig. 3 satisfies conditions (1) to (3). We evaluate the number of the occurrences of the case in Claim C by using an integral number ψ defined by

$$\psi = \sum_{s \in V} \frac{b'(s) - a'(s)}{\alpha_s}. \tag{4}$$

By (c2) and (3), $(b'(s) - a'(s))/\alpha_s \le 2n$ holds. We will show that ψ does not increase if no scaling factors are updated, and that either ψ or a certain scaling factor is strictly reduced except in one case, which does not occur consecutively in a sense.

We set $a'(s)$ and $b'(s)$ to be

$$a'(s) := x(s) - \alpha_s \lfloor (x(s) - a(s))/\alpha_s \rfloor,$$
$$b'(s) := x(s) + \alpha_s \lfloor (b(s) - x(s))/\alpha_s \rfloor, \tag{5}$$

just after an update of α_s. Obviously, these $a'(s)$, $b'(s)$ satisfy conditions (1) to (3). We always put $a'(s) := a(s)$ and $b'(s) := b(s)$ when $\alpha_s = 1$, i.e.,

$$\alpha_s = 1 \implies a'(s) = a(s) \text{ and } b'(s) = b(s). \tag{6}$$

For a fixed element $s \in V$, we describe how to modify $a'(s)$ and $b'(s)$ in CASE2. The modification in CASE3 is similar. Values $a'(s)$ and $b'(s)$ are modified when s is selected as either u or w in functions CASE2 or CASE3. In the sequel, updated values are denoted by $a(s)$, $b(s)$, $a'(s)$, $b'(s)$ and $x(s)$, and that old values are denoted by (old) $a(s)$, (old) $b(s)$, etc.

We first suppose that s is selected as u (e.g., see the modification from Fig. 3 to Fig. 4). We put $b'(s) := x(s)$ and do not change $a'(s)$. Obviously, (1) is preserved and

$$a'(s) - 2\alpha_s < a(s) \tag{7}$$
$$b(s) - b'(s) = \alpha_s - 1 \tag{8}$$

hold. By (7) and (8), we have

$$b'(s) - a'(s) + 2\alpha_s > b(s) - a(s) - \alpha_s + 1 = (q-1)\alpha_s + r + 1,$$

where $q = \lfloor (b(s) - a(s))/\alpha_s \rfloor$ and $r = (b(s) - a(s)) - q\alpha_s$. Since $b'(s) - a'(s) + 2\alpha_s$ is divisible by α_s, $a'(s)$ and $b'(s)$ satisfy (2). All inequalities in (3) other than $a'(s) \le x(s)$ are trivially satisfied. If $x(s) < a'(s)$ then $b(s) - a(s) < 2\alpha_s$ holds by (7), (8) and the equation $x(s) = b'(s) = a'(s) - \alpha_s$, and hence, α_s must be updated at least once at line $\ell 8$ (and $a'(s)$ and $b'(s)$ are updated by (5)); otherwise (3) is satisfied. Moreover, the integral number $(b'(s) - a'(s))/\alpha_s$ is reduced at least by one when α_s remains the same, by $b'(s) + \alpha_s = \lceil (\text{old}) \, x(s) \rceil \le \lceil (\text{old}) \, b'(s) \rceil$. We note that the number of marks in $[a(s), b(s)] \setminus [a'(s), b'(s)]$ is at most one if s is selected as u, because of (5), (7) and (8).

We next suppose that s is selected as w and $\alpha_u \le \alpha_w$ (e.g., see the modification from Fig. 3 to Fig. 5). We modify $a'(s)$ and $b'(s)$ as follows: $a'(s)$ is set to be as large as possible in such a way that $a'(s) - 2\alpha_s < a(s) \le a'(s) \le x(s)$ and $x(s) - a'(s)$ is divisible by α_s; $b'(s)$ is next set to be as large as possible so that $b'(s) \le b(s)$, $b'(s) - x(s)$ is divisible by α_s, and $b'(s) - a'(s)$ does not increase. If $b(s) < b'(s) + \alpha_s$, then we have $b(s) - a(s) < b'(s) - a'(s) + 3\alpha_s$ by $a'(s) - 2\alpha_s < a(s)$; otherwise, $b'(s) - a'(s) = (\text{old}) \, (b'(s) - a'(s))$ must hold by the definition of $b'(s)$. In either case, $a'(s)$ and $b'(s)$ satisfy (2). Suppose to the contrary that $b'(s) + 2\alpha_s \le b(s)$ holds. This inequality and (2) imply $a'(s) - \alpha_s < a(s)$. By $a'(s) - \alpha_s < a(s)$ and the definition of $a'(s)$, we have $a'(s) = x(s)$. Since (old) $a'(s) \le (\text{old}) \, x(s) < x(s) = a'(s)$ and $b'(s) - a'(s) = (\text{old}) \, (b'(s) - a'(s))$, we have (old) $b'(s) < b'(s) \le b(s) - 2\alpha_s$. This contradicts the fact that (old) $b'(s)$

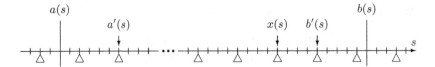

Fig. 3. A case where $\alpha_s = 4$

Fig. 4. A case where $s = u$ and $\alpha_s = 4$

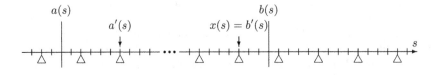

Fig. 5. A case where $s = w$, $\alpha_s = 4$ and $\alpha_u = 2$

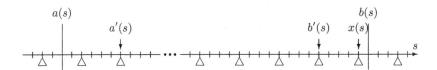

Fig. 6. A case where $s = w$, $\alpha_s = 4$ and $\alpha_u = 8$ (inadmissible case)

satisfies (3). Thus, $a'(s)$ and $b'(s)$ satisfy (1), (2) and inequalities in (3) except for $x(s) \leq b'(s)$. Moreover, $b'(s) - a'(s)$ does not increase. If $a'(s)$ and $b'(s)$ can satisfy an additional requirement $x(s) \leq b'(s)$, then (3) can be satisfied. In the other case, we have $\alpha_s > 1$ and $b'(s) + \alpha_s = x(s)$ because (1), (6) and $b'(s) < x(s) \leq b(s) < b'(s) + 2\alpha_s$ hold. Inequalities $x(s) - (n-1)(\alpha_u - 1) \leq a(s)$, $b(s) < b'(s) + 2\alpha_s$, $b'(s) + \alpha_s = x(s)$ and $\alpha_u \leq \alpha_s$ yield $b(s) - a(s) < n\alpha_s$, and hence, α_s must be updated at least once ($a'(s)$ and $b'(s)$ are updated by (5)).

We finally suppose that s is selected as w and $\alpha_u > \alpha_w$ (e.g., see the modification from Fig. 3 to Fig. 6). Since α_u is divisible by α_w, marks on the s-axis remain the same. In the same way as the case where $s = w$ and $\alpha_u \leq \alpha_w$, we can modify $a'(s)$ and $b'(s)$ so that $b'(s) - a'(s)$ does not increase and that (1), (2) and inequalities in (3) except for $x(s) \leq b'(s)$ hold. If $a'(s)$ and $b'(s)$ can be chosen to satisfy an additional requirement $x(s) \leq b'(s)$, then (3) can be satisfied. Here we call the other case an *inadmissible case*. Since marks on the s-axis remain the same, the above modification guarantees that the number of marks in $[(\text{old})a(s), (\text{old})b(s)] \setminus [(\text{old})a'(s), (\text{old})b'(s)]$ must be precisely two in the inadmissible case. We also have $x(s) = b'(s) + \alpha_s \leq b(s)$ because $b'(s) < x(s) \leq b(s) < b'(s) + 2\alpha_s$ and $b'(s) - x(s)$ is divisible by α_s. To satisfy (3), it suffices to increase $b'(s)$ by α_s. The number of marks in $[a(s), b(s)] \setminus [a'(s), b'(s)]$ is at most one after the inadmissible case, because $b(s) < b'(s) + \alpha_s$ holds.

By the above discussion, the integral number ψ defined in (4) does not increase if no scaling factors are updated. In particular, either ψ or a certain scaling factor is strictly reduced except in the inadmissible case. We now attach label w to each iteration. We claim that, for a fixed $s \in V$, the inadmissible case with the same scaling factor α_s does not occur consecutively in the subsequence of iterations labeled s. Hence, the case in Claim C occurs $O(n^2 \log(L/n))$ times.

We prove the last claim. Suppose, to the contrary, that the inadmissible case with the same scaling factor α_s occurs at iterations i and j ($i < j$) labeled s and that s is not selected as w at iterations $i+1$ to $j-1$. Although s may be selected as u at iterations $i+1$ to $j-1$ in CASE2 or CASE3, marks on the s-axis remain the same. After the modification at iteration i, $[a(s), b(s)] \setminus [a'(s), b'(s)]$ contains at most one mark and this must be preserved just before iteration j. However, this contradicts the fact that the inadmissible case occurs at iteration j.   □

By Lemmas 3 and 4, our main result is obtained.

Theorem 5. *Suppose that $f : \mathbb{Z}^V \to \mathbb{R} \cup \{+\infty\}$ is an M-convex function with a bounded effective domain and that $x^0 \in \text{dom} f$ is given. Algorithm* COORDINATEWISE_SCALING *finds a minimizer of f in* $O(Fn^3 \log(L/n))$ *time, where F is an upper bound of time to evaluate f and $L = \max\{|x(u) - y(u)| \mid u \in V, x, y \in \text{dom} f\}$.*

Proof. By Lemma 3, if COORDINATEWISE_SCALING terminates then it finds a minimizer of f. Lines $\ell 01$ to $\ell 03$ terminate in $O(Fn^2 \log L)$ time from Lemma 2. Lemma 4 says that the while loop of COORDINATEWISE_SCALING terminates in $O(n^2 \log(L/n))$ iterations. Since the value $f_a^b(x)$ is evaluated in $O(F)$ time for any $x \in \mathbb{Z}^V$ and function f_a^b is referred to at most $3n$ times per iteration, the total time complexity is $O(Fn^3 \log(L/n))$.   □

Remark 2. For the sake of simplicity, both u_1 and u_2 are computed before line $\ell 08$, and hence, f_a^b is evaluated $3|V'|$ times in each iteration. We can easily reduce the number of evaluations of f_a^b from $3|V'|$ to $2|V'|$ per iteration. We can also replace constant n with variable $|V'|$. These revision may improve execution time of our algorithm.

4   Concluding Remarks

In this paper, we proposed an $O(Fn^3 \log(L/n))$ time algorithm for M-convex function minimization. We finally note that the algorithm can be extended to a more general class of functions, called quasi M-convex functions.

Murota and Shioura [24] introduced the concept of quasi M-convex functions and generalized results on M-convexity to quasi M-convexity. A function f : $\mathbb{Z}^V \to \mathbb{R} \cup \{+\infty\}$ with dom $f \neq \emptyset$ is called *quasi M-convex* if it satisfies

(SSQM$^{\neq}$) $\forall x, y \in$ dom f with $f(x) \neq f(y)$, $\forall u \in$ supp$^+(x - y)$, $\exists v \in$ supp$^-(x - y)$:

(i) $f(x - \chi_u + \chi_v) - f(x) \geq 0 \Rightarrow f(y + \chi_u - \chi_v) - f(y) \leq 0$, and
(ii) $f(y + \chi_u - \chi_v) - f(y) \geq 0 \Rightarrow f(x - \chi_u + \chi_v) - f(x) \leq 0$.

Obviously, an M-convex function satisfies (SSQM$^{\neq}$). Murota and Shioura [24] show that Theorems 1 and 2 are extended to quasi M-convex functions. Furthermore, we can verify that Theorem 4 is also extended to quasi M-convex functions.

Let $\mathcal{C}$ be the class of functions f such that f satisfies (SSQM$^{\neq}$) and arg min f is an M-convex set. It is noted that the class $\mathcal{C}$ includes the semistrictly quasi M-convex functions due to [24], which are defined by the property obtained from (SSQM$^{\neq}$) by eliminating hypothesis $f(x) \neq f(y)$.

The key properties of M-convex functions stated in Propositions 1 and 2 are shared by functions in $\mathcal{C}$ as follows. Since the set of the minimizers of any function in $\mathcal{C}$ is an M-convex set, Proposition 1 is trivially extended to $\mathcal{C}$. For any $f \in \mathcal{C}$ and $a, b \in \mathbb{Z}^V$, if arg min $f \cap [a, b] \neq \emptyset$, then arg min f_a^b is an M-convex set and f_a^b belongs to $\mathcal{C}$, and hence, Proposition 2 is also extended to $\mathcal{C}$. Since our algorithm, COORDINATEWISE_SCALING, relies solely on the above properties and the fact that l and u with dom $f \subseteq [l, u]$ are efficiently computable, it also works with the same time complexity for a subclass of $\mathcal{C}$ such that l and u are efficiently computable.

Acknowledgement

The author thanks Kazuo Murota and Akiyoshi Shioura for their helpful comments and suggestions.

References

1. Cunningham, W. H. and Frank, A.: A Primal-dual Algorithm for Submodular Flows. Math. Oper. Res. **10** (1985) 251–262
2. Danilov, V., Koshevoy, G., Murota, K.: Discrete Convexity and Equilibria in Economies with Indivisible Goods and Money. Math. Social Sci. **41** (2001) 251–273
3. Dress, A. W. M., Wenzel, W.: Valuated Matroid: A New Look at the Greedy Algorithm. Appl. Math. Lett. **3** (1990) 33–35
4. Dress, A. W. M., Wenzel, W.: Valuated Matroids. Adv. Math. **93** (1992) 214–250
5. Edmonds, J.: Submodular Functions, Matroids and Certain Polyhedra. In: Guy, R., Hanani, H., Sauer, N., Schönheim, J. (eds.): Combinatorial Structures and Their Applications. Gordon and Breach, New York (1970) 69–87
6. Edmonds, J., Giles, R.: A Min-max Relation for Submodular Functions on Graphs. Ann. Discrete Math. **1** (1977) 185–204

7. Frank, A.: Generalized Polymatroids. In: Hajnal, A., Lovász, L., Sós, V. T. (eds.): Finite and Infinite Sets, I. North-Holland, Amsterdam (1984) 285–294
8. Frank, A., Tardos, É.: Generalized Polymatroids and Submodular Flows. Math. Program. **42** (1988) 489–563
9. Fujishige, S.: Lexicographically Optimal Base of a Polymatroid with Respect to a Weight Vector. Math. Oper. Res. **5** (1980) 186–196
10. Fujishige, S.: Submodular Functions and Optimization. Annals of Discrete Mathematics 47, North-Holland, Amsterdam (1991)
11. Groenevelt, H.: Two Algorithms for Maximizing a Separable Concave Function over a Polymatroid Feasible Region. European J. Oper. Res. **54** (1991) 227–236
12. Hochbaum, D. S.: Lower and Upper Bounds for the Allocation Problem and Other Nonlinear Optimization Problems. Math. Oper. Res. **19** (1994) 390–409
13. Iwata, S.: A Fully Combinatorial Algorithm for Submodular Function Minimization. J. Combin. Theory Ser. B (to appear)
14. Iwata, S., Shigeno, M.: Conjugate Scaling Algorithm for Fenchel-type Duality in Discrete Convex Optimization. SIAM J. Optim. (to appear)
15. Iwata, S., Fleischer, L., Fujishige, S.: A Combinatorial Strongly Polynomial Algorithm for Minimizing Submodular Functions. J. ACM **48** (2001) 761–777
16. Lovász, L.: Submodular Functions and Convexity. In: Bachem, A., Grötschel, M., Korte, B. (eds.): Mathematical Programming — The State of the Art. Springer-Verlag, Berlin (1983) 235–257
17. Moriguchi, S., Murota, K., Shioura, A.: Minimization of an M-convex Function with a Scaling Technique (in Japanese). IPSJ SIG Notes 2001–AL–76 (2001) 27–34
18. Moriguchi, S., Murota, K., Shioura, A.: Scaling Algorithms for M-convex Function Minimization. IEICE Trans. Fund. (to appear)
19. Murota, K.: Convexity and Steinitz's Exchange Property. Adv. Math. **124** (1996) 272–311
20. Murota, K.: Discrete Convex Analysis. Math. Program. **83** (1998) 313–371
21. Murota, K.: Submodular Flow Problem with a Nonseparable Cost Function. Combinatorica **19** (1999) 87–109
22. Murota, K.: Matrices and Matroids for Systems Analysis. Springer-Verlag, Berlin (2000)
23. Murota, K.: Discrete Convex Analysis (in Japanese). Kyoritsu Publ. Co., Tokyo (2001)
24. Murota, K., Shioura, A.: Quasi M-convex and L-convex functions: Quasi-convexity in discrete optimization. Discrete Appl. Math. (to appear)
25. Murota, K., Tamura, A.: New Characterizations of M-convex Functions and Their Applications to Economic Equilibrium Models with Indivisibilities. Discrete Appl. Math. (to appear)
26. Murota, K., Tamura, A.: Application of M-convex Submodular Flow Problem to Mathematical Economics. In: Eades, P., Takaoka, T. (eds.): Proceedings of 12th International Symposium on Algorithms and Computation, Lecture Notes in Computer Science, Vol. 2223. Springer-Verlag, Berlin Heidelberg New York (2001) 14–25
27. Rockafellar, R. T.: Convex Analysis. Princeton University Press, Princeton (1970)
28. Schrijver, A.: A Combinatorial Algorithm Minimizing Submodular Functions in Strongly Polynomial Time. J. Combin. Theory Ser. B **80** (2000) 346–355
29. Shioura, A.: Minimization of an M-convex Function. Discrete Appl. Math. **84** (1998) 215–220

The Quickest Multicommodity Flow Problem⋆

Lisa Fleischer[1] and Martin Skutella[2]

[1] Graduate School of Industrial Administration, Carnegie Mellon University
Pittsburgh, PA 15213, USA
lkf@andrew.cmu.edu
http://www.andrew.cmu.edu/user/lkf/
[2] Institut für Mathematik, MA 6–1, Technische Universität Berlin
Straße des 17. Juni 136, 10623 Berlin, Germany
skutella@math.tu-berlin.de
http://www.math.tu-berlin.de/~skutella/

Abstract. Traditionally, flows over time are solved in time-expanded networks which contain one copy of the original network for each discrete time step. While this method makes available the whole algorithmic toolbox developed for static flows, its main and often fatal drawback is the enormous size of the time-expanded network. In particular, this approach usually does not lead to efficient algorithms with running time polynomial in the input size since the size of the time-expanded network is only pseudo-polynomial.

We present two different approaches for coping with this difficulty. Firstly, inspired by the work of Ford and Fulkerson on maximal s-t-flows over time (or 'maximal dynamic s-t-flows'), we show that static, length-bounded flows lead to provably good multicommodity flows over time. These solutions not only feature a simple structure but can also be computed very efficiently in polynomial time.

Secondly, we investigate 'condensed' time-expanded networks which rely on a rougher discretization of time. Unfortunately, there is a natural tradeoff between the roughness of the discretization and the quality of the achievable solutions. However, we prove that a solution of arbitrary precision can be computed in polynomial time through an appropriate discretization leading to a condensed time expanded network of polynomial size. In particular, this approach yields a fully polynomial time approximation scheme for the quickest multicommodity flow problem and also for more general problems.

1 Introduction

We consider flow problems in networks with fixed capacities and transit times on the arcs. The transit time of an arc specifies the amount of time it takes for flow to travel from the tail to the head of that arc. In contrast to the classical case of static flows, a *flow over time* in such a network specifies a flow rate entering

---

⋆ Extended abstract; information on the full version of the paper can be obtained via the authors' WWW-pages.

W.J. Cook and A.S. Schulz (Eds.): IPCO 2002, LNCS 2337, pp. 36–53, 2002.

an arc for each point in time. In this setting, the capacity of an arc limits the rate of flow into the arc at each point in time.

Flows over time may be applied to various areas of operations research and have many real-world applications such as traffic control, evacuation plans, production systems, communication networks, and financial flows. Examples and further application can be found in the survey articles of Aronson [1] and Powell, Jaillet, and Odoni [18].

Flows over time have been introduced about forty years ago by Ford and Fulkerson [5,6]. They consider the problem of sending the maximal possible amount of flow from a source node s to a sink node t within a given time T. This problem can efficiently be solved by essentially one min-cost flow computation on the given network where transit times of arcs are interpreted as costs per unit of flow. Ford and Fulkerson show that an optimal solution to this min-cost flow problem can be turned into a maximal flow over time by first decomposing it into flows on paths. The corresponding flow over time starts to send flow on each path at time zero, and repeats each so long as there is enough time left in the T time units for the flow along the path to arrive at the sink. A flow over time featuring this structure is called *temporally repeated*.

A problem closely related to the problem of computing a maximal flow over time is the *quickest flow problem*: Send a given amount of flow from the source to the sink in the shortest possible time. This problem can be solved in polynomial time by incorporating the algorithm of Ford and Fulkerson in a binary search framework. Burkard, Dlaska, and Klinz [2] present a faster algorithm which even solves the quickest flow problem in strongly polynomial time.

A natural generalization of the quickest flow problem and the maximal flow problem considered by Ford and Fulkerson can be defined on networks with additional costs on the arcs. Klinz and Woeginger [14] show that the search for a quickest or a maximal flow over time with minimal cost cannot be restricted to the class of temporally repeated flows. In fact, adding costs has also a considerable impact on the complexity of these problems. Klinz and Woeginger prove NP-hardness results even for the special case of series parallel graphs. Moreover, they show that the problem of computing a maximal temporally repeated flow with minimal cost is strongly NP-hard.

Another generalization is the quickest transshipment problem: Given a vector of supplies and demands at the nodes, the task is to find a flow over time that zeroes all supplies and demands within minimal time. Hoppe and Tardos [13] show that this problem can still be solved in polynomial time. They introduce the use of *chain decomposable flows* which generalize the class of temporally repeated flows and can also be compactly encoded as a collection of paths. However, in contrast to temporally repeated flows, these paths may also contain backward arcs. Therefore, a careful analysis is necessary to show feasibility of the resulting flows over time.

All results mentioned so far work with a discrete time model, that is, time is discretized into steps of unit length. In each step, flow can be sent from a node v through an arc (v, w) to the adjacent node w, where it arrives $\tau_{(v,w)}$ time steps later; here, $\tau_{(v,w)}$ denotes the given integral transit time of arc (v, w). In particular, the time-dependent flow on an arc is represented by a time-indexed vector in this model. In contrast to this, in the continuous time model the flow on an arc e is a function $f_e : \mathbb{R}^+ \to \mathbb{R}^+$. Fleischer and Tardos [4] point out a strong connection between the two models. They show that many results and algorithms which have been developed for the discrete time model can be carried over to the continuous time model.

In the discrete time model, flows over time can be described and computed in *time-expanded networks* which were introduced by Ford and Fulkerson [5,6]. A time expanded network contains a copy of the node set of the underlying 'static' network for each discrete time step. Moreover, for every arc e in the static network with integral transit time τ_e, there is a copy between all pairs of time layers with distance τ_e in the time-expanded network. Unfortunately, due to the time expansion, the size of the resulting network is in general exponential in the input size of the problem. This difficulty has already been pointed out by Ford and Fulkerson[1]. On the other hand, the advantage of this approach is that it turns the problem of determining an optimal flow over time into a classical 'static' network flow problem on the time-expanded network. This problem can then be solved by well-known network flow algorithms. A main contribution of this paper is to provide a possibility to overcome the difficulties caused by the size of time-expanded networks while still maintaining the latter advantage.

A straightforward idea is to reduce the size of time-expanded networks by replacing the time steps of unit length by larger steps. In other words, applying a sufficiently rough discretization of time leads to a *condensed* time-expanded network of polynomial size. However, there is a tradeoff between the necessity to reduce the size of the time-expanded network and the desire to limit the loss of precision of the resulting flow model since the latter results in a loss of quality of achievable solutions.

We show that there is a satisfactory solution to this tradeoff problem. An appropriate choice of the step length leads to a condensed time-expanded network of polynomial size which still allows a $(1 + \varepsilon)$-approximate precision in time, for any $\varepsilon > 0$. This observation has potential applications for many problems involving flows over time. In particular, it yields a fully polynomial time approximation scheme (FPTAS) for the quickest multicommodity flow problem.

To the best of our knowledge, this is the first approximation result for the general time-dependent multicommodity flow problem. Moreover, the complexity of the quickest multicommodity flow problem is still open. It has neither been proved to be NP-hard nor is it known to be solvable in polynomial time. Apart from this, we believe that our result is also of interest for flow problems, like the quickest transshipment problem, which are known to be solvable in polynomial time. While the algorithm of Hoppe and Tardos [13] for the quickest transshipment problem relies on submodular function minimization, the use of condensed

---

[1] They use the time-expanded network only in the analysis of their algorithm. The optimality of the computed flow over time is proven by a cut in the time-expanded network whose capacity equals the value of the flow.

time-expanded networks leads to an FPTAS which simply consists of a series of max-flow computations.

Moreover, our approach also works in the setting with costs and we can give a bicriteria FPTAS[2] for the min-cost quickest multicommodity flow problem. Notice that already the single-commodity version of this problem is known to be NP-hard [14].

We next introduce a geometrically-condensed time-expanded network satisfying the following property: For any point in time θ, every unit of flow which arrives at its destination in a given flow at time θ, arrives there before time $(1+\varepsilon)\theta$ in a corresponding solution for this network. We use this to give the first FPTAS for the *earliest arrival flow problem* when there are multiple sources and a single sink. In contrast to Hoppe and Tardos' FPTAS for the single-source, single-sink problem where the amount of flow is approximately optimal at every moment of time [12], we obtain optimal flows in approximately optimal time.

While our analysis shows that condensed time-expanded networks lead to theoretically efficient (polynomial time) algorithms with provably good worst case performance, these algorithms can certainly not compete with methods, like the algorithm of Ford and Fulkerson, which solely work on the underlying static network. On the other hand, such methods are only known for restricted problems with one single commodity. For more general problems, like multicommodity flows over time, it is not even clear how to encode optimal solutions efficiently. Complex time dependency seems to be an inherent 'defect' of optimal solutions to these problems. Against this background, it is interesting to ask for provably good solutions featuring a reasonably simple structure.

Inspired by the work of Ford and Fulkerson on maximal s-t-flows over time, we show that static, *length-bounded* flows in the underlying static network lead to provably good multicommodity flows over time which, in addition, can be computed very efficiently. A length-bounded flow has a path decomposition where the length of each flow-carrying path is bounded. Based on such a path decomposition, we can construct a temporally repeated flow over time. Moreover, if we start with a maximal length-bounded flow, the resulting flow over time needs at most twice as long as a quickest flow over time. If one allows a $(1+\varepsilon)$-violation of the length bound, a maximal length-bounded flow can be computed efficiently in polynomial time. Therefore, this approach yields a $(2+\varepsilon)$-approximation algorithm for the quickest multicommodity flow problem with costs. In this context it is interesting to remember that the problem of computing a quickest temporally repeated flow with bounded cost is strongly NP-hard [14] and therefore does not allow an FPTAS, unless P=NP. We also present pathological instances which imply a lower bound on the worst case performance of our algorithm and show that the given analysis is tight.

The paper is organized as follows. In Sect. 2 we give a precise description of the problem under consideration and state basic properties of static (length-bounded) flows and flows over time. In Sect. 3 we present a $(2+\varepsilon)$-approximation

---

[2] A family of approximation algorithms with simultaneous performance ratio $1+\varepsilon$ for time and cost, for any $\varepsilon > 0$.

algorithm based on length-bounded flows. In Sect. 4 we introduce time-expanded networks and discuss the interconnection between flows over time and (static) flows in time-expanded networks. Our main result on condensed time expanded networks is presented in Sect. 5. Finally, in Sect. 6 and 7 we discuss extensions of this result to the corresponding min-cost flow problems and to earliest arrival flows. In this extended abstract, we omit all details in the last two sections due to space restrictions.

2   Preliminaries

We consider routing problems on a network $\mathcal{N} = (V, A)$. Each arc $e \in A$ has an associated *transit time* or *length* τ_e and a capacity u_e. An arc e from node v to node w is sometimes also denoted (v, w); in this case, we write $\mathrm{head}(e) = w$ and $\mathrm{tail}(e) = v$. There is a set of commodities $K = \{1, \ldots, k\}$ that must be routed through network $\mathcal{N}$. Each commodity is defined by a source-sink pair $(s_i, t_i) \in V \times V$, $i \in K$, and it is required to send a specified amount of flow d_i from s_i to t_i, called the demand. In the setting with costs, each arc e has associated cost coefficients $c_{e,i}$, $i \in K$, which determine the per unit cost for sending flow of commodity i through the arc.

2.1   Static Flows

A *static (multicommodity) flow* x on $\mathcal{N}$ assigns every arc-commodity pair (e, i) a non-negative flow value $x_{e,i}$ such that *flow conservation constraints*

$$\sum_{e \in \delta^-(v)} x_{e,i} \ - \sum_{e \in \delta^+(v)} x_{e,i} \ = \ 0 \qquad \text{for all } v \in V \setminus \{s_i, t_i\},$$

are obeyed for any commodity $i \in K$. Here, $\delta^+(v)$ and $\delta^-(v)$ denote the set of arcs e leaving node v ($\mathrm{tail}(e) = v$) and entering node v ($\mathrm{head}(e) = v$), respectively. The static flow x *satisfies the multicommodity demands* if

$$\sum_{e \in \delta^-(t_i)} x_{e,i} \ - \sum_{e \in \delta^+(t_i)} x_{e,i} \ \geq \ d_i \ ,$$

for any commodity $i \in K$. Moreover, x is called *feasible* if it obeys the *capacity constraints* $x_e \leq u_e$, for all $e \in A$, where $x_e := \sum_{i \in K} x_{e,i}$ is the total flow on arc e. In the setting with costs, the cost of a static flow x is defined as

$$c(x) \ := \ \sum_{e \in A} \sum_{i \in K} c_{e,i} x_{e,i} \ . \tag{1}$$

2.2   Flows over Time

In many applications of flow problems, static routing of flow as discussed in Sect. 2.1 does not satisfactorily capture the real structure of the problem since not only the amount of flow to be transmitted but also the time needed for the transmission plays an essential role.

A *(multicommodity) flow over time* f on $\mathcal{N}$ with time horizon T is given by a collection of Lebesgue-measurable functions $f_{e,i} : [0, T) \to \mathbb{R}^+$ where $f_{e,i}(\theta)$ determines the rate of flow (per time unit) of commodity i entering arc e at time θ. Transit times are fixed throughout, so that flow on arc e progresses at a uniform rate. In particular, the flow $f_{e,i}(\theta)$ of commodity i entering arc e at time θ arrives at $\mathrm{head}(e)$ at time $\theta + \tau_e$. Thus, in order to obey the time horizon T, we require that $f_{e,i}(\theta) = 0$ for $\theta \in [T - \tau_e, T)$.

In our model, we allow intermediate storage of flow at nodes. This corresponds to holding inventory at a node before sending it onward. Thus, the flow conservation constraints are integrated over time to prohibit deficit at any node:

$$
\sum_{e \in \delta^-(v)} \int_{\tau_e}^{\xi} f_{e,i}(\theta - \tau_e) \, d\theta \;-\; \sum_{e \in \delta^+(v)} \int_0^{\xi} f_{e,i}(\theta) \, d\theta \;\geq\; 0 \;, \tag{2}
$$

for all $\xi \in [0, T)$, $v \in V \setminus \{s_i\}$, and $i \in K$. Moreover, we require that equality holds in (2) for $\xi = T$ and $v \in V \setminus \{s_i, t_i\}$, meaning that no flow should remain in the network after time T.

The flow over time f satisfies the multicommodity demands if

$$
\sum_{e \in \delta^-(t_i)} \int_{\tau_e}^{T} f_{e,i}(\theta - \tau_e) \, d\theta \;-\; \sum_{e \in \delta^+(t_i)} \int_0^{T} f_{e,i}(\theta) \, d\theta \;\geq\; d_i \;, \tag{3}
$$

for any commodity i. Moreover, f is called feasible if it obeys the capacity constraints. Here, capacity u_e is interpreted as an upper bound on the rate of flow entering arc e, i. e., a capacity per unit time. Thus, the capacity constraints are $f_e(\theta) \leq u_e$, for all $\theta \in [0, T)$ and $e \in A$, where $f_e(\theta) := \sum_{i \in K} f_{e,i}(\theta)$ is the total flow into arc e at time θ.

For flows over time, a natural objective is to minimize the *makespan*: the time T necessary to satisfy all demands. In the *quickest (multicommodity) flow problem*, we are looking for a feasible flow over time with minimal time horizon T that satisfies the multicommodity demands.

In the setting with costs, the cost of a flow over time f is defined as

$$
c(f) \;:=\; \sum_{e \in A} \sum_{i \in K} c_{e,i} \int_0^{T} f_{e,i}(\theta) \, d\theta \;. \tag{4}
$$

The *quickest (multicommodity) flow problem with costs* is to find a feasible flow over time f with minimal time horizon T that satisfies the multicommodity demands and whose cost is bounded from above by a given *budget* B. A natural variant of this problem is to bound the cost for every single commodity i by a budget B_i, that is, $\sum_{e \in A} c_{e,i} \int_0^{T} f_{e,i}(\theta) \, d\theta \leq B_i$, for all $i \in K$. All of our results on the quickest multicommodity flow problem with costs work in this setting also.

2.3   Length-Bounded Static Flows

While static flows are not defined with reference to transit times, we are interested in static flows that suggest reasonable routes with respect to transit times.
To account for this, we consider decompositions of static flows into paths.

It is well known that in a static flow x the flow $(x_{e,i})_{e \in A}$ of any commodity $i \in K$ can be decomposed into the sum of flows on a set of s_i-t_i-paths and
flows on cycles. We denote the set of all s_i-t_i-paths by $\mathcal{P}_i$ and the flow value of
commodity i on path $P \in \mathcal{P}_i$ is denoted by $x_{P,i}$. Then, for each arc $e \in A$,

$$x_{e,i} \;=\; \sum_{\substack{P \in \mathcal{P}_i: \\ e \in P}} x_{P,i} \;.$$

We assume without loss of generality that there are no cycles in the flow decomposition; otherwise, the solution x can be modified by decreasing flow on those
cycles.

The static flow x is called T-*length-bounded* if the flow of every commodity $i \in K$ can be decomposed into the sum of flows on s_i-t_i-paths such that the
length $\tau(P) := \sum_{e \in P} \tau_e$ of any path $P \in \mathcal{P}_i$ with $x_{P,i} > 0$ is at most T.

While the problem of computing a feasible static flow that satisfies the multicommodity demands can be solved efficiently, it is NP-hard to find such a
flow which is in addition T-length-bounded, even for the special case of a single
commodity. This follows by a straightforward reduction from the NP-complete
problem PARTITION. On the other hand, the length-bounded flow problem can
be approximated within arbitrary precision in polynomial time. More precisely,
if there exists a feasible T-length-bounded static flow x which satisfies the multicommodity demands, then, for any $\varepsilon > 0$, we can compute a feasible $(1 + \varepsilon)T$-
length-bounded static flow x' of cost $c(x') \leq c(x)$ satisfying all demands with
running time polynomial in the input size and $1/\varepsilon$.

In order to prove this, we first formulate the problem of finding a feasible T-
length-bounded static flow as a linear program in path-variables. Let

$$\mathcal{P}_i^T \;:=\; \{P \in \mathcal{P}_i \mid \tau(P) \leq T\}$$

be the set of all s_i-t_i-paths whose lengths are bounded from above by T. The
cost of path $P \in \mathcal{P}_i$ is defined as $c_i(P) := \sum_{e \in P} c_{e,i}$. The length bounded flow
problem can then be written as:

$$\min \sum_{i \in K} \sum_{P \in \mathcal{P}_i^T} c_i(P) x_{P,i}$$

$$\text{s. t.} \quad \sum_{P \in \mathcal{P}_i^T} x_{P,i} \;\geq\; d_i \qquad\qquad \text{for all } i \in K,$$

$$\sum_{i \in K} \sum_{\substack{P \in \mathcal{P}_i^T: \\ e \in P}} x_{P,i} \;\leq\; u_e \qquad\qquad \text{for all } e \in A,$$

$$x_{P,i} \;\geq\; 0 \qquad\qquad \text{for all } i \in K,\, P \in \mathcal{P}_i^T.$$

Unfortunately, the number of paths in $\mathcal{P}_i^T$ and thus the number of variables in this linear program are in general exponential in the size of the underlying network $\mathcal{N}$. If we dualize the program we get:

$$\max \quad \sum_{i \in K} d_i z_i \;-\; \sum_{e \in A} u_e y_e$$

$$\text{s. t.} \quad \sum_{e \in P} (y_e + c_{e,i}) \;\geq\; z_i \qquad \text{for all } i \in K,\, P \in \mathcal{P}_i^T,$$

$$z_i, y_e \;\geq\; 0 \qquad \text{for all } i \in K,\, e \in A.$$

The corresponding separation problem can be formulated as a length-bounded shortest path problem: Find a shortest s_i-t_i-path P with respect to the arc weights $y_e + c_{e,i}$ whose length $\tau(P)$ is at most T, that is, $P \in \mathcal{P}_i^T$. While this problem is NP-hard [10], it can be solved approximately in the following sense: For any $\varepsilon > 0$, one can find in time polynomial in the size of the network $\mathcal{N}$ and $1/\varepsilon$ an s_i-t_i-path P with $\tau(P) \leq (1+\varepsilon)T$ whose length with respect to the arc weights $y_e + c_{e,i}$ is bounded from above by the length of a shortest path in $\mathcal{P}_i^T$ [11,15,17]. Using the equivalence of optimization and separation [9], this means for our problem that we can find in polynomial time an optimal solution to a modified dual program which contains additional constraints corresponding to paths of length at most $(1 + \varepsilon)T$. From this dual solution we get a primal solution which uses additional variables corresponding to those paths of length at most $(1 + \varepsilon)T$.

Notice that the method described above relies on the ellipsoid method and is therefore of rather restricted relevance for solving length-bounded flow problems in practice. However, the FPTASes developed in [8,3] for multicommodity flow problems can be generalized to the case of length-bounded flows: Those algorithms iteratively send flow on shortest paths with respect to some length function. In order to get a length-bounded solution, these shortest paths must be replaced by (up to a factor of $(1 + \varepsilon)$) length-bounded shortest paths.

3   A Simple $(2 + \varepsilon)$-Approximation Algorithm

In this section we generalize the basic approach of Ford and Fulkerson [5,6] to the case of multiple commodities and costs. However, in contrast to the algorithm of Ford and Fulkerson which is based on a (static) min-cost flow computation, the method we propose employs length-bounded static flows.

Any feasible flow over time f with time horizon T and cost at most B naturally induces a feasible static flow x on the underlying network $\mathcal{N}$ by averaging the flow on every arc over time, that is,

$$x_{e,i} \;:=\; \frac{1}{T} \int_0^T f_{e,i}(\theta)\, d\theta$$

for all $e \in A$ and $i \in K$. By construction, the static flow x is feasible and it satisfies the following three properties:

Fig. 1. An instance of the quickest multicommodity flow problem containing three commodities; commodities 1 and 3 have demand value 1, commodity 2 has demand value 2. The numbers at the arcs indicate the transit times; all arcs have unit capacity. A quickest flow with waiting at intermediate nodes allowed takes 3 time units and stores one unit of commodity 2 at the intermediate node $t_1 = s_3$ for two time units. However, if flow cannot be stored at intermediate nodes, an optimal solution takes time 4

(i) it is T-length-bounded;
(ii) it satisfies a fraction of $\frac{1}{T}$ of the demands covered by the flow over time f;
(iii) $c(x) = c(f)/T$.

Due to the fixed time horizon T, flow can only travel on paths of length at most T in f such that property (i) is fulfilled. Property (ii) follows from (3). Finally, property (iii) is a consequence of (1) and (4).

On the other hand, given an arbitrary feasible static flow x meeting requirements (i),(ii), and (iii), it can easily be turned into a feasible flow over time g meeting the same demands at the same cost as f within time horizon $2T$: Pump flow into every s_i-t_i-path P given by the length-bounded path decomposition of x at the corresponding flow rate $x_{P,i}$ for T time units; then wait for at most T additional time units until all flow has arrived at its destination. In particular, no flow is stored at intermediate nodes in this solution.

Lemma 1. *Allowing the storage of flow at intermediate nodes in $\mathcal{N}$ saves at most a factor of 2 in the optimal makespan. On the other hand, there are instances where the optimal makespan without intermediate storage is 4/3 times the optimal makespan with intermediate storage.*

Proof. The bound of 2 follows from the discussion above. In Fig. 1 we give an instance with a gap of 4/3 between the optimal makespan without storing and the optimal makespan with storing at intermediate nodes.   □

Notice that the gap of 4/3 is not an artifact of the small numbers in the instance depicted in Fig. 1. It holds for more general demands and transit times as well: For instance, scale all transit times and capacities of arcs by a factor of q and multiply all pairwise demands by a factor of q^2. The ratio of optimal makespans for the problems without to with storage is still 4/3.

In contrast to Ford and Fulkerson's temporally repeated flows, the flows over time resulting from length-bounded static flows described above, do not necessarily use flow-carrying paths as long as possible. However, we can easily enforce this property by scaling the flow rate $x_{P,i}$ on any path P by a factor $T/(2T - \tau(P)) \leq 1$ and sending flow into the path at this modified rate during the time interval $[0, 2T - \tau(P))$.

We can now state the main result of this section.

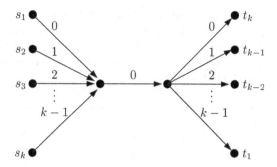

Fig. 2. An instance with k commodities showing that the analysis in the proof of Theorem 1 is tight. All arcs have unit capacity and transit times as depicted above. The demand value of every commodity is 1. A quickest flow needs $T^* = k$ time units. However, any static flow can satisfy at most a fraction of $1/k$ of the demands. In particular, the makespan of the resulting flow over time is at least $2k - 1$

Theorem 1. *For the quickest multicommodity flow problem with costs, there exists a polynomial time algorithm that, for any $\varepsilon > 0$, finds a solution of the same cost as optimal with makespan at most $2 + \varepsilon$ times the optimal makespan.*

Proof. Using binary search, we can guess the optimal makespan with precision $1 + \varepsilon/4$, that is, we get T with $T^* \leq T \leq (1+\varepsilon/4)T^*$. If we relax property (i) to allow flow on paths of length at most $(1 + \varepsilon/4)T \leq (1 + 3\varepsilon/4)T^*$, a feasible static flow meeting properties (i) to (iii) can be computed in polynomial time; see Sect. 2.3. This static flow can then be turned into a flow over time with makespan $(1 + 3\varepsilon/4)T^* + (1 + \varepsilon/4)T^* = (2 + \varepsilon)T^*$ as described above.   □

In Fig. 2 we present an instance which shows that the analysis in the proof of Theorem 1 is tight, that is, the performance guarantee of the discussed approximation algorithm is not better than 2.

4   Flows over Time and Time-Expanded Networks

Traditionally, flows over time are solved in a time-expanded network. Given a network $\mathcal{N} = (V, A)$ with integral transit times on the arcs and an integral time horizon T, the T-*time-expanded network* of $\mathcal{N}$, denoted $\mathcal{N}^T$ is obtained by creating T copies of V, labeled V_0 through V_{T-1}, with the θth copy of node v denoted v_θ, $\theta = 0, \ldots, T - 1$. For every arc $e = (v, w)$ in A and $0 \leq \theta < T - \tau_e$, there is an arc e_θ from v_θ to $w_{\theta+\tau_e}$ with the same capacity and cost as arc e. In addition, there is a *holdover arc* from v_θ to $v_{\theta+1}$, for all $v \in V$ and $0 \leq \theta < T-1$, which models the possibility to hold flow at node v.

Any flow in this time-expanded network may be taken by a flow over time of equal cost: interpret the flow on arc e_θ as the flow through arc $e = (v, w)$ that starts at node v in the time interval $[\theta, \theta + 1)$. Similarly, any flow over time completing by time T corresponds to a flow in $\mathcal{N}^T$ of the same value and cost obtained by mapping the total flow starting on e in time interval $[\theta, \theta+1)$ to flow on arc e_θ. More details can be found below in Lemma 2 (set $\Delta := 1$). Thus, we may solve any flow-over-time problem by solving the corresponding static flow problem in the time-expanded graph.

One problem with this approach is that the size of $\mathcal{N}^T$ depends linearly on T, so that if T is not bounded by a polynomial in the input size, this is not a polynomial-time method of obtaining the required flow over time. However, if all arc lengths are a multiple of $\Delta > 0$ such that $\lceil T/\Delta \rceil$ is bounded by a polynomial in the input size, then instead of using the T-time-expanded graph, we may rescale time and use a condensed time-expanded network that contains only $\lceil T/\Delta \rceil$ copies of V. Since in this setting every arc corresponds to a time interval of length Δ, capacities are multiplied by Δ. We denote this condensed time-expanded network by $\mathcal{N}^T/\Delta$, and the copies of V in this network by $V_{\rho\Delta}$ for $\rho = 0, \ldots, \lceil T/\Delta \rceil - 1$.

Lemma 2. *Suppose that all arc lengths are multiples of Δ and T/Δ is an integer. Then, any flow over time that completes by time T corresponds to a static flow of equal cost in $\mathcal{N}^T/\Delta$, and any flow in $\mathcal{N}^T/\Delta$ corresponds to a flow over time of equal cost that completes by time T.*

Proof. Given an arbitrary flow over time, a modified flow over time of equal value and cost can be obtained by averaging the flow value on any arc in each time interval $[\rho\Delta, (\rho+1)\Delta)$, $\rho = 0, \ldots, T/\Delta - 1$. This modified flow over time defines a static flow in $\mathcal{N}^T/\Delta$ in a canonical way. Notice that the capacity constraints are obeyed since the total flow starting on arc e in interval $[\rho\Delta, (\rho+1)\Delta)$ is bounded by Δu_e. The flow values on the holdover arcs are defined in such a way that flow conservation is obeyed in every node of $\mathcal{N}^T/\Delta$.

On the other hand, a static flow on $\mathcal{N}^T/\Delta$ can easily be turned into a flow over time. The static flow on an arc with tail in $V_{\rho\Delta}$ is divided by Δ and sent for Δ time units starting at time $\rho\Delta$. If the head of the arc is in $V_{\sigma\Delta}$ for $\sigma \geq \rho$, then the length of the arc is $(\sigma-\rho)\Delta$, and the last flow (sent before time $(\rho+1)\Delta$) arrives before time $(\sigma+1)\Delta$. Note that if costs are assigned to arcs of $\mathcal{N}^T/\Delta$ in the natural way, then the cost of the flow over time is the same as the cost of the corresponding flow in the time-expanded graph. □

If we drop the condition that T/Δ is integral, we get the following slightly weaker result.

Corollary 1. *Suppose that all arc lengths are multiples of Δ. Then, any flow over time that completes by time T corresponds to a static flow of equal value and cost in $\mathcal{N}^T/\Delta$, and any flow in $\mathcal{N}^T/\Delta$ corresponds to a flow over time of equal value that completes by time $T + \Delta$.*

5 An FPTAS for Multicommodity Flow over Time

Our FPTAS for flow over time will use a graph $\mathcal{N}^T/\Delta$ for an appropriately defined Δ. We show below that, even when all arc lengths are not multiples of Δ, for an appropriate choice of Δ that depends on ε we may round the lengths up to the nearest multiple of Δ, and suffer only a $1+\varepsilon$ factor increase in the makespan of our flow. Thus, our algorithm is simply to first round the arc lengths, construct the corresponding condensed time-expanded network $\mathcal{N}^T/\Delta$, solve the flow problem in this time-expanded graph, and then translate this solution into a flow over time. We show below that this natural algorithm yields an FPTAS for minimizing the makespan of flow over time problems.

In the last step of the sketched algorithm, we make use of the following straightforward observation which will also be employed at several points during the analysis of the algorithm.

Observation 1. *Any flow over time with time horizon T in a network with elongated arc lengths induces a flow over time satisfying the same demands within the same time horizon in the original network.*

5.1   Increasing the Transit Times

Our analysis starts with an optimal flow that completes by time T^*, and then shows how to modify this flow so that it completes by time $(1+\varepsilon)T^*$ in a network with elongated arc lengths. Throughout the proof we often use the following *freezing technique* for modifying a given flow over time in a fixed network $\mathcal{N}$: At some point in time θ, we '*freeze*' the flow in progress and '*unfreeze*' it later at time $\theta+\delta$, thereby increasing the completion time of the flow by at most δ. More formally, freezing a flow in progress during a time interval $[\theta, \theta+\delta)$ means that no new flow is sent onto any arc during this interval; instead, at time θ, every unit of flow which is currently traveling on some arc continues on its path at its regular pace until it reaches the next node and then rests there for δ time units before it continues its journey. The motivation for introducing such a freezing period is that it provides free capacity which can be used to send additional flow on an arc during the time interval $[\theta, \theta+\delta)$.

Let f^* be an optimal flow over time for network $\mathcal{N}$ with m arcs, n nodes, and with commodities $1 \leq i \leq k$ associated with source-sink node pairs (s_i, t_i) and flow demands d_i. Let T^* be the completion time of this flow. The following theorem contains the key result for the construction of our FPTAS.

Theorem 2. *Let $\varepsilon > 0$ and $\Delta \leq \frac{\varepsilon^2}{4k^2m^4}T^*$. Increasing the transit time of every arc by an additive factor of at most Δ increases the minimum time required to satisfy the multicommodity demands by at most εT^*.*

The rough idea of the proof is that, for each elongated arc e, its original transit time can be emulated by holding ready additional units of flow in a buffer at the head of the arc. Since the required amount of flow in this buffer depends on the total flow that is traveling on the arc at any moment of time, we first show that we can bound the maximal rate of flow into an arc, without too much increase in the makespan.

For an arbitrary flow over time f, consider the flow on arc e of commodity i. This flow travels from s_i to t_i on a set of paths $\mathcal{P}_{e,i}(f)$, all containing arc e. For

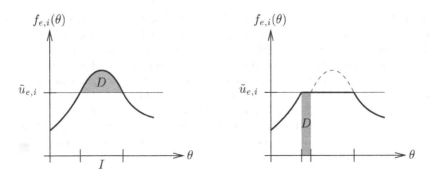

Fig. 3. Modification of the flow over time f: The flow of commodity i on arc e is truncated at $\tilde{u}_{e,i}$. In the modified solution, the extra D flow units (shaded area on the left hand side) are sent onto arc e while the flow in the remaining network is frozen (shaded area on the right hand side)

any path P, let $u(P)$ denote the minimum capacity of an arc $e' \in P$ and $|P|$ the number of arcs on P. Define $u_{e,i}(f) := \max_{P \in \mathcal{P}_{e,i}(f)} u(P)$.

Lemma 3. *For any $\varepsilon > 0$, there is a flow over time f in the network $\mathcal{N}$ that fulfills the following properties:*

(i) *It satisfies all demands by time $(1 + \varepsilon/2)T^*$.*
(ii) *For any arc e and any commodity i, the rate of flow onto arc e of commodity i is bounded by*

$$\tilde{u}_{e,i}(f) := \frac{2km^2}{\varepsilon} u_{e,i}(f)$$

at any time.

Proof. The optimal flow over time f^* obviously satisfies property (i). We start with the flow $f := f^*$ and carefully modify it until it also satisfies (ii). During this modification we only delay but never reroute flow such that $u_{e,i}(f)$ and $\tilde{u}_{e,i}(f)$ stay fixed and are therefore denoted by $u_{e,i}$ and $\tilde{u}_{e,i}$, respectively.

An illustration of the following modification is given in Fig. 3. In any time interval[3] I when f sends more than $\tilde{u}_{e,i}$ units of flow of commodity i onto arc e, we truncate the flow at $\tilde{u}_{e,i}$. In order to compensate for the resulting loss of

$$D := \int_I (f_{e,i}(\theta) - \tilde{u}_{e,i}) \, d\theta$$

flow units, we freeze the flow in $\mathcal{N}$ for $D/\tilde{u}_{e,i}$ time units at the beginning of time interval I and push D units of flow onto arc e (see the shaded area on the

---

[3] It follows from the discussion in Sect. 4 (see Lemma 2.) that there are only finitely many such intervals since we can assume without loss of generality that $f^*_{e,i}$ is a step function.

right hand side of Fig. 3). Afterwards, the flow is unfrozen again. Due to this freeze-unfreeze process, the D flow units arrive early (compared to the remaining flow) at the head of arc e and are stored there until it is time to send them onto another arc.

We repeat this freeze-unfreeze step for every commodity-arc pair. By construction, the resulting flow f fulfills property (ii) and satisfies all demands. It thus remains to show that f completes before time $(1 + \varepsilon/2)T^*$. Notice that the increase in the completion time of f compared to f^* is exactly the additional time added to the total flow schedule when the flow gets frozen due to some arc e and some commodity i. In the following we show that the total freezing time caused by arc e and commodity i is at most $\frac{\varepsilon}{2mk}T^*$, for any e and i.

Whenever the flow is frozen due to arc e and commodity i, the flow rate of commodity i onto arc e is exactly $\tilde{u}_{e,i}$; see Fig. 3. It therefore suffices to show that a total of at most $\tilde{u}_{e,i}\frac{\varepsilon}{2mk}T^* = mu_{e,i}T^*$ flow of commodity i is sent through arc e in the optimal solution f^* and thus also in f. Consider all flow of commodity i on paths in $\mathcal{P}_{e,i}$. If we choose for each path $P \in \mathcal{P}_{e,i}$ an arc with capacity $u(P) \leq u_{e,i}$, the total flow through the chosen arcs bounds the total flow of commodity i through arc e. Since we can choose at most m arcs and the makespan of f^* is T^*, this gives the desired bound of $mu_{e,i}T^*$ on the total flow of commodity i sent through arc e. This completes the proof.   □

With the aid of Lemma 3 we can now prove Theorem 2.

Proof (of Theorem 2). Let f be a flow over time in network $\mathcal{N}$ (with original transit times). We start by modifying f as described in the proof of Lemma 3 so that it fulfills properties (i) and (ii). Let $\mathcal{N}_\Delta$ be $\mathcal{N}$ modified so that the transit time of every arc is rounded up to an integral multiple of Δ. We show how the flow over time f can be modified to satisfy all demands in $\mathcal{N}_\Delta$ by time $(1+\varepsilon)T^*$. Although some flow units will be rerouted during this modification, the set of paths $\mathcal{P}_{e,i} := \mathcal{P}_{e,i}(f)$ remains unchanged, for any arc e and any commodity i. In particular, $u_{e,i}(f)$ and $\tilde{u}_{e,i}(f)$ stay fixed and are therefore denoted by $u_{e,i}$ and $\tilde{u}_{e,i}$, respectively.

The modification is done in m steps such that in each step the transit time of only one arc e is increased at the cost of increasing the makespan by at most $\frac{\varepsilon}{2m}T^*$. Thus, the total increase of the makespan of f after m steps is at most $\frac{\varepsilon}{2}T^*$. Together with the prior modifications discussed in Lemma 3, this implies that the resulting flow completes by time $(1 + \varepsilon)T^*$.

Each step has two phases. In the first phase, the transit time of arc e remains unchanged but the demand satisfied by f is increased to $d_i + \Delta\tilde{u}_{e,i}$, for all commodities i. Then, in the second phase, the extra $\Delta\tilde{u}_{e,i}$ units of flow from the first phase are used to emulate the original transit time τ_e on the elongated arc e of length $\tau_e + \Delta$. (It follows from Observation 1 that it suffices to consider the extreme case of increasing the transit time by exactly Δ).

Phase 1: For each commodity i, let $P_{e,i} \in \mathcal{P}_{e,i}$ be an s_i-t_i-path with capacity $u_{e,i}$. The additional $\Delta\tilde{u}_{e,i}$ units of flow are routed through path $P_{e,i}$: At time 0, we freeze the current flow throughout the network for $\Delta\frac{2km^2}{\varepsilon}$ time units and pump an extra $\Delta \frac{2km^2}{\varepsilon} u_{e,i} = \Delta \tilde{u}_{e,i}$ units of flow of commodity i into the first arc on this path. When this flow arrives at the next arc on path $P_{e,i}$, we again freeze the current flow for $\Delta \frac{2km^2}{\varepsilon}$ time units and send this flow onto the next arc, and so on.

Notice that the extra flow does not violate the capacity constraints since the capacity of any arc on path $P_{e,i}$ is at least $u_{e,i}$. Moreover, the extra units of flow arrive at their destination t_i before time

$$\tau(P_{e,i}) + \Delta \frac{2km^2}{\varepsilon} \leq T^* + \Delta \frac{2km^2}{\varepsilon} .$$

We add this flow to f. The makespan of f is increased by

$$|P_{e,i}| \Delta \frac{2km^2}{\varepsilon} \leq \Delta \frac{2km^3}{\varepsilon} \leq \frac{\varepsilon}{2km} T^* .$$

Repeating this for all commodities i increases the makespan by at most $\frac{\varepsilon}{2m} T^*$.

Phase 2: Now, we increase the length of arc e to $\tau_e + \Delta$ and modify the current flow over time f as follows. The point on arc e that is distance τ_e from tail(e) is called apex(e). For any commodity i, the first $\Delta \tilde{u}_{e,i}$ units of flow of commodity i traveling across arc e are stored in a special buffer as soon as they arrive at head(e). The flow stored in the buffer at head(e) can then be used to emulate the original transit time on arc e for any further unit of flow of commodity i.

An illustration of the following argument is given in Fig. 4. Any further unit of flow of commodity i arriving at apex(e) is instantly 'replaced' by a unit of flow from the buffer. Then, after Δ time units, when the flow has traveled from apex(e) to head(e), the buffer is refilled again. In other words, the flow in the buffer is treated as newly arriving flow at head(e), and the flow reaching the head enters the buffer. Thus, due to property (ii) from Lemma 3, the choice of buffer size $\Delta \tilde{u}_{e,i}$ ensures that the buffer is never empty.

Notice that this modification of f does not increase its makespan. The result of the modification is that exactly d_i units of flow of commodity i arrive at destination t_i and $\Delta \tilde{u}_{e,i}$ units of flow remain in the buffer at the head of arc e in the end. Since the latter effect is undesired, we revoke it as follows: The $\Delta \tilde{u}_{e,i}$ units of flow in the buffer are exactly those that arrive last at head(e). We can simply delete them from the entire solution, that is, we never send them from s_i into the network.   □

5.2   The FPTAS

As a consequence of Theorem 2 we can now give an FPTAS for the problem of computing a quickest multicommodity flow over time:

Input: A directed network $\mathcal{N}$ with non-negative integral transit times and capacities on the arcs, and commodities $1 \leq i \leq k$ associated with source-sink node pairs (s_i, t_i) and flow demands d_i; a number $\varepsilon > 0$.

Output: A multicommodity flow over time satisfying all demands.

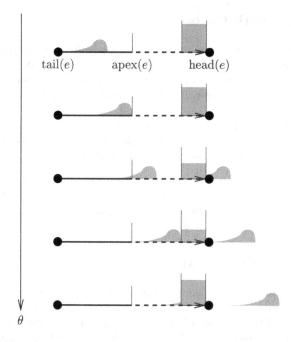

Fig. 4. In Phase 2, the flow in the buffer at the head of the elongated arc e is used to emulate its original length τ_e

Step 1. Compute a good lower bound L on the optimal makespan T^* such that L is at least a constant fraction of T^*.

Step 2. Set $\Delta := \frac{\varepsilon^2}{4m^4k^2}L$ and round the transit times up to the nearest multiple of Δ.

Step 3. Find the minimal $T = \ell\Delta$, $\ell \in \mathbb{N}$, such that there exists a feasible flow satisfying all demands in the condensed time-expanded network $\mathcal{N}^T/\Delta$.

Step 4. Output the flow over time that corresponds to the flow in $\mathcal{N}^T/\Delta$ from Step 3 (see Lemma 2 and Observation 1).

It follows from Theorem 2 and Corollary 1 that the above algorithm computes a solution with makespan $T \le (1 + \varepsilon)T^* + \Delta \le (1 + 2\varepsilon)T^*$. Moreover, it can be implemented to run in time polynomial in n, m, and $1/\varepsilon$: Step 1 can be done in polynomial time using the constant factor approximation algorithm from Sect. 3 (see Theorem 1). Step 2 is trivial and Step 3 can be done in polynomial time by binary search since $\ell \in O(k^2 m^4/\varepsilon^2)$ by Theorem 2 and choice of Δ. Here, Δ is chosen so that the size of the condensed time-expanded network $\mathcal{N}^T/\Delta$ is polynomial. Thus, Step 4 can also be done in polynomial time.

Theorem 3. *There is an FPTAS for the problem of computing a quickest multicommodity flow over time.*

6   Problems with Costs

Unfortunately, Theorem 3 cannot directly be generalized to the quickest multi-commodity flow problem with costs. The reason is that in our analysis we have to reroute flow in order to show that there exists a reasonably good solution in the condensed time-expanded network $\mathcal{N}^T/\Delta$ (see proof of Theorem 2). Since the thick paths $P_{e,i}$, which are used to route additional units of flow, might be relatively expensive, the modified flow in the network with increased transit times can possibly violate the given budget B. However, we can prove the following bicriteria result. We omit the proof in this extended abstract.

Theorem 4. *Given an instance of the quickest multicommodity flow problem with costs and $\varepsilon > 0$, one can compute in time polynomial in the input size and $1/\varepsilon$ a flow over time of cost at most $(1+\varepsilon)B$ whose makespan is within a factor of $1+\varepsilon$ of the optimal makespan.*

7   Earliest Arrival Flows

For a single commodity problem, an *earliest arrival flow* is a flow that simultaneously maximizes the amount of flow arriving at the sink before time θ, for all $\theta = 0, \ldots, T$. The existence of such a flow was first observed by Gale [7] for the case of a single source. Both Wilkinson [19] and Minieka [16] give equivalent pseudo-polynomial time algorithms for this case, and Hoppe and Tardos [12] describe an FPTAS for the problem. For multiple sources, an earliest arrival flow over time can be computed in the discrete time model by using lexicographically maximal flows in the time-expanded network [16]. However, due to the exponential size of the time-expanded network, this does not lead to an efficient algorithm for the problem.

Unfortunately, lexicographically maximal flows in *condensed* time-expanded networks do not necessarily yield approximate earliest arrival flows. One problem is that, in our analysis, the first units of flow on an arc are always used to fill the buffer at the head of the arc and are therefore 'lost'. As a consequence, the first units of flow that actually arrive at the sink might be pretty late.

Another problem arises due to the discretization of time itself. Although we can interpret a static flow in a time-expanded network as a continuous flow over time, in doing so, we only get solutions where the rate of flow arriving at the sink is constant (i.e., averaged) within each discrete time interval. While this effect is negligible for late intervals in time, it might well cause problems within the first time intervals. In the full version of this paper, we introduce a *geometrically-condensed time-expanded network* to surmount this difficulty and obtain the following result.

Theorem 5. *For any $\varepsilon > 0$, a $(1+\varepsilon)$-approximate earliest arrival flow in a network with multiple sources and a single sink can be computed in time polynomial in the input size and $1/\varepsilon$ by computing a lexicographically maximal flow in an appropriate geometrically condensed time-expanded network.*

References

1. J. E. Aronson. A survey of dynamic network flows. *Annals of Operations Research*, 20:1–66, 1989.
2. R. E. Burkard, K. Dlaska, and B. Klinz. The quickest flow problem. *ZOR – Methods and Models of Operations Research*, 37:31–58, 1993.
3. L. K. Fleischer. Approximating fractional multicommodity flows independent of the number of commodities. *SIAM Journal on Discrete Mathematics*, 13:505–520, 2000.
4. L. K. Fleischer and É Tardos. Efficient continuous-time dynamic network flow algorithms. *Operations Research Letters*, 23:71–80, 1998.
5. L. R. Ford and D. R. Fulkerson. Constructing maximal dynamic flows from static flows. *Operations Research*, 6:419–433, 1958.
6. L. R. Ford and D. R. Fulkerson. *Flows in Networks*. Princeton University Press, Princeton, NJ, 1962.
7. D. Gale. Transient flows in networks. *Michigan Mathematical Journal*, 6:59–63, 1959.
8. N. Garg and J. Könemann. Faster and simpler algorithms for multicommodity flow and other fractional packing problems. In *Proceedings of the 39th Annual IEEE Symposium on Foundations of Computer Science*, pages 300–309, Palo Alto, CA, 1998.
9. M. Grötschel, L. Lovász, and A. Schrijver. *Geometric Algorithms and Combinatorial Optimization*, volume 2 of *Algorithms and Combinatorics*. Springer, Berlin, 1988.
10. G. Handler and I. Zang. A dual algorithm for the constrained shortest path problem. *Networks*, 10:293–310, 1980.
11. R. Hassin. Approximation schemes for the restricted shortest path problem. *Mathematics of Operations Research*, 17:36–42, 1992.
12. B. Hoppe and É Tardos. Polynomial time algorithms for some evacuation problems. In *Proceedings of the 5th Annual ACM–SIAM Symposium on Discrete Algorithms*, pages 433–441, Arlington, VA, 1994.
13. B. Hoppe and É Tardos. The quickest transshipment problem. *Mathematics of Operations Research*, 25:36–62, 2000.
14. B. Klinz and G. J. Woeginger. Minimum cost dynamic flows: The series-parallel case. In E. Balas and J. Clausen, editors, *Integer Programming and Combinatorial Optimization*, volume 920 of *Lecture Notes in Computer Science*, pages 329–343. Springer, Berlin, 1995.
15. D. H. Lorenz and D. Raz. A simple efficient approximation scheme for the restricted shortest path problem. *Operations Research Letters*, 28:213–219, 2001.
16. E. Minieka. Maximal, lexicographic, and dynamic network flows. *Operations Research*, 21:517–527, 1973.
17. C. A. Phillips. The network inhibition problem. In *Proceedings of the 25th Annual ACM Symposium on the Theory of Computing*, pages 776–785, San Diego, CA, 1993.
18. W. B. Powell, P. Jaillet, and A. Odoni. Stochastic and dynamic networks and routing. In M. O. Ball, T. L. Magnanti, C. L. Monma, and G. L. Nemhauser, editors, *Network Routing*, volume 8 of *Handbooks in Operations Research and Management Science*, chapter 3, pages 141–295. North–Holland, Amsterdam, The Netherlands, 1995.
19. W. L. Wilkinson. An algorithm for universal maximal dynamic flows in a network. *Operations Research*, 19:1602–1612, 1971.

A New Min-Cut Max-Flow Ratio for Multicommodity Flows

Oktay Günlük

Math. Sci. Dept., IBM Research, Yorktown Heights, NY 10598
oktay@watson.ibm.com

Abstract. We present an improved bound on the min-cut max-flow ratio for multicommodity flow problems with specified demands. To obtain the numerator of this ratio, capacity of a cut is scaled by the demand that has to cross the cut. In the denominator, the maximum concurrent flow value is used. Our new bound is proportional to $\log(k^*)$ where k^* is the cardinality of the minimal vertex cover of the demand graph.

1 Introduction

In this paper we study the multicommodity flow problem on undirected graphs and present a new bound on the associated min-cut max-flow ratio. Starting with the pioneering work of Leighton and Rao [11] there has been ongoing research in the area of "approximate min-cut max-flow theorems" for multicommodity flows. Currently, the best known bound for this ratio is proportional to $\log(k)$ where k is the number of origin-destination pairs with positive demand. Our new bound is proportional to $\log(k^*)$ where k^* is the cardinality of the minimal vertex cover of the demand graph.

Throughout the paper, we assume that the input graph $G = (V, E)$ is connected and has positive capacity c_e on all edges $e \in E$. A (multicommodity) flow is considered feasible if for all edges $e \in E$ total load on e (regardless of the orientation) does not exceed c_e.

1.1 Single Commodity Flows

Given two special nodes s, $v \in V$, and a specified flow requirement $t \in \mathbb{R}_+$, the well-known min-cut max-flow theorem [5] implies that t units of flow can be routed from s to v if and only if *the minimum cut-capacity to cut-load ratio* ρ^* is at least 1, where

$$\rho^* = \min_{S \subset V : s \in S, v \notin S} \left\{ \frac{\sum_{e \in \delta(S)} c_e}{t} \right\}.$$

It is possible to generalize this result to a collection of flow requirements that share a common source node as follows: Given a source node s and a collection of sink nodes $v_q \in V \setminus \{s\}$ for $q \in Q$, it is possible to simultaneously route $t_q \in \mathbb{R}_+$ units of flow from s to v_q for all $q \in Q$ if and only if $\rho^* \geq 1$ where

$$\rho^* = \min_{S \subset V : s \in S} \left\{ \frac{\sum_{e \in \delta(S)} c_e}{\sum_{q \in Q : v_q \notin S} t_q} \right\}.$$

W.J. Cook and A.S. Schulz (Eds.): IPCO 2002, LNCS 2337, pp. 54–66, 2002.
© Springer-Verlag Berlin Heidelberg 2002

This observation is the main motivation behind our study as it shows that a min-cut max-flow relationship holds tight for network flow problems (with specified flow requirements) as long as the sink nodes share a common source node. Note that, since G is undirected, the min-cut max-flow relationship also holds tight when there is a single sink node and multiple source nodes.

1.2 Multicommodity Flows

A natural extension of this observation is to consider multicommodity flows, where a collection of pairs of vertices $\{s_q, v_q\}$, $q \in Q$ together with a flow requirement t_q for each pair is provided. Let the minimum cut-capacity to cut-load ratio for multicommodity flows be similarly defined as

$$\rho^* = \min_{S \subset V} \left\{ \frac{\sum_{e \in \delta(S)} c_e}{\sum_{q \in Q : |S \cap \{s_q, v_q\}| = 1} t_q} \right\}.$$

In the remainder of the paper we refer to ρ^* as the minimum cut ratio. Clearly, it is possible to simultaneously route t_q units of flow from s_q to v_q for all $q \in Q$, only if $\rho^* \geq 1$. But the converse is not true ([14], [16]) and a simple counter example is the complete bipartite graph $K_{2,3}$ with unit capacity edges and unit flow requirements between every pair of nodes that are not connected by an edge.

For multicommodity flows, *metric inequalities* provide the necessary and sufficient conditions for feasibility, (see [7], and [17]). More precisely, it is possible to simultaneously route t_q units of flow from s_q to v_q for all $q \in Q$, if and only if the edge capacities satisfy

$$\sum_{e \in E} w_e c_e \geq \sum_{q \in Q} dist(s_q, v_q) t_q$$

for all $w \in \mathbb{R}_+^{|E|}$, where $dist(u, v)$ denotes the shortest path distance from u to v using w as edge weights. The set of all important edge weights form a well-defined polyhedral cone. Notice that the above example with $K_{2,3}$ does not satisfy the metric inequality "generated" by $w_e = 1$ for all $e \in E$. It is easy to show that the condition $\rho^* \geq 1$ is implied by metric inequalities.

The *maximum concurrent flow* problem is the optimization version of the multicommodity flow feasibility problem, (see [20] and [14]). The objective now is to find the maximum value of κ such that κt_q units of flow can be simultaneously routed from s_q to v_q for all $q \in Q$. Note that κ can be greater than one.

For a given instance of the multicommodity flow problem, let κ^* denote the value of the maximum concurrent flow. Clearly κ^* can not exceed ρ^*. Our main result is the following reverse relationship

$$\kappa^* \geq \frac{1}{c \lceil \log k^* \rceil} \rho^* \tag{1}$$

where c is a constant and k^* is the cardinality of the minimal vertex cover of the demand graph. In other words, k^* is the size of the smallest set $K^* \subseteq V$ such that K^* contains at least one of s_q or v_q for all $q \in Q$. Throughout the paper, we assume that $k^* > 1$.

In literature, these bounds are often called "approximate min-cut max-flow theorems", as they relate the maximum (concurrent) flow of a multicommodity flow problem to the (scaled) capacity of the minimum cut. As discussed above, this bound is tight, i.e., $\rho^* = \kappa^*$, when $k^* = 1$.

1.3   Related Work

Starting with the pioneering work of Leighton and Rao [11] there has been ongoing interest in the area of approximate min-cut max-flow theorems. The first such result in [11] shows that the denominator in (1) is at most $O(\log |V|)$ when $t_q = 1$ for all $q \in Q$. Later Klein, Agrawal, Ravi and Rao [10] extend this result to general t_q and show that the bound is $O(\log C \log D)$ where D is the sum of (integral) demands (i.e. $D = \sum_{q \in Q} t_q$) and C is the sum of (integral) capacities (i.e. $C = \sum_{e \in E} c_e$). Tragoudas [21] has later improved this bound to $O(\log |V| \log D)$ and Garg, Vazirani and Yannakakis [6] has further improved it to $O(\log k \log D)$, where $k = |Q|$.

Plotkin and Tardos [19] present a bound of $O(\log^2 k)$ which is the first bound that does not depend on the input data. Finally Linial, London and Rabinovich [12] and Aumann and Rabani [1] independently improve this bound to $O(\log k)$.

Our result improves this best known bound to $O(\log k^*)$. To emphasize the difference between $O(\log k)$ and $O(\log k^*)$, we note that for an instance of the multicommodity flow problem with a single source node and $|V| - 1$ sink nodes, $k = |V| - 1$ whereas $k^* = 1$. In general, $k \geq k^* \geq k/|V|$.

2   Formulation

When formulating a multicommodity problem as a linear program, what is meant by a "commodity" can affect the size of the formulation significantly. This has been exploited by researchers interested in solving these linear programs (see, for example, [2] and [13]).

Given flow requirements t_q for pairs of vertices $\{s_q, v_q\}$, $q \in Q$, let T denote the corresponding traffic matrix, where, $T_{[k,j]} = \sum_{q \in Q \,:\, s_q = k,\, v_q = j} t_q$ for all $k, j \in V$. Let $K = \{k \in V \,:\, \sum_{j \in V} T_{[k,j]} > 0\}$ denote the set of nodes with positive supply. The original linear programming formulation of the maximum concurrent flow problem presented in Shahrokhi and Matula [20] defines a commodity for each $k \in K$ to obtain a formulation with $|K|$ commodities. Notice that this is significantly more "compact" than the natural formulation which would have $|Q|$ commodities.

To find the smallest set of commodities that would model the problem instance correctly, we do the following: Let $G^T = (V, E^T)$ denote the (undirected) demand graph where $E^T = \{\{i, j\} \in V \times V \,:\, T_{[i,j]} + T_{[j,i]} > 0\}$ and let $K^* \subseteq V$ be a minimal vertex cover of G^T. In other words, K^* is a smallest cardinality set that satisfies $\{i, j\} \cap K^* \neq \emptyset$ for all $\{i, j\} \in E^T$. We then modify the entries of the flow matrix T so that $T_{[k,j]} > 0$ only if $k \in K^*$. Note that this can be done without loss of generality since the capacity constraints in the formulation do not depend on the orientation of the flow.

We next present the resulting multicommodity formulation with $|K^*|$ commodities.

Maximize κ

Subject to

$$\sum_{v:\{j,v\}\in E} f_{jv}^k - \sum_{v:\{v,j\}\in E} f_{vj}^k + \kappa\, T_{[k,j]} \leq \quad 0 \quad \text{for all } j \in V,\ k \in K^*, j \neq k$$

$$\sum_{k\in K^*} \left(f_{jv}^k + f_{vj}^k \right) \qquad\qquad \leq c_{\{j,v\}} \text{ for all } \{j,v\} \in E$$

$$\kappa \text{ free, } f_{jv}^k \geq \quad 0 \quad \text{for all } k \in K^*, \{j,v\} \in E\ ,$$

where variable f_{vj}^k denotes the flow of commodity k from node i to node j, and variable κ denotes the value of the concurrent flow. Note that in this formulation (i) there are no flow balance equalities for the source nodes $k \in K^*$, (ii) the flow balance equalities for the remaining nodes are in inequality form, and (iii) there is no non-negativity requirement for κ. Also note that using an aggregate flow vector f, it is easy to find disaggregated flows for node pairs (k, j) with $T_{[k,j]} > 0$. The disaggregation, however, is not necessarily unique.

As a side remark, we note that it is possible to further aggregate this formulation to verify (or, falsify) that the maximum concurrent flow value κ^* is at least β for a fixed $\beta > 0$. This can be done by introducing a new demand node z^k for every $k \in K^*$ together with new edges (j, z^k) with capacity $\beta T_{[k,j]}$ for all $j \in V$. Let $\hat{\kappa}$ denote the value of the maximum concurrent flow for the transformed problem and note that $\beta \geq \hat{\kappa}$. If $\beta = \hat{\kappa}$, then $\kappa^* \geq \beta$ for the original problem. If, on the other hand, $\beta > \hat{\kappa}$ then $\beta > \kappa^*$ since κ^* can not exceed $\hat{\kappa}$. Note that κ^* can be strictly smaller than $\hat{\kappa}$.

The dual of the above linear program is:

Minimize $\displaystyle\sum_{\{j,v\}\in E} c_{\{j,v\}}\, w_{\{j,v\}}$

Subject to

$$\sum_{k\in K^*} \sum_{j\in V} T_{[k,j]}\, y_j^k \ = 1$$

$$\left.\begin{array}{l} y_v^k - y_j^k + w_{\{j,v\}} \ \geq 0 \\[2mm] y_j^k - y_v^k + w_{\{j,v\}} \ \geq 0 \end{array}\right\} \text{ for all } k \in K^*, \text{ and } \{j,v\} \in E$$

$$y_k^k \ = 0 \text{ for all } k \in K^*$$

$$y_j^k \ \geq 0 \text{ for all } j \in V,\ k \in K^* \text{ with } j \neq k$$

$$w_{\{j,v\}} \ \geq 0 \text{ for all } \{j,v\} \in E\ ,$$

where dual variables y_k^k, for $k \in K^*$ are included in the formulation even though there are no corresponding primal constraints. These variables are set to zero in a separate constraint. The main reason behind this is to obtain a dual formulation that would have an optimal solution that satisfies the following properties.

Proposition 1. *Let $[\bar{y}, \bar{w}]$ be an optimal solution to the dual problem. Let $\hat{y} \in \mathbb{R}^{|V| \times |V|}$ be the vector of shortest path distances using $\bar{w}$ as edge weights with y_j^k denoting distance from node k to j.*

(i) *For any $k \in K^*$ and $j \in V$, with $T_{[k,j]} > 0$, $\bar{y}_j^k$ is equal to $\hat{y}_j^k$.*

(ii) *For any $\{j, v\} \in E$, $w_{\{j,v\}}^-$ is equal to $\hat{y}_v^j$.*

Proof. (i) For any $k \to j$ path $P = \{\{k, v_1\}, \{v_1, v_2\}, \ldots, \{v_{|P|-1}, j\}\}$ we have $\sum_{e \in P} w_e \geq \bar{y}_j^k$, implying $\hat{y}_j^k \geq \bar{y}_j^k$. If $\hat{y}_j^k > \bar{y}_j^k$ for some $k \in K^*$, $j \in V$ with $T_{[k,j]} > 0$, we can write $\sum_{k \in K^*} \sum_{j \in V} T_{[k,j]} \, \hat{y}_j^k = \sigma > \sum_{k \in K^*} \sum_{j \in V} T_{[k,j]} \, \bar{y}_j^k = 1$. Which implies that a new solution, with an improved objective function value, can be constructed by scaling $[\hat{y}, \bar{w}]$ by $1/\sigma$, a contradiction.

(ii) Clearly, $w_{\{j,v\}}^- \geq \hat{y}_v^j$ for all $\{j, v\} \in E$. If the inequality is strict for some $e \in E$, then replacing $\bar{y}$ with $\hat{y}$ and all $w_{\{j,v\}}^-$ by $\hat{y}_v^j$ in the solution improves the objective function value, a contradiction (remember that $c_{\{j,v\}} > 0$ for all $\{j, v\} \in E$). Note that this replacement maintains feasibility since $\hat{y}$ denotes the shortest path distances and therefore $|\hat{y}_v^k - \hat{y}_j^k| \leq \hat{y}_v^j$ for all $k \in K^*$ and $j, v \in V$. □

As a side remark, we note that it is therefore possible to substitute some of the dual variables and consequently it is possible to combine some of the constraints in the primal formulation.

We next express the maximum concurrent flow value using shortest path distances with respect to $\bar{w}$.

Corollary 1. *Let κ^* be the optimal value of the primal (or, the dual) problem. Then,*

$$\kappa^* = \frac{\sum\limits_{\{j,v\} \in E} c_{\{j,v\}} \, dist(j, v)}{\sum\limits_{k \in K^*} \sum\limits_{v \in V} T_{[k,v]} \, dist(k, v)} \tag{2}$$

where $dist(j, v)$ denotes the shortest path distance from node j to node v with respect to some edge weight vector.

3   The Min-Cut Max-Flow Ratio

We next argue that there exists a mapping $\Phi : V \to \mathbb{R}_+^p$ for some p, such that $||\Phi(u) - \Phi(v)||_1$ is not very different from $dist(u, v)$ for node pairs $\{u, v\}$ that are of interest. We then substitute $||\Phi(u) - \Phi(v)||_1$ in place of $dist(u, v)$ in (2) and relate the new right hand side of (2) to the minimum cut ratio. More precisely, we show that for some $\alpha = O(\log|K^*|)$

$$\kappa^* \geq \frac{1}{\alpha} \times \frac{\sum_{\{u,v\}\in E} c_{\{u,v\}} \, ||\Phi(u) - \Phi(v)||_1}{\sum_{k\in K^*} \sum_{v\in V} T_{[k,v]} \, ||\Phi(k) - \Phi(v)||_1} \geq \frac{1}{\alpha} \rho^*.$$

3.1 Mapping the Nodes of the Graph with Small Distortion

Our approach follows general structure of the proof of a related result by Bourgain [3] that shows that any $n-$point metric space can be embedded into l_1 with logarithmic distortion. We state this result more precisely in Sect. 4.

Given an undirected graph $G = (V, E)$ with edge weights $w_e \geq 0$, for $e \in E$, let $d(u,v)$ denote the shortest path distance from $u \in V$ to $v \in V$ using w as edge weights. For $v \in V$ and $S \subseteq V$ let $d(v,S) = \min_{k\in S}\{d(v,k)\}$ and define $d(v,\emptyset) = \sigma = \sum_{u\in V}\sum_{j\in V} d(u,j)$. Furthermore, let $K \subseteq V$ with $|K| > 1$ be also be given.

For any $j,t \geq 1$, let Q_j^t be random subset of K such that members of Q_j^t are chosen independently and with equal probability $P(k \in Q_j^t) = 1/2^t$ for all $k \in K$. Note that for all $j \geq 1$, Q_j^t has an identical probability distribution and $E[|Q_j^t|] = |K|/2^t$. For $m = \lceil\log(|K|)\rceil$ and $L = 300 \cdot \lceil\log(|V|)\rceil$, define the following (random) mapping $\Phi^R : V \to \mathbb{R}_+^{mL}$

$$\Phi^R(v) = \frac{1}{L \cdot m} \left[d(v, Q_1^1), \ldots, d(v, Q_j^t), \ldots, d(v, Q_L^m) \right] .$$

Note that, by triangle inequality, $|d(u,S) - d(v,S)| \leq d(u,v)$ for any $S \subseteq V$, and therefore:

$$||\Phi^R(u) - \Phi^R(v)||_1 = \frac{1}{L \cdot m} \sum_{i=1}^{m}\sum_{j=1}^{L} \left| d(u, Q_j^i) - d(v, Q_j^i) \right|$$

$$\leq \frac{1}{L \cdot m} \cdot L \cdot m \cdot d(u,v) = d(u,v) \qquad (3)$$

for all $u, v \in V$. We next bound $||\Phi^R(u) - \Phi^R(v)||_1$ from below.

Lemma 1. *For all $u \in K$ and $v \in V$ and for some $\alpha = O(\log|K|)$ the following property*

$$||\Phi^R(u) - \Phi^R(v)||_1 \geq \frac{1}{\alpha} \cdot d(u,v)$$

holds simultaneously with positive probability.

Proof. See Appendix A.   □

An immediate corollary of this result is the existence of a (deterministic) mapping with at most $O(\log|K|)$ distortion.

Corollary 2. *There exists a collection of sets* $\bar{Q}_j^i \subseteq K$ *for* $m \geq i \geq 1$ *and* $L \geq j \geq 1$ *such that the corresponding mapping* $\Phi^D : V \to \mathbb{R}_+^{mL}$ *satisfies the following two properties:*

(i) $d(u,v) \geq \|\Phi^D(u) - \Phi^D(v)\|_1$ *for all* $u, v \in V$

(ii) $d(u,v) \leq \alpha \|\Phi^D(u) - \Phi^D(v)\|_1$ *for all* $u \in K$ *and* $v \in V$,

where $\alpha = c \log |K|$ *for some constant* c.

3.2   Bounding the Maximum Concurrent Flow Value

Combining Corollary 1 and Corollary 2, we now bound the maximum concurrent flow value as follows:

$$
\begin{aligned}
\kappa^* &= \frac{\sum_{\{u,v\}\in E} c_{\{u,v\}} \, dist(u,v)}{\sum_{k\in K^*} \sum_{v\in V} T_{[k,v]} \, dist(k,v)} \\[2mm]
&\geq \frac{\sum_{\{u,v\}\in E} c_{\{u,v\}} \, \|\Phi^D(u) - \Phi^D(v)\|_1}{\sum_{k\in K^*} \sum_{v\in V} T_{[k,v]} \, \alpha \, \|\Phi^D(k) - \Phi^D(v)\|_1} \\[2mm]
&= \frac{1}{\alpha} \times \frac{\sum_{i=1}^m \sum_{j=1}^L \left(\sum_{\{u,v\}\in E} c_{\{u,v\}} \, |d(u,\bar{Q}_j^i) - d(v,\bar{Q}_j^i)| \right)}{\sum_{i=1}^m \sum_{j=1}^L \left(\sum_{k\in K^*} \sum_{v\in V} T_{[k,v]} \, |d(k,\bar{Q}_j^i) - d(v,\bar{Q}_j^i)| \right)} \\[2mm]
&\geq \frac{1}{\alpha} \times \frac{\sum_{\{u,v\}\in E} c_{\{u,v\}} \, |d(u,Q^*) - d(v,Q^*)|}{\sum_{k\in K^*} \sum_{v\in V} T_{[k,v]} \, |d(k,Q^*) - d(v,Q^*)|}
\end{aligned}
\tag{4}
$$

for $Q^* = \bar{Q}_{j_*}^{i*}$ for some $m \geq i^* \geq 1$ and $L \geq j^* \geq 1$. Note that, we have essentially bounded maximum concurrent flow value (from below) by a collection of cut ratios. We next bound it by the minimum cut ratio.

First, we assign indices $\{1, 2, \ldots, |V|\}$ to nodes in V so that $d(v_p, Q^*) \geq d(v_{p-1}, Q^*)$ for all $|V| \geq p \geq 2$, and let $x_p = d(v_p, Q^*)$. Next, we define $|V|$ nested sets $S_p = \{v_j \in V : j \leq p\}$ and the associated cuts $C_p = \{\{u,v\} \in E : |\{u,v\} \cap S_p| = 1\}$ and $T_p = \{(k,v) \in K^* \times V : |\{k,v\} \cap S_p| = 1\}$. We can now rewrite the summations in (4) as follows:

$$
\frac{1}{\alpha} \times \frac{\sum_{\{v_i,v_j\}\in E} c_{\{v_i,v_j\}} \, |x_i - x_j|}{\sum_{v_i\in K^*} \sum_{v_j\in V} T_{[v_i,v_j]} \, |x_i - x_j|} = \frac{1}{\alpha} \times \frac{\sum_{p=1}^{|V|-1} (x_{p+1} - x_p) \sum_{\{u,v\}\in C_p} c_{\{u,v\}}}{\sum_{p=1}^{|V|-1} (x_{p+1} - x_p) \sum_{(k,v)\in T_p} T_{[k,v]}}
$$

$$
\geq \frac{1}{\alpha} \times \frac{\sum_{\{u,v\}\in C_{p*}} c_{\{u,v\}}}{\sum_{(k,v)\in T_{p*}} T_{[k,v]}} \geq \frac{1}{\alpha} \rho^*
$$

for some $p^* \in \{1, \ldots, |V| - 1\}$.

$$
\frac{1}{\alpha} \times \frac{\sum_{\{v_i,v_j\}\in E} c_{\{v_i,v_j\}} \, |x_i - x_j|}{\sum_{v_i\in K^*} \sum_{v_j\in V} T_{[v_i,v_j]} \, |x_i - x_j|} = \frac{1}{\alpha} \times \frac{\sum_{p=2}^{|V|} (x_p - x_{p-1}) \sum_{\{u,v\}\in C_p} c_{\{u,v\}}}{\sum_{p=2}^{|V|} (x_p - x_{p-1}) \sum_{(k,v)\in T_p} T_{[k,v]}}
$$

$$
\geq \frac{1}{\alpha} \times \frac{\sum_{\{u,v\}\in C_{p*}} c_{\{u,v\}}}{\sum_{(k,v)\in T_{p*}} T_{[k,v]}} \geq \frac{1}{\alpha} \rho^*
$$

for some $p^* \in \{1, \ldots, |V|\}$. We have therefore shown that:

Theorem 1. *Given a multicommodity problem, let κ^* denote the maximum concurrent flow value, ρ^* denote the minimum cut ratio and k^* denote the cardinality of the minimal vertex cover of the associated demand graph. If $k^* > 1$, then*

$$c \lceil \log k^* \rceil \geq \frac{\rho^*}{\kappa^*} \geq 1$$

for some constant c.

3.3   A Tight Example

We next show that there are problem instances for which the above bound on the min-cut max-flow ratio is tight, up to a constant. This result is a relatively straight forward extension of a similar result by Leighton and Rao [11].

Lemma 2. *For any given $n, k^* \in Z_+$ with $n \geq k^*$, it is possible to construct an instance of the multicommodity problem with n nodes and k^* (minimal) aggregate commodities such that*

$$\frac{\rho^*}{\kappa^*} \geq c \lceil \log k^* \rceil$$

for some constant c.

Proof. We start with constructing a bounded-degree expander graph G^{k^*} with k^* nodes and $O(k^*)$ edges. See, for example, [15] for a definition, and existence of constant degree expander graphs. As discussed in [11], these graphs (with unit capacity for all edges and unit flow requirement between all pairs of vertices) provide examples with $\rho^*/\kappa^* \geq c \lceil \log k^* \rceil$ for some constant c. Note that the demand graph is complete and therefore the minimal vertex cover has size k^*.

We next augment G^{k^*} by adding $n - k^*$ new vertices and $n - k^*$ edges. Each new vertex has degree one and is connected to an arbitrary vertex of G^{k^*}. The new edges are assigned arbitrary capacities. The augmented graph, with the original flow requirements, has n nodes and satisfies $\rho^*/\kappa^* \geq c \lceil \log k^* \rceil$.   □

4   Geometric Interpretation

Both of the more recent studies (namely; Linial, London and Rabinovich [12] and Aumann and Rabani [1]) that relate the min-cut max-flow ratio to the number of origin-destination pairs in the problem instance, take a geometric approach and base their results on the fact that a finite metric space can be mapped into a Euclidean space with logarithmic distortion. More precisely, they base their analysis on the following result (see [3], [12]) that shows that n points can be mapped from l_∞^n to l_1^p with $O(\log n)$ distortion (where l_b^a denotes $\mathbb{R}^a$ equipped with the norm $||x||_b = (\sum_{i=1}^{a} |x_i|^b)^{1/b}$).

Lemma 3 (Bourgain). *Given n points $x_1, \ldots, x_n \in \mathbb{R}^n$, there exists a mapping $\Phi : \mathbb{R}^n \to \mathbb{R}^p$, with $p = O(\log n)$, that satisfies the following two properties:*

(i) $\|x_i - x_j\|_\infty \geq \quad \|\Phi(x_i) - \Phi(x_j)\|_1 \quad$ *for all $i, j \leq n$*

(ii) $\|x_i - x_j\|_\infty \leq \alpha \|\Phi(x_i) - \Phi(x_j)\|_1 \quad$ *for all $i, j \leq n$*
where $\alpha = c \log n$ for some constant c. □

Using this result, it is possible to map the optimal dual solution of the disaggregated (one commodity for each source-sink pair) formulation to l_1^p with logarithmic distortion, see [12] and [1]. One can then show a $O(\log k)$ bound by using arguments similar to the ones presented in Sect. 3.2.

We next give a geometric interpretation of Corollary 2 in terms of mapping n points from l_∞^m to l_1^p with logarithmic distortion with respect to a collection of "seed" points.

Lemma 4. *Given n points $x_1, \ldots, x_n \in \mathbb{R}^m$, the first $t \leq n$ of which are special, $t > 1$, there exists a mapping $\Phi : \mathbb{R}^m \to \mathbb{R}^p$ with $p = O(\log n)$, that satisfies the following two properties:*

(i) $\|x_i - x_j\|_\infty \geq \quad \|\Phi(x_i) - \Phi(x_j)\|_1 \quad$ *for all $i, j \leq n$*

(ii) $\|x_i - x_j\|_\infty \leq \alpha \|\Phi(x_i) - \Phi(x_j)\|_1 \quad$ *for all $i \leq t, \ j \leq n$*
where $\alpha = c \log t$ for some constant c.

Proof. See Appendix B. □

5   Conclusion

In this paper we presented an improved bound on the min-cut max-flow ratio for the multicommodity flow problem. Our bound is motivated by a "compact" linear programming formulation based on the minimal vertex cover of the demand graph. This result suggest that the quality of the ratio depends on the demand graph in a more structural way than the size of the edge set (i.e. number of origin-destination pairs).

We also note that in a longer version of this paper, we use the same approach to show a similar bound on the maximum multicommodity flow problem.

Acknowledgment

We are very grateful to Gregory Sorkin for his helpful advise in formulating and presenting Lemma 1. We also thank Baruch Schieber and Maxim Sviridenko for fruitful discussions.

References

1. Y. Aumann and Y. Rabani: Approximate min-cut max-flow theorem and approximation algorithm. SIAM Journal on Computing **27** (1998) 291–301
2. D. Bienstock and O. Günlük: Computational experience with a difficult multicommodity flow problem. Mathematical Programming **68** (1995) 213–238

3. J. Bourgain: On Lipschitz embedding of finite metric spaces in Hilbert space. Israel Journal of Mathematics **52** (1985) 46–52

4. "P. Fishburn and P. Hammer: Bipartite dimensions and bipartite degrees of graphs. Discrete Math. **160** (1996) 127–148

5. L. R. Ford Jr. and D. R. Fulkerson: Flows in Networks. Princeton University Press (1962)

6. N. Garg, V.V. Vazirani and M. Yannakakis: Approximate max-flow min-(multi)cut theorems and their applications. in Proceedings of the 25th Annual ACM Symposium on Theory of Computing (1993) 698–707

7. M. Iri: On an extension of the max-flow min-cut theorem to multicommodity flows. Journal of the Operations Research Society of Japan **13** (1971) 129–135

8. W. Johnson and J. Lindenstrauss: Extensions of Lipschitz mappings into a Hilbert space. Contemporary Mathematics **26** (1984) 189–206

9. A. V. Karzanov: Polyhedra related to undirected multicommodity flows. Linear Algebra and its Applications **114** (1989) 293–328

10. P. Klein, S. Rao, A. Agrawal, and R. Ravi: An approximate max-flow min-cut relation for undirected multicommodity flow with applications. Combinatorica **15** (1995) 187–202

11. F.T. Leighton and S. Rao: An approximate max-flow min-cut theorem for uniform multicommodity flow problems with applications to approximation algorithms. in Proceedings of the 29th Annual IEEE Symposium on Foundations of Computer Science (1988) 422–431

12. N. Linial, E. London, and Y. Rabinovich (1995): The geometry of graphs and some of its algorithmic applications. Combinatorica **15** 215–245

13. C. Lund, S. Phillips and N. Reingold: Private communication

14. D. W. Matula: Concurrent flow and concurrent connectivity in graphs. in Graph Theory and its Applications to Algorithms and Computer Science Wiley New York (1985) 543–559

15. R. Motwani and P. Raghavan: Randomized Algorithms. Cambridge University Press Cambridge (1995)

16. H. Okamura and P. Seymour: Multicommodity flows in planar graphs. Journal of Comput. Theory – Ser. B **31** (1981) 75–81

17. K. Onaga, and O. Kakusho: On feasibility conditions of multicommodity flows in networks. IEEE Transactions on Circuit Theory CT-18 **4** (1971) 425–429

18. G. Pisier: The Volume of Convex Bodies and Banach Space Geometry. Cambridge University Press Cambridge (1989)

19. S. Plotkin and E. Tardos: Improved Bounds on the max-flow min-cut ratio for multicommodity flows. Proceedings 25'th Symposium on Theory of Computing (1993)

20. F. Shahrokhi and D. W. Matula: The maximum concurrent flow problem. Journal of Association for Computing Machinery **37** (1990) 318–334

21. S. Tragoudas: Improved approximations for the min-cut max-flow ratio and the flux. Mathematical Systems Theory **29** (1996) 157–167

Appendix A: Proof of Lemma 1

For any $v \in V$ and $\delta \geq 0$, let $B(v, \delta) = \{k \in K : d(v, k) \leq \delta\}$ and $B^o(v, \delta) = \{k \in K : d(v, k) < \delta\}$, respectively, denote the collection of members of K that lie within the closed and open balls around v. We next define a sequence of δ's for pairs of nodes.

For any fixed $u \in K$ and $v \in V$, let

$$t^*_{uv} = \max\left\{1, \left\lceil \log\left(max\{|B(u, d(u, v)/2)|, |B(v, d(u, v)/2)|\}\right)\right\rceil\right\}$$

and define

$$\delta^t_{uv} = \begin{cases} 0 & t = 0 \\ \max\{\delta \geq 0 : |B^o(u, \delta)| < 2^t \text{ and } |B^o(v, \delta)| < 2^t\} & t^*_{uv} > t > 0 \\ d(u, v)/2 & t = t^*_{uv} \end{cases}$$

We use the following three observations in the proof:

1. $m = \lceil\log(|K|)\rceil \geq t^*_{uv} > 0$,

2. $\max\{|B(u, \delta^t_{uv})|, |B(v, \delta^t_{uv})|\} \geq 2^t$ for all $t < t^*_{uv}$, and,

3. $\max\{|B^o(u, \delta^t_{uv})|, |B^o(v, \delta^t_{uv})|\} < 2^t$ for all $t \leq t^*_{uv}$.

For fixed $u \in K$, $v \in V$, and $t \geq 0$ such that $t < t^*_{uv}$, rename u and v as $z_{\max}$ and z_{other} so that $|B(z_{\max}, \delta^t_{uv})| \geq |B(z_{\text{other}}, \delta^t_{uv})|$. Using $\frac{1}{e} \geq (1 - \frac{1}{x})^x \geq \frac{1}{4}$, for any $x \geq 2$, we can write the following for any Q^{t+1}_j for $L \geq j \geq 1$:

$$P\left(Q^{t+1}_j \cap B(z_{\max}, \delta^t_{uv}) = \emptyset\right) = \left(1 - 2^{-(t+1)}\right)^{|B(z_{\max}, \delta^t_{uv})|}$$
$$\leq \left(1 - 2^{-(t+1)}\right)^{2^t} \leq e^{-\frac{1}{2}},$$

$$P\left(Q^{t+1}_j \cap B^o(z_{\text{other}}, \delta^{t+1}_{uv}) = \emptyset\right) = \left(1 - 2^{-(t+1)}\right)^{|B^o(z_{\text{other}}, \delta^{t+1}_{uv})|}$$
$$\geq \left(1 - 2^{-(t+1)}\right)^{2^{t+1}} \geq \frac{1}{4}.$$

Notice that $Q^{t+1}_j \cap B(z_{\max}, \delta^t_{uv}) \neq \emptyset$ implies that $d(z_{\max}, Q^{t+1}_j) \leq \delta^t_{uv}$, and similarly, $Q^{t+1}_j \cap B^o(z_{\text{other}}, \delta^{t+1}_{uv}) = \emptyset$ implies that $d(z_{\text{other}}, Q^{t+1}_j) \geq \delta^{t+1}_{uv}$. Using the independence of the two events (since the two balls are disjoint) we can now write:

$$P\left(Q^{t+1}_j \cap B(z_{\max}, \delta^t_{uv}) \neq \emptyset, Q^{t+1}_j \cap B^o(z_{\text{other}}, \delta^{t+1}_{uv}) = \emptyset\right) \geq \left(1 - e^{-\frac{1}{2}}\right) \cdot \frac{1}{4} \geq \frac{1}{11}$$

and therefore,

$$P\left(|d(z_{\text{other}}, Q^{t+1}_j) - d(z_{\max}, Q^{t+1}_j)| \geq \delta^{t+1}_{uv} - \delta^t_{uv}\right) \geq \frac{1}{11}$$

or, equivalently,

$$P\left(\left|d(u, Q_j^{t+1}) - d(v, Q_j^{t+1})\right| \geq \delta_{uv}^{t+1} - \delta_{uv}^t\right) \geq \frac{1}{11}$$

for all $t < t_{uv}^*$.

Let X_{uv}^{tj} be a random variable taking value 1 if $\left|d(u, Q_j^{t+1}) - d(v, Q_j^{t+1})\right| \geq \delta_{uv}^{t+1} - \delta_{uv}^t$, and 0 otherwise. Note that for any fixed $u \in K$ and $v \in V$ if $\sum_{j=1}^L X_{uv}^{tj} \geq L/22$ (that is, at least one-half the expected number) for all $t < t_{uv}^*$, then we can write:

$$\|\Phi^R(u) - \Phi^R(v)\|_1 = \frac{1}{L \cdot m} \sum_{i=1}^m \sum_{j=1}^L \left|d(u, Q_j^i) - d(v, Q_j^i)\right|$$

$$\geq \frac{1}{L \cdot m} \sum_{i=1}^{t_{uv}^*} \frac{L}{22} \left(\delta_{uv}^i - \delta_{uv}^{i-1}\right) \quad = \quad \frac{1}{22m} \left(\delta_{uv}^{t_{uv}^*} - \delta_{uv}^0\right)$$

$$= \frac{d(u, v)}{44m} \ .$$

To this end, we first use the Chernoff bound (see for example [15], Chapter 4) to claim that

$$P\left(\sum_{j=1}^L X_{uv}^{tj} < \frac{1}{2} \times \frac{L}{11}\right) < e^{-\frac{1}{2} \times \frac{1}{4} \times \frac{L}{11}} \ = \ e^{-\frac{L}{88}}$$

for any $u \in K$, $v \in V$ and $t < t_{uv}^*$, which, in turn, implies that

$$P\left(\sum_{j=1}^L X_{uv}^{tj} < \frac{L}{22} \text{ for some } u \in K,\ v \in V, t < t_{uv}^*\right) < |K||V| \lceil \log(|K|) \rceil e^{-L/88}$$

where the right hand side of the inequality is less than 1 for $L \geq 88(3 \cdot \log(|V|)$. Therefore, with positive probability, $\sum_{j=1}^L X_{uv}^{tj} \geq \frac{L}{22}$ for all $u \in K$, $v \in V$ and $t < t_{uv}^*$, which implies that, with positive probability,

$$\|\Phi^R(u) - \Phi^R(v)\|_1 \geq \frac{d(u, v)}{44m}$$

for all $u \in K$, $v \in V$. $\square$

Appendix B: Proof of Lemma 4

Let $G = (V, E)$ be a complete graph with n nodes where each node v_i is associated with point x_i for $i = 1, \ldots, n$. For $e = \{v_i, v_j\} \in E$, let $w_e = \|x_i - x_j\|_\infty$ be the edge weight. Furthermore, let $d(v_i, v_j)$ denote the shortest path length between nodes $v_i, v_j \in V$ using w as edge weights. Note that

$$\|x_i - x_j\|_\infty \leq \|x_i - x_k\|_\infty + \|x_k - x_j\|_\infty$$

for any $i, j, k \leq n$ and therefore $d(v_i, v_j) = ||x_i - x_j||_\infty$ for all $i, j \leq n$. We can now use Corollary 2 to show the existence of a mapping $\Phi' : \mathbb{R}^m \to \mathbb{R}^q$ with $q = O(\log n \log t)$ that satisfies the desired properties.

To decrease the dimension of the image space, we scale Φ' by $\sqrt{Lm}$ to map the points $x_1, \ldots, x_n$ to l_2^q with $c' \log t$ distortion. More precisely, we use $\Phi'' : \mathbb{R}^m \to \mathbb{R}^q$ where $\Phi''(x) = \sqrt{Lm}\, \Phi'(x)$. It is easy to see that:

$$(i) \quad ||\Phi''(x_i) - \Phi''(x_j)||_2 \leq \sqrt{(1/Lm) \sum_{k=1}^{m} \sum_{q=1}^{L} d(v_i, v_j)^2} \;=\; d(v_i, v_j)$$
$$= ||x_i - x_j||_\infty \; ,$$

$$(ii) \quad ||\Phi''(x_i) - \Phi''(x_j)||_2 \geq ||\Phi'(x_i) - \Phi'(x_j)||_1 \;\geq\; c' \log t \, d(v_i, v_j)$$
$$= c' \log t \, ||x_i - x_j||_\infty \; .$$

We can now use the following two facts (also used in [12]), to reduce the dimension of the image space to $O(\log n)$: (i) For any $q \in Z_+$, n points can be mapped from l_2^q to l_2^p, where $p = O(\log n)$ with constant distortion (see [8]), and (ii) For any $p \in Z_+$, l_2^p can be embedded in l_1^{2p} with constant distortion (see [18], Chapter 6). □

Improved Rounding Techniques for the MAX 2-SAT and MAX DI-CUT Problems*

Michael Lewin, Dror Livnat, and Uri Zwick

School of Computer Science, Tel-Aviv University, Tel-Aviv 69978, Israel
{mikel,dror,zwick}@tau.ac.il

Abstract. Improving and extending recent results of Matuura and Matsui, and less recent results of Feige and Goemans, we obtain improved approximation algorithms for the MAX 2-SAT and MAX DI-CUT problems. These approximation algorithms start by solving semidefinite programming relaxations of these problems. They then *rotate* the solution obtained, as suggested by Feige and Goemans. Finally, they round the rotated vectors using random hyperplanes chosen according to *skewed* distributions. The performance ratio obtained by the MAX 2-SAT algorithm is at least 0.940, while that obtained by the MAX DI-CUT algorithm is at least 0.874. We show that these are essentially the best performance ratios that can be achieved using any combination of pre-rounding rotations and skewed distributions of hyperplanes, and even using more general families of rounding procedures. The performance ratio obtained for the MAX 2-SAT problem is fairly close to the inapproximability bound of about 0.954 obtained by Håstad. The performance ratio obtained for the MAX DI-CUT problem is very close to the performance ratio of about 0.878 obtained by Goemans and Williamson for the MAX CUT problem.

1 Introduction

In a seminal paper, Goemans and Williamson [5] used semidefinite programming to obtain a 0.878-approximation algorithm for the MAX CUT and MAX 2-SAT problems, and a 0.796-approximation algorithm for the MAX DI-CUT problem. These algorithms solve semidefinite programming relaxations of the problems, and then round the obtained solution using random hyperplanes. The normal vector of this random hyperplane is uniformly distributed over the n-dimensional sphere. Feige and Goemans [3] improved the last two of these results by obtaining a 0.931-approximation algorithm for the MAX 2-SAT problem, and a 0.859-approximation algorithms for the MAX DI-CUT problem. (A tighter analysis of these algorithms is presented in Zwick [12].) The improved approximation ratios are obtained by *rotating* the vectors obtained as the solutions of the semidefinite programming relaxations before rounding them using a random hyperplane. The normal of the random hyperplane used is again uniformly distributed over the $(n+1)$-dimensional sphere. In their paper, Feige and Goemans [3] suggest several

---

* This research was supported by the ISRAEL SCIENCE FOUNDATION (grant no. 246/01).

other techniques that may be used to obtain improved approximation algorithms for these problems. One of these suggestions is that the normal of the random hyperplane used to round the vectors should be chosen according to a distribution that is *not* uniform on the $(n + 1)$-dimensional sphere, but rather *skewed* toward, or away, from v_0. (In the relaxation of the MAX 2-SAT problem, the vector v_0 is used to represent the value `false`.) Feige and Goemans [3], however, do not explore this possibility. The first to explore this possibility were Matuura and Matsui [8,9]. Overcoming technical difficulties that stand in the way of employing the skewed distribution idea, they obtain in [8] a 0.863-approximation algorithm for the MAX DI-CUT problem, and in [9] a 0.935-approximation algorithm for the MAX 2-SAT problem. In [9] they also show, partially relying on results of Mahajan and Ramesh [10], that these algorithms can be derandomized.

Matuura and Matsui [8,9] obtained their approximation algorithms by using skewed distributions taken from a specific, easy to work with, family of skewed distributions. We conduct a *systematic search* for skewed distributions that yield the *best* performance ratios for the MAX 2-SAT and MAX DI-CUT problems. As a result, we obtain further improved approximation ratios, namely 0.937 for MAX 2-SAT and 0.866 for MAX DI-CUT. These are essentially the best performance ratios that can be used by using skewed distribution of hyperplanes on their own.

We obtain further improved results, however, by *combining* the skewed hyperplane technique with pre-rounding rotations. For the MAX 2-SAT problem we get an approximation ratio of at least 0.940. For the MAX DI-CUT problem we get an approximation ratio of at least 0.874. These are essentially the best ratios that can be obtained using *any* combination of pre-rounding rotations and skewed distributions of hyperplanes, and even using more general families of rounding procedures.

Håstad [6], using gadgets of Trevisan *et al.* [11], showed that for any $\epsilon > 0$, if there is a $\left(\frac{21}{22} + \epsilon\right)$-approximation algorithm for the MAX 2-SAT problem, or a $\left(\frac{12}{13} + \epsilon\right)$-approximation algorithm for the MAX DI-CUT problem, then P=NP. (Note that $\frac{21}{22} \simeq 0.95454$ and that $\frac{12}{13} \simeq 0.92308$. There are still gaps, therefore, between the best approximation and inapproximability ratios for the MAX 2-SAT and MAX DI-CUT problems. The gap for the MAX 2-SAT problem, between 0.940 and 0.954, is quite small. The gap for the MAX DI-CUT is larger. But, the ratio of 0.874 we obtain for the MAX DI-CUT problem is very close to the best ratio, of about 0.878, obtained by Goemans and Williamson [5] for the MAX CUT problem, which may be seen as a subproblem of the MAX DI-CUT problem.

The rest of this paper is organized as follows. In the next section we formally define the MAX 2-SAT and MAX DI-CUT problems and describe their semidefinite programming relaxations. We also briefly sketch there the approximation algorithms of [5] and [3]. In Sect. 3 we discuss families of rounding procedures that can be used to round configuration of vectors obtained by solving semidefinite programming relaxations. In Sect. 4 we explain how to compute certain probabilities that appear in the analysis of approximation algorithms that use rounding procedures from the families defined in Sect. 3. In Sect. 5 we describe the computation of lower bounds on the performance ratios achieved by such algorithms. In Sect. 6 we describe the essentially best approximation algorithms that can be obtained using rounding procedures from the family $\mathcal{SKEW}$. In Sect. 7 we describe our best approximation algorithms. These algorithms use rounding procedures from the family $\mathcal{ROT+SKEW}$. In Sect. 8 we show that these algorithms are essentially the best algorithms that can be obtained using rounding procedures from $\mathcal{ROT+SKEW}$. We end in Sect. 9 with some concluding remarks.

2   Preliminaries

An instance of MAX 2-SAT in the Boolean variables $x_1, \ldots, x_n$ is composed of a collection of *clauses* $C_1, \ldots, C_m$ with non-negative weights $w_1, \ldots, w_m$ assigned to them. Each clause C_i is of the form z_1 or of the form $z_1 \vee z_2$ where each z_j is either a variable x_k or its negation $\bar{x}_k$. (Each such z_j is called a *literal*.) The goal is to assign the variables $x_1, \ldots, x_n$ Boolean values 0 or 1 so that the total weight of the satisfied clauses is maximized. A clause is satisfied if at least one of the literals appearing in it is assigned the value 1. (If a variable is assigned the value 0 then its negation is assigned the value 1, and vice versa.)

It is convenient to let $x_{n+i} = \bar{x}_i$, for $1 \leq i \leq n$, and also to let $x_0 = 0$. Each clause is then of the form $x_i \vee x_j$ where $0 \leq i, j \leq 2n$. An instance of MAX 2-SAT can then be encoded as an array (w_{ij}), where $0 \leq i, j \leq 2n$, where w_{ij} is interpreted as the weight of the clause $x_i \vee x_j$. The goal is then to assign the variables $x_1, \ldots, x_{2n}$ Boolean values, such that $x_{n+i} = \bar{x}_i$ and such that the total weight of the satisfied clauses is maximized.

The semidefinite programming relaxation of the MAX 2-SAT problem used by Feige and Goemans [3] is given in Fig. 1. In this relaxation, a unit vector v_i is assigned to each literal x_i, where $0 \leq i \leq 2n$. As $x_0 = 0$, the vector v_0 corresponds to the constant 0 (`false`). To ensure consistency, we require $v_i \cdot v_{n+i} = -1$, for $1 \leq i \leq n$. As the value of a solution of the SDP program depends only on the inner products between the vectors, we can assume, with out loss of generality, that $v_0 = (1, 0, \ldots, 0) \in \mathbb{R}^{n+1}$. Note that as the 'triangle constraints' holds for any $1 \leq i, j \leq 2n$, and as $v_i = -v_{n+i}$, for $1 \leq i \leq n$, we get that if v_0, v_i and v_j appear in a feasible solution of the SDP, then we have:

$$v_0 \cdot v_i + v_0 \cdot v_j + v_i \cdot v_j \geq -1 \quad,$$
$$-v_0 \cdot v_i - v_0 \cdot v_j + v_i \cdot v_j \geq -1 \quad,$$
$$-v_0 \cdot v_i + v_0 \cdot v_j - v_i \cdot v_j \geq -1 \quad,$$
$$v_0 \cdot v_i - v_0 \cdot v_j - v_i \cdot v_j \geq -1 \quad.$$

We can strengthen the relaxation by adding the constraints $v_i \cdot v_j + v_i \cdot v_k + v_j \cdot v_k \geq -1$, for any $1 \leq i, j, k \leq 2n$, but we do not know how to take advantage of these additional constraints in our analysis, so we do not know whether this helps.

An instance of MAX DI-CUT is a directed graph $G = (V, E)$, with non-negative weights assigned to its edges. The goal is to find a partition of V into

$$\text{Max } \frac{1}{4} \sum_{i,j} w_{ij}(3 - v_0 \cdot v_i - v_0 \cdot v_j - v_i \cdot v_j)$$

$$
\begin{aligned}
v_0 \cdot v_i + v_0 \cdot v_j + v_i \cdot v_j &\geq -1 \quad , \quad 1 \leq i, j \leq 2n \\
v_i \cdot v_{n+i} &= -1 \quad , \quad 1 \leq i \leq n \\
v_i \in \mathbb{R}^{n+1} , \; v_i \cdot v_i &= 1 \quad , \quad 0 \leq i \leq 2n
\end{aligned}
$$

Fig. 1. A semidefinite programming relaxation of MAX 2-SAT

$$\text{Max } \frac{1}{4} \sum_{i,j} w_{ij}(1 + v_0 \cdot v_i - v_0 \cdot v_j - v_i \cdot v_j)$$

$$
\begin{aligned}
v_0 \cdot v_i + v_0 \cdot v_j + v_i \cdot v_j &\geq -1 \quad , \quad 1 \leq i, j \leq n \\
-v_0 \cdot v_i - v_0 \cdot v_j + v_i \cdot v_j &\geq -1 \quad , \quad 1 \leq i, j \leq n \\
-v_0 \cdot v_i + v_0 \cdot v_j - v_i \cdot v_j &\geq -1 \quad , \quad 1 \leq i, j \leq n \\
v_0 \cdot v_i - v_0 \cdot v_j - v_i \cdot v_j &\geq -1 \quad , \quad 1 \leq i, j \leq n \\
v_i \in \mathbb{R}^{n+1} , \; v_i \cdot v_i &= 1 \quad , \quad 0 \leq i \leq n
\end{aligned}
$$

Fig. 2. A semidefinite programming relaxation of MAX DI-CUT

$(S, V - S)$, where $S \subseteq V$, such that the total weight of the edges that cross this cut, i.e., edges $u \to v$ such that $u \in S$ but $v \notin S$, is maximized.

It is convenient to assume that $V = \{1, 2, \ldots, n\}$. An instance of MAX DI-CUT can then be represented as an array (w_{ij}), where $1 \leq i, j \leq n$, where w_{ij} is interpreted as the weight of the edge $i \to j$. Note that $w_{ij} \geq 0$, for every $1 \leq i, j \leq n$. We can also assume that $w_{ii} = 0$, for $1 \leq i \leq n$.

The semidefinite programming relaxation of the MAX DI-CUT problem used by Feige and Goemans [3] is given in Fig. 2. Here, there is a unit vector v_i corresponding to each vertex $1 \leq i \leq n$ of the graph, and there is also a vector v_0, corresponding to the side S of the directed cut. Again, we can add more triangle constraints, but it is not clear whether this helps.

The approximation algorithms of Goemans and Williamson [5] solve weakened versions of the SDP relaxations, they do not use the triangle constraints, and then *round* the resulting solution using a random hyperplane. More specifically, they choose a random vector $r = (r_0, r_1, \ldots, r_n)$ according a standard $(n+1)$-dimensional normal distribution. The random variables r_i, for $0 \leq i \leq n$, are i.i.d. random variables distributed according to the standard normal distribution. The normalized vector $r/\|r\|$ is then uniformly distributed on S^n, the unit sphere in $\mathbb{R}^{n+1}$. For a MAX 2-SAT instance, the literal x_i is assigned the value 0 if and only if v_i and v_0 lie on the same side of the hyperplane whose normal is r, or equivalently, if $sgn(v_i \cdot r) = sgn(v_0 \cdot r)$. For a MAX DI-CUT instance, a vertex i is placed in S if and only if v_i and v_0 lie on the same side of the random hyperplane.

The approximation algorithms of Feige and Goemans [3] solve the SDP relaxations given in Fig. 1 and 2. Before rounding the resulting vectors, they rotate them using a rotation function $f : [0, \pi] \to [0, \pi]$ that satisfies $f(\pi - \theta) = \pi - f(\theta)$, for $0 \le \theta \le \pi$. The vector v_i, forming an angle $\theta_i = \arccos(v_0 \cdot v_i)$ with v_0, is rotated, in the 2-dimensional plane spanned by v_0 and v_i, into a vector v_i' whose angle $\theta_i' = \arccos(v_0 \cdot v_i')$ with v_0 satisfies $\theta_i' = f(\theta_{0i})$.

3   Families of Rounding Procedures

Feige and Goemans [3] suggest the following very general family of rounding procedures:

$\mathcal{GEN}$:   Let $F : \mathbb{R}^2 \times [-1, 1] \to \{0, 1\}$ be an arbitrary function. Let $r = (r_0, r_1, \ldots, r_n)$ be a standard $(n+1)$-dimensional random variable. For $1 \le i \le n$, let $x_i \gets F(v_0 \cdot r, v_i \cdot r, v_0 \cdot v_i)$.

For consistency, the function F should satisfy $F(x, -y, -z) = 1 - F(x, y, z)$ for every $x, y \in \mathbb{R}$ and $z \in [-1, 1]$. In the spirit of the RPR^2 rounding procedure of Feige and Langberg [4], this can be further generalized to:

$\mathcal{GEN}^+$:   Let $F : \mathbb{R}^2 \times [-1, 1] \to [0, 1]$ be an arbitrary function. Let $r = (r_0, r_1, \ldots, r_n)$, where $r_0, r_1, \ldots, r_n$ are independent standard normal variables. For $1 \le i \le n$, let $x_i \gets 1$, independently with probability $F(v_0 \cdot r, v_i \cdot r, v_0 \cdot v_i)$.

The basic random hyperplane rounding technique of Goemans and Williamson [5] is of course a very special case of general family with

$$F(v_0 \cdot r, v_i \cdot r, v_0 \cdot v_i) = [(v_0 \cdot r)(v_i \cdot r) \le 0],$$

where $[a \le b]$ gets the value 1 if $a \le b$, and 0, otherwise. We can assume, without loss of generality that $v_0 = (1, 0, \ldots, 0)$, and that $r_0 \ge 0$. The expression for F is then further simplified to

$$F(v_0 \cdot r, v_i \cdot r, v_0 \cdot v_i) = [(v_i \cdot r) \le 0].$$

To obtain their improved MAX 2-SAT and MAX DI-CUT algorithms, Feige and Goemans [3] do not use the full generality of the $\mathcal{GEN}$ family. They use, instead the following more restricted family:

$\mathcal{ROT}$:   Let $f : [0, \pi] \to [0, \pi]$ be an arbitrary *rotation* function. Let $\theta_i = \arccos(v_0 \cdot v_i)$ be the angle between v_0 and v_i. Rotate v_i into a vector v_i' that forms an angle of $\theta_i' = f(\theta_i)$ with v_0. Round the vectors $v_0, v_1', \ldots, v_n'$ using the random hyperplane technique of Goemans and Williamson, i.e., let $x_i \gets [(v_0 \cdot r)(v_i' \cdot r) \le 0]$, for $1 \le i \le n$.

For consistency, the function f should satisfy $f(\pi - \theta) = \pi - f(\theta)$. Feige and Goemans [3] also suggest, but do not analyze, the following *skewed hyperplane* rounding technique:

$\mathcal{SKEW}$:   Let $r = (r_0, r_1, \ldots, r_n)$ be a *skewed* normal vector, i.e., $r_1, r_2, \ldots, r_n$ are still independent standard normal variables, but r_0 is chosen according to a different distribution. Again, let $x_i \leftarrow [(v_0 \cdot r)(v_i \cdot r) \leq 0]$, for $1 \leq i \leq n$.

It may seem, at first sight, that $\mathcal{SKEW}$ is not a sub-family of $\mathcal{GEN}$, as r_0 in $\mathcal{SKEW}$ is not a standard normal variable. We will see shortly, however, that $\mathcal{SKEW}$ is indeed a sub-family of $\mathcal{GEN}$, and it is not necessary, therefore, to further generalize $\mathcal{GEN}$ by allowing r_0 to be chosen according to a non-normal distribution.

Matuura and Matsui [8,9] take up the challenge of analyzing this technique, in a slightly different form. They obtain some improved results but do not consider *all* rounding procedures from this family, and hence do not exploit the full power of $\mathcal{SKEW}$. In Sect. 6 we obtain the essentially best algorithms for the MAX 2-SAT and MAX DI-CUT problems that can be obtained using rounding procedures for $\mathcal{SKEW}$.

We obtained further improved algorithms, however, by considering combinations of pre-rounding rotations and skewed hyperplane distributions. We denote the family of rounding procedures containing all these combinations by $\mathcal{ROT}+\mathcal{SKEW}$. In Sect. 7 we obtain the essentially best algorithms for the MAX 2-SAT and MAX DI-CUT problems that can be obtained using rounding procedures from this family. These are the best approximation algorithms that we get for these two problems. We have not been able to improve these algorithms using more general rounding techniques. Extensive numerical experiments that we have conducted suggest that, at least for the MAX 2-SAT problem, no further improvements are possible, even using rounding procedures from $\mathcal{GEN}^+$.

Our next task will be to determine the form of the function $F(x, y, z)$ that corresponds to rounding procedures from the family $\mathcal{ROT}+\mathcal{SKEW}$. Before continuing, it would be convenient to present an equivalent formulation of $\mathcal{GEN}$ (and $\mathcal{GEN}^+$). We assume, as stated, that $v_0 = (1, 0, \ldots, 0)$. Any vector $r = (r_0, r_1, \ldots, r_n) \in \mathbb{R}^{n+1}$ can thus be written as

$$r = r_0 v_0 + r^\perp,$$

where, $r^\perp = r - r_0 v_0 = (0, r_1, \ldots, r_n)$. Clearly, $v_0 \cdot r^\perp = 0$. We can therefore replace the three parameters $v_0 \cdot r$, $v_i \cdot r$ and $v_0 \cdot v_i$, by the three parameters r_0, $v_i \cdot r^\perp$ and $v_0 \cdot v_i$. As $v_i \cdot r = r_0 + v_i \cdot r^\perp$, we can easily move from one set of parameters to the other. Instead of using an arbitrary function $F(v_0 \cdot r, v_i \cdot r, v_0 \cdot v_i)$, we can, therefore, use an arbitrary function $G(r_0, v_i \cdot r^\perp, v_0 \cdot v_i)$. Note that r_0 is now an argument of G. By applying an appropriate function on r_0 we can get a random variable distributed according to any desired distribution.

Let v_i' be the vector obtained by rotating the vector v_i, with respect to v_0, using the rotation function f. Let $g(x) = f(\arccos x)$, for $-1 \leq x \leq 1$. Also let $x_i = v_0 \cdot v_i$. As v_i' lies in the 2-dimensional space spanned by v_0 and v_i, we get that $v_i' = \alpha v_0 + \beta v_i$, for some $\alpha, \beta \in \mathbb{R}$. By the definition of the rotation process, we have $\beta \geq 0$. We can get two equations for α and β:

$$v_i' \cdot v_i' = (\alpha v_0 + \beta v_i) \cdot (\alpha v_0 + \beta v_i) = \alpha^2 + \beta^2 + 2\alpha\beta x_i = 1 ,$$
$$v_0 \cdot v_i' = v_0 \cdot (\alpha v_0 + \beta v_i) = \alpha + \beta x_i = g(x_i) .$$

Solving these equations, keeping in mind that $\beta \geq 0$, we get that:

$$\alpha = g(x_i) - x_i\sqrt{\frac{1-g^2(x_i)}{1-x_i^2}} = \cos\theta_i' - \cot\theta_i \sin\theta_i' ,$$
$$\beta = \sqrt{\frac{1-g^2(x_i)}{1-x_i^2}} = \frac{\sin\theta_i'}{\sin\theta_i} .$$

As

$$v_i' \cdot r = (\alpha v_0 + \beta v_i) \cdot (r_0 v_0 + r^\perp) = (\alpha + \beta x)r_0 + \beta v_i \cdot r^\perp ,$$

the condition $v_i' \cdot r \leq 0$ of $\mathcal{ROT}+\mathcal{SKEW}$ is equivalent to:

$$v_i \cdot r^\perp \leq -\frac{\alpha + \beta x_i}{\beta}r_0 = -\cot\theta_i' \sin\theta_i\, r_0 = -\cot f(\theta_i)\sin\theta_i\, r_0 .$$

We can also define $v_i^\perp = v_i - (v_0 \cdot v_i)$. Clearly $v_0 \cdot v_i^\perp = 0$, and therefore $v_i \cdot r^\perp = v_i^\perp \cdot r^\perp$. The above condition is, therefore, also equivalent to the condition $v_i^\perp \cdot r^\perp \leq -\cot f(\theta_i)\sin\theta_i\, r_0$.

This leads us to the definition of the following family of rounding procedures that properly contains $\mathcal{ROT}+\mathcal{SKEW}$:

$\mathcal{THRESH}$:   Let $r = (r_0, r_1, \ldots, r_n)$, where $r_1, r_2, \ldots, r_n$ are independent standard normal variables, and r_0 is an independent random variable distributed according to an arbitrary distribution. (Without loss of generality, r_0 assumes only nonnegative values.) Let $r^\perp = (0, r_1, \ldots, r_n)$. Let $T : \mathbb{R} \times [-1, 1] \to \mathbb{R}$ be a *threshold* function. Let $x_i \leftarrow [v_i \cdot r^\perp \leq T(r_0, v_0 \cdot v_i)]$, for $1 \leq i \leq n$.

From the above discussion, we see that $\mathcal{ROT}+\mathcal{SKEW}$ is just the sub-family of $\mathcal{THRESH}$ obtained by requiring the threshold function T to be of the form $T(r_0, x_i) = r_0 S(x_i)$, i.e., multiplicative in r_0. (We simply take $S(x) = -\cot f(\arccos x)\sqrt{1-x^2}$, or $f(\theta) = \arctan(-\frac{\sin\theta}{S(\cos\theta)})$.)

Interestingly, our best rounding procedures for the MAX 2-SAT and the MAX DI-CUT problems choose r_0 according to a very simple distribution. They just pick a *fixed* value for r_0! They fall, therefore, to the following sub-family of $\mathcal{THRESH}$:

$\mathcal{THRESH}^-$:   Let $r = (0, r_1, \ldots, r_n)$, where $r_1, r_2, \ldots, r_n$ are independent standard normal variables. Let $S : [-1, 1] \to \mathbb{R}$ be a *threshold* function. Let $x_i \leftarrow [v_i \cdot r \leq S(v_0 \cdot v_i)]$, for $1 \leq i \leq n$.

In rounding procedures from $\mathcal{THRESH}^-$, the normal vector r is chosen in the n-dimensional space orthogonal to v_0 and $r^\perp = r$. $\mathcal{THRESH}^-$ is the sub-family of $\mathcal{ROT}+\mathcal{SKEW}$ in which r_0 assumes a fixed value. Note that the exact value assumed by r_0 is not really important. Let us represent a rounding procedure from $\mathcal{ROT}+\mathcal{SKEW}$ using a pair (a, f), where a is the fixed value assumed by r_0, and f is the rotation function used. Then, it is easy to check that the rounding procedure (a_1, f_1) is equivalent to the rounding procedure (a_2, f_2), where $\cot f_2(\theta) = \frac{a_1}{a_2} \cot f_1(\theta)$. Thus, the choice of the fixed value assumed by r_0 is, in fact, arbitrary. In our best rounding procedures, the rotation functions used are linear. The exact value assumed by r_0 does play a role in that case. Note that the exact values assumed by r_0 are also important when no rotation function is used, as is the case in Sect. 6.

Our best algorithms for the MAX 2-SAT and MAX DI-CUT problems use rounding procedures from the family $\mathcal{ROT}+\mathcal{SKEW}$. Our MAX 2-SAT algorithm, however, is essentially the best algorithm that can be obtained using any rounding procedure from $\mathcal{THRESH}$. Our MAX DI-CUT algorithm is essentially the best algorithm that can be obtained using any rounding procedure from $\mathcal{THRESH}^-$. At least for the MAX 2-SAT problem, we believe that no better rounding procedure can be found even in $\mathcal{GEN}^+$, the most general family of rounding procedures defined above. (This belief is supported by some numerical experimentation. We will elaborate on this point in the full version of the paper.)

4   Computing Probabilities

The analysis of most semidefinite programming based approximation algorithm is *local* in nature. In the analysis of the MAX 2-SAT and MAX DI-CUT algorithms that are presented in the next sections, we only have to evaluate the probability that a certain clause is satisfied (in the case of MAX 2-SAT) or that a certain edge is cut (in the case of MAX DI-CUT). Let $prob(v_1, v_2)$, where $v_1, v_2 \in \mathbb{R}^{n+1}$ be the probability that the corresponding variables x_1 and x_2 both receive the value 1 when the vectors v_1 and v_2 are rounded using the chosen rounding procedure. Then, the probability that a 2-SAT clause is satisfied is $1 - prob(-v_1, -v_2)$ and the probability that a directed edge is cut is $prob(-v_1, v_2)$.

To evaluate such a probability, it is enough to consider the projection of r on the three dimensional space spanned by v_0, v_1 and v_2. Let $r' = (r_0, r'_1, r'_2)$ be the projection of r on this space. It is easy to see that r_0, r'_1, r'_2 are still independent random variables, and that r'_1 and r'_2 are standard normal variables. We may assume, therefore, without loss of generality, that v_0, v_1 and v_2 lie in $\mathbb{R}^3$. We may also assume that $v_1 = (a_0, a_1, 0)$ and $v_2 = (b_0, b_1, b_2)$, for some $a_0, a_1, b_2 \in \mathbb{R}^+$ and $b_0, b_1 \in \mathbb{R}$. We also let $r = (r_0, r_1, r_2)$ and $r^\perp = (0, r_1, r_2)$.

Suppose now that the vectors v_1 and v_2 are rounded using a rounding procedure from $\mathcal{GEN}^+$, the most general family of rounding procedures considered in the previous section. Let $G : \mathbb{R}^2 \times [-1, 1] \to [0, 1]$ be the rounding function used. By definition, we then have:

$$prob(v_1, v_2) =$$
$$\int_0^\infty \left[\int_{-\infty}^\infty \int_{-\infty}^\infty G(x, a_1 y, a_0) G(x, b_1 y + b_2 z, b_0) \phi(y) \phi(z) \, dy \, dz \right] f(x) \, dx \ ,$$

where $\phi(y) = \frac{1}{\sqrt{2\pi}} e^{-x^2/2}$ is the density function of a standard normal variable, and $f(x)$ is the density function of the random variable r_0. If r_0 is a discrete random variable that assumes the value s_k with probability p_k, for $1 \leq k \leq \ell$, then the outer integral is replaced by a sum:

$$prob(v_1, v_2) =$$

$$\sum_{k=1}^{\ell} p_k \left[\int_{-\infty}^{\infty} \int_{-\infty}^{\infty} G(s_k, a_1 y, a_0) G(s_k, b_1 y + b_2 z, b_0) \phi(y) \phi(z) dy dz \right].$$

If the rounding procedure belongs to $\mathcal{THRESH}$, so that $G(r_0, v_i \cdot r^\perp, v_0 \cdot v_i) = [v_i \cdot r^\perp \leq T(r_0, v_0 \cdot v_i)]$, then we have a further simplification:

$$prob(v_1, v_2) = \sum_{k=1}^{\ell} p_k \left[\iint_{\substack{a_1 y \leq T(s_k, a_0) \\ b_1 y + b_2 z \leq T(s_k, b_0)}} \phi(y)\phi(z) dy dz \right].$$

Note that the conditions $a_1 y \leq T(s_k, a_0)$ and $b_1 y + b_2 z \leq T(s_k, b_0)$, which are equivalent to the conditions $v_1 \cdot r^\perp \leq T(r_0, v_0 \cdot v_1)$ and $v_2 \cdot r^\perp \leq T(r_0, v_0 \cdot v_2)$, define a wedge in the y-z plane, the plane perpendicular to v_0. (Note that $T(s_k, a_0)$ and $T(s_k, b_0)$ here are constant.) This wedge integral can be evaluated numerically. It is not difficult to reduce it to single integration over the normal distribution function $\Phi(x) = \frac{1}{\sqrt{2\pi}} \int_{-\infty}^{x} \phi(y) dy$.

When the rounding procedure is from the family $\mathcal{SKEW}$, or $\mathcal{SKEW+ROT}$, further simplifications are possible, and $prob(v_1, v_2)$ can be evaluated using a single integration, even if r_0 is a continuous random distribution. More details would appear in the full version of the paper.

5   Computing Performance Ratios

In this section we describe the standard way in which we get a lower bound on the performance ratio achieved by our algorithms. For concreteness, we consider the MAX 2-SAT problem. The details for the MAX DI-CUT problem are analogous.

Let $P(v_i, v_j) = 1 - prob(-v_i, -v_j)$ be the probability that the clause $x_i \vee x_j$ is satisfied by the assignment produced by the chosen rounding procedure. Let

$$V(v_i, v_j) = \tfrac{1}{4}(3 - v_0 \cdot v_i - v_0 \cdot v_j - v_i \cdot v_j)$$

be the contribution of this clause to the value of the semidefinite program. As shown by Goemans and Williamson [5], the performance ratio of the MAX 2-SAT algorithm is at least

$$\alpha = \min_{(v_i, v_j) \in \Omega} \frac{P(v_i, v_j)}{V(v_i, v_j)},$$

where Ω is the set of all $(v_i, v_j) \in \mathbb{R}^3 \times \mathbb{R}^3$ that satisfy all the constraints of the semidefinite program. We evaluated such minimums numerically, using the optimization toolbox of Matlab.

To obtain an upper bound the performance ratio that can be obtained using *any* rounding procedure from a given family, we select a finite set of configurations $C = \{(v_1^i, v_2^i) \mid 1 \leq i \leq t\}$. We then estimate, numerically, that maximum, over all rounding procedures from the given family, of the minimum

$$\bar{\alpha} = \min_{i=1}^{t} \frac{P(v_1^i, v_2^i)}{V(v_1^i, v_2^i)} .$$

This bounding technique is implicit in Feige and Goemans [3] and is described explicitly in Zwick [12].

The computation of α, for a given rounding procedure, and of $\bar{\alpha}$, for a given family of rounding procedure, are very difficult global optimization problems. We have made extensive experiments to verify the numerical bounds that we state in this paper. However, we cannot claim them as theorems. It is possible, at least in principle, to obtain rigorous proofs for our claims using a tool such as $\mathcal{R}eal\mathcal{S}earch$ (see [13]), but this would require a tremendous amount of work.

As mentioned, our best algorithms for the MAX 2-SAT and MAX DI-CUT problems simply set r_0 to be a constant. This may seem very surprising. We next try to explain why this is not so surprising after all.

Let C be a set of t configurations used to obtain an upper bound of $\bar{\alpha}$ on the performance ratio that can be achieved using rounding procedures from a given family. Consider now the search for rounding procedures that actually achieve a performance ratio of at least $\bar{\alpha}$ on each one of the t configurations of the set. Such optimal rounding procedures, for C, may choose r_0 according to an arbitrary continuous distribution. However, we can discretize the problem and assume that r_0 assumes only ℓ different values, where ℓ may be a huge number. If ℓ is large enough, we loose very little in this process. Let us assume that r_0 assumes the value s_i with probability p_i, for $1 \leq i \leq \ell$.

Let R_{ij} be the ratio obtained for the j-th configuration when r_0 assumes the value s_i. To find the optimal probabilities in which the values $s_1, s_2, \ldots, s_\ell$ should be chosen, we solve the following linear program:

$$\max \alpha$$
$$\text{s.t. } \alpha \leq \textstyle\sum_{i=0}^{\ell} R_{ij} p_i \, , \quad 1 \leq j \leq t$$
$$\textstyle\sum_{i=0}^{\ell} p_i = 1$$
$$p_i \geq 0 \qquad\qquad , \quad 1 \leq i \leq \ell .$$

This linear program has a basic optimal solution with at most t of the p_i's are non-zero. (Note that there are $\ell + 1$ variables and only $t + 1$ inequalities other than the inequalities $p_i \geq 0$, for $1 \leq i \leq \ell$. Thus, at least $\ell - t$ of these inequalities must be satisfied with equality.)

Thus, we see that when C is of size t, it is not necessary to let r_0 assume more than t different values. The sets of configurations that we use to bound the performance ratio that can be achieved are very small. For MAX 2-SAT, for example, we use only two configurations.

Table 1. (a) The probability distribution of r_0 used to obtain a 0.937-approximation algorithm for the MAX 2-SAT problem. (b) The probability distribution of r_0 used to obtain a 0.866-approximation algorithm for the MAX DI-CUT problem

<table>
<tr><td colspan="2">(a)</td></tr>
<tr><td>prob.</td><td>value</td></tr>
<tr><td>0.0430</td><td>$N(0,1)$</td></tr>
<tr><td>0.0209</td><td>0.145</td></tr>
<tr><td>0.0747</td><td>0.345</td></tr>
<tr><td>0.3448</td><td>0.755</td></tr>
<tr><td>0.5166</td><td>1.635</td></tr>
</table>

<table>
<tr><td colspan="2">(b)</td></tr>
<tr><td>prob.</td><td>value</td></tr>
<tr><td>0.0714</td><td>$N(0,1)$</td></tr>
<tr><td>0.0644</td><td>0.170</td></tr>
<tr><td>0.3890</td><td>0.640</td></tr>
<tr><td>0.4752</td><td>1.520</td></tr>
</table>

6   Algorithms from $\mathcal{SKEW}$

In this section we describe the essentially best approximation algorithms for the MAX 2-SAT and MAX DI-CUT problems that can be obtained using rounding procedures from $\mathcal{SKEW}$.

6.1   A MAX 2-SAT Algorithm

To specify our algorithms, we only need to specify how the random variable r_0 is chosen. The probability distribution of r_0 used by our MAX 2-SAT algorithm is given in Table 1(a). The distribution of r_0 is very discrete in nature. With a probability of more than $\frac{1}{2}$, namely 0.5166, it assumes the fixed value 1.635. It assumes each of three other values with smaller probabilities. Only with the relatively small probability of 0.0430 it is chosen according to the standard normal distribution. (If we want r_0 to assume only nonnegative values, we take, with this probability, the absolute value of a standard normal variable.) This small probability of assuming a standard normal distribution allows the algorithm to easily achieve good approximation ratios on clauses which contribute very small values to the overall value of the semidefinite program.

6.2   A MAX DI-CUT Algorithm

The probability distribution of r_0 used by our MAX DI-CUT algorithm is given in Table 1(b). Again, the distribution of r_0 is very discrete in nature. With a probability of almost $\frac{1}{2}$, namely 0.4752, it assumes the fixed value 1.52. It assumes only two other values with smaller probabilities. Only with the relatively small probability of 0.0714, r_0 is chosen according to the standard normal distribution which again is needed to handle edges which have small values in the semidefinite solution.

Numerical calculations shows that the performance ratio achieved using this rounding procedure is at least 0.866. Again, more details would appear in the full version of the paper.

The approximation algorithm for MAX DI-CUT presented here, as most other MAX DI-CUT approximation algorithms, is also an approximation algorithm for MAX 2-AND.

7   Algorithms from $\mathcal{ROT}+\mathcal{SKEW}$

In this section we describe our best approximation algorithms for the MAX 2-SAT and MAX DI-CUT problems. These algorithms use rounding procedures from the $\mathcal{ROT}+\mathcal{SKEW}$ family. These algorithms are essentially the best algorithms that can be obtained using rounding procedures from this family, and even from larger families.

7.1   A MAX 2-SAT Algorithm

To specify a rounding procedure from the $\mathcal{ROT}+\mathcal{SKEW}$ family, we have to specify a rotation function f and the distribution of r_0. The rounding procedure used by our best MAX 2-SAT algorithm has a very simple representation. The 'random' variable r_0 assumes the value 2 and the rotation function f is linear:

$$f(\theta) \;=\; 0.58831458\,\theta + 0.64667394\,.$$

Numerical calculations show that the performance ratio achieved by this algorithm is at least 0.9401.

The worst performance ratio of about 0.9401 is achieved on two configurations. It is convenient to represent a configuration (v_1, v_2) using a triplet $(\theta_{01}, \theta_{02}, \theta_{12})$, where $\theta_{ij} = \arccos(v_i \cdot v_j)$ is the angle between v_i and v_j, for $0 \leq i < j \leq 2$. The angles corresponding to the first of the these two worst configurations (v_1, v_2) are roughly $(1.401, 1.401, 2.294)$. The second configuration is then just $(-v_1, -v_2)$, which is roughly represented by the triplet $(1.741, 1.741, 2.294)$.

7.2   A MAX DI-CUT Algorithm

The rounding procedure used by our best MAX DI-CUT algorithm has again a fairly simple representation. The 'random' variable r_0 assumes the constant value 2.5. The rotation function f is now a piecewise linear function defined by the points given in Table 2.

Numerical calculations show that the performance ratio achieved by this algorithm is at least 0.8740.

8   Upper Bounds on Achievable Ratios

In this section we explain our claim that the approximation algorithm presented in Sect. 7.1 is essentially the best MAX 2-SAT algorithm that can be obtained using a rounding procedure from $\mathcal{ROT}+\mathcal{SKEW}$, or even from $\mathcal{THRESH}$. (We also believe that it is essentially the best rounding procedure even from $\mathcal{GEN}^+$,

Table 2. The piecewise linear rotation function used for the 0.8740 algorithm for the MAX DI-CUT problem

θ	$f(\theta)$
0	0.88955971
1.24676999	1.42977923
1.30701991	1.45654677
1.36867009	1.48374672
1.39760143	1.49640918
1.44728317	1.51791721
1.50335282	1.54210593
$\pi/2$	$\pi/2$

but our evidence in support of this claim is less conclusive at this stage.) We also explain our claim that the approximation algorithm presented in Sect. 7.2 is essentially the best approximation algorithm that can be obtained using a rounding procedure from $\mathcal{ROT}+\mathcal{SKEW}$, or even from $\mathcal{THRESH}^-$.

We would like to stress that these claims are based on difficult numerical computations, carried out using the optimization toolbox of Matlab. We do not claim, therefore, that they are theorems.

Another point that should be explained is the following. Our analysis of the various rounding procedures is very local in nature and it produces only *lower bounds* on the actual performance ratios. We do not know whether these lower bounds are tight. The local analysis of the MAX CUT algorithm of Goemans and Williamson [5] was shown to be tight by Karloff [7] (see also [1] and [2]). It is not known whether the analysis of the MAX 2-SAT and MAX DI-CUT algorithms of Feige and Goemans [3] (see also [12]) is tight. When we say here that a certain rounding procedure is essentially optimal, what we actually mean is that no better *lower bound* on the performance ratio can be obtained, using the local form of analysis, for any other rounding procedure from the family.

8.1   Upper Bound for MAX 2-SAT

Let $C = \{(v_1, v_2), (-v_1, -v_2)\}$ be the set of the two worst configurations for the MAX 2-SAT algorithm of Sect. 7.1. To show that no rounding procedure from $\mathcal{THRESH}$ can be used to obtain a better approximation algorithm for MAX 2-SAT, it is enough to show that no rounding procedure can achieve a better performance ratio even on the set of these two configurations.

A rounding procedure from $\mathcal{THRESH}$ is defined using a threshold function $T : \mathbb{R} \times [-1, 1] \to \mathbb{R}$ and a distribution for r_0. To ensure that the literals x_i and $\bar{x}_i$ are assigned opposite values, this function should satisfy $T(x, -y) = -T(x, y)$. As the set C contains only two configurations, it follows from the discussion in Sect. 5 that there is an optimal rounding procedure for it in which r_0 assumes only two different values. Let us denote these two values by s_1 and s_2. Let p be the probability that r_0 gets the value s_1.

Let $\Theta_1 = (\theta, \theta, \theta_{12})$ and $\Theta_2 = (\pi - \theta, \pi - \theta, \theta_{12})$ be the triplets corresponding to the configurations (v_1, v_2) and $(-v_1, -v_2)$. (We have $\theta \simeq 1.401$ and $\theta_{12} = \arccos(2\cos\theta - 1) \simeq 2.294$.) To determine the behavior of this rounding procedure on the configurations (v_1, v_2) and $(-v_1, -v_2)$ we only need to specify the values of the thresholds $T(s_1, \cos\theta)$ and $T(s_2, \cos\theta)$. (Note that $T(s_i, -\cos\theta) = -T(s_i, -\cos\theta)$, for $i = 1, 2$.) The search for the best rounding procedure from $\mathcal{THRESH}$ is therefore reduced to the search for the optimal setting of the three parameters $T(s_1, \cos\theta)$, $T(s_2, \cos\theta)$ and p. (Note that the values of s_1 and s_2 are not important.) We have numerically searched for the best setting of these three parameters and could not improve on the result obtained using the algorithm of Sect. 7.1. This algorithm is, therefore, the best algorithm that can be obtained using *any* rounding procedure from $\mathcal{THRESH}$.

A rounding procedure from $\mathcal{GEN}^+$ is characterized by a function $G : \mathbb{R}^2 \times [-1, 1] \to [0, 1]$ that satisfies $G(x, -y, -z) = 1 - G(x, y, z)$. As before, we may assume that an optimal rounding procedure for C assigns r_0 only two distinct values s_1 and s_2. To determine the behavior of the rounding procedure on the configurations (v_1, v_2) and $(-v_1, -v_2)$ we have to specify p, the probability that r_0 assumes the value s_1, and the two univariate functions $g_1(x) = G(s_1, x, \cos\theta)$ and $g_2(x) = G(s_2, x, \cos\theta)$. We have attempted to improve the performance ratio obtained for C using functions $g_1(x)$ and $g_2(x)$ that are piecewise linear. The attempt failed. This seems to suggest that no improvement can be obtained using rounding procedures from $\mathcal{GEN}^+$. (We plan to elaborate on this point in the full version of the paper.)

8.2   Upper Bound for MAX DI-CUT

The upper bound on the performance ratio that can be obtained for the MAX DI-CUT problem using rounding procedures from the family $\mathcal{THRESH}^-$ is obtained using similar ideas. Things are more complicated here, however, as we need to use more configurations. To obtain an upper bound that almost matches the performance of the MAX DI-CUT algorithm of Sect. 7.2 we need to use a collection of six configurations.

For every $k \geq 1$, Zwick [12] describes a family C_k of $k + 1$ configurations that are parametrized by k angles $\theta_1, \theta_2, \ldots, \theta_k$. For example, for $k = 6$, we have $C_6 = \{\Theta_{17}, \Theta_{26}, \Theta_{35}, \Theta_{44}, \Theta_{43}, \Theta_{52}, \Theta_{61}\}$, where $\Theta_{ij} = (\theta_i, \theta_j, \arccos(\theta_i + \theta_j - 1))$, for $1 \leq i, j \leq 7$, and $\theta_7 = \frac{\pi}{2}$. To obtain our best upper bounds we used the values

$$\theta_1 = 1.24677 \,,\ \theta_2 = 1.30703 \,,\ \theta_3 = 1.36867 \,,$$
$$\theta_4 = 1.39760 \,,\ \theta_5 = 1.44728 \,,\ \theta_6 = 1.50335 \,.$$

To specify a rounding procedure from $\mathcal{THRESH}^-$ we only need to specify a threshold function $S : [-1, 1] \to \mathbb{R}$. To determine the behavior of this rounding procedure on the configurations of C_6, we only need to specify six numbers, namely $S(\cos\theta_i)$, for $1 \leq i \leq 6$. We have numerically searched for the best setting of these six parameters and could not improve on the result obtained using the algorithm of Sect. 7.2. This algorithm is, therefore, the best algorithm that can be obtained using *any* rounding procedure from $\mathcal{THRESH}^-$. We hope to determine, in the near future, whether any improvements are possible using rounding procedures from $\mathcal{THRESH}$.

9   Concluding Remarks

We presented improved approximation algorithms for the MAX 2-SAT and MAX DI-CUT problems using combinations of pre-rounding rotations and skewed hyperplane rounding. We showed that the performance ratios achieved by these algorithms are essentially the best ratios that can be achieved using this technique. Obtaining further improved approximation ratios would require additional ideas.

Our investigation also sheds some new light on the nature a general family of rounding procedures suggested by Feige and Goemans [3].

Acknowledgement

We would like to thank Uri Feige and Michael Langberg for stimulating discussions on skewed hyperplane rounding techniques, and Shiro Matuura and Tomomi Matsui for sending us a copy of [9].

References

1. N. Alon and B. Sudakov. Bipartite subgraphs and the smallest eigenvalue. *Combinatorics, Probability & Computing*, 9:1–12, 2000.
2. N. Alon, B. Sudakov, and U. Zwick. Constructing worst case instances for semidefinite programming based approximation algorithms. *SIAM Journal on Discrete Mathematics*, 15:58–72, 2002.
3. U. Feige and M.X. Goemans. Approximating the value of two prover proof systems, with applications to MAX-2SAT and MAX-DICUT. In *Proceedings of the 3rd Israel Symposium on Theory and Computing Systems, Tel Aviv, Israel*, pages 182–189, 1995.
4. U. Feige and M. Langberg. The RPR^2 rounding technique for semidefinite programs. In *Proceedings of the 28th International Colloquium on Automata, Languages and Programming, Crete, Greece*, pages 213–224, 2001.
5. M.X. Goemans and D.P. Williamson. Improved approximation algorithms for maximum cut and satisfiability problems using semidefinite programming. *Journal of the ACM*, 42:1115–1145, 1995.
6. J. Håstad. Some optimal inapproximability results. In *Proceedings of the 29th Annual ACM Symposium on Theory of Computing, El Paso, Texas*, pages 1–10, 1997. Full version available as E-CCC Report number TR97-037.
7. H. Karloff. How good is the Goemans-Williamson MAX CUT algorithm? *SIAM Journal on Computing*, 29:336–350, 1999.
8. S. Matuura and T. Matsui. 0.863-approximation algorithm for MAX DICUT. In *Approximation, Randomization and Combinatorial Optimization: Algorithms and Techniques, Proceedongs of APPROX-RANDOM'01, Berkeley, California*, pages 138–146, 2001.

9. S. Matuura and T. Matsui. 0.935-approximation randomized algorithm for MAX 2SAT and its derandomization. Technical Report METR 2001-03, Department of Mathematical Engineering and Information Physics, the University of Tokyo, Japan, September 2001.

10. S. Mahajan and H. Ramesh. Derandomizing approximation algorithms based on semidefinite programming. *SIAM Journal on Computing*, 28:1641–1663, 1999.

11. L. Trevisan, G.B. Sorkin, M. Sudan, and D.P. Williamson. Gadgets, approximation, and linear programming. *SIAM Journal on Computing*, 29:2074–2097, 2000.

12. U. Zwick. Analyzing the MAX 2-SAT and MAX DI-CUT approximation algorithms of Feige and Goemans. Submitted for publication, 2000.

13. U. Zwick. Computer assisted proof of optimal approximability results. In *Proceedings of the 13th Annual ACM-SIAM Symposium on Discrete Algorithms, San Francisco, California*, pages 496–505, 2002.

Finding the Exact Integrality Gap
for Small Traveling Salesman Problems*

Sylvia Boyd and Geneviève Labonté

University of Ottawa, Ottawa, Canada

Abstract. The Symmetric Traveling Salesman Problem (STSP) is to
find a minimum weight Hamiltonian cycle in a weighted complete graph
on n nodes. One direction which seems promising for finding improved
solutions for the STSP is the study of a linear relaxation of this problem
called the Subtour Elimination Problem (SEP). A well known conjecture
in combinatorial optimization says that the integrality gap of the SEP is
4/3 in the metric case. Currently the best upper bound known for this
integrality gap is 3/2.
Finding the exact value for the integrality gap for the SEP is difficult
even for small values of n due to the exponential size of the data involved.
In this paper we describe how we were able to overcome such difficulties
and obtain the exact integrality gap for all values of n up to 10. Our
results give a verification of the 4/3 conjecture for small values of n, and
also give rise to a new stronger form of the conjecture which is dependent
on n.

1   Introduction

Given the complete graph $K_n = (V, E)$ on n nodes with non-negative edge costs
$c \in \mathbb{R}^E$, $c \neq 0$, the *Symmetric Traveling Salesman Problem* (henceforth STSP)
is to find a Hamiltonian cycle (or *tour*) in K_n of minimum cost. When the costs
satisfy the triangle inequality, i.e. when $c_{ij} + c_{jk} \geq c_{ik}$ for all $i, j, k \in V$, we call
the problem the *metric* STSP.

For any edge set $F \subseteq E$ and $x \in \mathbb{R}^E$, let $x(F)$ denote the sum $\sum_{e \in F} x_e$.
For any node set $W \subset V$, let $\delta(W)$ denote $\{uv \in E : u \in W, v \notin W\}$. Let
$\mathbf{S} = \{S \subset V, 3 \leq |S| \leq n - 3\}$. An integer linear programming (ILP) formulation
for the STSP is as follows:

$$\text{minimize } cx \tag{1}$$
$$\text{subject to: } x(\delta(v)) = 2 \quad \text{for all } v \in V, \tag{2}$$
$$x(\delta(S)) \geq 2 \quad \text{for all } S \in \mathbf{S}, \tag{3}$$
$$0 \leq x_e \leq 1 \quad \text{for all } e \in E, \tag{4}$$
$$x \quad \text{integer.} \tag{5}$$

We use *TOUR* to denote the optimal solution value for ILP (1). If we drop
the integer requirement (5) from the above ILP, we obtain a linear programming

---

* This research was partially supported by grants from the Natural Sciences and Engineering Research Council of Canada.

W.J. Cook and A.S. Schulz (Eds.): IPCO 2002, LNCS 2337, pp. 83–92, 2002.
© Springer-Verlag Berlin Heidelberg 2002

(LP) relaxation of the STSP called the *Subtour Elimination Problem* (SEP). We use *SUBT* to denote the optimal solution value for the SEP. The associated poly-tope, which is denoted by S^n, is the set of all vectors x satisfying the constraints of the SEP, i.e.

$$S^n = \{x \in \mathbb{R}^E : x \text{ satisfies } (2), (3), (4)\} \ .$$

Note that despite the fact that there are an exponential number of constraints (3), the SEP can be solved in polynomial time, since there is an exact polynomial-time separation algorithm for each of its constraints [7].

The STSP is known to be NP-hard, even in the metric case [8]. One approach taken for finding reasonably good solutions is to look for a k-approximation algorithm for the problem, i.e. try to find a heuristic for the STSP which finds a tour which is guaranteed to be of cost at most $k * (TOUR)$ for some constant $k \geq 1$. Currently the best k-approximation algorithm known for the metric STSP is Christofides algorithm [6] for which $k = 3/2$. Surprisingly, no one has been able to improve upon this algorithm in over two decades. Note that for general costs there does not exist a k-approximation algorithm unless $P = NP$ [8].

A related approach for finding improved STSP solutions is to study the *in-tegrality gap* α for the SEP, which is the worst-case ratio between $TOUR$ and $SUBT$, i.e.

$$\alpha = \max_{c \geq 0} \left(\frac{TOUR}{SUBT} \right) .$$

Value α gives one measure of the quality of the lower bound provided by the SEP for the STSP. Moreover, a constructive proof for value α would provide an α-approximation algorithm for the STSP.

It is known that for the metric STSP, α is at most $3/2$ ([10], [11]), however, no example for which this ratio comes close to $3/2$ has yet been found. In fact, a well-known conjecture states the following:

Conjecture 1. For the metric STSP, the integrality gap for the SEP is $4/3$.

If Conjecture 1 is true, then it is best possible, for consider the metric STSP example with costs c shown in Fig. 1(a). In the figure, imagine the three hori-zontal paths each have length k, and let the cost c_{uv} for edges $uv \in E$ not shown be the cost of a cheapest u to v path in the figure. The optimal solution x^* for the SEP for this set of costs is shown in Fig. 1(b), where $x_e^* = \frac{1}{2}$ for the dashed edges e, 1 for the solid edges e, and 0 for all other edges. The optimal tour is shown in Fig. 1(c). Thus this set of costs gives the ratio

$$\frac{TOUR}{SUBT} = \frac{4k + 6}{3k + 6}$$

which tends to $4/3$ as $k \to \infty$.

Note that not much progress has been made on Conjecture 1 in the past decade. It has been verified for a very special class of cost functions in [2], and recent results which are implied by the correctness of the conjecture can be found in [1] and [4].

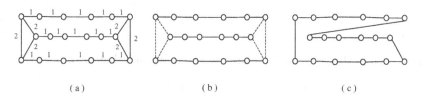

(a)   (b)   (c)

Fig. 1. An example for which $\alpha = 4/3$

In this paper we examine the problem of finding the exact integrality gap for the metric SEP for fixed values of n, where n is small. Note that many models for this problem are too complex and too large to be practical to solve. In Sect. 2 we describe how we were able to overcome such difficulties by instead solving a series of reasonably-sized LPs based on a related problem for a subset of the vertices of the subtour elimination polytope S^n. In Sect. 3 we describe how we found the necessary set of vertices for our method, and also describe some new structural properties for the vertices of S^n which we needed to use in order to make our ideas work. Finally, in Sect. 4 we report our results, i.e. we give the exact integrality gap for all values of n up to 10 which we found using our method. These results not only provide a verification of Conjecture 1 for $n \leq 10$, but also give rise to a new stronger form of the conjecture which is dependent on n. Such a formulation of the conjecture could be useful in an inductive proof approach for Conjecture 1.

2   Finding the Integrality Gap for Fixed n

For the metric STSP, we wish to solve the problem of finding the integrality gap for the SEP when n, the number of nodes in the complete graph $K_n = (V, E)$, is fixed in size. We denote this integrality gap by α_n, i.e.

$$\alpha_n = \max_{c \geq 0} \left(\frac{TOUR}{SUBT} \right)$$

for all metric STSPs on n nodes. Note that it is known that the SEP and the STSP are equivalent problems for $n \leq 5$, and thus $\alpha_n = 1$ for $n \leq 5$.

We first note that for a particular metric STSP, if we divide all the edge costs $c \in \mathbb{R}^E$ by the optimum tour value $TOUR$, then the new costs also satisfy the triangle inequality, and the ratio $TOUR/SUBT$ remains unchanged. Note the new value of $TOUR$ using this new cost function will be 1. Thus to solve our problem, it is sufficient to only consider metric cost functions c for which the minimum tour value $TOUR$ is 1, and we have

$$\alpha_n = \max_{\substack{c \geq 0 \\ TOUR=1 \text{ for } c}} \left(\frac{1}{SUBT} \right),$$

or equivalently,

$$\frac{1}{\alpha_n} = \min_{\substack{c \geq 0 \\ TOUR=1 \text{ for } c}} (SUBT) \ . \tag{6}$$

This leads to the following quadratic programming model of the problem:

$$\text{minimize } cx \tag{7}$$

$$\text{subject to: } x \text{ satisfies constraints (2)-(4)} \tag{8}$$

$$x(T) \geq 1 \qquad \text{for all } T \in \mathbf{T}, \tag{9}$$

$$c_{ij} + c_{jk} \geq c_{ik} \qquad \text{for all } i, j, k \in V, \tag{10}$$

$$c_e \geq 0 \qquad \text{for all } e \in E, \tag{11}$$

where $\mathbf{T}$ represents all the 0-1 incidence vectors of tours of K_n. Constraints (8) simply ensure that vector x is in the subtour elimination polytope S^n, while constraints (9) and (10) ensure that the minimum tour value is 1 for c and our cost vector c is metric. Note that if the optimal solution for (7) is k, then α_n has value $1/k$. Also note that this model has an exponential number of constraints of types (8) and (9).

As the model above proved impractical to solve even for small n using the tools available to us, we tried several other models in an attempt to get rid of the quadratic objective function. One such model was a binary integer programming model with an explosive number of both constraints and variables. We were able to use this model to find α_n for $n = 6, 7$ and 8, but only through a great deal of exhaustive work used to reduce the problem size by fixing different sets of variables. It was clear that this method was impractical for $n > 8$.

We then found a way of modeling our problem as a series of LPs of a more reasonable size (although still exponential) as follows. Let $\mathbf{X} = \{x_{(1)}, x_{(2)}, \ldots, x_{(p)}\}$ be a complete list of all the vertices of the subtour elimination polytope S^n. We know from polyhedral theory that for every cost function $c \in \mathbb{R}^E$, there exists at least one vertex x^* in $\mathbf{X}$ such that c is minimized over S^n at x^*, i.e. such that $SUBT = cx^*$. This provides a way of breaking our set of cost functions c into different sets (although these sets will not be disjoint). For each vertex $x_{(i)}$, let

$$C_i = \{c \in \mathbb{R}^E : c \text{ is minimized over } S^n \text{ at } x_{(i)}\} \ . \tag{12}$$

So now we need to solve the following problem for each C_i, where $x_{(i)}$ is fixed and is no longer a variable vector:

$$(\min cx_{(i)} : c \text{ satisfies constraints (9), (10) and (11)}, c \in C_i) \ . \tag{13}$$

If we let OPT_i represent the optimal value of (13), then from (6) we have

$$\frac{1}{\alpha_n} = \min_{1 \leq i \leq p} (OPT_i). \tag{14}$$

Note that everything in model (13) is linear except the condition that $c \in C_i$. This condition can also be represented by a set of linear constraints using duality theory, as we explain below.

The *dual* LP of the SEP has variables $y \in \mathbb{R}^V$, $u \in \mathbb{R}^E$, and $d \in \mathbb{R}^\mathbf{S}$ and is defined as follows:

$$\text{maximize } 2 \cdot y - 1 \cdot u + 2 \cdot d \tag{15}$$

$$\text{subject to: } y_i + y_j - u_{ij} + \sum(d_S : S \in \mathbf{S}, ij \in \delta(S)) \le c_{ij} \quad \forall \, ij \in E, \tag{16}$$

$$u_e \ge 0 \quad \forall \, e \in E, \tag{17}$$

$$d_S \ge 0 \quad \forall \, S \in \mathbf{S}. \tag{18}$$

Given a feasible solution x^* for the SEP, a feasible solution (y, u, d) for the dual LP (15) is said to *satisfy the complementary slackness conditions for x^** if it satisfies the following:

$$y_i + y_j - u_{ij} + \sum(d_S : S \in \mathbf{S}, ij \in \delta(S)) = c_{ij} \quad \forall \, ij \in E, x_{ij}^* > 0, \tag{19}$$

$$u_e = 0 \quad \forall \, e \in E, x_e^* < 1, \tag{20}$$

$$d_S = 0 \quad \forall \, S \in \mathbf{S}, x^*(\delta(S)) > 2. \tag{21}$$

It follows from the theory of duality and complementary slackness that cost function $c \in C_i$ if and only if there exists a feasible solution (y, u, d) for the dual LP (15) which satisfies the complementary slackness conditions (19), (20) and (21) for $x_{(i)}$. Thus for each C_i we can find OPT_i by solving the following LP, which has variables c and (y, u, d):

$$OPT_i = (\min cx_{(i)} : c \text{ satisfies (9) - (11) , and}$$
$$(y, u, d) \text{ satisfy (16) - (21) for } x_{(i)}). \tag{22}$$

Note that (22) has an exponential number of variables and constraints. Nevertheless, for small values of n we were able to solve (22) using CPLEX in a reasonable amount of time. For example, for $n = 10$, each of these LPs took approximately one to two minutes to solve.

3   Finding the Necessary Vertices for S^n

We need to avoid solving the LP (22) for OPT_i for every vertex $x_{(i)} \in \mathbf{X}$ as the size of $\mathbf{X}$ can be very large even for small n. Luckily we were able to find a much smaller subset of vertices which was sufficient for finding α_n.

Given a vertex $x \in S^n$, the *weighted support graph* $G_x = (V_x, E_x)$ of x is the subgraph of K_n induced by the edge set $E_x = \{e \in E : x_e > 0\}$, with edge weights x_e for $e \in \mathbb{R}^{E_x}$. Observe that many of the vertices of S^n will have isomorphic weighted support graphs (such as all the vertices representing tours), and thus will have the same objective value OPT_i for LP (22). This observation led to the following steps for reducing the number of vertices we needed to consider:

Step 1 Generate the set of all the vertices of S^n. To do this we used a software package called PORTA (POlyhedron Representation Transformation Algorithm) [5], a program which, given an LP, generates all the vertices for the corresponding polyhedron.

Step 2 Reduce the set of vertices from Step 1 by finding the vertices within our set with non-isomorphic weighted support graphs. To do this we used a software package called nauty [9]. Note that this package deals with unweighted graphs, but we were able to transform our weighted graphs such that this package could be used for our purposes.

The above method worked for finding α_n for $n = 6, 7$, however for $n = 8$, PORTA was unable to generate all of the vertices for S^n even after running for days. Thus for $n \geq 8$ we needed to exploit some properties of the vertices of S^n in order to aid PORTA in Step 1. These properties are described in the three theorems below. The proofs of these theorems use ideas similar to those found in [3].

Theorem 1. *Let x be a vertex of S^n. Then the number of edges in the support graph G_x of x is at most $2n - 3$.*

Proof. This result follows from a stronger result in [3] which says that the number of edges in G_x is at most $2n - k - 3$, where k is the number of nodes in G_x which have degree 3 and for which none of the corresponding incident edges e have value $x_e = 1$.   □

Theorem 2. *Consider $x \in \mathbb{R}^E$ such that for some node v we have $x_{uv} = x_{vw} = 1$. Let $\hat{x}$ be the vector indexed by the edge set of $K_n \setminus \{v\}$ defined by*

$$\hat{x}_e = \begin{cases} 1 & \text{if } e = uw, \\ x_e & \text{for all other edges } e \ . \end{cases} \tag{23}$$

Then $\hat{x}$ is a vertex of S^{n-1} if and only if x is a vertex of S^n.

Proof. We begin by showing that $x \in S^n$ if and only if $\hat{x} \in S^{n-1}$.

Suppose that $x \in S^n$. The fact that x satisfies $x(\delta(\{u, v, w\})) \geq 2$ implies that $x_{uw} = 0$. Therefore, since $\hat{x}_{uw} = 1$, the sum of the x_e values for edges incident with u and edges incident with w did not change when $\hat{x}$ was created. Thus the constraints (2) are satisfied by $\hat{x}$ for nodes u and w. For all other constraints (2), (3) and (4) for the SEP, it is clear that if x satisfies them, then so does $\hat{x}$. Hence $\hat{x} \in S^{n-1}$.

Suppose that $\hat{x} \in S^{n-1}$. Vector x is essentially formed from $\hat{x}$ by taking edge uw of value 1 in the support graph of $\hat{x}$ and adding a node v to split it into two edges uv and vw, each with value $x_e = 1$. Thus $x(\delta(v)) = 2$ for vector x. For any $S \in \mathbf{S}$ such that $v \in S$ and $u, w \notin S$, we have $x(\delta(S)) \geq x_{uv} + x_{vw} = 2$, and thus all such cut constraints are satisfied by x. For all other constraints (2), (3) and (4) for the SEP, it is clear that if $\hat{x}$ satisfies them, then so does x. Hence $x \in S^n$.

Now suppose that $x \in S^n$ and x is not a vertex of S^n. This implies that x can be expressed as a convex combination of k distinct points of S^n, $k \geq 2$. For each of these k points x', we must have $x'_{uv} = x'_{vw} = 1$, since $x_{uv} = x_{vw} = 1$. So for each point x' we can form a corresponding point $\hat{x}'$ which will necessarily be in S^{n-1} by the discussion above. By doing this we obtain $\hat{x}$ as a convex combination of k distinct points of S^{n-1}, $k \geq 2$. This shows that $\hat{x}$ is also not a vertex of S^{n-1}.

In a similar manner, we can show that if $\hat{x} \in S^{n-1}$ is not a vertex of S^{n-1} then x is not a vertex of S^n. This completes the proof.   □

Theorem 3. *Let x be a vertex of S^n, $n \geq 5$. Then the maximum degree in the support graph G_x of x is $n - 3$.*

Proof. Suppose the theorem is not true, and for some vertex x of S^n there is a node of degree greater than $n - 3$ in its support graph. Consider a counter example G_x with the number of nodes n as small as possible. We know that the theorem is true for $n = 5$ since all vertices of S^5 correspond to tours. Hence we can assume that $n \geq 6$.

Suppose that G_x has a node v of degree 2, which implies that we have $x_{uv} = x_{vw} = 1$. Thus by Theorem 2 we could shrink this vertex x to a vertex $\hat{x}$ of S^{n-1} whose support graph would also contain a node of degree greater than $n - 3$. This contradicts the fact that G_x was a smallest counter example.

If G_x has no nodes of degree 2, then

$$2|E(G_x)| = \sum (d_x(v) : v \in V) \geq 3(n - 1) + (n - 2) = 4n - 5 \ ,$$

where $d_x(v)$ denotes the degree of node v in G_x. However, by Theorem 1 we have that $2|E(G_x)| \leq 4n - 6$, so again we have a contradiction.   □

We used Theorems 1 and 3 to reduce the work for PORTA by first generating all the non-isomorphic graphs on n nodes that contain $2n - 3$ edges and have a maximum degree of $n - 3$ (i.e. all the possible non-isomorphic support graphs for vertices of S^n), and then for each of these graphs G' we ran PORTA on the SEP problem constraints with the added constraints $x_e = 0$ for each of the edges e not appearing in G'.

Theorem 2 implies that we can obtain any vertex of S^n which has a node of degree 2 in its support graph by subdividing a 1-edge in the support graph of a vertex of S^{n-1}. Thus we obtained all the vertices with a node of degree 2 in the support graph directly from our list of non-isomorphic vertices for S^{n-1}, and only used PORTA to search for vertices for which all nodes of the corresponding support graph have degree at least 3. This meant that we could add the assumption that our subgraphs for the support graphs have a minimum degree of 3 in the above.

To summarize, we found α_n as follows:

Step 1' Find all the non-isomorphic graphs on n nodes that contain $3n - 2$ edges, and for which all nodes have a maximum degree of $n - 3$ and a minimum degree of 3.

Step 2' For each graph G' from Step 1', use PORTA to find all vertices of the polytope defined by the constraints of the SEP plus the constraints $x_e = 0$ for all edges not appearing in G'. (Note that this polytope is a face of S^n.) Let **A** represent the union of all of these vertices found.

Table 1. Integrality gap results

n	Number of vertices	α_n
6	2	10/9
7	3	9/8
8	13	8/7
9	38	7/6
10	421	20/17

Step 3′ Find all the vertices of S^n which have a node of degree 2 in the support graph by directly generating them from the list of non-isomorphic vertices created for S^{n-1}. Let **B** represent this set of vertices.

Step 4′ Use the nauty software package to find the vertices in $\mathbf{A} \cup \mathbf{B}$ with non-isomorphic weighted support graphs.

Step 5′ For each vertex $x_{(i)}$ in the set of vertices from Step 4′, solve the LP (22) to find OPT_i. Let the minimum value for OPT_i found be k. Then by (14), α_n is $1/k$.

4   Results and a New Stronger Conjecture

Using our method, we were able to find the exact value of α_n for all values of n up to 10. The results are shown in Table 1. In column 2 of the table we list the number of non-isomorphic vertices of S^n we found. Note that this also represents the number of LPs (22) that we needed to solve for each value of n. Column 3 contains the value of α_n, which is clearly less than 4/3 for all $n \le 10$.

It surprised us that for each value of n, there was a unique vertex of S^n that gave the maximum ratio α_n. Even more surprising was the definite pattern that these vertices formed. The unique vertex x^* of S^n which gave the maximum ratio for each value of n is shown in Fig. 2, where $x_e^* = 1$ for solid edges e, $x_e^* = \frac{1}{2}$ for dashed edges e, and $x_e^* = 0$ for all the other edges. Note that the costs c which gave α_n are shown on the edges, and for edges not shown the cost c_{ij} was the cost of a minimum cost i to j path using the costs shown.

By extrapolating the pattern of the vertices shown in Fig. 2, we developed the following conjecture for the value of α_n.

Conjecture 2. For all integers $n \ge 3$,

$$\alpha_n = \begin{cases} \dfrac{4n+6}{3n+9} & \text{if } n = 0 \text{ (mod 3)}, \\[2mm] \dfrac{4\lfloor \frac{n}{3} \rfloor^2 + 2\lfloor \frac{n}{3} \rfloor - 2}{3\lfloor \frac{n}{3} \rfloor^2 + 3\lfloor \frac{n}{3} \rfloor - 2} & \text{if } n = 1 \text{ (mod 3)}, \\[2mm] \dfrac{4\lfloor \frac{n}{3} \rfloor^2 + 2\lfloor \frac{n}{3} \rfloor - 4}{3\lfloor \frac{n}{3} \rfloor^2 + 3\lfloor \frac{n}{3} \rfloor - 4} & \text{if } n = 2 \text{ (mod 3)}. \end{cases} \quad (24)$$

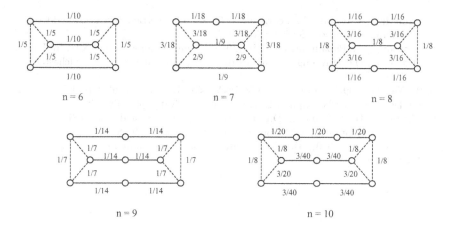

Fig. 2. Vertices and costs which give α_n

Note that in the formulas in this new conjecture, α_n approaches $4/3$ as $n \to \infty$. Thus the correctness of Conjecture 2 would imply that Conjecture 1 is true. Also note that for each n we have found a set of costs for which the value in Conjecture 2 is tight, i.e. for which $TOUR/SUBT = \alpha_n$ as defined by the formula in Conjecture 2.

Acknowledgments

We would like to thank Lucia Moura and Marc Raaphorst for their helpful suggestions on creating non-isomorphic subgraphs with certain degree properties.

References

1. Boyd, S., Carr. R. (1999): A new bound for the ratio between the 2-matching problem and its linear programming relaxation, Math. Prog. Series A **86**, 499-514.
2. Boyd, S., Carr, R. (2000): Finding low cost TSP and 2-matching solutions using certain half-integer subtour vertices, Technical Report TR-96-12, University of Ottawa, Ottawa.
3. Boyd, S., Pulleyblank, W.R. (1991): Optimizing over the subtour polytope of the traveling salesman problem, Math. Prog. **49**, 163-187.
4. Carr, R., Ravi, R. (1998): A new bound for the 2-edge connected subgraph problem, Proceedings of the Conference on Integer Programming and Combinatorial Optimization (IPCO'98).
5. Christof, T., Löbel, A, Stoer, M. (1997): PORTA, A POlyhedron Representation Transformation Algorithm,
   http://www.zib.de/Optimization/Software/Porta/index.html
6. Christofides, N. (1976): Worst case analysis of a new heuristic for the traveling salesman problem, Report 388, Graduate School of Industrial Administration, Carnegie Mellon University, Pittsburgh.

7. Grötschel, M., Lovasz, L., Schrijver, A. (1988): Geometric Algorithms and Combinatorial Optimization, Springer-Verlag, Berlin.

8. Johnson, D.S., Papadimitriou, C.H. (1985): Computational Complexity, In: Lawler et al, eds., The Traveling Salesman Problem , John Wiley & Sons, Chichester, 37-85.

9. McKay, B. (1991): nauty User's Guide (Version 1.5), Technical Report TR-CS-90-02, Department of Computer Science, Australia National University.

10. Shmoys, D.B., Williamson, D.P. (1990): Analyzing the Held-Karp TSP bound: A monotonicity property with application, Inf. Process. Lett. **35**, 281-285.

11. Woolsey, L.A. (1980): Heuristic analysis, linear programming and branch and bound, Math. Prog. Study **13**, 121-134.

Polynomial-Time Separation
of Simple Comb Inequalities

Adam N. Letchford[1] and Andrea Lodi[2]

[1] Department of Management Science, Lancaster University,
Lancaster LA1 4YW, England
A.N.Letchford@lancaster.ac.uk
[2] D.E.I.S., University of Bologna,
Viale Risorgimento 2, 40136 Bologna, Italy
alodi@deis.unibo.it

Abstract. The *comb* inequalities are a well-known class of facet-inducing inequalities for the Traveling Salesman Problem, defined in terms
of certain vertex sets called the *handle* and the *teeth*. We say that a
comb inequality is *simple* if the following holds for each tooth: either
the intersection of the tooth with the handle has cardinality one, or the
part of the tooth outside the handle has cardinality one, or both. The
simple comb inequalities generalize the classical *2-matching* inequalities
of Edmonds, and also the so-called *Chvátal comb* inequalities.

In 1982, Padberg and Rao [29] gave a polynomial-time algorithm for *separating* the 2-matching inequalities – i.e., for testing if a given fractional
solution to an LP relaxation violates a 2-matching inequality. We extend
this significantly by giving a polynomial-time algorithm for separating
the simple comb inequalities. The key is a result due to Caprara and
Fischetti.

1 Introduction

The famous *Symmetric Traveling Salesman Problem* (STSP) is the $\mathcal{NP}$-hard
problem of finding a minimum cost Hamiltonian cycle (or *tour*) in a complete
undirected graph. The most successful optimization algorithms at present (e.g.,
Padberg & Rinaldi [31], Applegate, Bixby, Chvátal & Cook [1]), are based on
an integer programming formulation of the STSP due to Dantzig, Fulkerson &
Johnson [7], which we now describe.

Let G be a complete graph with vertex set V and edge set E. For each
edge $e \in E$, let c_e be the cost of traversing edge e. For any $S \subset V$, let $\delta(S)$
(respectively, $E(S)$), denote the set of edges in G with exactly one end-vertex
(respectively, both end vertices) in S. Then, for each $e \in E$, define the 0-1
variable x_e taking the value 1 if e is to be in the tour, 0 otherwise. Finally let
$x(F)$ for any $F \subset E$ denote $\sum_{e \in F} x_e$. Then the formulation is:

W.J. Cook and A.S. Schulz (Eds.): IPCO 2002, LNCS 2337, pp. 93–108, 2002.
© Springer-Verlag Berlin Heidelberg 2002

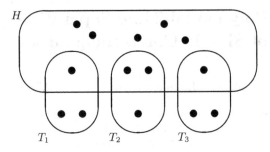

Fig. 1. A comb with three teeth

Minimize $\sum_{e \in E} c_e x_e$

Subject to:

$$x(\delta(\{i\})) = 2 \qquad \forall i \in V, \tag{1}$$

$$x(E(S)) \leq |S| - 1 \qquad \forall S \subset V : 2 \leq |S| \leq |V| - 2, \tag{2}$$

$$x_e \geq 0 \qquad \forall e \in E, \tag{3}$$

$$x \in Z^{|E|} . \tag{4}$$

Equations (1) are called *degree equations*. The inequalities (2) are called *subtour elimination constraints* (SECs) and the inequalities (3) are simple non-negativity conditions. Note that an SEC with $|S| = 2$ is a mere upper bound of the form $x_e \leq 1$ for some edge e.

The convex hull in $\mathbb{R}^{|E|}$ of vectors satisfying (1) - (4) is called a *Symmetric Traveling Salesman Polytope*. The polytope defined by (1) - (3) is called a *Subtour Elimination Polytope*. These polytopes are denoted by STSP(n) and SEP(n) respectively, where $n := |V|$. Clearly, STSP(n) $\subseteq$ SEP(n), and containment is strict for $n \geq 6$.

The polytopes STSP(n) have been studied in great depth and many classes of valid and facet-inducing inequalities are known; see the surveys by Jünger, Reinelt & Rinaldi [19,20] and Naddef [22]. Here we are primarily interested in the *comb* inequalities of Grötschel & Padberg [15,16], which are defined as follows. Let $t \geq 3$ be an odd integer. Let $H \subset V$ and $T_j \subset V$ for $j = 1, \ldots, t$ be such that $T_j \cap H \neq \emptyset$ and $T_j \setminus H \neq \emptyset$ for $j = 1, \ldots, t$, and also let the T_j be vertex-disjoint. (See Fig. 1 for an illustration.) The comb inequality is:

$$x(E(H)) + \sum_{j=1}^{t} x(E(T_j)) \leq |H| + \sum_{j=1}^{t} |T_j| - \lceil 3t/2 \rceil . \tag{5}$$

The set H is called the *handle* of the comb and the T_j are called *teeth*.

Comb inequalities induce facets of STSP(n) for $n \geq 6$ [15,16]. The validity of comb inequalities in the special case where $|T_j \cap H| = 1$ for all j was proved by Chvátal [6]. For this reason inequalities of this type are sometimes referred to as *Chvátal comb* inequalities. If, in addition, $|T_j \setminus H| = 1$ for all j, then the inequalities reduce to the classical *2-matching* inequalities of Edmonds [8].

In this paper we are concerned with a class of inequalities which is intermediate in generality between the class of comb inequalities and the class of Chvátal comb inequalities. For want of a better term, we call them *simple* comb inequalities, although the reader should be aware that the term *simple* is used with a different meaning in Padberg & Rinaldi [30], and with yet another meaning in Naddef & Rinaldi [23,24].

Definition 1. *A comb (and its associated comb inequality) will be said to be simple if, for all j, either $|T_j \cap H| = 1$ or $|T_j \setminus H| = 1$ (or both).*

So, for example, the comb shown in Fig. 1 is simple because $|T_1 \cap H|$, $|T_2 \setminus H|$ and $|T_3 \cap H|$ are all equal to 1. Note however that it is not a Chvátal comb, because $|T_2 \cap H| = 2$.

For a given class of inequalities, a *separation algorithm* is a procedure which, given a vector $x^* \in \mathbb{R}^{|E|}$ as input, either finds an inequality in the class which is violated by x^*, or proves that none exists (see Grötschel, Lovász & Schrijver [14]). A desirable property of a separation algorithm is that it runs in polynomial time.

In 1982, Padberg & Rao [29] discovered a polynomial-time separation algorithm for the *2-matching* inequalities. First, they showed that the separation problem is equivalent to the problem of finding a *minimum weight odd cut* in a certain weighted labeled graph. This graph has $\mathcal{O}(|E^*|)$ vertices and edges, where $E^* := \{e \in E : x_e^* > 0\}$. Then, they proved that the desired cut can be found by solving a sequence of $O(|E^*|)$ max-flow problems. Using the well-known *pre-flow push* algorithm (Goldberg & Tarjan [12]) to solve the max-flow problems, along with some implementation tricks given in Grötschel & Holland [13], the Padberg-Rao separation algorithm can be implemented to run in $\mathcal{O}(n|E^*|^2 \log(n^2/|E^*|))$ time, which is $\mathcal{O}(n^5)$ in the worst case, but $\mathcal{O}(n^3 \log n)$ if the support graph is sparse. (The *support graph*, denoted by G^*, is the subgraph of G induced by E^*.)

In Padberg & Grötschel [28], page 341, it is conjectured that there also exists a polynomial-time separation algorithm for the more general *comb* inequalities. This conjecture is still unsettled, and in practice many researchers resort to *heuristics* for comb separation (see for example Padberg & Rinaldi [30], Applegate, Bixby, Chvátal & Cook [1], Naddef & Thienel [25]). Nevertheless, some progress has recently been made on the theoretical side. In chronological order:

- Carr [5] showed that, for a *fixed* value of t, the comb inequalities with t teeth can be separated by solving $\mathcal{O}(n^{2t})$ maximum flow problems, i.e., in $\mathcal{O}(n^{2t+1}|E^*|\log(n^2/|E^*|))$ time using the pre-flow push algorithm.
- Fleischer & Tardos [9] gave an $\mathcal{O}(n^2 \log n)$ algorithm for detecting *maximally violated* comb inequalities. (A comb inequality is maximally violated if it is violated by $\frac{1}{2}$, which is the largest violation possible if $x^* \in SEP(n)$.) However this algorithm only works when G^* is planar.
- Caprara, Fischetti & Letchford [3] showed that the comb inequalities were contained in a more general class of inequalities, called $\{0, \frac{1}{2}\}$-*cuts*, and showed how to detect maximally violated $\{0, \frac{1}{2}\}$-cuts in $\mathcal{O}(n^2|E^*|)$ time.

- Letchford [21] defined a different generalization of the comb inequalities, called *domino-parity* inequalities, and showed that the associated separation problem can be solved in $\mathcal{O}(n^3)$ time when G^* is planar.
- Caprara & Letchford [4] showed that, if the *handle H* is fixed, then the separation problem for $\{0, \frac{1}{2}\}$-cuts can be solved in polynomial time. They did not analyze the running time, but the order of the polynomial is likely to be very high.

In this paper we make another step forward in this line of research, by proving the following theorem:

Theorem 1. *Simple comb inequalities can be separated in polynomial time, provided that $x^* \in SEP(n)$.*

This is a significant extension of the Padberg-Rao result. The proof is based on some results of Caprara & Fischetti [2] concerning $\{0, \frac{1}{2}\}$-cuts, together with an 'uncrossing' argument which enables one to restrict attention to a small (polynomial-sized) collection of *candidate teeth*.

The structure of the paper is as follows. In Sect. 2 we summarize the results given in [2] about $\{0, \frac{1}{2}\}$-cuts and show how they relate to the simple comb inequalities. In Sect. 3 the uncrossing argument is given. In Sect. 4 we describe the separation algorithm and analyze its running time, which turns out to be very high at $\mathcal{O}(n^9 \log n)$. In Sect. 5, we show that the running time can be reduced to $\mathcal{O}(n^3 |E^*|^3 \log n)$, and suggest ways in which it could be reduced further. Conclusions are given in Sect. 6.

2   Simple Comb Inequalities as $\{0, \frac{1}{2}\}$-Cuts

As mentioned above, we will need some definitions and results from Caprara & Fischetti [2]. We begin with the definition of $\{0, \frac{1}{2}\}$-*cuts*:

Definition 2. *Given an integer polyhedron $P_I := conv\{x \in Z_+^q : Ax \le b\}$, where A is a $p \times q$ integer matrix and b is a column vector with p integer entries, a $\{0, \frac{1}{2}\}$-cut is a valid inequality for P_I of the form*

$$\lfloor \lambda A \rfloor x \le \lfloor \lambda b \rfloor, \tag{6}$$

where $\lambda \in \{0, \frac{1}{2}\}^p$ is chosen so that λb is not integral.

(Actually, Caprara & Fischetti gave a more general definition, applicable when variables are not necessarily required to be non-negative; but the definition given here will suffice for our purposes. Also note that an equation can easily be represented by two inequalities.)

Caprara and Fischetti showed that many important classes of valid and facet-inducing inequalities, for many combinatorial optimization problems, are $\{0, \frac{1}{2}\}$-cuts. They also showed that the associated separation problem is strongly $\mathcal{NP}$-hard in general, but polynomially solvable in certain special cases. To present these special cases, we need two more definitions:

Definition 3. *The* mod-2 support *of an integer matrix A is the matrix obtained by replacing each entry in A by its parity (0 if even, 1 if odd).*

Definition 4. *A $p \times q$ binary matrix A is called an* edge-path incidence matrix *of a tree (EPT for short), if there is a tree T with p edges such that each column of A is the characteristic vector of the edges of a path in T.*

The main theorem in [2] is then the following:

Theorem 2 (Caprara & Fischetti [2]). *The separation problem for $\{0, \frac{1}{2}\}$-cuts for a system $Ax \leq b$ can be solved in polynomial time if the mod-2 support of A, or its transpose, is EPT.*

Now let us say that the ith inequality in the system $Ax \leq b$ is *used* if its multiplier is non-zero, i.e., if $\lambda_i = \frac{1}{2}$. Moreover, we will also say that the non-negativity inequality for a given variable x_i has been *used* if the i-th coefficient of the vector λA is fractional. The reasoning behind this is that rounding down the coefficient of x_i on the left hand side of (6) is equivalent to adding one half of the non-negativity inequality $x_i \geq 0$ (written in the reversed form, $-x_i \leq 0$).

Given these definitions, it can be shown that:

Proposition 1 (Caprara & Fischetti [2]). *Let $x^* \in \mathbb{R}^q_+$ be a point to be separated. Then a $\{0, \frac{1}{2}\}$-cut is violated by x^* if and only if the sum of the slacks of the inequalities used, computed with respect to x^*, is less than 1.*

Under the (reasonable) assumption that $Ax^* \leq b$, all slacks are non-negative and Proposition 1 also implies that the slack of each inequality used must be less than 1.

The reason that these results are of relevance is that comb inequalities can be derived as $\{0, \frac{1}{2}\}$-cuts from the degree equations and SECs; see Caprara, Fischetti & Letchford [3] for details. However, we have not been able to derive a polynomial-time separation algorithm for comb inequalities (simple or not) based on this observation alone. Instead we have found it necessary to consider a certain *weakened version* of the SECs, presented in the following lemma:

Lemma 1. *For any $S \subset V$ such that $1 \leq |S| \leq |V| - 2$, and any $i \in V \setminus S$, the following* tooth *inequality is valid for STSP(n):*

$$2x(E(S)) + x(E(i : S)) \leq 2|S| - 1, \tag{7}$$

where $E(i : S)$ denotes the set of edges with i as one end-node and the other end-node in S.

Proof: The inequality is the sum of the SEC on S and the SEC on $S \cup \{i\}$.

We will call i the 'root' of the tooth and S the 'body'.

The next proposition shows that, if we are only interested in *simple* comb inequalities, we can work with the tooth inequalities instead of the (stronger) SECs:

Proposition 2. *Simple comb inequalities can be derived as $\{0, \frac{1}{2}\}$-cuts from the degree equations (1) and the tooth inequalities (7).*

Proof: First, sum together the degree equations for all $i \in H$ to obtain:

$$2x(E(H)) + x(\delta(H)) \leq 2|H|. \tag{8}$$

Now suppose, without loss of generality, that there is some $1 \leq k \leq t$ such that $|T_j \cap H| = 1$ for $j = 1, \ldots, k$, and $|T_j \setminus H| = 1$ for $k+1, \ldots, t$. For $j = 1, \ldots, k$, associate a tooth inequality of the form (7) with tooth T_j, by setting $\{i\} := T_j \cap H$ and $S := T_j \setminus H$. Similarly, for $j = k+1, \ldots, t$, associate a tooth inequality with tooth T_j, by setting $\{i\} := T_j \setminus H$ and $S := T_j \cap H$. Add all of these tooth inequalities to (8) to obtain:

$$2x(E(H)) + x(\delta(H)) + \sum_{j=1}^{k} \left(2x(E(T_j \setminus H)) + x(E(T_j \cap H : T_j \setminus H))\right)$$

$$+ \sum_{j=k+1}^{t} \left(2x(E(T_j \cap H)) + x(E(T_j \cap H : T_j \setminus H))\right) \leq 2|H| + 2\sum_{j=1}^{t} |T_j| - 3t.$$

This can be re-arranged to give:

$$2x(E(H)) + 2\sum_{j=1}^{t} x(E(T_j)) + x\left(\delta(H) \setminus \bigcup_{j=1}^{t} E(T_j \cap H : T_j \setminus H)\right)$$

$$\leq 2|H| + 2\sum_{j=1}^{t} |T_j| - 3t.$$

Dividing by two and rounding down yields (5).

Although not crucial to the remainder of this paper, it is interesting to note that the SECs (2) can themselves be regarded as 'trivial' $\{0, \frac{1}{2}\}$-cuts, obtained by dividing a single tooth inequality by two and rounding down. Indeed, as we will show in the full version of this paper, any $\{0, \frac{1}{2}\}$-cut which is derivable from the degree equations and tooth inequalities is either an SEC, or a simple comb inequality, or dominated by these inequalities.

3   Candidate Teeth: An Uncrossing Argument

Our goal in this paper is to apply the results of the previous section to yield a polynomial-time separation algorithm for simple comb inequalities. However, a problem which immediately presents itself is that there is an exponential number of tooth inequalities, and therefore the system $Ax \leq b$ defined by the degree and tooth inequalities is of exponential size.

Fortunately, Proposition 1 tells us that we can restrict our attention to tooth inequalities whose slack is less than 1, without losing any violated $\{0, \frac{1}{2}\}$-cuts. Such tooth inequalities are polynomial in number, as shown in the following two lemmas.

Lemma 2. *Suppose that $x^* \in SEP(n)$. Then the number of sets whose SECs have slack less than $\frac{1}{2}$ is $\mathcal{O}(n^2)$, and these sets can be found in $\mathcal{O}(n|E^*|(|E^*| + n \log n))$ time.*

Proof: The degree equations can be used to show that the slack of the SEC on a set S is less than $\frac{1}{2}$ if and only if $x^*(\delta(S)) < 3$. Since the minimum cut in G^* has weight 2, we require that the cut-set $\delta(S)$ has a weight strictly less than $\frac{3}{2}$ times the weight of the minimum cut. It is known (Hensinger & Williamson [18]) that there are $\mathcal{O}(n^2)$ such sets, and that the algorithm of Nagamochi, Nishimura & Ibaraki [26] finds them in $\mathcal{O}(n|E^*|(|E^*| + n \log n))$ time.

Lemma 3. *Suppose that $x^* \in SEP(n)$. Then the number of distinct tooth inequalities with slack less than 1 is $\mathcal{O}(n^3)$, and these teeth can be found in $\mathcal{O}(n|E^*|(|E^*| + n \log n))$ time.*

Proof: The slack of the tooth inequality is equal to the slack of the SEC for S plus the slack of the SEC for $\{i\} \cup S$. For the tooth inequality to have slack less than 1, the slack for at least one of these SECs must be less than $\frac{1}{2}$. So we can take each of the $\mathcal{O}(n^2)$ sets mentioned in Lemma 2 and consider them as candidates for either S or $\{i\} \cup S$. For each candidate, there are only n possibilities for the root i. The time bottleneck is easily seen to be the Nagamochi, Nishimura & Ibaraki algorithm.

Now consider the system of inequalities $Ax \leq b$ formed by the degree equations and the $\mathcal{O}(n^3)$ tooth inequalities mentioned in Lemma 3. If we could show that the mod-2 support of A (or its transpose) is always an EPT matrix, then we would be done. Unfortunately this is *not* the case. (It is easy to produce counter-examples even for $n = 6$.)

Therefore we must use a more involved argument if we wish to separate simple comb inequalities via $\{0, \frac{1}{2}\}$-cut arguments. It turns out that the key is to pay special attention to tooth inequalities whose slack is strictly less than $\frac{1}{2}$. This leads us to the following definition and lemma:

Definition 5. *A tooth in a comb is said to be* light *if the slack of the associated tooth inequality is less than $\frac{1}{2}$. Otherwise it is said to be* heavy.

Lemma 4. *If a simple comb inequality is violated by a given $x^* \in SEP(n)$, then at most one of its teeth can be heavy.*

Proof: If two of the teeth in a comb are heavy, the slacks of the associated tooth inequalities sum to at least $\frac{1}{2} + \frac{1}{2} = 1$. Then, by Proposition 1, the comb inequality is not violated.

We now present a slight refinement of the light/heavy distinction.

Definition 6. *For a given $i \in V$, a vertex set $S \subset V \setminus \{i\}$ is said to be i-*light *if the tooth inequality with root i and body S has slack strictly less than $\frac{1}{2}$. Otherwise it is said to be i-*heavy.

Now we invoke our uncrossing argument. The classical definition of *crossing* is as follows:

Definition 7. *Two vertex sets* $S_1, S_2 \subset V$ *are said to* cross *if each of the four sets* $S_1 \cap S_2$, $S_1 \setminus S_2$, $S_2 \setminus S_1$ *and* $V \setminus (S_1 \cup S_2)$ *is non-empty.*

However, we will need a slightly modified definition:

Definition 8. *Let* $i \in V$ *be a fixed root. Two vertex sets* $S_1, S_2 \subset V \setminus \{i\}$ *are said to* i-cross *if each of the four sets* $S_1 \cap S_2$, $S_1 \setminus S_2$, $S_2 \setminus S_1$ *and* $V \setminus (S_1 \cup S_2 \cup \{i\})$ *is non-empty.*

Theorem 3. *Let* $i \in V$ *be a fixed root. If* $x^* \in SEP(n)$, *it is impossible for two* i-light *sets to* i-cross.

Proof: If we sum together the degree equations (1) for all $i \in S_1 \cap S_2$, along with the SECs on the four vertex sets $i \cup S_1 \cup S_2$, $S_1 \setminus S_2$, $S_2 \setminus S_1$ and $S_1 \cap S_2$, then (after some re-arranging) we obtain the inequality:

$$x^*(E(i:S_1)) + 2x^*(E(S_1)) + x^*(E(i:S_2)) + 2x^*(E(S_2)) \leq 2|S_1| + 2|S_2| - 3. \quad (9)$$

On the other hand, the sum of the tooth inequality with root i and body S_1 and the tooth inequality with root i and body S_2 is:

$$x^*(E(i:S_1)) + 2x^*(E(S_1)) + x^*(E(i:S_2)) + 2x^*(E(S_2)) \leq 2|S_1| + 2|S_2| - 2. \quad (10)$$

Comparing (10) and (9) we see that the sum of the slacks of these two tooth inequalities is at least 1. Since $x^* \in SEP(n)$, each of the individual slacks is non-negative. Hence at least one of the slacks must be $\geq \frac{1}{2}$. That is, at least one of S_1 and S_2 is i-heavy.

Corollary 1. *For a given root* i, *there are only* $\mathcal{O}(n)$ i-light *vertex sets.*

Corollary 2. *The number of distinct tooth inequalities with slack less than* $\frac{1}{2}$ *is* $\mathcal{O}(n^2)$.

The following lemma shows that we can eliminate half of the i-light sets from consideration.

Lemma 5. *A tooth inequality with root* i *and body* S *is equivalent to the tooth inequality with root* i *and body* $V \setminus (S \cup \{i\})$.

Proof: The latter inequality can be obtained from the former by subtracting the degree equations for the vertices in S, and adding the degree equations for the vertices in $V \setminus (S \cup \{i\})$.

Therefore, for any i, we can pick any arbitrary vertex $j \neq i$ and eliminate all i-light sets containing j, without losing any violated $\{0, \frac{1}{2}\}$-cuts. It turns out that the remaining i-light sets have an interesting structure:

Lemma 6. *For any* i *and any arbitrary* $j \neq i$, *the* i-light *sets which do not contain* j *are 'nested'. That is, if* S_1 *and* S_2 *are* i-light *sets which do not contain* j, *then either* S_1 *and* S_2 *are disjoint, or one is entirely contained in the other.*

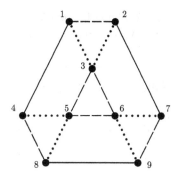

Fig. 2. A fractional point contained in SEP(9)

Proof: If S_1 and S_2 did not meet this condition, then they would i-cross.

We close this section with a small example.

Example: Fig. 2 shows the support graph G^* for a vector x^* which lies in SEP(9). The solid lines, dashed lines and dotted lines show edges with $x_e^* = 1$, $2/3$ and $1/3$, respectively. The 1-light sets are $\{2\}, \{4\}, \{3, \ldots, 9\}$ and $\{2, 3, 5, 6, 7, 8, 9\}$; the 3-light sets are $\{5\}, \{6\}, \{5, 6\}, \{1, 2, 4, 7, 8, 9\}, \{1, 2, 4, 5, 7, 8, 9\}$ and $\{1, 2, 4, 6, 7, 8, 9\}$. The 1-heavy sets include, for example, $\{3\}, \{4, 8\}$ and $\{3, 5, 6\}$. The 3-heavy sets include, for example, $\{1\}, \{2\}$ and $\{1, 4\}$. The reader can easily identify light and heavy sets for other roots by exploiting the symmetry of the fractional point.

4   Separation

Our separation algorithm has two stages. In the first stage, we search for a violated simple comb inequality in which all of the teeth are light. If this fails, then we proceed to the second stage, where we search for a violated simple comb inequality in which one of the teeth is heavy. Lemma 4 in the previous section shows that this approach is valid.

The following theorem is at the heart of the first stage of the separation algorithm:

Theorem 4. *Let $Ax \leq b$ be the inequality system formed by the degree equations (written in less-than-or-equal-to form), and, for each i, the tooth inequalities corresponding to the i-light sets forming a nested family obtained as in the previous section. Then the mod-2 support of the matrix A is an EPT matrix.*

The proof of this theorem is long and tedious and will be given in the full version of the paper. Instead, we demonstrate that the theorem is true for the fractional point shown in Fig. 2. It should be obvious how to proceed for other fractional points.

Example (Cont.): Consider once again the fractional point shown in Fig. 2. There are four i-light sets for each of the roots $1, 2, 4, 7, 8, 9$ and six i-light sets for the remaining three roots. Applying Lemma 5 we can eliminate half of these from consideration. So suppose we choose:

- 1-light sets: $\{2\}, \{4\}$;
- 2-light sets: $\{1\}, \{7\}$;
- 3-light sets: $\{5\}, \{6\}, \{5,6\}$;
- 4-light sets: $\{1\}, \{8\}$;
- 5-light sets: $\{3\}, \{6\}, \{3,6\}$;
- 6-light sets: $\{3\}, \{5\}, \{3,5\}$;
- 7-light sets: $\{2\}, \{9\}$;
- 8-light sets: $\{4\}, \{9\}$;
- 9-light sets: $\{7\}, \{8\}$.

This leads to 21 light tooth inequalities in total. However, there are some duplicates: a tooth inequality with root i and body $\{j\}$ is identical to a tooth inequality with root j and body $\{i\}$ (in both cases the inequality is a simple upper bound, $x_{ij} \leq 1$). In fact there are only 12 distinct inequalities, namely:

$$2x_{35} + x_{36} + x_{56} \leq 3 \tag{11}$$

$$2x_{36} + x_{35} + x_{56} \leq 3 \tag{12}$$

$$2x_{56} + x_{35} + x_{36} \leq 3, \tag{13}$$

plus the upper bounds on $x_{12}, x_{14}, x_{27}, x_{35}, x_{36}, x_{48}, x_{56}, x_{79}$ and x_{89}. Therefore the matrix A has 36 columns (one for each variable), and 21 rows (12 tooth inequalities plus 9 degree equations). In fact we can delete the columns associated with variables which are zero at x^*. This leaves only 15 columns; the resulting matrix A is as follows (the first three rows correspond to (11) - (13), the next nine to simple upper bounds, and the final nine to degree equations):

$$
\begin{pmatrix}
0\,0\,0\,0\,0\,2\,1\,0\,0\,1\,0\,0\,0\,0\,0 \\
0\,0\,0\,0\,0\,1\,2\,0\,0\,1\,0\,0\,0\,0\,0 \\
0\,0\,0\,0\,0\,1\,1\,0\,0\,2\,0\,0\,0\,0\,0 \\
1\,0\,0\,0\,0\,0\,0\,0\,0\,0\,0\,0\,0\,0\,0 \\
0\,0\,1\,0\,0\,0\,0\,0\,0\,0\,0\,0\,0\,0\,0 \\
0\,0\,0\,0\,1\,0\,0\,0\,0\,0\,0\,0\,0\,0\,0 \\
0\,0\,0\,0\,0\,1\,0\,0\,0\,0\,0\,0\,0\,0\,0 \\
0\,0\,0\,0\,0\,0\,1\,0\,0\,0\,0\,0\,0\,0\,0 \\
0\,0\,0\,0\,0\,0\,0\,0\,1\,0\,0\,0\,0\,0\,0 \\
0\,0\,0\,0\,0\,0\,0\,0\,0\,1\,0\,0\,0\,0\,0 \\
0\,0\,0\,0\,0\,0\,0\,0\,0\,0\,0\,0\,0\,1\,0 \\
0\,0\,0\,0\,0\,0\,0\,0\,0\,0\,0\,0\,0\,0\,1 \\
1\,1\,1\,0\,0\,0\,0\,0\,0\,0\,0\,0\,0\,0\,0 \\
1\,0\,0\,1\,1\,0\,0\,0\,0\,0\,0\,0\,0\,0\,0 \\
0\,1\,0\,1\,0\,1\,1\,0\,0\,0\,0\,0\,0\,0\,0 \\
0\,0\,1\,0\,0\,0\,0\,1\,1\,0\,0\,0\,0\,0\,0 \\
0\,0\,0\,0\,1\,0\,1\,0\,1\,1\,0\,0\,0\,0\,0 \\
0\,0\,0\,0\,0\,0\,1\,0\,0\,1\,0\,1\,1\,0\,0 \\
0\,0\,0\,1\,0\,0\,0\,0\,0\,0\,1\,0\,1\,0 \\
0\,0\,0\,0\,0\,0\,0\,1\,0\,1\,0\,0\,0\,1 \\
0\,0\,0\,0\,0\,0\,0\,0\,0\,0\,0\,1\,1\,1
\end{pmatrix}
$$

The mod-2 support of this matrix is indeed EPT; this can be checked by reference to the tree given in Fig. 3. The edge associated with the ith degree equation is labeled d_i. The edge associated with the upper bound $x_{ij} \leq 1$ is labeled u_{ij}. Finally, next to the remaining three edges, we show the root and body of the corresponding light tooth inequality.

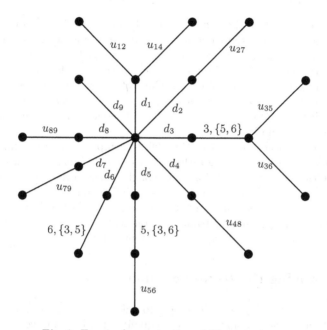

Fig. 3. Tree in demonstration of Theorem 4

Corollary 3. *If $x^* \in SEP(n)$, then a violated simple comb inequality which uses only light teeth can be found in polynomial time, if any exists.*

Proof: In the previous section we showed that the desired tooth inequalities are polynomial in number and that they can be found in polynomial time. The necessary system $Ax \leq b$ and its associated EPT matrix can easily be constructed in polynomial time. The result then follows from Theorem 2.

Example (Cont.): Applying stage 1 to the fractional point shown in Fig. 2, we find a violated simple comb inequality with $H = \{1,2,3\}$, $T_1 := \{1,4\}$, $T_2 = \{2,7\}$ and $T_3 = \{3,5,6\}$. The inequality is

$$x_{12} + x_{13} + x_{23} + x_{14} + x_{27} + x_{35} + x_{36} + x_{56} \leq 5,$$

and it is violated by one-third.

Now we need to deal with the second stage, i.e., the case where one of the teeth involved is heavy. From Lemma 3 we know that there are $\mathcal{O}(n^3)$ candidates for this heavy tooth. We do the following for each of these candidates: we take the collection of $\mathcal{O}(n^2)$ light tooth inequalities, and eliminate the ones whose teeth have a non-empty intersection with the heavy tooth under consideration. As we will show in the full version of this paper, the resulting modified matrix is still an EPT matrix, and the Caprara-Fischetti procedure can be repeated. This proves Theorem 1.

Now let us analyze the running time of this separation algorithm. First we consider stage 1. A careful reading of Caprara & Fischetti [2] shows that to separate $\{0, \frac{1}{2}\}$-cuts derived from a system $Ax \leq b$, where A is a $p \times q$ matrix, it is necessary to compute a minimum weight odd cut in a labeled weighted graph with $p - 1$ vertices and $p + q'$ edges, where q' is the number of variables which are currently non-zero. In our application, p is $\mathcal{O}(n^2)$, because there are $\mathcal{O}(n^2)$ light tooth inequalities; and q' is $\mathcal{O}(|E^*|)$. Therefore the graph concerned has $\mathcal{O}(n^2)$ vertices and $\mathcal{O}(n^2)$ edges.

Using the Padberg-Rao algorithm [29] to compute the minimum weight odd cut, and the pre-flow push algorithm [12] to solve the max-flow problems, stage 1 takes $\mathcal{O}(n^6 \log n)$ time. This is bad enough, but an even bigger running time is needed for stage 2, which involves essentially repeating the procedure used in stage 1 $\mathcal{O}(n^3)$ times. This leads to a running time of $\mathcal{O}(n^9 \log n)$ for stage 2, which, though polynomial, is totally impractical.

In the next section, we explore the potential for reducing this running time.

5   Improving the Running Time

To improve the running time, it suffices to reduce the number of 'candidates' for the teeth in a violated simple comb inequality. To this end, we now describe a simple lemma which enables us to eliminate teeth from consideration. We will then prove that, after applying the lemma, only $\mathcal{O}(|E^*|)$ *light* teeth remain.

Lemma 7. *Suppose a violated $\{0, \frac{1}{2}\}$-cut can be derived using the tooth inequality with root i and body S. If there exists a set $S' \subset V \setminus \{i\}$ such that*

- $E(i : S) \cap E^* = E(i : S') \cap E^*,$
- $|S'| - 2x^*(E(S')) - x^*(E(i : S')) \leq |S| - 2x^*(E(S)) - x^*(E(i : S)),$

then we can obtain a $\{0, \frac{1}{2}\}$-cut violated by at least as much by replacing the body S with the body S' (and adjusting the set of used non-negativity inequalities accordingly).

Proof: By Proposition 1, we have to consider the net change in the sum of the slacks of the *used* inequalities. The second condition in the lemma simply says that the slack of the tooth inequality with root i and body S' is not greater than the slack of the tooth inequality with root i and body S. Therefore replacing S with S' causes the sum of the slacks to either remain the same or decrease. Now we consider the *used* non-negativity inequalities. The only variables to receive an *odd* coefficient in a tooth inequality with root i and body S are those which correspond to edges in $E(i : S)$, and a similar statement holds for S'. So, for the edges in $E(i : (S \setminus S') \cup (S' \setminus S))$, the non-used non-negativity inequalities must now be used and vice-versa. But this has no effect on the sum of the slacks, because $E(i : (S \setminus S') \cup (S' \setminus S)) \subset E \setminus E^*$ by assumption and the slack of a non-negativity inequality for an edge in $E \setminus E^*$ is zero. Hence, the total sum of slacks is either unchanged or decreased and the new $\{0, \frac{1}{2}\}$-cut is violated by at least as much as the original.

A naive implementation of this idea runs in $\mathcal{O}(n^3|E^*|)$ time. The next theorem shows that, after it is applied, only $\mathcal{O}(|E^*|)$ *light* tooth inequalities remain.

Theorem 5. *After applying the elimination criterion of Lemma 7, at most $4|E^*| - 2n$ light tooth inequalities remain.*

Proof: Recall that, for any root i, the i-light sets form a nested family. Let S, S' be two distinct i-light sets remaining after the elimination criterion has been applied. If $S' \subset S$, then $E(i : (S \setminus S')) \cap E^* \neq \emptyset$. It is then easy to check that the total number of i-light sets is at most $2(d_i^* - 1)$, where d_i^* denotes the degree of i in G^*. So the total number of light tooth inequalities is at most $2 \sum_{i \in V} (d_i^* - 1) = 4|E^*| - 2n$.

The effect on the running time of the overall separation algorithm is as follows. The graphs on which the minimum weight odd cuts must be computed now have only $\mathcal{O}(|E^*|)$ vertices and edges. Each odd cut computation now only involves the solution of $\mathcal{O}(|E^*|)$ max-flow problems. Thus, again using the pre-flow push max-flow algorithm, the overall running time of the separation algorithm is reduced from $\mathcal{O}(n^9 \log n)$ to $\mathcal{O}(n^3|E^*|^3 \log n)$. In practice, G^* is very sparse, so that this is effectively $\mathcal{O}(n^6 \log n)$.

It is natural to ask whether the running time can be reduced further. In our opinion further reduction is possible. First, we believe that a more complicated argument can reduce the number of *light* teeth to only $\mathcal{O}(n)$, and the number of *heavy* teeth to $\mathcal{O}(n|E^*|)$. Moreover, empirically we have found that, for each root i, it is possible to partition the set of i-heavy sets into $\mathcal{O}(d_i^*)$ nested families, each containing $\mathcal{O}(d_i^*)$ members (where d_i^* is as defined in the proof of Theorem 5). If this result could be proven to hold in general, we would be need to perform only $\mathcal{O}(|E^*|)$ minimum odd cut computations in stage 2. These changes would reduce the overall running time to only $\mathcal{O}(n^2|E^*|^2 \log(n^2/|E^*|))$, which leads us to hope that a practically useful implementation is possible. Further progress on this issue (if any) will be reported in the full version of the paper.

6   Concluding Remarks

We have given a polynomial-time separation algorithm for the simple comb inequalities, thus extending the result of Padberg and Rao [29]. This is the latest in a series of positive results concerned with comb separation (Padberg & Rao [29], Carr [5], Fleischer & Tardos [9], Caprara, Fischetti & Letchford [3], Letchford [21], Caprara & Letchford [4]).

A number of open questions immediately spring to mind. The main one is, of course, whether there exists a polynomial-time separation algorithm for *general* comb inequalities, or perhaps a generalization of them such as the *domino-parity* inequalities [21]. We believe that any progress on this issue is likely to come once again from the characterization of comb inequalities as $\{0, \frac{1}{2}\}$-cuts. Moreover, since separation of $\{0, \frac{1}{2}\}$-cuts is equivalent to the problem of finding a minimum weight member of a binary clutter [2], it might be necessary to draw on some of the deep decomposition techniques which are used in the theory of binary clutters and binary matroids (compare [2] with Grötschel & Truemper [17]).

Another interesting question is concerned with *lower bounds*. Let us call the lower bound obtained by optimizing over SEP(n) the *subtour bound*. The subtour bound is typically good in practice, and a widely held conjecture is that, when the edge costs satisfy the triangle inequality, the subtour bound is at least 3/4 of the optimal value. Moreover, examples are known which approach this value arbitrarily closely, see Goemans [11]. Now, using the results in this paper, together with the well-known equivalence of separation and optimization [14], we can obtain a stronger lower bound by also using the simple comb inequalities. Let us call this the *simple comb bound*. A natural question is whether the worst-case ratio between the simple comb bound and the optimum is greater than 3/4. Unfortunately, we know of examples where the ratio is still arbitrarily close to 3/4. Details will be given in the full version of the paper.

Finally, we can also consider special classes of graphs. For a given graph G, let us denote by SC(G) the polytope defined by the degree equations, the SECs, and the non-negativity and simple comb inequalities. (Now we only define variables for the edges in G.) Let us say that a graph G is *SC-perfect* if SC(G) is an integral polytope. Clearly, the TSP is polynomially-solvable on SC-perfect graphs. It would be desirable to know which graphs are SC-perfect, and in particular whether the class of SC-perfect graphs is closed under the taking of minors (see Fonlupt & Naddef [10]). Similarly, let us say that a graph is *SC-Hamiltonian* if SC(G) is non-empty. Obviously, every Hamiltonian graph is SC-Hamiltonian, but the reverse does not hold. (The famous *Peterson graph* is SC-Hamiltonian, but not Hamiltonian.) It would be desirable to establish structural properties for the SC-Hamiltonian graphs, just as Chvátal [6] did for the so-called *weakly Hamiltonian* graphs.

References

1. D. Applegate, R.E. Bixby, V. Chvátal & W. Cook (1995) *Finding cuts in the TSP (a preliminary report)*. Technical Report 95–05, DIMACS, Rutgers University, New Brunswick, NJ.
2. A. Caprara & M. Fischetti (1996) $\{0, \frac{1}{2}\}$-Chvátal-Gomory cuts. *Math. Program.* 74, 221–235.
3. A. Caprara, M. Fischetti & A.N. Letchford (2000) On the separation of maximally violated mod-k cuts. *Math. Program.* 87, 37–56.
4. A. Caprara & A.N. Letchford (2001) On the separation of split cuts and related inequalities. To appear in *Math. Program.*

5. R. Carr (1997) Separating clique trees and bipartition inequalities having a fixed number of handles and teeth in polynomial time. *Math. Oper. Res.* 22, 257–265.
6. V. Chvátal (1973) Edmonds polytopes and weakly Hamiltonian graphs. *Math. Program.* 5, 29–40.
7. G.B. Dantzig, D.R. Fulkerson & S.M. Johnson (1954) Solution of a large-scale traveling salesman problem. *Oper. Res.* 2, 393–410.
8. J. Edmonds (1965) Maximum matching and a polyhedron with 0-1 vertices. *J. Res. Nat. Bur. Standards* 69B, 125–130.
9. L. Fleischer & É. Tardos (1999) Separating maximally violated comb inequalities in planar graphs. *Math. Oper. Res.* 24, 130–148.
10. J. Fonlupt & D. Naddef (1992) The traveling salesman problem in graphs with excluded minors. *Math. Program.* 53, 147–172.
11. M.X. Goemans (1995) Worst-case comparison of valid inequalities for the TSP. *Math. Program.* 69, 335–349.
12. A.V. Goldberg & R.E. Tarjan (1988) A new approach to the maximum flow problem. *J. of the A.C.M.* 35, 921–940.
13. M. Grötschel & O. Holland (1987) A cutting plane algorithm for minimum perfect 2-matching. *Computing* 39, 327–344.
14. M. Grötschel, L. Lovász & A.J. Schrijver (1988) *Geometric Algorithms and Combinatorial Optimization.* Wiley: New York.
15. M. Grötschel & M.W. Padberg (1979) On the symmetric traveling salesman problem I: inequalities. *Math. Program.* 16, 265–280.
16. M. Grötschel & M.W. Padberg (1979) On the symmetric traveling salesman problem II: lifting theorems and facets. *Math. Program.* 16, 281–302.
17. M. Grötschel & K. Truemper (1989) Decomposition and optimization over cycles in binary matroids. *J. Comb. Th. (B)* 46, 306–337.
18. M.R. Hensinger & D.P. Williamson (1996) On the number of small cuts in a graph. *Inf. Proc. Lett.* 59, 41–44.
19. M. Jünger, G. Reinelt, G. Rinaldi (1995) The traveling salesman problem. In M. Ball, T. Magnanti, C. Monma & G. Nemhauser (eds.). *Network Models*, Handbooks in Operations Research and Management Science, 7, Elsevier Publisher B.V., Amsterdam, 225–330.
20. M. Jünger, G. Reinelt & G. Rinaldi (1997) The traveling salesman problem. In M. Dell'Amico, F. Maffioli & S. Martello (eds.) *Annotated Bibliographies in Combinatorial Optimization.* Chichester, Wiley, 199-221.
21. A.N. Letchford (2000) Separating a superclass of comb inequalities in planar graphs. *Math. Oper. Res.* 25, 443–454.
22. D. Naddef (2001) *Polyhedral theory and branch-and-cut algorithms for the symmetric TSP.* In G. Gutin & A. Punnen (eds.), *The Traveling Salesman Problem and its Variations.* Kluwer Academic Publishers, 2002 (to appear).
23. D. Naddef & G. Rinaldi (1991) The symmetric traveling salesman polytope and its graphical relaxation: composition of valid inequalities. *Math. Program.* 51, 359–400.
24. D. Naddef & G. Rinaldi (1993) The graphical relaxation: a new framework for the symmetric traveling salesman polytope. *Math. Program.* 58, 53–88.
25. D. Naddef & S. Thienel (1998) Efficient separation routines for the symmetric traveling salesman problem I: general tools and comb separation. *Working paper*, LMC-IMAG, Grenoble.
26. H. Nagamochi, K. Nishimura and T. Ibaraki (1997) Computing all small cuts in undirected networks. *SIAM Disc. Math.* 10, 469–481.
27. G.L. Nemhauser and L.A. Wolsey (1988) *Integer and Combinatorial Optimization.* New York: Wiley.

28. M.W. Padberg & M. Grötschel (1985) Polyhedral computations. In E.L. Lawler, J.K. Lenstra, A.H.G. Rinnooy Kan & D.B. Schmoys (Eds.) *The Traveling Salesman Problem*. John Wiley & Sons, Chichester.

29. M.W. Padberg & M.R. Rao (1982) Odd minimum cut-sets and b-matchings. *Math. Oper. Res.* 7, 67–80.

30. M.W. Padberg & G. Rinaldi (1990) Facet identification for the symmetric traveling salesman polytope. *Math. Program.* 47, 219–257.

31. M.W. Padberg & G. Rinaldi (1991) A branch-and-cut algorithm for the resolution of large-scale symmetric traveling salesman problems. *SIAM Rev.* 33, 60–100.

A New Approach to Cactus Construction Applied to TSP Support Graphs

Klaus M. Wenger

Institute of Computer Science
University of Heidelberg
Im Neuenheimer Feld 368
69120 Heidelberg, Germany
`Klaus.Wenger@Informatik.Uni-Heidelberg.DE`

Abstract. We present a novel approach to the construction of the cactus representation of all minimum cuts of a graph. The representation is a supergraph that stores the set of all mincuts compactly and mirrors its structure. We focus on support graphs occurring in the branch-and-cut approach to traveling salesman, vehicle routing and similar problems in a natural way. The ideas presented also apply to more general graphs. Unlike most previous construction approaches, we do not follow the Karzanov-Timofeev framework or a variation of it. Our deterministic algorithm is based on inclusion-minimal mincuts. We use Fleischer's approach [J. Algorithms, 33(1):51-72, 1999], one of the fastest to date, as benchmark. The new algorithm shows an average speed-up factor of 20 for TSP-related support graphs in practice. We report computational results. Compared to the benchmark, we reduce the space required during construction for n-vertex graphs with m edges from $\mathcal{O}(n^2)$ to $\mathcal{O}(m)$.

1 Introduction

We define a *graph* as a pair consisting of a finite set of *vertices* and a set of *edges* which are two-point subsets of the vertex set. This excludes multiple edges and loops $\{v, v\} = \{v\}$. The vertices of an edge are called its *ends*. A graph is called *undirected* to emphasize that edge $\{u, v\}$ equals $\{v, u\}$. A *weighted graph* has *edge weights* assigned to its edges. Here, weights are in $\mathrm{I\!R}^+$. In a *connected graph* there is a path of edges between two different vertices. We call a graph that has a subset of $\mathrm{I\!N}$ assigned to each vertex a *supergraph*.

A *cut* in a graph is a non-empty proper vertex subset. A cut is different from its complement called *complementary cut*. The cardinality of a cut is called its *size*. In a weighted graph, the *weight of a cut* is the sum of weights of edges with exactly one end in the cut. Regard an unweighted graph as a weighted graph with unit weights. A weighted graph has a finite number of cuts and hence cuts of minimum weight. These are called *mincuts* or *global minimum weight cuts*.

There are several algorithms computing mincuts. For an overview and computational studies for weighted graphs see [6,17]. Network reliability and augmentation are fields of application for mincuts. Another major field is cutting-plane

W.J. Cook and A.S. Schulz (Eds.): IPCO 2002, LNCS 2337, pp. 109–126, 2002.
© Springer-Verlag Berlin Heidelberg 2002

generation in branch-and-cut approaches to combinatorial optimization problems. Further applications are mentioned in [18,29]. A suitable representation of all mincuts of a graph in terms of storage space and access time is often desirable. This paper deals with the construction of such a representation.

The input of our algorithm is a connected undirected graph $G = (V, E)$ with strictly positive edge weights and a special property: Every vertex is a mincut. For arbitrary edge weights the mincut problem is equivalent to the $\mathcal{NP}$-hard maxcut problem of finding a cut of maximum weight. The special property is not artificial. Such graphs naturally appear in the cutting-plane or branch-and-cut approach to traveling salesman, vehicle routing and related problems.

A graph weighted with an optimal solution x to the Linear Program (LP) relaxation (possibly tightened by cutting-planes) of the classical 0/1 Integer LP (ILP) model [9] of the symmetric Traveling Salesman Problem (TSP) satisfies:

(a) every vertex has weight 2 (*degree equations*) and

(b) every cut has weight at least 2 (*subtour elimination constraints*).

Remove edges with weight 0 to obtain a valid input graph. A graph with edge weights in $]0, 1]$ satisfying (a) and (b) is called a *TSP support graph*. For some purposes, directed support graphs carrying LP solutions associated with the asymmetric TSP can be transformed to undirected counterparts [5] which are actually TSP support graphs. By transforming the multisalesmen problem to the symmetric TSP, the Capacitated Vehicle Routing Problem (CVRP) [33] can be modelled as symmetric TSP with additional constraints [32]. Therefore, TSP support graphs also are encountered in connection with CVRPs.

After scaling edge weights, every graph with the properties of G above is a TSP support graph with mincut weight $\lambda = 2$. Essentially, the valid input graphs for our algorithm are the TSP support graphs. We rely on no further characteristics of the symmetric TSP. We do not require that the edge weight vector x is an extreme point of the *subtour elimination polytope* (SEP) consisting of all points in the unit cube satisfying (a) and (b); we only require that x is within the SEP. This requirement is common [5,12,20,25,30]. For an n-vertex graph, we do not require $x \notin \mathrm{STSP}(n)$ where $\mathrm{STSP}(n)$ denotes the polytope associated with the symmetric TSP on n vertices. Support graphs of CVRPs can have weight vectors x within the corresponding TSP polytope.

Because of the strong interest in both traveling salesman [1,16,30] and vehicle routing [33] problems, TSP support graphs G deserve special treatment. We have implemented our algorithmic cactus construction idea for such graphs.

The output of our algorithm is a cactus of G. More precisely: It is the unique canonical cactus representation $\mathcal{H}(G)$ of the set $\mathcal{M}$ of all mincuts of G. It can be viewed as a condensed list of G's mincuts. If $n = |V|$, then G has $\mathcal{O}(n^2)$ mincuts [3,10,18]. An explicit list requires $\mathcal{O}(n^3)$ space. $\mathcal{H}(G)$ is a supergraph with linear space requirement in n showing the inclusion- and intersection-structure of $\mathcal{M}$. Every $M \in \mathcal{M}$ can be extracted from $\mathcal{H}(G)$ in time $\mathcal{O}(|M|)$. Dinitz, Karzanov and Timofeev [10] reported this elegant representation of mincuts in 1976.

Karzanov and Timofeev [19] presented the first polynomial algorithm to build a cactus of a graph. The subsequent approaches [8,11,18,22,27] generalized and improved their framework. We abandon it. The idea of the new deterministic algorithm is: repeatedly find a basic mincut M in G, build the corresponding part of $\mathcal{H}(G)$ and shrink M. By our definition, a *basic mincut* does not properly contain any non-trivial mincut. We call a single-vertex mincut *trivial*.

A (u, v)-mincut is a mincut containing vertex u but not v. A (U, v)-cut contains vertex set U but not v. We call a (u, v)-mincut M *degenerate* if M or $\overline{M}$ is trivial. We use the preflow-push mincut algorithm of Hao and Orlin (H-O) [14] to decide if there is a non-degenerate (u, v)-mincut for $u \neq v$. We can detect all edges $\{u, v\}$ for which there is a non-degenerate (u, v)-mincut within a single H-O call. By removing these edges, we can simultaneously find the material basic mincuts of a TSP support graph that have more than 2 vertices.

Cutting-plane generation procedures are crucial to solving hard combinatorial optimization problems to optimality [1,30]. Several such procedures [5,12,25,34] can benefit from our algorithm. During the cutting-plane phase of a sub-problem corresponding to a node of the problem tree in a branch-and-cut framework [15], one may want to repeatedly construct cacti of support graphs.

Section 2 summarizes the cactus representation of mincuts. In Sect. 3 we comment on previous cactus construction approaches and on the space requirement of different algorithms. Our new approach is detailed in Sect. 4. We mention some applications of cacti to cutting-plane generation in Sect. 5. The suitability of our new cactus construction algorithm for TSP support graphs is demonstrated by computational experiments in Sect. 6. Section 7 concludes the paper.

2   The Cactus of a Weighted Graph

2.1   Definition and Properties

We define a *cactus* as a weighted graph in which every edge is within at most one cycle (Figs. 1,3). There are two types of edges: *cycle edges* are in one cycle, *tree edges* are in none. Tree edges carry twice the weight of cycle edges in a cactus. Cacti we encounter are connected. For cacti we say *node* instead of vertex.

Let a cactus be given. Map the set of vertices $V = \{1, \ldots, n\}$ of G to the node set of this cactus. Then a subset (possibly $\emptyset$) of V is associated with every node of the cactus. Let C be a cut in the cactus. Denote the vertices of G mapped to a node in C by $V(C)$ and say that $V(C)$ is *induced* by C. A cactus node inducing $\emptyset$ is called *empty* (Fig. 3). If $\mathcal{M} = \{V(C) \mid C \text{ is a mincut of the cactus}\}$, then the cactus as a weighted supergraph is called a *cactus representation of $\mathcal{M}$* or a *cactus of G* (Fig. 1). Every mincut of a cactus of G induces a mincut of G and every mincut of G can be obtained at least once in this way.

The mincuts of a cactus are known. They can all be obtained by removing a tree edge or two different edges of the same cycle which splits a cactus into two connected components. The nodes of each component form a mincut of the cactus. By traversing a component and collecting the vertices assigned to nodes of a mincut of a cactus of G, we quickly get the induced mincut $M \in \mathcal{M}$ of G in time $\mathcal{O}(|M|)$. Assign the mincut weight λ of G to tree edges and $\lambda/2$ to cycle edges to give a cactus of G the mincut weight λ.

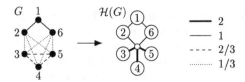

Fig. 1. A TSP support graph and the canonical cactus representation of its mincuts

We call a cycle with k edges a k-*cycle*. Removing the edges of a k-cycle splits a cactus of G into k connected components with node sets C_i where $i = 1, \ldots, k$. The vertex sets $V(C_i) \in \mathcal{M}$ are called the *beads* of the cycle.

A cactus of G with at most $2n$ nodes [21] actually exists [10,27]. The number of edges is also $\mathcal{O}(n)$ [10]. This means the representation is compact.

2.2   Uniqueness of the Representation

A graph can have more than one cactus. A k-*way cutnode* splits a cactus into k connected components when removed together with its incident edges.

One source of ambiguity is redundancy. Contracting a tree edge T incident to an empty 2-way cutnode does not reduce the number of mincuts represented but results in a smaller cactus having fewer empty nodes. A cactus is called *normal* [21] or *simple* [8] if neither operation O_1 nor O_2 in Fig. 2 can be applied. A mincut of G is induced by at most two mincuts of a simple cactus of G [8]. Only empty nodes of degree 4 common to exactly 2 cycles can be the reason for mincuts to be induced twice.

Fig. 2. The two simplifying operations [8,21]

The second source of ambiguity are empty 3-way cutnodes. The same mincuts are induced if we replace them by 3-cycles of empty nodes [8]. We say that a cactus is of *cycle type* [21,23] if it does not contain any empty 3-way cutnodes.

If a cactus of G is normal and of cycle type, then it is unique [21], denoted by $\mathcal{H}(G)$ and called *canonical* [8,11]. We directly construct $\mathcal{H}(G)$.

2.3   Structure and Cactus Representation of $\mathcal{M}$

Two cuts are said to *cross* if their union does not cover V, if they intersect and if none is contained in the other. A k-cycle with $k > 3$ in $\mathcal{H}(G)$ represents crossing mincuts (Fig. 3). Let G be weighted by a vector x indexed by the elements of E. Let $U, W \subset V$. For $U \cap W = \emptyset$ define $(U:W) = (W:U)$ as the set of edges with one end in U and one in W. Write $\delta(U)$ for $(U:\overline{U})$. Write $x(F)$ for $\sum_{e\in F\subseteq E} x_e$ and $x(U:W)$ for $x((U:W))$. Since the following well known lemma [3,10,13] is fundamental to this paper and our algorithm, we give a proof.

Lemma 1 (Crossing Mincuts). *If $A, B \in \mathcal{M}$ cross, then*

1. *$M_1 = A \cap \overline{B}, M_2 = A \cap B, M_3 = \overline{A} \cap B, M_4 = \overline{A \cup B}$ are mincuts,*
2. *$x(M_1 : M_3) = x(M_2 : M_4) = 0$ and*
3. *$x(M_1 : M_2) = x(M_2 : M_3) = x(M_3 : M_4) = x(M_4 : M_1) = \lambda/2$.*

Proof. We have $2\lambda = x(\delta(A)) + x(\delta(B))$ since A, B are mincuts. Repeatedly use $x(U:W) = x(U:W_1) + x(U:W_2)$ if $\{W_1, W_2\}$ is a partition of W. We arrive at $2\lambda = x(\delta(M_1)) + x(\delta(M_3)) + 2x(M_2:M_4) = x(\delta(M_2)) + x(\delta(M_4)) + 2x(M_1:M_3)$. Since $x(\delta(M_i)) \geq \lambda$ we get claim 2. We now see that $x(\delta(M_i)) > \lambda$ is impossible which proves 1. If $M_i \cup M_j \in \mathcal{M}$, then $\lambda = x(\delta(M_i \cup M_j)) = x(\delta(M_i)) + x(\delta(M_j)) - 2x(M_i:M_j) = 2\lambda - 2x(M_i : M_j)$. This proves the last claim 3. □

This implies that the union and intersection of crossing mincuts is again a mincut. From Lemma 1 we also see that there can be no edges in G between non-consecutive beads of a k-cycle of $\mathcal{H}(G)$. The weight in G between two consecutive beads of a cycle of $\mathcal{H}(G)$ is $\lambda/2$. Lemma 1 implies this for $k > 3$ and the statement obviously also holds for $k = 3$.

A *partition* of V is a set of pairwise disjoint non-empty subsets of V whose union is V. A tupel $(V_1, \ldots, V_k)$ is called an *ordered partition* if the V_i form a partition. We get a *circular partition* if we locate the V_i around a circle so that V_1 follows V_k. A circular partition of *size* $k \geq 3$ is called (maximal) *mincut circular partition* (MCCP) if the union of any $1 \leq r < k$ consecutive V_i is in $\mathcal{M}$ and either M or $\overline{M}$ is contained in some V_i for any $M \in \mathcal{M}$ that cannot be obtained as the union of consecutive V_i of the circular partition. If G has crossing mincuts, then G has a MCCP (Circular Partition Lemma [2,3,8,10]).

A MCCP of size $k > 3$ is represented by a k-cycle in every cactus of G [8]. In a cactus of cycle type this also holds for $k = 3$. The MCCPs of G are in 1-1 correspondence with the cycles of $\mathcal{H}(G)$. The V_i are the beads.

Two MCCPs $(U_1, \ldots, U_k)$ and $(V_1, \ldots, V_l)$ are called *equal* if they are equal as partitions. They are called *compatible* if there is a unique i^* and j^* such that $V_j \subset U_{i^*}$ for $j \neq j^*$ and $U_i \subset V_{j^*}$ for $i \neq i^*$. Two different MCCPs of G are compatible (proof in [11,12]). This, too, is reflected in $\mathcal{H}(G)$ (Fig. 3).

2.4   Cacti of TSP Support Graphs

The statements above also hold for cacti of graphs without the property that every vertex is a mincut. Such cacti may have more than one vertex assigned to a node – also to a cutnode. There is a mincut separating two different vertices of a graph exactly if these vertices are assigned to different nodes of a cactus of the graph. We call a non-cutnode or trivial mincut in a cactus a *leaf*.

Lemma 2. *A cactus of an n-vertex TSP support graph has n leaves, each leaf has exactly one vertex assigned to it and all cutnodes are empty.*

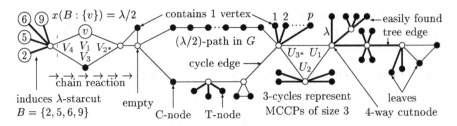

Fig. 3. Annotated cactus $\mathcal{H}(G)$ of a TSP support graph G (not shown)

Assume that vertex i of G is assigned to node i of $\mathcal{H}(G)$. Then nodes $i > n$ of $\mathcal{H}(G)$ are empty. Knowing λ does not give a big lead. Computing λ is a lot faster in practice than constructing $\mathcal{H}(G)$.

3   Previous Approaches and Memory Requirement

From Lemma 1 and $x_e > 0$ we see [19]: All (S,t)-mincuts M for $S \subset V$ and $t \notin S$ adjacent to some vertex in S have the form $M = V_1 \cup \ldots \cup V_r$ ($1 \le r < k$). Here $(V_1, \ldots, V_k)$ is an ordered partition (*chain*) with $S \subset V_1$ and $t \in V_k$. Let $(v_1, \ldots, v_n)$ be an *adjacency order* of V, that is v_i ($2 \le i \le n$) is adjacent to some v_j ($1 \le j < i$). The chains representing the nested (non-crossing) $(\{v_1, \ldots, v_{i-1}\}, v_i)$-mincuts ($2 \le i \le n$) form a so-called *chain representation* of $\mathcal{M}$. The Karzanov-Timofeev (K-T) framework [19] merges chains into cacti. For a corrected K-T framework for weighted graphs taking $\mathcal{O}(n^2 m \log(n^2/m))$ time and $\mathcal{O}(m)$ space see De Vitis [8]. Randomization in [18] speeds up chain computation and needs $\mathcal{O}(n^2)$ space. A randomized cactus construction algorithm represents with high probability all $M \in \mathcal{M}$.

In a variation of the K-T framework, the $(\{u_1, \ldots, u_{j-1}\}, u_j)$-mincuts are not always nested since (u_j) is not then an adjacency order. Examples are [2,22,27] where [2,27] may omit mincuts of G in the cactus. The asymptotic run-times of [11,22,23] are comparable and faster than [8] by a factor of approximately n where [23] does not follow K-T but decomposes G into subgraphs. The smaller the number γ of cycles in $\mathcal{H}(G)$, the faster the algorithm of Nagamochi, Nakao and Ibaraki [23]. The number of cycles in $\mathcal{H}(G)$ is $\mathcal{O}(n)$. The run-time of [11] and [23] is $\mathcal{O}(nm \log(n^2/m))$ and $\mathcal{O}(nm + n^2 \log n + \gamma m \log n)$, respectively.

Our benchmark algorithm by Fleischer [11] uses Hao-Orlin [14] to build $n-1$ representations of $(\{u_1, \ldots, u_{j-1}\}, u_j)$-mincuts where (u_j) cannot be guaranteed to be an adjacency order [14]. The main contribution of [11] is a transformation of these representations into $n - 1$ chains. This takes $\mathcal{O}(n^2)$ space since each of the $n - 1$ representations takes $\mathcal{O}(n)$ space and they have to be simultaneously accessible for the transformation. Following K-T, the chain representation of $\mathcal{M}$ is then merged to obtain $\mathcal{H}(G)$. We do not follow K-T.

The algorithm detailed in the next section takes $\mathcal{O}(m)$ space. As with [8,22], we only have to deal with a single max-(S,t)-flow residual network at a time.

4   The New Cactus Construction Algorithm

4.1   The Three Types of Basic Mincuts

We call leaves incident to a tree edge *T-nodes* and the others *C-nodes* (Fig. 3). Let $E(U) = \{\{v, w\} \in E \mid v, w \in U\}$. Then $U \in \mathcal{M} \Leftrightarrow x(E(U)) = \lambda(|U| - 1)/2$.

Denote the set of basic mincuts of G by $\mathcal{B}$. There are three *types*. The trivial mincuts are the $B \in \mathcal{B}$ of size 1. The $e \in E$ with $x_e = \lambda/2$ are the $B \in \mathcal{B}$ of size 2. Figure 1 shows that basic mincuts of size 2 can intersect and can cross. Lemma 3 establishes the third type. It follows from the definition of basic mincuts and the way in which mincuts of a cactus can be obtained.

Lemma 3. *Every basic mincut B of G with $|B| \geq 3$ is induced by a mincut of $\mathcal{H}(G)$ consisting of an empty node and $|B|$ adjacent T-nodes (Fig. 3).*

Example 1. As an extreme example, let $\mathcal{H}(G)$ be a central empty node with n ($n \geq 4$) adjacent T-nodes. Then G has exactly $2n$ mincuts, they are all basic and n of them have size $n - 1$. Basic mincuts B with $|B| \geq 3$ can intersect.

We call a vertex set *starset* if it is induced by an empty node and all of its adjacent T-nodes of the canonical cactus of a graph. Figure 3 shows how a starset with p vertices can be induced. A starset is not always a cut. Example 1 shows that a starset can be the entire set of vertices of G. The annotated empty node in Fig. 3 shows that a starset can be empty. We call a starset which is a mincut a *λ-starcut*. Every $B \in \mathcal{B}$ with $|B| \geq 3$ is a λ-starcut of G. The graph G can have trivial λ-starcuts (see Fig. 3). However:

Remark 1. Since we use cacti of cycle type, a λ-starcut cannot have size 2.

Shrinking a vertex set U means deleting $E(U)$, identifying vertices in U and replacing resulting multiple edges with a single edge weighted by the total weight of the multiple edges. Denote the result by G/U. Let u be the new vertex replacing U. Substitute u in a mincut of G/U by U to get a mincut of G *found in G/U*. Propositions 1,2 motivate our algorithm and imply its correctness.

Proposition 1. *If B is a λ-starcut of G, then all $M \in \mathcal{M}$ except the trivial mincuts in B are found in G/B. $\mathcal{H}(G)$ can be constructed from G, B and $\mathcal{H}(G/B)$.*

Proof. The first claim follows from the definition of λ-starcuts and Lemma 1. To see the second, let b be the vertex of G/B obtained by shrinking the mincut B in G. Replace the leaf of $\mathcal{H}(G/B)$ inducing $\{b\}$ by an empty node with $|B|$ adjacent T-nodes. Map B injectively to the new T-nodes to obtain $\mathcal{H}(G)$. □

Proposition 2. *Let B be a basic mincut of G having size 2. Then $\mathcal{H}(G)$ can be quickly constructed from G, B and the smaller cactus $\mathcal{H}(G/B)$.*

Proof. Let b be obtained by shrinking $B = \{u, v\}$. Let b' be the node of $\mathcal{H}(G/B)$ inducing $\{b\}$. If b' is a T-node, then let a' be its adjacent node, remove b' with its incident tree edge, create nodes u' and v' containing u and v respectively, join a', u', v' by cycle edges to get a 3-cycle. There are 3 cases if b' is a C-node:

$$b' \qquad x(V_i : \{u\}) = \lambda/2 \qquad x(V_i : \{v\}) = \lambda/2$$
$$- V_i -\!\!\fbox{b}\!\!- \;\longrightarrow\; - V_i -\!\!\fbox{u}\!\!-\!\!\fbox{v}\!\!- \quad \text{or} \quad - V_i -\!\!\fbox{v}\!\!-\!\!\fbox{u}\!\!- \quad \text{or} \quad - V_i$$

Let V_i be a bead next to bead $\{b\}$ of the cycle of $\mathcal{H}(G/B)$ to which b' belongs. Using Lemma 1 in G we can decide in $\mathcal{O}(m)$ time which case applies. The third case applies if neither the first nor the second applies. The weight in G between consecutive beads of a cycle of $\mathcal{H}(G)$ has to be $\lambda/2$. $\qquad\square$

4.2 How to Find λ-Starcuts

Algorithm NEWCACTUS below has to find λ-starcuts of a TSP support graph G' as a sub-problem. The idea of how to find them is to "remove" edges from G' in such a way that the λ-starcuts are among the vertex sets of connected components of the remaining graph. We use the Hao-Orlin algorithm [14] to find the edges that we have to remove. Note that shrinking λ-starcuts of a TSP support graph can generate new λ-starcuts (see Fig. 6).

Replacing the edges of an n'-vertex TSP support graph G' with pairs of anti-parallel arcs results in a directed network. Hao-Orlin computes max-(S_i, t_i)-flow $(1 \leq i < n')$ residual networks R_i where $S_{i+1} = S_i \cup \{t_i\}$, S_1 is an arbitrary vertex and the *sinks* t_i are chosen by H-O. Every sink vertex is a trivial mincut. The (S_i, t_i)-mincuts of G' are in 1-1 correspondence with the closed vertex sets containing S_i in the residual network R_i, see Picard and Queyranne [28]. The *closure* of U is the smallest vertex set containing U and having no arc leaving it. The so-called dormant vertices [14] at the end of max-flow i form the inclusion-minimal (S_i, t_i)-mincut, see [11]. The residual networks R_i $(1 \leq i < n')$ implicitly represent all mincuts of G'. Lemmata 4,5 are the key to finding λ-starcuts.

Lemma 4. *If B is a λ-starcut of G' with vertices $u \neq v$ in B, then there are only the two degenerate (u, v)-mincuts in G'.*

Proof. Assume that there is a non-degenerate (u, v)-mincut M in G'. Since B is a basic mincut, M cannot be contained in B. If $\overline{B} \subset M$, then $\overline{M}$ is a non-trivial mincut properly contained in B; this contradicts the assumption that B is basic. If $\overline{B} \not\subset M$, then Lemma 1 can be applied to the crossing mincuts B and M. If the mincut $B \cap M \subset B$ is non-trivial, then we have the same contradiction. If $B \cap M$ is trivial, then the mincut $B \cap \overline{M} \subset B$ is non-trivial (see Remark 1) and we again have the contradiction to the assumtion that B is basic. The lemma can also be seen by considering the mincut of $\mathcal{H}(G')$ that induces B. $\qquad\square$

Lemma 5. *Let B be a λ-starcut of G' with $|B| \leq n' - 2$. Let $u \in B$ and $v \notin B$. Then G' has a non-degenerate (u, v)-mincut M.*

Proof. If $|B| > 1$, then let $M := B$. If $B = \{b\}$, then B is induced by a cut of $\mathcal{H}(G')$ consisting of an empty k-way cutnode z with $k > 3$ and its single adjacent T-node t. Removing z and its incident edges splits $\mathcal{H}(G')$ into k mincuts. Only b is assigned to the mincut $\{t\}$ inducing B. The other $k-1$ mincuts of $\mathcal{H}(G')$ induce $k - 1$ mincuts of G' each having at least 2 vertices. Let M be the complement of the mincut out of these $k - 1$ mincuts of G' that contains v. $\qquad\square$

The graph $(M, E'(M))$ is connected for every mincut M of G'. In particular, $(B, E'(B))$ is connected for a λ-starcut B in G'. In order to find λ-starcuts of G' simultaneously, color black each edge of G' and then color white each edge $\{u, v\}$ for which there is a non-degenerate (u, v)-mincut. The desired λ-starcuts are among the vertex sets of connected components if only black edges are taken into account (*black-connected* components).

This method finds all λ-starcuts with at most $n' - 2$ vertices. If we have the case of Example 1, then the method stops with all edges of the connected graph G' colored black so that we end up with a single black-connected component equal to G'. The method colors white all edges incident to a trivial λ-starcut (Lemma 5). We do not want to find trivial λ-starcuts.

Remark 2. This method also finds subsets of starsets U where U has cut weight greater than λ. The reason why only subsets are found is that the induced subgraph $(U, E'(U))$ of G' is not always connected if U is not a mincut.

How can we decide whether there is a non-degenerate (u, v)-mincut in G'? Let H-O compute one residual network R_i at a time (only $\mathcal{O}(m)$ space requirement) and compute the strongly connected components (SCCs) of R_i. If we shrunk the SCCs, we would get so-called DAGs (*directed acyclic graphs*) representing the (S_i, t_i)-mincuts in R_i [28]. However, we do not need the DAGs.

Lemma 6. *The residual network R_i represents a (u, v)-mincut of G' $\Leftrightarrow$ the vertices u and v of G' are in different SCCs of R_i.*

Proof. Let $(C_1, \ldots, C_r)$ be a reverse topological order of the SCCs of R_i. If there is an arc in R_i from C_j to C_k, then $k < j$. We have $S_i \subset C_1$ and $C_r = \{t_i\}$. "$\Rightarrow$": A closed vertex set in R_i cannot contain a non-empty proper subset of a SCC. Hence, it is not possible that u and v are in the same SCC. "$\Leftarrow$": Assume that u is in C_j and w.l.o.g. v is in the SCC C_l with $l > j$. The closed vertex sets $C_1 \cup \ldots \cup C_k$ with $j \leq k < l$ are desired (u, v)-mincuts of G'. There can be even more since the reverse topological order is not unique. $\square$

We want to know whether R_i represents a non-degenerate (u, v)-mincut M. Using the notation of the proof above, if $u \in C_1$ or $v \in C_r$ we have to distinguish some straightforward cases in order to check whether M can be chosen so that neither M nor $\overline{M}$ is trivial. Note that $|C_1|$ can be 1 and $|C_r|$ is always 1.

4.3   The Algorithm: Shrinking Basic Mincuts Only

We repeatedly shrink basic mincuts. They have three properties: (i) we can efficiently find them, (ii) shrinking them ultimately results in a 2-vertex graph with an obvious cactus, (iii) shrinking them can be reversed (Propositions 1,2).

In parallel with shrinking, NEWCACTUS constructs $\mathcal{H}(G)$ from the leaves inwards. Think of $p(v)$ as pointing to parts of the gradually constructed $\mathcal{H}(G)$. If $p(v)$ points to one node, then to a leaf (1) or to a cutnode (10). Otherwise it points to the two ends of a cycle segment of $\mathcal{H}(G)$. The algorithm is intuitive if the part of $\mathcal{H}(G)$ pointed to by $p(v)$ is considered to be "contained" in vertex v of G'. Once the leaves have been created in (1), only empty nodes are created.

Algorithm: NEWCACTUS

Input: n-vertex TSP support graph $G = (V, E)$ weighted with $x \in]0, 1]^E$

Output: the unique canonical cactus $\mathcal{H}(G)$ of G

(1)   create cactus nodes $c_1, \ldots, c_n$ and assign vertex $i \in V = \{1, \ldots, n\}$ to c_i

(2)   let G' be a copy of G and color all edges of G' white

(3)   associate the set $p(v) = \{c_v\}$ with every vertex v of G'

(4)   if there are only two vertices u and v left in G'

(5)   finish $\mathcal{H}(G)$ by joining the constructed parts of $\mathcal{H}(G)$ (Fig. 4)

(6)   stop.

(7)   search for a basic mincut B with $|B| \geq 2$ in G' (exploit edge colors)

(8)   if the search was successful

(9)   if $B = \{u, v\}$ then construct part of a cycle of $\mathcal{H}(G)$ (Fig. 5)

(10)   if $|B| \geq 3$ then create a new empty node z, let $r_1 = r_2 = z$ and

(11)   for all $v \in B$: join the nodes $p(v)$ by (tree/cycle) edges to z

(12) $G' := G'/B$, let v be the resulting vertex, $p(v) := \{r_1, r_2\}$

(13)   go to (4)

(14) color all edges of the current graph G' black

(15) for all max-(S_i, t_i)-flow residual networks R_i (Sect. 4.2, call Hao-Orlin)

(16)   for all black edges $e = \{u, v\}$ of G'

(17)   if there is a non-degenerate (u, v)-mincut in R_i: color edge e white

(18) if G' equals a single black-connected component

(19)   for all vertices v of G': join nodes $p(v)$ to a new central empty node z

(20)   stop.

(21) go to (7)

If $|p(v)| = 2$ in (11) or (19), then the two elements are the end nodes of a cycle segment of $\mathcal{H}(G)$ and have to be joined to z by two cycle edges. This *closes* the cycle. If $|p(v)| = 1$, then the node has to be joined to z by a tree edge.

For the search in (7), consider the black-connected components (Sect. 4.2) if there is no $\lambda/2$-edge in G' and if there is no λ-starcut of size 3. In order to reduce the size of the H-O input in (15), search λ-starcuts of size 3 seperately without relying upon black-connected components. This can be done quickly since they contain at least one edge of weight $\lambda/3$.

If shrinking B in (12) produces new basic mincuts that can be easily found (size 2 or 3), we say that it *ignites* a *chain reaction* (Fig. 3). Large parts of $\mathcal{H}(G)$ can often be constructed in this way with little effort.

Algorithm NEWCACTUS correctly constructs $\mathcal{H}(G)$ of G. Nothing can stop G' from getting smaller until only 2 vertices are left. Therefore, the algorithm stops after a finite number of steps. No superfluous empty nodes are created. That is, the resulting cactus representation is normal. Whenever a $\lambda/2$-edge is encountered, a part of a cycle is constructed. This guarantees that we construct a cactus of cycle type. The construction step (11) and Figs. 4,5 ensure that every mincut of the resulting cactus induces a mincut of the input graph. Every mincut $M \in \mathcal{M}$ of the input graph G is represented. Shrinking basic mincuts ensures that we do not destroy mincuts that we have not yet represented.

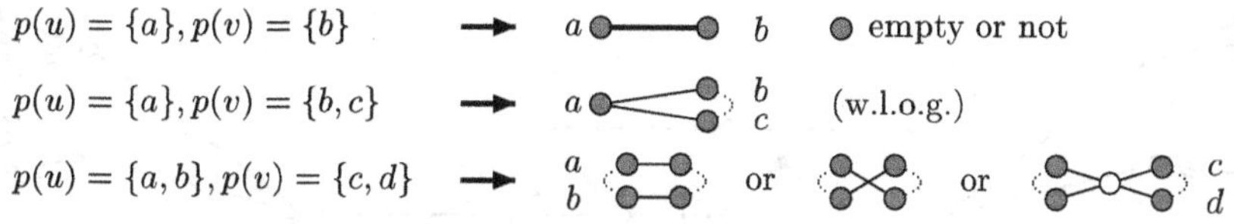

$p(u) = \{a\}, p(v) = \{b\}$ $\longrightarrow$ $a \bullet\!\!-\!\!\bullet\, b$ $\bullet$ empty or not

$p(u) = \{a\}, p(v) = \{b, c\}$ $\longrightarrow$ $a \bullet\!\!\!<\!\!\!^{\bullet\, b}_{\bullet\, c}$ (w.l.o.g.)

$p(u) = \{a, b\}, p(v) = \{c, d\}$ $\longrightarrow$ $^{a}_{b}\,$⟨$\bullet\!\!-\!\!\bullet$⟩   or   ⟨$\bullet\!\!\times\!\!\bullet$⟩   or   ⟨$\bullet\!\!\!>\!\!\!-\!\!\!<\!\!\!\bullet$⟩ $^{c}_{d}$

Fig. 4. Finish $\mathcal{H}(G)$. Use Lemma 1 in G in order to make decisions

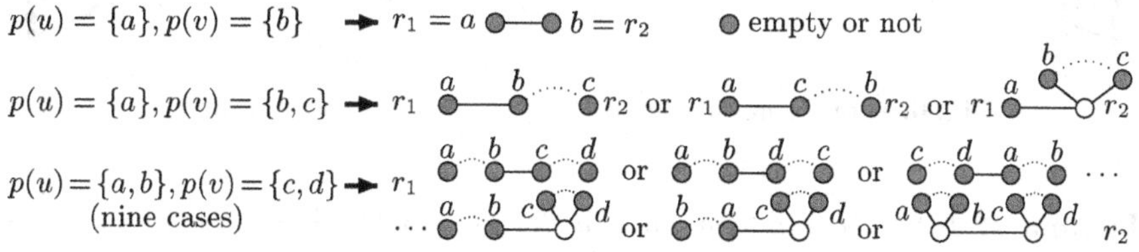

$p(u) = \{a\}, p(v) = \{b\}$ $\longrightarrow$ $r_1 = a \;\bullet\!\!-\!\!\bullet\; b = r_2$ $\bullet$ empty or not

$p(u) = \{a\}, p(v) = \{b, c\}$ $\longrightarrow$ $r_1\; \overset{a}{\bullet}\!\!-\!\!\overset{b}{\bullet}\cdots\overset{c}{\bullet} r_2$ or $r_1\; \overset{a}{\bullet}\!\!-\!\!\overset{c}{\bullet}\cdots\overset{b}{\bullet} r_2$ or $r_1\; \overset{a}{\bullet}\!\!-\!\!\!\!<\!\!^{\overset{b}{}\cdots\overset{c}{}}_{} r_2$

$p(u) = \{a, b\}, p(v) = \{c, d\}$ $\longrightarrow$ $r_1\; \overset{a}{\bullet}\,\overset{b}{\bullet}\!\!-\!\!\overset{c}{\bullet}\,\overset{d}{\bullet}$ or $\overset{a}{\bullet}\,\overset{b}{\bullet}\!\!-\!\!\overset{d}{\bullet}\,\overset{c}{\bullet}$ or $\overset{c}{\bullet}\,\overset{d}{\bullet}\!\!-\!\!\overset{a}{\bullet}\,\overset{b}{\bullet}\cdots$
(nine cases) $\quad\cdots\; \overset{a}{\bullet}\,\overset{b}{\bullet}\,\overset{c}{\bullet}\!\!>\!\!\overset{}{\circ}d$ or $\overset{b}{\bullet}\,\overset{a}{\bullet}\,\overset{c}{\bullet}\!\!>\!\!\circ d$ or $a\,\overset{}{\circ}\!\!<\!\!^{b\, c}\!\!>\!\!\circ d$ r_2

Fig. 5. Constructing a cycle segment of $\mathcal{H}(G)$. Compare Proposition 2

4.4   Beyond Shrinking Basic Mincuts

Consider the cactus in Fig. 6. Let q be small and let p be very large. Since $\mathcal{H}(G)$ does not contain any cycle edge with two non-empty end nodes, G does not contain any $\lambda/2$-edge. Further, G does not have any mincut of size 3. The search in (7) fails. There are no black-connected components yet. H-O is called in order to compute such components. The 8 outer λ-starcuts of size 4 are then found in (7), the corresponding parts of $\mathcal{H}(G)$ are constructed in (11), and the λ-starcuts are shrunken in (12). No chain reaction is ignited. We are stuck with 2 new λ-starcuts of size 4 in G'. We have to call H-O again. Unfortunately, the very large starset of size p is still in G' and H-O has to work through it again. In order to avoid this, we also shrink subsets of starsets of G' (see Remark 2). The corresponding parts of $\mathcal{H}(G)$ are constructed analoguous to λ-starcuts. Shrunken graphs are no longer TSP support graphs, finding λ-starcuts gets more difficult and we have to be more careful when joining parts of $\mathcal{H}(G)$. H-O is burdened only once with large starsets if they are shrunken as soon as they are detected.

If H-O is called repeatedly, then the input graph size of H-O usually decreases very fast in practice. We need pathological examples if we want to witness the opposite. Can the input graph size of H-O be guaranteed to decrease by a constant factor for repeated calls? Let p in Fig. 6 be small. After H-O has been called for the first time, the size of G' is decreased by a certain number of vertices by shrinking the outer λ-starcuts. Enhancing the number q of triangles shows that the factor by which the input graph size decreases can be made arbitrarily bad for two consecutive H-O calls. After the second H-O call, chain reactions are ignited in Fig. 6 and $\mathcal{H}(G)$ can be quickly completed without a third call.

We describe a further operation that reduces G' so that the input graph size for repeated H-O calls decreases. The induced subgraphs of special kinds of starsets are connected. These starsets are entirely detected by our λ-starcut search method. Compare Remark 2.

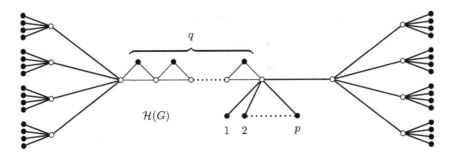

Fig. 6. A problematic cactus $\mathcal{H}(G)$ of a TSP support graph G

Lemma 7. *Let M be a mincut and S a starset with $M \cap S = \emptyset$. Let $N := M \cup S$ be a mincut. Let $|S| \geq 3$. Then the subgraph induced by S is connected.*

Proof. Assume $(S, E(S))$ is not connected. Then there exist $A, B \subset S$ with $A, B \neq \emptyset$, $A \cup B = S$, $A \cap B = \emptyset$ and $x(A : B) = 0$. We have $x(A : \overline{N}) \geq \lambda - x(A : M)$ and $x(B : \overline{N}) \geq \lambda - x(B : M)$. Since A or B is a subset of the starset S with more than 1 vertex, at least one of the two inequalities above is strict ($>$). We have $\lambda = x(\delta(M)) = x(M : \overline{N}) + x(M : A) + x(M : B)$. Further, $\lambda = x(\delta(N)) = x(\overline{N} : M) + x(\overline{N} : A) + x(\overline{N} : B)$. We arrive at $\lambda \geq x(M : A) + x(M : B) > \lambda$. A contradiction. $\square$

We have $x(M : S) > \lambda/2$. Assume that starset S has already been shrunken. If we shrink the mincut M in (12), we generate an edge e with $x_e > \lambda/2$. We link the parts of $\mathcal{H}(G)$ representing S and M. If shrinking S resulted in vertex s and shrinking M in vertex v, then we have to join the nodes $p(v)$ to the empty node $p(s)$ just like in (11) or (20). We then shrink e.

Applying these reduction criteria to G', we can see that $O(\log n)$ is an upper bound for the number of H-O calls in the complete algorithm. An H-O call is only needed when we get stuck. This bound is only reached for pathological graphs in our testbed. Further, the input graph size for H-O is decreasing very rapidly in practice. We conjecture that the time complexity of our complete algorithm equals that of a single H-O call for the original graph G. We have to leave a formal worst case analysis of the run-time for future work.

5   Cacti and Cutting-Plane Generation

All of the following applications benefit from a fast construction of $\mathcal{H}(G)$.

5.1   The Literature

We quickly summarize some cutting-plane generation procedures from the literature that rely on the cactus of a TSP support graph G weighted with a fractional LP solution x or that can benefit from such a representation of mincuts.

Fleischer and Tardos [12]: An important class of facet inducing inequalities for the symmetric TSP are the *comb inequalities* defined on a *handle* $H \subset V$ and an odd number r of pairwise disjoint *teeth* $T_i \subset V$ that intersect H but are not contained in H. The tight triangular (TT) form [24] of comb inequalities is:

$$x(\delta(H)) + \sum_{i=1}^{r} x(\delta(T_i)) \geq 3r + 1 \ . \tag{1}$$

Both the variable vector and weight vector are denoted by x. The algorithm in [12] searches for a comb inequality which is maximally violated by the LP solution x. The maximal violation is 1. Either a violated comb inequality is returned or a guarantee is given that no comb inequality is violated by 1. The input graph G has to be planar. The algorithm exploits the structure of $\mathcal{M}$ represented by $\mathcal{H}(G)$. The run-time is $\mathcal{O}(n + T_{\mathcal{H}(G)})$ where $T_{\mathcal{H}(G)}$ denotes the time needed to construct $\mathcal{H}(G)$. Obviously, the run-time is dominated by cactus construction. The faster the construction of $\mathcal{H}(G)$, the faster [12] in practice. For finding maximally violated combs in planar graphs, [12] is faster than [5].

Caprara, Fischetti and Letchford [5]: Let $\min\{c^t x \mid Ax \leq b, x \text{ integer}\}$ be an ILP with A, b integer. The TSP and many other problems can be written in this form. For k integer, a *mod-k cutting-plane* has the form $\ell^t A x \leq \lfloor \ell^t b \rfloor$ where ℓ has components in $\{0, 1/k, \ldots, (k-1)/k\}$, $\ell^t A$ is integer and $\lfloor \cdot \rfloor$ denotes rounding down. It is well known that comb inequalities are mod-2 cuts [5]. For the TSP, the class of mod-k cuts is considerably larger than the class of comb inequalities and it also contains inequalities that do not induce facets. In [5] an algorithm is presented that finds a maximally violated mod-k cut in polynomial time if such a cut exists. The maximal violation is $(k-1)/k$. Here, not the TT form of inequalities is used. At the heart of the algorithm is the solution of a system of linear equations over the finite field $\mathbb{F}_k$. It is enough to consider k prime. For the asymmetric and symmetric TSP the number of variables in this system can be reduced from $\mathcal{O}(n^2)$ to $\mathcal{O}(n)$ if one has $\mathcal{H}(G)$ at hand, see [5].

Naddef and Thienel [25]: In some heuristics to detect (not necessarily maximally) violated comb inequalities [25] and other handle- and tooth-based violated inequalities [26], one requires suitable candidate sets for handles and teeth. These sets can, for instance, be grown in several ways from seed vertex sets by adding further vertices while trying to keep the cut weight low [25]. Equation (1) shows that the smaller the weights of the candidate sets for H and T_i, the better the chances of violating (1). Good teeth are teeth with low weight. As a starting point for the topic of near-minimum cuts see Benczúr [2] and Karger [18].

The cactus $\mathcal{H}(G)$ can be regarded as a container storing $\mathcal{O}(n^2)$ ideal candidate sets with minimal weight in only $\mathcal{O}(n)$ space. Since $\mathcal{H}(G)$ also mirrors the intersection-structure of $\mathcal{M}$, it is more than a simple container. Sets $S, S_1, S_2 \in \mathcal{M}$ with $S = S_1 \cup S_2$ and $S_1 \cap S_2 = \emptyset$ can be easily extracted from $\mathcal{H}(G)$ if they exist. See also [1]. Consider a cycle of $\mathcal{H}(G)$ and let S, S_1, S_2 be the union of appropriate consecutive beads of this cycle. In [25] the sets S_i are considered to be ideal seed sets for growing a handle H.

Some heuristic cutting-plane generation procedures tend to perform better if G has many 1-paths. Generating 1-edges by shrinking appropriate mincuts in G helps in [25,26] to find violated inequalities in the shrunken graph that could not be found in the original TSP support graph G. These cuts are then lifted. Using $\mathcal{H}(G)$ it is easy to find two mincuts whose shrinking results in a 1-edge: The weight between two consecutive beads of a cycle is 1.

5.2   PQ-Trees

Cacti are not the only way to represent mincuts. PQ-trees, introduced by Booth and Lueker [4], are an alternative. The algorithm in [12] uses a PQ-tree representation of $\mathcal{M}$ in addition to $\mathcal{H}(G)$ and is inspired by ideas from [1]. Applegate, Bixby, Chvátal and Cook [1] use PQ-trees to represent cuts of weight 2. In the graphs they consider these cuts are not necessarily mincuts. They use PQ-trees to find violated comb inequalities. A PQ-tree representing $\mathcal{M}$ can be derived from $\mathcal{H}(G)$ in time $\mathcal{O}(n)$. Therefore, we regard our cactus construction algorithm as part of a new and fast way to construct a PQ-tree representing $\mathcal{M}$.

5.3   Collections of Disjoint Mincuts

Consider the problem of finding a collection of pairwise disjoint $M_i \in \mathcal{M}$ that has to satisfy additional requirements. Examples of requirements are: (i) the number r of mincuts in the collection has to be odd (think of teeth), (ii) r is fixed or (iii) shrinking the M_i in G has to result in an n'-vertex TSP support graph with n' fixed. We might also be interested in more than one such collection. Further, we might be more selective and accept only a subset $D \subset \mathcal{M}$ as domain for the M_i. Examples: (i) only consider mincuts of size at least 2, (ii) do not consider mincuts that constitute only a small part of a long 1-path or (iii) ignore mincuts that contain a depot in a CVRP. Let us assume that we use an enumeration framework to find the collections. On top of testing the additional requirements, we then have to test whether an $M \in D$ considered for inclusion in a collection intersects some mincut that is already in the collection. This can be done using an array in time $\mathcal{O}(|M|)$. However, this is too slow for many applications.

We give the idea of a fast test based on $\mathcal{H}(G)$. For $M \in D$ we select one of the (at most 2) mincuts of $\mathcal{H}(G)$ that induce M (Sect. 2.2) and denote it by M'. We assume that D does not contain complementary cuts. We order the mincuts in D in non-increasing size. Whenever we consider a mincut for inclusion in a collection, its size is not greater than the smallest mincut already in the collection. When we include a mincut M in our collection, we *lock* M' in $\mathcal{H}(G)$. That is, we mark the nodes in M' as occupied. If M' is obtained by removing two edges of cycle c of $\mathcal{H}(G)$, define $Q(M')$ as the intersection of M' with the nodes of cycle c. If M' is obtained by removing a tree edge t, define $Q(M')$ as the end of t lying in M'. Note that $|Q(M')| \leq |M| \leq |M'|$ and usually $Q(M')$ is considerably smaller than M and M'. The fast cactus-based test ([34]) is:

Table 1. Construction CPU *time* in seconds, *avg* stands for the arithmetic mean

shrunken graph 1-paths	type	Time BEN avg	max	Time NEW avg	max	Speed-Up avg	max
yes	1	5	48	0.2	1.2	21	61
yes	2	17	146	0.6	4.4	19	40
no	1	117	1172	0.3	1.6	237	899
no	2	126	1235	0.7	4.5	119	298

Proposition 3. *Let pairwise disjoint mincuts M_i' $(i \in I)$ of $\mathcal{H}(G)$ be locked in $\mathcal{H}(G)$. Let $M \in D \subset \mathcal{M}$ be induced by the mincut M' of $\mathcal{H}(G)$. If $|M| \leq |M_i|$ $(i \in I)$ where $M_i = V(M_i') \in D$, then: M intersects some M_i exactly if $Q(M')$ contains a locked node.*

In [34] Proposition 3 is used to enumerate all collections of pairwise disjoint mincuts that shrink G (short 1-paths) down to exactly n' vertices. Using the above cactus-based test, this enumeration takes seconds where a naive approach takes hours [34]. Exploiting the known facet structure of low-dimensional TSP polytopes $STSP(n')$, the n'-vertex graphs are used to generate cutting-planes which can be lifted then. Pioneering results in this spirit are reported in [7].

6 Computational Results

We compare the run-time of the new algorithm (NEW) with that of Fleischer's [11,34] which we use as benchmark (BEN). We implemented both the algorithm described in Sect. 4 and the algorithm [11]. We do not have any other cactus implementation available. Both C-implementations use the same subroutines and were prepared carefully from scratch. The run-time of our Hao-Orlin [14] implementation [34] is $\mathcal{O}(n^2\sqrt{m})$ since we use highest level pushing. We used gcc 2.95.3 with optimization -O3 on a 450 Mhz SUN sparc with 4 GB memory.

We generated two TSP support graphs for each of 32 TSPLIB [31] instances in the range of 1,000 to 20,000 cities. For type 1, only subtour elimination constraints were separated. In addition, simple combs were heuristically separated [30] for type 2. For the instances with 2,000 to 20,000 cities (Fig. 7), cactus construction for graphs of type 1 was always faster than for type 2.

The run-time of both algorithms was measured for these 64 original graphs and for a further 64 graphs obtained by shrinking paths of 1-edges in the original graphs to a single 1-edge (Table 1). Whether parts of a graph can be shrunken [30] prior to cactus construction depends on the application. Naddef and Thienel [25] report a situation in which they prefer not to make long 1-paths shorter than 4 edges. Shrinking only simple 1-paths leaves all interesting parts of a graph for the algorithms to work through. The speed-up factor of 20 given in the abstract is the arithmetic mean for shrinking of 1-paths in advance before cactus construction. Table 1 also shows speed-up factors for unmodified TSP support graphs as they are produced by a branch-and-cut framework.

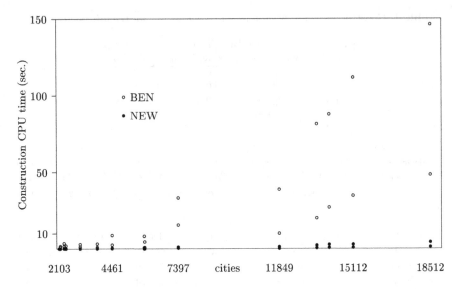

Fig. 7. Time required for big TSP instances; 1-paths shrunken in advance

For the testbed of 128 TSP support graphs, the number of Hao-Orlin calls was usually 1 but ranged between 0 and 2. We also generated TSP support graphs whose cacti are full k-ary trees of height h. The *height* is the maximum level of a node where the root has level 0. For $k \geq 5$ the number of H-O calls needed is h. Speed-up factors for these pathological graphs, which are unfavourable for the new approach in the sense that many H-O calls are needed, are at least 2.

7   Summary and Conclusion

We have presented a new kind of construction algorithm for cactus representations of mincuts with a view to cutting-plane generation. The new algorithmic idea is intuitive and differs completely from previous approaches. We construct the canonical cactus of a TSP support graph from the leaves inwards.

Fleischer's deterministic approach to build cacti from Hao-Orlin [11] is among the fastest proposed so far. We used this state-of-the-art approach for weighted graphs as benchmark. For TSP support graphs, the algorithm detailed here outperforms the benchmark in terms of construction time (Table 1, Fig. 7).

While the cactus of an n-vertex graph takes $\mathcal{O}(n)$ space to store, the memory requirement during construction can differ substantially between algorithms. For the larger testgraphs in our testbed, the benchmark algorithm takes GBytes where the new algorithm takes MBytes. This corresponds to the reduction in space requirement from $\mathcal{O}(n^2)$ to $\mathcal{O}(m)$.

References

1. D. Applegate, R. Bixby, V. Chvátal, and W. Cook. Finding Cuts in the TSP (A preliminary report). Technical Report 95-05, Center for Discrete Mathematics & Theoretical Computer Science, DIMACS, 1995.
   Available at http://dimacs.rutgers.edu/TechnicalReports.

2. A. A. Benczúr. *Cut structures and randomized algorithms in edge-connectivity problems*. Department of Mathematics, MIT, 1997.
   Available at http://theses.mit.edu.

3. R. E. Bixby. The Minimum Number of Edges and Vertices in a Graph with Edge Connectivity n and m n-Bonds. *Networks*, 5:253–298, 1975.

4. K. S. Booth and G. S. Lueker. Testing for the consecutive ones property, interval graphs, and graph planarity using PQ-tree algorithms. *Journal of Computer Systems and Science*, 13:335–379, 1976.

5. A. Caprara, M. Fischetti, and A. N. Letchford. On the separation of maximally violated mod-k cuts. *Math. Program. Series A*, 87(1):37–56, January 2000.

6. C. S. Chekuri, A. V. Goldberg, D. R. Karger, M. S. Levine, and C. Stein. Experimental Study of Minimum Cut Algorithms. In *Proceedings of the 8th Annual ACM-SIAM Symposium on Discrete Algorithms (SODA '97)*, pages 324–333, 1997.

7. T. Christof and G. Reinelt. Combinatorial Optimization and Small Polytopes. *Top*, 4(1):1–64, 1996. ISSN 1134-5764.

8. A. De Vitis. The cactus representation of all minimum cuts in a weighted graph. Technical Report 454, IASI, Viale Manzoni 30, 00185 Roma, Italy, May 1997. Write to biblio@iasi.rm.cnr.it.

9. G. Dantzig, R. Fulkerson, and S. Johnson. Solution of a large-scale traveling-salesman problem. *Journal of the ORSA*, 2:393–410, 1954.

10. E. A. Dinits, A. V. Karzanov, and M. V. Lomonosov. On the structure of a family of minimal weighted cuts in a graph. In A. A. Fridman, editor, *Studies in Discrete Optimization*, pages 290–306. Nauka, Moscow, 1976. Original article in Russian. English translation from www.loc.gov or www.nrc.ca/cisti.

11. L. Fleischer. Building Chain and Cactus Representations of All Minimum Cuts from Hao-Orlin in the Same Asymptotic Run Time. *J. Algorithms*, 33(1):51–72, October 1999. A former version appeared in LNCS 1412, pages 294–309, 1998.

12. L. Fleischer and É. Tardos. Separating maximally violated comb inequalities in planar graphs. *Mathematics of Operations Research*, 24(1):130–148, February 1999. A former version appeared in LNCS 1084, pages 475–489, 1996.

13. R. E. Gomory and T. C. Hu. Multi-terminal network flows. *J. Soc. Indust. Appl. Math.*, 9(4):551–570, December 1961.

14. J. Hao and J. B. Orlin. A Faster Algorithm for Finding the Minimum Cut in a Directed Graph. *J. Algorithms*, 17:424–446, 1994.

15. M. Jünger and D. Naddef, editors. *Computational Combinatorial Optimization: Optimal or Provably Near-Optimal Solutions*, volume 2241 of *LNCS*. Springer, 2001.

16. M. Jünger, G. Reinelt, and G. Rinaldi. The Traveling Salesman Problem. In M. O. Ball, T. L. Magnanti, C. L. Monma, and G. L. Nemhauser, editors, *Network Models*, volume 7 of *Handbooks in Operations Research and Management Science*, chapter 4, pages 225–330. Elsevier, 1995.

17. M. Jünger, G. Rinaldi, and S. Thienel. Practical Performance of Efficient Minimum Cut Algorithms. *Algorithmica*, 26:172–195, 2000.

18. D. R. Karger and C. Stein. A new approach to the minimum cut problem. *Journal of the ACM*, 43(4):601–640, July 1996.

19. A. V. Karzanov and E. A. Timofeev. Efficient algorithm for finding all minimal edge cuts of a nonoriented graph. *Cybernetics*, 22:156–162, 1986. Translated from *Kibernetika*, 2:8–12, 1986.

20. A. N. Letchford. Separating a superclass of comb inequalities in planar graphs. *Mathematics of Operations Research*, 25(3):443–454, August 2000.

21. H. Nagamochi and T. Kameda. Canonical Cactus Representation for Minimum Cuts. *Japan Journal of Industrial and Applied Mathematics*, 11(3):343–361, 1994.

22. H. Nagamochi and T. Kameda. Constructing cactus representation for all minimum cuts in an undirected network. *Journal of the Operations Research Society of Japan*, 39(2):135–158, 1996.

23. H. Nagamochi, Y. Nakao, and T. Ibaraki. A Fast Algorithm for Cactus Representations of Minimum Cuts. *Japan Journal of Industrial and Applied Mathematics*, 17(2):245–264, June 2000.

24. D. Naddef and G. Rinaldi. The crown inequalities for the symmetric traveling salesman polytope. *Mathematics of Operations Research*, 17(2):308–326, May 1992.

25. D. Naddef and S. Thienel. Efficient Separation Routines for the Symmetric Traveling Salesman Problem I: General Tools and Comb Separation. Technical report, ENSIMAG (France), Universität zu Köln (Germany), 1999. To Appear in Math. Program.

26. D. Naddef and S. Thienel. Efficient Separation Routines for the Symmetric Traveling Salesman Problem II: Separating Multi Handle Inequalities. Technical report, ENSIMAG (France), Universität zu Köln (Germany), January 2001. To Appear in Math. Program.

27. D. Naor and V. V. Vazirani. Representing and Enumerating Edge Connectivity Cuts in $\mathcal{RNC}$. *LNCS 519*, pp. 273–285, 1991.

28. J.-C. Picard and M. Queyranne. On the structure of all minimum cuts in a network and applications. *Math. Program. Study*, 13:8–16, 1980.

29. J.-C. Picard and M. Queyranne. Selected applications of minimum cuts in networks. *INFOR, Can. J. Oper. Res. Inf. Process.*, 20(4):394–422, November 1982.

30. M. Padberg and G. Rinaldi. Facet identification for the symmetric traveling salesman polytope. *Math. Program.*, 47:219–257, 1990.

31. G. Reinelt. TSPLIB - A Traveling Salesman Problem Library. *ORSA Journal on Computing*, 3(4):376–384, Fall 1991.
    http://www.informatik.uni-heidelberg.de/groups/comopt/software/TSPLIB95

32. T. K. Ralphs, L. Kopman, W. R. Pulleyblank, and L. E. Trotter Jr. On the Capacitated Vehicle Routing Problem. Submitted to Math. Program., 2000. Available at www.branchandcut.org/VRP.

33. P. Toth and D. Vigo, editors. *The Vehicle Routing Problem*. Monographs on Discrete Mathematics and Applications. SIAM, 2002. ISBN 0-89871-498-2.

34. K. M. Wenger. Kaktus-Repräsentation der minimalen Schnitte eines Graphen und Anwendung im Branch-and-Cut Ansatz für das TSP. Diplomarbeit, Institut für Informatik, Ruprecht-Karls-Universität Heidelberg, Im Neuenheimer Feld 368, 69120 Heidelberg, Germany, 1999.

Split Closure and Intersection Cuts*

Kent Andersen, Gérard Cornuéjols, and Yanjun Li

Graduate School of Industrial Administration
Carnegie Mellon University, Pittsburgh, PA, USA
{kha,gc0v,yanjun}@andrew.cmu.edu

Abstract. In the seventies, Balas introduced intersection cuts for a
Mixed Integer Linear Program (MILP), and showed that these cuts can
be obtained by a closed form formula from a basis of the standard linear programming relaxation. In the early nineties, Cook, Kannan and
Schrijver introduced the split closure of an MILP, and showed that the
split closure is a polyhedron. In this paper, we show that the split closure can be obtained using only intersection cuts. We give two different
proofs of this result, one geometric and one algebraic. Furthermore, the
result is used to provide a new proof of the fact that the split closure
is a polyhedron. Finally, we extend the result to more general two-term
disjunctions.

1 Introduction

In the seventies, Balas showed how a cone and a disjunction can be used to
derive a cut [1] for a Mixed Integer Linear Program (MILP). In that paper, the
cone was obtained from an optimal basis to the standard linear programming
relaxation. The cut was obtained by a closed form formula and was called the
intersection cut.

Later in the seventies, Balas generalized the idea to polyhedra [2]. It was
demonstrated that, given a polyhedron, and a valid but violated disjunction, a
cut could be obtained by solving a linear program. The idea was further expanded
in the early nineties, where Cook, Kannan and Schrijver [4] studied split cuts
obtained from two-term disjunctions that are easily seen to be valid for an MILP.
The intersection of all split cuts is called the split closure of an MILP. Cook,
Kannan and Schrijver proved that the split closure of an MILP is a polyhedron.

Any basis of the constraint matrix describing the polyhedron can be used,
together with a disjunction, to derive an intersection cut, i.e. the basis used does
not have to be optimal or even feasible. A natural question is how intersection
cuts relate to disjunctive cuts obtained from polyhedra. This question was an-
swered by Balas and Perregaard [3] for the 0-1 disjunction for Mixed Binary
Linear Programs. The conclusion was that any disjunctive cut obtained from a
polyhedron and the 0-1 disjunction is identical to, or dominated by, intersection
cuts obtained from the 0-1 disjunction and bases of the constraint matrix describing the polyhedron. We generalize this result from 0-1 disjunctions to more

---

* Supported by NSF grant DMI-0098427 and ONR grant N00014-97-1-0196.

general two-term disjunctions. This is the main result of this paper. We provide two different proofs, one geometric and one algebraic. A consequence is a new proof of the fact that the split closure is a polyhedron.

We consider the Mixed Integer Linear Program (MILP):

$$(\text{MILP}) \qquad min\{c^T x : Ax \le b, x_j \text{ integer}, j \in N_I\},$$

where $c \in \mathbb{R}^n$, $b \in \mathbb{R}^m$, $N_I \subseteq N := \{1, 2, \dots, n\}$ and A is an $m \times n$ matrix. r denotes the rank of A. LP is the Linear Programming problem obtained from MILP by dropping the integrality conditions on x_j, $j \in N_I$. P_I and P denote the sets of feasible solutions to MILP and LP respectively. $M := \{1, 2, \dots, m\}$ is used to index the rows of A. a_i, for $i \in M$, denotes the ith row of A. We assume $a_i \ne 0_n$ for all $i \in M$. Given $S \subseteq M$, $r(S)$ denotes the rank of the sub-matrix of A induced by the rows in S ($r(M) = r$). Furthermore, for $S \subseteq M$, the relaxation of P obtained by dropping the constraints indexed by $M \setminus S$ from the description of P, is denoted $P(S)$, i.e. $P(S) := \{x \in \mathbb{R}^n : (a_i)^T x \le b_i, \forall i \in S\}$ ($P(M) = P$). A *basis* of A is an n-subset B of M, such that the vectors a_i, $i \in B$, are linearly independent. Observe that, if $r < n$, A does not have bases. $\mathcal{B}_k^*$, where k is a positive integer, denotes the set of k-subsets S of M, such that the vectors a_i, $i \in S$, are linearly independent ($\mathcal{B}_n^*$ denotes the set of bases of A).

The most general two-term disjunction considered in this paper is an expression D of the form $D^1 x \le d^1 \vee D^2 x \le d^2$, where $D^1 : m_1 \times n$, $D^2 : m_2 \times n$, $d^1 : m_1 \times 1$ and $d^2 : m_2 \times 1$. The set of points in $\mathbb{R}^n$ satisfying D is denoted F_D. The set $conv(P \cap F_D)$ is called the *disjunctive set* defined by P and D in the remainder. In addition, given a subset S of the constraints, the set $conv(P(S) \cap F_D)$ is called the disjunctive set defined by S and D. Finally, given a basis B in $\mathcal{B}_r^*$, the set $conv(P(B) \cap F_D)$ is called a *basic* disjunctive set.

An important two-term disjunction, in the context of an MILP, is the *split disjunction* $D(\pi, \pi_0)$ of the form $\pi^T x \le \pi_0 \vee \pi^T x \ge \pi_0 + 1$, where $(\pi, \pi_0) \in \mathbb{Z}^{n+1}$ and $\pi_j = 0$ for all $j \notin N_I$. For the split disjunction, a complete description of the basic disjunctive sets is available as follows. Given a set B in $\mathcal{B}_r^*$, the basic disjunctive set defined by B and $D(\pi, \pi_0)$ is the set of points in $P(B)$ that satisfy the intersection cut derived from B and $D(\pi, \pi_0)$ (Lemma 1). Let $\Pi^n(N_I) := \{(\pi, \pi_0) \in \mathbb{Z}^{n+1} : \pi_j = 0, j \notin N_I\}$. The *split closure* of MILP, denoted by SC, is defined to be the intersection of the disjunctive sets defined by P and $D(\pi, \pi_0)$ over all disjunctions (π, π_0) in $\Pi^n(N_I)$. Similarly, given $S \subseteq M$, SC(S), is defined to be the intersection of the disjunctive sets defined by $P(S)$ and $D(\pi, \pi_0)$ over all disjunctions (π, π_0) in $\Pi^n(N_I)$. A *split cut* is a valid inequality for SC.

The first contribution in this paper is a theorem (Theorem 1) stating that the split closure of MILP can be written as the intersection of the split closures of the sets $P(B)$ over all sets B in $\mathcal{B}_r^*$. (i.e. $SC = \bigcap_{B \in \mathcal{B}_r^*} SC(B)$). We prove this theorem by proving that the disjunctive set defined by P and $D(\pi, \pi_0)$, for a split disjunction $D(\pi, \pi_0)$, can be written as the intersection of the basic disjunctive sets (i.e. $conv(P \cap F_{D(\pi, \pi_0)}) = \bigcap_{B \in \mathcal{B}_r^*} conv(P(B) \cap F_{D(\pi, \pi_0)})$). We provide both a geometric and an algebraic proof of this result. The result implies that both the disjunctive set defined by P and $D(\pi, \pi_0)$ and the split closure of MILP can be obtained using only intersection cuts. This generalizes a result of Balas and Perregaard showing the same result for the disjunction $x_j \leq 0 \vee x_j \geq 1$ for Mixed Binary Linear Programs [3]. (In fact, in that paper, it was assumed that $r = n$, whereas the theorem presented here does not have this assumption). Furthermore, the result leads to a new proof of the fact that the split closure is a polyhedron (Theorem 3).

The second contribution in this paper is a theorem (Theorem 6) stating that the disjunctive set defined by P and a general two-term disjunction D can be written as the intersection of disjunctive sets defined by D and $(r + 1)$-subsets S of M (i.e. $conv(P \cap F_D) = \bigcap_{S \in \mathcal{C}_1^*} conv(P(S) \cap F_D)$, where $\mathcal{C}_1^*$ is some family of $(r + 1)$-subsets of M to be defined later). The theorem implies that any valid inequality for the disjunctive set defined by P and D is identical to, or dominated by, inequalities derived from the disjunction D and $(r+1)$-subsets of the constraints describing the polyhedron. Furthermore, in the special case where $r = n$, we show that it is enough to consider a certain family $\mathcal{C}_2^*$ of n-subsets of the constraints describing the polyhedron.

The rest of this paper is organized as follows. In Sect. 2, we consider basic disjunctive sets for split disjunctions. In Sect. 3, the characterization of the split closure in terms of intersection cuts is presented, and a geometric argument for the validity of the result is presented. In Sect. 4 we give a new proof of the fact that the split closure is a polyhedron. In Sect. 5, we generalize the results of Sect. 3 to more general two-term disjunctions. We also give an example showing that other extensions are incorrect. The arguments used in this section are mostly algebraic. In fact, Sect. 3 and Sect. 5 could be read independently.

2   A Complete Description of the Basic Disjunctive Set for Split Disjunction

In this section, we describe the set $P(B)$ for B in $\mathcal{B}_r^*$ as the translate of a cone, and use this cone together with a split disjunction to derive an intersection cut. The intersection cut is then used to characterize the basic disjunctive set obtained from the split disjunction.

Let $B \in \mathcal{B}_r^*$ be arbitrary. The set $P(B)$ was defined using half-spaces in the introduction. We now give an alternative description. Let $\bar{x}(B)$ satisfy $a_i^T \bar{x}(B) = b_i$ for all $i \in B$. Furthermore, let $L(B) := \{x \in \mathbb{R}^n : a_i^T x = 0, \forall i \in B\}$. Finally, let $r^i(B)$, where $i \in B$, be a solution to the system $(a_{k.})^T r^i(B) = 0$, for $k \in B \setminus \{i\}$, and $(a_{i.})^T r^i(B) = -1$. We have:

$$P(B) = \bar{x}(B) + L(B) + Cone\left(\{r^i(B) : i \in B\}\right), \tag{1}$$

where $Cone\left(\{r^i(B) : i \in B\}\right) := \{x \in \mathbb{R}^n : x = \sum_{i \in B} \lambda_i r^i(B), \lambda_i \geq 0, i \in B\}$ denotes the cone generated by the vectors $r^i(B)$, $i \in B$. Observe that the vectors $r^i(B)$, $i \in B$, are linearly independent.

Let $(\pi, \pi_0) \in \mathbb{Z}^{n+1}$. Assume that all points y in $\bar{x}(B) + L(B)$ violate the disjunction $D(\pi, \pi_0)$ (Lemma 1 below shows that this is the only interesting case). Observe that this implies that the linear function $\pi^T x$ is constant on $\bar{x}(B) + L(B)$. The intersection cut defined by B and $D(\pi, \pi_0)$ can now be described. Define $\epsilon(\pi, B) := \pi^T \bar{x}(B) - \pi_0$ to be the amount by which the points in $\bar{x}(B) + L(B)$ violate the first term in the disjunction. Also, for i in B, define:

$$\alpha_i(\pi, B) := \begin{cases} -\epsilon(\pi, B)/(\pi^T r^i(B)) & \text{if } \pi^T r^i(B) < 0, \\ (1 - \epsilon(\pi, B))/(\pi^T r^i(B)) & \text{if } \pi^T r^i(B) > 0, \\ \infty & \text{otherwise.} \end{cases} \tag{2}$$

The interpretation of the numbers $\alpha_i(\pi, B)$, for $i \in B$, is the following. Let $x^i(\alpha, B) := \bar{x}(B) + \alpha r^i(B)$, where $\alpha \in \mathbb{R}_+$, denote the half-line starting in $\bar{x}(B)$ in the direction $r^i(B)$. The value $\alpha_i(\pi, B)$ is the smallest value of $\alpha \in \mathbb{R}_+$, such that $x^i(\alpha, B)$ satisfies the disjunction $D(\pi, \pi_0)$, i.e. $\alpha_i(\pi, B) = \inf\{\alpha \geq 0 : x^i(\alpha, B) \in F_{D(\pi,\pi_0)}\}$. Given the numbers $\alpha_i(\pi, B)$ for $i \in B$, the intersection cut associated with B and $D(\pi, \pi_0)$ is given by:

$$\sum_{i \in B} (b_i - a_{i.}^T x)/\alpha_i(\pi, B) \geq 1 . \tag{3}$$

The validity of this inequality for $Conv(P(B) \cap F_{D(\pi,\pi_0)})$ was proven by Balas [1]. In fact, we have:

Lemma 1. *Let $B \in \mathcal{B}_r^*$ and $D(\pi, \pi_0)$ be a split disjunction, where $(\pi, \pi_0) \in \mathbb{Z}^{n+1}$.*

(i) *If $\pi^T x \notin]\pi_0, \pi_0 + 1[$, for some $x \in \bar{x}(B) + L(B)$, then $Conv(P(B) \cap F_{D(\pi,\pi_0)}) = P(B)$.*

(ii) *If $\pi^T x \in]\pi_0, \pi_0 + 1[$, for all $x \in \bar{x}(B) + L(B)$, then $Conv(P(B) \cap F_{D(\pi,\pi_0)}) = \{x \in P(B) : (3)\}$.*

3   Split Closure Characterization

In this section, we give a geometric proof of the following theorem, which characterizes the split closure in terms of certain basic subsets of the constraints.

Theorem 1.

$$SC = \bigcap_{B \in \mathcal{B}_r^*} SC(B) . \tag{4}$$

We prove this result in the following lemmas and corollaries. Let $\alpha^T x \leq \beta$ and $\alpha^T x \geq \psi$ be inequalities, where $\alpha \in \mathbb{R}^n$ and $\beta < \psi$. When $\alpha \neq 0_n$, $\alpha^T x \leq \beta$ and $\alpha^T x \geq \psi$ represent two non-intersecting half-spaces. We have the following key lemma:

Lemma 2. *Assume P is full-dimensional. Then:*

$$Conv((P \cap \{x : \alpha^T x \leq \beta\}) \cup (P \cap \{x : \alpha^T x \geq \psi\})) =$$
$$\bigcap_{B \in \mathcal{B}_r^*} Conv((P(B) \cap \{x : \alpha^T x \leq \beta\}) \cup (P(B) \cap \{x : \alpha^T x \geq \psi\})) . \tag{5}$$

Proof. The following notation will be convenient. Define $P_1 := P \cap \{x : \alpha^T x \leq \beta\}$ and $P_2 := P \cap \{x : \alpha^T x \geq \psi\}$. Furthermore, given a set $B \in \mathcal{B}_r^*$, let $P_1(B) := P(B) \cap \{x : \alpha^T x \leq \beta\}$ and $P_2(B) := P(B) \cap \{x : \alpha^T x \geq \psi\}$.

When $\alpha = 0_n$, no matter what the values of β and ψ are, at least one of P_1 and P_2 is empty (Notice that we always have $\beta < \psi$). If both are empty, then the lemma holds trivially by $\emptyset = \emptyset$. If one is not empty, then $Conv(P_1 \cup P_2) = P$ and, similarly, $\bigcap_{B \in \mathcal{B}_r^*} Conv((P_1(B) \cup P_2(B))) = \bigcap_{B \in \mathcal{B}_r^*} P(B) = P$. The last equality is due to the assumption $a_i. \neq 0_n$ for $i \in M$. Therefore, we assume $\alpha \neq 0_n$ in the rest of the proof.

Because $P \subseteq P(B)$ for any $B \in \mathcal{B}_r^*$, it is clear that $Conv(P_1 \cup P_2) \subseteq \bigcap_{B \in \mathcal{B}_r^*} Conv((P_1(B) \cup P_2(B)))$. Therefore, we only need to show the other direction of the inclusion.

Observe that it suffices to show that any valid inequality for $Conv(P_1 \cup P_2)$ is valid for $Conv(P_1(B) \cup P_2(B))$ for at least one $B \in \mathcal{B}_r^*$. Now, let $\delta^T x \leq \delta_0$ be a valid inequality for $Conv(P_1 \cup P_2)$. Clearly, we can assume $\delta^T x \leq \delta_0$ is facet-defining for $Conv(P_1 \cup P_2)$. This is clearly true for the valid inequalities of P, since we can always choose a $B \in \mathcal{B}_r^*$, by applying the techniques of linear programming, such that the valid inequality of P is valid for $P(B)$. So we may assume that the inequality is valid for $Conv(P_1 \cup P_2)$ but not valid for P.

Case 1. $P_2 = \emptyset$.
Since $P_2 = \emptyset$, $Conv(P_1 \cup P_2) = P_1 = P \cap \{x : \alpha^T x \leq \beta\}$. Hence $\alpha^T x \leq \beta$ is a valid inequality for $Conv(P_1 \cup P_2)$. We just want to show that it is also valid for $Conv((P(B) \cap \{x : \alpha^T x \leq \beta\}) \cup (P(B) \cap \{x : \alpha^T x \geq \psi\}))$ for some $B \in \mathcal{B}_r^*$. Because $P_2 = \emptyset$ and P is full-dimensional, applying the techniques of linear programming shows that the value $\tilde{\gamma} = \max\{\gamma : P \cap \{x : \alpha^T x = \gamma\} \neq \emptyset\}$ specifies $B \in \mathcal{B}_r^*$ such that $P(B) \cap \{x : \alpha^T x \geq \psi\} = \emptyset$ and $\tilde{\gamma} = \max\{\gamma : P(B) \cap \{x : \alpha^T x = \gamma\} \neq \emptyset\}$, where $\tilde{\gamma} < \psi$. We have $Conv((P(B) \cap \{x : \alpha^T x \leq \beta\}) \cup (P(B) \cap \{x : \alpha^T x \geq \psi\})) = P(B) \cap \{x : \alpha^T x \leq \beta\}$ for this particular B. Therefore, $\alpha^T x \leq \beta$ is valid for $Conv((P(B) \cap \{x : \alpha^T x \leq \beta\}) \cup (P(B) \cap \{x : \alpha^T x \geq \psi\}))$.

Case 2. P_1 and P_2 are full-dimensional.
Consider an arbitrary facet F of $Conv(P_1 \cup P_2)$ which does not induce a valid inequality for P. We are going to use the corresponding F-defining inequality (half-space) and F-defining equality (hyperplane). Our goal is to show that the F-defining inequality is valid for $Conv(P(B) \cap \{x : \alpha^T x \leq \beta\}) \cup (P(B) \cap \{x : \alpha^T x \geq \psi\}))$ for some $B \in \mathcal{B}_r^*$.

Let $F_1 := F \cap P_1$ and $F_2 := F \cap P_2$. Since the F-defining inequality is valid for P_1 and P_2 but not valid for P, we can deduce $F_1 \subseteq \{x \in \mathbb{R}^n | \alpha^T x = \beta\}$ and $F_2 \subseteq \{x \in \mathbb{R}^n : \alpha^T x = \psi\}$. F is the convex combination of F_1 and F_2, where F_1 is a k-dimensional face of P_1 and F_2 is an m-dimensional face of P_2. Since F is of dimension $n - 1$, we have $0 \leq k \leq n - 2$ and $n - k - 2 \leq m \leq n - 2$.

The intersection of the F-defining hyperplane with $\alpha^T x = \beta$ (or $\alpha^T x = \psi$) defines a $(n - 2)$-dimensional affine subspace which contains F_1 (or F_2). Therefore, $Aff(F_2)$ contains two affine subspaces S_1 and S_2 of dimensions $n-k-2$ and $m - (n - k - 2)$ respectively, where S_1 is orthogonal to $Aff(F_1)$ and S_2 is parallel to $Aff(F_1)$. In other words, $Aff(F_2)$ can be orthogonally decomposed into two affine subspaces S_1 and S_2 such that $S_1 \cap S_2$ has an unique point x_0, $(x_1 - x_0)^T (x_2 - x_0) = 0$ for any $x_1 \in S_1$ and $x_2 \in S_2$, and for some $x_3 \in \text{Aff}(F_1)$ we have $\{x_i - x_0 : x_i \in S_2\} \subseteq \{x_i - x_3 : x_i \in \text{Aff}(F_1)\}$ and $(x_1 - x_0)^T (x_4 - x_3) = 0$ for any $x_1 \in S_1$ and $x_4 \in \text{Aff}(F_1)$.

There exist $n - k - 1$ constraints of P such that the corresponding hyperplanes, together with $\alpha^T x = \beta$, define $\text{Aff}(F_1)$. Let these hyperplanes be $A_{(n-k-1) \times n} x = b_1$. Similarly, there are $n - m - 1$ constraints of P such that the corresponding hyperplanes, together with $\alpha^T x = \psi$, define $\text{Aff}(F_2)$. Let them be $A_{(n-m-1) \times n} x = b_2$. From the discussion in the previous paragraphs, one can easily see that the equations $A_{(n-k-1) \times n} x = 0$, $A_{(n-m-1) \times n} x = 0$ have solution space $\{x_i - x_0 : x_i \in S_2\}$ with dimension $m - (n - k - 2)$. Since $m - (n - k - 2) \equiv n - [(n - m - 1) + (n - k - 1)]$, the matrix $\begin{pmatrix} A_{(n-k-1) \times n} \\ A_{(n-m-1) \times n} \end{pmatrix}$ is full row-rank. Because the rank of A is r, $(n - k - 1) + (n - m - 1) \le r$. This allows us to choose another $r - (n - k - 1) - (n - m - 1)$ constraints $A_{[r - (n-k-1) - (n-m-1)] \times n} x \le b_3$ from $Ax \le b$, together with $A_{(n-k-1) \times n} x \le b_1$ and $A_{(n-m-1) \times n} x \le b_2$, to construct a $B \in \mathcal{B}_r^*$ such that the F-defining inequality is valid for $\text{Conv}(P(B) \cap \{x : \alpha^T x \le \beta\}) \cup (P(B) \cap \{x : \alpha^T x \ge \psi\}))$.

Case 3. $P_1 \ne \emptyset$, $P_2 \ne \emptyset$, and some of P_1 and P_2 is not full-dimensional.
Instead of considering the inequalities $\alpha^T x \le \beta$ and $\alpha^T x \ge \psi$, we construct two inequalities $\alpha^T x \le \beta + \epsilon$ and $\alpha^T x \ge \psi - \epsilon$, where ϵ is an arbitrarily small positive number satisfying $\beta + \epsilon < \psi - \epsilon$. Let $P_1^\epsilon := P \cap \{x : \alpha^T x \le \beta + \epsilon\}$ and $P_2^\epsilon := P \cap \{x : \alpha^T x \ge \psi - \epsilon\}$. Since $P_1 \ne \emptyset$ and $P_2 \ne \emptyset$, P_1^ϵ and P_2^ϵ are full-dimensional polyhedra.

Because $|\mathcal{B}_r^*|$ is finite and B and P are closed sets in $\mathbb{R}^n$, by the definition of the Conv operation and the result proved in Case 2, we have

$$\bigcap_{B \in \mathcal{B}_r^*} \text{Conv}((P(B) \cap \{x : \alpha^T x \le \beta\}) \cup (P(B) \cap \{x : \alpha^T x \ge \psi\}))$$

$$= \bigcap_{B \in \mathcal{B}_r^*} \lim_{\epsilon \to 0^+} \text{Conv}((P(B) \cap \{x : \alpha^T x \le \beta + \epsilon\}) \cup (P(B) \cap \{x : \alpha^T x \ge \psi - \epsilon\}))$$

$$= \lim_{\epsilon \to 0^+} \bigcap_{B \in \mathcal{B}_r^*} \text{Conv}((P(B) \cap \{x : \alpha^T x \le \beta + \epsilon\}) \cup (P(B) \cap \{x : \alpha^T x \ge \psi - \epsilon\}))$$

$$= \lim_{\epsilon \to 0^+} \text{Conv}((P \cap \{x : \alpha^T x \le \beta + \epsilon\}) \cup (P \cap \{x : \alpha^T x \ge \psi - \epsilon\}))$$

$$= \text{Conv}((P \cap \{x : \alpha^T x \le \beta\}) \cup (P \cap \{x : \alpha^T x \ge \psi\})) .$$
□

From Lemma 2 we immediately have the following:

Corollary 1. Let $S \subset \mathbb{R}^{n+2}$ be a set of $(\alpha, \beta, \psi) \in \mathbb{R}^{n+2}$ such that $\alpha \in \mathbb{R}^n$ and $\beta < \psi$. When P is full-dimensional,

$$\bigcap_{(\alpha, \beta, \psi) \in S} \text{Conv}((P \cap \{x : \alpha^T x \le \beta\}) \cup (P \cap \{x : \alpha^T x \ge \psi\})) =$$

$$\bigcap_{B \in \mathcal{B}_r^*} \bigcap_{(\alpha, \beta, \psi) \in S} \text{Conv}((P(B) \cap \{x : \alpha^T x \le \beta\}) \cup (P(B) \cap \{x : \alpha^T x \ge \psi\})) . \quad (6)$$

By choosing $S = \{(\pi, \pi_0, \pi_0 + 1) : (\pi, \pi_0) \in Z^{n+1}\}$ we get:

Corollary 2. *Equation (4) holds when P is full-dimensional.*

Now assume that P is not full-dimensional. We first consider the case when P is empty:

Lemma 3. *Equation (4) holds when P is empty.*

Proof. There always exist $\bar{S} \subseteq M$ and $\bar{i} \in M$, where $\bar{i} \notin \bar{S}$, such that $\bar{S}$ contains $\bar{B} \in \mathcal{B}_r^*$ and $\{x \in \mathbb{R}^n : a_{i.}^T x \leq b_i, \ i \in \bar{S}\} \neq \emptyset$ and $\{x \in \mathbb{R}^n : a_{i.}^T x \leq b_i, \ i \in \bar{S} \cup \{\bar{i}\}\} = \emptyset$. Actually, $\bar{S}$ and $\bar{i}$ can be chosen by an iterative procedure. In iteration k $(k \geq 0)$, an $i_k \in M_k \subseteq M$ is chosen such that $M_k \backslash \{i_k\}$ contains a $B \in \mathcal{B}_r^*$, where $M_0 := M$. If $\{x \in \mathbb{R}^n : a_{i.}^T x \leq b_i, \ i \in M_k \backslash \{i_k\}\} \neq \emptyset$, then $\bar{S} := M_k \backslash \{i_k\}$ and $\bar{i} := i_k$. Otherwise, let $M_{k+1} := M_k \backslash \{i_k\}$ and proceed until finally we obtain $\bar{S}$ and $\bar{i}$. The fact that $\{x \in \mathbb{R}^n : a_{i.}^T x \leq b_i, \ i \in \bar{B}\} \neq \emptyset$ for any $\bar{B} \in \mathcal{B}_r^*$ ensures the availability of $\bar{S}$ and $\bar{i}$.

By applying the techniques of linear programming, we see that $\bar{S}$ contains $\bar{B}^* \in \mathcal{B}_r^*$ such that $\{x \in \mathbb{R}^n : a_{i.}^T x \leq b_i, \ i \in \bar{B}^* \cup \{\bar{i}\}\} = \emptyset$. It is possible to choose $\tilde{i} \in \bar{B}^*$ such that $\tilde{B}^* := (\bar{B}^* \backslash \{\tilde{i}\}) \cup \{\bar{i}\} \in \mathcal{B}_r^*$. Then $\{x \in \mathbb{R}^n : a_{i.}^T x \leq b_i, \ i \in \bar{B}^*\} \cap \{x \in \mathbb{R}^n : a_{i.}^T x \leq b_i, \ i \in \tilde{B}^*\} = \emptyset$.

Because $P = \emptyset$, $\mathrm{SC} = \emptyset$. $\mathrm{SC}(\bar{B}^*) \cap \mathrm{SC}(\tilde{B}^*) = \emptyset$ follows $\mathrm{SC}(\bar{B}^*) \subseteq \{x \in \mathbb{R}^n : a_{i.}^T x \leq b_i, \ i \in \bar{B}^*\}$, $\mathrm{SC}(\tilde{B}^*) \subseteq \{x \in \mathbb{R}^n : a_{i.}^T x \leq b_i, \ i \in \tilde{B}^*\}$ and $\{x \in \mathbb{R}^n : a_{i.}^T x \leq b_i, \ i \in \bar{B}^*\} \cap \{x \in \mathbb{R}^n : a_{i.}^T x \leq b_i, \ i \in \tilde{B}^*\} = \emptyset$. Therefore, $\mathrm{SC} = \emptyset = \mathrm{SC}(\bar{B}^*) \cap \mathrm{SC}(\tilde{B}^*) = \bigcap_{B \in \mathcal{B}_r^*} \mathrm{SC}(B)$. $\square$

In the remainder of this section, we assume that P is non-empty and not full-dimensional. Let $\mathit{Aff}(P)$ denote the affine hull of P. When P is not full-dimensional, it is full-dimensional in $\mathit{Aff}(P)$. Let $l := \dim(\mathit{Aff}(P)) < n$ denote the dimension of P. Also, $M^= := \{i \in M : a_{i.}^T x = b_i, \ \forall x \in P\}$ denotes the constraints describing P satisfied with equality by all points in P. Since $\dim(P) = l$, there exists a set $B \in \mathcal{B}_{n-l}^*$, such that $\mathit{Aff}(P) = \{x \in \mathbb{R}^n : a_{i.}^T x = b_i, i \in B\}$. $\mathcal{S}^= := \{B \in \mathcal{B}_{n-l}^* : B \subseteq M^=\}$ denotes the set of all such sets B. The following properties are needed:

Lemma 4. *Assume P is non-empty and not full-dimensional. Then:*

(i) $\mathit{Aff}(P) = P(M^=)$.

(ii) *Let $i^* \in M^=$ be arbitrary. The linear program $\min\{a_{i^*}^T x : x \in P(M^= \backslash \{i^*\})\}$ is bounded, the optimal objective value is b_{i^*} and the set of optimal solutions is $\mathit{Aff}(P)$.*

(iii) *There exist $B' \in \mathcal{S}^=$ and $i' \in M^= \backslash B'$ such that $\mathit{Aff}(P) = P(B' \cup \{i'\})$.*

(iv) *Let i' and B' be as in (iii). There exists $i'' \in B'$ such that $B'' := (B' \backslash \{i''\}) \cup \{i'\} \in \mathcal{S}^=$ and $\mathit{Aff}(P) = P(B') \cap P(B'')$.*

Proof. The correctness of (i), (ii) and (iv) is easy to check. So next we just prove (iii).

Because of (i), we have $|M^=| \geq n - l + 1$. Choose $i' \in M^=$ such that $(M^= \backslash \{i'\}) \cap \mathcal{B}_{n-l}^* \neq \emptyset$. So (ii) is true for $i^* = i'$. Since $(M^= \backslash \{i'\}) \cap \mathcal{B}_{n-l}^* \neq \emptyset$, the optimal dual solution of $\min\{a_{i.}^T x : a_{i.} \leq b_i, \ i \in M^= \backslash \{i'\}\}$ specifies a $B' \in (M^= \backslash \{i'\}) \cap \mathcal{B}_{n-l}^*$ such that $\min\{a_{i'}^T x : a_{i.}^T x \leq b_i, \ i \in B'\} = b_{i'}$ with optimal solution set $\mathit{Aff}(P)$. Therefore, $\mathit{Aff}(P) = \{x \in \mathbb{R}^n : a_{i.}^T x \leq b_i, \ i \in B' \cup \{i'\}\}$. $\square$

Let B' and B'' be as in Lemma 4. The sets B' and B'' might not be of cardinality r, i.e. B' (B'') might not be a maximal subset of M such that the vectors $a_{i.}$, for $i \in B'$ ($i \in B''$), are linearly independent. Define $\gamma := r - (n - l)$. It follows that $0 \leq \gamma \leq l$. $\mathcal{B}_\gamma^A := \{B \in \mathcal{B}_\gamma^* : B \cup B' \in \mathcal{B}_r^*\}$ ($= \{B \in \mathcal{B}_\gamma^* : B \cup B'' \in \mathcal{B}_r^*\}$) denotes the family of γ-subsets B in $\mathcal{B}_\gamma^*$ such that $B' \cup B$ ($B'' \cup B$) is an r-subset in $\mathcal{B}_r^*$. Also, $\mathcal{B'}_r^* := \{B \in \mathcal{B}_r^* : B \supseteq B'\}$ and $\mathcal{B''}_r^* := \{B \in \mathcal{B}_r^* : B \supseteq B''\}$ denotes the families of r-subsets in $\mathcal{B}_r^*$ containing B' and B'' respectively. The following is immediate from the definitions of $\mathcal{B}_\gamma^A$, $\mathcal{B'}_r^*$ and $\mathcal{B''}_r^*$:

Lemma 5. *There is a one-to-one mapping from $\mathcal{B}_\gamma^A$ to $\mathcal{B'}_r^*$, and there is a one-to-one mapping from $\mathcal{B}_\gamma^A$ to $\mathcal{B''}_r^*$.*

We are now able to finish the proof of Theorem 1:

Lemma 6. *Equation (4) holds when P is non-empty and not full-dimensional in $\mathbb{R}^n$.*

Proof. P is full-dimensional in $Aff(P)$. If there exists $(\pi, \pi_0) \in \mathbb{Z}^{n+1}$ such that $Aff(P)$ is between the hyperplanes $\pi^T x = \pi_0$ and $\pi^T x = \pi_0 + 1$ and $Aff(P)$ does not intersect them, then Lemma 6 is trivially true with $\emptyset = \emptyset$. Otherwise, we only need to consider $(\pi, \pi_0) \in \mathbb{Z}^{n+1}$ such that both hyperplanes $\pi^T x = \pi_0$ and $\pi^T x = \pi_0 + 1$ intersect $Aff(P)$ and neither of them contains $Aff(P)$. Denote the set of these (π, π_0) by S_A.

Now we have $\bigcap_{(\pi,\pi_0) \in \mathbb{Z}^{n+1}} Conv(P \cap F_{D(\pi,\pi_0)}) = \bigcap_{(\pi,\pi_0) \in S_A} Conv((P \cap \{x \in Aff(P) : \pi^T x \leq \pi_0\}) \cup (P \cap \{x \in Aff(P) : \pi^T x \geq \pi_0 + 1\}))$. Applying Corollary 1 to the affine subspace $Aff(P)$, we see that the latter is equal to $\bigcap_{B \in \mathcal{B}_\gamma^A} \bigcap_{(\pi,\pi_0) \in S_A} Conv(((P(B) \cap Aff(P)) \cap \{x \in Aff(P) : \pi^T x \leq \pi_0\}) \cup ((P(B) \cap Aff(P)) \cap \{x \in Aff(P) : \pi^T x \geq \pi_0 + 1\}))$. By Lemma 4(iv) and Lemma 5, for any $P(B)$, where $B \in \mathcal{B}_\gamma^A$, there always exist $P(\tilde{B}')$ and $P(\tilde{B}'')$, where $\tilde{B}' \in \mathcal{B'}_r^*$ and $\tilde{B}'' \in \mathcal{B''}_r^*$, such that $\tilde{B}' = B' \cup B$, $\tilde{B}'' = B'' \cup B$ and $P(B) \cap Aff(P) = P(\tilde{B}') \cap P(\tilde{B}'')$. Therefore, $\bigcap_{B \in \mathcal{B}_\gamma^A} \bigcap_{(\pi,\pi_0) \in S_A} Conv(((P(B) \cap Aff(P)) \cap \{x \in Aff(P) : \pi^T x \leq \pi_0\}) \cup ((P(B) \cap Aff(P)) \cap \{x \in Aff(P) : \pi^T x \geq \pi_0 + 1\})) \supseteq \bigcap_{B \in \mathcal{B}_r^*} \bigcap_{(\pi,\pi_0) \in \mathbb{Z}^{n+1}} Conv(P(B) \cap F_{D(\pi,\pi_0)})$, which implies $\bigcap_{(\pi,\pi_0) \in \mathbb{Z}^{n+1}} Conv(P \cap F_{D(\pi,\pi_0)}) \supseteq \bigcap_{B \in \mathcal{B}_r^*} \bigcap_{(\pi,\pi_0) \in \mathbb{Z}^{n+1}} Conv(P(B) \cap F_{D(\pi,\pi_0)})$. Because $P \subseteq P(B)$ for any $B \in \mathcal{B}_r^*$, it is easy to obtain

$$\bigcap_{(\pi,\pi_0) \in \mathbb{Z}^{n+1}} Conv(P \cap F_{D(\pi,\pi_0)}) \subseteq \bigcap_{B \in \mathcal{B}_r^*} \bigcap_{(\pi,\pi_0) \in \mathbb{Z}^{n+1}} Conv(P(B) \cap F_{D(\pi,\pi_0)}).$$

The lemma is proved. □

Theorem 1 is implied by Corollary 2, Lemma 3 and Lemma 6. In fact, the proofs allow us to extend Theorem 1 to arbitrary subsets of $\Pi^n(N_I)$:

Theorem 2. *Assume $S \subseteq \Pi^n(N_I)$. Then:*

$$\bigcap_{(\pi,\pi_0) \in S} Conv(P \cap F_{D(\pi,\pi_0)}) = \bigcap_{B \in \mathcal{B}_r^*} \bigcap_{(\pi,\pi_0) \in S} Conv(P \cap F_{D(\pi,\pi_0)}). \tag{7}$$

4   Polyhedrality of Split Closure

In this section, we assume $A \in \mathbb{Q}^{m \times n}$ and $b \in \mathbb{Q}^n$, i.e. that P is a rational polyhedron. Cook, Kannan and Schrijver proved the following result [4]:

Theorem 3. *The split closure of MILP is a polyhedron.*

We will give a new proof of this result using the characterization of the split closure obtained in the previous section. Let $\tilde{C} := \tilde{x} + Cone\left(\{\tilde{r}_i : i = 1, 2, \ldots, q\}\right)$ be (the translate of) a cone with apex $\tilde{x} \in \mathbb{Q}^n$ and q linearly independent extreme ray vectors $\{\tilde{r}_i\}_{i=1}^q$, where $q \leq n$ and $\tilde{r}_i \in \mathbb{Z}^n$ for $1 \leq i \leq q$. The following lemma plays a key role in proving Theorem 3.

Lemma 7. *The split closure of $\tilde{C}$ is a polyhedron.*

Proof. Suppose that the disjunction $\pi^T x \leq \pi_0$ and $\pi^T x \geq \pi_0 + 1$ induces a split cut that is not valid for some part of C. Then it must be not valid for $\tilde{x}$ either. So we know $\pi_0 < \pi^T \tilde{x} < \pi_0 + 1$, i.e. the point $\tilde{x}$ is between the two hyperplanes $\pi^T x = \pi_0$ and $\pi^T x = \pi_0 + 1$.

Choose an extreme ray generated by vector $\tilde{r}_i$ and assume that the hyperplane $\pi^T x = \pi_0$ intersects the extreme ray at $\tilde{x} + \tilde{\alpha}_i \tilde{r}_i$, where $\tilde{\alpha}_i > 0$ ($\tilde{\alpha}_i = +\infty$ is allowed). Then $\tilde{\alpha}_i = \frac{\epsilon}{-\pi^T \tilde{r}_i}$ can be easily calculated, where $\pi^T \tilde{r}_i \leq 0$, $\epsilon :=$ $\pi^T \tilde{x} - \pi_0$ and $0 < \epsilon < 1$.

We claim that $\tilde{\alpha}_i$ is either $+\infty$ or bounded above by 1. When $\pi^T \tilde{r}_i = 0$, $\tilde{\alpha}_i = +\infty$, in which case the hyperplane $\pi^T x = \pi_0$ is parallel to the vector $\tilde{r}_i$. When $\pi^T \tilde{r}_i < 0$, $0 < \tilde{\alpha}_i < +\infty$, which means that the hyperplane intersects the ray at some finite point. In this case, because π and $\tilde{r}_i$ are in $\mathbb{Z}^n$, we know $-\pi \tilde{r}_i \geq 1$. Hence, $\tilde{\alpha}_i = \frac{\epsilon}{-\pi^T \tilde{r}_i} \leq \epsilon < 1$.

Let $\tilde{x} = (\tilde{x}_1, \tilde{x}_2, \cdots, \tilde{x}_n)^T \in \mathbb{Q}^n$. Let g be the least common multiple of all the denominators of $\tilde{x}_1, \tilde{x}_2, \cdots, \tilde{x}_n$. Noticing the fact that $\tilde{\alpha}_i = \frac{\epsilon}{-\pi^T \tilde{r}_i}$, it follows that $\tilde{\alpha}_i$ can be expressed as $\frac{p}{wg}$, where $p, w \in \mathbb{Z}_+$ and $0 < p < g$.

By the following claim, we actually prove the lemma.

Claim. There is only a finite number of undominated split cuts for cone C.

Proof of claim. By induction on m, the number of extreme rays of C.

When $q = 1$, $C = \{x \in \mathbb{R}^n : x = \tilde{x} + \alpha_1 \tilde{r}_1, \alpha_1 \geq 0\}$. The case $\tilde{\alpha}_1 = +\infty$ does not yield a split cut, so $\tilde{\alpha}_1$ is bounded above by 1 for every split cut. Note that the maximum value of $\tilde{\alpha}_1$ is reachable, because $\tilde{\alpha}_1$ has the form of $\frac{p}{wg}$ as mentioned above. Let $\pi^* x = \pi_0^*$ be the hyperplane for which $\tilde{\alpha}_1$ reaches its maximum. Then the split cut $\pi^* x \leq \pi_0^*$ is an undominated split cut of C.

Assume that the claim is true for $q = k < n$. Let us consider the case of $q = k + 1$.

Let $C_i := \{x \in \mathbb{R}^n : x = \tilde{x} + \sum_{j \neq i} \alpha_j \tilde{r}_j, \alpha_j \geq 0, 1 \leq j \leq k + 1, j \neq i\}$, where $1 \leq i \leq k + 1$. Each C_i is a polyhedral cone with apex $\tilde{x}$ and k linearly independent extreme rays. By induction hypothesis, there is only a finite number of undominated split cuts for each cone C_i. Among those points obtained by intersecting the undominated split cuts for C_i with the extreme ray generated by $\tilde{r}_i$, let z_i be the closest point to $\tilde{x}$.

Now we claim that any undominated split cut of C cannot intersect the extreme ray generated by $\tilde{r}_i$ ($1 \leq i \leq k+1$) at a point which is closer to $\tilde{x}$ than z_i. Otherwise, let us assume that there is an undominated split cut $\mathcal{H}$ of C which intersects the extreme ray generated by $\tilde{r}_i$ at a point $\tilde{z}_i$ that is between $\tilde{x}$ and z_i. By the definition of z_i, the cut (when restricted to C_i) must be dominated by a cut of C_i, say $\mathcal{H}'$. Wlog, assume that the cut $\mathcal{H}'$ is an undominated cut for C_i. So $\mathcal{H}'$ must intersect the extreme ray generated by $\tilde{r}_i$ at z_i' that is not between $\tilde{x}$ and z_i. But now $\mathcal{H}'$ dominates $\mathcal{H}$, a contradiction to the choice of $\mathcal{H}$.

We know that the intersection point of any undominated split cut with the extreme ray of $\tilde{r}_i$ ($1 \leq i \leq k+1$) is either at infinity or between z_i and $\tilde{x} + \tilde{r}_i$. Since $\tilde{\alpha}_i = \frac{p}{wg}$, there are only finitely many points between z_i and $\tilde{x} + \tilde{r}_i$ that could be the intersections of the split cuts with the extreme ray. Therefore, we see that the claim is true when $q = k+1$. $\square$

Let $B \in \mathcal{B}_r^*$ be arbitrary. From Sect. 2 we know that $P(B)$ can be written as $P(B) = \bar{x}(B) + L(B) + C(B)$, where $C(B) := Cone\left(\{r^i(B) : i \in B\}\right)$ and $\{r^i(B) : i \in B\} \subseteq \mathbb{Z}^n$ are linearly independent (by scaling). The following lemmas are straightforward:

Lemma 8. *Let $(\pi, \pi_0) \in \mathbb{Z}^{n+1}$ and $B \in \mathcal{B}_r^*$ be arbitrary. If an inequality is valid for $Conv\left(P(B) \cap F_{D(\pi, \pi_0)}\right)$ but not valid for $P(B)$, then $\pi^T = \sum_{i \in B} \alpha_i a_{i\cdot}^T$, where $\alpha_i \in \mathbb{R}$ ($i \in B$).*

Define $S_B := \{\pi \in \mathbb{Z}^n : \pi^T = \sum_{i \in B} \alpha_i a_{i\cdot}^T, \ \alpha_i \in \mathbb{R}, \ i \in B\}$. From Lemma 8 we have:

Lemma 9. *Let $B \in \mathcal{B}_r^*$, $(\pi, \pi_0) \in S_B \times \mathbb{Z}$.*

(i) *Assume there exists a facet F of $Conv\left(P(B) \cap F_{D(\pi, \pi_0)}\right)$, which is not a facet of $P(B)$. Then F is unique and there exists a unique facet $\tilde{F}$ of $Conv\left((\bar{x}(B) + C(B)) \cap F_{D(\pi, \pi_0)}\right)$, which is not a facet of $\bar{x}(B) + C(B)$. Furthermore $F = L(B) + \tilde{F}$.*

(ii) *Assume there exists a facet $\tilde{F}$ of $Conv\left((\bar{x}(B) + C(B)) \cap F_{D(\pi, \pi_0)}\right)$, which is not a facet of $\bar{x}(B) + C(B)$. Then $\tilde{F}$ is unique and there exists a unique facet F of $Conv\left(P(B) \cap F_{D(\pi, \pi_0)}\right)$, which is not a facet of $P(B)$. Furthermore $\tilde{F} = F \cap (\bar{x}(B) + C(B))$.*

The following result is implied by Lemma 7 and Lemma 9:

Lemma 10. *$SC(B)$, where $B \in \mathcal{B}_r^*$, is a polyhedron.*

Now, Theorem 3 follows from Theorem 1 and Lemma 10.

5 Disjunctive Sets Derived from Polyhedra and Two-Term Disjunctions

In this section, two decomposition results for the set $conv(P \cap F_D)$ are presented. The first decomposition result (Theorem 6) states that $conv(P \cap F_D)$ can be written as the intersection of sets $conv(P(T) \cap F_D)$ over sets $T \in C_1^*$, where C_1^* is some family of $(r + 1)$-subsets of M. Furthermore, in the special case where $r = n$, we show that it is enough to consider r-subsets T of M. The second result strengthens the first result for split disjunctions $D(\pi, \pi_0)$, and states that the set $conv(P \cap F_{D(\pi,\pi_0)})$ can be written as the intersection of the sets $conv(P(B) \cap F_{D(\pi,\pi_0)})$ for $B \in \mathcal{B}_r^*$. We start by proving the following:

Theorem 4. *Let $S \subseteq M$ be non-empty. If S satisfies $|S| \geq r(S) + 2$, then*

$$conv(P(S) \cap F_D) = \bigcap_{i \in S} conv(P(S \setminus \{i\}) \cap F_D) . \tag{8}$$

Furthermore, (8) remains true if $r(S) = n$ and $|S| = n + 1$.

One direction is easy to prove:

Lemma 11. *Let $S \subseteq M$ be non-empty. Then,*

$$conv(P(S) \cap F_D) \subseteq \bigcap_{i \in S} conv(P(S \setminus \{i\}) \cap F_D) . \tag{9}$$

Proof. Clearly $P(S) \subseteq P(S \setminus \{i\})$ for all $i \in S$. Intersecting with F_D on both sides gives $P(S) \cap F_D \subseteq P(S \setminus \{i\}) \cap F_D$ for all $i \in S$. Convexifying both sides results in $Conv(P(S) \cap F_D) \subseteq Conv(P(S \setminus \{i\}) \cap F_D)$ for all $i \in S$, and finally, since this holds for all $i \in S$, the result follows. $\qquad\square$

The proof of the other direction involves the idea introduced by Balas [2] of lifting the set $Conv(P(S) \cap F_D)$ onto a higher dimensional space. Specifically, $Conv(P(S) \cap F_D)$ can be described as the projection of the set, described by the following constraints, onto the space of x-variables:

$$x = x^1 + x^2, \tag{10}$$
$$a_{i.}^T x^1 \leq b_i \lambda^1, \qquad \forall i \in S, \tag{11}$$
$$a_{i.}^T x^2 \leq b_i \lambda^2, \qquad \forall i \in S, \tag{12}$$
$$\lambda^1 + \lambda^2 = 1, \tag{13}$$
$$D^1 x^1 \leq d^1 \lambda^1, \tag{14}$$
$$D^2 x^2 \leq d^2 \lambda^2, \tag{15}$$
$$\lambda^1, \lambda^2 \geq 0 . \tag{16}$$

The description (10)-(16) can be projected onto the (x, x^1, λ^1)-space by using constraints (10) and (13). By doing this, we arrive at the following characterization of $Conv(P(S) \cap F_D)$. Later, the constraints below will be used in the formulation of an LP problem. Therefore, we have written the names of the corresponding dual variables next to the constraints:

$$-\lambda^1 b_i + a_{i.}^T x^1 \leq 0, \qquad \forall i \in S, \qquad (u_i) \qquad (17)$$

$$-\lambda^1 b_i + a_{i.}^T x^1 \leq b_i - a_{i.}^T x, \qquad \forall i \in S, \qquad (v_i) \qquad (18)$$

$$\lambda^1 \leq 1, \qquad (w_0) \qquad (19)$$

$$-\lambda^1 d^1 + D^1 x^1 \leq 0_{m_1}, \qquad (u^0) \qquad (20)$$

$$\lambda^1 d^2 - D^2 x^1 \leq d^2 - D^2 x, \qquad (v^0) \qquad (21)$$

$$\lambda^1 \geq 0 . \qquad (t_0) \qquad (22)$$

Consider now relaxing constraints (20) and (21), i.e. replacing (20) and (21) by the following constraints:

$$-\lambda^1 d^1 + D^1 x^1 - s1_{m_1} \leq 0_{m_1}, \qquad (u^0) \qquad (23)$$

$$\lambda^1 d^2 - D^2 x^1 - s1_{m_2} \leq d^2 - D^2 x, \qquad (v^0) \qquad (24)$$

$$s \geq 0 . \qquad (t_1) \qquad (25)$$

Now, the problem of deciding whether or not a given vector $x \in \mathbb{R}^n$ belongs to $Conv(P(S) \cap F_D)$ can be decided by solving the following linear program, which will be called $P_{LP}(x, S)$ in the following:

$$\max \ -s$$
$$s.t. \quad (17) - (19), \ (22) \ \text{and} \ (23) - (25) . \qquad (P_{LP}(x, S))$$

Observe that $P_{LP}(x, S)$ is feasible if and only if $x \in P(S)$, and that $P_{LP}(x, S)$ is always bounded above by zero. Finally, note that $x \in Conv(P(S) \cap F_D)$ if and only if $P_{LP}(x, S)$ is feasible and bounded, and there exists an optimal solution in which the variable s has the value zero.

The other direction is proved with the aid of the problem $P_{LP}(x, S)$ and its dual $D_{LP}(x, S)$. Suppose $S \subseteq M$, satisfies $S \neq \emptyset$, $|S| \geq r(S) + 1$ and that $\bar{x} \in \bigcap_{i \in S} Conv(P(S \setminus \{i\}) \cap F_D)$. Then $\bar{x} \in \bigcap_{i \in S} P(S \setminus \{i\})$, and since $|S| \geq 2$, we have $\bar{x} \in P(S)$. Hence $P_{LP}(\bar{x}, S)$ is feasible and bounded if S satisfies $|S| \geq r(S) + 1$ and $\bar{x} \in \bigcap_{i \in S} Conv(P(S \setminus \{i\}) \cap F_D)$. In the case where $P_{LP}(\bar{x}, S)$ is feasible and bounded, $(\bar{x}^1, \bar{\lambda}^1, \bar{s})$ denotes an optimal basic feasible solution to $P_{LP}(\bar{x}, S)$ and $(\bar{u}, \bar{v}, \bar{u}^0, \bar{v}^0, \bar{w}_0, \bar{t}_0, \bar{t}_1)$ denotes a corresponding optimal basic feasible solution to $D_{LP}(\bar{x}, S)$.

For $u^0 \geq 0_{m_1}$, $u_i \geq 0$ for $i \in S$ and $j \in N$, define the quantities $\alpha_j^1(S, u, u^0) := \sum_{i \in S} u_i a_{i,j} + (u^0)^T D_{.j}^1$ and $\beta^1(S, u, u^0) := \sum_{i \in S} u_i b_i + (u^0)^T d^1$, where $D_{.j}^1$ denotes the jth column of D^1. The inequality $(\alpha^1(S, u, u^0))^T x \leq \beta^1(S, u, u^0)$ is valid for $\{x \in P(S) : D^1 x \leq d^1\}$. Similarly, for $v^0 \geq 0_{m_2}$, $v_i \geq 0$ for $i \in S$ and $j \in N$, defining the quantities $\alpha_j^2(S, v, v^0) := \sum_{i \in S} v_i a_{i,j} + (v^0)^T D_{.j}^2$ and $\beta^2(S, v, v^0) := \sum_{i \in S} v_i b_i + (v^0)^T d^2$, gives the inequality $(\alpha^2(S, v, v^0))^T x \leq \beta^2(S, v, v^0)$, which is valid for $\{x \in P(S) : D^2 x \leq d^2\}$. With these quantities, the dual $D_{LP}(\bar{x}, S)$ of $P_{LP}(\bar{x}, S)$ can be formulated as follows:

$$\min \quad \beta^2(S, v, v^0) - (\alpha^2(S, v, v^0))^T \bar{x} + w_0$$

$$\text{s.t.} \quad \alpha_j^1(S, u, u^0) - \alpha_j^2(S, v, v^0) = 0, \qquad \forall j \in N, \qquad (x_j^1) \qquad (26)$$

$$\beta^2(S, v, v^0) - \beta^1(S, u, u^0) + w_0 - t_0 = 0, \qquad (\lambda^1) \qquad (27)$$

$$1_{m_1}^T u^0 + 1_{m_2}^T v^0 + t_1 = 1, \qquad (s) \qquad (28)$$

$$u^0 \geq 0_{m_1}, \qquad (29)$$

$$v^0 \geq 0_{m_2}, \qquad (30)$$

$$w_0, t_0, t_1 \geq 0, \qquad (31)$$

$$u_i, v_i \geq 0 \qquad \forall i \in S . \qquad (32)$$

Lemma 12. *Let $S \subseteq M$ be arbitrary. Suppose $\bar{x} \in P(S) \setminus Conv(P(S) \cap F_D)$. Then $\bar{u}^0 \neq 0_{m_1}$ and $\bar{v}^0 \neq 0_{m_2}$.*

Proof. Let $\bar{x}$ be as stated, and suppose first that $\bar{v}^0 = 0_{m_2}$. The inequality $(\alpha^2(S, \bar{v}, 0_{m_2}))^T x \leq \beta^2(S, \bar{v}, 0_{m_2})$ is valid for $P(S)$. However, since the optimal objective value to $D_{LP}(\bar{x}, S)$ is negative, and $\bar{w}_0 \geq 0$, we have $\beta^2(S, \bar{v}, 0_{m_2}) - (\alpha^2(S, \bar{v}, 0_{m_2}))^T \bar{x} < 0$, which contradicts the assumption that $\bar{x} \in P(S)$.

Now suppose $\bar{u}_0 = 0_{m_1}$. The inequality $(\alpha^1(S, \bar{u}, 0_{m_1}))^T x \leq \beta^1(S, \bar{u}, 0_{m_1})$ is valid for $P(S)$, but $\beta^1(S, \bar{u}, 0_{m_1}) - (\alpha^1(S, \bar{u}, 0_{m_1}))^T \bar{x} \leq \beta^2(S, \bar{v}, \bar{v}^0) + w_0 - (\alpha^2(S, \bar{v}, \bar{v}^0))^T \bar{x} < 0$, where we have used (20) and (21). This contradicts the assumption that $\bar{x} \in P(S)$. $\qquad \square$

The next lemma is essential for the proof of the converse direction of Theorem 4:

Lemma 13. *Let $S \subseteq M$, and let $T \subseteq S$ satisfy $|T| \geq 2$. Also, suppose $\bar{x} \in \bigcap_{i \in T} Conv(P(S \setminus \{i\}) \cap F_D)$ and $\bar{x} \notin Conv(P(S) \cap F_D)$. Then $D_{LP}(\bar{x}, S)$ is feasible and bounded. Furthermore, $\bar{u}_i > 0$ or $\bar{v}_i > 0$ for all $i \in T$.*

Proof. Let S, T and $\bar{x}$ be as stated. The fact that $D_{LP}(\bar{x}, S)$ is feasible and bounded follows from the facts $\bar{x} \in \bigcap_{i \in T} Conv(P(S \setminus \{i\}) \subseteq P(S)$ and $|T| \geq 2$.

Now, suppose $\bar{u}_{i'} = 0$ and $\bar{v}_{i'} = 0$ for some $i' \in T$. Then the problem $(D'_{LP}(\bar{x}, S \setminus \{i'\}))$, obtained from $D_{LP}(\bar{x}, S)$ by eliminating $u_{i'}$, $v_{i'}$ and the normalization constraint:

$$\min \quad \beta^2(S \setminus \{i'\}, v, v^0) - (\alpha^2(S \setminus \{i'\}, v, v^0))^T \bar{x} + w_0$$

$$\text{s.t.} \quad \alpha_j^1(S \setminus \{i'\}, u, u^0) - \alpha_j^2(S \setminus \{i'\}, v, v^0) = 0, \qquad \forall j \in N, \quad (x_j^1)$$

$$\beta^2(S \setminus \{i'\}, v, v^0) - \beta^1(S \setminus \{i'\}, u, u^0) + w_0 - t_0 = 0, \qquad (\lambda^1)$$

$$u^0 \geq 0_{m_1}, \ v^0 \geq 0_{m_2}, \ w_0, t_0 \geq 0, \ u_i, v_i \geq 0, \ \forall i \in S \setminus \{i'\},$$

is unbounded (since $\bar{x} \notin Conv(P(S) \cap F_D)$). This means that the dual of $D'_{LP}(\bar{x}, S \setminus \{i'\})$, the problem $P'_{LP}(\bar{x}, S \setminus \{i'\})$, is infeasible:

$$\max \quad 0_n^T x^1 + 0\lambda^1$$

$$\text{s.t.} \quad -\lambda^1 b_i + a_{i.}^T x^1 \le 0, \qquad \forall i \in S \setminus \{i'\}, \tag{u_i}$$

$$\lambda^1 b_i - a_{i.}^T x^1 \le b_i - a_{i.}^T \bar{x}, \qquad \forall i \in S \setminus \{i'\}, \tag{v_i}$$

$$\lambda^1 \le 1, \tag{w_0}$$

$$-\lambda^1 d^1 + D^1 x^1 \le 0_{m_1}, \tag{u^0}$$

$$\lambda^1 d^2 - D^2 x^1 \le d^2 - D^2 \bar{x}, \tag{v^0}$$

$$\lambda^1 \ge 0 . \tag{t_0}$$

However, these constraints are the conditions that must be satisfied for $\bar{x}$ to be in $Conv(P(S \setminus \{i'\}) \cap F_D)$, which is a contradiction. □

With the above lemmas, we are ready to prove the converse of Theorem 4:

Lemma 14. *Let $S \subseteq M$, and suppose that either $|S| \ge r(S) + 2$ or $r(S) = n$ and $|S| = n + 1$. Then*

$$Conv(P(S) \cap F_D) \supseteq \bigcap_{i \in S} Conv(P(S \setminus \{i\}) \cap F_D) . \tag{33}$$

Proof. Let $\bar{x} \in \bigcap_{i \in S} Conv(P(S \setminus \{i\}) \cap F_D)$, and suppose $\bar{x} \notin Conv(P(S) \cap F_D)$. Define $B_u := \{i \in S : u_i \text{ basic }\}$ and $B_v := \{i \in S : v_i \text{ basic }\}$ to be the set of basic u's and v's respectively in the solution $(\bar{u}, \bar{v}, \bar{u}^0, \bar{v}^0, \bar{w}_0, \bar{t}_0, \bar{t}_1)$ to $D_{LP}(\bar{x}, S)$. From Lemma 13 and the fact that a variable with positive value, that does not have an upper bound, is basic, it follows that $(B_u \cup B_v) = S$.

The feasible set for the problem $D_{LP}(\bar{x}, S)$ is of the form $\{y \in \mathbb{R}^{n'} : Wy = w_0, y \ge 0_{n'}\}$, where W and w_0 are of suitable dimensions. The column of W corresponding to the variable u_i, $i \in S$, is given by $[a_{i.}^T, -b_i, 0]^T$. Similarly, the column of W corresponding to the variable v_i, $i \in S$, is given by $[-a_{i.}^T, b_i, 0]^T$. Since for all $i \in S$, either u_i or v_i is basic, the vectors $[a_{i.}^T, -b_i]^T$, $i \in S$, are linearly independent. Clearly, there can be at most $r(S) + 1$ of these. Hence $|S| = r(S) + 1$. This excludes the case $|S| \ge r(S) + 2$, so we must have $r(S) = n$ and $|S| = n + 1$.

The number of basic variables in the solution $(\bar{u}, \bar{v}, \bar{u}^0, \bar{v}^0, \bar{w}_0, \bar{t}_0, \bar{t}_1)$ to $D_{LP}(\bar{x}, S)$ is at most $n + 2$, since the number of basic variables is bounded by the number of equality constraints in $D_{LP}(\bar{x}, S)$. The number of basic variables among the variables u_i and v_i, $i \in S$, is $|S| = n + 1$. However, according to Lemma 12, at least two of the variables in $(\bar{u}^0, \bar{v}^0)$ are basic, which gives a total of $n + 3$ basic variables – a contradiction. □

Now, we strengthen Theorem 4 for the case where $|S| \ge r(S) + 2$. Let $\bar{I}(S)$ be the set of constraints $i \in S$ for which $r(S \setminus \{i\}) = r(S)$, i.e. $\bar{I}(S) := \{i \in S : r(S) = r(S \setminus \{i\})\}$. We have:

Theorem 5. *Let $S \subseteq M$ satisfy $|S| \ge r(S) + 2$. Then*

$$Conv(P(S) \cap F_D) = \bigcap_{i \in \bar{I}(S)} Conv(P(S \setminus \{i\}) \cap F_D) . \tag{34}$$

Like in Theorem 4, and with the same proof, one direction of Theorem 5 is easy. For the converse, observe that $|S| \geq r(S) + 2$ implies $|\bar{I}(S)| \geq 2$. It also implies $\bigcap_{i \in \bar{I}(S)} Conv(P(S \setminus \{i\}) \cap F_D) \subseteq P(S)$. From Lemma 13, we have:

Corollary 3. *Suppose $|S| \geq r(S) + 2$, $\bar{x} \in \bigcap_{i \in \bar{I}(S)} Conv(P(S \setminus \{i\}) \cap F_D)$ and $\bar{x} \notin Conv(P(S) \cap F_D)$. Then $\bar{u}_i > 0$ or $\bar{v}_i > 0$ for all $i \in \bar{I}(S)$.*

We can now prove the converse of Theorem 5:

Lemma 15. *Let $S \subseteq M$ satisfy $|S| \geq r(S) + 2$. Then:*

$$Conv(P(S) \cap F_D) \supseteq \bigcap_{i \in \bar{I}(S)} Conv(P(S \setminus \{i\}) \cap F_D) . \qquad (35)$$

Proof. Let $\bar{x} \in \bigcap_{i \in \bar{I}(S)} Conv(P(S \setminus \{i\}) \cap F_D)$, and suppose $\bar{x} \notin Conv(P(S) \cap F_D)$. Observe that it suffices to prove that the vectors $[a_{i.}^T, -b_i]$, $i \in S$, are linearly independent, since that would contradict $|S| \geq r(S)+2$. Suppose they are linearly dependent, and let $\bar{i} \in S$ satisfy $[a_{i.}^T, -b_{\bar{i}}] = \sum_{i \in S \setminus \{\bar{i}\}} \mu_i [a_{i.}^T, -b_i]$, where μ_i, for $i \in S \setminus \{\bar{i}\}$, are scalars, not all zero. Suppose first that $\bar{i} \in S \setminus \bar{I}(S)$. Then we have $a_{\bar{i}.} \in span(\{a_{i.} : i \in S \setminus \{\bar{i}\})$. However, that implies $r(S) = r(S \setminus \{\bar{i}\})$, which is a contradiction. Hence, we must have $\bar{i} \in \bar{I}(S)$ and $\mu_i = 0$ for all $i \in S \setminus \bar{I}(S)$. However, according to Corollary 3, the vectors $[a_{i.}^T, -b_i]$, $i \in \bar{I}(S)$, are linearly independent. $\square$

Applying Theorem 5 iteratively, and Theorem 4 for sets of size $n + 1$, we get the following:

Theorem 6. *Suppose $|M| = m \geq r + 1$. Also, define $C_1^* := \{S \subseteq M : |S| = r + 1 \text{ and } r(S) = r\}$ and $C_2^* := \{S \subseteq M : |S| = r \text{ and } (r(S) = r \vee r(S) = r - 1)\}$. We have,*

$$Conv(P \cap F_D) = \bigcap_{T \in C_1^*} Conv(P(T) \cap F_D) . \qquad (36)$$

Furthermore, if $r = n$,

$$Conv(P \cap F_D) = \bigcap_{T \in C_2^*} Conv(P(T) \cap F_D) . \qquad (37)$$

The example in Figure 1 demonstrates that the assumption $|S| \geq r(S) + 2$ is necessary for (35) to hold. In this example, P has 3 constraints $a_i^T x \leq b_i$, $i = 1, 2, 3$, and D is a two-term disjunction involving the 2 constraints $D^1 x \leq d^1$ and $D^2 x \leq d^2$. C^1 denotes the cone defined by $a_1^T x \leq b_1$ and $a_3^T x \leq b_3$, and C^2 denotes the cone defined by $a_2^T x \leq b_2$ and $a_3^T x \leq b_3$. We clearly see $Conv(P(M) \cap F_D) \neq \bigcap_{i \in \bar{I}(M)} Conv(P(M \setminus \{i\}) \cap F_D)$.

The example does not exclude, however, that (35) is true for sets S satisfying $|S| = r(S) + 1$ for the special case of a split disjunction. In fact, the example suggests that it is also necessary for the disjunction to be a split disjunction. In

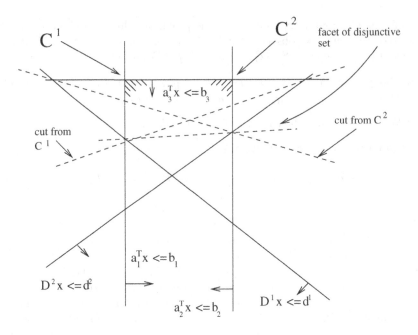

Fig. 1. Example of two-term disjunction

the following, we will prove that (35) remains valid for $|S| = r(S) + 1$ for the special case of a split disjunction.

Let $D(\pi, \pi_0)$ be a split disjunction. The problem $P_{LP}(\bar{x}, S)$, for a split disjunction $D(\pi, \pi_0)$, which will be called $P_{LP}^S(\bar{x}, S)$ in the following, is obtained from the problem $P_{LP}(\bar{x}, S)$ by replacing (23) and (24) with:

$$-\lambda^1 \pi_0 + \pi^T x^1 - s \leq 0, \qquad\qquad (u^0) \qquad (38)$$

$$-\lambda^1 (\pi_0 + 1) + \pi^T x^1 - s \leq -(\pi_0 + 1) + \pi^T \bar{x} . \qquad (v^0) \qquad (39)$$

The dual of $P_{LP}^S(\bar{x}, S)$ is the problem $D_{LP}^S(\bar{x}, S)$ defined as:

$$\min \ \sum_{i \in S} v_i(b_i - a_{i\cdot}^T \bar{x}) + w_0 + v^0(\pi^T \bar{x} - (\pi_0 + 1)) \qquad (40)$$

$$\text{s.t.} \ \sum_{i \in S} a_{i\cdot}(u_i - v_i) + \pi(u_0 + v_0) = 0_n, \qquad (x^1) \qquad (41)$$

$$\sum_{i \in S} b_i(v_i - u_i) - \pi_0(u_0 + v_0) - v^0 + w_0 - t_0 = 0, \quad (\lambda^1) \qquad (42)$$

$$u^0 + v^0 + t_1 = 1, \qquad\qquad\qquad (s) \qquad (43)$$

$$u^0, v^0, w_0, t_0, t_1 \geq 0, \qquad\qquad\qquad (44)$$

$$u_i, v_i \geq 0, \quad \forall i \in S . \qquad\qquad\qquad (45)$$

The solution $(\bar{u}, \bar{v}, \bar{u}^0, \bar{v}^0, \bar{w}_0, \bar{t}_0, \bar{t}_1)$ to $D_{LP}^S(\bar{x}, S)$, for the case where $|S| = r(S) + 1$, can be characterized as follows (see also Lemma 2 in [3]):

Lemma 16. *Suppose* $|S| = r(S) + 1$, $\bar{x} \in \bigcap_{i \in \bar{I}(S)} Conv(P(S \setminus \{i\}) \cap F_{D(\pi,\pi_0)})$ *and* $\bar{x} \notin Conv(P(S) \cap F_{D(\pi,\pi_0)})$. *Let* $B_u := \{i \in S : u_i \text{ basic }\}$ *and* $B_v := \{i \in S : v_i \text{ basic }\}$ *be the set of basic u's and v's respectively in the solution* $(\bar{u}, \bar{v}, \bar{u}^0, \bar{v}^0, \bar{w}_0, \bar{t}_0, \bar{t}_1)$ *to* $D^S_{LP}(\bar{x}, S)$. *Then* $B_u \cap B_v = \emptyset$, $r(S) = n$, $|B_u \cup B_v| = n$, *and the vectors* $a_{i\cdot}$, $i \in B_u \cup B_v$, *are linearly independent.*

Proof. As mentioned earlier, the feasible set for $D^S_{LP}(\bar{x}, S)$ can be written as $\{y \in \mathbb{R}^{n'} : Wy = z_0, y \geq 0_{n'}\}$, where W and z_0 are of suitable dimensions.

We first argue that the variables w_0, t_0 and t_1 are non-basic in the solution $(\bar{u}, \bar{v}, \bar{u}^0, \bar{v}^0, \bar{w}_0, \bar{t}_0, \bar{t}_1)$ to $D^S_{LP}(\bar{x}, S)$. t_1 is clearly non-basic, since s is basic in $P^S_{LP}(\bar{x}, S)$. From Lemma 12 it follows that both u^0 and v^0 are basic. The column corresponding to u^0 is $[\pi, -\pi_0, 1]$ and the column corresponding to v^0 is $[\pi, -(\pi_0 + 1), 1]$. Subtracting the column corresponding to v_0 from the column corresponding to u_0 gives e_{n+1}, i.e. the $(n+1)^{th}$ unit vector in $\mathbb{R}^{n+2}$. Since this is exactly the column corresponding to w_0 and $-t_0$, and since basic columns must be linearly independent, it follows that w_0 and t_0 are both non-basic.

As argued earlier, not both v_i and u_i, $i \in S$, can be in the basis, since their corresponding columns in W are multiples of each other. Hence $B_u \cap B_v = \emptyset$. Now, since $(\bar{u}, \bar{v}, \bar{u}^0, \bar{v}^0, \bar{w}_0, \bar{t}_0, \bar{t}_1)$ is a basic solution to $D^S_{LP}(\bar{x}, S)$, the solution to the system:

$$\sum_{i \in B_u} a_{i\cdot} u_i - \sum_{i \in B_v} a_{i\cdot} v_i + \pi = 0_n, \tag{46}$$

$$\sum_{i \in B_v} b_i v_i - \sum_{i \in B_u} b_i u_i - \pi_0 - v^0 = 0, \tag{47}$$

$$u_0 + v_0 = 1 \tag{48}$$

is unique. The system (46)-(48) is of the form $W^B y = z_0^B$. The number of rows of W^B (and z_0^B) is $n+2$, and the number of columns is $|B_u \cup B_v| + 2$. All columns of W^B are linearly independent. If $|B_u \cup B_v| + 2 < n + 2$, multiple solutions would exist. Hence, we must have $|B_u \cup B_v| = n$.

Now suppose the vectors $a_{i\cdot}$, $i \in B_u \cup B_v$, are linearly dependent. Then there exists a non-zero solution (u^*, v^*) to the system $\sum_{i \in B_u} a_{i\cdot} u_i^* - \sum_{i \in B_v} a_{i\cdot} v_i^* = 0_n$. Define the scalars $u_i(\delta) := \bar{u}_i + \delta u_i^*$ for $i \in B_u$ and $v_i(\delta) := \bar{v}_i + \delta v_i^*$ for $i \in B_v$, where $\delta \in \mathbb{R}$. We have that $(u(\delta), v(\delta), u_0, v_0)$ satisfies (46)-(48) if and only if $u_0 + v_0 = 1$ and $\bar{v}_0 + \delta(\sum_{i \in B_v} b_i v_i^* - \sum_{i \in B_u} b_i u_i^*) - v_0 = 0$. Defining $v_0(\delta) := \bar{v}_0 + \delta(\sum_{i \in B_v} b_i v_i^* - \sum_{i \in B_u} b_i u_i^*)$ and $u_0(\delta) := 1 - v_0(\delta)$, we have that $(u(\delta), v(\delta), u_0(\delta), v_0(\delta))$ satisfies (46)-(48). Since u_i^* for $i \in B_u$ and v_i^* for $i \in B_v$ are not all zero, and none of the vectors $a_{i\cdot}$, $i \in M$, are zero vectors, there must exist $\delta^* \in \mathbb{R}$ such that $(u(\delta^*), v(\delta^*), u_0(\delta^*), v_0(\delta^*))$ is a different solution to (46)-(48) than $(\bar{u}, \bar{v}, \bar{u}_0, \bar{v}_0)$, a contradiction. $\square$

From the above lemma, we immediately have the desired extension of Theorem 5 for the split disjunction:

Lemma 17. *Suppose* $S \subseteq M$ *satisfies* $|S| = r(S) + 1$. *Then:*

$$Conv(P(S) \cap F_{D(\pi,\pi_0)}) = \bigcap_{i \in \bar{I}(S)} Conv(P(S \setminus \{i\}) \cap F_{D(\pi,\pi_0)}). \tag{49}$$

Proof. Suppose $\bar{x} \in \bigcap_{i \in \bar{I}(S)} Conv(P(S \setminus \{i\}) \cap F_{D(\pi,\pi_0)})$ and $\bar{x} \notin Conv(P(S) \cap F_{D(\pi,\pi_0)})$. Let B_u and B_v and $(\bar{u}, \bar{v}, \bar{u}^0, \bar{v}^0, \bar{w}_0, \bar{t}_0, \bar{t}_1)$ be as in Lemma 6. We have $B_u \cap B_v = \emptyset$, $r(S) = n$, $|B_u \cup B_v| = n$, and the vectors $a_{i\cdot}$, $i \in B_u \cup B_v$, are linearly independent. Let $\{\bar{i}\} = S \setminus (B_u \cup B_v)$. We can not have $\bar{i} \in \bar{I}(S)$, since by Corollary 1, that would imply $\bar{u}_{\bar{i}} > 0$ or $\bar{v}_{\bar{i}} > 0$, which contradicts $\bar{i} \notin B_u \cup B_v$. Hence, we must have $\bar{i} \in S \setminus \bar{I}(S)$. But that means that $\bar{i}$ is in every basis of S, which contradicts $\bar{i} \notin B_u \cup B_v$. $\square$

From Theorem 5 and Lemma 17, we get the following:

Theorem 7.

$$Conv(P \cap F_{D(\pi,\pi_0)}) = \bigcap_{T \in \mathcal{B}_r^*} Conv(P(T) \cap F_{D(\pi,\pi_0)}) . \tag{50}$$

By intersection over all possible split disjunctions, and interchanging intersections, we get Theorem 1 of Sect. 3.

References

1. Balas, E.: Intersection cuts – a new type of cutting planes for integer programming. Operations Research **19** (1971) 19–39.
2. Balas, E.: Disjunctive programming. Annuals of Discrete Mathematics **5** (1979) 3–51.
3. Balas, E., Perregaard, M.: A precise correspondence between lift-and-project cuts, simple disjunctive cuts, and mixed integer Gomory cuts for 0-1 programming. Technical Report MSRR-631, GSIA, Carnegie Mellon University, 2000.
4. Cook, W., Kannan, R., Schrijver, A.: Chvátal closures for mixed integer programs. Mathematical Programming **47** (1990) 155–174.

An Exponential Lower Bound on the Length of Some Classes of Branch-and-Cut Proofs*

Sanjeeb Dash

Princeton University, Princeton NJ 08544, USA
sanjeebd@princeton.edu

Abstract. Branch-and-cut methods are among the more successful techniques for solving integer programming problems. They can also be used to prove that all solutions of an integer program satisfy a given linear inequality. We examine the complexity of branch-and-cut proofs in the context of 0-1 integer programs. We prove an exponential lower bound on the length of branch-and-cut proofs in the case where branching is on the variables and the cutting planes used are lift-and-project cuts (also called simple disjunctive cuts by some authors), Gomory-Chvátal cuts, and cuts arising from the N_0 matrix-cut operator of Lovász and Schrijver. A consequence of the lower-bound result in this paper is that branch-and-cut methods of the type described above have exponential running time in the worst case.

1 Introduction

Branch-and-cut methods for integer programs are currently the most important techniques for solving a wide range of integer programming problems. The two main ingredients of branch-and-cut methods, namely linear programming based branch-and-bound and cutting planes, were developed in the late 50's but were used successfully in combination only during the last decade.

Branch-and-cut methods employ a variety of cutting planes, some of which are specific to the problem class (e.g., comb inequalities for TSP instances) and also some that are defined for all integer programs, e.g., Gomory-Chvátal cutting planes (see [11]), disjunctive cuts [4] and Gomory mixed integer cuts [29]. In the case of 0-1 integer programs, the matrix cuts of Lovász and Schrijver [38] and the lift-and-project cuts of Balas, Ceria and Cornuéjols [5] define general families of cutting planes, i.e., they can be used for any 0-1 integer program. Branch-and-cut methods also utilize different branching strategies. The effectiveness of lift-and-project cuts (which are a class of facial disjunctive cuts) in a branch-and-cut framework is studied in [5], while [7] illustrates that Gomory mixed-integer cuts can be useful in branch-and-cut algorithms. The matrix cuts of Lovász and Schrijver have not been studied in much detail.

The integer programming problem (also the 0-1 variant) is NP-hard; no polynomial-time branch-and-cut algorithm is likely to exist. Some special classes

---

* This work was supported by ONR Grant N00014-98-1-0014.

W.J. Cook and A.S. Schulz (Eds.): IPCO 2002, LNCS 2337, pp. 145–160, 2002.

of branch-and-cut algorithms are known to have exponential time-complexity. Jeroslow [33] showed that any branch-and-bound procedure for the 0-1 integer program $\max\{x_1 \mid x_1 + \cdots + x_n = n/2, x_1, \ldots, x_n \in \{0,1\}\}$ where n is an odd integer, requires at least $2^{(n+1)/2}$ iterations to solve it (i.e., to establish infeasibility). Chvátal [12] exhibited a class of 0-1 knapsack problems such that every branch-and-bound procedure requires an exponential number of iterations for almost every problem in this class. For 0-1 integer programs, every cutting-plane algorithm which uses only Gomory-Chvátal cuts requires exponential time in the worst case; this follows from Pudlák's [39] results on the complexity of cutting-plane proofs. We extend Pudlak's idea and prove exponential lower bounds on the complexity of a large class of branch-and-cut algorithms for 0-1 integer programs, via complexity results for branch-and-cut proofs.

Chvátal [11] introduced the notion of a *cutting-plane proof* and developed this idea further in [13] and [14]; see also [43]. Examples of such proofs can be found in [30]. It is easy to extend the notion of cutting-plane proofs to *branch-and-cut proofs* which we study in this paper. We refer to the cutting-plane proofs studied in Chvátal [11] as Gomory-Chvátal cutting-plane proofs or as *G-C proofs*. We describe cutting-plane proofs which use lift-and-project cuts and matrix cuts in addition to Gomory-Chvátal cuts and call these *L-M-G proofs*. If in addition branching on variables is used, we refer to such proofs as *L-M-G branch-and-cut proofs*.

Pudlák [39] gave a class of infeasible 0-1 integer programs such that every G-C proof of infeasibility has exponential length. This non-trivial result uses bounds from monotone circuit complexity theory (see also Cook and Haken [17]) and the interpolation technique from proof complexity theory. A discussion of cutting-plane proofs in the context of complexity theory can also be found in [20], [19], [40] and [9]. Lovász suggested the study of proofs using matrix cuts, and Pudlák [40] presents some results on such proofs, and mentions some open problems. We prove an exponential lower bound for the length of L-M-G branch-and-cut proofs. Thus branch-and-cut algorithms which use branching on the variables and the cuts above have exponential worst-case complexity.

The paper is organized as follows. In Sect. 2, we describe the different cuts we use in this paper, and in Sect. 3 we discuss the notion of branch-and-cut proofs. In Sect. 4, we discuss interpolation and its usefulness in establishing lower bounds on the lengths of cutting-plane proofs. We also answer a question posed by Pudlák [40]. In Sect. 5, the concept of monotone interpolation is discussed, and an exponential lower bound on the length of lift-and-project cutting-plane proofs (or simple disjunctive cutting-plane proofs) is proved. This result is used to obtain the main result of this paper, an exponential lower bound on the length of some classes of branch-and-cut proofs for 0-1 integer programs.

2   Some Classes of Cutting Planes

Two classes of cutting planes available for general integer programs which can be used for 0-1 problems are Gomory-Chvátal cutting planes and the disjunctive cuts of Balas [4]. The disjunctive cuts we consider in this paper are called simple disjunctive cuts in [24]. Balas, Ceria and Cornuéjols [5] consider lift-and-project cuts for 0-1 problems which are identical to simple disjunctive cuts. Lift-and-project cuts form a subclass of the matrix cuts of Lovász and Schrijver [38].

Let $Q_n = [0,1]^n$ be the 0-1 cube in $\mathbb{R}^n$. If the dimension is obvious from the context, we denote the 0-1 cube by Q. Let $a_i^T x \leq b_i$ $(i = 1, \ldots, m)$ be a system of rational linear inequalities in $\mathbb{R}^n$ (we generally assume linear inequalities are rational). Assume that the inequalities $0 \leq x_j \leq 1$ $(j = 1, \ldots, n)$ are included in the above system. Let $P \subseteq Q$ be defined by

$$P = \{x \mid a_i^T x \leq b_i, \ i = 1, \ldots, m\} \ , \tag{1}$$

and let P_I stand for the convex hull of 0-1 points in P.

If $c^T x \leq d$ is a linear inequality valid for P and c is integral, then $c^T x \leq \lfloor d \rfloor$ is valid for all 0-1 points in P, and is called a *Gomory-Chvátal cutting plane* for P (abbreviated as a *G-C cut*). The *Chvátal closure* of P is the set of points satisfying all G-C cuts for P and is denoted by P'.

In what follows, we write $a^T x \leq b$ as $b - a^T x \geq 0$. All points in P satisfy

$$(b_i - a_i^T x) x_j \geq 0, \quad i = 1, \ldots, m, j = 1, \ldots, n \ ,$$
$$(b_i - a_i^T x)(1 - x_j) \geq 0, \ i = 1, \ldots, m, j = 1, \ldots, n \ , \tag{2}$$

obtained by multiplying the inequalities in (1) with the inequalities defining Q. In addition, 0-1 points in P satisfy

$$x_j^2 - x_j = 0, \text{ for } j = 1, \ldots, n \ . \tag{3}$$

Adding non-negative multiples of the inequalities in (2) and arbitrary multiples of (3) yields inequalities valid for all 0-1 points in P. A linear inequality of this form is a cutting plane for P, and is called an N-cut.

Formally, an inequality $c^T x \leq d$ or $d - c^T x \geq 0$ is called an N-*cut* for P if

$$d - c^T x = \sum_{i,j} \alpha_{ij}(b_i - a_i^T x)x_j + $$
$$\sum_{i,j} \beta_{ij}(b_i - a_i^T x)(1 - x_j) + \tag{4}$$
$$\sum_j \lambda_j(x_j^2 - x_j),$$

where $\alpha_{ij} \geq 0, \beta_{ij} \geq 0$ and $\lambda_j \in \mathbb{R}$ for $i = 1, \ldots, m, j = 1, \ldots, n$. A weakening of N-cuts, called N_0-*cuts*, can be obtained if in (4) we insist that $x_i x_j$ and $x_j x_i$ are distinct terms, for all i, j with $i \neq j$. A *lift-and-project cut* for P with respect to a variable x_k is a linear inequality of the form

$$\alpha(b_1 - a_1^T x)x_k + \beta(b_2 - a_2^T x)(1 - x_k) + \lambda(x_k^2 - x_k) \ , \tag{5}$$

where $a_1^T x \leq b_1$ and $a_2^T x \leq b_2$ are valid for P, λ is some real number and α and β are non-negative reals. Thus a lift-and-project cut is a special type of N_0-cut.

N-cuts and N_0-cuts are examples of matrix cuts (see [38]). Stronger matrix cuts, called N_+-cuts, are defined in [38]. An inequality $d - c^T x \geq 0$ is called an N_+-*cut* for P if it is formed by adding $n + 1$ squares of linear functions (i.e., terms of the form $(g_k + h_k^T x)^2$) to the sum in (4).

Adding all cuts of one of the types listed above to P yields a closed convex set which contains P_I and is strictly contained in P (unless $P = P_I$). The sets $N(P)$ and $N_+(P)$ are defined as follows:

$$N(P) \text{ is the set of points satisfying all } N\text{-cuts for } P , \tag{6}$$

$$N_+(P) \text{ is the set of points satisfying all } N_+\text{-cuts for } P . \tag{7}$$

$N_0(P)$ is defined similarly in terms of N_0-cuts; these sets have alternative projection representations, see [38]. $N(P)$ and $N_0(P)$ are polytopes, whereas $N_+(P)$ is generally non-polyhedral. P_k is the polytope obtained from P by adding all lift-and-project cuts with respect to the variable x_k. Observe that any inequality valid for $N(P)$ is an N-cut for P; this is true even if $N(P) = \emptyset$. The operators N_0, N_+ and P_k have similar properties.

The matrix-cut operators can be iterated to obtain approximations of P_I which are strictly contained in P (if $P \neq P_I$). Let $N^0(P) = P$ and $N^{t+1}(P) = N(N^t(P))$ if t is a non-negative integer. Let $N_0^t(P)$ and $N_+^t(P)$ be similarly defined. Lovász and Schrijver proved the following important results.

Theorem 1. *[38] Let $P \subseteq Q_n$ be a polytope. Then $N_0^n(P) = P_I$.*

Theorem 2. *[38] Let $P = \{x \mid Ax \leq b\}$ be a polytope contained in Q_n. For any fixed value of t, it is possible to optimize linear functions over both $N_0^t(P)$ and $N^t(P)$ in time bounded by a polynomial function of the encoding size of $Ax \leq b$.*

The first theorem remains true if N_0 is replaced by N or N_+, as both of these operators yield smaller convex sets than N_0. In the case of N_+, it is possible to approximate the maximum or minimum of a linear function over $N_+(P)$ to within a prescribed error tolerance in polynomial time.

The Chvátal closure can be iterated to obtain P_I from P; see [11], [43], and [26]. Lift-and-project cuts can also be generated iteratively. In fact, Theorem 1 follows from the following result of Balas [4].

Theorem 3. *[5] If $i_1, i_2, \ldots, i_n$ is any permutation of $\{1, 2, \ldots, n\}$ and P is a polytope in Q_n, then $(\cdots ((P_{i_1})_{i_2}) \cdots)_{i_n} = P_I$.*

The next result can be found in Cook and Dash [22]; an analogous property holds in the case of the Chvátal operator. Lemma 4 will be necessary for our lower-bound proof.

Lemma 4. *[22] If F is a face of a polytope $P \subseteq Q$, then $N(F) = N(P) \cap F$. This equation is also valid for the N_+ and N_0 operators.*

Lemma 4 is useful in many contexts. Let F be a face of a polytope P and let $c^T x \leq d$ be an N-cut for F. From Lemma 4, $N(F)$ is a face of $N(P)$. As $N(P)$ is a polytope, we can "rotate" $c^T x \leq d$ to get an inequality $(c')^T x \leq d'$ valid for $N(P)$, and hence an N-cut for P, such that

$$F \cap \{x \mid c^T x \leq d\} = F \cap \{x \mid (c')^T x \leq d'\} . \tag{8}$$

Compare this with Lemma 6.33 in Cook, Cunningham, Pulleyblank, and Schrijver [21] (the same result for P'). We may not be able to "rotate" $c^T x \leq d$ in the case of $N_+(P)$ as it may not be a polytope. Lemma 4 also implies:

$$F \text{ is a face of } Q \Rightarrow N(P \cap F) = N(P) \cap F . \tag{9}$$

The following two results give useful properties of $N_0(P)$.

Lemma 5. *[5,38] Assume that $a^T x \leq b$ is valid for $P \cap \{x \mid x_i = 0\}$ and $P \cap \{x \mid x_i = 1\}$, where P is a polytope, and $1 \leq i \leq n$. Then $a^T x \leq b$ is valid for P_i and $N_0(P)$.*

Proof. It is easy to see that there are numbers α and β, such that

$$a^T x - \alpha x_i \leq b \text{ is valid for } P,$$
$$a^T x - \beta(1 - x_i) \leq b \text{ is valid for } P.$$

Multiplying the first inequality by $1 - x_i$ and the second by x_i, replacing x_i^2 by x_i and adding, we see that $a^T x \leq b$ is a lift-and-project cut with respect to the variable x_i. $\square$

Lemma 6. *[5,38] If $P \subseteq Q_n$ is a polytope, then $P_i = conv((P \cap \{x \mid x_i = 0\}) \cup (P \cap \{x \mid x_i = 1\}))$ for $i = 1, \ldots, n$, and $N_0(P) = \cap_i P_i$.*

We now state some well-known properties of (symmetric) positive semidefinite matrices which we will need; see Horn and Johnson [32]. We denote the fact that a matrix A is positive semidefinite by $A \succeq 0$. A *principal submatrix* of a matrix is a square submatrix obtained by deleting some rows and the corresponding columns from the matrix. If $A \succeq 0$, then every principal submatrix of A is positive semidefinite. If a matrix A has a block-diagonal decomposition, then $A \succeq 0 \Leftrightarrow$ every block is positive semidefinite. For example,

$$\text{if } A = \begin{bmatrix} A_1 & 0 \\ 0 & A_2 \end{bmatrix}, \text{ then } A \succeq 0 \Leftrightarrow A_1 \succeq 0, A_2 \succeq 0 . \tag{10}$$

Proposition 7. *(Schur complements) Let A be a non-singular matrix, and let B and C be matrices.*

$$\text{If } D = \begin{bmatrix} A & B \\ B^T & C \end{bmatrix}, \text{ then } D \succeq 0 \Leftrightarrow C - B^T A^{-1} B \succeq 0 .$$

The following result is crucial for the exponential bounds we prove later on.

Lemma 8. *Let $P_1 = \{(x,y) \mid Ax \le e\}$ and $P_2 = \{(x,y) \mid By \le f\}$ be two polytopes contained in Q. Then $N_+(P_1 \cap P_2) = N_+(P_1) \cap N_+(P_2)$. An identical result holds for N and N_0.*

Proof. For any two polytopes P_1 and P_2, it is true that $N_+(P_1 \cap P_2) \subseteq N_+(P_1) \cap N_+(P_2)$. Let P_1 and P_2 satisfy the conditions of the lemma. To prove the reverse inclusion, assume that

$$z = \begin{pmatrix} x \\ y \end{pmatrix} \in N_+(P_1) \cap N_+(P_2) \ . \tag{11}$$

Let $\bar{z} = \begin{pmatrix} 1 \\ z \end{pmatrix}$. Then there are symmetric matrices $X \in M_+(P_1)$ and $Y \in M_+(P_2)$ ($M_+(P)$ and $M(P)$ are defined in [38]), such that $\bar{z} = Xe_0 = Ye_0$, where

$$X = \begin{pmatrix} 1 & x^T & y^T \\ x & X_{11} & X_{12}^T \\ y & X_{12} & X_{22} \end{pmatrix} \text{ and } Y = \begin{pmatrix} 1 & x^T & y^T \\ x & Y_{11} & Y_{12}^T \\ y & Y_{12} & Y_{22} \end{pmatrix}. \tag{12}$$

Also $X, Y \succeq 0$. As X and Y are positive semidefinite,

$$\begin{pmatrix} 1 & x^T \\ x & X_{11} \end{pmatrix} \succeq 0 \text{ and } \begin{pmatrix} 1 & y^T \\ y & Y_{22} \end{pmatrix} \succeq 0 \ ; \tag{13}$$

the above matrices are principal submatrices of X and Y. We can conclude that $X_{11} - xx^T$ and $Y_{22} - yy^T$ are both positive semidefinite. This is true because of Proposition 7 (set $A = 1$ in Proposition 7).

Now, let Z be the matrix defined by

$$Z = \begin{pmatrix} 1 & x^T & y^T \\ x & X_{11} & xy^T \\ y & yx^T & Y_{22} \end{pmatrix}. \tag{14}$$

It is not difficult to verify that Z is contained in $M(P_1 \cap P_2)$. Also, observe that $Z - \bar{z}\bar{z}^T$ is a block-diagonal matrix with non-zero blocks $X_{11} - xx^T$ and $Y_{22} - yy^T$. Therefore $Z - \bar{z}\bar{z}^T$ is positive semidefinite by (10); this implies that $Z \succeq 0$. Since $\bar{z} = Ze_0$, we have shown that $z \in N_+(P_1 \cap P_2)$, and the result follows for the semidefinite operator.

Observe that in (12), if we start out with X in $M(P_1)$ and Y in $M(P_2)$, then Z yields the result for $N(P)$. Let X and Y belong to $M_0(P_1)$ and $M_0(P_2)$ respectively. Then X and Y are as in (12), except that they are non-symmetric and X_{12}^T is replaced by X_{21}, and Y_{12}^T by Y_{21}. The matrix Z above, formed from X and Y, belongs to $M_0(P_1 \cap P_2)$, and the result for $N_0(P)$ follows.   □

3   Cutting-Plane Proofs and Branch-and-Cut Proofs

Traditionally, the phrase "cutting-plane proof" refers to a proof using G-C cuts. We refer to such proofs as *Gomory-Chvátal cutting-plane proofs* (or as *G-C proofs* for shortness sake). By a cutting-plane proof, we mean one which uses any of the cutting planes discussed in Sect. 2.

Let $Ax \leq b$ denote the following linear system in $\mathbb{R}^n$:

$$a_i^T x \leq b_i \ (i = 1, \ldots, m) \ . \tag{15}$$

Assume $c^T x \leq d$ is valid for all 0-1 solutions of $Ax \leq b$. An N-cutting-plane proof of $c^T x \leq d$ from $Ax \leq b$ is a sequence,

$$a_{m+k}^T x \leq b_{m+k} \ (k = 1, \ldots, M) \ , \tag{16}$$

with $c^T x \leq d$ the last inequality in the sequence, and a collection of numbers

$$\alpha_{jl}^k, \beta_{jl}^k \geq 0 \ (j = 1, \ldots, m + k - 1, l = 1, \ldots, n) \tag{17}$$

such that, for $k = 1, \ldots, M$ $a_{m+k}^T x \leq b_{m+k}$ is derived as in (4) using α_{jl}^k and β_{jl}^k from

$$a_j^T x \leq b_j \ (j = 1, \ldots, m + k - 1) \ . \tag{18}$$

Informally, an inequality in the sequence is an N-cut for the previous inequalities in the sequence. The *length* of the cutting-plane proof is M and its *size* is the sum of the sizes of the inequalities and numbers $\alpha_{jl}^k, \beta_{jl}^k$ in the proof. (The size of a proof is the number of bits required to write it down). If an inequality belongs to or is implied by $Ax \leq b$, we say it has an N-cutting-plane proof of length 0 from $Ax \leq b$. We can assume that inequalities in (15) and (16) integral. Proofs using disjunctive cuts, N_0-cuts, or N_+-cuts are defined in a similar fashion. An N-cutting-plane proof will be abbreviated as an N-*proof*. We analogously define N_+-*proofs*.

If an inequality $c^T x \leq d$ has an N-proof from P, then $c^T x \leq d$ is valid for P_I. Conversely, an inequality valid for P_I has an N-proof from P; this follows from Theorem 1 and the fact that $N(P)$ is a polytope whenever P is. If P_I is empty, we refer to a cutting-plane proof of $0^T x \leq -1$ as a *cutting-plane proof of infeasibility*.

We can use both G-C cuts and N_+-cuts in a cutting-plane proof; such a proof will be called an N_*-proof. We also define $N_\#$-proofs; each cut is either an N-cut or a G-C cut for the previous ones.

The branch-and-cut method can be used to prove that a given inequality is satisfied by all 0-1 solutions of (15); this yields the notion of a "branch-and-cut proof". The length of the proof is defined as the sum of the number of cuts in the proof and the number of times we branch on a variable. We will mainly deal with *branch-and-cut proofs of infeasibility* where $0^T x \leq -1$ is the last inequality in each branch.

4   Interpolation and Cutting-Plane Proofs

An important goal in the theory of propositional proofs is to establish exponential worst-case complexity for proofs in various proof systems. Achieving this goal for all proof systems would show $P \neq NP$. See Beame and Pitassi [8], and also Pudlák [40], for recent surveys on propositional proof complexity. Some methods used in this endeavour are the *bottleneck counting method* of Haken [31], the *restriction method* of Ajtai [1], and the *interpolation method*. Krajíček [35,36] proposed the idea of using *effective interpolation* to establish lower bounds on the lengths of proofs in different proof systems. This idea was first used in [42] and [9]. Pudlák [39] derived exponential lower bounds for lengths of G-C proofs using interpolation. We now discuss this technique.

Assume that the following linear system is infeasible:

$$Ax + Cz \leq e \ ,$$

$$By + Dz \leq f \ , \tag{19}$$

$$x, y, z \text{ are 0-1} \ .$$

Then $0^T x + 0^T y + 0^T z \leq -1$ has a G-C proof $\mathcal{P}$ from (19). Let z' denote some 0-1 assignment to z. The system

$$Ax \leq e - Cz' \ ,$$

$$By \leq f - Dz' \ , \tag{20}$$

$$x, y \text{ are 0-1} \ ,$$

obtained from (19), is still infeasible. Now, $\mathcal{P}$ can be modified to a proof $\mathcal{P}'$ of infeasibility of (20), with the same length (let $\mathcal{P}_i$ stand for the ith inequality in $\mathcal{P}$):

$$\text{if } \mathcal{P}_i \text{ is } a^T x + b^T y + c^T z \leq d \text{ then } \mathcal{P}'_i \text{ is } a^T x + b^T y \leq d - c^T z' \ . \tag{21}$$

In (20) we have two linear systems, with no variables in common, and at least one of the two is infeasible. Which of the two systems is infeasible can decided by examining $\mathcal{P}'$. Pudlák [39] showed:
Given any z', it is possible to construct in polynomial time (in the size of $\mathcal{P}$), two G-C proofs, one involving x alone and the other involving only y, such that either the last inequality in the first proof is $0^T x \leq -1$ or the last in the second proof is $0^T y \leq -1$.
This yields a polynomial-time algorithm $\mathcal{F}_{\mathcal{P}}(z)$ for each composite G-C proof $\mathcal{P}$, which takes as input z' and decides which of the two systems in (20) is infeasible. This is called *effective interpolation* and $\mathcal{F}_{\mathcal{P}}(z)$ is called an *interpolating algorithm*.

A strong lower bound can sometimes be derived for the complexity of the interpolating algorithm $\mathcal{F}_{\mathcal{P}}$; this immediately yields a lower bound for the length of $\mathcal{P}$. We discuss this later. First we give Pudlák's effective interpolation result for G-C proofs [39, Theorem 3] in Proposition 9.

Assume that the following system, in n variables and m inequalities, is infeasible:

$$Ax \leq e, \quad x \text{ is 0-1} \ , \tag{22}$$

$$By \leq f, \quad y \text{ is 0-1} \quad, \tag{23}$$

and assume that x and y have no variables in common. For convenience, we assume that the initial inequalities in every cutting-plane proof of infeasibility from the above system are precisely the inequalities in the system.

Proposition 9. *[39] Let $\mathcal{R}$ be a G-C proof of $0^T x + 0^T y \leq -1$ from (22) and (23). In polynomial time (in the size of $\mathcal{R}$), a G-C proof of infeasibility of either (22), or of (23), can be constructed from $\mathcal{R}$.*

Proof. Let $a_i^T x + b_i^T y \leq d_i$ be the ith inequality in $\mathcal{R}$ and call this $\mathcal{R}_i$. Now, $\mathcal{R}_1, \ldots, \mathcal{R}_m$ are just (22) and (23) and $\mathcal{R}_k$ (for some k) is precisely $0^T x + 0^T y \leq -1$. We can assume that $\mathcal{R}$ has integral inequalities. We say that $\mathcal{R}_i$ is derived from $\mathcal{R}_1, \ldots, \mathcal{R}_{i-1}$ if

$$a_i^T x + b_i^T y = \sum_j \lambda_{ij}(a_j^T x + b_j^T y) \text{ and } d_i = \left\lfloor \sum_j \lambda_{ij} d_j \right\rfloor, \tag{24}$$

where $\lambda_{ij} \geq 0$ for $j = 1, \ldots, i-1$.

We construct a sequence of inequalities $\mathcal{S}$ involving only x, and another sequence $\mathcal{T}$, involving only y, such that $\mathcal{S}_i$ and $\mathcal{T}_i$ together imply $\mathcal{R}_i$. Let I_i stand for $\{1, \ldots, i-1\}$. For $i = 1, \ldots, m$, if $\mathcal{R}_i$ involves only x, then set $\mathcal{S}_i$ to $\mathcal{R}_i$ and $\mathcal{T}_i$ to $0^T y \leq 0$, otherwise set $\mathcal{S}_i$ to $0^T x \leq 0$ and $\mathcal{T}_i$ to $\mathcal{R}_i$. Define subsequent terms of $\mathcal{S}$ and $\mathcal{T}$ as follows: for $i = m+1, \ldots, k$, if $\mathcal{R}_i$ is derived from $\mathcal{R}_j$ ($j \in I_i$) with the numbers $\lambda_{ij} \geq 0$ ($j \in I_i$), then let $\mathcal{S}_i$ be derived from $\mathcal{S}_j$ ($j \in I_i$) and let $\mathcal{T}_i$ be derived from $\mathcal{T}_j$ ($j \in I_i$), with the same numbers λ_{ij}. If the right-hand sides of $\mathcal{S}_i$ and $\mathcal{T}_i$ are g_i and h_i respectively, we can conclude that

$$\mathcal{S}_i \equiv a_i^T x \leq g_i \text{ and } \mathcal{T}_i \equiv b_i^T y \leq h_i \text{ with } g_i + h_i \leq d_i. \tag{25}$$

Therefore the last inequalities in $\mathcal{S}$ and $\mathcal{T}$ are, respectively, $0^T x \leq g_k$ and $0^T y \leq h_k$. Since $d_k = -1$, one of g_k and h_k is at most -1, and we have a G-C proof of infeasibility of either (22) or (23). This is a polynomial-time construction. $\square$

Suppose $\mathcal{P}$ is a G-C proof of $0^T x + 0^T y + 0^T z \leq -1$ from (19). The interpolating algorithm $\mathcal{F}_\mathcal{P}(z)$ mentioned above broadly does the following. If z' is a 0-1 assignment to z, then $\mathcal{F}_\mathcal{P}$ first computes (20). Then it computes one of the two proofs in Proposition 9, say $\mathcal{S}$. If in $\mathcal{S}_k$ the right-hand side $g_k \leq -1$, $\mathcal{F}_\mathcal{P}$ returns 0 to indicate that $Ax \leq e - Cz'$ is infeasible, else it knows that $h_k \leq -1$, and it returns 1. We need not compute $\mathcal{T}$ or $\mathcal{P}'$. In fact, $\mathcal{F}_\mathcal{P}$ only needs to compute $e - Cz'$ and the numbers g_i to decide which of the two systems in (20) is infeasible.

To obtain a similar property for N-proofs, Pudlák proved the following result.

Proposition 10. *[40] Let $\mathcal{R}$ be an $N_\#$-proof of $0^T x \leq -1$ from (22) and (23). In polynomial time (in the size of $\mathcal{R}$), an $N_\#$-proof of infeasibility of either (22), or of (23), can be constructed from $\mathcal{R}$.*

We now proceed to extend the above result to N_+-proofs (more precisely N_*-proofs) thus answering a question which is mentioned as being unsolved by Pudlák; see[40, page 11].

Proposition 11. *Let $\mathcal{R}$ be an N_*-proof of $0^T x \leq -1$ from (22) and (23). In polynomial time (in the size of $\mathcal{R}$), an N_*-proof of infeasibility of either (22), or of (23), can be constructed from $\mathcal{R}$.*

Proof. We can assume that all inequalities in $\mathcal{R}$ are integral and N_+-cuts for the previous ones in $\mathcal{R}$. G-C cuts can be handled as in Proposition 9.

Let $\mathcal{R}$ be as in Proposition 9 (only $\mathcal{R}$ is an N_+-proof of infeasibility). We construct two N_+-proofs $\mathcal{S}$ and $\mathcal{T}$ with the same length as $\mathcal{R}$. Let the first m terms in $\mathcal{S}$ and $\mathcal{T}$ be as in Proposition 9. By definition, if $\mathcal{R}_j$ is $a_j^T x + b_j^T y \leq d_j$ and $i = m + 1$, then for all $j < i$

$$\mathcal{S}_j \equiv a_j^T x \leq g_j \quad \text{and} \quad \mathcal{T}_j \equiv b_j^T y \leq h_j \quad \text{with} \quad g_j + h_j \leq d_j \ . \tag{26}$$

Assume that we have obtained $\mathcal{S}_j$ and $\mathcal{T}_j$ satisfying (26) for all $j < i$, where i is some number greater than m. Let P_1 be the polytope defined by $\mathcal{S}_1, \ldots, \mathcal{S}_{i-1}$. Similarly, let P_2 be defined by $\mathcal{T}_1, \ldots, \mathcal{T}_{i-1}$. Now, if P is defined by $\mathcal{R}_1, \ldots, \mathcal{R}_{i-1}$, then $P_1 \cap P_2 \subseteq P$.

It follows that $a_i^T x + b_i^T y \leq d_i$ is valid for $N_+(P_1 \cap P_2)$. By Lemma 8 and Carathéodory's Theorem, we have

$$a_i^T x + b_i^T y = \sum_{j \in J} \alpha_j p_j^T x + \sum_{k \in K} \beta_k q_k^T y \quad \text{and} \quad \sum_{j \in J} \alpha_j r_j + \sum_{k \in K} \beta_k s_k \leq d_i \ , \tag{27}$$

where $p_j^T x \leq r_j$ ($j \in J$) are $N_+(P_1)$-cuts and $q_k^T x \leq s_k$ ($k \in K$) are $N_+(P_2)$-cuts and $\alpha_j \geq 0$ and $\beta_k \geq 0$ for all $j \in J$ and $k \in K$. Adding separately the $N_+(P_1)$-cuts and the $N_+(P_2)$-cuts, we see that there are there are real numbers g_i' and h_i' such that

$$a_i^T x \leq g_i' \text{ is an } N_+(P_1)\text{-cut and } b_i^T y \leq h_i' \text{ is an } N_+(P_2)\text{-cut} \ , \tag{28}$$

and $g_i' + h_i' \leq d_i$. The numbers g_i' and h_i' can be computed as

$$g_i' = \max\{a_i^T x \mid (x, y) \in N_+(P_1)\}, \quad h_i' = \max\{b_i^T y \mid (x, y) \in N_+(P_2)\} \ . \tag{29}$$

To get the ith terms of $\mathcal{S}$ and $\mathcal{T}$, we compute g_i' and h_i' as in (29), by solving two semidefinite programs (see Theorem 2 and the discussion following it). As these are not necessarily integers, we round them down to get g_i and h_i; we have completed the construction of the ith terms in $\mathcal{S}$ and $\mathcal{T}$. Repeating this process we get, as the last terms in $\mathcal{S}$ and $\mathcal{T}$, $0^T x \leq g_k$ and $0^T x \leq h_k$ where at least one of g_k or h_k is bounded above by -1.   □

In the proof above, if we replace $N_+(P)$ by $N(P)$, we get Pudlák's result, Proposition 10, by combining with Proposition 9. We can also use $N_0(P)$ instead of $N_+(P)$ and get a result analogous to Proposition 10 in the case where we have proofs using N_0-cuts and G-C cuts.

5   Monotone Interpolation

Let $f : \mathbb{R}^n \to \mathbb{R}$ be a real-valued non-decreasing function, that is, if $x \leq y$ with x, y in $\mathbb{R}^n$, then $f(x) \leq f(y)$. Such a function is also called a *monotone function*. We use the term "monotone computation" to refer to an application of a monotone function to given inputs. Examples of monotone unary and binary functions (referred to as *monotone operations*) are

$$tx, \quad r + x, \quad x + y, \quad \lfloor x \rfloor, \quad thr(x, -1) \tag{30}$$

where t is a non-negative number, x and y are real variables, and r is a real number; $thr(x, -1)$ is a *threshold function* which returns 0, if $x \leq -1$, and returns 1 otherwise.

Consider a fixed system in (19), and the interpolating algorithm $\mathcal{F}_{\mathcal{P}}$ mentioned in the context of G-C proofs. Once we have computed $e - Cz'$ in (20), all further computations to obtain the numbers g_i in Proposition 9 are monotone. We can perform a sequence of monotone operations, with precisely the ones in (30), to get g_i, for $i = 1, \ldots, k$ and then apply $thr(x, -1)$ to g_k to get $\mathcal{F}_{\mathcal{P}}(z')$ (which is 0 or 1). Further, if all coefficients in C are negative, then it is possible to compute $e - Cz'$ and therefore $\mathcal{F}_{\mathcal{P}}(z')$ by monotone operations only; the number of monotone operations is polynomially bounded by the length of the proof $\mathcal{P}$. This is an example of *monotone interpolation*. This result is a crucial step in Pudlák's proof of an exponential lower bound on the length of G-C proofs.

Razborov [41], and independently Andreev [3], proved super-polynomial lower bounds on the lengths of sequences of monotone boolean operations ('and' and 'or') required to compute certain monotone boolean functions. See also [2] and [10]. Pudlák [39] and Cook and Haken [17] extended the above results to the case where arbitrary monotone operations are used in computing monotone boolean functions. The following follows from their results.

Proposition 12. *[39] There is a polynomial function $p(n)$ and an exponential function $e(n)$, such that for every positive integer n, there is a system of inequalities $\mathcal{I}_n$ with the properties:*

(i) $\mathcal{I}_n$ has the same form as (19) with $C \leq 0$ and has no 0-1 solution,
(ii) $\mathcal{I}_n$ has at most $p(n)$ inequalities and variables,
(iii) Any sequence of monotone operations which has z as input and computes 0 if $Ax \leq e - Cz'$ is infeasible and 1 otherwise has length at least $e(n)$.

Pudlák combined the monotone interpolation result mentioned above along with Proposition 12 to obtain an exponential lower bound on the length of G-C proofs. To get similar bounds for L-M-G proofs we first prove an effective interpolation result for N_+-cuts.

Theorem 13. *Assume that (19) is infeasible and assume that $C \leq 0$. If $\mathcal{P}$ is an N_*-proof of infeasibility of (19), then there is an algorithm $\mathcal{F}_{\mathcal{P}}(z)$ with the following properties:*

(i) *if z' is any 0-1 assignment to z, then $\mathcal{F}_\mathcal{P}$ computes 0 if $Ax \leq e - cz'$ is infeasible, and 1 otherwise;*

(ii) *$\mathcal{F}_\mathcal{P}$ performs only monotone computations;*

(iii) *The number of monotone computation steps performed by $\mathcal{F}_\mathcal{P}$ is bounded above by a polynomial function of the length of $\mathcal{P}$.*

Proof. Let $\mathcal{P}$ be an N_*-proof of $0^T x + 0^T y + 0^T z \leq -1$ from (19). By Lemma 4, if $\mathcal{P}'$ is defined by (21) then $\mathcal{P}'$ is an N_*-proof of the infeasibility of (20). Now, let $\mathcal{R}$ denote $\mathcal{P}'$; as in Proposition 11, obtaining the N_*-proof $\mathcal{S}$ suffices to decide whether $Ax \leq e - cz'$ is feasible or not. We proceed as in the case of G-C proofs and first compute $e - Cz'$ and then the numbers $g_1, g_2, \ldots, g_k$ described in Proposition 11, and finally $thr(g_k, -1)$ to get $\mathcal{F}_\mathcal{P}(z')$.

If $C \leq 0$ then $e - Cz'$ can be obtained by monotone operations. The numbers $g_1, g_2, \ldots, g_k$ can be obtained by monotone computations. To see this, observe that the following computation

$$\max\{a_i^T x \mid x \in N_+(P)\},$$

$$P = \{x \in Q \mid a_1^T x \leq g_1, \ldots, a_{i-1}^T x \leq g_{i-1}\} \ , \tag{31}$$

performed in (29), is monotone in the inputs $g_1, g_2, \ldots, g_{i-1}$. If the numbers $g_1, \ldots, g_{i-1}$ are increased, then P is larger, and so is $N_+(P)$, and the maximum in (31) increases.

Finally, assume (19) has m inequalities and n variables. Computing $e - Cz'$ requires at most $2mn$ monotone operations. The numbers $g_1, g_2, \ldots, g_k$ each require a monotone computation of the form (31). Hence, the number of monotone computation steps is at most $2nL$, where L is the length of $\mathcal{P}$ (here $m \leq L$). $\square$

Theorem 13 does not yield an exponential lower bound on the length of N_*-proofs as all monotone computations do not have a bounded number of inputs (they are not monotone operations, for example). But we can prove:

Theorem 14. *Any lift-and-project cutting-plane proof of infeasibility of the system $\mathcal{I}_n$ given in Proposition 12 has exponential length (in n).*

Proof. Let $\mathcal{O}$ be a lift-and-project cutting-plane proof of length L of the infeasibility of $\mathcal{I}_n$ for a given integer n. Every $\mathcal{O}_i$ is a lift-and-project cut (say with respect to the variable x_k; this variable changes with i) for $\mathcal{O}_1, \ldots, \mathcal{O}_{i-1}$. Therefore $\mathcal{O}_i$ is equal to $\alpha q_1 + \beta q_2 + \lambda(x_k^2 - x_k)$ for some $\alpha, \beta \geq 0$ and some λ, where q_1 and q_2 are inequalities implied by $\mathcal{O}_1, \ldots, \mathcal{O}_{i-1}$. Define a sequence $\mathcal{P}$ of length at most $3L$ by

$$\mathcal{P}_{3i-2} \equiv q_1, \ \mathcal{P}_{3i-1} \equiv q_2, \ \mathcal{P}_{3i} \equiv \mathcal{O}_i \ . \tag{32}$$

As lift-and-project cuts are also N_0-cuts, $\mathcal{P}$ is an N_0-cutting-plane proof of $0^T x + 0^T y + 0^T z \leq -1$ with the special property that each inequality is either a non-negative linear combination of $p(n) + 1$ previous inequalities in $\mathcal{P}$ ($\mathcal{I}_n$ has at most $p(n)$ variables) or is an N_0-cut derived from at most two previous inequalities.

If in Theorem 13, we replace N_* by N_0, we get an interpolating algorithm $\mathcal{F}_\mathcal{P}$ which performs only monotone operations. To see this, let z' be some 0-1 assignment to z, and let $\mathcal{R} \equiv \mathcal{P}'$ where $\mathcal{P}'$ is defined as in (21). As in the proof of Proposition 11 define the proofs $\mathcal{S}$ and $\mathcal{T}$ from $\mathcal{R}$ with the following property. If $\mathcal{R}_i = \lambda_1 \mathcal{R}_1 + \cdots + \lambda_{i-1} \mathcal{R}_{i-1}$ for non-negative numbers $\lambda_1, \ldots, \lambda_{i-1}$, then $\mathcal{S}_i$ equals $\lambda_1 \mathcal{S}_1 + \cdots + \lambda_{i-1} \mathcal{S}_{i-1}$ and $\mathcal{T}_i$ equals $\lambda_1 \mathcal{T}_1 + \cdots + \lambda_{i-1} \mathcal{T}_{i-1}$. If $\mathcal{R}_i$ is the inequality $a_i^T x + b_i^T y \leq d_i$ and is an N_0-cut derived from two previous inequalities, say $\mathcal{R}_j$ and $\mathcal{R}_l$, then $\mathcal{S}_i$ is the inequality $a_i^T x \leq g_i$ such that

$$g_i = \max\{a_i^T x \mid x \in N_0(P)\},$$

$$P = \{x \in Q \mid a_j^T x \leq g_j, \, a_l^T x \leq g_l\} \ . \tag{33}$$

Similarly $\mathcal{T}_i$ is the inequality $b_i^T y \leq h_i$ derived as an N_0-cut from $b_j^T y \leq h_j$ and $b_l^T y \leq h_l$; also $g_i + h_i \leq d_i$. Hence, if $\mathcal{P}$ has k inequalities, then either $g_k \leq -1$ or $h_k \leq -1$. Also, $g_1, g_2, \ldots, g_k$ can be computed by monotone operations only. Either g_i is the non-negative linear combination of $p(n) + 1$ previous numbers from $g_1, \ldots, g_{i-1}$ (this requires $2(p(n) + 1)$ monotone operations) or g_i is computed as in (33), which is a monotone operation. The result follows. $\square$

The next result implies that any N_0-cutting-plane proof of the infeasibility of $\mathcal{I}_n$, given in Proposition 12, must have exponential length.

Lemma 15. *Let $\mathcal{P}$ be an N_0-proof of $c^T x \leq d$, from some polytope in Q_n, of length L. There is a lift-and-project cutting-plane proof of $c^T x \leq d$ of length at most $(n + 1)L$.*

Proof. Let $a^T x \leq b$ be an inequality in the proof. Let $P = \{x \mid Ax \leq b\}$ be the polytope defined by the inequalities used in deriving $a^T x \leq b$ as an N_0-cut. By Lemma 6, $N_0(P) = \cap_i P_i$, where P_i is the lift-and-project operator with respect to the variable x_i. Therefore $N_0(P)$ is completely defined by the inequalities defining the polytopes P_i. Hence, by Carathéodory's Theorem, $a^T x \leq b$ is a nonnegative linear combination of n inequalities $g_1^T x \leq h_1, \ldots, g_n^T x \leq h_n$, where for $i = 1, \ldots, n$, $g_i^T x \leq h_i$ is valid for some P_k ($1 \leq k \leq n$). It follows that $a^T x \leq b$ is a nonnegative linear combination of n lift-and-project cuts for P. Let $\mathcal{R}$ be a sequence of inequalities such that $a^T x \leq b$ in $\mathcal{P}$ is replaced by the inequalities $g_1^T x \leq h_1, \ldots, g_n^T x \leq h_n$ and $a^T x \leq b$ (in order). In $\mathcal{R}$, every inequality is either a nonnegative linear combination of previous inequalities or a lift-and-project cut derived from previous inequalities; hence $\mathcal{R}$ is a lift-and-project cutting-plane proof of length at most $(n + 1)L$. $\square$

Combining Proposition 9, Proposition 12, Theorem 14 and Lemma 15, we get the following theorem.

Theorem 16. *Let $\mathcal{I}_n$ be as in Proposition 12. Then any cutting-plane proof of infeasibility which uses only N_0-cuts, lift-and-project cuts and G-C cuts must have exponential length, i.e., length at least $e(n)/h(n)$ where $e(n)$ is the exponential function in Proposition 12, and $h(n)$ is some polynomial function of n.*

To get an exponential lower bound on the length of branch-and-cut proofs for $\mathcal{I}_n$, we prove the following lemma.

Lemma 17. *Let $P \subseteq Q_n$ be a polytope which does not contain 0-1 points. Let $\mathcal{P}$ be an L-M-G branch-and-cut proof of $0^T x \leq -1$ from P, with m cutting planes and k branches. Then there is an L-M-G cutting-plane proof of $0^T x \leq -1$ with length exactly $m + k$.*

Proof. As discussed earlier, every lift-and-project cut is an N_0-cut; so we can assume that the cuts in the proof are either N_0-cuts or Gomory-Chvátal cuts. Assume $\mathcal{P}$ has exactly one branch and s inequalities in the left branch and t inequalities in the right branch. Further assume that the very first step in the branch-and-cut proof consists of branching on the variable x_1. In the the left branch, we impose the condition $x_1 = 0$ and in the right branch the condition $x_1 = 1$.

Let the inequalities in the left branch be $a_1^T x \leq b_1, a_2^T x \leq b_2, \ldots$. Let $a_1^T x \leq b_1$ be an N_0-cut for P and the inequality $x_1 = 0$. We can assume that $a_1^T x \leq b_1$ is an N_0-cut for $P \cap \{x \mid x_1 = 0\}$, i.e., $a_1^T x \leq b_1$ is an $N_0(P \cap \{x \mid x_1 = 0\})$-cut. From Lemma 4 and the subsequent discussion, we know that $a_1^T x \leq b_1$ can be "rotated" to get an N_0-cut for P of the form $a_1^T x + \alpha_1 x_1 \leq b_1$ for some number α_1. To continue this process, for $i = 2, \ldots, s$ do the following: if $a_i^T x \leq b_i$ is an N_0-cut from P, $x_1 = 0$, and the inequalities $a_1^T x \leq b_1, \ldots, a_{i-1}^T x \leq b_{i-1}$, let P_{i-1} equal P intersected with the inequalities $a_1^T x + \alpha_1 x_1 \leq b_1, \ldots, a_{i-1}^T x + \alpha_{i-1} x_1 \leq b_{i-1}$. Then $a_i^T x \leq b_i$ is an $N_0(P_{i-1} \cap \{x \mid x_1 = 0\})$-cut and can be rotated to get an $N_0(P_{i-1})$-cut of the form $a_i^T x + \alpha_i x_1 \leq b_i$. This rotation process can also be performed if $a_i^T x \leq b_i$ is a Gomory-Chvátal cut. Similarly an inequality $c_i^T \leq d_i$ in the right branch gets rotated to $c_i^T x + \beta_i(1 - x_1) \leq d_i$ for some β_i.

The last inequality in the left branch is mapped to $0^T x + \alpha_s x_1 \leq -1$ and the last inequality in the right branch is mapped to $0^T x + \beta_t(1 - x_1) \leq -1$. Multiplying the first of these two inequalities by $(1 - x_1)$ and the second by x_1, replacing x_1^2 by x_1, and adding, we get the inequality $0^T x \leq -1$ as an N_0-cut. Observe that we have removed a branch and replaced it by an N_0-cut, to get a cutting-plane proof of infeasibility of length $s + t + 1$.

If a branch-and-cut proof has many branches, we can start from the lowermost branches, and recursively eliminate the branches by adding an extra N_0-cut for every branch eliminated. This completes the proof. □

The exponential lower bounds on the length of L-M-G branch-and-cut proofs follows immediately from Theorem 16.

References

1. Ajtai, M.: The complexity of the pigeonhole principle. Combinatorica **14** (1994) 417–433
2. Alon, N., Boppana, R.: The monotone circuit complexity of Boolean functions. Combinatorica **7** (1987) 1–22

3. Andreev, A.E.: On a method for obtaining lower bounds for the complexity of individual monotone functions. Soviet Mathematics - Doklady **31** (1985) 530–534

4. Balas, E.: Disjunctive programming. Annals of Discrete Mathematics **5** (1979) 3–51

5. Balas, E., Ceria, S., Cornuéjols, G.: A lift-and-project cutting plane algorithm for mixed 0-1 programs. Mathematical Programming **58** (1993) 295–324

6. Balas, E., Ceria, S., Cornuéjols, G.: Mixed 0-1 programming by lift-and-project in a branch-and-cut framework. Management Science **42** (1996) 1229–1246

7. Balas, E., Ceria, S., Cornuéjols, G., Natraj, G.: Gomory cuts revisited. Operations Research Letters **19** (1996) 1–9

8. Beame, P., Pitassi, T.: Propositional proof complexity: past, present, and future. Bulletin of the European Association for Theoretical Computer Science **65** (1989) 66–89

9. Bonet, M., Pitassi, T., Raz, R.: Lower bounds for cutting planes proofs with small coefficients. The Journal of Symbolic Logic **62** (1997) 708–728.

10. Boppana, R., Sipser, M.: The complexity of finite functions. In: Handbook of Theoretical Computer Science, Volume A: Algorithms and Complexity. MIT Press/Elsevier (1990) 757–804

11. Chvátal, V.: Edmonds polytopes and a hierarchy of combinatorial problems. Discrete Mathematics **4** (1973) 305–337

12. Chvátal, V.: Hard knapsack problems. Operations Research **28** (1980) 1402–1411

13. Chvátal, V.: Cutting-plane proofs and the stability number of a graph. Report Number 84326-OR. Insitut für Ökonometrie und Operations Research, Universität Bonn, Bonn. (1984)

14. Chvátal, V.: Cutting planes in combinatorics. European Journal of Combinatorics **6** (1985) 217-226

15. Chvátal, V., Cook, W., Hartmann, M.: On cutting plane proofs in combinatorial optimization. Linear Algebra and its Applications **114/115** (1989) 455–499

16. Chvátal, V., Szemerédi, E.: Many hard examples for resolution. Journal of the Association for Computing Machinery **35** (1988) 759–768

17. Cook, S.A., Haken, A.: An exponential lower bound for the size of monotone real circuits. Journal of Computer and System Sciences **58** (1999) 326–335

18. Cook, S.A., Reckhow, R.A.: The relative efficiency of propositional proof systems. Journal of Symbolic Logic **44** (1979) 36–50

19. Cook, W.: Cutting-plane proofs in polynomial space. Mathematical Programming **47** (1990) 11–18

20. Cook, W., Coullard, C.R., Turán, Gy.: On the complexity of cutting-plane proofs. Discrete Applied Mathematics **18** (1987) 25–38

21. Cook, W., Cunningham, W., Pulleyblank, W., Schrijver, A.: Combinatorial Optimization. Wiley, New York (1998)

22. Cook, W., Dash, S.: On the matrix-cut rank of polyhedra. Mathematics of Operations Research **26** (2001) 19–30

23. Cook, W., Hartmann, M.: On the complexity of branch and cut methods for the traveling salesman problem. In: Seymour, P.D., Cook, W. (eds.): Polyhedral Combinatorics, Vol. 1, DIMACS Series in Discrete Mathematics and Theoretical Computer Science (1990) 75–82

24. G. Cornuéjols and Li, Y.: Elementary closures for integer programs. Operations Research Letters **28** (2001) 1–8

25. Dash, S.: On the matrix cuts of Lovász and Schrijver and their use in integer programming. Ph.D. Thesis. Rice University, Houston, Texas (2001). Available as: Technical Report TR01-08. Rice University (2001)

26. Eisenbrand, F., Schulz, A.S.: Bounds on the Chvátal rank of polytopes in the 0/1-cube. In: Cornuéjols, G., Burkard, R.E., Woeginger, G.J. (eds.): Integer Programming and Combinatorial Optimization, 7th International IPCO Conference. Springer, Berlin (1999) 137–150
27. Goemans, M.X., Tunçel, L.: When does the positive semidefiniteness constraint help in lifting procedures. Mathematics of Operations Research **26** (2001) 796–815
28. Gomory, R.E.: Outline of an algorithm for integer solutions to linear programs. Bulletin of the American Mathematical Society **64** (1958) 275–278
29. Gomory, R.E.: An algorithm for the mixed integer problem. RM-2597, The Rand Corporation (1960)
30. Grötschel, M., Pulleyblank, W.R.: Clique tree inequalities and the symmetric traveling salesman problem. Mathematics of Operations Research **11** (1986) 1–33
31. Haken, A.: The intractability of resolution. Theoretical Computer Science **39** (1985) 297–308
32. Horn, R.A., Johnson, C.R.: Matrix Analysis. Cambridge University Press (1985)
33. Jeroslow, R.G.: Trivial integer programs unsolvable by branch-and-bound. Mathematical Programming **6** (1974) 105–109
34. Jeroslow, R.G.: A cutting-plane game for facial disjunctive programs. SIAM Journal on Control and Optimization **18** (1980) 264–281
35. Krajíček, J.: Lower bounds to the size of constant-depth propositional proofs. Journal of Symbolic Logic **59** (1994) 73–86
36. Krajíček, J.: Interpolation theorems, lower bounds for proof systems and independence results for bounded arithmetic. Journal of Symbolic Logic **62** (1997) 457–486
37. Lovász, L.: Stable sets and polynomials. Discrete Mathematics **124** (1994) 137–153
38. Lovász, L., Schrijver, A.: Cones of matrices and set-functions and 0-1 optimization. SIAM Journal on Optimization **1** (1991) 166–190
39. Pudlák, P.: Lower bounds for resolution and cutting plane proofs and monotone computations. Journal of Symbolic Logic **62** (1997) 981–998
40. Pudlák, P.: On the complexity of propositional calculus. In: Sets and Proofs, Invited papers from Logic Colloquium 1997. Cambridge University Press (1999) 197–218
41. Razborov, A.A., Lower bounds for the monotone complexity of some boolean functions. Dokladi Akademii Nauk SSSR **281** (1985) 798–801 (in Russian). English translation in: Soviet Mathematics - Doklady **31** (1985) 354–357
42. Razborov, A.A., Unprovability of lower bounds on the circuit size in certain fragments of bounded arithmetic. Izvestiiya of the RAN **59** (1995) 201–224
43. Schrijver, A.: On cutting planes. In: Deza, M., Rosenberg, I.G. (eds.): Combinatorics 79 Part II, Annals of Discrete Mathematics 9. North Holland, Amsterdam (1980) 291–296
44. Schrijver, A.: Theory of Linear and Integer Programming. Wiley, Chichester (1986)
45. Sherali, H.D., Adams, W.P.: A hierarchy of relaxations between the continuous and convex representations for zero-one programming problems. SIAM Journal on Discrete Mathematics **3** (1990) 411–430

Lifted Inequalities for 0-1 Mixed Integer Programming: Basic Theory and Algorithms*

Jean-Philippe P. Richard[1], Ismael R. de Farias, Jr.[2], and George L. Nemhauser[3]

[1] School of Industrial and Systems Engineering, Georgia Institute of Technology, Atlanta, GA 30332-0205, USA
jpacc@isye.gatech.edu
[2] Center for Operations Research and Econometrics, 34 Voie du Roman Pays, 1348 Louvain-La-Neuve, Belgium
defarias@core.ucl.ac.be
[3] School of Industrial and Systems Engineering, Georgia Institute of Technology, Atlanta, GA 30332-0205, USA
george.nemhauser@isye.gatech.edu

Abstract. We study the mixed 0-1 knapsack polytope, which is defined by a single knapsack constraint that contains 0-1 and bounded continuous variables. We develop a lifting theory for the continuous variables. In particular, we present a pseudo-polynomial algorithm for the sequential lifting of the continuous variables. We introduce the concept of superlinear inequalities and show that our lifting scheme can be significantly simplified for them. Finally, we show that superlinearity results can be generalized to nonsuperlinear inequalities when the coefficients of the continuous variables lifted are large.

1 Introduction

Lifted cover inequalities, derived from 0-1 knapsack inequalities, have proven to be a useful family of cuts for solving 0-1 integer programs by branch-and-cut algorithms. The idea was first proposed by Crowder, Johnson and Padberg [5], with the theoretical foundation coming from the studies of the 0-1 knapsack polytope by Balas [3], Balas and Zemel [2], Hammer, Johnson and Peled [15] and Wolsey [23]. A computational study is presented in Gu, Nemhauser and Savelsbergh [13]. In order to extend these ideas to the mixed integer case, in which continuous variables are present as well, it is necessary to develop a lifting theory for the continuous variables. In this paper we study the polytope generated by a mixed 0-1 knapsack inequality with an arbitrary number of bounded continuous variables and we develop a lifting theory for them.

Although we are not aware of any previous study of the mixed 0-1 knapsack polytope, valid inequalities and facets for related polyhedra have been known for quite some time. For example, there are the mixed integer cuts introduced by Gomory [11], the MIR inequalities introduced by Nemhauser and Wolsey

---

* This research was supported by NSF grants DMI-0100020 and DMI-0121495.

W.J. Cook and A.S. Schulz (Eds.): IPCO 2002, LNCS 2337, pp. 161–175, 2002.

[17], and the mixed disjunctive cuts introduced by Balas, Ceria and Cornuéjols [1]. More closely related to our study is the "0-1 knapsack polyhedron with a single nonnegative continuous variable" introduced by Marchand and Wolsey [16]. However, as we discuss in the next section, our polytope yields more general results with respect to generating cuts.

Given $M = \{1, \ldots, m\}$, $N = \{1, \ldots, n\}$, the sets of integers $\{a_1, \ldots, a_m\}$, $\{b_1, \ldots, b_n\}$, and the integer d, let

$$S = \{(x, y) \in \{0, 1\}^m \times [0, 1]^n \mid \sum_{j \in M} a_j x_j + \sum_{j \in N} b_j y_j \leq d\}.$$

We define the mixed 0-1 knapsack polytope as $PS = conv(S)$. Note that the choice of 1 as an upper bound for the (bounded) continuous variables is without loss of generality, since they can be rescaled. We assume throughout the paper that

Assumption 1. $M \neq \emptyset$ and $N \neq \emptyset$.

Assumption 2. $0 < a_j \leq d$ $\forall j \in M$, and $0 < b_j \leq d$ $\forall j \in N$.

If Assumption 1 is not satisfied, PS is either trivial ($M = \emptyset$) or has already been studied ($N = \emptyset$). Clearly, there is no loss of generality in Assumption 2.

We now give a few results about PS and its inequality description. Whenever it is clear from the context, we use the term facet to denote the corresponding inequality as well.

Proposition 1. *PS is full-dimensional.*   □

Some inequalities of the linear description of PS are easy to explain. We give their characterization in the following proposition.

Proposition 2. *The inequality*

$$x_i \geq 0 \tag{1}$$

is a facet of PS $\forall i \in M$. *The inequality*

$$x_i \leq 1 \tag{2}$$

is a facet of PS for $i \in M$ *iff* $\max\{a_j \mid j \in M - \{i\}\} + a_i \leq d$ *and* $a_i < d$. *The inequality*

$$y_i \geq 0 \tag{3}$$

is a facet of PS $\forall i \in N$. *The inequality*

$$y_i \leq 1 \tag{4}$$

is a facet of PS for $i \in N$ *iff* $\max\{a_j \mid j \in M\} + b_i \leq d$.   □

We call the inequalities of Proposition 2 trivial. Another trivial inequality that may or may not be a facet of PS is the initial knapsack inequality. It does not seem that there exists a simple characterization of when this inequality is a facet except for particular cases, as for example, when the coefficients of the continuous variables are big (see Marchand and Wolsey [16] and Sect. 2). Sufficient conditions can also be derived from Theorem 4 presented later. In the remainder of the paper, we characterize some families of nontrivial facets of PS that can be obtained by sequential lifting. The nontrivial facets of PS satisfy the following proposition.

Proposition 3. *Assume that $\sum_{j \in M} \alpha_j x_j + \sum_{j \in N} \beta_j y_j \le \delta$ is a facet of PS that is not a multiple of (1) or (3). Then $\alpha_j \ge 0$ for $j \in M$, $\beta_j \ge 0$ for $j \in N$, and $\delta > 0$.*   □

In Sect. 2 we discuss why the polytope we study is more general than the polyhedron of Marchand and Wolsey. In Sect. 3 we develop sequential lifting schemes for continuous variables. For the sequential lifting of continuous variables fixed at 0 ("lifting from 0"), we show that the lifting coefficients are 0 almost always (Theorem 2). Finally, we consider the sequential lifting of continuous variables fixed at 1 ("lifting from 1"), which is much more interesting and difficult, and we give a pseudo-polynomial algorithm for it (Theorem 5). In Sect. 4 we present a way to alleviate some of the computational burden associated with our lifting scheme by imposing structure on the inequality to be lifted and we introduce the concept of superlinear lifting (Theorem 7). In Sect. 5 we present another way to reduce the amount of computation in our algorithm by imposing restrictions on the coefficients of the continuous variables to be lifted (Theorem 11). We conclude in Sect. 6 with some discussion on how our lifting theory can be used algorithmically.

2   A Related Polyhedron

Let

$$T = \{(x, s) \in \{0, 1\}^m \times [0, \infty) \mid \sum_{j \in M} a_j x_j \le d + s\} .$$

Marchand and Wolsey [16] studied the 0-1 knapsack polyhedron with a single nonnegative continuous variable $PT = conv(T)$. The polytope PS and the polyhedron PT are similar. However, as we show in this section, because PS is bounded and has more than one continuous variable, it is possible to derive from it certain cuts for 0-1 mixed integer programming that cannot be derived from PT. We first show that PS with $|N| = 1$ is more general than PT because it is bounded. Then we show that PS with more than one continuous variable is more general than the polytope obtained by aggregating them into a single continuous variable.

When $|N| = 1$, there exists a natural transformation that converts a polytope of the form PS into a polyhedron of the form PT and vice-versa. The simple addition of the constraint $s \leq \mu$ transforms PT into PS after adequately scaling and complementing the continuous variable. We call the resulting polytope $PS' = conv\{(x, y) \in \{0, 1\}^m \times [0, 1] \mid \sum_{j \in M} a_j x_j + \mu y \leq d + \mu\}$, where $y = \frac{\mu - s}{\mu}$. When μ is large enough, the facets of PS' are in one-to-one correspondence with the facets of PT (if we remove the facet $y \geq 0$ from the linear description of PS').

Now consider a transformation from PS with a single continuous variable to PT. If we scale and complement the variable y, defining $s = b - by$, we obtain an equivalent polytope where $0 \leq s \leq b$. Suppose we relax the upper bound on s (i.e. we relax the lower bound on y). Then we obtain a polytope of the form PT that we call $PT' = conv\{(x, s) \in \{0, 1\}^m \times [0, \infty) \mid \sum_{j \in M} a_j x_j \leq d - b + s\}$. In general, there is no bijection between the facets of PS and those of PT', although valid inequalities for PT' can be turned into valid inequalities of PS by substituting $s = b - by$. However, even when starting from a facet of PT', this procedure is not guaranteed to generate a facet of PS as the following example illustrates.

Example 1. Consider the polytope PS defined by

$$12x_1 + 8x_2 + 8x_3 + 7x_4 + 5y \leq 24 .$$

Let $s = 5 - 5y$. Introduce s in the previous inequality and relax its upper bound. The inequality becomes

$$12x_1 + 8x_2 + 8x_3 + 7x_4 \leq 19 + s,$$

which is in the format studied by Marchand and Wolsey. The inequality

$$8x_1 + 4x_2 + 8x_3 + 7x_4 \leq 15 + s$$

is a facet of the polytope PT'. Substituting back $s = 5 - 5y$, we obtain

$$8x_1 + 4x_2 + 8x_3 + 7x_4 + 5y \leq 20 .$$

This inequality is not a facet of PS since it defines a face whose dimension is only 3. □

Now we show that we need to consider several continuous variables to describe the most general results about facets. Consider the polytopes $PS_1 = conv\{(x, y) \in \{0, 1\}^m \times [0, 1] \mid \sum_{j \in M} a_j x_j + by \leq d\}$ and $PS_n = conv\{(x, y) \in \{0, 1\}^m \times [0, 1]^n \mid \sum_{j \in M} a_j x_j + \sum_{j \in N} b_j y_j \leq d\}$. Note that any polytope of the form PS_n can be turned into a polytope of the fom PS_1 by defining $y = \frac{\sum_{j \in N} b_j y_j}{b}$ and $b = \sum_{j \in N} b_j$. If facets of PS_1 are known, the substitution $y = \sum_{j \in N} b_j y_j$ will turn them into valid inequalities for PS_n. There are two main issues regarding this aggregation procedure. The first is that the inequality generated is not necessarily a facet of PS_n. The second is that there are many facets of PS_n that cannot be obtained in this way. The following example illustrates this last observation.

Example 2. Consider the polytope PS defined by

$$30x_1 + 25x_2 + 23x_3 + 20x_4 + 18x_5 + 17x_6 + 13x_7 + 12x_8 + 16y_1 + 7y_2 \leq 103 \ .$$

The linear description of this polytope was obtained using Porta [4] and contains 3114 inequalities. The following five

$$
\begin{array}{llllllllllll}
2x_1 + & 2x_2 + & 2x_3 + & 2x_4 + & x_5 + & x_6 + & x_7 + & x_8 & & & \leq & 8, & (5) \\
4x_1 + & 4x_2 + & 4x_3 + & 4x_4 + & 4x_5 + & 4x_6 + & 4x_7 & & +16y_1 & & \leq & 16, & (6) \\
6x_1 + & 4x_2 + & 4x_3 + & 4x_4 + & 2x_5 + & 2x_6 + & 2x_7 + & 2x_8 & & +7y_2 & \leq & 23, & (7) \\
16x_1 + & 16x_2 + & 12x_3 + & 12x_4 + & 12x_5 + & 8x_6 + & 8x_7 + & 8x_8 & +16y_1 & +7y_2 & \leq & 71, & (8) \\
20x_1 + & 20x_2 + & 15x_3 + & 15x_4 + & 15x_5 + & 10x_6 + & 10x_7 + & 10x_8 & +48y_1 & +7y_2 & \leq & 115, & (9)
\end{array}
$$

are some of them. $\qquad\square$

Only inequalities (5) and (8) can be derived by the above procedure. But inequalities (6) and (7) of Example 2 suggest a more general aggregation technique to obtain facets of PS_n. Here, we choose $J \subseteq N$ and define $y = \frac{\sum_{j \in J} b_j y_j}{b}$ where $b = \sum_{j \in J} b_j$. PS_n can be converted into PS_1 by dropping all the variables y_j for $j \in N \backslash J$ and replacing $\sum_{j \in J} b_j y_j$ with by. If facets of PS_1 are known, the substitution $y = \frac{\sum_{j \in J} b_j y_j}{b}$ will turn them into valid inequalities of PS_n. The disadvantages of this aggregation procedure are identical to the ones described for the initial one. The inequalities generated are not necessarily facets and there are some facets of PS_n that cannot be explained in this way. Inequality (9) of Example 2 illustrates this last disadvantage. We note, however, that aggregation can be helpful practically since many facets of PS_n can be obtained in this way. Therefore we will refer to inequalities that can be scaled in such a way that coefficients of the continuous variables are either 0 or the given coefficient of the knapsack inequality as having the *ratio property*.

3   Lifting Theory for Continuous Variables

Describing facets of high dimensional polytopes can be a difficult task. One alternative is to reduce the number of variables by fixing some of them at their upper or lower bounds to obtain a polytope for which at least one nontrivial inequality is known. Once such an inequality is available, it is converted progressively into facets of higher dimensional polytopes by the mechanism known as lifting. Lifting was introduced in the context of the group problem by Gomory [12]. Its computational possibilites were emphasized by Padberg [18] and the approach was generalized by Wolsey [24], Zemel [25] and Balas and Zemel [2]. Although lifting techniques have been studied extensively for 0-1 variables, see Gu [14], there has been limited study of lifting for continuous variables (de Farias [10], de Farias, Johnson and Nemhauser [6,8,7] and de Farias and Nemhauser [9] are exceptions). The purpose of this section is to develop algorithms for it. In principle we should investigate the lifting of continuous variables that are fixed at any value within the interval $[0, 1]$. However, we have shown (see Richard, de Farias and Nemhauser [20]) that when S is defined by only one 0-1 mixed knapsack inequality, the inequalities obtained by fixing continuous variables at fractional values are only rarely different from the ones that can be generated from the fixing of continuous variables at 0 and 1. Thus we focus only on the lifting of continuous variables fixed to 0 and 1. For the lifting of continuous variables from 0, we show that the lifting coefficients are almost always equal to 0. We also describe when they are not equal to 0 and how to obtain the lifting coefficients in this case. Finally, we study the lifting of continuous variables from 1 and we show that it is much richer and more difficult than the lifting from 0.

3.1   General Lifting Results

Given M_0, M_1, two disjoint subsets of M, and N_0, N_1 two disjoint subsets of N, we define

$$S(M_0, M_1, N_0, N_1) = S \cap \{ (x, y) \in \{0, 1\}^m \times [0, 1]^n \, | \, x_j = 0 \, \forall j \in M_0,$$
$$x_j = 1 \, \forall j \in M_1, \, y_j = 0 \, \forall j \in N_0, \, y_j = 1 \, \forall j \in N_1\},$$

and $PS(M_0, M_1, N_0, N_1) = conv(S(M_0, M_1, N_0, N_1))$. Whenever it is clear from the context, we abbreviate $S(M_0, M_1, N_0, N_1)$ as S^*, $PS(M_0, M_1, N_0, N_1)$ as PS^*, $M - (M_0 \cup M_1)$ as M^*, $N - (N_0 \cup N_1)$ as N^*, and $d - \sum_{j \in M_1} a_j - \sum_{j \in N_1} b_j = d^*$. We also let $m^* = |M^*|$ and $n^* = |N^*|$. Note that PS^* is defined by the constraint

$$\sum_{j \in M^*} a_j x_j + \sum_{j \in N^*} b_j y_j \leq d^*. \tag{10}$$

We represent nontrivial valid inequalities of PS^* by

$$\sum_{j \in M^*} \alpha_j x_j + \sum_{j \in N^*} \beta_j y_j \leq \delta^*. \tag{11}$$

We assume throughout the paper that:

Assumption 3. $i \in M_0$ whenever $a_i > d^*$.

Note that, as with Assumption 2, there is no loss of generality in Assumption 3 and it implies that PS^* is full-dimensional. As a result, it is simpler to lift (11).

In the remainder of this paper, we assume that $r_1, \dots, r_s$ are distinct elements of N_0 and N_1 and that we wish to sequentially lift the corresponding variables $y_{r_1}, \dots, y_{r_s}$ in (11). We denote the associated lifting coefficients by $\beta_{r_1}, \dots, \beta_{r_s}$. Lemma 1 establishes the lifting procedure, see [10,24] for a proof. In the lemma, i indicates the value we lift from, i.e. 0 or 1. Given a polytope Q we let $V(Q)$ be the set of its extreme points.

Lemma 1. *Let (11) be a valid (resp. facet-defining) inequality of PS^*. Let $i \in \{0,1\}$ and let $r_1, \ldots, r_s \in N_i$. Define $\tilde{N}_i^q = N_i - \{r_1, \ldots, r_q\}$ and $\tilde{N}_{1-i}^q = N_{1-i}$ for $q = 1, \ldots, s$. Then,*

$$\sum_{j \in M^*} \alpha_j x_j + \sum_{j \in N^*} \beta_j y_j + \sum_{t=1}^{s} \beta_{r_t} y_{r_t} \leq \delta^* + \sum_{t=1}^{s} i\beta_{r_t} \tag{12}$$

is a valid (resp. facet-defining) inequality for $PS(M_0, M_1, \tilde{N}_0^s, \tilde{N}_1^s)$, where β_{r_q} is the optimal value of $L_i(q)$:

$$\beta_{r_q} = (-1)^i \min \, (-1)^i \frac{\sum_{j \in M^*} \alpha_j x_j + \sum_{j \in N^*} \beta_j y_j + \sum_{t=1}^{q-1} \beta_{r_t}(y_{r_t} - i) - \delta^*}{i - y_{r_q}}$$

$$s.t. \ (x,y) \in V(PS(M_0, M_1, \tilde{N}_0^q, \tilde{N}_1^q)) \ and \ y_{r_q} \in \left[\frac{1-i}{b_{r_q}}, 1 - \frac{i}{b_{r_q}} \right]$$

for $q = 1, \ldots, s$. □

Regardless of whether the lifting coefficients of the continuous variables satisfy the ratio property, they satisfy the following relations.

Theorem 1. *Let $r \in N_0$ and $s \in N_1$ be distinct.*

1. *If y_{r_1} and y_{r_2} are lifted from 0 in (11) then $\frac{\beta_{r_1}}{b_{r_1}} \geq \frac{\beta_{r_2}}{b_{r_2}}$.*
2. *If y_{r_1} and y_{r_2} are lifted from 1 in (11) then $\frac{\beta_{r_1}}{b_{r_1}} \leq \frac{\beta_{r_2}}{b_{r_2}}$.*
3. *If y_r is lifted from 0 and y_s is lifted from 1 in (11) then $\frac{\beta_r}{b_r} \leq \frac{\beta_s}{b_s}$ (note that the lifting order is irrelevant in this case).* □

3.2 Lifting Continuous Variables from 0 (L_0)

We show that the lifting coefficients of the continuous variables fixed at 0 are almost always zero. We also develop a pseudo-polynomial algorithm when this is not the case.

Theorem 2. *Assume that (11) is a facet of PS^* that is not a multiple of (10). Then, when lifting continuous variables from 0 in (11), we have that $\beta_{r_1} = \cdots = \beta_{r_s} = 0$.* □

We now consider the lifting of continuous variables from 0 in an inequality that is a multiple (10). First we consider the case with $N^* \neq \emptyset$.

Theorem 3. *Let (10) be a facet of PS^* with $N^* \neq \emptyset$. When lifting continuous variables from 0 in (10) we have that $\beta_{r_q} = b_{r_q}$ for $q = 1, \ldots, s$.* □

Now we consider the case where $N^* = \emptyset$. Given a polytope PS^* with $N^* = \emptyset$, let

$$\sigma = d^* - \max \sum_{j \in M^*} a_j x_j$$

$$s.t. \sum_{j \in M^*} a_j x_j \leq d^* - 1, \tag{13}$$

$$x_j \in \{0,1\} \, \forall j \in M^* \ .$$

Theorem 4. *Let (10) be a facet of PS^* with $N^* = \emptyset$. When lifting continuous variables from 0 in (10), we have that*

1. *If $b_{r_1} \geq \sigma$ then $\beta_{r_q} = b_{r_q}$ for $q = 1, \ldots, s$*
2. *If $b_{r_1} < \sigma$ then $\beta_{r_1} = \sigma$ and $\beta_{r_q} = 0$ for $q = 2, \ldots, s$.*   □

Theorem 4 leads to a simple pseudo-polynomial algorithm to perform the lifting from 0 of continuous variables when (11) is a multiple of (10): we compute σ by dynamic programming, and then we use Theorem 4 to deduce the lifting coefficients.

3.3   Lifting Continuous Variables from 1 (L_1)

In Lemma 1, we described a formal way to lift continuous variables from 1. In order to turn it into a practical scheme, we define the function

$$\Lambda(w) = \min \sum_{j \in M^*} a_j x_j + \sum_{j \in N^*} b_j y_j - d^*$$

$$s.t. \sum_{j \in M^*} \alpha_j x_j + \sum_{j \in N^*} \beta_j y_j = \delta^* + w$$

$$x_j \in \{0, 1\} \, \forall j \in M, y_j \in [0, 1] \, \forall j \in N.$$

The domain of the function Λ is

$$\mathbb{W} = \{w \in \mathbb{R} \mid \exists (x, y) \in \{0, 1\}^{m^*} \times [0, 1]^{n^*} \text{ s.t. } \sum_{j \in M^*} \alpha_j x_j + \sum_{j \in N^*} \beta_j y_j = \delta^* + w\}.$$

We say that (11) is satisfied at equality at least once, abbreviated SEO, if there is at least one $(x^*, y^*) \in S^*$ such that $\sum_{j \in M^*} \alpha_j x_j^* + \sum_{j \in N^*} \beta_j y_j^* = \delta^*$. For $A \subseteq \mathbb{R}$, we define $A_+ = \{x \in A \mid x > 0\}$.

We focus on the case $N^* = \emptyset$, i.e. we will consider the lifting of continuous variables from 1 with respect to

$$\sum_{j \in M^*} \alpha_j x_j \leq \delta^* \tag{14}$$

(which is purely 0-1 and SEO). The reason for restricting ourselves to this case is that Λ is a discrete function and it is simpler to obtain the lifting coefficients. The case $N^* \neq \emptyset$ can be treated similarly, see [20].

Let $\mathbb{W}^q = \{w \in \mathbb{W} \mid \Lambda(w) \leq \sum_{j=1}^q b_{r_j}\}$, $\mathbb{S}^q = \mathbb{W}^q \backslash \mathbb{W}^{q-1}$, and $\mathbb{T}^q = \{w \in \mathbb{S}^q \mid \frac{\beta_{r_{q-1}}}{b_{r_{q-1}}} < \frac{w - \sum_{i=1}^{q-1} b_{r_i}}{\Lambda(w) - \sum_{i=1}^{q-1} b_{r_i}}\}$ for $q \in \{1, \ldots, s\}$.

Theorem 5. *When lifting continuous variables from 1 in (14), we have that*

1. *If $\mathbb{T}^q = \emptyset$ then $\frac{\beta_{r_q}}{b_{r_q}} = \frac{\beta_{r_{q-1}}}{b_{r_{q-1}}}$*

2. *If $\mathbb{T}^q \neq \emptyset$ then $\frac{\beta_{r_q}}{b_{r_q}} = \max\{\frac{w - \sum_{j=1}^{q-1} b_{r_j}}{\Lambda(w) - \sum_{j=1}^{q-1} b_{r_j}} \mid w \in \mathbb{T}^q\}$*

for $q \in \{1, \ldots, s\}$, where we define $\frac{\beta_{r_0}}{b_{r_0}} = 0$.   □

Table 1 gives an algorithm that computes the lifting coefficients of continuous variables fixed at 1. The algorithm requires $\mathbb{W}_+$, the function Λ and the initial coefficients b_i of the continuous variables to be lifted. It outputs θ_i and the desired lifting coefficients β_i are computed as $\beta_i = \theta_i b_i$ for $i = 1, \ldots, m$.

Table 1. Algorithm for the lifting of continuous variables from 1

Lift$(\mathbb{W}, \Lambda, b)$

$\theta_0 = 0,\ \beta^{old} = 0,\ b^{old} = 0$

For $i = 1 : m$

$\qquad b^{new} = b^{old} + b_i$
$\qquad \theta_i = \theta_{i-1}$

$\qquad$ **For Each** $w \in \mathbb{W}$ s.t. $b^{old} < \Lambda(w) \le b^{new}$
$\qquad\qquad \theta_i = \max\{\theta_i, \frac{w - \beta^{old}}{\Lambda(w) - b^{old}}\}$
$\qquad$ **End For**

$\qquad b^{old} = b^{new}$
$\qquad \beta^{old} = \beta^{old} + \theta_i b_i$

End For

Let w_1 and w_2 be distinct points of $\mathbb{W}_+$. We say that w_1 is dominated by w_2 if $w_2 > w_1$ and $\Lambda(w_1) \ge \Lambda(w_2)$. It can be shown that it suffices in the algorithm of Table 1 to consider the nondominated points in the domain of the function Λ. This result is important because it allows us to compute the sets $\mathbb{T}^q$ of Theorem 5 without performing any sorting of the ordinates of Λ. From this observation, we obtain the complexity of the algorithm.

Theorem 6. *The algorithm presented in Table 1 runs in time $O(m + |\mathbb{W}_+|)$.* □

This algorithm is pseudo-polynomial when the inequalities we lift have no particular structure. However, for structured inequalities like covers, it is polynomial since $|\mathbb{W}_+|$ is polynomially bounded in m.

Example 2 (continued). Consider the polytope PS presented in Example 2. Let $N_1 = \{1, 2\}$. The inequality

$$4x_1 + 4x_2 + 3x_3 + 3x_4 + 3x_5 + 2x_6 + 2x_7 + 2x_8 \le 12 \qquad (15)$$

is valid and defines a facet of $PS(\emptyset, \emptyset, \emptyset, N_1)$. In fact,

$$4x_2 + 3x_3 + 3x_4 + 3x_5 + 2x_6 + 2x_7 + 2x_8 \le 8$$

is a Chvátal-Gomory inequality with weight $\frac{1}{6}$ in $PS(\emptyset, \{1\}, \emptyset, N_1)$ and can be extended into a valid inequality of $PS(\emptyset, \emptyset, \emptyset, N_1)$ by lifting x_1, leading to (15). It can be shown that (15) is a facet of the associated polytope. So assume that we want to lift from 1 the continuous variables y_2 and y_1 in this order (i.e. $r_1 = 2$ and $r_2 = 1$) using Theorem 5. We first need to compute the function Λ associated with (15). Its values are represented in Table 2.

Table 2. Function Λ associated with (15)

w	1	2	3	4	5	6	7	8	9	10	11
$\Lambda(w)$	5	8	18	25	31	38	48	55	61	∞	78

We compute the first lifting coefficient using Theorem 5. We have that $\mathbb{T}^1 = \{1\}$. Therefore $\frac{\beta_{r_1}}{b_{r_1}} = \max\{\frac{w}{\Lambda(w)} \mid w \in \mathbb{T}^1\} = \frac{1}{5}$. Next, we compute $\mathbb{S}^2 = \{2,3\}$ and $\mathbb{T}^2 = \{2\}$. It follows that

$$\frac{\beta_{r_2}}{b_{r_2}} = \max\{\frac{w - \frac{7}{5}}{\Lambda(w) - 7} \mid w \in \mathbb{T}^2\} = \frac{3}{5}.$$

This shows that the inequality

$$4x_1 + 4x_2 + 3x_3 + 3x_4 + 3x_5 + 2x_6 + 2x_7 + 2x_8 + \frac{48}{5}y_1 + \frac{7}{5}y_2 \le 12 + \frac{48}{5} + \frac{7}{5}$$

is a facet of PS. In fact, this inequality is (9) from Example 2.   □

4   Superlinear Lifting

Our algorithm is not polynomial because $|\mathbb{W}_+|$ can be large. There are three directions we can consider to diminish the impact of this pseudo-polynomiality. The first is to consider families of inequalities for which $|\mathbb{W}_+|$ is polynomially bounded (like covers). The second is to determine conditions that would make the function Λ yield an easier lifting process (and a simpler algorithm than the one presented before). The third is to impose conditions on the coefficients of the variables to be lifted so that the lifting algorithm can be simplified. Each of these three approaches is fruitful in its own way. We will describe the second one in this section, leaving the third one for Sect. 5, see [21] for details.

Definition 1. *The function $\Lambda(w)$ associated with (14) in PS^* and the inequality (14) are said to be superlinear if for $w \ge w^*$, $w^*\Lambda(w) \ge w\Lambda(w^*)$ with $w^* = \max \operatorname{argmin}\{\Lambda(w) \mid w \in \mathbb{W}_+\}$. We call w^* the superlinearity point.*

Example 2 (continued). Inequality (15) is not superlinear. In fact, we can see from Table 2 that $w^* = 1$ and $\Lambda(1) = 5$. In order for the function Λ to be

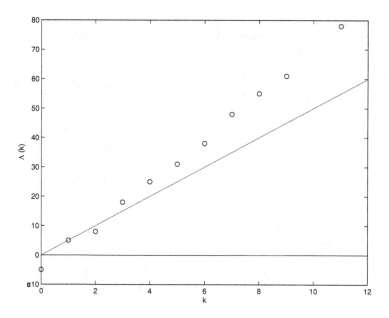

Fig. 1. Function Λ associated with (15)

superlinear, we should have $\Lambda(w) \geq w\Lambda(1)$ which is not true when $w = 2$. In Fig. 1, we see that Λ is not superlinear because the point $(2, \Lambda(2))$ lies beneath the line containing the points $(0, 0)$ and $(w^*, \Lambda(w^*))$. □

Superlinear inequalities play a pivotal role in our study since when continuous variables are lifted from 1, the inequality generated always satisfies the ratio property. Moreover, once their superlinearity point is known, the lifting from 1 of any continuous variable becomes trivial. Let $p = \max\{i \in \{0, \dots, s\} \mid \sum_{j=1}^{i} b_{r_j} < \Lambda(w^*)\}$.

Theorem 7. *If (14) is superlinear, then in the lifting of continuous variables from 1 we obtain* $\frac{\beta_{r_i}}{b_{r_i}} = 0$ *for* $i = 1, \dots, p$ *and* $\frac{\beta_{r_i}}{b_{r_i}} = \frac{w^*}{\Lambda(w^*) - \sum_{j=1}^{p} b_{r_j}}$ *for* $i = p+1, \dots, s$. □

Note that Theorem 7 can be used to generate facets of PS. First, we fix all the continuous variables to 0 or 1, then we generate a facet of the resulting pure 0-1 knapsack polytope that we finally lift with respect to the continuous variables. This scheme is restricted by Assumption 3 so that, after we perform the lifting of continuous variables, we are left with a facet of $PS(M_0, \emptyset, \emptyset, \emptyset)$ where $M_0 = \{i \in M \mid a_i \geq d^*\}$. Surprisingly, there exists a closed form expression for lifting the members of M_0 that is presented in the next theorem. For $a \in \mathbb{R}$, let $(a)^+ = \max\{a, 0\}$.

Theorem 8. *Let (N_0, N_1) be a partition of N, $N_1 = \{1, \ldots, n_1\}$ and $M_0 = \{i \in M \mid a_i \geq d - \sum_{j \in N_1} a_j\}$. Assume that (14) is superlinear and is not a multiple of (2) and that $p < n_1$. Then*

$$\sum_{j \in M^*} \alpha_j x_j + \sum_{j=p+1}^{n_1} \bar{\theta} b_{r_j} y_{r_j} + \sum_{j \in M_0} (\delta^* + \bar{\theta}(a - d^* - b)^+) x_j \leq \delta^* + \sum_{j=p+1}^{n_1} \bar{\theta} b_{r_j} \quad (16)$$

is a facet of $PS(\emptyset, \emptyset, N_0, \emptyset)$ where $b = \sum_{j=1}^{p} b_{r_j}$ and $\bar{\theta} = \frac{w^*}{\Lambda(w^*)-b}$. $\qquad\square$

The practical relevance of Theorem 8 depends on our ability to find families of pure 0-1 facets that are superlinear. We will show next that covers are superlinear. Let $P = conv\{x \in \{0,1\}^m \mid \sum_{j \in M} a_j x_j \leq d\}$. Let (C_1, U, C_2) be a partition of M and C_1 be a minimal cover in $P(U, C)$ where $x_i = 0$ for $i \in U$ and $x_i = 1$ for $i \in C_2$. We say that the inequality

$$\sum_{j \in C_1} x_j + \sum_{j \in U} \alpha_j x_j + \sum_{j \in C_2} \alpha_j x_j \leq |C_1| - 1 + \sum_{j \in C_2} \alpha_j \quad (17)$$

is a partitioned cover inequality based on (C_1, U, C_2) if it is obtained from the inequality $\sum_{j \in C_1} x_j \leq |C_1| - 1$ by sequentially lifting the members of U from 0 and then the members of C_2 from 1.

Theorem 9. *Let (17) be a facet of P that is a partitioned cover based on (C_1, U, C_2). Then $\Lambda(w) \geq w\Lambda(1)$ for $w \in \mathbb{W}_+$.* $\qquad\square$

We already know, from the algorithm presented in Table 1, how to lift 0-1 covers in polynomial time. Together with Theorem 8, Theorem 9 provides a faster technique to construct facets of PS based on 0-1 covers. We illustrate this technique in the next example.

Example 2 (continued). Let $M_0 = \{3, 4\}$, $M_1 = \{1, 2\}$, $N_0 = \{1\}$ and $N_1 = \{2\}$. The cover inequality $x_5 + x_6 + x_7 + x_8 \leq 3$ is a facet of PS^*. It can be turned into the partitioned cover facet of $PS(\emptyset, \emptyset, N_0, N_1)$

$$3x_1 + 2x_2 + 2x_3 + 2x_4 + x_5 + x_6 + x_7 + x_8 \leq 8$$

by lifting the variables x_3, x_4, x_1 and x_2 in this order. By Theorem 9, this inequality is superlinear and we have also that $\Lambda(1) = 2$. Therefore,

$$3x_1 + 2x_2 + 2x_3 + 2x_4 + x_5 + x_6 + x_7 + x_8 + \frac{7}{2}y_2 \leq 8 + \frac{7}{2}$$

is a facet of PS, which is inequality (7). $\qquad\square$

We have shown that the lifting of continuous variables from 1 in a superlinear inequality always leads to an inequality that satisfies the ratio property. Moreover, as stated in the next theorem, these are the only inequalities for which this can be guaranteed.

Theorem 10. *Assume (14) is not superlinear. Let $\bar{\Lambda} = \min\{\Lambda(w) \mid w^*\Lambda(w) < w\Lambda(w^*)$ and $w \in \mathbb{W}_+\}$ and $\bar{w} = \max argmin\{\Lambda(w) \mid w^*\Lambda(w) < w\Lambda(w^*)$ and $w \in \mathbb{W}_+\}$. If $b_{r_1} = \Lambda(w^*)$ and $b_{r_2} = \bar{\Lambda} - \Lambda(w^*)$ then $0 < \frac{\beta_{r_1}}{b_{r_1}} < \frac{\beta_{r_2}}{b_{r_2}}$.* $\qquad\square$

5 Pseudo-superlinear Lifting

Although lifted nonsuperlinear inequalities do not necessarly have the ratio property, we would like to understand when they do. Refer to Table 2 and the lifting of (15). The only reason we obtain two different lifting ratios is because b_{r_1} is too small which forces the point $(2, \Lambda(2))$ out of $\mathbb{S}^1$. This suggests that the ratio property can possibly be salvaged if the first continuous variable lifted has a big coefficient.

Definition 2. *The pseudo-superlinearity ratio of the function $\Lambda(w)$ associated with (14) in PS^* and the inequality (14) is $\max\{\frac{w}{\Lambda(w)} \mid w \in \mathbb{W}_+\}$. Their pseudo-superlinearity point $\tilde{w}$ is $\min\{w \in \mathbb{W}_+ \mid \frac{w}{\Lambda(w)} = \tilde{\theta}\}$.*

The pseudo-superlinearity point can be seen to be a substitute for the superlinearity point when the inequality is not superlinear. This statement is supported by the following characterization of superlinear inequalities.

Proposition 4. *$\Lambda(w)$ is superlinear if and only if $w^* = \tilde{w}$.* $\qquad\square$

Therefore, we can generalize Theorem 8 using $\tilde{w}$ as a substitue for w^*.

Theorem 11. *Let (N_0, N_1) be a partition of N and $M_0 = \{i \in M \mid a_i \geq d - \sum_{j \in N_1} a_j\}$. Assume that (14) is not a multiple of (2), $\sum_{j=1}^{p+1} b_{r_j} \geq \Lambda(\tilde{w})$ and $p < n_1$. Then (16) is a facet of $PS(\emptyset, \emptyset, N_0, \emptyset)$ where $b = \sum_{j=1}^{p} b_{r_j}$, $\tilde{\theta} = \max\{\frac{w}{\Lambda(w)-b} \mid w \in \mathbb{W}_+ \cap [w^*, \tilde{w}]\}$ and $\bar{w} = \min\{w \in \mathbb{W}_+ \cap [w^*, \tilde{w}] \mid \frac{w}{\Lambda(w)-b}\}$.* $\qquad\square$

Note that Theorem 11 is a result about asymptotic lifting. In fact, if we assume that $b_{r_1} \geq \Lambda(\tilde{w})$, we can conclude that $\frac{\beta_{r_i}}{b_{r_i}} = \tilde{\theta}$ for $i = 1, \ldots, s$. On the other hand, if the inequality we start from is superlinear, we have $w^* = \bar{w}$ and Theorem 11 reduces to Theorem 8 since the condition $\sum_{j=1}^{p+1} b_{r_j} \geq \Lambda(\tilde{w})$ becomes void and $\tilde{\theta} = \frac{w^*}{\Lambda(w^*)-b}$. Next, we present an example of the asymptotic version of Theorem 11.

Example 2 (continued). Consider inequality (15), which we already observed is not superlinear. We can easily establish that its pseudo-superlinearity ratio is $\tilde{\theta} = \frac{1}{4}$ and that $\Lambda(\tilde{w}) = 8$. Now suppose we want to sequentially lift from 1 the continuous variables y_1 and y_2 in this order (i.e. $r_1 = 1$ and $r_2 = 2$). Since $b_{r_1} \geq \Lambda(\tilde{w})$, we can apply Theorem 11 to obtain that inequality (8) from Example 2,

$$4x_1 + 4x_2 + 3x_3 + 3x_4 + 3x_5 + 2x_6 + 2x_7 + 2x_8 + \frac{1}{4}(16y_1 + 7y_2) \leq 12 + \frac{23}{4}$$

is a facet of PS. $\qquad\square$

Note that sometimes it is easy to find the pseudo-superlinearity ratio. As an example, the weight inequalities introduced by Weismantel [22] have $\tilde{\theta} = 1$. The lifting of continuous variables is therefore easy for them provided that the coefficients of the variables to be lifted are large. Large coefficients arise, for example, in the transformation of PT into PS presented in Sect. 2. This simple observation gives rise to the following refinement of a theorem by Marchand and Wolsey [16]. Let M_0, M_1 be nonintersecting subsets of M and define $PT(M_0, M_1) = \{x \in \{0,1\}^m \mid \sum_{j \in M} a_j x_j \leq d, x_j = 0 \,\forall j \in M_0, x_j = 1 \,\forall j \in M_1\}$.

Theorem 12. *Let $M_0 = \{i \in M \mid a_i \geq d\}$. Assume (14) is a facet of $PT(M_0, \emptyset)$ which is not a multiple of (2). Then the inequality*

$$\sum_{j \in M^*} \frac{\alpha_j}{\tilde{\theta}} x_j + \sum_{j \in M_0} (\frac{\delta^*}{\tilde{\theta}} + (a_j - d^*)^+) x_j \leq \frac{\delta^*}{\tilde{\theta}} + s$$

is a facet of PT.   □

Note that $\tilde{\theta} = 1$ for the weight inequalities and $\tilde{\theta} = \frac{1}{A(1)}$ for partitioned covers.

6   Concluding Remarks and Further Research

We investigated the lifting of continuous variables for the mixed 0-1 knapsack polytope. Although our algorithm is in general only pseudo-polynomial, we showed that it is polynomial for a particular class of well known and extensively used facets of the 0-1 knapsack polytope and we showed how to decrease its computational burden by introducing the concept of superlinearity. We also have shown how the lifting is made easier when the continuous variables have large coefficients. Currently, we are evaluating the practical impact of these results in the context of cutting plane algorithms. Our mixed integer cuts are not only strong but robust in the sense that a violated mixed integer cut of the type given in this paper can always be found when a basic solution to the linear programming relaxation violates integrality. This observation leads to a simplex-based algorithm for which finite convergence can be proved [19]. We are currently testing the significance of using these inequalities in a branch-and-cut algorithm designed to solve general 0-1 mixed integer programs.

References

1. E. Balas, S. Ceria, and G. Cornuéjols. A lift-and-project cutting plane algorithm for mixed 0-1 programs. *Mathematical Programming*, 58:295–324, 1993.
2. E. Balas and E. Zemel. Facets of the knapsack polytope from minimal covers. *SIAM Journal on Applied Mathematics*, 34:119–148, 1978.
3. E. Balas. Facets of the knapsack polytope. *Mathematical Programming*, 8:146–164, 1975.
4. T. Christof and A. Löbel. PORTA : A POlyhedron Representation Transformation Algorithm. http://www.zib.de/Optimization/Software/Porta/, 1997.
5. H.P. Crowder, E.L. Johnson, and M.W. Padberg. Solving large-scale zero-one linear programming problems. *Operations Research*, 31:803–834, 1983.

6. I.R. de Farias, Johnson E.L., and Nemhauser G.L. A polyhedral study of the cardinality constrained knapsack problem. Technical Report 01-05, Georgia Institute of Technology, 2001.

7. I.R. de Farias, E.L. Johnson, and G.L. Nemhauser. A generalized assignment problem with special ordered sets : A polyhedral approach. *Mathematical Programming*, 89:187–203, 2000.

8. I.R. de Farias, E.L. Johnson, and G.L. Nemhauser. Facets of the complementarity knapsack polytope. *To appear in Mathematics of Operations Research*.

9. I.R. de Farias and G.L. Nemhauser. A family of inequalities for the generalized assignment polytope. *Operations Research Letters*, 29:49–51, 2001.

10. I.R. de Farias. *A Polyhedral Approach to Combinatorial Problems*. PhD thesis, School of Industrial and Systems Engineering, Georgia Institute of Technology, 1995.

11. R.E. Gomory. An algorithm for the mixed integer problem. Technical Report RM-2597, RAND Corporation, 1960.

12. R.E. Gomory. Some polyhedra related to combinatorial problems. *Linear Algebra and Its Applications*, 2:451–558, 1969.

13. Z. Gu, G.L. Nemhauser, and M.W.P. Savelsbergh. Lifted cover inequalities for 0-1 integer programs: Computation. *INFORMS Journal on Computing*, 10:427–437, 1998.

14. Z. Gu. *Lifted Cover Inequalities for 0-1 and Mixed 0-1 Integer Programs*. PhD thesis, School of Industrial and Systems Engineering, Georgia Institute of Technology, 1995.

15. P.L. Hammer, E.L. Johnson, and U.N. Peled. Facets of regular 0-1 polytopes. *Mathematical Programming*, 8:179–206, 1975.

16. H. Marchand and L.A. Wolsey. The 0-1 knapsack problem with a single continuous variable. *Mathematical Programming*, 85:15–33, 1999.

17. G.L. Nemhauser and L.A. Wolsey. A recursive procedure for generating all cuts for 0-1 mixed integer programs. *Mathematical Programming*, 46:379–390, 1990.

18. M.W. Padberg. On the facial structure of set packing polyhedra. *Mathematical Programming*, 5:199–215, 1973.

19. J.-P. P. Richard, I.R. de Farias, and G.L. Nemhauser. A simplex based algorithm for 0-1 mixed integer programming. Technical Report 01-09, Georgia Institute of Technology, 2001.

20. J.-P. P. Richard, I.R. de Farias, and G.L. Nemhauser. Lifted inequalities for 0-1 mixed integer programming : Basic theory and algorithms. Technical report, Georgia Institute of Technology, (in preparation).

21. J.-P. P. Richard, I.R. de Farias, and G.L. Nemhauser. Lifted inequalities for 0-1 mixed integer programming : Superlinear lifting. Technical report, Georgia Institute of Technology, (in preparation).

22. R. Weismantel. On the 0/1 knapsack polytope. *Mathematical Programming*, 77:49–68, 1997.

23. L.A. Wolsey. Faces for a linear inequality in 0-1 variables. *Mathematical Programming*, 8:165–178, 1975.

24. L.A. Wolsey. Facets and strong valid inequalities for integer programs. *Operations Research*, 24:367–372, 1976.

25. E. Zemel. Lifting the facets of zero-one polytopes. *Mathematical Programming*, 15:268–277, 1978.

On a Lemma of Scarf

Ron Aharoni[1,*] and Tamás Fleiner[2,**]

[1] Technion, Haifa
`ra@techunix.technion.ac.il`
[2] Alfréd Rényi Institute of Mathematics Hungarian Academy of Sciences,
and Eötvös Loránd University, Operations Research Department
`fleiner@renyi.hu`

Abstract. The aim of this note is to point out some combinatorial applications of a lemma of Scarf, proved first in the context of game theory. The usefulness of the lemma in combinatorics has already been demonstrated in [1], where it was used to prove the existence of fractional kernels in digraphs not containing cyclic triangles. We indicate some links of the lemma to other combinatorial results, both in terms of its statement (being a relative of the Gale-Shapley theorem) and its proof (in which respect it is a kin of Sperner's lemma). We use the lemma to prove a fractional version of the Gale-Shapley theorem for hypergraphs, which in turn directly implies an extension of this theorem to general (not necessarily bipartite) graphs due to Tan [12]. We also prove the following result, related to a theorem of Sands, Sauer and Woodrow [10]: given a family of partial orders on the same ground set, there exists a system of weights on the vertices, which is (fractionally) independent in all orders, and each vertex is dominated by them in one of the orders.

1 Introduction

A famous theorem of Gale and Shapley [5] states that given a bipartite graph and, for each vertex v, a linear order $\leq_v$ on the set of edges incident with v, there exists a stable matching. Here, a matching M is called *stable* if for every edge $e \notin M$ there exists an edge in M meeting e and beating it in the linear order of the vertex at which they are incident. (The origin of the name "stable" is that in such a matching no non-matching edge poses a reason for breaking marriages: for every non-matching edge, at least one of its endpoints prefers its present spouse to the potential spouse provided by the edge.) Alternatively, a stable matching is a kernel in the line graph of the bipartite graph, where the edge connecting two vertices (edges of the original graph) is directed from the larger to the smaller, in the order of the vertex of the original graph at which they meet.

* Research supported by the fund for the promotion of research at the Technion and by a grant from the Israel Science Foundation.
** Part of the research was done at the Centrum voor Wiskunde en Informatica (CWI) and during two visits to the Technion, Haifa. Research was supported by the Netherlands Organization for Scientific Research (NWO) and by OTKA T 029772.

W.J. Cook and A.S. Schulz (Eds.): IPCO 2002, LNCS 2337, pp. 176–187, 2002.

It is well known that the theorem fails for general graphs, as shown by the following simple example: let G be an undirected triangle on the vertices u, v, w, and define: $(u, v) >_u (w, u)$, $(v, w) >_v (u, v)$, $(w, u) >_w (v, w)$. But the theorem is true for general graphs if one allows fractional matchings, as follows easily from a result of Tan [12] (see Theorem 3 below). For example, in the example of the triangle one could take the fractional matching assigning each edge the weight $\frac{1}{2}$: each edge is then dominated at some vertex by edges whose sum of weights is 1 (for example, the edge (u, v) is dominated in this way at v).

The notions of stable matchings and fractional stable matchings can be extended to hypergraphs. A *hypergraphic preference system* is a pair $(H, \mathcal{O})$, where $H = (V, E)$ is a hypergraph, and $\mathcal{O} = \{\leq_v : v \in V\}$ is a family of linear orders, $\leq_v$ being an order on the set $D(v)$ of edges containing the vertex v. If H is a graph we call the system a *graphic* preference system.

A set M of edges is called a *stable matching* with respect to the preference system if it is a matching (that is, its edges are disjoint) and for every edge e there exists a vertex $v \in e$ and an edge $m \in M$ containing v such that $e \leq_v m$.

Recall that a function w assigning non-negative weights to edges in H is called a *fractional matching* if $\sum_{v \in h} w(h) \leq 1$ for every vertex v. A fractional matching w is called *stable* if every edge e contains a vertex v such that $\sum_{v \in h, e \leq_v h} w(h) = 1$.

As noted, by a result of Tan every graphic preference system has a fractional stable matching. Does this hold also for general hypergraphs? The answer is yes, and it follows quite easily from a result of Scarf [11]. This result is the starting point of the present paper. It was originally used in the proof of a better known theorem in game theory, and hence gained the name "lemma". Its importance in combinatorics has already been demonstrated in [1], where it was used to prove the existence of a fractional kernel in any digraph not containing a cyclic triangle.

Scarf's lemma is intriguing in that it seems unrelated to any other body of knowledge in combinatorics. In accord, its proof appears to be of a new type. The aim of this paper is to bring it closer to the center of the combinatorial scene. First, by classifying it as belonging to the Gale-Shapley family of results. Second, by pointing out its similarity (in particular, similarity in proofs) to results related to Brouwer's fixed point theorem.

In [4], it was noted that the Gale-Shapley theorem is a special case of a result of Sands, Sauer and Woodrow [10] on monochromatic paths in edge two-colored digraphs. This result can also be formulated in terms of dominating antichains in two partial orders (see Theorem 7 below). We shall use Scarf's lemma to prove a fractional generalization of this "biorder" theorem to an arbitrary number of partial orders.

In [4], a matroidal version of the Gale-Shapley theorem was proved, for two matroids on the same ground set. Using Scarf's lemma, we prove a fractional version of this result to arbitrarily many matroids on the same ground set.

We finish the introduction with stating Scarf's lemma. (Apart from the original paper, a proof can also be found in [1]. The basic ideas of the proof are mentioned in the last section of the present paper).

Theorem 1 (Scarf [11]). *Let $n < m$ be positive integers, b a vector in $\mathrm{I\!R}_+^n$. Also let $B = (b_{i,j}), C = (c_{i,j})$ be matrices of dimensions $n \times m$, satisfying the following three properties: the first n columns of B form an $n \times n$ identity matrix (i.e. $b_{i,j} = \delta_{i,j}$ for $i, j \in [n]$), the set $\{x \in \mathrm{I\!R}_+^n : Bx = b\}$ is bounded, and $c_{i,i} < c_{i,k} < c_{i,j}$ for any $i \in [n]$, $i \neq j \in [n]$ and $k \in [m] \setminus [n]$.*

Then there is a nonnegative vector x of $\mathrm{I\!R}_+^m$ such that $Bx = b$ and the columns of C that correspond to $\mathrm{supp}(x)$ form a dominating set, that is, for any column $i \in [m]$ there is a row $k \in [n]$ of C such that $c_{k,i} \leq c_{k,j}$ for any $j \in \mathrm{supp}(x)$.

As we shall see in the last section, under the assumption that the columns of B are in general position and the entries in each row of C are different the proof of the lemma yields a stronger fact, namely that there exists an odd number of vectors x as in the lemma.

2   Some Applications

In this section we study some extensions of the stable marriage theorem.

Theorem 2 (Gale-Shapley [5]). *If $(G, \mathcal{O})$ is a graphic preference model and graph $G = (V, E)$ is bipartite then there exists a stable matching.*   □

As we mentioned in the introduction, in nonbipartite graphic preference models Theorem 2 is not true. The first algorithm to decide the existence of a stable matching in this case is due to Irving [6]. Later on, based on Irving's proof, Tan gave a compact characterization of those models that contain a stable matching [12]. In what follows, we formulate Tan's theorem.

In a graphic preference model $(G, \mathcal{O})$, a subset $C = \{c_1, c_2, \ldots, c_k\}$ of E is a *preference cycle* if $c_1 <_{v_1} c_2 <_{v_2} c_3 <_{v_3} \cdots <_{v_{k-1}} c_k <_{v_k} c_1$ for different vertices $v_1, v_2, \ldots, v_k$ of V. A preference cycle C is *odd* if $|C|$ is odd, otherwise C is *even*. A *stable partition* of model $(G, \mathcal{O})$ is a subset S of E with the following properties.

1. Any component of S is either a cycle or an edge, and
2. each cycle component of S is a preference cycle, and
3. for any edge e of $E \setminus S$ there is a vertex v covered by S and incident with e such that $e <_v s$ for any edge s of S incident with v.

It is easy to see that a stable partition with no cycle component is a stable matching.

Let us define a *stable half-matching* as a stable fractional matching x so that $2x$ is an integral vector. Clearly, for a graphic preference model, the support of a stable half-matching is a stable partition. Also, if S is a stable partition then x_S is a stable half matching, where

$$x_S(e) = \begin{cases} 0 & \text{if } e \notin S, \\ 1 & \text{if } e \text{ is an edge-component of } S, \\ \frac{1}{2} & \text{if } e \text{ belongs to a cycle-component of } S. \end{cases}$$

Theorem 3 (Tan, [12]). *Any graphic preference model has a stable partition.*

Note that a weaker version of Theorem 3 can be proved the following way: Define model $\mathcal{M}' = (G', \mathcal{O}')$ by $G' = (V', E')$, $V' := \{v_m, v_w : v \in V\}$, $E' := \{u_m v_w, u_w v_m : uv \in E\}$ and $u_m v_w <_{u_m} u_m v'_w$ iff $v_m u_w <_{u_w} v'_m u_w$ iff $uv <_u uv'$. That is, we introduce a bipartite preference model by duplicating the original one. According to Theorem 2, there is a stable matching M in $\mathcal{M}'$. Define $S := \{uv : u_m v_w \in M \text{ or } v_m u_w \in M\}$. It is straightforward to check that S satisfies the first two requirements in the definition of a stable partition, but instead of 3., we have the weaker property

3'. For any edge e of $E \setminus S$ there are two edges s_1 and s_2 of S such that $s_1 \leq_v e$ and $s_2 \leq_u e$ for some $u, v \in V$. If s_1 is an edge-component of S then $s_1 = s_2$ is allowed, otherwise $s_1 \neq s_2$.

However, the above weak version of Theorem 3 does not help us to decide the existence of a stable matching in a model. In particular, the following fact is not true with the weaker notion of stable partition.

Observation 4 *If S is a stable partition for graphic preference model $\mathcal{M} = (G, \mathcal{O})$ and there are no odd preference cycles in S then there exists a stable matching of $\mathcal{M}$.*

An immediate consequence of Observation 4 and Theorem 3 is that if a model $\mathcal{M}$ is free of odd preference cycles then there is a stable matching of $\mathcal{M}$.

Proof. Throw away each second edge of each cycle in S. By the definition of a stable partition, what is left from S after these deletions is exactly a stable matching of $\mathcal{M}$. $\square$

So if stable partition S does not contain an odd cycle then we immediately see a stable matching. On the other hand, an odd cycle in S means that no stable matching exists in $\mathcal{M}$. More specifically, there is the following theorem.

Theorem 5 (Tan [12]). *Let $\mathcal{M} = (G, \mathcal{O})$ be a graphic preference model and S be a stable partition of $\mathcal{M}$. If there is an odd cycle C in S then C is present in each stable partition of $\mathcal{M}$.*

To prove Theorem 3, we justify the promised fractional version of the Gale-Shapley theorem for hypergraphs.

Theorem 6. *Any hypergraphic preference system has a fractional stable matching.*

Proof. Let $(H, \mathcal{O})$ be a hypergraphic preference system, where $H = (V, E)$ and $\mathcal{O} = \{\leq_v : v \in V\}$. Let B be the incidence matrix of H, with the identity matrix adjoined to it at its left. Let C' be a $V \times E$ matrix satisfying the following two conditions:

(1) $c'_{v,e} < c'_{v,f}$ whenever $v \in e \cap f$ and $e <_v f$,
(2) $c'_{v,f} < c'_{v,e}$ whenever $v \in f \setminus e$.

Let C be obtained from C' by adjoining to it on its left a matrix so that C satisfies the conditions of Theorem 1. Let x be a vector as in Theorem 1 for B and C, where b is taken as the all $1's$ vector $\mathbf{1}$. Define $x' = x|_E$, namely the restriction of x to E. Clearly, x' is a fractional matching. To see that it is dominating, let e be an edge of H. By the conditions on x, there exists a vertex v such that $c_{v,e} \leq c_{v,j}$ for all $j \in \text{supp}(x)$. Since $c_{v,v} < c_{v,e}$ it follows that $v \notin \text{supp}(x)$. Since $Bx = \mathbf{1}$ it follows that $\text{supp}(x)$ contains an edge f containing v (otherwise $(Bx)_v = 0$). Since $c_{v,f} \geq c_{v,e}$ it follows by condition (2) above that $v \in e$. The condition $(Bx)_v = 1$ now implies that e is dominated by x at v. □

In fact, the vector x' can be assumed to be a vertex of the fractional matching polytope of H. To see this, write $x' = \sum \alpha_i y_i$, where $\alpha_i > 0$ for all i, $\sum \alpha_i = 1$ and the y_i's are vertices of the fractional matching polytope. Then each y_i must be a fractional stable matching. It is well known (see e.g. [8]) that the vertices of the fractional matching polytope of a graph are half integral, that is, they have only $0, \frac{1}{2}, 1$ coordinates. This yields Theorem 3. Next we give a direct proof of this fact.

Proof (of Theorem 3). Let $\mathcal{M} = (G, E)$ be a graphic preference model, x be a fractional stable matching for $\mathcal{M}$ that exists by Theorem 6 and define $S := \text{supp}(x)$. We shall prove that S is a stable partition. By the stability of x, we can orient each edge e of E so that the corresponding arc $\mathbf{e}$ points to a vertex v such that

$$\sum_{e \leq_v f} x(f) = 1 . \tag{1}$$

Let $D = (V, A)$ be the resulted digraph. From (1), it follows that if $\mathbf{e}, \mathbf{f} \in S$ then $\mathbf{e}$ and $\mathbf{f}$ have different endvertices. Also, if $\mathbf{e}, \mathbf{f}$ is a directed path for some $e, f \in S$ then $x(f) < 1$ as x is a fractional matching. Then (1) yields that there is an edge $g \in S$ such that $\mathbf{e}, \mathbf{f}, \mathbf{g}$ is a directed path. These two properties of S imply that the components of S correspond to disjoint edges and directed cycles in D. Condition 3. in the definition of a stable partition holds for S because if edge e of $E \setminus S$ is oriented as $\mathbf{e} = uv$ then $e <_v s$ for any $s \in S$ because of (1). So $S = \text{supp}(x)$ is indeed a stable partition of $\mathcal{M}$. □

For the sake of completeness we finish this section by proving Theorem 5.

Proof (of Theorem 5). Introduce preference model $\mathcal{M}' = (G', \mathcal{O}')$ by $G' = (V, E')$, $E' := \{e^u, e^v : e = uv \in E, e^u \text{ and } e^v \text{ are parallel to } e\}$, $\mathcal{O}' := \{<'_v : v \in V\}$, where

$$e^u <'_v f^w \text{ iff } (e <_v f \text{ or } (e = f, u = v, w \neq v)) .$$

(We duplicate all edges, and extend the order to the duplicates in a natural way, so that we only have to take extra care for the relation of the two copies of the same edge.)

Observe that if $C = \{e_1, e_2, \ldots, e_k\}$ is a preference cycle of model $\mathcal{M}$ such that $e_{i+1} <_{v_i} e_i$ (i is modulo k) then $C' := \{e_1^{v_1}, e_2^{v_2}, \ldots, e_k^{v_k}\}$ is a corresponding preference cycle of model $\mathcal{M}'$. So for any stable partition S of model $\mathcal{M}$ there is a corresponding subset S' of E' defined by

$$S' := \bigcup \{ C' : C \text{ is a cycle component of } S \} \cup$$
$$\cup \{ e^u, e^v : e = uv \text{ is an edge component of } S \} .$$

Observe that S' is a stable partition of $\mathcal{M}'$ and each component of S' is a preference cycle of $\mathcal{M}'$.

Let S and T be stable partitions of model $\mathcal{M}$ so that C is an odd cycle component of S. For the corresponding stable partitions S' and T' of $\mathcal{M}'$ we get

$$|S'|+|T'| \leq |\{(s,t_1,t_2,v): s \in S', t_1, t_2 \in T', v \in V, t_1 \neq t_2, s \leq'_v t_1, s \leq'_v t_2\}| \ (2)$$
$$+|\{(t,s_1,s_2,v): t \in T', s_1, s_2 \in S', v \in V, s_1 \neq s_2, t \leq'_v s_1, t \leq'_v s_2\}| \ (3)$$
$$\leq 2|V(S') \cap V(T')| \leq |V(S')| + |V(T')| = |S'| + |T'| . \qquad (4)$$

The inequality in (2) is true because of property 3. of stable partitions S' and T' (note that both S' and T' are 2-regular subgraphs of (V, E')). The inequality between (3) and (4) holds because the contribution of each vertex that is covered by S' and T' is at most two. So there must be equality throughout (2) to (4). This means that $V(S) = V(S') = V(T') = V(T)$ (i.e. any two stable partitions cover the same set of vertices) and that every vertex v of $V(S')$ contributes exactly two at (2,3).

From this latter property, it follows that there is no vertex v of $V(S')$ that is incident with exactly three edges of $S' \cup T'$. This is because the contribution of v can only be two, if these three edges are e, f and g so that $e <'_v f$, $e <'_v g$ and $e \in S' \cap T'$. But if we follow the two preference cycle components of S' and T' starting at common edge e then we shall find a vertex $u \neq v$ of $V(S')$ so that u is incident with exactly three edges of $S' \cup T'$, and the common edge of S' and T' is the $<'_u$-maximal of the three. The contribution to (2,3) of vertex u is only one, hence the degrees in $S' \cup T'$ can only be 0, 2 and 4. It also follows that if S' and T' share an edge e then the cycle component containing e is the the same for S' and T'.

So assume that the odd cycle component C' of S' is not a component of T'. This means that each vertex v of C' is incident with exactly four edges of $S' \cup T'$. As the contribution of v in (2,3) is exactly two, we have only two possibilities for vertex v. Either the two $<'_v$-smaller edges of these four edges belong to S' and the two $<'_v$-bigger edges belong to T' (in which case we say that v is an S'-vertex) or vice versa, when v is a T'-vertex. From property 3. of S' it follows that for no edge $e = uv$ of C' it can happen that both u and v are S'-vertices. As C' is an odd cycle, it means that there must be an edge $e = uv$ so that both u and v are T'-vertices. But then the inequality in (2) is strict at e.

The contradiction shows that $C' \subseteq T'$, hence C is a component of T. $\qquad \square$

3   Dominating Antichains and Matroid-Kernels

In [10], there was proved a generalization of the Gale-Shapley theorem by Sands *et al.* Its original formulation was in terms of paths in digraphs whose edges are two-colored. But at its core is a fact about pairs of partial orders.

Let V be a finite ground set and $\leq_1$ and $\leq_2$ be two partial orders on V. A *dominating common antichain of* $\leq_1$ *and* $\leq_2$ is a subset A of V such that A is an antichain in both partial orders and for any element v of V there is an element a in A with $v \leq_1 a$ or $v \leq_2 a$.

Theorem 7 (see [4,3]). *For any two partial orders $\leq_1$ and $\leq_2$ on the same finite ground set V, there exists a dominating antichain of $\leq_1$ and $\leq_2$.*   □

The Gale-Shapley theorem is obtained by applying this theorem to the two orders on the edge set of the bipartite graph, each being obtained by taking the (disjoint) union of the linear orders induced by the vertices in one side of the graph.

The theorem is false for more than two partial orders. But a fractional version is true. For given partial orders $\leq_1, \leq_2, \ldots, \leq_k$ on a ground set V, a nonnegative vector x of $\mathbb{R}_+^V$ is called a *fractional dominating antichain* if x is a *fractional antichain* (i.e. $\sum_{c \in C} x(c) \leq 1$ for any chain C of any of the partial orders $\leq_i$) and x is a *fractional upper bound* for any element of V, that is for each element v of V there is a chain $v = v_0 \leq_i v_1 \leq_i v_2 \leq_i \ldots \leq_i v_l$ of some partial order $\leq_i$ with $\sum_{j=0}^{l} x(v_j) = 1$. Note that if a fractional dominating antichain x happens to be integral then it is the characteristic vector of a dominating antichain.

Theorem 8. *Any finite set $\leq_1, \leq_2, \ldots, \leq_k$ of partial orders on the same ground set V has a fractional dominating antichain.*

Proof. For each $i \leq k$ let $\mathcal{D}_i$ be the set of maximal chains in the partial order $\leq_i$. Let $\mathcal{J} = \bigcup_{i \leq k} \{i\} \times \mathcal{D}_i$ (that is, $\mathcal{J}$ is the union, with repetition, of the families $\mathcal{D}_i$).

Let B' be the $V \times \mathcal{J}$ incidence matrix of the chains of $\mathcal{J}$ (that is, for $v \in V$ and a maximal chain D in $\leq_i$, the $(v, (i, D))$ entry of B' is 1 if $v \in D$, otherwise it is 0). Let $B := [I_n, B']$ be obtained by adding an $n \times n$ identity matrix I_n in front of B'.

Next we define a $V \times \mathcal{J}$ matrix C'. For $v \in V$ and $j = (i, D) \in \mathcal{J}$ define $C'_{v,j}$ as $|D| + 1$ if $v \notin D$, and as the height of v in D in the order $\leq_i$ if $v \in D$. Append now on the left of C' a matrix so that the resulting matrix C satisfies the conditions of Theorem 1.

Applying Theorem 1 to the above matrices B, C and the all 1's vector $b = \mathbf{1}_n$, we get a nonnegative vector $x \in \mathbb{R}^{\mathcal{J} \cup V}$. Let x' be the restriction of x to $\mathbb{R}^V$. As $B \cdot x = b = \mathbf{1}$, we have $B' \cdot x' \leq \mathbf{1}$, meaning that x' is a fractional antichain. The domination property of x implies that for any element v of V there is a chain D of some partial order $\leq_i$ such that for any element u from $D \cap \text{supp}(x)$ we have $v \leq_i u$. Since $c_{(i,D),(i,D)}$ is smallest in row (i, D) of C, it follows that the column (i, D) of C does not belong to $\text{supp}(x)$. The equality $(Bx)_{(i,D)} = 1$ thus means that $\sum_{d \in D} x(d) = 1$, showing that $\sum_{d \in D, d \geq_i v} x(d) = 1$. This proves the fractional upper bound property of x. $\qquad\square$

Our last application is a generalization of a matroid version of the Gale-Shapley theorem.

An *ordered matroid* is a triple $\mathcal{M} = (E, \mathcal{C}, \leq)$ such that $(E, \mathcal{C})$ is a matroid and $\leq$ is a linear order on E. For two ordered matroids $\mathcal{M}_1 = (E, \mathcal{C}_1, \leq_1)$ and $\mathcal{M}_2 = (E, \mathcal{C}_2, \leq_2)$ on the same ground set, a subset K of E is an $\mathcal{M}_1\mathcal{M}_2$-*kernel*, if K is independent in both matroids $(E, \mathcal{C}_1)$ and $(E, \mathcal{C}_2)$, and for any element e in $E \setminus K$ there is a subset C_e of K and an index $i = 1, 2$ so that

$$\{e\} \cup C_e \in \mathcal{C}_i \text{ and } e \leq_i c \text{ for any } c \in C_e .$$

Theorem 9 (see [4,3]). *For any pair $\mathcal{M}_1$, $\mathcal{M}_2$ of ordered matroids there exists an $\mathcal{M}_1\mathcal{M}_2$-kernel.* $\qquad\square$

Let $\mathcal{M}_1, \mathcal{M}_2, \ldots, \mathcal{M}_k$ be ordered matroids on the same ground set E, where $\mathcal{M}_i = (E, \mathcal{C}_i, \leq_i)$. A vector $x \in \mathbb{R}_+^E$ is called a *fractional kernel for matroids* $\mathcal{M}_1, \mathcal{M}_2, \ldots, \mathcal{M}_k$ if it satisfies the following two properties:

(1) x is *fractionally independent*, namely $\sum_{e \in E'} x(e) \leq r_i(E')$ for any subset E' of E, where r_i is the rank function of the matroid $\mathcal{M}_i$.

(2) every element e of E is *fractionally optimally spanned* in one of the matroids, namely there exists a subset E' of E and a matroid $\mathcal{M}_i$, such that $e \leq_i e'$ for any $e' \in E'$, and $\sum_{e \in E'} x(e) = r_i(E' \cup \{e\})$.

Note that a fractional matroid-kernel for two matroids that happens to be integral is a matroid kernel.

Theorem 10. *Every family $\mathcal{M}_i = (E, \mathcal{C}_i, \leq_i)$ $(i = 1, 2, \ldots, k)$ of ordered matroids has a fractional kernel.*

Proof. Let B' be a matrix whose rows are indexed by pairs (i, F), where $1 \leq i \leq k$ and $F \subseteq E$, and whose columns are indexed by E, the $((i, F), e)$ entry being 1 if $e \in F$, 0 otherwise. Let $B := [I, B']$.

Define matrix C' on the same row and column sets as those of B', by letting its $((i, F), e)$ entry be the height of e in $\leq_i$ if $e \in F$ and $|F| + 1$ otherwise. Append an appropriate matrix on the left of C', so as to get a matrix C as in Theorem 1. Let b be the vector on E defined by $b_{(i,F)} := r_i(F)$.

Apply Theorem 1 to B, C and b. Let x be the vector whose existence is guaranteed in the theorem and x' be the restriction of x to E. We claim that x' is a fractional kernel for our matroids. As $Bx = b$ and both B and x are nonnegative, we have $B'x' \leq b$. In other words, x' is fractionally independent. The domination property of $\operatorname{supp}(x)$ yields that for any element e of E there is a subset F and a matroid $\mathcal{M}_i$ such that we have

$$e \leq_i f \qquad \text{for any element } f \text{ of } F' := F \cap \operatorname{supp}(x) . \tag{5}$$

Since $c_{(i,F),(i,F)}$ is smallest in row (i, F) of C, column (i, F) does not belong to $\operatorname{supp}(x)$. Thus $(Bx)_{(i,F)} = r_i(F)$ implies

$$r_i(F') \geq \sum_{f \in F'} x'(f) = \sum_{f \in F} x(f) = r_i(F) \geq r_i(F') .$$

In particular, $F' \neq \emptyset$, hence (5) and the definition of $\leq_i$ shows that $e \in F$, and this proves the optimal spanning property of x'. □

We finish this section by pointing out a difference between Theorem 8 and Theorem 10. Namely, we show that Theorem 10 (the fractional version of Theorem 9), together with the well-known fact about the integrality of the matroid intersection polytope implies Theorem 9. The proof is analogous to the method of Aharoni and Holzman in [1]. There, they proved the existence of an integral kernel for any normal orientation of any perfect graph from the existence of a so called strong fractional kernel and from the linear description of the independent set polytope of perfect graphs. On the other hand, there is no similar polyhedral argument to deduce Theorem 7 from Theorem 8.

The polyhedral result that we need here is the following theorem of Edmonds.

Theorem 11 (Edmonds [2]). *If $\mathcal{M}_1 = (E, \mathcal{C}_1)$ and $\mathcal{M}_2 = (E, \mathcal{C}_2)$ are matroids on the same ground set then*

$$\text{conv}\{\chi^I : I \text{ is independent both in } \mathcal{M}_1 \text{ and in } \mathcal{M}_2\} =$$
$$\{x \in \mathbb{R}^E : \mathbf{0} \leq x, \sum_{f \in F} x(f) \leq r_i(F) \text{ for any } i \in \{1, 2\} \text{ and } F \subseteq E\} .$$

□

Proof (Alternative proof of Theorem 9.). By Theorem 10, there is a fractional kernel x for ordered matroids $\mathcal{M}_1$ and $\mathcal{M}_2$. As x is fractionally independent, Theorem 11 implies that $x = \sum_{j=1}^{l} \lambda_j \chi^{I_j}$ is a convex combination of the characteristic vectors of common independent sets of $\mathcal{M}_1$ and $\mathcal{M}_2$. We claim that any of the I_j-s is an $\mathcal{M}_1 \mathcal{M}_2$-kernel.

As x is a fractionally optimally spanning set, for any element e of E there is an index $i \in \{1, 2\}$ and a subset E' of E such that $w_i(e') \leq w_i(e)$ for any element e' of E and

$$\sum_{j=1}^{l} \lambda_j |I_j \cap E'| = \sum_{e \in E'} x(e) = r_i(E' \cup \{e\}) \geq \sum_{e \in E' \cup \{e\}} x(e) = \sum_{j=1}^{l} \lambda_j |I_j \cap (E' \cup \{e\})| .$$

This means that each common independent set I_j intersects E' in $r_i(E' \cup \{e\})$ elements, that is each I_j spans e. □

In contrast to the above argument, there is no polyhedral proof for Theorem 7 from Theorem 8 along similar lines. Namely, it can happen that for two partial orders $<_1$ and $<_2$ on the same ground set, a fractional dominating antichain is not a convex combination of dominating antichains. Figure 1 shows the Hasse diagrams of two partial orders on four elements. As any two elements of the common ground set are comparable in one of the partial orders, a dominating antichain contains exactly one element. However, it is easy to check that the all-$\frac{1}{3}$ vector is a fractional dominating antichain of total weight $\frac{4}{3}$.

Fig. 1. Hasse diagrams of the counterexample partial orders

4 A Link with a Theorem of Shapley

A *simplicial complex* is a non-empty family $\mathcal{C}$ of subsets of a finite ground set such that $A \subset B \in \mathcal{C}$ implies $A \in \mathcal{C}$. Members of $\mathcal{C}$ are called *simplices* or *faces*. Let us call a simplicial complex *manifold-like* if, denoting its rank by n (that is, the maximum cardinality of a simplex in it is $n+1$), every face of cardinality n in it is contained in two faces of cardinality $n+1$. The *dual* $\mathcal{D}^*$ of a complex $\mathcal{D}$ is the set of complements of its simplices. Just like in the case of complexes, members of a dual complex are also called *faces*.

Lemma 1. *If $\mathcal{C}, \mathcal{D}$ are two manifold-like complexes on the same ground set X, then the number of maximum cardinality faces of $\mathcal{C}$ that are also minimum cardinality faces of $\mathcal{D}^*$ is even.*

Proof. Let $\mathcal{C}_{max}$ be the family of faces of $\mathcal{C}$ of maximum cardinality and $\mathcal{D}^*_{min}$ be the set of faces of $\mathcal{D}^*$ of minimum cardinality. We may clearly assume that these two cardinalities are equal, as otherwise the lemma claims the triviality that zero is an even number. Fix an element x of X, and define an auxiliary digraph $\boldsymbol{G}$ on $\mathcal{C}_{max} \cup \mathcal{D}^*_{min}$ by drawing an arc from $C \in \mathcal{C}_{max}$ to $D \in \mathcal{D}^*_{min}$ if $D \setminus C = \{x\}$.

Let $D \in \mathcal{D}^*_{min}$. If $x \notin D$ or $D \setminus \{x\} \not\subset C$ then no arc enters D in $\boldsymbol{G}$. Otherwise, as $\mathcal{C}$ is manifold-like, there are exactly two different members C_1, C_2 of $\mathcal{C}_{max}$ of the form $C_i = D \setminus \{x\} \cup \{y_i\}$ for some different elements y_1, y_2 of X. If $y_1 \neq x \neq y_2$ then the in-degree of D in $\boldsymbol{G}$ is two and D is not a member of $\mathcal{C}_{max}$. Else D has in-degree exactly one, and $x \in D \in \mathcal{C}_{max} \cap \mathcal{D}^*_{min}$.

Similarly, let $C \in \mathcal{C}_{max}$. If $x \in C$ or $C \cup \{x\} \not\subset \mathcal{D}^*$ then no arc of $\boldsymbol{G}$ leaves C. Otherwise, $\mathcal{D}$ being manifold-like, there are exactly two members D_1, D_2 of $\mathcal{D}^*_{min}$ of the form $D_i = C \cup \{x\} \setminus \{y_i\}$ for some different elements y_1, y_2 of X. If $y_1 \neq x \neq y_2$ then the out-degree of C is exactly two and C is not a member of $\mathcal{D}^*_{min}$. Else the out-degree of C in $\boldsymbol{G}$ is exactly one and $x \notin C \in \mathcal{C}_{max} \cap \mathcal{D}^*_{min}$.

Let G be the underlying undirected graph of $\boldsymbol{G}$. The above argument shows that a vertex v of G has degree zero or two if $v \in \mathcal{C}_{max} \triangle \mathcal{D}^*_{min}$ and v has degree one if $v \in \mathcal{C}_{max} \cap \mathcal{D}^*_{min}$. As the number of odd degree vertices of a finite graph is even, the lemma follows.   □

What examples are there of manifold-like complexes? Of course, a triangulation of a closed manifold is of this sort. (We call this complex a *manifold-complex*.) Another well known example of a dual manifold-like complex is the *cone complex*: let X be a set of vectors in $\mathbb{R}^n$, and b a vector not lying in the positive cone spanned by any $n - 1$ elements of X. Consider the set $\mathcal{C}^* := \{A \subseteq X : b \in \text{cone}(A)\}$. It is a well known fact from linear programming that if $b \in \text{cone}(A)$, where $A \subset X$, $|A| = n$ and $z \in X \setminus A$, then there exists a unique element $a \in A$ such that $b \in \text{cone}(A \cup \{z\} \setminus \{a\})$. That is, $\mathcal{C}^*$ is indeed a dual manifold-like complex.

A third example of a manifold-like complex is the *domination complex*. Let C be a matrix as in Theorem 1 with the additional property that in each row of C all entries are different. Then it is not difficult to check that the family of dominating column sets together with the extra member $[n]$ is a manifold-like complex. (For the details, see [1].)

Lemma 1 directly implies a generalization of Sperner's lemma.

Lemma 2. *Let the vertices of a triangulation T of a closed n-dimensional manifold be labeled with vectors from $\mathbb{R}^{n+1}$. Let $b \in \mathbb{R}^{n+1}$ be a vector that does not belong to the cone spanned by fewer than $n + 1$ labels. Then there are an even number of simplices S of the triangulation with the property that b is in the cone spanned by the vertex-labels of S.*

Proof. Let $\mathcal{C}$ be the manifold complex of T. Define family $\mathcal{D}^*$ by

$$\mathcal{D}^* := \{A \subseteq V(T) : b \in \text{cone}(L(A))\},$$

where $V(T)$ is the set of vertices of triangulation T and for subset A of $V(T)$, $L(A)$ denotes the set of labels on the vertices of A. By the condition on the vertex-labels, $\mathcal{D}^*$ is a cone complex. Clearly, the common members of $\mathcal{C}$ and $\mathcal{D}^*$ are minimum cardinality faces of $\mathcal{C}$ and maximum cardinality faces of $\mathcal{D}^*$. By Lemma 1, there is an even number of common members of $\mathcal{C}$ and $\mathcal{D}^*$, and these common members are exactly those simplices of T whose labels contain b in their cone.   □

Sperner's lemma is obtained from Lemma 2 by taking the n-sphere S^n as the closed manifold, by choosing the vertex-labels from the standard unit vectors $(0, 0, \ldots, 1, \ldots, 0)$ and by fixing $b = \mathbf{1}$ (the all 1's vector). This is not the standard way the lemma is stated, but is well known to be equivalent to it, see e.g. [7]. The more general Lemma 2 is undoubtedly known, but we do not know a reference to it. Shapley [9] proved it for the case that the labels are $0, 1$ vectors, but his proof works also for general vectors.

Next we apply Theorem 1 to prove Scarf's lemma. The proof is essentially the same as in [1].

Proof (of Theorem 1). By slightly changing vector b and the entries of matrix C, we can construct vector b' and matrix C' with the following properties. No $n - 1$ columns of B span a cone that contains b' and if n columns of B span a cone that contains b' then this cone also contains b. For C' we require that in each row of C' all entries are different, and if $c_{ij} < c_{ik}$ then for the corresponding C'-entires the same holds: $c'_{ij} < c'_{ik}$.

Define family $\mathcal{D}^*$ on $[m]$ by $X \in \mathcal{D}^*$ if and only if $\text{cone}_B(X)$ (the cone of those columns of B that are indexed by X) contain b'. Then $\mathcal{D}^*$ is a cone complex, by the choice of b'. Let $\mathcal{C}$ be the domination complex defined by C'. By the choice of B, b' and C', any common member of $\mathcal{C}$ and $\mathcal{D}^*$ is a maximum cardinality face of $\mathcal{C}$ and a minimum cardinality face of $\mathcal{D}^*$. So Lemma 1 implies that there is an even number of such common faces. But a common face of $\mathcal{C}$ and $\mathcal{D}^*$ is either $[n]$ or it corresponds to a dominating set of C' (which is also a dominating set of C) and to a column set of B that contains b' (hence b as well) in its cone. As $[n]$ is indeed a common face of $\mathcal{C}$ and $\mathcal{D}^*$, we get that there exists a common face of the second type. $\square$

Shapley's theorem can be proved via Brouwer's fixed point theorem (which is also easily implied by it). This, and the similarity between its proof and the proof of Scarf's lemma, suggests that perhaps there is a fixed point theorem related to the latter. A supporting fact is that in [4] there was given a proof of Gale-Shapley's theorem using the Knaster-Tarski fixed point theorem for lattices.

Acknowledgement

We are indebted to Ron Holzman and to Bronislaw Wajnryb for several helpful discussions.

References

1. R. Aharoni and R. Holzman. Fractional kernels in digraphs. *J. Combin. Theory Ser. B*, 73(1):1–6, 1998.
2. J. Edmonds. Submodular functions, matroids, and certain polyhedra. In *Combinatorial Structures and their Applications (Proc. Calgary Internat. Conf., Calgary, Alta., 1969)*, pages 69–87. Gordon and Breach, New York, 1970.
3. T. Fleiner. A fixed-point approach to stable matchings and some applications. submitted to *Mathematics of Operations Research* and EGRES Technical Report TR-2001-01, http://www.cs.elte.hu/egres, 2001 March
4. T. Fleiner. Stable and crossing structures, August, 2000. PhD dissertation, http://www.renyi.hu/~fleiner.
5. D. Gale and L.S. Shapley. College admissions and stability of marriage. *Amer. Math. Monthly*, 69(1):9–15, 1962.
6. R. W. Irving. An efficient algorithm for the "stable roommates" problem. *J. Algorithms*, 6(4):577–595, 1985.
7. L. Lovász, L. Matroids and Sperner's lemma, *European J. Combin.* 1 (1980), no. 1, 65–66.
8. L. Lovász and M. D. Plummer. *Matching theory*. North-Holland Publishing Co., Amsterdam, 1986. Annals of Discrete Mathematics, 29.
9. L.S Shapley, On balanced games without side payments, in T.C. Hu and S.M. Robinson, eds., *Mathematical Programming*, Academic Press, NY (1973), 261–290.
10. B. Sands, N. Sauer, and R. Woodrow. On monochromatic paths in edge-coloured digraphs. *J. Combin. Theory Ser. B*, 33(3):271–275, 1982.
11. H. E. Scarf. The core of an N person game. *Econometrica*, 35:50–69, 1967.
12. J. J.M. Tan. A necessary and sufficient condition for the existence of a complete stable matching. *J. Algorithms*, 12(1):154–178, 1991.

A Short Proof
of Seymour's Characterization of the Matroids
with the Max-Flow Min-Cut Property

Bertrand Guenin

Department of Combinatorics and Optimization, Faculty of Mathematics,
University of Waterloo, Waterloo, ON N2L 3G1, Canada
bguenin@math.uwaterloo.ca *

Abstract. Seymour proved that the set of odd circuits of a signed binary matroid (M, Σ) has the Max-Flow Min-Cut property if and only if it does not contain a minor isomorphic to $(M(K_4), E(K_4))$. We give a shorter proof of this result.

1   Introduction

The matroids considered in this paper are all binary. A *signed matroid* is a pair (M, Σ) where M is a matroid and $\Sigma \subseteq E(M)$. A subset X of elements of M is called *odd* (resp. *even*) if $|X \cap \Sigma|$ is odd (resp. even). We denote the set of odd circuits of (M, Σ) by $\mathcal{C}(M, \Sigma)$. We say that $\Sigma' \subseteq E(M)$ is a *signature* of (M, Σ) if $\mathcal{C}(M, \Sigma) = \mathcal{C}(M, \Sigma')$. Consider weights $w \in Z_+^{E(M)}$. We say that a subset $\mathcal{P}$ of $\mathcal{C}(M, \Sigma)$ is a *w-packing* (of odd circuits) if, for every element e of M, at most w_e circuits of $\mathcal{P}$ use e. A subset B of $E(M)$ is a *cover* of (M, Σ) if every odd circuit of (M, Σ) contains some element of B. It is straightforward to show that (inclusion-wise) minimal covers are signatures. Evidently, for every w-packing $\mathcal{P}$ and every cover B we must have $w(B) \geq |\mathcal{P}|$. If equality holds we say that (M, Σ) *packs with respect to weights* w. When $w_e = 1$ for all $e \in E(M)$ then a w-packing is called a *packing* and we say that (M, Σ) *packs* if it packs with respect to w. A signed matroid (M, Σ) has the *Max-Flow Min-Cut property* if it packs with respect to all non-negative integral weights w.

Let $e \in E(M)$. The *deletion* $(M, \Sigma) \backslash e$ of (M, Σ) is defined as $(M \backslash e, \Sigma - \{e\})$. The *contraction* $(M, \Sigma)/e$ of (M, Σ) is defined as follows: if $e \notin \Sigma$ then $(M, \Sigma)/e := (M/e, \Sigma)$; if $e \in \Sigma$ and e is not a loop then there exists a cocircuit D of M with $e \in D$ and $(M, \Sigma)/e := (M/e, \Sigma \triangle D)$. A *minor* of (M, Σ) is any signed matroid which can be obtained by a sequence of deletions and contractions. We say that (M, Σ) is *isomorphic* to (M', Σ') if after relabeling elements of M' we have $\mathcal{C}(M, \Sigma) = \mathcal{C}(M', \Sigma')$. It is easy to see that the Max-Flow Min-Cut property is closed under taking minors. We denote by $M(K_4)$ the graphic matroid of the complete graph on four vertices. Note that $(M(K_4), E(K_4))$ does not pack, as

* This work supported by a grant from the Natural Sciences and Engineering Research Council of Canada.

there are no two disjoint odd circuits and no edge intersects all odd circuits. Thus the essence of the next theorem is the "if" direction.

Main Theorem. *(Seymour [2]) A signed binary matroid (M, Σ) has the Max-Flow Min-Cut property if and only if it has no minor that is isomorphic to $(M(K_4), E(K_4))$.*

2   A Short Proof

Some of the ideas in this proof were used to give a characterization of evenly-bipartite graphs [1]. That paper also contains a proof for the graphic case of Seymour's theorem. The presentation of the proof below follows closely the presentation of that proof.

Proof (of the main theorem). Let (M_0, Σ_0) be a minor-minimal signed binary matroid which does not have the Max-Flow Min-Cut property. Let e_0 be an element of M_0. Choose $w \in Z_+^{E(M_0)}$ such that (M_0, Σ_0) does not pack with respect to w and such that it minimizes $w(E(M_0)) - \frac{3}{2}w_{e_0}$. Let (M, Σ) be obtained by replacing each element $f \in E(M)$ by w_f parallel elements (where f and its copies belong to Σ if and only if $f \in \Sigma_0$). Then (M, Σ) does not pack. Let e be one of the copies of e_0. Choose a set $\mathcal{F}$ of odd circuits of (M, Σ) such that:

(1) $\{C - \{e\} : C \in \mathcal{F}\}$ are disjoint.
(2) $|\mathcal{F}|$ is maximum with respect to (1).
(3) $|\{C : e \in C \in \mathcal{F}\}|$ is minimum with respect to (1) and (2).

Let $(\mathcal{F}_e, \mathcal{F}_{\bar{e}})$ be the partition of $\mathcal{F}$ into circuits containing e and not containing e respectively.

Claim 1. $|\mathcal{F}_e| = 2$.

Proof. Choices (1) and (2) for $\mathcal{F}$ imply that $\{C - \{e\} : C \in \mathcal{F}\}$ is a maximum packing of $(M, \Sigma)/e$. Because of the choice of (M_0, Σ_0), $(M, \Sigma)/e$ packs. Thus there exists a cover B of $(M, \Sigma)/e$ with $|B| = |\mathcal{F}|$. Since B is also a cover of (M, Σ) and since (M, Σ) does not pack, $|\mathcal{F}_e| \geq 2$. Suppose $|\mathcal{F}_e| > 2$. Let (M_1, Σ_1) be obtained by adding an element e_1 parallel to e (with $\Sigma_1 = \Sigma$ if $e \notin \Sigma$ and $\Sigma_1 = \Sigma \cup \{e_1\}$ otherwise). By the choice of w, (M_1, Σ_1) packs and let $\mathcal{F}_1, B_1$ be the corresponding packing and cover with $|\mathcal{F}_1| = |B_1|$. Let $\mathcal{F}'$ be obtained by replacing the circuit C of $\mathcal{F}_1$ using e_1, by $C - \{e_1\} \cup \{e\}$. Since $2 = |\{C : e \in C \in \mathcal{F}'\}| < |\mathcal{F}_e|$, we must have $|\mathcal{F}'| < |\mathcal{F}|$. Thus $|B_1| < |\mathcal{F}|$. Since B_1 intersects all circuits of $\mathcal{F}$, $e \in B_1$. Since e_1 is parallel to e, $e_1 \in B_1$. But the packing consisting of the set of circuits of $\mathcal{F}_1$ avoiding e_1 together with the cover $B_1 - \{e_1\}$ imply that (M, Σ) pack, a contradiction. □

Let C_1, C_2 denote the circuits in $\mathcal{F}_e$. Note $C_1 \cap C_2 = \{e\}$. Recall that in a binary matroid, every cycle is the union of disjoint circuits.

Claim 2. *There are no odd cycles of (M, Σ) in $C_1 \triangle C_2$.*

Proof. Suppose for a contradiction, there exists an odd cycle $C \subseteq C_1 \triangle C_2$. Then C and $C_1 \triangle C_2 \triangle C$ contain odd circuits, say S and S' respectively. But then $\{S, S'\} \cup \mathcal{F}_{\bar{e}}$ contradicts choice (3) for $\mathcal{F}$ since $e \notin S \cup S'$. $\square$

Let (M', Σ') be a minor of (M, Σ) where $E(M) - E(M') \subseteq C_1 \triangle C_2$. Let $C \subseteq C_1 \cup C_2$ be an odd circuit of (M', Σ'). We say that B is a *good cover* of C if it is a cover of (M', Σ') and $|B - C| = |\mathcal{F}_{\bar{e}}|$. We say that B is a *small cover* if it is a cover of (M', Σ') and $|B - \{e\}| = |\mathcal{F}_{\bar{e}}|$. Since circuits of $\mathcal{F}_{\bar{e}}$ are odd circuits of (M', Σ') it follows that if B is a good cover of C (resp. if B is a small cover) then $B - C$ (resp. $B - \{e\}$) intersects every odd circuit of $\mathcal{F}_{\bar{e}}$ exactly once.

Claim 3. *Every odd circuit C of (M, Σ) included in $C_1 \cup C_2$ has a good cover.*

Proof. Let $\mathcal{F}'$ be a maximum packing of $(M, \Sigma) \backslash C$. Observe that if $|\mathcal{F}'| > |\mathcal{F}_{\bar{e}}|$ then $\mathcal{F}' \cup \{C\}$ violates either choice (2) or (3) for $\mathcal{F}$. Thus $|\mathcal{F}'| = |\mathcal{F}_{\bar{e}}|$. It is easy to see that (M_0, Σ_0) has no odd loops. Hence, neither does (M, Σ) and $|C| \geq 2$. It follows from the choice of w that $(M, \Sigma) \backslash C$ packs. Hence there exists a cover B' of $(M, \Sigma) \backslash C$ with $|B'| = |\mathcal{F}_{\bar{e}}|$. Then $B := B' \cup C$ is a good cover of C. $\square$

Claim 4. *For every element $f \neq e$ there exists a minimum cover B of (M, Σ) with $f \in B$. Moreover, B intersects every odd cycle included in $C_1 \cup C_2$ exactly once.*

Proof. Because of the choice of w, $(M, \Sigma) \backslash f$ packs and let $\mathcal{F}', B'$ be the corresponding packing and cover with $|\mathcal{F}'| = |B'|$. Since (M, Σ) does not pack, the size of the minimum cover of (M, Σ) is at least $|\mathcal{F}'| + 1 = |B'| + 1$. Thus $B := B' \cup \{f\}$ is a minimum cover of (M, Σ). Suppose for a contradiction there exists an odd cycle $C \subseteq C_1 \cup C_2$ with $|B \cap C| > 1$. As B is minimal, it is a signature, thus $|B \cap C|$ is odd, hence at least 3. Because B intersects all circuits of $\mathcal{F}_{\bar{e}}$, $|B| \geq |\mathcal{F}_{\bar{e}}| + 3 > |\mathcal{F}|$. Since $(M, \Sigma)/e$ packs and since $\{C - \{e\} : C \in \mathcal{F}\}$ is a maximum packing of $(M, \Sigma)/e$, minimum covers of $(M, \Sigma)/e$ have cardinality $|\mathcal{F}|$. Hence, minimum covers of (M, Σ) have cardinality at most $|\mathcal{F}|$. But then B is not a minimum cover, a contradiction. $\square$

Let (M', Σ') be a minor of (M, Σ) which is minimal and satisfies the following properties:

(1) $E(M) - E(M') \subseteq C_1 \triangle C_2$.
(2) There exists odd cycles $C_1', C_2' \subseteq C_1 \cup C_2$ of (M', Σ') such that $\{e\} = C_1' \cap C_2'$.
(3) Every odd circuit $C \subseteq C_1' \cup C_2'$ of (M', Σ') has a good cover.
(4) (M', Σ') has no small cover.
(5) For all $f \in C_1' \triangle C_2'$, there exists $f' \in C_1' \triangle C_2'$ such that $\{f, f'\}$ intersects every odd cycle of (M', Σ') included in $C_1' \cup C_2'$ exactly once.

We claim that (M, Σ) satisfies properties (1)-(5). (1) is trivial; for (2) choose $C_1' = C_1, C_2' = C_2$; (3) holds because of Claim 3; (4) is satisfied since (M, Σ) does not pack. Let $f \in C_1 \triangle C_2$. Claim 4 implies that f is contained in a minimum cover B which intersects odd cycles included in $C_1 \cup C_2$ exactly once. Then C_1, C_2 imply that B contains exactly two elements in $C_1 \triangle C_2$. Hence (5) holds. Thus (M', Σ') is well defined.

Claim 5. *The only odd cycles of* (M', Σ') *included in* $C'_1 \cup C'_2$ *are* C'_1 *and* C'_2.

Proof.

Fact (a). Each odd cycle in $C'_1 \cup C'_2$ is a circuit. In particular C'_1, C'_2 are odd circuits.

Proof (of fact). Otherwise there exists an odd cycle $C \subseteq C'_1 \cup C'_2$ which is not a circuit. Partition C into an even cycle C_{even} and an odd circuit C_{odd}. Consider (5) and choose $f \in C_{even}$. Since C_{odd} is odd, we must have $f' \in C_{odd}$, but then $|\{f, f'\} \cap C| = 2$, a contradiction. □

Suppose for a contradiction, there exists an odd cycle $C \subseteq C'_1 \cup C'_2$ distinct from C'_1, C'_2. We know from Claim 2 that $e \in C$. Note that Fact (a) implies that C and $\bar{C} := C'_1 \triangle C'_2 \triangle C$ are odd circuits. Define:

$$P_1 := C'_1 \cap C - \{e\} \qquad\qquad Q_1 := C'_1 \cap \bar{C} - \{e\}$$
$$P_2 := C'_2 \cap \bar{C} - \{e\} \qquad\qquad Q_2 := C'_2 \cap C - \{e\}$$

Note $(P_1, Q_1, \{e\})$ partitions C'_1 and $(P_2, Q_2, \{e\})$ partitions C'_2. Since $P_1 \cup P_2 = C'_2 \triangle C, Q_1 \cup Q_2 = C'_1 \triangle C$ it follows that $P_1 \cup P_2$ and $Q_1 \cup Q_2$ are even cycles. Odd circuits $C'_1, C'_2, C, \bar{C}$ imply,

Fact (b). For f, f' as in (5) either one element is in P_1 the other in P_2, or one is in Q_1 the other in Q_2.

Fact (c). Let S, S' be odd circuits of (M', Σ') which are included in $C'_1 \cup C'_2$ and which are disjoint in $P_1 \cup P_2$. Then $P_1 \cup P_2 \subseteq S \cup S'$.

Proof (of fact). Suppose for a contradiction $(P_1 \cup P_2) - (S \cup S') \neq \emptyset$. Consider f, f' as in (5) and choose $f \in P_i - S - S'$ for some $i \in \{1, 2\}$. Fact (b) implies $f' \in P_{3-i}$. Since S, S' are disjoint in $P_1 \cup P_2$, $\{f, f'\}$ intersects at most one of S, S', a contradiction. □

Case 1: There exists $P \subseteq P_1 \cup P_2, Q \subseteq Q_1 \cup Q_2$ such that $P \cup Q \cup \{e\}$ is an odd circuit of (M', Σ') such that for all covers B' of (M', Σ'), $|B' - P - \{e\}| > |\mathcal{F}_{\bar{e}}|$.

We may assume, after relabeling, that $P \cup Q \cup \{e\}$ corresponds to C'_1; that the odd cycles $(P \cup Q \cup \{e\}) \triangle C'_1 \triangle C'_2, (P \cup Q \cup \{e\}) \triangle (Q_1 \cup Q_2)$ correspond respectively to C'_2 and C; and that $P_1 = P, Q_1 = Q$. Let $(M'', \Sigma'') := (M', \Sigma') \backslash P_1/P_2$. We will show that (M'', Σ'') satisfies conditions (1)-(5) thereby contradicting the minimality of (M', Σ'). Clearly (1) holds. (2) is satisfied since $C''_1 := Q_1 \cup \{e\} = \bar{C} - P_2$ and $C''_2 := Q_2 \cup \{e\} = C'_2 - P_2$ are odd cycles of (M'', Σ''). Let S be any odd circuit of (M'', Σ'') included in $C''_1 \cup C''_2$. Then there exists an odd circuit $S' \subseteq S \cup P_2$ in (M', Σ'). Since C'_1 and S' are disjoint in $P_1 \cup P_2$, Fact (c) implies that $S' = S \cup P_2$. Fact (a) implies that $\bar{S} := (P_1 \cup P_2) \triangle S' = S \cup P_1$ is an odd circuit of (M', Σ'). From (3) we know that there exists a cover B' of (M', Σ') such that $|B' - \bar{S}| = |(B' - P_1) - S| = |\mathcal{F}_{\bar{e}}|$. Then $B' - P_1$ is a good cover of S

in (M'', Σ''). Thus (3) holds. Note (4) holds by hypothesis (Case 1). Finally, (5) follows from Fact (b).

Case 2: For all $P \subseteq P_1 \cup P_2, Q \subseteq Q_1 \cup Q_2$ such that $P \cup Q \cup \{e\}$ is an odd circuit of (M', Σ') there exists a cover B' of (M', Σ') such that $|B' - P - \{e\}| = |\mathcal{F}_{\bar{e}}|$.

Let $(M'', \Sigma'') := (M', \Sigma')/(Q_1 \cup Q_2)$. We will show that (M'', Σ'') satisfies conditions (1)-(5) thereby contradicting the minimality of (M', Σ'). Clearly (1) holds. (2) is satisfied since $C_1'' := P_1 \cup \{e\} = C_1' - Q_1$ and $C_2'' := P_2 \cup \{e\} = C_2' - Q_2$ are odd cycles of (M'', Σ''). Consider an odd circuit included in $C_1'' \cup C_2''$ of (M'', Σ''). It is of the form $P \cup \{e\}$ where $P \subseteq P_1 \cup P_2$. Then there is an odd circuit $P \cup Q \cup \{e\}$ of (M', Σ') where $Q \subseteq Q_1 \cup Q_2$. By hypothesis (Case 2) there is a cover B' such that $|B' - P - \{e\}| = |\mathcal{F}_{\bar{e}}|$. Then B' is disjoint from $Q_1 \cup Q_2$. Hence, it is a good cover of $P \cup \{e\}$ in (M'', Σ''). Thus (3) holds. (4) is satisfied because (M'', Σ'') is a contraction minor of (M', Σ'). Finally, (5) follows from Fact (b).   □

Let C_1', C_2' be the odd circuits given in (2). Let B_1, B_2 be the good covers of C_1', C_2' given by (4). Note $e \in B_1 \cap B_2$.

Claim 6. *There exists an odd circuit S of (M', Σ') such that $S \cap B_i \subseteq C_i' - \{e\}$ for $i = 1, 2$.*

Proof. Let $T := ((B_1 \cup B_2) - C_1' - C_2') \cup \{e\}$. Suppose for a contradiction, T is not a cover. Then there exists a minimal cover $B \subseteq T$. For each $C \in \mathcal{F}_{\bar{e}}$ we have $|B \cap C| \leq 2$. Since B is a signature of (M', Σ'), $|B \cap C|$ is odd, hence equal to 1. But then B is a small cover, a contradiction to (5). Therefore, T is not a cover, i.e. $E(M) - T$ contains an odd circuit S.   □

Choose S in Claim 6 so that $|S - C_1' - C_2'|$ is minimized. Let $(M'', \Sigma'') := (M', \Sigma') \backslash (E(M') - C_1' - C_2' - S)$. For $i = 1, 2$, let $B_i' \subseteq C_i'$ be the minimal cover of (M'', Σ'') corresponding to B_i. Note $|C_i' \cap B_i'|$ is odd, thus in particular $|C_i'| \geq 3$.

Claim 7. *There are no two disjoint odd circuits in (M'', Σ'').*

Proof. Let $E' := S - C_1' - C_2'$. Consider an odd circuit C of (M'', Σ'') distinct from C_1', C_2'. Then Claim 5 implies $C \cap E' \neq \emptyset$. Suppose $e \notin C$. Then $E' \subseteq C$, for otherwise we would have chosen C instead of S. Moreover, (for $i = 1, 2$) $C \cap C_i' \neq \emptyset$ since $C \cap B_i' \neq \emptyset$.   □

Note $\{C_1', C_2', S\}$ implies that covers of (M'', Σ'') have cardinality at least two. Thus (M'', Σ'') does not pack and by minimality we must have $(M, \Sigma) = (M'', \Sigma'')$. Since $(M, \Sigma)/f$ packs for all $f \in E(M)$ and since (M, Σ) has no two disjoint odd circuits, M has no parallel elements f, f' where both $f, f' \in \Sigma$ or both $f, f' \notin \Sigma$. Hence, $w_f = 1$ for every $f \in E(M_0)$ and $(M_0, \Sigma_0) = (M, \Sigma)$. Therefore element e plays the same role as any other element of M. Thus for each $f \in E(M)$ we have odd circuits, say C_1^f, C_2^f, which intersect exactly in f. Define a graph G as follows: $V(G) := E(M)$ and $(f, f') \in E(G)$ if and only if $\{f, f'\}$ is a cover of (M, Σ).

Claim 8. *Edges of G form a perfect matching. Moreover, for each $(f, f') \in E(G), \{f, f'\} = E(M) - (C_1^f \triangle C_2^f)$.*

Proof. Claim 4 and the fact that every element plays the same role implies that every element of (M, Σ) is in a cover of cardinality 2. Thus, every $f \in E(M)$ has degree at least one in G. Let f be any element of M. Odd circuits C_1^f, C_2^f imply (minimum covers are signatures) that all edges of G with endpoint f have an endpoint in $E(M) - (C_1^f \cup C_2^f)$, and conversely, all edges with an endpoint in $E(M) - (C_1^f \cup C_2^f)$ have endpoint f. It follows that for each $f \in E(M)$ all its neighbors in G have degree one. Therefore, edges of G form a perfect matching. Finally, let $f', f'' \in E(M) - (C_1^f \cup C_2^f)$, then both (f, f') and (f, f'') are edges of G. Since $E(G)$ is a matching $f' = f''$. Thus $E(M) - (C_1^f \cup C_2^f) = \{f'\}$.   □

Let $e \in E(M)$. Recall $|C_1^e|, |C_2^e| \geq 3$. Claim 8 implies that we have elements $f, h \in C_1^e - \{e\}, f', h' \in C_2^e - \{e\}$, where $(f, f'), (h, h')$ are independent edges of G. It follows from Claim 8 that $\{f, h, f', h'\} = C_1^f \triangle C_2^f \triangle C_1^h \triangle C_2^h$. Thus $C_1^e \triangle \{f, h, f', h'\}$ is an odd cycle. It follows from Claim 5 that $C_1^e \triangle \{f, h, f', h'\} = C_2^e$. Hence, $|C_1^e| = |C_2^e| = 3$. Let $e' \in E(M)$ be such that $\{e, e'\}$ is a cover. Claim 8 implies that $E(M) = \{e, e', f, f', h, h'\}$. Then $C_1^{e'} = \{e', h, f'\}$ and $C_2^{e'} = \{e', f, h'\}$. Since $C_1^e, C_2^e, C_1^{e'}, C_2^{e'}$ are also covers of (M, Σ), the only odd circuits of (M, Σ) are $C_1^e, C_2^e, C_1^{e'}, C_2^{e'}$, i.e. (M, Σ) is isomorphic to $(M(K_4), E(K_4))$.

Acknowledgments

I would like to thank Lex Schrijver for suggesting a shorter ending to the proof (Claim 8 and the following paragraph). A similar argument was used in Seymour's original proof. I would also like to thank Bert Gerards for his comments.

References

1. J. Geelen and B. Guenin. Packing odd circuits in Eulerian graphs. *To appear in J. Comb. Theory Ser. B.*
2. P.D. Seymour. The Matroids with the Max-Flow Min-Cut property. *J. Comb. Theory Ser. B,* 23:189–222, 1977.

Integer Programming
and Arrovian Social Welfare Functions

Jay Sethuraman[1], Chung-Piaw Teo[2], and Rakesh V. Vohra[3]

[1] IEOR Department, Columbia University, New York, NY 10027, USA
jay@ieor.columbia.edu
[2] Department of Decision Sciences, National University of Singapore,
Singapore 117591
bizteocp@nus.edu.sg
[3] Department of Managerial Economics and Decision Sciences, Kellogg Graduate
School of Management, Northwestern University, Evanston IL 60208, USA
r-vohra@nwu.edu

Abstract. We formulate the problem of deciding which preference domains admit a non-dictatorial Arrovian Social Welfare Function as one of verifying the feasibility of an integer linear program. Many of the known results about the presence or absence of Arrovian social welfare functions, impossibility theorems in social choice theory, and properties of majority rule etc., can be derived in a simple and unified way from this integer program. We characterize those preference domains that admit a non-dictatorial, neutral Arrovian social welfare Function and give a polyhedral characterization of Arrovian social welfare functions on single-peaked domains.

1 Introduction

The Old Testament likens the generations of men to the leaves of a tree. It is a simile that applies as aptly to the literature inspired by Arrow's impossibility theorem [2]. Much of it is devoted to classifying those preference domains that admit or exclude the existence of a non-dictatorial Arrovian social welfare function (ASWF)[1]. We add another leaf to that tree. Specifically, we formulate the problem of deciding which preference domains admit a non-dictatorial Arrovian social welfare function as one of verifying the feasibility of an integer linear program. Many of the known results about the presence or absence of Arrovian social welfare functions, impossibility theorems in social choice theory, properties of the majority rule etc., can be derived in a simple and unified way from this integer program. The integer program also leads to some interesting new results such as (a) a characterization of preference domains that admit a non-dictatorial, neutral Arrovian social welfare function; and (b) a polyhedral characterization of Arrovian social welfare Functions on single-peaked domains.

---

[1] An ASWF is a social welfare function that satisfies the axioms of the Impossibility theorem.

W.J. Cook and A.S. Schulz (Eds.): IPCO 2002, LNCS 2337, pp. 194–211, 2002.
© Springer-Verlag Berlin Heidelberg 2002

Let $\mathcal{A}$ denote the set of alternatives (at least three). Let Σ denote the set of all transitive, antisymmetric and total binary relations on $\mathcal{A}$. An element of Σ is a preference ordering. The set of admissible preference orderings for members of a society of n-agents (voters) will be a subset of Σ and denoted Ω. Let Ω^n be the set of all n-tuples of preferences from Ω, called *profiles*. An element of Ω^n will typically be denoted as $\mathbf{P} = (\mathbf{p_1}, \mathbf{p_2}, \ldots, \mathbf{p_n})$, where $\mathbf{p_i}$ is interpreted as the preference ordering of agent i. (In the language of Le Breton and Weymark [7], we assume the "common preference domain" framework; this assumption can be relaxed, see Sect. 2.) An n-person social welfare function is a function $f : \Omega^n \to \Sigma$. Thus for any $\mathbf{P} \in \Omega^n$, $f(\mathbf{P})$ is an ordering of the alternatives. We write $xf(\mathbf{P})y$ if x is ranked above y under $f(\mathbf{P})$. An n-person *Arrovian social welfare function* (ASWF) on Ω is a function $f : \Omega^n \to \Sigma$ that satisfies the following two conditions:

1. **Unanimity:** If for $\mathbf{P} \in \Omega^n$ and some $x, y \in \mathcal{A}$ we have $x\mathbf{p_i}y$ for all i then $xf(\mathbf{P})y$.
2. **Independence of Irrelevant Alternatives:** For any $x, y \in \mathcal{A}$ suppose $\exists \mathbf{P}, \mathbf{Q} \in \Omega^n$ such that $x\mathbf{p_i}y$ if an only if $x\mathbf{q_i}y$ for $i = 1, \ldots, n$. Then $xf(\mathbf{P})y$ if an only if $xf(\mathbf{Q})y$.

The first axiom stipulates that if all voters prefer alternative x to alternative y, then the social welfare function f must rank x above y. The second axiom states that the ranking of x and y in f is not affected by how the voters rank the other alternatives. An obvious social welfare function that satisfies the two conditions is the *dictatorial rule*: rank the alternatives in the order of the preferences of a particular voter (the dictator). Formally, an ASWF is *dictatorial* if there is an i such that $f(\mathbf{P}) = \mathbf{p_i}$ for all $\mathbf{P} \in \Omega^n$. An ordered pair $x, y \in \mathcal{A}$ is called *trivial* if $x\mathbf{p}y$ for all $\mathbf{p} \in \Omega$. In view of unanimity, any ASWF must have $xf(\mathbf{P})y$ for all $\mathbf{P} \in \Omega^n$ whenever x, y is a trivial pair. If Ω consists only of trivial pairs then distinguishing between dictatorial and non-dictatorial ASWF's becomes nonsensical, so we assume that Ω contains at least one non-trivial pair. The domain Ω is *Arrovian* if it admits a non-dictatorial ASWF.

The main contributions of this paper are summarized below.

- We provide an integer linear programming formulation of the problem of finding an n-person ASWF. For each Ω we construct a set of linear inequalities with the property that every feasible 0-1 solution corresponds to an n-person ASWF.
- When restricted to the class of neutral ASWF's the integer program yields a simple and easily checkable characterization of domains that admit neutral, non-dictatorial ASWF's. This result contains as a special case the results of Sen [14] and Maskin [8] about the robustness of the majority rule.
- For the case when Ω is single-peaked, we show that the polytope defined by the set of linear inequalities is *integral*: the vertices of the polytope correspond to ASWF's and every ASWF corresponds to a vertex of the polytope. This gives the first characterization of ASWF's on this domain. The same proof technique yields a characterization of the generalized majority rule on single peaked domains, originally due to Moulin [10].

- We show that the computational complexity of deciding whether a domain is Arrovian depends critically on the way the domain is described. We propose a graph-theoretical method to identify stronger linear inequalities for ASWF's. For cases with a small number of alternatives (3 or 4), our approach is able to characterize the polytope of all ASWF's. Thus for *any* Ω and any set of alternatives size at most 4 we characterize the polyhedral structure of all ASWF's.

2   The Integer Program

Denote the set of all ordered pairs of alternatives by $\mathcal{A}^2$. Let E denote the set of all agents, and S^c denote $E \setminus S$ for all $S \subseteq E$.

To construct an n-person ASWF we exploit the independence of irrelevant alternatives condition. This allows us to specify an ASWF in terms of which ordered pair of alternatives a particular subset, S, of agents is decisive over.

Definition 1. *For a given ASWF f, a subset S of agents is* weakly decisive for x over y *if whenever all agents in S rank x over y and all agents in S^c rank y over x, the ASWF f ranks x over y.*

Since this is the only notion of decisiveness used in the paper, we omit the qualifier 'weak' in what follows.

For each non-trivial element $(x, y) \in \mathcal{A}^2$, we define a 0-1 variable as follows:

$$d_S(x, y) = \begin{cases} 1, & \text{if the subset } S \text{ of agents is decisive for } x \text{ over } y; \\ 0, & \text{otherwise.} \end{cases}$$

If $(x, y) \in \mathcal{A}^2$ is a trivial pair then by default we set $d_S(x, y) = 1$ for all $S \neq \emptyset$.

Given an ASWF f, we can determine the associated d variables as follows: for each $S \subseteq E$, and each non-trivial pair (x, y), pick a $\mathbf{P} \in \Omega^n$ in which agents in S rank x over y, and agents in S^c rank y over x; if $x f(\mathbf{P}) y$, set $d_S(x, y) = 1$, else set $d_S(x, y) = 0$.

In the rest of this section, we identify some conditions satisfied by the d variables associated with an ASWF f.

Unanimity: To ensure unanimity, for all $(x, y) \in \mathcal{A}^2$, we must have

$$d_E(x, y) = 1. \tag{1}$$

Independence of Irrelevant Alternatives: Consider a pair of alternatives $(x, y) \in \mathcal{A}^2$, a $\mathbf{P} \in \Omega^n$, and let S be the set of agents that prefer x to y in $\mathbf{P}$. (Thus, each agent in S^c prefers y to x in $\mathbf{P}$.) Suppose $x f(\mathbf{P}) y$. Let $\mathbf{Q}$ be any other profile such that all agents in S rank x over y and all agents in S^c rank y over x. By the independence of irrelevant alternatives condition $x f(\mathbf{Q}) y$. Hence the set S is decisive for x over y. However, had $y f(\mathbf{P}) x$ a similar argument would imply that S^c is decisive for y over x. Thus, for all S and $(x, y) \in \mathcal{A}^2$, we must have

$$d_S(x, y) + d_{S^c}(y, x) = 1. \tag{2}$$

A consequence of Eqs. (1) and (2) is that $d_\emptyset(x, y) = 0$ for all $(x, y) \in \mathcal{A}^2$.

Transitivity: To motivate the next class of constraints, it is useful to consider the majority rule. If the number n of agents is odd, the majority rule can be described using the following variables:

$$d_S(x, y) = \begin{cases} 1, & \text{if} \quad |S| > n/2, \\ 0, & \text{otherwise.} \end{cases}$$

These variables satisfy both (1) and (2). However, if Ω admits a *Condorcet* triple (e.g., $\mathbf{p_1}, \mathbf{p_2}, \mathbf{p_3} \in \Omega$ with $x\mathbf{p_1}y\mathbf{p_1}z$, $y\mathbf{p_2}z\mathbf{p_2}x$, and $z\mathbf{p_3}x\mathbf{p_3}y$), then such a rule does not always produce an *ordering* of the alternatives for each preference profile. Our next constraint (*cycle elimination*) is designed to exclude this and similar possibilities.

Let A, B, C, U, V, and W be (possibly empty) *disjoint* sets of agents whose union includes all agents. For each such partition of the agents, and any triple x, y, z,

$$d_{A \cup U \cup V}(x, y) + d_{B \cup U \cup W}(y, z) + d_{C \cup V \cup W}(z, x) \leq 2, \tag{3}$$

where the sets satisfy the following conditions (hereafter referred to as conditions (*)):

$$A \neq \emptyset \text{ only if there exists } \mathbf{p} \in \Omega, x\mathbf{p}z\mathbf{p}y,$$
$$B \neq \emptyset \text{ only if there exists } \mathbf{p} \in \Omega, y\mathbf{p}x\mathbf{p}z,$$
$$C \neq \emptyset \text{ only if there exists } \mathbf{p} \in \Omega, z\mathbf{p}y\mathbf{p}x,$$
$$U \neq \emptyset \text{ only if there exists } \mathbf{p} \in \Omega, x\mathbf{p}y\mathbf{p}z,$$
$$V \neq \emptyset \text{ only if there exists } \mathbf{p} \in \Omega, z\mathbf{p}x\mathbf{p}y,$$
$$W \neq \emptyset \text{ only if there exists } \mathbf{p} \in \Omega, y\mathbf{p}z\mathbf{p}x.$$

The constraint ensures that on any profile $\mathbf{P} \in \Omega^n$, the ASWF f does not produce a ranking that "cycles".

Theorem 1. *Every feasible integer solution to (1)-(3) corresponds to an ASWF and vice-versa.*

Proof. Given an ASWF, it is easy to see that the corresponding d vector satisfies (1)-(3). Now pick any feasible solution to (1)-(3) and call it d. To prove that d gives rise to an ASWF, we show that for every profile of preferences from Ω, d generates an ordering of the alternatives. Unanimity and Independence of Irrelevant Alternatives follow automatically from the way the d_S variables are used to construct the ordering.

Suppose d does not produce an ordering of the alternatives. Then, for some profile $\mathbf{P} \in \Omega^n$, there are three alternatives x, y and z such that d ranks x over y, y over z and z over x. For this to happen there must be three non-empty sets H, I, and J such that

$$d_H(x, y) = 1, \quad d_I(y, z) = 1, \quad d_J(z, x) = 1,$$

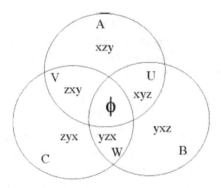

Fig. 1. The sets and the associated orderings

and for the profile **P**, agent i ranks x over y (resp. y over z, z over x) if and only if i is in H (resp. I, J). Note that $H \cup I \cup J$ is the set of all agents, and $H \cap I \cap J = \emptyset$.

Let

$$A \leftarrow H \setminus (I \cup J), B \leftarrow I \setminus (H \cup J), C \leftarrow J \setminus (H \cup I),$$

$$U \leftarrow H \cap I, V \leftarrow H \cap J, W \leftarrow I \cap J.$$

Now A (resp. B, C, U, V, W) can only be non-empty if there exists **p** in Ω with x**pzpy** (resp. y**pxpz**, z**pypx**, x**pypz**, z**pxpy**, y**pzpx**).

In this case constraint (3) is violated since

$$d_{A \cup U \cup V}(x,y) + d_{B \cup U \cup W}(y,z) + d_{C \cup V \cup W}(z,x) = d_H(x,y) + d_I(y,z) + d_J(z,x) = 3.$$

□

For the case $n = 2$, constraint (3) can be simplified as follows: (i) if for some **p, q** $\in \Omega$ and $x, y, z \in \mathcal{A}$, we have x**pypz** and y**qzqx**, then

$$d_S(x,y) \leq d_S(x,z), \tag{4}$$

$$d_S(z,x) \leq d_S(y,x); \tag{5}$$

and (ii) if for some **p** $\in \Omega$ and $x, y, z \in \mathcal{A}$, we have x**pypz**, then

$$d_S(x,y) + d_S(y,z) \leq 1 + d_S(x,z), \tag{6}$$

$$d_S(z,y) + d_S(y,x) \geq d_S(z,x). \tag{7}$$

These inequalities, discovered earlier by Kalai and Muller [5], are called *decisiveness implications*. Thus, Constraints (3) generalize the decisiveness implication conditions to $n \geq 3$. We will sometimes refer to (1)-(3) as IP.

General Domains. The IP characterization obtained above can be generalized to the case in which the domain of preferences for each voter is non-identical. In general, let D be the domain of profiles over alternatives. In this case, for each set S, the d_S variables need not be well-defined for each pair of alternatives x, y, if there is no profile in which all agents in S (resp. S^c) rank x over y (resp. y over x). d_S is thus only defined for (x, y) if such profiles exist. Note that $d_S(x, y)$ is well-defined if and only if $d_{S^c}(y, x)$ is well-defined. With this proviso inequalities (1) and (2) remains valid. We only need to modify (3) to the following:

> Let A, B, C, U, V, and W be (possibly empty) *disjoint* sets of agents whose union includes all agents. For each such partition of the agents, and any triple x, y, z,

$$d_{A \cup U \cup V}(x, y) + d_{B \cup U \cup W}(y, z) + d_{C \cup V \cup W}(z, x) \leq 2, \qquad (8)$$

where the sets satisfy the following conditions (hereafter referred to as condition (**)):

$$A \neq \emptyset \text{ only if there exists } \mathbf{p_i}, i \in A, \text{ with } x\mathbf{p_i}z\mathbf{p_i}y,$$
$$B \neq \emptyset \text{ only if there exists } \mathbf{p_i}, i \in B, \text{ with } y\mathbf{p_i}x\mathbf{p_i}z,$$
$$C \neq \emptyset \text{ only if there exists } \mathbf{p_i}, i \in C, \text{ with } z\mathbf{p_i}y\mathbf{p_i}x,$$
$$U \neq \emptyset \text{ only if there exists } \mathbf{p_i}, i \in U, \text{ with } x\mathbf{p_i}y\mathbf{p_i}z,$$
$$V \neq \emptyset \text{ only if there exists } \mathbf{p_i}, i \in V, \text{ with } z\mathbf{p_i}x\mathbf{p_i}y,$$
$$W \neq \emptyset \text{ only if there exists } \mathbf{p_i}, i \in W, \text{ with } y\mathbf{p_i}z\mathbf{p_i}x.$$
$$\text{and } (\mathbf{p_1}, \ldots, \mathbf{p_n}) \in D.$$

The following theorem is immediate from our discussion. We omit the proof.

Theorem 2. *Every feasible integer solution to (1), (2) and (8) corresponds to an ASWF on domain D and vice-versa.*

This yields a new characterization of non-dictatorial profile domains D, and can be used to obtain a simple proof of a result due to Fishburn and Kelly [4] on super non-Arrovian domains; we state this result without proof.

A domain D is called *super non-Arrovian* if it is non-Arrovian and every domain D' containing D is also non-Arrovian. Furthermore, if d_S is well defined for every pair of alternatives x, y and every S, we say that the domain D satisfies the *near-free doubles* condition.

Theorem 3 (Fishburn and Kelly [4]). *A domain D is super-non-Arrovian if and only if it is non-Arrovian and satisfies the near-free doubles condition.*

3   Applications

Arrow's Theorem. Our first use of IP is to provide a simple proof of Arrow's theorem.

Theorem 4 (Arrow's Impossibility theorem). *When $\Omega = \Sigma$, the 0-1 solutions to the IP correspond to dictatorial rules.*

Proof: When $\Omega = \Sigma$, we know from constraints (4-5) and the existence of all possible triples that $d_S(x, y) = d_S(y, z) = d_S(z, u)$ for all alternatives x, y, z, u. We will thus write d_S in place of $d_S(x, y)$ in the rest of the proof.

We show first that $d_S = 1 \Rightarrow d_T = 1$ for all $S \subset T$. Suppose not. Let T be the set containing S with $d_T = 0$. Constraint (2) implies $d_{T^c} = 1$. Choose $A = T \setminus S$, $U = T^c$ and $V = S$ in (3). Then, $d_{A \cup U \cup V} = d_E = 1$, $d_{B \cup U \cup W} = d_{T^c} = 1$ and $d_{C \cup V \cup W} = d_S = 1$, which contradicts (3).

The same argument implies that $d_T = 0 \Rightarrow d_S = 0$ whenever $S \subset T$. Note also that if $d_S = d_T = 1$, then $S \cap T \neq \emptyset$, otherwise the assignment $A = (S \cup T)^c, U = S, V = T$ will violate the cycle elimination constraint. Furthermore, $d_{S \cap T} = 1$, otherwise the assignment $A = (S \cup T)^c, U = T \setminus S, V = S \setminus T, W = S \cap T$ will violate the cycle elimination constraint. Hence there exists a minimal set S^* with $d_{S^*} = 1$ such that all T with $d_T = 1$ contains S^*. We show that $|S^*| = 1$. If not there will be $j \in S$ with $d_j = 0$, which by (2) implies $d_{E \setminus \{j\}} = 1$. Since $d_{S^*} = 1$ and $d_{E \setminus \{j\}} = 1$, $d_{E \setminus \{j\} \cap S^*} \equiv d_{S^* \setminus \{j\}} = 1$, contradicting the minimality of S^*. $\qquad\square$

Born Loser rule. For subsequent applications we introduce the *born loser* rule. For each j, we define the *born loser* rule with respect to j (denoted by B_j) in the following way: (i) set $d_E^{B_j}(x, y) = 1$ for every $x, y \in \mathcal{A}^2$; (ii) set $d_\emptyset^{B_j}(x, y) = 0$ for every $x, y \in \mathcal{A}^2$; and (iii) for every non-trivial pair (x, y), and for any $S \neq \emptyset, E$, $d_S^{B_j}(x, y) = 0$ if $S \ni j$, $d_S^{B_j}(x, y) = 1$ otherwise.

Theorem 5. *For any j and $n > 2$, the born loser rule B_j is a non-dictatorial n-person ASWF if and only if for all x, y, z, there do not exist $\mathbf{p_1}, \mathbf{p_2}, \mathbf{p_3}$ in Ω with*

$$x\mathbf{p_1}z\mathbf{p_1}y, \quad x\mathbf{p_2}y\mathbf{p_2}z, \quad z\mathbf{p_3}x\mathbf{p_3}y,$$

Proof. It is clear that by definition, d^{B_j} satisfies (1, 2). To see that it satisfies (3), observe that in every partition of the agents, one of the sets obtained must contain j. Say $j \in A \cup U \cup V$. If $d_{A \cup U \cup V}^{B_j}(x, y) = 0$, then (3) is clearly valid. So we may assume that $d_{A \cup U \cup V}^{B_j}(x, y) = 1$. This happens only when $A \cup U \cup V = E$ (or if (x, y) is trivial, which in turns imply that all the other sets are empty). We may assume $U, V \neq \emptyset$ and $j \in A$, otherwise (3) is clearly valid. But according to condition (*), this implies existence of $\mathbf{p_1}, \mathbf{p_2}, \mathbf{p_3}$ in Ω with

$$x\mathbf{p_1}z\mathbf{p_1}y, \quad x\mathbf{p_2}y\mathbf{p_2}z, \quad z\mathbf{p_3}x\mathbf{p_3}y,$$

which is a contradiction.

So, d^{B_j} satisfies (1-3) and hence corresponds to an ASWF. When $n > 2$, B_j is clearly non-dictatorial. $\qquad\square$

Anonymous and Neutral Rules. Two additional conditions that are sometimes imposed on an ASWF are anonymity and neutrality. An ASWF is called *anonymous* if its ranking over pairs of alternatives remains unchanged when the labels of the agents are permuted. Hence $d_S(x, y) = d_T(x, y)$ for all $(x, y) \in \mathcal{A}^2$

whenever $|S| = |T|$. In particular a dictatorial rule is not anonymous. An ASWF is called *neutral* if its ranking over any pair of alternatives depends only on the pattern of agents' preferences over that pair, not on the alternatives' labels. Neutrality implies that $d_S(x, y) = d_S(a, b)$ for any $(x, y), (a, b) \in \mathcal{A}^2$. Thus the value of $d_S(\cdot, \cdot)$ is determined by S alone. When anonymity and neutrality are combined, $d_S(\cdot, \cdot)$ is determined by $|S|$ alone. In such a case, we write d_S as d_r where $r = |S|$. If n is even, it is not possible for an anonymous ASWF to be neutral because Eq. (2) cannot be satisfied for $|S| = n/2$. The IP (1)-(3) can be used to derive a number of old and new results regarding anonymous and neutral ASWF's in a unified way; we state these results next.

Recall that Ω admits a *Condorcet* triple if there are x, y and $z \in \mathcal{A}$ and $\mathbf{p_1}$, $\mathbf{p_2}$ and $\mathbf{p_3} \in \Omega$ such that $x\mathbf{p_1}y\mathbf{p_1}z$, $y\mathbf{p_2}z\mathbf{p_2}x$, and $z\mathbf{p_3}x\mathbf{p_3}y$.

The following results are well known and follow directly from the IP characterization:

Theorem 6. *(Sen [14]) For an odd number of agents, the majority rule is an ASWF on Ω if and only if Ω does not contain a Condorcet triple.*

Theorem 7. *(Maskin [8]) Suppose there are at least 3 agents. If Ω admits an anonymous, neutral ASWF, then Ω has no Condorcet triples.*

Theorem 8. *(Maskin [8]) Suppose that g is anonymous, neutral, satisfies unanimity and independence of irrelevant alternatives, and is not the majority rule. Then there exists a domain Ω on which g is not an ASWF but the majority rule is.*

The next result, which is new, shows that checking whether Ω admits a neutral, non-dictatorial ASWF reduces to checking whether the majority rule or the born loser rule is an ASWF on that domain. Notice that no parity assumption on the number of voters is needed.

Theorem 9. *For $n \geq 3$, a domain Ω admits a neutral, non-dictatorial ASWF if an only if the majority rule or the born loser rule is an ASWF on Ω.*

Proof. If either the majority rule or the born loser rule is an ASWF on Ω, Ω clearly admits a neutral, non-dictatorial ASWF. Suppose then Ω admits a neutral, non-dictatorial ASWF, but neither the majority rule nor the born loser rule is an ASWF on Ω. Since the majority rule is not an ASWF, Ω admits a Condorcet triple $\{a, b, c\}$. Since the born loser rule is not an ASWF on Ω, by corollary 1 there exist $\mathbf{p_1}, \mathbf{p_2}, \mathbf{p_3}$ in Ω and $x, y, z \in \mathcal{A}$ with

$$x\mathbf{p_1}z\mathbf{p_1}y, \quad x\mathbf{p_2}y\mathbf{p_2}z, \quad z\mathbf{p_3}x\mathbf{p_3}y.$$

We will need the existence of these orderings to construct a partition of the agents that satisfies the cycle elimination constraints. The proof will mimic the proof of Arrow's theorem (Theorem 4) given earlier.

Neutrality implies that $d_S(x, y) = d_S(y, z) = d_S(z, u)$ for all alternatives x, y, z, u. We will thus write d_S in place of $d_S(x, y)$ in the rest of the proof.

First, $d_S = 1 \Rightarrow d_T = 1$ for all $S \subset T$. Suppose not. Let T be the set containing S with $d_T = 0$. Constraint (2) implies $d_{T^c} = 1$. Choose $A = T \setminus S$, $U = T^c$ and $V = S$ in (3). We can do this because of $\mathbf{p_1}, \mathbf{p_2}, \mathbf{p_3}$. Then, $d_{A \cup U \cup V} = d_E = 1$, $d_{B \cup U \cup W} = d_{T^c} = 1$ and $d_{C \cup V \cup W} = d_S = 1$, which contradicts (3).

The same argument implies that $d_T = 0 \Rightarrow d_S = 0$ whenever $S \subset T$. Note also that if $d_S = d_T = 1$, then $S \cap T \neq \emptyset$, otherwise the assignment $A = (S \cup T)^c, U = S, V = T$ will violate the cycle elimination constraint.

Next we show that $d_{S \cap T} = 1$. Suppose not. Consider the assignment $U = E \setminus S$, $V = S \setminus T$ and $W = S \cap T$. We can choose such a partition because $\{a, b, c\}$ form a Condorcet triple. For this specification, $d_{A \cup U \cup V} = d_{E \setminus \{S \cap T\}} = 1$. Since $T \subset B \cup U \cup W$, $d_{B \cup U \cup W} = 1$ and $d_{C \cup V \cup W} = d_S = 1$, which contradicts (3).

Hence there exists a minimal set S^* such that $d_{S^*} = 1$ and all T with $d_T = 1$ contains S^*. We show that $|S^*| = 1$. If not there will be $j \in S$ with $d_j = 0$, and hence $d_{E \setminus \{j\} \cap S^*} = 1$, contradicting the minimality of S^*.   □

A simple consequence of this result is the following theorem due to Kalai and Muller [5]. The proof is new.

Theorem 10. *A non-dictatorial solution to (1, 2, 4 - 7) exists for the case $n = 2$ agents if and only if a non-dictatorial solution to (1-3) exists for any n.*

Proof. Given a 2 person non-dictatorial AWSF, we can build an ASWF for the n-person case by focusing only on the preferences submitted by the first two voters and ranking the alternatives using the 2-person ASWF. This is clearly a non-dictatorial ASWF for the n-person case. Hence we only need to give a proof of the converse.

Let d^* be a non-dictatorial solution to (1-3). Suppose d does not imply a neutral ASWF. Then there is a set of agents S such that $d_S^*(x, y)$ is non-zero for some but not all $(x, y) \in \mathcal{A}^2$. Hence, $d_1 = d_S^*, d_2 = d_{S^c}^*$ would be a non-dictatorial solution to (1, 2, 4-7).

Suppose then d implies a neutral ASWF. By the previous theorem we can choose d to be either the majority rule or the born loser rule. In the first case, we can build a 2 person ASWF by using a dummy voter with a fixed ordering from Ω and using the (3 person) majority rule. In the second case, we can build a 2 person ASWF by adding a dummy born loser.   □

The following refinement to Maskin's result also follows directly from Theorem 9.

Theorem 11. *Let the number of agents be odd. Suppose Ω does not contain any Condorcet triples, and suppose there exist $\mathbf{p_1}, \mathbf{p_2}, \mathbf{p_3}$ in Ω and $x, y, z \in \mathcal{A}$ with*

$$x \mathbf{p_1} z \mathbf{p_1} y, \quad x \mathbf{p_2} y \mathbf{p_2} z, \quad z \mathbf{p_3} x \mathbf{p_3} y.$$

*Then, the majority rule is the **only** anonymous, neutral ASWF on Ω.*

Proof.(Sketch) From the proof to Theorem 9, we know that if d_S corresponds to a neutral ASWF, and if there exist $\mathbf{p_1}, \mathbf{p_2}, \mathbf{p_3}$ in Ω and $x, y, z \in \mathcal{A}$ with $x \mathbf{p_1} z \mathbf{p_1} y$, $x \mathbf{p_2} y \mathbf{p_2} z$, $z \mathbf{p_3} x \mathbf{p_3} y$, then d_S is monotonic. i.e., $d_S \leq d_T$ if $S \subset T$. By May's Theorem, it has to be the majority rule since the majority rule is the only ASWF that is anonymous, neutral and monotonic.   □

Single-peaked Domains. The domain Ω is single-peaked with respect to a linear ordering $\mathbf{q}$ over $\mathcal{A}$ if $\Omega \subseteq \{\mathbf{p} \in \Sigma :$ for every triple (x, y, z) if $x\mathbf{q}y\mathbf{q}z$ then it is **not** the case that $x\mathbf{p}y$ and $z\mathbf{p}y\}$. The class of single-peaked preferences has received a great deal of attention in the literature. Here we show how the IP can be used to characterize the class of ASWF's on single peaked domains. We prove that the constraints (1–3), along with the non-negative constraints on the d variables, are *sufficient* to characterize the convex hull of the 0-1 solutions.

Theorem 12. *When Ω is single-peaked the set of non-negative solutions satisfying (1-3) is an integral polytope. All ASWF's are extreme point solutions of this polytope.*

Proof. (Sketch) It suffices to prove that every fractional solution satisfying (1-3) can be written as a convex combination of 0-1 solutions satisfying the same set of constraints. Let $\mathbf{q}$ be the linear ordering with respect to which Ω is single-peaked.

Let $d_S(\cdot)$ be a (possibly) fractional solution to the linear programming relaxation of (1-3). We round the solution d to the 0-1 solution d' in the following way:

- Generate a random number Z uniformly between 0 and 1.
- For $a, b \in \mathcal{A}$ with $a\mathbf{q}b$, and $S \subset E$, then
  - $d'_S(a, b) = 1$, if $d_S(a, b) > Z$, 0 otherwise;
  - $d'_S(b, a) = 1$, if $d_S(b, a) \geq 1 - Z$, 0 otherwise.

The 0-1 solution d'_S generated in the above manner clearly satisfies constraints (1). To verify that it satisfies constraint (2), consider a set $T \subseteq E$, an arbitrary pair of alternatives a, b, and suppose without loss of generality $a\mathbf{q}b$. From the linear programming relaxation, we know that either $d_T(a, b) > Z$ or $d_T(b, a) \geq 1 - Z$ (since the two variables add up to 1), but not both. Thus, exactly one of $d'_T(a, b)$ or $d'_T(b, a)$ is set to 1.

We show next that all the constraints in (3) are satisfied by the solution $d'_S(\cdot)$. Consider three alternatives a, b, c, and constraint (3) (with a, b, c replacing the role of x, y, z) can be re-written as:

$$d_{A \cup U \cup V}(a, b) + d_{B \cup U \cup W}(b, c) + d_{C \cup V \cup W}(c, a) \leq 2.$$

Suppose $a\mathbf{q}b\mathbf{q}c$. Then in constraints (3), by the single-peakedness property, we must have $A = V = \emptyset$. In this case, the constraint reduces to $d_U(a, b) + d_{B \cup U \cup W}(b, c) + d_{C \cup W}(c, a) \leq 2$.

We need to show that $d'_U(a, b) + d'_{B \cup U \cup W}(b, c) + d'_{C \cup W}(c, a) \leq 2$. By choosing the sets in constraints (3) in a different way, with $U' \leftarrow U$, $B' \leftarrow B$, $W' \leftarrow W \cup C$, $C' \leftarrow \emptyset$, we have a new inequality $d_{U'}(a, b) + d_{B' \cup U' \cup W'}(b, c) + d_{C' \cup W'}(c, a) \leq 2$, which is equivalent to $d_U(a, b) + 1 + d_{C \cup W}(c, a) \leq 2$. Hence we must have $d_U(a, b) + d_{C \cup W}(c, a) \leq 1$. Note that since $a\mathbf{q}b$ and $b\mathbf{q}c$, our rounding scheme ensures that $d'_U(a, b) + d'_{C \cup W}(c, a) \leq 1$. Hence $d'_U(a, b) + d'_{B \cup U \cup W}(b, c) + d'_{C \cup W}(c, a) \leq 2$.

To finish the proof, we need to show that constraint (3) holds for different orderings of a, b and c under $\mathbf{q}$; the above argument can be easily extended to handle all these cases to show that constraint (3) is valid. Integrality of the polytope follows directly from this rounding method. We omit the details here.

$\square$

The argument above shows the set of ASWF's on single-peaked domains (wrt $\mathbf{q}$) has a property similar to the generalized median property of the stable marriage problem (see Teo and Sethuraman [15]).

Theorem 13. *Let $f_1, f_2, \ldots, f_N$ be distinct ASWF's for the single-peaked domain Ω (with respect to $\mathbf{q}$). Define a function $F_k : \Omega^n \to \Sigma$ with the property:*

> *The set S under F_k is decisive for x over y if*
>   *$x\mathbf{q}y$, and S is decisive for x over y for at least $k + 1$ of the ASWF*
>   *f_i's; or*
>   *$y\mathbf{q}x$, and S is decisive for x over y for at least $N - k$ of the ASWF*
>   *f_i's.*
> *Then F_k is also an ASWF.*

One consequence of Theorem 13 is that when Ω is single-peaked, it is Arrovian, since the dictatorial ASWF's can be used to construct non-dictatorial ASWF in the above manner. For instance, consider the case $n = 2$. Let f_1 and f_2 be the dictatorial rule associated with agent's 1 and 2 respectively. The function F_1 constructed above reduces to the following ASWF:

> If $x\mathbf{q}y$, the social welfare function ranks x above y if and only if both agents prefer x over y.
> If $y\mathbf{q}x$, the social welfare function ranks y above x if and only if none of the agents prefer x above y.

Generalized Majority Rule. Moulin [10] has introduced a generalization of the majority rule called the generalized majority rule. A Generalized majority rule (GMR) M for n agents is of the following form:

- Add n-1 dummy agents, each with a fixed preference drawn from Ω.
- x is ranked above y under M if and only if the majority (of real and dummy agents) prefer x to y.

Each instance of a GMR can be described algebraically as follows. Fix a profile $\mathbf{R} \in \Omega^{n-1}$ and let $R(x, y)$ be the number of orderings in $\mathbf{R}$ where x is ranked above y. Given any profile $\mathbf{P} \in \Omega^n$, GMR ranks x above y if the number of agents who rank x above y under $\mathbf{P}$ is at least $n - R(x, y)$. To check that GMR is an ASWF on single peaked domains, set

$$g_S(x, y) = 1 \text{ iff } |S| \geq n - R(x, y)$$

and zero otherwise. It is easy to check that g satisfies (1)-(3) when Ω is single peaked.

GMR has two important properties. The first is that it is anonymous and second that it is monotonic.

Definition 2. *An ASWF is monotonic if when one switches from the profile* $\mathbf{P}$ *to* $\mathbf{Q}$ *by raising the ranking of* $x \in A$ *for at least one agent, then* $f(\mathbf{Q})$ *will not rank* x *lower than it is in* $f(\mathbf{P})$.

Theorem 14 (Moulin). *An ASWF that is anonymous and monotonic on a single-peaked domain* Ω *must be a generalized majority rule*

Proof. Let d_S be a solution to (1)-(3), corresponding to an anonymous and monotonic ASWF on the domain Ω. Let $\mathbf{q}$ be the underlying order of alternatives. For each $(x, y) \in A^2$, by anonymity, $d_S(x, y)$ depends only on the cardinality of S. Monotonicity implies $d_S(x, y) \leq d_T(x, y)$ if $S \subseteq T$. Thus

$$d_S(x, y) = 1 \text{ if and only if } |S| \geq e(x, y)$$

for some number $e(x, y)$. To complete the proof we need to determine a profile $\mathbf{R} \in \Omega^{n-1}$ such that

$$n - R(x, y) = e(x, y) \; \forall (x, y) \in A^2.$$

Since $d_S(x, y) + d_{S^c}(y, x) = 1$, we have

$$e(x, y) + e(y, x) = n + 1$$

for all (x, y) and (y, x). Note that $e(x, y) \geq 1$ and $e(x, y) \leq n$. Furthermore, if $x\mathbf{q}y\mathbf{q}z$, then by (4) and (5), $d_S(x, y) \leq d_S(x, z)$ and hence $e(x, y) \geq e(x, z)$. Similarly, we have $e(y, x) \leq e(z, x)$, $e(z, y) \geq e(z, x)$ and $e(x, z) \geq e(y, z)$.

We use the geometric construction used in the earlier proof to construct the profile $\mathbf{R} \in \Omega^{n-1}$.

- To each (x, y) such that $x\mathbf{q}y$, associate the interval $[0, e(x, y)]$ and label it $l(x, y)$.
- To each (x, y) such that $y\mathbf{q}x$, associate the interval $[n + 1 - e(x, y), n + 1]$ and label it $l(x, y)$.

We construct preferences in $\mathbf{R}$ in the following way:

- For each $k = 1, 2, \ldots, n - 1$, if $l(x, y)$ covers the point $k + 0.5$, then the kth dummy voter ranks y over x. Otherwise the dummy voter ranks x over y.

Since the intervals $l(x, y)$ and $l(y, x)$ are disjoint and cover $[0, n+1]$ the procedure is well-defined. If $R(x, y)$ is the number of dummy voters who rank x above y in this construction it is easy to see that $n - R(x, y) = e(x, y)$, which is what we need. It remains then to to show that the profile constructed is in Ω^{n-1}.

Claim. The procedure returns a linear ordering of the alternatives.

Proof. Suppose otherwise and consider three alternatives x, y, z where the procedure (for some dummy voter) ranks x above y, y above z and z above x. Hence the intervals $l(x, y)$, $l(y, z)$ and $l(z, x)$ do not cover the point $k + 0.5$. From symmetry, it suffices to consider the following two cases:

- Case 1. Suppose $x\mathbf{q}y\mathbf{q}z$. Since $l(x,z)$ covers the point $k+0.5$ and $e(x,y) \geq e(x,z)$, $l(x,y)$ must cover the point $k+0.5$, a contradiction.
- Case 2. Suppose $y\mathbf{q}x\mathbf{q}z$. Now, there exists $\mathbf{p}$ and $\mathbf{p}'$ in Ω with $z\mathbf{p}x\mathbf{p}y$ and $x\mathbf{p}'y\mathbf{p}'z$, hence $l(z,x) \geq l(z,y)$. This is impossible as $l(z,y)$ covers the point $k+0.5$ but $l(z,x)$ does not.

Hence the ordering constructed is a linear order.   □

Claim. The linear orderings constructed for the dummy voters correspond to orderings from Ω.

Proof. If not there exist k and $x\mathbf{q}y\mathbf{q}z$ with the kth dummy voter ranking y below x and z. i.e. $l(x,y)$ does not cover the point $k+0.5$ and $l(y,z)$ does. Hence $e(x,y) < e(y,z)$. Now, using $x\mathbf{q}y\mathbf{q}z$, we have

$$d_S(x,y) \leq d_S(x,z), d_S(x,z) \leq d_S(y,z).$$

So

$$e(x,y) \geq e(x,z), e(x,z) \geq e(y,z),$$

which is a contradiction.   □

Muller-Satterthwaite Theorem. A *social choice function* maps profiles of preferences into a single alternative. These are objects that have received as much attention as social welfare functions. It is therefore natural to ask if the integer programming approach described above can be used to obtain results about social choice functions. Up to a point, yes. The difficulty is that knowing what alternative a social choice function will pick from a set of size two, does not, in general, allow one to infer what it will choose when the set of alternatives is extended by one. However, given the additional assumptions imposed upon a social choice function one can surmount this difficulty. We illustrate how with an example.

The analog of Arrow's impossibility theorem for social choice functions is the Muller-Satterthwaite theorem [11]. The counterpart of Unanimity and the Independence of Irrelevant Alternatives condition for social choice functions are called *pareto optimality* and *monotonicity*. To define them, denote the preference ordering of agent i in profile $\mathbf{P}$ by $\mathbf{p_i}$.

1. **Pareto Optimality:** Let $\mathbf{P} \in \Omega^n$ such that $x\mathbf{p}y$ for all $\mathbf{p} \in \mathbf{P}$. Then $f(\mathbf{P}) \neq y$.
2. **Monotonicity:** For all $x \in \mathcal{A}$, $\mathbf{P}, \mathbf{Q} \in \Omega^n$ if $x = f(\mathbf{P})$ and $\{y : x\mathbf{p_i}y\} \subseteq \{y : x\mathbf{q_i}y\}$ $\forall i$ then $x = f(\mathbf{Q})$.

We call a social choice function that satisfies pareto-optimality and monotonicity an Arrovian social choice function (ASCF).

Theorem 15 (Muller-Satterthwaite). *When $\Omega = \Sigma$, all ASCF's are dictatorial[2].*

---

[2] The more well known result about strategy proof social choice functions is due to Gibbard [3] and Satterthwaite [13]. It is a consequence of Muller-Satterthwaite [11].

Proof: For each subset S of agents and ordered pair of alternatives (x, y), denote by $[S, x, y]$ the set of all profiles where agents in S rank x first and y second, and agents in S^c rank y first and x second. By the hypothesis on Ω this collection is well defined.

For any profile $\mathbf{P} \in [S, x, y]$ it follows by pareto optimality that $f(\mathbf{P}) \in \{x, y\}$. By monotonicity, if $f(\mathbf{P}) = x$ for one such profile $\mathbf{P}$ then $f(\mathbf{P}) = x$ for all $\mathbf{P} \in [S, x, y]$.

Suppose then for all $\mathbf{P} \in [S, x, y]$ we have $f(\mathbf{P}) \neq y$. Let $\mathbf{Q}$ be any profile where all agents in S rank x above y, and all agents in S^c rank y above x. We show next that $f(\mathbf{Q}) \neq y$ too.

Suppose not. That is $f(\mathbf{Q}) = y$. Let $\mathbf{Q}'$ be a profile obtained by moving x and y to the top in every agents ordering but preserving their relative position within each ordering. So, if x was above y in the ordering under $\mathbf{Q}$, it remains so under $\mathbf{Q}'$. Similarly if y was above x. By monotonicity $f(\mathbf{Q}') = y$. But monotonicity with respect to $\mathbf{Q}'$ and $\mathbf{P} \in [S, x, y]$ implies that $f(\mathbf{P}) = y$ a contradiction.

Hence, if there is one profile in which all agents in S rank x above y, and all agents in S^c rank y above x, and y is not selected, then all profiles with such a property will not select y. This observation allows us to describe ASCF's using the following variables.

For each $(x, y) \in \mathcal{A}^2$ define a 0-1 variable as follows:

- $g_S(x, y) = 1$ if when all agents in S rank x above y and all agents in S^c rank y above x then y is never selected,
- $g_S(x, y) = 0$ otherwise.

If E is the set of all candidates we set $g_E(x, y) = 1$ for all $(x, y) \in \mathcal{A}^2$. This ensures pareto optimality.

Consider a $\mathbf{P} \in \Omega^n$, $(x, y) \in \mathcal{A}^2$ and subset S of agents such that all agents in S prefer x to y and all agents in S^c prefer y to x. Then, $g_S(x, y) = 0$ implies that $g_{S^c}(y, x) = 1$ to ensure a selection. Hence for all S and $(x, y) \in \mathcal{A}^2$ we have

$$g_S(x, y) + g_{S^c}(y, x) = 1 \ . \tag{9}$$

We show that the variables g_S satisfy the cycle elimination constraints. If not there exists a triple $\{x, y, z\}$, and set A, B, C, U, V, W such that the cycle elimination constraint is violated. Consider the profile $\mathbf{P}$ where each voter ranks the triple $\{x, y, z\}$ above the rest, and with the ordering of x, y, z depending on whether the voter is in A, B, C, U, V or W. Since $g_{A \cup U \cup V}(x, y) = 1$, $g_{B \cup U \cup W} = 1$, and $g_{C \cup V \cup W} = 1$, none of the alternatives x, y, z is selected for the profile $\mathbf{P}$. This violates pareto optimality, a contradiction.

Hence g_S satisfies constraints (1-3). Since $\Omega = \Sigma$, by Arrow's Impossibility Theorem, g_S corresponds to a dictatorial solution. $\qquad\qquad\square$

4   Decomposability, Complexity and Valid Inequalities

A domain is called *decomposable* if and only if there is a non-trivial solution (not all 1's or all 0's) to the system of inequalities (1, 2, 4–7) for the case

$n = 2$. The main result of [5] (cf. Theorem 10) can be phrased as follows: *the domain Ω is non-dictatorial if and only if it is decomposable.* This result allows one to formulate the problem of deciding whether Ω is arrovian as an integer program involving a number of variables and constraints that is polynomial in $|\mathcal{A}|$. However, the set $\mathcal{A}$ is not the only input to the problem. The preference domain Ω is also an input. If Ω is specified by the set of permutations it contains, and if it has exponentially many permutations (say $O(2^{|\mathcal{A}|})$), the the straight forward input model needs at least $O(2^{|\mathcal{A}|})$ bits. Recall the number of decision variables for the integer program for 2-person ASWF's is polynomial in $|\mathcal{A}|$. Furthermore, the time complexity of verifying the existence of triplets in Ω can trivially be performed in time $O(n^3 2^{|\mathcal{A}|})$. Hence the decision version of the decomposability conditions can be solved in time polynomial in the size of the input.

Suppose, however, instead of listing the elements of Ω, we prescribe a polynomial time oracle to check membership in Ω. The complexity issue of deciding whether the domain is decomposable now depends on how we encode the membership oracle, and not on the number of elements in Ω. In this model, we exhibit an example to show that checking whether a triplet exists in Ω is already NP-hard.

Let G be a graph with vertex set V. Let Ω_G consist of all orderings of V that correspond to a Hamiltonian path in G. Given any triple $(u, v, w) \in V$, the problem of deciding if G admits a Hamiltonian path in which u precedes v precedes w is NP-complete[3]. Hence the problem of deciding whether there is a preference ordering $\mathbf{p}$ in Ω with $u\mathbf{p}v\mathbf{p}w$ is already NP-complete.

Thus, given an Ω specified by hamiltonian paths, it is already NP-hard just to write down the set of inequalities specified by the decomposability conditions!

One way to by-pass the above difficulties is to focus on ordering on triplets that are realized by some preferences in Ω. The input to the complexity question is thus the set of orderings on triplets ($O(n^3)$ size) that are admissible in Ω. We will focus on this input model for the rest of the paper.

Ignore, for the moment, inequalities of types (6) and (7). The constraint matrix associated with the inequalities of types (1, 2, 4, 5) and $0 \leq d(x, y) \leq 1 \; \forall (x, y) \in \mathcal{A}^2$ is totally unimodular. This is because each inequality can be reduced to one that contains at most two coefficients of opposite sign and absolute value of 1 [4]. Hence the extreme points are all 0-1. If one or more of these extreme points was different from the all 0's solution and all 1's solution we would know that Ω is Arrovian. If the only extreme points were the all 0's solution and all 1's solution that would imply that Ω is not Arrovian.

Thus difficulties with determining the existence of a feasible 0-1 solution different from the all 0's and all 1's solution have to do with the inequalities of the form (6) and (7). Notice that any admissible ordering (by Ω) of three

---

[3] If not, we can apply the algorithm for this problem thrice to decide if G admits a Hamiltonian cycle.

[4] It is well known that such matrices are totally unimodular. See for example, Theorem 11.12 in [1].

alternatives gives rise to an inequality of types (6) and (7). However some of them will be redundant. Constraints (6, 7) are not redundant only when they are obtained from a triplet (x, y, z) with the property:

There exists $\mathbf{p}$ such that $x\mathbf{p}y\mathbf{p}z$ but no $\mathbf{q} \in \Omega$ such that $y\mathbf{q}z\mathbf{q}x$ or $z\mathbf{q}x\mathbf{q}y$.

Such a triplet is called an *isolated triplet*.

Call the inequality representation of Ω, by inequalities of types (1, 2, 4, 5), the *unimodular representation* of Ω. Note that all inequalities in the unimodular representation are of the type $d(x, u) \leq d(x, v)$ or $d(u, x) \leq d(v, x)$. *Furthermore, $d(x, u) \leq d(x, v)$ and $d(u, y) \leq d(v, y)$ appear in the representation only if there exist $\mathbf{p}, \mathbf{q}$ with $u\mathbf{p}x$ and $v\mathbf{p}x$ and $x\mathbf{q}u$ and $x\mathbf{q}v$.*

This connection allows us to provide a graph-theoretic representation of the unimodular representation of Ω as well as a graph-theoretic interpretation of when Ω is not Arrovian.

With each *non-trivial* element of $\mathcal{A}^2$ we associate a vertex. If in the unimodular representation of Ω there is an inequality of the form $d_1(a, b) \leq d_1(x, y)$ where (a, b) and $(x, y) \in \mathcal{A}^2$ then insert a *directed* edge from (a, b) to (x, y). Call the resulting directed graph D^Ω.

If (x, y) is a trivial pair (and hence $(x, y) \notin D^\Omega$), then $d_1(x, y)$ is automatically fixed at 1, and $d_1(y, x)$ fixed at 0. An inequality of the form $d_1(x, y) \leq d_1(x, z)$ (or $d_1(z, y)$) *cannot* appear in the unimodular representation, for any alternative z in $\mathcal{A}$. Otherwise there must be some $\mathbf{p} \in \Omega$ with $y\mathbf{p}x$. Similarly, if (x, y) is trivial, $d_1(y, x) \geq d_1(z, x)$ (or $d_1(y, z)$) *cannot* appear in the unimodular representation, for any alternative z in $\mathcal{A}$. Thus fixing the values of $d_1(x, y)$ and $d_1(y, x)$ arising from a trivial pair (x, y) does not affect the value of $d_1(a, b)$ for $(a, b) \in D^\Omega$.

A subset S of vertices in D^Ω is *closed* if there is no edge directed out of S. That is, there is no directed edge with its tail incident to a vertex in S and its head incident to a vertex outside S. Notice that $d_1(x, y) = 1 \; \forall (x, y) \in S$ and 0 otherwise (and together with those arising from the trivial pairs) is a feasible 0-1 solution to the unimodular representation of Ω if S is closed. Hence every closed set in D^Ω corresponds to a feasible 0-1 solution to the unimodular representation. The converse is also true.

Theorem 16. *If D^Ω is strongly connected then Ω is non-Arrovian.*

Proof. The set of all vertices of D^Ω is clearly a closed set. The solution corresponding to this closed set is the ASWF where agent 1 is the dictator. The empty set of vertices is closed and this corresponds to agent 2 being the dictator. If D^Ω is strongly connected[5], these are the only closed sets in the graph. Since any ASWF must correspond to some closed set in D^Ω, we conclude that Ω is non-Arrovian. □

We note that verifying whether a directed graph is strongly connected can be done efficiently. See [1] for details. Note also that if Ω does not contain any isolated triplets, then Ω is Arrovian if and only D^Ω is not strongly connected.

---

[5] A directed graph is strongly connected if there is a directed cycle through every pair of vertices.

We describe next a sequential lifting method to derive valid inequalities for the problem to strengthen the LP formulation, using the directed graph D^{Ω} defined previously. We say that the node u *dominates* the node v if there is a directed path in D^{Ω} from v to u (i.e. $d(u) \geq d(v)$).

Sequential Lifting Method:

– For each isolated triplet (x, y, z), we have the inequality

$$1 + d(x, z) \geq d(x, y) + d(y, z). \tag{10}$$

– Let $D(x, y)$ (and resp. $D(y, z)$) denote the set of nodes in D^{Ω} that are dominated by the node (x, y) (resp. (y, z)) in D^{Ω}.
– For each node (a, b) in D^{Ω}, if

$$u \in D(a, b) \cap D(x, y) \neq \emptyset, \ v \in D(a, b) \cap D(y, z) \neq \emptyset,$$

then the constraint arising from the isolated triplet can be augmented by the following valid inequalities:

$$d(a, b) + d(x, z) \geq d(u) + d(v). \tag{11}$$

To see the validity of the above constraint, note that by the definition of domination, we have $d(x, y) \geq d(u), d(y, z) \geq d(v), d(a, b) \geq d(u), d(a, b) \geq d(v)$. If $d(a, b) = 0$, then $d(u) = d(v) = 0$ and hence (11) is trivially true. If $d(a, b) = 1$, then (11) follows from (10).

We have successfully verified that the sequential lifting method finds the convex hull of the set of all ASWF's whenever the number of alternatives is at most four. A natural question is if whether the sequential lifting method will gives rise to all facets even for the case $|\mathcal{A}| \geq 5$; we do not yet know, although we suspect the answer to be negative.

5   Conclusions

In this paper, we study the connection between Arrow's Impossibility Theorem and Integer Programming. We show that the set of ASWF's can be expressed as integer solutions to a system of linear inequalities. Many of the well known results connected to the impossibility theorem are direct consequences of the Integer Program. Furthermore, the polyhedral structure of the IP formulation warrants further study in its own right. We have initiated the study on this class of polyhedra by characterizing the polyhedral structure of ASWF's on single peaked domain. We have also demonstrated by an extensive computational experiment that the sequential lifting method proposed in this paper can be used to obtain the complete polyhedral description of ASWF's when the number of alternatives is small. Several interesting problems still remain:

1. Given a domain Ω specified by certain membership oracle, is it possible to check for existence of non-dictatorial ASWF's in polynomial time? Is the problem in the class NP?

2. The LP relaxation of our proposed IP formulation characterizes the ASWF's for single peaked domain. What are the domains that can be characterized by the LP relaxation given by the sequential lifting method?

3. Can the conditions for ASCF's be written down as a system of integer linear inequalities?

We leave the above questions for future research.

References

1. Ahuja, R., T. L. Magnanti and J. B. Orlin. (1993). **Network Flows: Theory, Applications and Algorithms**, Prentice-Hall, Englewood Cliffs, New Jersey, U.S.A.

2. Arrow, K. J. (1963). **Social Choice and Individual Values**, Wiley, New York.

3. Gibbard, A. (1973). "Manipulation of Voting Schemes: A general result", *Econometrica*, 41, 587-602.

4. P. C. Fishburn and J. S. Kelly (1997), Super Arrovian domains with strict preferences, SIAM J. Disc. Math., 11, pp. 83-95.

5. Kalai, E. and E. Muller. (1977). "Characterization of Domains Admitting Non-dictatorial Social Welfare Functions and Non-manipulable Voting Procedures", *J. of Economic Theory*, 16, 2, 457-469.

6. Kalai, E., E. Muller and M. Satterthwaite. (1979). "Social Welfare Functions when Preferences are Convex, Strictly Monotonic and Continuous", *Public Choice*, 34, 87-97.

7. Le Breton, M. and J. Weymark. (1996). "An Introduction to Arrovian Social Welfare Functions on Economic and Political Domains". In **Collective Decision Making: Social Choice and Political Economy**, edited by Norman Schofield, Kluwer Academic Publishers, Boston.

8. Maskin, E. (1995), "Majority Rule, Social Welfare Functions, and Game Forms", in K. Basu, P.K. Pattanaik and K. Suzumura, eds., **Choice, Welfare and Development** (Oxford: Clarendon Press).

9. Moulin, H. (1980). "On Strategy-Proofness and Single Peakedness", *Public Choice*, 35, 437-455.

10. Moulin, H. (1984). "Generalized Condorcet Winners for Single Peaked and Single Plateau Preferences", *Social Choice and Welfare*, 1, 127-147.

11. Muller, E. and M. Satterthwaite. (1977). "The Equivalence of Strong Positive Association and Strategy-Proofness", *Journal of Economic Theory*, 14, 412-418.

12. Nemhauser, G. and L. Wolsey. (1998). **Integer and Combinatorial Optimization**, John Wiley, New York.

13. Satterthwaite, M. (1975). "Strategy-Proofness and Arrow's Conditions: Existence and correspondence theorems for voting procedures and social welfare functions", *J. of Economic Theory*, 10, 187-217.

14. Sen, A. (1966). "A Possibility Theorem on Majority Decisions", *Econometrica*, vol. 34, Issue 2, 491-499.

15. Teo, C.P. and J. Sethuraman (1998) "Geometry of fractional stable matchings and its applications", *Mathematics of Operations Research*, Vol 23, Number 4, 874-891.

Integrated Logistics: Approximation Algorithms Combining Facility Location and Network Design*

Ramamoorthi Ravi and Amitabh Sinha

GSIA, Carnegie Mellon University, Pittsburgh, PA 15213-3890, USA
{ravi,asinha}@andrew.cmu.edu

Abstract. We initiate a study of the approximability of integrated logistics problems that combine elements of facility location and the associated transport network design.

In the simplest version, we are given a graph $G = (V, E)$ with metric edge costs c, a set of potential facilities $\mathcal{F} \subseteq V$ with nonnegative facility opening costs ϕ, a set of clients $D \subseteq V$ (each with unit demand), and a positive integer u (cable capacity). We wish to open facilities and construct a network of cables, such that every client is served by some open facility and all cable capacities are obeyed. The objective is to minimize the sum of facility opening and cable installation costs. With only one zero-cost facility and infinite u, this is the *Steiner tree* problem, while with unit capacity cables this is the *Uncapacitated Facility Location* problem. We give a $(\rho_{ST} + \rho_{UFL})$-approximation algorithm for this problem, where ρ_P denotes any approximation ratio for problem P.

For an extension when the facilities don't have costs but no more than p facilities may be opened, we provide a bicriteria approximation algorithm that has total cost at most $\rho_{p-MEDIAN} + 2$ times the minimum but opens up to $2p$ facilities.

Finally, for the general version with k different types of cables, we extend the techniques of [Guha, Meyerson, Munagala, STOC 2001] to provide an $O(k)$ approximation.

1 Introduction

A ubiquitous problem faced by every corporation which manufactures and sells products to a geographically spread-out market is the following: Where should the factories be built, and how should the finished goods be transported to the markets, so as to minimize costs? Earlier work on facility location problems and network design problems have sought to address these two questions independently. In this paper, we initiate an integrated study of the overall problem; We define and study some simple versions of problems that combine the two objectives, and provide polynomial time approximation algorithms for them.

* Supported in part by a research grant from the Carnegie Bosch Institute, CMU, and by an NSF grant CCR-0105548.

W.J. Cook and A.S. Schulz (Eds.): IPCO 2002, LNCS 2337, pp. 212–229, 2002.

Consider the following scenario: A multinational corporation wishes to enter a promising new geographic market, characterized by demand at each city. It has also identified potential locations of its manufacturing facilities, and the associated costs. Suppose the shipping of the goods (from the facilities to the cities) is to be outsourced to a transport company. This transport company has only one type of truck, with a large capacity. For each truck, the transport company charges at a fixed rate per mile, and offers no discount in case the truck is not utilized to full capacity. The overall logistics problem facing the corporation is to decide on the location of its manufacturing facilities, and a shipping plan of the finished goods to each city, so that the total demand at each city is met and the total cost is minimized. Assume for the sake of simplicity that facilities have no capacity limitations.

If the facility location costs were not an issue (e.g., if the company had already decided where to open its facilities), the problem becomes a single sink edge installation problem [12] (If several facilities are open, they can all be identified into a single sink node). If the transport company charged in proportion to the amount shipped instead of the (discrete) number of trucks used, the problem becomes the uncapacitated facility location problem [21]. Both these problems have been well studied in the past. However, to the best of our knowledge, there has been no effort to study the problem in an integrated way that would allow one to exploit the savings that may result from making both decisions in a coordinated way to reduce the total cost of location and transportation. Our paper addresses this gap, and provides approximation algorithms for some simple versions of the integrated problem.

The first problem we consider is exactly as defined above. We call this the *capacitated cable facility location* problem (CCFL for short). A variant of this problem is the *median* version. Here there are no facility location costs, but we are not allowed to open more than p facilities. We call this the capacitated cable p-median problem (CCpM). Finally, we study an extension of CCFL where, for example, the transport company may provide a range of truck types, each with a different capacity and cost. We call this the *k-cable facility location* problem (KCFL for short). We note for all of these problems, the assumption that the edge-lengths obey the triangle inequality is without loss of generality, since we may use the metric completion of the costs in running our algorithms and replace solution edges by the corresponding shortest paths in the underlying graph of the same total length.

All three problems generalize known NP-hard problems, and hence are NP-hard. We provide polynomial time approximation algorithms for these problems.

1.1   Previous Work

While this is a first attempt to combine the facility location and transport network design objectives, a lot of work has been done on each of the individual problems. The uncapacitated facility location problem has been the focus of much attention in recent years [3,5,6,7,9,13,14,16,17,21,22]. Shmoys, Tardos and Aardal [21] provided the first $O(1)$ approximation algorithm for the uncapacitated facility location problem. They used LP-rounding, thus also showing that the integrality gap of their IP formulation is 4. The bound on the integrality gap was improved to 3 by a primal-dual algorithm due to Jain and Vazirani [14]. An alternative IP formulation was recently provided by Jain, Mahdian and Saberi [13], with integrality gap 1.61. The current best known approximation algorithm is a 1.52 approximation due to Mahdian, Ye and Zhang [17].

Charikar, Guha, Shmoys and Tardos [7] gave the first constant factor approximation algorithm for the p-median problem with metric costs. A local search technique by Korupulu, Plaxton and Rajaraman [16] provided an improved approximation, and this was further improved by Arya, Garg, Khandekar, Meyerson, Munagala and Pandit [3] to a factor of $3 + \epsilon$, which is the best known at present.

Cable installation problems have also received a lot of attention in the recent past [2,4,8,11,12,20,23]. Hassin, Ravi and Salman [12] provide a constant factor approximation for the single sink single cable version of the cable installation problem; we use their method as a subroutine. A constant factor approximation for the multiple cable single sink edge installation problem was first provided by Guha, Meyerson and Munagala [11]; we use this method also in our solution to the general problem. Recently, Talwar [23] showed that the IP formulation of this problem has a constant integrality gap, thus also providing an improved approximation factor for the multiple cable single sink edge installation problem.

1.2 Our Results

The CCFL problem with unit demands at the clients generalizes both the Steiner tree (ST) and uncapacitated facility location (UFL) problems. In the next section (Sect. 2), we present a $\rho_{ST} + \rho_{UFL}$ approximation algorithm for CCFL, where ρ_P is any approximation factor achievable for the problem P. We do this by carefully combining solutions to appropriately set up Steiner tree and UFL problems that capture two natural lower bounds for our problem. With the current best approximation factors, this is a 3.07-approximation algorithm. We also present an integer programming formulation of the problem, and show that its integrality gap is no more than the sum of the integrality gaps of natural formulations of the Steiner Tree and UFL problems. Again, with the current best results, this gap is less than 5. For the case where clients have arbitrary demands and the entire demand for a client must be served by the same facility, we provide a $\rho_{ST} + 2\rho_{UFL}$ approximation, which is currently at most 4.59 (Sect. 2.7).

For CCpM, in Sect. 3, we provide a bicriteria approximation that delivers a solution of cost at most $(\rho_{p-MEDIAN} + 2)$ times the optimum while opening up to $2p$ medians. Again, our method combines approximate solutions to a corresponding p-median problem and a 2-approximation for a newly defined p-Steiner forest problem. With the current best approximation factor for the p-median problem, this is a $(5 + \epsilon, 2)$ bicriteria approximation algorithm for the (total cost, number of medians) problem.

Finally, in Sect. 4, we study the KCFL problem where k different cable types (or truck sizes) are available to us, each with a different cost and capacity. We provide an $O(k)$ approximation for this problem, by extending and adapting the algorithm of Guha et al [11] to incorporate the choices for facility location.

2   The Capacitated-Cable Facility Location Problem

2.1   Problem Definition

The *capacitated-cable facility location* problem (CCFL) is defined as follows. We are given an undirected graph $G = (V, E)$. There is a weight function on the edges, $c : E \to \mathbb{R}^+$, which satisfies the triangle inequality. The *clients* (markets) consist of a subset of nodes, $D \subset V$. The set of *potential facilities*, $\mathcal{F} \subset V$, is also part of the input. Each potential facility $j \in \mathcal{F}$ has a *facility opening cost* of ϕ_j. We are also given an integer $u > 0$, which is the *capacity* of the cable type available to us.

Each client has a demand of one unit, which needs to be serviced by routing one unit of flow from it to some *open* facility. On any edge, we are only allowed to install integral amounts of the cable. If we install z_e copies of the cable on edge e, we can route uz_e units of flow through it, and it costs us $c_e z_e$. Hence our total cost is the cost of all cables installed plus the cost of all the facilities we have opened. The objective of CCFL is to open facilities and install cables connecting clients to open facilities such that no capacity constraint is violated, all clients are served, and the total cost is minimized.

2.2   Hardness and Relation to Other Problems

If there is only a single potential facility ($|\mathcal{F}| = 1$) and u is infinity, then the problem reduces to the *Steiner tree* problem. If there is a single facility and $1 < u < \infty$, CCFL is the *single-sink, single-cable edge installation* problem. If $u = 1$ but $|\mathcal{F}| > 1$, CCFL is the *uncapacitated facility location* problem. All these problems have been studied in the past, and all three are known to be MAX-SNP-hard. Hence CCFL is also MAX-SNP-hard.

2.3   Lower Bounds

We begin with two lemmas which provide lower bounds to an optimal solution of CCFL.

Lemma 1. *Consider a UFL (uncapacitated facility location) instance defined as follows. The set of clients and potential facilities remain the same as in the CCFL instance, but for all edges e, we set the edge cost to be c_e/u. Then the cost of an optimal solution to this UFL instance is a lower bound on the optimal solution to CCFL.*

Proof. Consider the optimal solution to CCFL. In the UFL instance, open all facilities which were opened by CCFL. Every client in CCFL is able to send one unit of flow to an open facility. Construct these flow paths. Now for each client, assign it to the facility it is assigned to in CCFL. The cost of this assignment is at most $\frac{1}{u}$ of that of the flow path used by this client in the CCFL solution, by triangle inequality. This constitutes a feasible solution to UFL, of cost no more than that of the CCFL solution. Hence an optimal UFL solution has cost at most that of the optimal CCFL solution.

Lemma 2. *Consider the graph* $G' = (V \cup \{r\}, E \cup E')$ *where* $E' = \{(j, r) : j \in \mathcal{F}\}$ *and* $c_{(j,r)} = \phi_j$. *Define the set of terminals to be* $R = D \cup \{r\}$. *Then the cost of a minimum Steiner tree in* G' *is a lower bound on the optimal solution of CCFL.*

Proof. Consider the optimal solution to CCFL. The set of edges in the CCFL solution, along with the edges (j, r) such that facility j is opened in the CCFL solution, constitutes a Steiner tree in G' of the same cost as the CCFL solution. Dropping all but one copy of edges which have multiplicity more than 1 in the CCFL solution only reduces the cost. Hence an optimal Steiner tree must cost no more than the optimal CCFL solution.

We use the two lower bounds in Lemma 1 and Lemma 2 (and approximation algorithms for these two problems) to build our solution. We use a flow rerouting algorithm introduced by Hassin, Ravi and Salman [12] to efficiently construct our solution.

2.4   Algorithm

We first run approximation algorithms for an uncapacitated facility location instance and a Steiner tree instance by transforming our problem as described in Lemmas 1 and 2. We then merge the two solutions to obtain a feasible solution of cost no more than the sum of these two approximate solutions (in a *Merge phase*).

To carry out the Merge phase, we adapt a re-routing algorithm described in [12]. We first open all facilities identified by the earlier two phases. Consider the subtrees associated with the facilities opened in the Steiner tree phase. If such a subtree has at most u clients, this subtree along with the facility it is attached to is a feasible solution, without adding any additional copies of the cable.

On the other hand, a subtree that has more than u clients is not feasible right away, since more cables have to be installed along the tree to route all the demand in this overloaded subtree. This is where we use the UFL solution - we clump the demands in these overloaded subtrees into subtrees which are disjoint with respect to edge capacities such that each new subtree has exactly u clients (with one remaining subtree with at most u clients attached to the facility opened in the original overloaded subtree). The fact that such a clumping is possible was proved in [12]; we describe it in detail in Algorithm 1 and prove it in Lemma 3. For each such clump, we use the UFL solution to select the client which is closest to an open facility in the UFL solution, and install one cable from this client to its nearest open facility. The idea is that since each client can pay a $1/u$ fraction of the cable cost to the facility assigned to it by the UFL phase, u such clients in a clump can together pay for one full cable from a client to an open facility if this distance is the cheapest distance among these u clients. In order to achieve this, we need to re-route flow from the $u - 1$ other clients to our selected client in a clump. However, this rerouting takes place along the original Steiner tree solution at no extra cost since the subtrees obtained in the clumping are disjoint with respect to edge capacities. We finally prune the solution by getting rid of unused facilities and cables.

The algorithm is formally described in Fig. 1.

2.5   Analysis

Lemma 3. (due to Hassin, Ravi and Salman [12]) *The solution produced by our algorithm is feasible for CCFL.*

Proof. In the demand routing phase, client demands from a subtree that is not fully served in an iteration may be reassigned in a later iteration. In particular, let's say part of the subtree's demand is routed to a picked client (say i_w^1) in a sibling subtree using upward flow on its parent arc. In the next iteration of the While loop (Steps 21-29), suppose one of the unsatisfied clients (say i_w^2) in this subtree is part of a picked pair. Now, flow from sibling subtrees in this iteration may be routed into it using a downward flow on the same parent arc. However, by standard flow cancellation arguments, no cable is used in both an upward and a downward direction. This flow cancellation implicitly reassigns the clients from the subtree initially assigned to to i_w^1 to i_w^2, and instead redirects the appropriate demand from elsewhere headed for i_w^2 in the second iteration to i_w^1.

The flow cancellation only reduces flow in the upward direction. If any cable has an upward flow, this flow has value at most $u - 1$, and this may potentially be cancelled by downward flow when a client in the subtree below it is part of a picked pair. Downward flow is assigned to any cable at most once, and the quantity of flow assigned is at most $u - 1$. After such an assignment, all the clients in this subtree are deleted from further consideration.

We have argued that both the underlying UFL instance and the associated Steiner tree problem are lower bounds for our CCFL instance. Hence the facilities opened by these two phases can be paid for by these two lower bounds.

The cables purchased by the Steiner tree phase can be paid for by the Steiner tree lower bound. We also install fresh cables in the Merge phase. Each cable has exactly u demand flowing through it. Each of the terminals which use this cable were assigned a facility whose distance is at least the length of the cable in the UFL phase. Hence we can charge the cost of this cable to the cost in the UFL solution.

Recall that ρ_{ST} and ρ_{UFL} denote the currently best known approximation ratios for the Steiner tree and UFL problems respectively. We have the following theorem.

Theorem 1. Algorithm CCFL *is a* $\rho_{ST} + \rho_{UFL}$ *approximation algorithm for CCFL.*

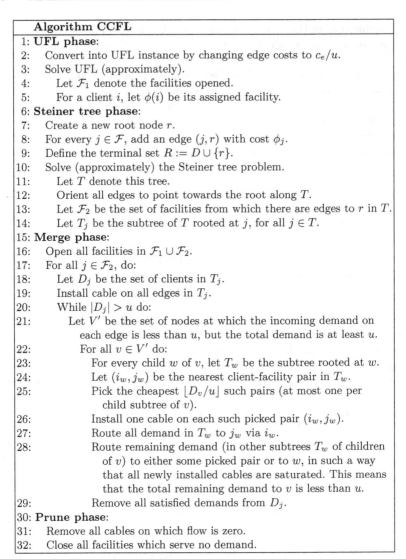

Algorithm CCFL
1: **UFL phase:**
2:     Convert into UFL instance by changing edge costs to c_e/u.
3:     Solve UFL (approximately).
4:         Let $\mathcal{F}_1$ denote the facilities opened.
5:         For a client i, let $\phi(i)$ be its assigned facility.
6: **Steiner tree phase:**
7:     Create a new root node r.
8:     For every $j \in \mathcal{F}$, add an edge (j, r) with cost ϕ_j.
9:     Define the terminal set $R := D \cup \{r\}$.
10:     Solve (approximately) the Steiner tree problem.
11:         Let T denote this tree.
12:         Orient all edges to point towards the root along T.
13:         Let $\mathcal{F}_2$ be the set of facilities from which there are edges to r in T.
14:         Let T_j be the subtree of T rooted at j, for all $j \in T$.
15: **Merge phase:**
16:     Open all facilities in $\mathcal{F}_1 \cup \mathcal{F}_2$.
17:     For all $j \in \mathcal{F}_2$, do:
18:         Let D_j be the set of clients in T_j.
19:         Install cable on all edges in T_j.
20:         While $|D_j| > u$ do:
21:             Let V' be the set of nodes at which the incoming demand on each edge is less than u, but the total demand is at least u.
22:             For all $v \in V'$ do:
23:                 For every child w of v, let T_w be the subtree rooted at w.
24:                 Let (i_w, j_w) be the nearest client-facility pair in T_w.
25:                 Pick the cheapest $\lfloor D_v/u \rfloor$ such pairs (at most one per child subtree of v).
26:                 Install one cable on each such picked pair (i_w, j_w).
27:                 Route all demand in T_w to j_w via i_w.
28:                 Route remaining demand (in other subtrees T_w of children of v) to either some picked pair or to w, in such a way that all newly installed cables are saturated. This means that the total remaining demand to v is less than u.
29:                 Remove all satisfied demands from D_j.
30: **Prune phase:**
31:     Remove all cables on which flow is zero.
32:     Close all facilities which serve no demand.

Fig. 1. Algorithm for CCFL

Proof. This follows from Lemmas 1, 2 and 3.

The current best approximation algorithm for the Steiner tree problem is the one by Robins and Zelikovsky [19], which achieves an approximation factor of 1.55. Mahdian, Ye and Zhang's algorithm [17] is the current best approximation for UFL, with a performance ratio of 1.52. With these values for ρ_{ST} and ρ_{UFL}, Theorem 1 gives a 3.07 approximation.

2.6 IP Formulation and Its Gap

There is a natural integer programming formulation of CCFL, which we describe next. We show that the techniques used in our approximation algorithm described above extend to providing a constant factor rounding algorithm for the linear relaxation of the IP formulation.

IP_{CCFL} is an integer program formulation of CCFL. Variable y_j is an indicator variable which is 1 iff facility j is opened. The number of copies of the cable on edge e is counted by z_e. Finally, f_e^i is the flow of the demand from client i along edge e. For a vertex set S, define $\delta^+(S) = \{(u,v) \in E : u \in S, v \notin S\}$. Define $\delta^-(S) = \delta^+(V \setminus S)$, and for a vertex v, define $\delta^+(v) = \delta^+(\{v\})$. The first four constraints are standard flow conservation and capacity constraints. The last is a connectivity constraint which strengthens the linear relaxation of IP_{CCFL} – these constraints enforce that for any set S containing a client, either it must contain a facility or it must have at least one cable leaving the set (to connect a client within it to a facility in the solution).

$$\min \quad \sum_j \phi_j y_j + \sum_e z_e c_e \qquad (IP_{CCFL})$$

$$\sum_{e \in \delta^+(i)} f_e^i - \sum_{e \in \delta^-(i)} f_e^i \geq 1 \quad \forall i \in D$$

$$\sum_i f_e^i - z_e u \leq 0 \quad \forall e \in E$$

$$\sum_{e \in \delta^-(j)} f_e^i - \sum_{e \in \delta^+(j)} f_e^i \leq y_j \quad \forall i \in D, \forall j \in \mathcal{F}$$

$$\sum_{e \in \delta^+(v)} f_e^i - \sum_{e \in \delta^-(v)} f_e^i = 0 \quad \forall i \in D, \forall v \in V \setminus (\mathcal{F} \cup \{i\})$$

$$\sum_{e \in \delta^+(S)} z_e + \sum_{j \in S} y_j \geq 1 \quad \forall S \subseteq V : S \cap D \neq \emptyset$$

$$z_e, y_j, f_e^i \qquad \text{non-negative integers} \quad.$$

Let gap_{ST} and gap_{UFL} denote the currently known upper bounds on the integrality gap of the undirected cut formulation of Steiner tree problem (See e.g., [1]) and the standard IP formulation of the uncapacitated facility location problem (due to Balinski [5]) respectively, that are obtainable by LP rounding algorithms.

Theorem 2. *The integrality gap of IP_{CCFL} is no more than $gap_{ST} + gap_{UFL}$.*

Proof. Consider an optimal solution to the linear relaxation (denoted LP_{CCFL}) of IP_{CCFL}. The linear relaxation of our UFL instance described in Lemma 1 is exactly LP_{CCFL} with the last constraint ignored. Hence an optimal solution to LP_{CCFL} costs no less than the solution to our UFL instance. If we use the rounding algorithm with gap gap_{UFL} in the UFL phase of our algorithm, this solution costs no more than gap_{UFL} times the value of an optimal solution to LP_{CCFL}.

Similarly, the linear relaxation of our Steiner tree instance described in Lemma 1 is identical to LP_{CCFL} if we ignore the first four (flow) constraints. Hence the value of an optimal solution of the Steiner tree relaxation is a lower bound on the cost of the optimal solution of LP_{CCFL}. Therefore the cost of the Steiner tree phase of our algorithm (using the rounding algorithm with ratio gap_{ST}) is no more than gap_{ST} times the value of an optimal solution to LP_{CCFL}.

The current best bounds on gap_{UFL} and gap_{ST} are 3 [14] and $(2 - \frac{2}{|D|+1})$ [1] respectively. Hence the integrality gap of this formulation of CCFL is less than 5. Note that there is an alternative IP formulation for UFL due to Jain et al. [13] with a lower integrality gap, but this formulation does not fit our framework.

2.7   Non-uniform Demands

The above algorithm generalizes directly to the case of non-uniform demands at the clients, provided we are allowed to split the demand at each client to different facilities. If the demands are unsplittable, the problem becomes more interesting. However, Hassin et al. [12] showed how their (single sink) problem can be solved in the unsplittable demand case with a slight increase in the approximation ratio. Clients which have more than u demand can be sent directly to their nearest facilities, incurring an additional factor of at most 2. To assign the remaining clients, we proceed as before. We now aggregate demands to lie between u and $2u$, and now use the UFL bound at most twice[1]. Hence the approximation ratio for this problem is at most $2\rho_{UFL} + \rho_{ST} = 4.59$.

3   The Capacitated Cable p-Median Problem

3.1   Problem Definition

The *capacitated cable p-median* problem (CCpM) is a minor variant of CCFL. Facilities can be opened for free in this version, but we are not allowed to open more than p facilities (called *medians* in this context). Everything else is as in CCFL. An (α, β) approximation for CCpM consists of a solution which uses βp medians and costs no more than α times the best possible solution which uses no more than p medians.

We first consider a simplified *spanning* version where every node in the graph is a client node and also an eligible median. Let $\rho_{p-MEDIAN}$ denote the best known approximation factor for the p-median problem. We provide a $(\rho_{p-MEDIAN}+1, 2)$-approximation for this restricted spanning version of CCpM.

In Sect. 3.4 we extend it to the case where every node may be a client, a potential median, both, or neither, and provide a $(\rho_{p-MEDIAN} + 2, 2)$ approximation.

---

[1] A slight refinement is possible here since only the cable costs of the UFL solution are used twice while the facility costs are charged at most once in the resulting merged solution.

3.2   Overview of Our Approach

Our approach is essentially the same as before, with appropriate modifications. The proof of the following lemma is identical to the proof of Lemma 1.

Lemma 4. *Consider a p-median instance as follows. The set of clients and potential facilities remain the same as in the CCpM instance, but for all edges e, we set the edge cost to be c_e/u. Then an optimal solution to this p-median instance is a lower bound on the optimal solution to CCpM.*

Definition 1. *Given a graph $G = (V, E)$ with edge costs, the p-spanning forest problem is to find a minimum cost forest with at most p trees.*

Lemma 5. *The p-spanning forest problem can be solved optimally in polynomial time.*

Proof. A minimum spanning tree with the $p - 1$ heaviest edges deleted can be verified to be an optimal solution to the p-spanning forest problem.

Lemma 6. *A minimum cost p-spanning forest in G is a lower bound on the optimal solution of CCpM.*

Proof. Consider the optimal solution to CCpM, and delete all but one copy of edges which have multiplicity more than one. This constitutes a p-spanning forest in G of no greater cost than the CCpM solution.

Our algorithm is now straightforward. We solve the above two problems on our input instance (The p-median instance can only be solved approximately, to a factor $\rho_{p-MEDIAN} = 3 + \epsilon$ [3]). For each tree in the p-spanning forest solution, designate any node as its median. We then reroute exactly as described in the *Merge* phase of *Algorithm CCFL*.

3.3   Analysis

Lemma 3 continues to hold and ensures that we do not violate any capacities in the solution we construct. Lemmas 4 and 6 bound the cost of the two stages of our solution. However, since each of our phases chooses p medians, we may end up with a solution which has as many as $2p$ medians.

Theorem 3. *There is a $(\rho_{p-MEDIAN} + 1, 2)$ approximation algorithm for the spanning version of CCpM.*

3.4   Unrestricted Version of CCpM

We now relax the simplification that every node is a client as well as a possible median. In the unrestricted case, a node may be a client, a possible median, both, or neither (Steiner node). As before, let $\mathcal{F} \subseteq V$ denote the set of possible medians. We can continue to use the $(3 + \epsilon)$ approximation for p-median for the p-median phase. However, our new p-*Steiner* forest problem is as follows. We wish to compute a minimum cost forest which has at most p trees, such that every client is in some tree, and each tree has at least one possible median.

The following is an integer program formulation of p-Steiner forest. Let y_r be an indicator variable indicating whether or not we designate node r to be a median. Let z_e denote whether or not we pick edge e, and for any $S \subseteq V$, let $\delta(S) = \{(u,v) \in E : u \in S, v \notin S\}$. Let $y(S) = \sum_{r \in S \cap \mathcal{F}} y_r$.

$$\min \sum_{e \in E} c_e z_e \qquad (IP_{pSF})$$

$$z(\delta(S)) + y(S) \geq 1 \qquad \forall S \subseteq V, S \cap D \neq \emptyset$$

$$\sum_{r \in \mathcal{F}} y_r \leq p$$

$$z, y \text{ non-negative integers} \quad .$$

The dual of the linear relaxation of IP_{pSF} is given below:

$$\max \sum_{S \subseteq V, S \cap D \neq \emptyset} u_S - \lambda p \qquad (DP_{pSF})$$

$$\sum_{S : e \in \delta(S)} u_S \leq c_e \qquad \forall e \in E$$

$$\sum_{S : r \in S} u_S \leq \lambda \qquad \forall r \in \mathcal{F}$$

$$u, \lambda \geq 0 \quad .$$

The following is a corollary of a lemma about the integrality ratio of the Steiner tree IP formulation, proved in [1] (See also [18]).

Corollary 1. *The primal-dual algorithm of [1] applied to IP_{pSF} provides a polynomial time 2 approximation for the p-Steiner forest problem.*

Proof. (Sketch) In the absence of the second constraint, IP_{pSF} is exactly the undirected Steiner tree IP formulation. The dual of its linear relaxation is exactly DP_{pSF} with the second constraint ignored.

Consider the primal-dual algorithm for Steiner tree of [1]. We can run the same algorithm here to search for a *locally optimal* dual solution. In the algorithm, we raise the value of dual variables corresponding to *minimally violated sets* simultaneously. A *violated* set is a subset of the vertex set which is either not connected by edges selected in the primal solution, or does not contain a median. If necessary, we simultaneously raise λ, as long as the number of connected components in the primal is more than p. We also construct a primal solution alongside, driven by the complementary slackness conditions. Since all variables are being raised simultaneously, we can define a notion of "time", such that the dual variables are being raised at the rate of one unit per unit time. Whenever two minimally violated sets have duals large enough to satisfy a constraint to equality, we add the tightened edge between them to our primal solution, and replace the two sets by their union (which is a new minimally violated set).

Clearly, while there are more than p components the total dual value is increasing. A locally optimal solution is obtained when there are at most p components, each containing a potential median. By imposing an ordering to break ties, we can find a locally maximal dual solution and a corresponding primal solution which has at most p connected components, each with at least one median. Note that in the case when λ is set to a positive value, the algorithm stops when there are exactly p components.

Lemma 5.3 in [1] proves that at time t, the cost of any tree constructed in the primal is no more than twice the total dual collected minus twice the time t. Clearly λ is never more than the final value of the time t, since λ is also raised at the same rate of one per unit time. Therefore, the cost of each component is at most twice the value of the duals collected by moats within it minus twice λ. Recall that when λ is positive, there are exactly p components in the solution. Summing this inequality over the (upto) p components, we get that the primal solution has cost at most twice the value of this locally maximal dual solution to DP_{pSF}.

Theorem 4. *There is a polynomial time $(\rho_{p-MEDIAN}+2,2)$ bicriteria approximation algorithm for the general version of the capacitated-cable p-median problem.*

Proof. This follows from Lemmas 4, 3 and Corollary 1.

The current best value of $\rho_{p-MEDIAN}$ is $3 + \epsilon$ [3]. Hence our algorithm is a $(5 + \epsilon, 2)$ bicriteria approximation to the general cable-capacitated p-median problem.

4   The Multiple-Cable Facility Location Problem

We now consider an extension of CCFL. Instead of just one cable type (or truck type), we have a suite of k cable types. Cable type i has fixed cost σ_i and variable (per unit) cost δ_i. That is, for using one copy of cable type i on edge e and transporting f_e flow through it, our cost is $(\sigma_i + f_e\delta_i)c_e$. This is equivalent to cables having fixed costs and capacities (within an approximation factor of two – see [8,11,23]), which is why we call it a generalization of CCFL. We call this problem the *k-cable facility location* problem, or KCFL.

In KCFL, if $|\mathcal{F}| = 1$, then the problem reduces to the *single sink edge installation* problem. Guha, Meyerson and Munagala provided a constant factor approximation algorithm for this problem [11]. In this section, we show how their algorithm can be adapted to incorporate many facilities. We will closely follow their paper, and many lemmas will not be proved here because they require only minor or notational changes from their paper. We use GMM to denote their paper and algorithm.

4.1   Algorithm

We begin with a brief overview of the GMM algorithm for the single sink edge installation problem. It can be shown that there is a near-optimal solution which is a tree, and where the flow path from any node to the sink uses cables in non-decreasing order of fixed costs. GMM build this tree in a bottom-up fashion. They start with the set of all clients, and build the first layer using a Steiner tree. They are able to charge this cost to a connectivity component of the cost of an optimal solution. They then find points of aggregation of sufficient demand along this tree, and solve a *lower bounded facility location* problem[2] to send some of that demand to the root. The rest of the demand is *aggregated* at certain points. The amount of aggregation is such that it justifies the use of the next type of cable, which has a higher fixed cost but a lower incremental cost. This process is repeated for all the cable types. The details may be read in their paper [11].

For our purposes, we begin by pre-processing the cables as in GMM. We order the cables in order of increasing fixed cost, and then retain cables such that for any i, we have $\sigma_i \leq \alpha \sigma_{i+1}$ and $\delta_{i+1} \leq \alpha \delta_i$, for some predefined constant $\alpha \in (0, \frac{1}{2})$. We also define b_i to be such that $\sigma_{i+1} + \delta_{i+1} b_i = 2\alpha(\sigma_i + \delta_i b_i)$. Intuitively, b_i is the demand quantity at which the cost of using cable types i and $i + 1$ are (almost) the same. We also define $u_i = \sigma_i/\delta_i$. Again, u_i is the point in building the part of the solution using cable type i when the problem shifts its focus from being a Steiner-tree-like problem to being a Shortest-path-aggregation-like problem, since after this threshold, the routing cost of any cable can pay for the fixed cost of installing it on any edge with this much flow.

We are now ready to state our algorithm. In the following, the changes we introduce are italicized, while the rest is the original GMM algorithm.

Iterate over cable types in order of increasing fixed cost:

1. **Steiner tree:** *Augment the graph by adding a "sink" and connecting it to every facility with edge cost equal to the facility cost.* All other edges have cost σ_i. Construct an approximately optimal Steiner tree with the terminals being the sources and the newly added sink. Walking along this tree, identify edges which have u_i demand and "cut" the tree at these edges.

2. **Consolidate:** For every tree in the forest created in the preceding step *not attached to a facility paid for by the Steiner tree*, transfer the total demand in the tree back to one of its sources with probability proportional to the demand at that source. *Route the demands in the trees attached to a paid facility via the tree using the current cable type.*

3. **Shortest path tree:** Set up a lower bounded facility location instance as follows. On every node, the lower bound is b_i and cost is 0. *On facility nodes, the lower bound is 0 and cost is the facility cost.* Edge costs are now δ_i per unit length. Solve this LBFL problem approximately. *Demand at nodes assigned to facility nodes in this solution are routed to them using the current cable type.*

---

[2] The lower bounded facility location can be solved to within a constant factor, see [10,15].

4. **Aggregate:** *Consider only nodes which are not assigned to facilities opened by the previous step.* For each such node, as in Step 2, transfer the total demand in that component of the solution back to one of the sources with probability proportional to the demand at that source.

In each stage, we also make available all facilities opened at all previous stages, at cost zero. In the last stage, the Shortest path tree step has no b_i since there are no more cable types. Hence all the demand aims to reach only facilities. There is no Aggregate step for the last cable type, since all the demand has reached open facilities.

4.2   Properties of a Near-Optimum Solution

If the set of facilities to be opened has already been decided for us, then the problem reduces to the single-sink edge installation problem. This is achieved by identifying all opened facilities into a new node called "sink", and updating the metric appropriately. This transformation allows us to show the existence of a near-optimum solution to KCFL which satisfies the properties mentioned in Theorem 5 below. Our algorithm also constructs a solution which satisfies these properties, and we compare ourselves with this near-optimal solution.

Theorem 5. (Similar to Theorem 3.2 in GMM) *There exists a solution to KCFL which uses cable type $i + 1$ on a link only if at least b_i demand is being routed across that link, and which routes all demand which entered a node using cable i, out of that node using cables i and $i + 1$. This solution pays at most $\frac{2}{\alpha} + 1$ times the optimum.*

Proof. Consider an optimum solution of KCFL. Take its set of open facilities, and identify them to a sink. The resulting solution is now a (possibly sub-optimal) single-sink solution. GMM show that there is a near-optimal solution to this single-sink instance which obeys the properties enumerated in the theorem. Hence we can transform our KCFL-optimal solution to a solution which satisfies these properties. We next reverse our "identification-of-facilities" operation, that is, we "separate" the facilities. This yields a solution to KCFL which is not too far from the optimum, and which satisfies the properties mentioned in the theorem.

The next two lemmas can be proved using nothing more than the definition of u_i and b_i.

Lemma 7. (Lemmas 3.3 and 3.4 in GMM) *For all i, we have $u_i \le b_i \le u_{i+1}$.*

Define $f_i(D) = \sigma_i + \delta_i(D)$ to be the per-unit-distance cost of routing D units of flow on cable type i.

Lemma 8. (Lemma 3.5 in GMM) *For any i and $D \ge b_i$, we have $f_{i+1}(D) \le 2\alpha f_i(D)$.*

4.3   Analysis

Our analysis will follow and make some modifications to the analysis in the paper by Guha, Meyerson and Munagala. Let d_v be the original demand of node v. Let D_v^i be the demand at node v at stage i in the algorithm.

Let T_i, P_i and N_i be the Steiner tree, Shortest path and consolidation-aggregation step costs respectively, at iteration i. Also, let T_i^I, T_i^F and T_i^ϕ denote the incremental (variable), fixed and facility opening cost components respectively of the Steiner tree step at iteration i. P_i^I, P_i^F and P_i^ϕ are similarly defined for the Shortest path step. The following lemmas are proved for the single sink version in GMM, and can be adapted to hold in our setting with facility costs too.

Lemma 9. (Lemma 4.1 in GMM)[3] *At the end of every consolidation and aggregation step, $E[D_v^i]$ at each node which has not yet been connected to an open facility is d_v.*

Proof. The main idea is that at the end of every consolidation and aggregation step, we re-route any demand which has not yet been connected to client nodes with probability proportional to the original demand at each node. Some nodes are connected to open facilities, and no demand is ever re-routed to any such node. Hence for any client node which still hasn't been connected to an open facility, its expected demand at any stage $E[D_v^i]$ is exactly d_v.

Since the consolidation and aggregation steps occur only after all excess demand has been removed, the total demand at each such step is no more than the cable capacity. Hence, we have the following.

Lemma 10. (Lemma 4.2 in GMM) *At every i, we have $E[N_i] \leq T_i + P_i$.*

Proof. The consolidation step occurs only along the Steiner tree which has just been built. It uses no extra edges, and hence costs no more than T_i. Similarly, the aggregation step costs no more than P_i. Hence the total consolidation-aggregation cost at step i, $E[N_i]$, is no more than $T_i + P_i$.

Lemma 11. (Lemmas 4.4 and 4.8 in GMM) *At every i, we have $P_i^F \leq P_i^I$ and $T_i^I \leq T_i^F$.*

Proof. At the start of every Shortest path tree stage, we know that the demand at any active node is at least u_i, since this is guaranteed by the preceding Steiner tree step. By the definition of u_i, the fixed cost at the shortest-path step is no more than the incremental cost, hence $P_i^F \leq P_i^I$.

Similarly, at the start of every Steiner tree step we have at least b_i demand at every active node. Again, using the definition of b_i, we can prove that $T_i^I \leq T_i^F$.

---

[3] Some of the results from GMM have been reworded for contextual clarity, while others have been extended by us to incorporate facility costs.

Now let C_i^* be the total cost of cables of type i in the near-optimum solution. Define $C^* = \sum_i C_i^*$ to be the total cable cost of the near-optimum solution. Let ϕ^* be the cost of the facilities opened by the near-optimum solution. Our main result is a consequence of the following two theorems.

Theorem 6. (Lemma 4.3 in GMM) *For every i, we have $E[P_i^I + P_i^\phi] = O(C^* + \phi^*)$.*

Proof Sketch. Consider the near-optimum solution, and replace all cables of type less than i by cable type i. For each $j < i$, the incremental cost of the new solution is a small fraction (α^{i-j}) of the incremental cost of the optimal solution's cable-j portion. The set of facilities opened in this new solution, combined with the new cables, constitutes a feasible solution for our problem of cost no more than $\phi^* + \frac{C^*}{(1-\alpha)}$. Since the lower bounded facility location problem can be solved within a constant factor ([10], [15]), the theorem follows.

Theorem 7. *(Lemmas 4.6 and 4.7 in GMM) For every i, we have $E[T_i^F + T_i^\phi] = O(C^* + \phi^*)$.*

Proof Sketch. Consider the near-optimum solution, and consider only those nodes which are candidate terminals in our stage i. Since it already has sufficient demand (b_i), thanks to stage $i - 1$, the expected cost incurred when we modify the near optimum solution to install a type i cable on the path from this node to an open facility is small, that is, within a constant of the original cost of the near-optimum solution.

This solution is a candidate solution for the Steiner tree stage, and hence a lower bound. Since we can find a Steiner tree within a constant factor of optimum [1], we are done.

Theorem 8. *There is an $O(k)$ approximation algorithm for KCFL.*

Proof. This follows from Lemma 10, Lemma 11, Theorem 6 and Theorem 7.

The precise details of the single sink edge installation problem can be read in the paper by Guha, Meyerson and Munagala [11]. They are able to prove much stronger versions of Theorems 6 and 7, which allows them to obtain an $O(1)$ approximation for their problem. However, they do not address facility costs. We have shown that we can incorporate facility costs, but we are only able to prove a weaker bound on the cost at each stage. We note that in our solution, the cable costs continue to meet the bounds proved in GMM, but the facility costs do not. In other words, the cable cost of our solution to KCFL is in fact within a constant factor of the best possible. It is an intriguing open question to bound the facility costs to within a constant as well.

5   Open Questions

Our approximation algorithm for CCpM provides a solution which uses twice as many medians as we are allowed to. An algorithm which is *uni-criterion*, that is, obeys the median restriction exactly, would be very desirable. Similarly, our approach does not work for the median version of KCFL, as it would open $O(kp)$ medians. Again, an approximation algorithm that opens $O(p)$ medians (independent of k) would be interesting.

Recently, Talwar [23] showed that the integrality gap of a natural IP formulation of the k-cable single sink edge installation problem [8] is a constant. Consequently, he provides a better approximation for this problem than GMM [11]. The KCFL problem studied by us can be modeled as an integer program which combines the IP studied by Talwar and IP_{CCFL}. It remains open whether this IP (for KCFL) can be rounded within a constant factor in polynomial time, thus providing an alternate approach to obtain an improved approximation algorithm for KCFL.

References

1. Agrawal, A., Klein, P., Ravi, R.: When trees collide: An approximation algorithm for the generalized Steiner problem on networks. SIAM J. Computing, 24:440-456, 1995.
2. Andrews, M., Zhang, L.: The access network design problem. Proc. of the 39th Ann. IEEE Symp. on Foundations of Computer Science, pages 42-49, 1998.
3. Arya, V., Garg, N., Khandekar, R., Pandit, V., Meyerson, A., Munagala, K.: Local search heuristic for k-median and facility location problems. Proc. 33rd ACM Symposium on the Theory of Computing, pages 21-29, 2001.
4. Awerbuch, B., Azar, Y.: Buy at bulk network design. Proc. 38th Ann. IEEE Symposium on Foundations of Computer Science, pages 542-547, 1997.
5. Balinski, M.L.: On finding integer solutions to linear programs. Proc. IBM Scientific Computing Symposium on Combinatorial Problems, pages 225-248, 1966.
6. Charikar, M., Guha, S.: Improved combinatorial algorithms for the facility location and k-median problems. Proc. 40th Ann. IEEE Symposium on Foundations of Computer Science, pages 378-388, 1999.
7. Charikar, M., Guha, S., Shmoys, D., Tardos, E.: A constant-factor approximation algorithm for the k-median problem. Proc. 31st ACM Symposium on Theory of Computing, pages 1-10, 1999.
8. Garg, N., Khandekar, R., Konjevod, G., Ravi, R., Salman, F.S., Sinha, A.: On the integrality gap of a natural formulation of the single-sink buy-at-bulk network design problem. Proc. 8th Conference on Integer Programming and Combinatorial Optimization, pages 170-184, 2001.
9. Guha, S., Khuller, S.: Greedy strikes back: Improved facility location algorithms. Proc. 9th ACM-SIAM Symposium on Discrete Algorithms, pages 649-657, 1998.
10. Guha, S., Meyerson, A., Munagala, K.: Hierarchical placement and network design problems. Proc. 41st Ann. IEEE Symposium on Foundations of Computer Science, pages 603-612, 2000.
11. Guha, S., Meyerson, A., Munagala, K.: A constant factor approximation for the single sink edge installation problems. Proc. 33rd ACM Symposium on the Theory of Computing, pages 383-388, 2001.
12. Hassin, R., Ravi, R., F.S. Salman, F.S.: Approximation algorithms for a capacitated network design problem. Approximation Algorithms for Combinatorial Optimization, Springer-Verlag Lecture Notes in Computer Science 1913, pages 167-176, 2000.

13. Jain, K., Mahdian, M., Saberi, A.: A new greedy approach for facility location problems. To appear in Proc. 34th ACM Symposium on the Theory of Computing, 2002.

14. Jain, K., Vazirani, V.V.: Primal-dual approximation algorithms for metric facility location and k-median problems. Proc. 40th Ann. IEEE Symposium on Foundations of Computer Science, pages 2-13, 1999.

15. Karger, D., Minkoff, M.: Building Steiner trees with incomplete global knowledge. Proc. 41st Ann. IEEE Symposium on Foundations of Computer Science, pages 613-623, 2000.

16. Korupulu, M., Plaxton, C., Rajaraman, R.: Analysis of a local search heuristic for facility location problems. Proc. 9th ACM-SIAM Symposium on Discrete Algorithms, pages 1-10, 1998.

17. Mahdian, M., Ye, Y., Zhang, J.: A 1.52 approximation algorithm for the uncapacitated facility location problem. Manuscript, 2001.

18. Ravi, R.: A primal-dual approximation algorithm for the Steiner Forest Problem. Information Processing Letters 50(4): 185-190, 1994.

19. Robins, G., Zelikovsky, A.: Improved Steiner tree approximation in graphs. Proc. 10th ACM-SIAM Symposium on Discrete Algorithms, pages 770-779, 1999.

20. Salman, F.S., Cheriyan, J., Ravi, R., Subramanian, S.: Buy-at-bulk network design: Approximating the single-sink edge installation problem. SIAM Journal on Optimization, 11:3, 595-610, 2000.

21. Shmoys, D., Tardos, E., Aardal, K.: Approximation algorithms for the facility location problem. Proc. 29th ACM Symposium on the Theory of Computing, pages 265-274, 1997.

22. Sviridenko, M.: A 1.582 approximation algorithm for the metric uncapacitated facility location problem. Proc. 9th Conference on Integer Programming and Combinatorial Optimization (2002), LNCS 2337, Springer Verlag, this volume.

23. Talwar, K.: Single sink buy-at-bulk LP has constant integrality gap. Proc. 9th Conference on Integer Programming and Combinatorial Optimization (2002), LNCS 2337, Springer Verlag, this volume.

The Minimum Latency Problem Is NP-Hard
for Weighted Trees

René Sitters

Department of Mathematics and Computer Science
Technische Universiteit Eindhoven
P.O.Box 513, 5600 MB Eindhoven, The Netherlands
r.a.sitters@tue.nl

Abstract. In the *minimum latency problem* (MLP) we are given n points $v_1, \ldots, v_n$ and a distance $d(v_i, v_j)$ between any pair of points. We have to find a tour, starting at v_1 and visiting all points, for which the sum of arrival times is minimal. The arrival time at a point v_i is the traveled distance from v_1 to v_i in the tour. The minimum latency problem is MAX-SNP-hard for general metric spaces, but the complexity for the problem where the metric is given by an edge-weighted tree has been a long-standing open problem. We show that the minimum latency problem is NP-hard for trees even with weights in $\{0, 1\}$.

1 Introduction

Given n points $v_1, \ldots, v_n$ and a distance between any pair of points, the *minimum latency problem* asks for a tour π, starting at v_1 and visiting all points, for which the sum of the arrival times $d_\pi(v_1, v_i)$ is minimum, where the arrival time is defined as the traveled distance from v_1 to v_i in tour π. The minimum latency problem has been well-studied in operations research, where it is also known as the *traveling repairman problem* and the *delivery man problem*. Unlike the traveling salesman problem, where the objective is minimizing maximum arrival time and therefore is server oriented, the MLP is client oriented. Interesting applications of the MLP are diskhead scheduling and searching information in a network [4],[12] (for example in the world wide web), where the MLP can be used to minimize expected search time.

The MLP was proven to be NP-complete for general metric spaces by Sahni and Gonzalez [16]. In fact both the traveling salesman problem and the MLP are MAX-SNP-hard for general metric spaces, but the minimum latency problem has a reputation for being much harder than the traveling salesman problem. The first constant-factor approximation algorithm for general metric spaces was a 144-approximation algorithm given by Blum et al. [6]. A 7.18-approximation algorithm follows from a result of Goemans and Kleinberg [11] and an improved version of Garg's algorithm [10] for the k-MST problem. The improvement is mentioned in the papers by Arora and Karakostas [3] and Chudak et al. [7]. We

W.J. Cook and A.S. Schulz (Eds.): IPCO 2002, LNCS 2337, pp. 230–239, 2002.
© Springer-Verlag Berlin Heidelberg 2002

refer to the papers of Goemans and Kleinberg [11], Arora and Karakostas [2], Ausiello et al. [4] and Lenstra et al. [13] for a recent overview of the MLP and related problems.

2   The Minimum Latency Problem for Trees

The problems in which the metric is defined by points on the line or an edge-weighted tree are denoted by the *line-MLP* and the *tree-MLP* respectively. The line-MLP can be solved in $O(n^2)$ time by dynamic programming, as was shown by Afrati et al. [1]. Recently a linear time algorithm has been proposed by García et al. [8]. Notice that the traveling salesman problem is trivial even for edge-weighted trees. However, for the tree-MLP no exact polynomial time algorithm is known. Classifying the complexity of the MLP for edge-weighted trees has been mentioned as an open problem in many papers [3,4,5,6,11,12,15,19,20]. For example Goemans and Kleinberg [11] write that 'the MLP is not known to be NP-hard in weighted trees, so it is worth considering whether it could be solved optimally.' The first attempt to solve the tree-MLP was made by Minieka [14] who gave an $O(n^{k+1})$ algorithm, where n is the number of vertices and k is the number of leaves. Koutsoupias et al. mention that the dynamic programming solution for the line-MLP can be extended to a $O(n^k)$ dynamic programming algorithm for the tree-MLP. Blum et al. [6] show that dynamic programming gives an $O(n^2)$ algorithm for the tree-MLP on trees with diameter at most 3. Wu [20] puts the dynamic programming approach for the MLP in a more general framework. If the tree is unweighted, then a tour is optimal if and only if it is a depth-first search. Proofs have been given by several authors [6], [15].

Goemans and Kleinberg [11] gave a 3.59-approximation algorithm for the tree-MLP, improving the 8-approximation algorithm given by Blum et al. [6]. Arora and Karakostas [2] give a quasipolynomial-time approximation scheme for weighted trees and for Euclidean spaces with fixed dimension. The performance of depth-first search algorithms has been studied by Webb [19].

3   Implications and Open Problems

A problem almost equivalent to the MLP is the graph searching problem (GSP), in which an arbitrary probability distribution on the nodes is given and we want to minimize the sum of all arrival times multiplied by their probability. If we imagine that the probabilities represent a probability that an object is hidden at the vertex, then the optimal tour minimizes the expected search time. From Theorem 1 it follows immediately that the GSP is strongly NP-hard for edge-weighted trees. Moreover, Corollary 1 implies that the GSP is NP-hard for trees with unit edge-lengths.

Another interesting problem is finding the search ratio of a graph, introduced by Koutsoupias et al. [12]. Again one has to find an object hidden in a vertex not known to the server, but instead of a distribution on the vertices one is facing an adversary who chooses at which vertex it hides the object. The search ratio is then defined as the optimal competitive ratio, where the competitive ratio is the distance traveled by the server divided by the length of the shortest path from the origin to the object. Koutsoupias et al. [12] show that both the randomized and the deterministic version are MAX-SNP-hard for general graphs. They note that computing the (randomized) search ratio for trees is a 'surprisingly tough problem' but show that, if one can solve the GSP for a class of graphs in polynomial time, then by duality one can solve the randomized search ratio problem for that same class of graphs. Our result excludes this tool for finding a polynomial time algorithm for the randomized search ratio problem for trees. However, polynomial time approximation algorithms for the MLP may be transferable to the randomized search ratio problem [12].

The MLP is NP-hard for the metric with distances 1 and 2, but the complexity for trees with edge distance 1 and 2 does not follow from our result. We conjecture that this problem is polynomially solvable.

The 3.59-approximation algorithm of Goemans and Kleinberg gives the current best ratio for the tree-MLP. However this algorithm hardly uses characteristics of the tree-metric. We conjecture that a significant improvement of this ratio is possible.

The value of the optimal MLP-tour clearly depends on the given starting vertex. To find the best starting vertex we could repeatedly run an algorithm that solves the MLP, each time using a different starting vertex. However, Sichmi-Levi and Berman [17] notice that for general metric spaces one run suffices. Wu [20] gives an $O(n^2)$ algorithm for this problem on the line, and Minieka [15] notices that for unweighted trees any vertex at the end of a longest path can be taken as the best starting vertex. It is unknown to us whether, for weighted trees and for general metric spaces, there is a more subtle way than solving an MLP instance to find the best starting vertex.

Koutsoupias et al. [12] mention that the MLP is conjectured to be NP-hard even for *caterpillars* (a path with edges sticking out). This conjecture remains unresolved.

In the forthcoming paper by Lenstra et al. [13] the authors define a large class of so called dial-a-ride problems. In this more general framework servers have to transport items from a source to a destination in a metric space. The class contains about 8,000 problems, from which all but 72 have been classified as NP-hard or solvable in polynomial time[1]. As one of the most interesting open problems, the authors mention the minimum latency problem with release dates and where the metric is the real line. This problem was already mentioned as an interesting open problem by Tsitsiklis [18].

4   Proof of NP-Hardness

We define an instance of the tree-MLP as a tree with root r and weights on the edges and the vertices. Notice that we can polynomially reduce any such instance to an instance with unit vertex-weights by replacing a vertex with weight w by

---

[1] By February 2002.

w vertices at distance zero from one another. Of course this only applies if all vertex-weights are polynomially bounded. For clarity we shall use the terms edge *length* and vertex *weight*. We will use the term *total completion time* for the sum of the weighted completion times. We assume that the weight of the starting vertex is zero and that the server always ends its tour in the origin.

To facilitate the exposition we assume that the server travels at unit speed such that distance traveled and elapsed time are equal. Given an instance of the MLP on a tree, let T be an optimal tour and let $t_0 = 0, t_1, \ldots, t_k$ be the moments at which the server is in the origin. Notice that t_k is equal to the length of T since we assumed that the server ends in the origin. Let T_i be the subtour between time t_{i-1} and t_i, $1 \le i \le k$, and let $|T_i|$ resp. W_i be the length resp. total weight of this subtour. Then the we have the following lemma.

Lemma 1. *For any optimal tour T the following holds.*

(i) $\frac{W_1}{|T_1|} \ge \frac{W_2}{|T_2|} \ge \cdots \ge \frac{W_k}{|T_k|}$.

(ii) *If $\frac{W_i}{|T_i|} = \frac{W_j}{|T_j|}$ for some $i, j \in \{1, \ldots, k\}$, then the total completion time remains the same if the subtours T_i and T_j swap their position on T.*

Proof. (i) We use a simple interchange argument. Assume that for some i $\frac{W_i}{|T_i|} < \frac{W_{i+1}}{|T_{i+1}|}$. If we change the order of the subtours T_i and T_{i+1}, then the increase in the total completion time is $W_i|T_{i+1}| - W_{i+1}|T_i| < 0$. Now (ii) follows directly from the proof of (i). □

To prove NP-hardness for the tree-MLP we give a reduction from the 3-Partition problem, which was proven to be NP-hard in the strong sense by Garey and Johnson [9].

3-Partition

Instance: A multiset of natural numbers $A = \{p_1, p_2 \ldots, p_{3n}\}$, with $P/4 < p_i < P/2$ for all $i \in \{1, \ldots, 3n\}$, and a number P such that $p_1 + \cdots + p_{3n} = nP$.
Question: Is it possible to partition the set A in n sets $A_1, \ldots, A_n$ such that $\sum_{p_i \in A_j} p_i = P$ for all $j \in \{1, \ldots, n\}$?

Theorem 1. *The minimum latency problem is strongly NP-hard for weighted trees.*

Proof. Given an instance of 3-Partition with the notation as described above, we define $a_i = Kp_i$ for all $i \in \{1, \ldots, 3n\}$ and $Q = KP$, where $K = 2Pn^4$. We define a tree on $3n(n + 2) + 1$ vertices as follows.

For each $i \in \{1, \ldots, 3n\}$ we construct a path $(r, v_{i1}, v_{i2}, \ldots, v_{in}, z_i)$. All these paths start in the root of the tree, which is appointed as the origin of the MLP-instance. To each of these paths an extra vertex u_i is attached through the edge (v_{i1}, u_i) (see Fig. 1). For the definition of the lengths of the edges in this tree we introduce the numbers m and l_i, $i \in 1, \ldots, 3n$, and choose their value appropriately later. The lengths of the edges are:

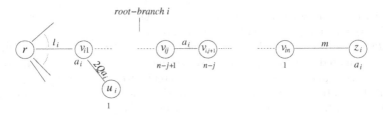

Fig. 1. Sketch of the MLP instance

$$
\begin{aligned}
d(r, v_{i1}) &= l_i, & i &= 1, \ldots, 3n; \\
d(v_{ij}, v_{i,j+1}) &= a_i, & i &= 1, \ldots, 3n, j = 1, \ldots, n-1; \\
d(v_{i1}, u_i) &= 2Qa_i, & i &= 1, \ldots, 3n; \\
d(v_{in}, z_i) &= m, & i &= 1, \ldots, 3n.
\end{aligned}
$$

The weights on the vertices are:

$$
\begin{aligned}
w(v_{i1}) &= a_i, & i &= 1, \ldots, 3n; \\
w(v_{ij}) &= n - j + 1, & i &= 1, \ldots, 3n, j = 2, \ldots, n; \\
w(z_i) &= a_i, & i &= 1, \ldots, 3n; \\
w(u_i) &= 1, & i &= 1, \ldots, 3n.
\end{aligned}
$$

We call the subtree rooted at r and constituted by the path $(r, v_{i1}, \ldots, v_{in}, z_i)$ and the edge (v_{i1}, u_i) the *root-branch* i, $i = 1, \ldots, 3n$.

If the values m and l_i are chosen appropriately large it is easy to see that an optimal tour satisfies the following properties for all i in $\{1, \ldots, 3n\}$:

(a) each edge (r, v_{i1}) is traversed exactly once in each direction;

(b) vertex u_i is visited before vertex z_i.

Moreover, we can choose l_i, $i = 1, \ldots, 3n$, such that for all i and h in $\{1, \ldots, 3n\}$

(c) $\frac{W_i}{L_i} = \frac{W_h}{L_h}$, where W_i and L_i are, respectively, the sum of all vertex-weights and the sum of all edge-lengths of root-branch i.

With respect to (c) we note that choosing $l_i = M(2a_i + n(n-1)/2 + 1) - (2Q + n - 1)a_i - m$ yields $\frac{W_i}{L_i} = M$, for all $i \in \{1, \ldots, 3n\}$, where M is a large number. It is intuitively clear that we can bound the numbers m and M by a polynomial in the size of the 3-Partition instance. A proof for this is submitted at the end of this proof.

By (a) and (b) there are only n different ways for the server to traverse root-branch i: for $k = 2, \ldots, n$, we call the subtour $(r, v_{i1}, \ldots, v_{ik}, v_{i,k-1}, \ldots, v_{i1}, u_i, v_{i1}, \ldots, v_{in}, z_i, v_{in}, \ldots, v_{i1}, r)$ the k-tour. The tour $(r, v_{i1}, u_i, v_{i1}, \ldots, v_{in}, z_i, v_{in}, \ldots, v_{i1}, r)$ is a 1-tour. Consider an optimal tour. Renumber the root branches according to the k-tour by which they are traversed: root branches $1, \ldots, i_1$ are traversed by a 1-tour, $i_1 + 1, \ldots, i_2$ by a 2-tour, etc. until $i_{n-1} + 1, \ldots, i_n = 3n$ being the root branches traversed by an n-tour. We know from Lemma 1(i) and (c) that in an optimal tour all root-branches $1, \ldots, i_1$ are served first followed by the root branches $i_1 + 1, \ldots, i_2$, etc.

Now consider the tour in which all root-branches are traversed by a 1-tour. We will compare the optimal tour with this tour. Due to Lemma 1(ii) and (c) we may assume that the order in which the root-branches are served is the same for both tours.

If root-branch i is served by a k-tour, then compared to serving this root-branch by a 1-tour, this will reduce the completion time of all vertices v_{ij} by $2 \leq j \leq k$ by $4Qa_i$ and increase the completion time of all vertices v_{ij} by $k+1 \leq j \leq n$ and vertices u_i and z_i by $2(k-1)a_i$. It also increases the completion time of all vertices in root-branches h for $h \geq i+1$ by $2(k-1)a_i$.

Thus the total decrease of serving according to the optimal solution instead of the all 1-tour solution is

$$\sum_{k=2}^{n} \sum_{i=i_{k-1}+1}^{3n} 4(n-k+1)Qa_i = \sum_{k=2}^{n} \left(4(n-k+1)Q \sum_{i=i_{k-1}+1}^{3n} a_i \right). \quad (1)$$

Let us now first study the total increase R due to a delay of all vertices except the vertices z_i and v_{i1}, $i = 1, \ldots, 3n$. There are $3n^2$ of them and their total weight is $3n(\frac{1}{2}n(n-1)+1) < 2n^3$. The delay for each of them is at most $2(n-1)Q$. Therefore, $R < 4Qn^4$.

To study the total increase due to a delay of the vertices v_{i1} and z_i, $i = 1, \ldots, 3n$, consider a root branch h that is served by a k-tour, i.e., $i_k \leq h < i_{k+1}$. Compared to serving it by a 1-tour, the vertex v_{h1} is not delayed, but the vertex z_h is delayed by an amount $2(k-1)a_h$, as noted before, which multiplied by its weight gives an increase in objective value of $2(k-1)a_h^2$. Moreover, all vertices v_{i1} and z_i for $i > h$ receive a delay of $2(k-1)a_h$ by this k-tour. Thus, the total increase by this k-tour due to delay of vertices v_{i1} and z_i, $i = h, \ldots, 3n$ amounts to

$$2(k-1)a_h \left(a_h + \sum_{i=h+1}^{3n} 2a_i \right).$$

The total increase of the optimal solution over the all 1-tour solution is then given by

$$R + \sum_{k=2}^{n} \sum_{h=i_{k-1}+1}^{i_k} 2(k-1)a_h \left(a_h + \sum_{i=h+1}^{3n} 2a_i \right) = R + \sum_{k=2}^{n} 2 \left(\sum_{i=i_{k-1}+1}^{3n} a_i \right)^2. \quad (2)$$

Let C^{OPT} and C be the total completion times, respectively, for the optimal tour and the all 1-tour solution. Combining (1) and (2) we obtain

$$C^{OPT} = C + R + \sum_{k=2}^{n} \left(2 \left(\sum_{i=i_{k-1}+1}^{3n} a_i \right)^2 - 4Q(n-k+1) \sum_{i=i_{k-1}+1}^{3n} a_i \right). \quad (3)$$

For each $k \in \{2, \ldots, n\}$, writing $b_k = \sum_{i=i_{k-1}+1}^{3n} a_i$, the term between the large brackets in (3) becomes $2b_k^2 - 4Q(n-k+1)b_k$, which is minimal if $b_k = (n-k+1)Q$.

If the 3-Partition instance has a yes-answer, implying that a perfect partition exists, then we can construct a tour $\tilde{T}$ for which $\tilde{b}_k = (n-k+1)Q$ for all $k \in \{2,\ldots,n\}$ (with $\tilde{b}_k$ defined as b_k with respect to $\tilde{T}$), which inserted in (3) yields:

$$
\begin{aligned}
C^{OPT} &= C + R + \sum_{k=2}^{n} (2b_k^2 - 4Q(n-k+1)b_k) \\
&\leq C + R + \sum_{k=2}^{n} (2\tilde{b}_k^2 - 4Q(n-k+1)\tilde{b}_k) \\
&= C + R - 2Q^2 \sum_{k=2}^{n} (n-k+1)^2 \\
&= C + R - \tfrac{1}{3}Q^2 n(n-1)(2n-1) \\
&< C + 4Qn^4 - \tfrac{1}{3}Q^2 n(n-1)(2n-1) \\
&= C + 2K^2 - \tfrac{1}{3}Q^2 n(n-1)(2n-1),
\end{aligned}
$$

using $Q = KP$ and $K = 2Pn^4$ in the second term of the last line.

Notice that b_k is a multiple of K for any $k \in \{2,\ldots,n\}$. Therefore standard calculus tells us that

$$
\sum_{k=2}^{n} 2b_k^2 - 4Q(n-k+1)b_k < -\frac{1}{3}Q^2 n(n-1)(2n-1) + 2K^2
$$

if and only if

$$
b_k = (n-k+1)Q, \text{ for all } k \in \{2,\ldots,n\}.
$$

Therefore if no perfect partition exists then

$$
\begin{aligned}
C^{OPT} &= C + R + \sum_{k=2}^{n} (2b_k^2 - 4Q(n-k+1)b_k) \\
&\geq C - \tfrac{1}{3}Q^2 n(n-1)(2n-1) + 2K^2 .
\end{aligned}
$$

We conclude that the 3-Partition instance has a yes-answer if and only if

$$
C^{OPT} < C - \frac{1}{3}Q^2 n(n-1)(2n-1) + 2K^2,
$$

which completes the proof.

In the remaining part of the proof we show that the numbers m and M can be bounded by a polynomial in the size of the 3-Partition instance, and still satisfy the conditions that each edge (r, v_{i1}) is traversed exactly once in each direction, and that vertex u_i is visited before vertex z_i.

Assume that the optimal tour traverses the edge (r, v_{i1}) more than once in each direction. For large enough M we can adjust the optimal tour to obtain a tour with smaller total completion time as follows. When the optimal tour visits vertex v_{i1} for the first time we continue this tour by visiting all the remaining vertices of root branch i in the same order as they are visited in the optimal tour. Next, we continue the optimal tour, leaving out the visits to root branch i (except for vertex r). For any vertex, the delay in completion time in this new tour is at most $2((2n - 2 + 2Q)a_i + m)$. On the other hand, for at least one of the vertices of root branch i, the completion time is decreased by at least $2l_i$. Therefore, the total decrease in the total completion time is at least $2l_i - 2((2n - 2 + 2Q)a_i + m)W$, where $W = \frac{3}{2}n^2(n - 1) + 3n + 2nQ$ is the total weight of the tree. Now it easy to see that we can choose M such that $l_i > ((2n - 2 + 2Q)a_i + m)W$ for all $i \in \{1, \ldots, 3n\}$ and M is bounded by a polynomial in the size of the 3-Partition instance and in m.

Now assume that the optimal tour traverses the edges (r, v_{i1}) $(i = 1 \ldots 3n)$ exactly once in each direction but vertex z_i is visited before vertex u_i for some $i \in \{1, \ldots, 3n\}$. If we change the optimal tour by traversing root branch i according to a 1-tour, then the completion time changes only for the vertices in root branch i. The decrease in completion time for vertex u_i is exactly $2(m + (n - 1)a_i)$, and the increase for the vertices $v_{i2}, \ldots, v_{in}, z_i$ is exactly $4Qa_i$. Therefore, the total decrease becomes $2(m + (n - 1)a_i) - 4Qa_i(n(n - 1)/2 + a_i)$ which is a positive number for $m > 2Qa_i(n(n - 1)/2 + a_i) - (n - 1)a_i$. □

We can strengthen Theorem 1 a little. We can reduce the MLP with polynomially bounded lengths and weights to the MLP with all vertex-weights 1 and all edge-lengths either 0 or 1.

Corollary 1. *The MLP is NP-hard for trees where all edge-lengths are either 0 or 1.*

Proof. Let I be an instance of the tree-MLP with vertices $v_1, \ldots, v_n$ and edges $e_1, \ldots, e_m$. Let w_i $(i \in \{1, \ldots, n\})$ and l_j $(j \in \{1, \ldots, m\})$ be the weight of vertex v_i and the length of edge l_j respectively. From I we define an instance I' of the tree-MLP with all vertex weights 1 and edge-lengths 0 and 1.

We define weight $w'_i = Kw_i$ on vertex v_i, $i = 1, \ldots, n$, where K is a large number that we choose appropriately later. As mentioned before, vertex-weights may be regarded as an equivalent number of points at distance zero from each other. For edge e_j with length l_j incident to vertices u and v we insert $l_j - 1$ extra vertices $v_{j1}, \ldots, v_{j,l_j-1}$ and the path $u, v_{j1}, \ldots, v_{j,l_j-1}, v$ containing l_j edges of length 1. The inserted vertices receive weight 1 and are referred to as intermediate vertices. We do this for all edges of I.

Choose $K > 2L(n + L - m)^2$, with L the sum of the lengths of all edges of I. We claim that, for any positive integer C, there exists a solution with total completion time $C(I) < C$ for I if and only if there exists a solution with total completion time $C(I') < KC$.

If there exists a tour T' for I' with $C(I') < KC$, then the tour for I that follows from T' in the obvious way has total completion time less than C. On the other hand, if there exists a tour T for I with $C(I) = C - 1$, then the tour T' for I' that follows from T in the obvious way has objective $K(C - 1) + D$, where D is the total completion time of the intermediate vertices. It suffices to show that $D < K$.

The number of vertices of I' is $n + L - m$. Notice that $2L(n + L - m)$ is an upper bound on the length of the tour that returns to the origin every time a new vertex has been visited, which in its turn is clearly an upper bound on the length of T'. Therefore, $D \leq 2L(n + L - m)^2 < K$. $\qquad\qquad\square$

The following statement is an obvious corollary of the above results. It makes a clear distinction between edge-weighted and vertex-weighted trees.

Corollary 2. *The MLP is strongly NP-hard for*

(i) vertex-weighted trees, i.e. all edges have length 1,

(ii) edge-weighted trees, where all vertices have weight 1 and the edge-lengths have integer values larger than or equal to 1.

Proof. The first follows directly from Corollary 1 and the observation that we can replace any vertex of weight w by w vertices at distance zero from one another. The second follows from Corollary 1 as well. Given an instance I of the tree-MLP with all edge-lengths either 0 or 1 (and all vertex-weights 1 by the definition of the tree-MLP), we define an instance I' of the tree-MLP by replacing an edge of length 0 by an edge of length 1, and replacing an edge of length 1 by an edge of length K, where K is some large number. It is easy to verify that if we choose $K \geq 2n^2$, then a tour is optimal for I if and only if it is optimal for I'. $\qquad\qquad\square$

References

1. F. Afrati, S. Cosmadakis, C. Papadimitriou, G. Papageorgiou, N. Papakostanti-nou, The complexity of the traveling repairman problem, *RAIRO Informatique Theorique et Applications* 20 (1986), 79–87.

2. S. Arora, G. Karakostas, Approximation schemes for minimum latency problems, *Proceedings of the 31st ACM Symposium on Theory of Computing* (Atlanta, 1999), 688–693.

3. S. Arora, G. Karakostas, A 2+ϵ approximation for the k-MST problem, *Proceedings of the 11th SIAM Symposium on Discrete Algorithms (SODA)* (San Francisco, 2000), 754–759.

4. G. Ausiello, S. Leonardi, A. Marchetti-Spaccamela, On salesmen, repairmen, spi-ders, and other traveling agents, *Proceedings of the 4th Italian Conference on Algorithms and Complexity* (Rome, Italy, 2000), Lecture Notes in Computer Science 1767, Springer, Berlin (2000), 1–16.

5. I. Averbakh, O. Berman, Sales-delivery man problems on treelike networks, *Networks* 25 (1995), 45–58.

6. A. Blum, P. Chalasani, D. Coppersmith, W. Pulleyblank, P. Raghavan, M. Sudan, The minimum latency problem, *Proceedings of the 26th ACM Symposium on the Theory of Computing* (Montreal, Quebec, Canada, 1994), 163-171.

7. F.A. Chudak, T. Roughgarden, D.P. Williamson, Approximate k-MSTs and k-Steiner trees via the primal-dual method and Lagrangean relaxation, *Proceedings of the 8th International Conference on Integer Programming and Combinatorial Optimization (IPCO)* (Utrecht, The Netherlands, 2001), Lecture Notes in Computer Science 2081, Springer, Berlin (2001), 60-70.

8. A. García, P. Jodrá, J. Tejel, A note on the traveling repairman problem, *Pre-publications del seminario mathematico* 3, University of Zaragoza, Spain (2001).

9. M.R. Garey, D.S. Johnson, Complexity results for multiprocessor scheduling under resource constraints, *SIAM Journal of Computing 4* (1975), 397–411.

10. N. Garg, A 3-approximation for the minimum tree spanning k vertices, *Proceedings of the 37th Annual IEEE Symposium on Foundations of Computer Science* (Burlington, VT, USA, 1996), IEEE Computer Society (1996), 302-309.

11. M. Goemans, J. Kleinberg, An improved approximation ratio for the minimum latency problem, *Mathematical Programming* 82 (1998), 111–124.

12. E. Koutsoupias, C. Papadimitriou, M. Yannakakis, Searching a fixed graph, *Proceedings of the 23rd International Colloquium on Automata, Languages, and Programming* (Paderborn, Germany, 1996), Lecture Notes in Computer Science 1099, Springer (1996), 280–289.

13. J.K. Lenstra, W.E. de Paepe, J. Sgall, R.A. Sitters, L. Stougie, Computer-aided complexity classification of dial-a-ride problems, to be published.

14. E. Minieka, The delivery man problem on a tree network, paper presented at the 1984 ORSA/TIMS San Francisco Meeting (1984).

15. E. Minieka, The delivery man problem on a tree network, *Annals of Operations Research* 18 (1989), 261–266.

16. S. Sahni, T. Gonzalez, P-complete approximation problems, *Journal of the Association for Computing Machinery* 23 (1976), 555–565.

17. D. Simchi-Levi, O. Berman, Minimizing the total flow time of n jobs on a network, *IIE Transactions* 23 (1991), 236–244.

18. J.N. Tsitsiklis, Special cases of traveling salesman and repairman problems with time windows, *Networks* 22 (1992), 263–282.

19. I.R. Webb, Depth-first solutions for the deliveryman problem on tree-like networks: an evaluations using a permutation model, *Transportation Science* 30 (1996), 134–147.

20. B.Y. Wu, Polynomial time algorithms for some minimum latency problems, *Information Processing Letters* 75 (2000), 225–229.

An Improved Approximation Algorithm for the Metric Uncapacitated Facility Location Problem

Maxim Sviridenko

IBM T. J. Watson Research Center, Yorktown Heights,
P.O. Box 218, NY 10598, USA
sviri@us.ibm.com
http://www.research.ibm.com/people/s/sviri/sviridenko.html

Abstract. We design a new approximation algorithm for the metric uncapacitated facility location problem. This algorithm is of LP rounding type and is based on a rounding technique developed in [5,6,7].

1 Introduction

In the uncapacitated facility location problem we are given a set of potential facility locations $\mathcal{F}$ and a set of demand points $\mathcal{D}$. Each point $i \in \mathcal{F}$ has an associated opening cost f_i. Once facility point i has been opened it can provide an unlimited amount of certain commodity. A client $j \in \mathcal{D}$ has a demand of commodity that must be shipped from one of the opened facilities. We assume that transportation costs of demand transferred from facility point i to demand point j are proportional to the distance c_{ij} between them. We assume that c is a metric, i.e. $c_{ii} = 0$ for all $i \in \mathcal{F}$, $c_{ij} = c_{ji}$ for all $i, j \in \mathcal{D} \cup \mathcal{F}$ and $c_{ij} + c_{jk} \geq c_{ik}$ for all $i, j, k \in \mathcal{D} \cup \mathcal{F}$. The goal is to determine a subset of potential facility points that minimizes the sum of total opening costs of facilities at these locations and total service costs of all demand points from opened facilities.

This problem can be written as the following mixed integer program

$$\min \sum_{i \in \mathcal{F}} \sum_{j \in \mathcal{D}} c_{ij} x_{ij} + \sum_{i \in \mathcal{F}} f_i y_i, \tag{1}$$

$$\sum_{i \in \mathcal{F}} x_{ij} = 1, \ j \in \mathcal{D}, \tag{2}$$

$$x_{ij} \leq y_i, \ i \in \mathcal{F}, j \in \mathcal{D}, \tag{3}$$

$$x_{ij} \geq 0, \ i \in \mathcal{F}, j \in \mathcal{D}, \tag{4}$$

$$y_i \in \{0, 1\}, \ i \in \mathcal{F} . \tag{5}$$

The uncapacitated facility location problem (UFLP) (another popular name is the simple plant location problem) is one of the most studied and known optimization problems. There are hundreds of papers on different aspects of this problem (see the survey chapter of Cornuejols, Nemhauser & Wolsey in [31] and short survey paper on polynomially solvable subcases of the problem by Ageev & Beresnev [3]). In this paper we are mainly interested in approximation

W.J. Cook and A.S. Schulz (Eds.): IPCO 2002, LNCS 2337, pp. 240–257, 2002.

algorithms with proven performance guarantees. We say that an algorithm has performance guarantee $\rho \geq 1$ if it always delivers a solution within a factor of ρ of optimum. We also call such an algorithm as ρ-*approximation algorithm.*

The first approximation algorithm for UFLP was built by Cornuejols, Fisher & Nemhauser [16]. They obtain $(1 - e^{-1})$-approximation algorithm for the maximization variant of UFLP. Recently, Ageev & Sviridenko [4] obtained an 0.828-approximation algorithm for this problem. The first approximation algorithm for the minimization version of the general UFLP (c might not satisfy the triangle inequality in this case) was built by Hochbaum [21]. Her algorithm has performance guarantee at most $\ln n + 1$. Slightly better algorithm was recently discovered by Young [33]. And it is easy to prove by a reduction from the set cover problem that the general UFLP is at least as hard to approximate as the set cover problem. On the other side Feige [17] proved that for any $\varepsilon > 0$ there is no $(1 - \varepsilon) \ln n$-approximation algorithm for the set cover problem unless $NP \not\subseteq DTIME(n^{\log \log n})$.

The most interesting special case of the general UFLP is *the metric UFLP.* In this problem distance function c satisfies the triangle inequality. The first approximation algorithm for this problem was obtained by Shmoys, Tardos & Aardal [32] using clustering and filtering techniques due to Lin & Vitter [26]. The performance guarantee of their algorithm is 3.16. Guha & Khuller [18] obtain 2.408-approximation algorithm by combining Shmoys, Tardos & Aardal's algorithm with the greedy algorithm. They also proved that there is no an approximation algorithm with performance guarantee better than 1.463 unless $NP \not\subseteq DTIME(n^{\log \log n})$ by the reduction from the set cover problem. Chudak & Shmoys [12,14] improved the Shmoys, Tardos & Aardal's algorithm by using dependent randomized rounding and information about the dual linear program to the linear programming relaxation of the problem (1)-(5). Their original performance guarantee [12] was $1 + 2e^{-1} \approx 1.736$ but using some additional tricks they slightly improved it to 1.732 [14]. Charikar & Guha [10] obtained 1.728-approximation algorithm by combining the primal-dual algorithm, the greedy algorithm and the Chudak & Shmoys's algorithm. Jain, Mahdian and Saberi [22] obtained 1.61-approximation algorithm using technique developed by Mahdian et al. [27]. Recently, combining few different approaches Mahdian, Ye & Zhang [28] obtained 1.52-approximation algorithm.

There are also few very interesting approximation algorithm for the metric UFLP based on different ideas with worse performance guarantees: local search algorithms was analyzed by Korupolu et al. [25], Arya et al. [9] and Charikar & Guha [10], primal-dual algorithms was studied by Jain & Vazirani [23] and Charikar & Guha [10]. Mettu & Plaxton [29] analyzed an algorithm very similar with Jain & Vazirani's one but their analysis doesn't use the primal-dual technique. The advantage of such algorithms is that all of them are combinatorial and very fast. All these techniques, developed originally for UFLP, was applied to different generalizations of this problem like capacitated facility location problem, multilevel facility location problem and others [32,23,15,13,1,25,30].

Several authors considered the metric UFLP in special metric spaces. Arora et al. [8] obtain a polynomial time approximation scheme (PTAS) for the UFLP on the plane, later Kolliopoulos & Rao [24] improved running time of this PTAS and they also generalized the result to the Euclidean spaces of constant dimension. Ageev [2] obtain a PTAS for the UFLP on planar graphs for instances of UFLP satisfying some technical condition.

The main result of this paper is an 1.582-approximation algorithm for the metric UFLP. The result is based on few main ideas. We use clustering technique developed in works [26,32,12,14]. We also use rounding technique developed in [5,6,7] for the problems with cardinality constraint, these technique allows us to derandomize randomized algorithms without actually constructing them. For example, assume that you want to construct a randomized algorithm which delivers an approximate solution with some expected value which is an analytical expression but you don't know how to do that. In some cases it is enough to know the analytical expression for the expectation and have some concavity property for this expression to construct a deterministic algorithm giving a solution of value no more than expected one.

2   Structure Lemma

Consider a linear programming relaxation of the problem (1)-(5) obtained by relaxing the integrality constraints (5) with the constraints

$$y_i \geq 0, i \in \mathcal{F} \ . \tag{6}$$

Let (x^*, y^*) be an optimal solution of the linear programming relaxation and let LP^* be the value of this solution. The following lemma is an extension of Lemma 4 in [14].

Lemma 1. *Suppose (x, y) is a feasible solution to the linear program (1)-(4),(6) for a given instance of the uncapacitated facility location problem $\mathcal{I}$ and let $\alpha \geq 1$ be some fixed constant. Then we can find, in polynomial time, an equivalent instance $\tilde{\mathcal{I}}$ and a feasible solution $(\tilde{x}_{ij}), (\tilde{y}_i)$ such that both solutions have the same fractional service and facility costs. The new instance $\tilde{\mathcal{I}}$ differs only by replacing each facility location by at most $\lceil \alpha \rceil |\mathcal{D}| + 1$ copies of the same location and the new feasible solution $(\tilde{x}_{ij}), (\tilde{y}_i)$ has three additional properties:*
1. $\tilde{x}_{ij} = \tilde{y}_i$ for all $i \in \tilde{\mathcal{F}}$, $j \in \mathcal{D}$ such that $\tilde{x}_{ij} > 0$,
2. $\tilde{y}_i \leq 1/\alpha, i \in \tilde{\mathcal{F}}$,
3. For any demand point j there is a permutation π^j of facility points $\tilde{\mathcal{F}}$ and facility point $\pi^j_{s(j)}$ such that $c_{\pi_1 j} \leq c_{\pi_2 j} \leq \ldots \leq c_{\pi_n j}$ (we omit superscript j from π^j) and $\sum_{i=1}^{s(j)} \tilde{y}_{\pi_i} = 1/\alpha$.

Proof. We start with $\tilde{\mathcal{F}} = \mathcal{F}$ and $(\tilde{x}_{ij}), (\tilde{y}_i) = (x^*_{ij}), (y^*_i)$ then on each iteration we will change an instance of the problem by adding new copies for some facilities, deleting the old ones and changing the current feasible solution.

1. Pick any facility i which violate the first property, i.e. there is a demand point j such that $0 < \tilde{x}_{ij} < \tilde{y}_i$. Let j_0 be a demand point with smallest value of $\tilde{x}_{ij}$ among all demand points with $\tilde{x}_{ij} > 0$. Instead of facility i create two new facilities i_1 and i_2 with the same location and opening cost and set $\tilde{y}_{i_1} = \tilde{x}_{ij_0}$ and $\tilde{y}_{i_2} = \tilde{y}_i - \tilde{x}_{ij_0}$. Then after that set $\tilde{x}_{i_1j} = \tilde{x}_{ij_0}$ and $\tilde{x}_{i_2j} = \tilde{x}_{ij} - \tilde{x}_{ij_0}$ for all demand points j such that $\tilde{x}_{ij} > 0$. After that repeat the process until the solution satisfies property 1 of the lemma. During the process, each facility in our original instance can be copied at most $|D| - 1$ times.

2. The second property is proved in a similar way. If $\tilde{y}_i > 1/\alpha$ we define a set S_i of at most $\lceil \alpha \rceil$ copies of facility i and corresponding variables $\tilde{y}_k, k \in S_i$ such that $\sum_{k \in S_i} \tilde{y}_k = \tilde{y}_i$ and $0 \leq \tilde{y}_k \leq 1/\alpha$, we also split the demand $\tilde{x}_{ij}$ between new facilities for any demand point j with $\tilde{x}_{ij} > 0$.

3. For each demand point $j \in \mathcal{D}$ fix any permutation π^j of the new set of facilities such that $c_{\pi_1 j} \leq c_{\pi_2 j} \leq \ldots \leq c_{\pi_n j}$. Let $k = \pi_{s(j)}$ be the facility point with smallest index $s(j)$ such that $\sum_{i=1}^{s(j)} \tilde{y}_{\pi_i} \geq 1/\alpha$. If $\sum_{i=1}^{s(j)} \tilde{y}_{\pi_i} > 1/\alpha$ then we define two new facilities k' and k'' instead of facility k. The facility assignment variables are defined as follows $\tilde{y}_{k''} = \sum_{i=1}^{s(j)} \tilde{y}_{\pi_i} - 1/\alpha$ and $\tilde{y}_{k'} = \tilde{y}_k - \tilde{y}_{k''}$. We also modify the permutation π^j: $\pi_t^j = \pi_t^j$ for $t = 1, \ldots, s(j) - 1$, $\pi_{s(j)}^j = k'$, $\pi_{s(j)+1}^j = k''$ and $\pi_t^j = \pi_{t-1}^j$ for $t = s(j) + 2, \ldots, n + 1$. All demands assigned to the facility k are splited now between two new facilities k' and k''. All other permutations $\pi^w, w \neq j$ are also modified accordingly. If $k = \pi_u^w$ then $\pi_t^w = \pi_t^w$ for $t = 1, \ldots, u - 1$, $\pi_u^w = k'$, $\pi_{u+1}^w = k''$ and $\pi_t^w = \pi_{t-1}^w$ for $t = u + 2, \ldots, n + 1$. We repeat this procedure for each demand point j.

So, after application of this lemma we have a new solution for the instance of the problem where we allow to locate few identical facilities at the same point. New solution has the same cost and any integral solution produced by the rounding procedure for the new instance can be easily transformed into a feasible solution of original problem with the same or smaller value of objective function (we just delete redundant copies of the same facility from approximate solution).

3   Approximation Algorithm: General Idea

On the first step of the algorithm we solve linear programming relaxation (1)-(4),(6).

On the second step we choose a parameter $\alpha \in [1, +\infty)$ from some probability distribution $g(\alpha)$ defined later (actually, $g(\alpha)$ will have nonzero value only when $\alpha \in [1, 1.71566]$). After that we construct a new instance $(\tilde{x}, \tilde{y})$ of the problem by applying the algorithm from the Lemma 1 and define a new solution $\bar{y}_i = \alpha \tilde{y}_i, i \in \tilde{\mathcal{F}}$ which is feasible by the second property from the Lemma 1. For convenience, we redefine $\mathcal{F} = \tilde{\mathcal{F}}$. New demand assignment variables $\bar{x}_{ij}$ are defined in an obvious way by using third property from the Lemma 1: $\bar{x}_{ij} = \bar{y}_i$ for closest $s(j)$ facilities and $\bar{x}_{ij} = 0$ for all other facilities, where $\pi_{s(j)}^j$ is a facility defined in the third property in the Lemma 1.

On the third step we define a set of d nonoverlapping clusters $K_1, \ldots, K_d$ (i.e. $K_p \cap K_q = \emptyset$ for any $p, q = 1, \ldots, d$) where each cluster is a set of facilities. Each cluster $K = K_1, \ldots, K_d$ satisfies the following equality $\sum_{i \in K} \bar{y}_i = 1$. The exact description of the clustering procedure will appear in the next section.

On the fourth step for each demand point j we define a function $F_j(z_1, \ldots, z_n, \alpha)$ of $n = |\mathcal{F}|$ variables $z_i \in [0, 1]$ and variable $\alpha \geq 1$. Note that the definition of the functions depends on a fractional solution $(\bar{x}, \bar{y})$ and clustering defined on the third step. These functions satisfy the following properties for any fractional solution $z_1, \ldots, z_n$ such that $\sum_{i \in K} z_i = 1$ for any cluster $K = K_1, \ldots, K_d$:

1. $\int_1^{+\infty} \left(\sum_{j \in \mathcal{D}} F_j(\bar{y}_1, \ldots, \bar{y}_n, \alpha) + \sum_{i \in \mathcal{F}} f_i \bar{y}_i \right) g(\alpha) d\alpha \leq 1.582 LP^*$
2. for any cluster $K = K_1, \ldots, K_d$ with at least one fractional facility in it there exist two fractional facilities $p, q \in K$ such that the functions $\varphi_j(\varepsilon, p, q) = F_j(z_1, \ldots, z_p + \varepsilon, \ldots, z_q - \varepsilon, \ldots, z_n, \alpha)$ are concave functions of variable $\varepsilon \in [-\min\{z_p, 1 - z_q\}, \min\{1 - z_p, z_q\}]$
3. the functions $\kappa_j(\varepsilon, p) = F_j(z_1, \ldots, z_p + \varepsilon, \ldots, z_n, \alpha)$ are concave functions of variable $\varepsilon \in [-z_p, 1 - z_p]$ for any facility point $p \in \mathcal{F} \setminus \cup_{i=1}^d K_i$
4. for any integral solution $z_1, \ldots, z_n$ such that there is at least one integrally open facility in each cluster $K_1, \ldots, K_d$ the value of $F_j(z_1, \ldots, z_n, \alpha)$ is an upper bound on the distance from the demand point j to the closest open facility in the solution $z_1, \ldots, z_n$

The firth step of approximation algorithm is the so-called "pipage rounding". Let $z_1, \ldots, z_n$ be a current fractional solution. In the beginning of the rounding procedure $z_i = \bar{y}_i, i \in \mathcal{F}$ and at the end $z_1, \ldots, z_n$ is an approximate fractional solution with exactly one integrally open facility in each cluster $K = K_1, \ldots, K_d$. During the rounding we always maintain the property that $\sum_{i \in K} z_i = 1$ for any cluster K.

If there is a cluster K with a fractional facility inside then we choose two fractional facilities p and q from cluster K such that all functions $\varphi_j(\varepsilon, p, q)$ are concave (Property 2). Let $z_i(\varepsilon) = z_i$ if $i \in \mathcal{F} \setminus \{p, q\}$, $z_p(\varepsilon) = z_p + \varepsilon$ and $z_q(\varepsilon) = z_q - \varepsilon$. Let $\varepsilon_1 = \min\{z_p, 1 - z_q\}$ and $\varepsilon_2 = \min\{1 - z_p, z_q\}$. The function

$$G(\varepsilon) = \sum_{j \in \mathcal{D}} \varphi_j(\varepsilon, p, q) + \sum_{i \in \mathcal{F}} f_i z_i(\epsilon)$$

is a concave function of variable ε on the interval $[-\varepsilon_1, \varepsilon_2]$ and therefore it is minimized when $\varepsilon = -\varepsilon_1$ or $\varepsilon = \varepsilon_2$, in particular $\min\{G(-\varepsilon_1), G(\varepsilon_2)\} \leq G(0)$. Let $\varepsilon^* \in \{-\varepsilon_1, \varepsilon_2\}$ be the value of ε minimizing the function $G(\varepsilon)$. We define $z_i = z_i(\varepsilon^*)$ and repeat the "pipage" step for the new solution z until all clusters contain a facility integrally assigned to it. Note that on each pipage step the number of fractional variables is decreasing.

The sixth and the last step is the final rounding. In the end of the previous pipage step we had a fractional solution such that each cluster K contains a facility i with $z_i = 1$ and $\sum_{i \in K} z_i = 1$. We apply a simpler rounding procedure since we do not have fractional facilities inside of clusters.

We choose any fractional facility p. Let $z_i(\varepsilon) = z_i$ if $i \in \mathcal{F} \setminus \{p\}$ and $z_p(\varepsilon) = z_p + \varepsilon$. Let $\varepsilon_1 = z_p$ and $\varepsilon_2 = 1 - z_p$. The function

$$G(\varepsilon) = \sum_{j \in \mathcal{D}} \kappa_j(\varepsilon, p) + \sum_{i \in \mathcal{F}} f_i z_i(\epsilon)$$

is concave (Property 3) and therefore it is minimized on the endpoints of the interval $[-\varepsilon_1, \varepsilon_2]$. Let $\varepsilon^* \in \{-\varepsilon_1, \varepsilon_2\}$ be the value of ε minimizing the function $G(\varepsilon)$. We define $z_i = z_i(\varepsilon^*)$ for $i \in \mathcal{F}$ and repeat rounding step again.

In the end of the rounding procedure we obtain an integral solution $z_1, \ldots, z_n$ such that

$$\sum_{j \in \mathcal{D}} F_j(z_1, \ldots, z_n, \alpha) + \sum_{i \in \mathcal{F}} f_i z_i \le \sum_{j \in \mathcal{D}} F_j(\bar{y}_1, \ldots, \bar{y}_n, \alpha) + \sum_{i \in \mathcal{F}} f_i \bar{y}_i \ .$$

Moreover, Property 4 guarantees that $F_j(z_1, \ldots, z_n, \alpha)$ is an upper bound on the distance from the demand point j to the closest open facility in the solution $z_1, \ldots, z_n$ and therefore by Property 1 the solution $z_1, \ldots, z_n$ is an 1.582-approximation solution for the metric uncapacitated facility location problem.

4   Approximation Algorithm: Details

In this section we describe the clustering procedure and give the definition for functions F_j then in the next few sections we show that F_j satisfy properties (1)-(4).

Let $N^\alpha(j) = \{i \in \mathcal{F} | \bar{x}_{ij} > 0\}$, i.e. $N^\alpha(j)$ is a set of facilities fractionally serving j in the modified fractional solution, let $R_j(\alpha) = \max_{i \in N^\alpha(j)} c_{ij}$ be a radius of $N^\alpha(j)$. Let $C_j(\alpha) = \sum_{i \in N^\alpha(j)} c_{ij} \bar{x}_{ij}$ be a service cost of demand point j in the modified fractional solution $(\bar{x}, \bar{y})$. Notice that $C_j(\alpha)$ is a nonincreasing function of α and in general $C_j(\alpha)$ can be smaller then $C_j(1) = \sum_{i \in N^1(j)} c_{ij} x_{ij}^*$. Let $D' = \emptyset$ and let $j_1 \in \mathcal{D}$ be a demand point such that $R_{j_1}(\alpha) + C_{j_1}(\alpha) = \min_{j \in \mathcal{D}} (R_j(\alpha) + C_j(\alpha))$. Add j_1 to D'. Delete j_1 and all demand points j such that $N^\alpha(j) \cap N^\alpha(j_1) \neq \emptyset$ from $\mathcal{D}$ (we will call such demand points as *demand points assigned to cluster center j_1*). Delete the set of facility points $N^\alpha(j_1)$ from $\mathcal{F}$ and repeat the clustering procedure until $\mathcal{D}$ is empty. This clustering procedure defines a set of clusters $K_1 = N^\alpha(j_1), \ldots, K_d = N^\alpha(j_d)$ and a set of cluster centers $D' = \{j_1, \ldots, j_d\}$. Notice that by construction $N^\alpha(j) \cap N^\alpha(j') = \emptyset$ for any two different cluster centers $j, j' \in D'$ and by the third property from the Lemma 1 $\sum_{i \in K} \bar{y}_i = 1$ for each cluster $K = K_1, \ldots, K_d$.

We now define functions $F_j(z_1, \ldots, z_n, \alpha)$ for all demand points j. Variable z_i always represents the fraction of facility assigned to facility location $i \in \mathcal{F}$. Originally, i.e. in the moment when we define functions F_j $z_i = \bar{y}_i = \alpha \tilde{y}_i$. Assume that demand point j was assigned to some cluster center j_k. For each demand point j let π^j be the permutation on the set of all facility location points $\mathcal{F}$ such that $c_{\pi_1 j} \le c_{\pi_2 j} \le \ldots \le c_{\pi_n j}$ (we omit superscript j from π^j)

and $N_j^1 = \cup_{k=1}^m \{\pi_k\}$ where $m = |N_j^1|$. The function $F_j(z_1, \ldots, z_n, \alpha)$ is defined as follows: if $N_{j_k}^\alpha \subseteq N_j^1$ then $F_j(z_1, \ldots, z_n, \alpha) =$

$$c_{\pi_1 j} z_{\pi_1} + c_{\pi_2 j}(1 - z_{\pi_1}) z_{\pi_2} + \ldots + c_{\pi_m j} \left(\prod_{k=1}^{m-1} (1 - z_{\pi_k}) \right) z_{\pi_m} + R_j(1) \prod_{k=1}^m (1 - z_{\pi_k}),$$

if $N_{j_k}^\alpha \setminus N_j^1 \neq \emptyset$ then $F_j(z_1, \ldots, z_n, \alpha) =$

$$c_{\pi_1 j} z_{\pi_1} + c_{\pi_2 j}(1 - z_{\pi_1}) z_{\pi_2} + \ldots + c_{\pi_m j} \left(\prod_{k=1}^{m-1} (1 - z_{\pi_k}) \right) z_{\pi_m} +$$

$$\prod_{k=1}^m (1 - z_{\pi_k}) \left(c_{j j_k} + \frac{\sum_{i \in N_{j_k}^\alpha \setminus N_j^1} c_{i j_k} z_i}{\sum_{i \in N_{j_k}^\alpha \setminus N_j^1} z_i} \right), \tag{7}$$

where $c_{j j_k}$ is a distance between demand point j and cluster center j_k.

The intuition behind this definitions is that if $N_{j_k}^\alpha \setminus N_j^1 \neq \emptyset$ then F_j is an upper bound for the expectation of the distance from demand point j to the closest open facility in the following random process. Open a facility at the facility point $i \in N_j^1$ at random independently with probability z_i, if we didn't open a facility in N_j^1 during this process then open one facility from $N_{j_k}^\alpha \setminus N_j^1$ at random such that facility i is chosen with probability $z_i / \sum_{i \in N_{j_k}^\alpha \setminus N_j^1} z_i$. If $N_{j_k}^\alpha \subseteq N_j^1$ then noticing that $c_{\pi_m j} = R_j(1)$ we can define the similar random process. Open a facility at the facility point π_k at random independently with probability z_{π_k} for $k = 1, \ldots, m-1$, if this random process didn't open any facility we open a facility at the point π_m with probability 1.

The reason we use a deterministic rounding instead of randomized one is that we don't know how to construct a random process which guarantee the above expectations for all demand points. The randomized rounding due to Chudak & Shmoys [12,14] gives similar expectations but they are a little bit different. Instead of distance from demand point j to the facility location point π_k they have some average distance from j to the set of facilities which belong to the same cluster (see [12,14] for details).

5   Proof of Properties

To prove that our approximation algorithm has performance guarantee 1.582 we need to prove four properties for functions F_j.

Theorem 1. *The function $F_j(z_1, \ldots, z_n, \alpha)$ satisfies Property 2.*

Proof. If there is a cluster k with some fractional facilities in it then we chose two facilities p, q to be fractional facilities in the cluster $K = N_{j_k}^\alpha$ with longest distances to the center j_k, i.e. $c_{p j_k}$ and $c_{q j_k}$ are the biggest among all distances between j_k and other facilities in $N_{j_k}^\alpha$. We prove that for this choice of p, q

the second derivative of $\varphi_j(\varepsilon, p, q)$ is always negative for $\varepsilon \in [-\min\{z_p, 1 - z_q\}, \min\{1 - z_p, z_q\}]$ and therefore $\varphi_j(\varepsilon, p, q)$ is a concave function on this interval for every $j \in \mathcal{F}$. See next section for details.

Property 3 follows from the fact that functions $\kappa_j(\varepsilon, p)$ are linear in ε for any index $p \in N_j^1$. We now prove Property 4. Let q be a closest to j opened facility. We know that $q \in N_j^1 \cup N_{j_k}^\alpha$ since there is always an opened facility in $N_{j_k}^\alpha$. If $q \in N_j^1$ then $z_p = 0$ for all facilities p such that $\pi^{-1}(p) < \pi^{-1}(q)$, therefore there is only one non-zero term in the expression for $F_j(z_1, \ldots, z_n, \alpha)$ and this term is equal to c_{qj}. If $q \in N_{j_k}^\alpha \setminus N_j^1$ then $F_j(z_1, \ldots, z_n, \alpha) = c_{jj_k} + c_{qj_k} \geq c_{qj}$ by the triangle inequality.

To prove Property 1 we define the density function $g(x)$ as follows

$$g(x) = \begin{cases} \frac{Ke^x}{x^3}, & \text{if } 1 \leq x \leq B, \\ 0, & \text{otherwise,} \end{cases}$$

where e is the base of the natural logarithm, $B = 1.71566$ and

$$K = \left(\int_1^B \frac{e^x}{x^3} dx \right)^{-1} \approx 0.839268$$

is a constant normalizing $g(x)$.

Theorem 2.

$$\int_1^{+\infty} \left(\sum_{j \in \mathcal{D}} F_j(\bar{y}_1, \ldots, \bar{y}_n, \alpha) + \sum_{i \in \mathcal{F}} f_i \bar{y}_i \right) g(\alpha) d\alpha \leq$$

$$\left(\max \left\{ K + \int_1^B \frac{B(e^{-\alpha/B} - e^{-\alpha})}{B - 1} g(\alpha) d\alpha, \right. \right.$$

$$\int_1^B B(1 - e^{-\alpha/B}) g(\alpha) d\alpha + \int_1^B \alpha e^{-\alpha} g(\alpha) d\alpha,$$

$$\left. \left. \int_1^B \alpha g(\alpha) d\alpha \right\} + \int_1^B e^{-\alpha} g(\alpha) d\alpha \right) LP^* = \rho LP^*$$

Proof. The proof of this theorem is rather complicated from technical viewpoint, so we put it into Sect. 7 and 8. It consists of many auxiliary lemmas on properties of F_j and the proof that $g(x)$ is a solution of certain integral equation of Volterra type.

We compute each quantity in the above expression and the performance guarantee ρ of our algorithm numerically by *Mathematica*

$$\rho \approx \max\{1.30404, 1.25777, 1.30404\} + 0.27707 = 1.58111 .$$

Note that our algorithm can be simply derandomized using the discretizing of the probability space and choosing the best parameter α from the discrete probability space. The simplest way to do it is to split interval $[1, B]$ into sufficiently many equally spaced intervals and then run the algorithm on each of the end-points of subintervals. Using this approach we are loosing small constant ε in our performance guarantee which can be made arbitrarily small by decreasing the length of subintervals.

6   Property 2

Proof of Theorem 1. Assume that we have the cluster $K = K_1, \ldots, K_d$ with at least one fractional facility (then we must have at least two fractional facilities due to the property that $\sum_{i \in K} z_i = 1$) and center j_k. Choose two fractional facilities $p, q \in K$ with biggest distances c_{pj_k} and c_{qj_k} among all fractional facilities in K.

Fix $j \in \mathcal{F}$. We consider now few cases. Assume that j was assigned to the cluster center j_k. If $N_{j_k}^\alpha \subseteq N_j^1$ then it is easy to see that the function $\varphi_j(\varepsilon, p, q)$ is a quadratic polynomial in ε. Moreover we will prove now that it has nonpositive main coefficient and therefore the function $\varphi_j(\varepsilon, p, q)$ is concave in this case.

W.l.o.g. we assume that facility p is closer to j then facility q, i.e. $\pi^{-1}(p) < \pi^{-1}(q)$, therefore we can explicitly write the main coefficient in quadratic polynomial $\varphi_j(\varepsilon, p, q)$ as

$$c_{qj} \left(\prod_{k \in \{1, \ldots, \pi^{-1}(q)-1\} \setminus \{\pi^{-1}(p)\}} (1 - z_{\pi_k}) \right) \varepsilon^2 -$$

$$\sum_{t=\pi^{-1}(q)+1}^{m} c_{\pi_t j} \left(z_{\pi_t} \prod_{k \in \{1, \ldots, t-1\} \setminus \{\pi^{-1}(p), \pi^{-1}(q)\}} (1 - z_{\pi_k}) \right) \varepsilon^2 -$$

$$R_j(1) \left(\prod_{k \in \{1, \ldots, m\} \setminus \{\pi^{-1}(p), \pi^{-1}(q)\}} (1 - z_{\pi_k}) \right) \varepsilon^2 \leq$$

$$c_{qj} \varepsilon^2 \left(\prod_{k \in \{1, \ldots, \pi^{-1}(q)-1\} \setminus \{\pi^{-1}(p)\}} (1 - z_{\pi_k}) \right) \times$$

$$\left(1 - \sum_{t=\pi^{-1}(q)+1}^{m} \left(z_{\pi_t} \prod_{k=\pi^{-1}(q)+1}^{t-1} (1 - z_{\pi_k}) \right) - \prod_{k=\pi^{-1}(q)+1}^{m} (1 - z_{\pi_k}) \right) = 0 \ .$$

If $N_{j_k}^\alpha \setminus N_j^1 \neq \emptyset$ then if both $p, q \in N_{j_k}^\alpha \setminus N_j^1$ then $\varphi_j(\varepsilon, p, q)$ is a linear function since only the last term in (7) is nonconstant and $+\varepsilon$ and $-\varepsilon$ are canceled in denominator of

$$\frac{\sum_{i \in N_{j_k}^\alpha \setminus N_j^1} c_{ij_k} z_i(\varepsilon)}{\sum_{i \in N_{j_k}^\alpha \setminus N_j^1} z_i(\varepsilon)} \ .$$

If both $p, q \in N_j^1$ then we claim again that $\varphi_j(\varepsilon, p, q)$ is a quadratic polynomial in ε with nonpositive main coefficient since $z_i(\varepsilon) = z_i$ for $i \in N_{j_k}^\alpha \setminus N_j^1$. The proof is very similar to the case when $N_{j_k}^\alpha \subseteq N_j^1$. We just should use the fact that

$$c_{jj_k} + \frac{\sum_{i \in N_{j_k}^\alpha \setminus N_j^1} c_{ij_k} z_i(\varepsilon)}{\sum_{i \in N_{j_k}^\alpha \setminus N_j^1} z_i(\varepsilon)} = \frac{\sum_{i \in N_{j_k}^\alpha \setminus N_j^1} (c_{ij_k} + c_{jj_k}) z_i}{\sum_{i \in N_{j_k}^\alpha \setminus N_j^1} z_i} \geq$$

$$\frac{\sum_{i \in N_{j_k}^\alpha \setminus N_j^1} c_{ij} z_i}{\sum_{i \in N_{j_k}^\alpha \setminus N_j^1} z_i} \geq c_{\pi_m j} .$$

The last inequality follows from the fact that $c_{ij} \geq c_{\pi_m j}$ for any $i \in N_{j_k}^\alpha \setminus N_j^1$. Note that this case covers the case when j was not assigned to the cluster center j_k but $p, q \in N_j^1$ if N_j^1 contains only p or only q and j was not assigned to the cluster center j_k then $\varphi_j(\varepsilon, p, q)$ is a linear function in ε.

We now consider the most difficult case when $q \in N_j^1 \cap N_{j_k}^\alpha$ and $p \in N_{j_k}^\alpha \setminus N_j^1$. In this case the nonlinear term of $\varphi_j(\varepsilon, p, q)$ is

$$\prod_{k=1}^{m} (1 - z_{\pi_k}(\varepsilon)) \left(\frac{\sum_{i \in N_{j_k}^\alpha \setminus N_j^1} c_{ij_k} z_i(\varepsilon)}{\sum_{i \in N_{j_k}^\alpha \setminus N_j^1} z_i(\varepsilon)} \right) . \tag{8}$$

If number of remaining fractional facilities in the set $N_{j_k}^\alpha \setminus N_j^1$ is exactly one then we assume that

$$\frac{\sum_{i \in N_{j_k}^\alpha \setminus N_j^1} c_{ij_k} z_i(\varepsilon)}{\sum_{i \in N_{j_k}^\alpha \setminus N_j^1} z_i(\varepsilon)} = c_{pj_k} .$$

The reason we stress it out is that we can have a division by zero during rounding if we would treat the above expression as function of ε. Therefore $\varphi_j(\varepsilon, p, q)$ is a linear function in this case.

So without loss of generality we consider the case when there are at least two fractional facilities in the set $N_{j_k}^\alpha \setminus N_j^1$. Dividing (8) by the positive constant factor $c_{pj_k} \times \prod_{k \in \{1,\dots,m\} \setminus \{\pi^{-1}(q)\}} (1 - z_{\pi_k}(\varepsilon)) = c_{pj_k} \times \prod_{k \in \{1,\dots,m\} \setminus \{\pi^{-1}(q)\}} (1 - z_{\pi_k})$ we obtain that nonlinear term (8) is proportional to

$$(1 - z_q + \varepsilon) \left(\frac{\varepsilon + \sum_{i \in N_{j_k}^\alpha \setminus N_j^1} c_{ij_k} z_i / c_{pj_k}}{\varepsilon + \sum_{i \in N_{j_k}^\alpha \setminus N_j^1} z_i} \right) .$$

Let $A = 1 - z_q$, $B = \sum_{i \in N_{j_k}^\alpha \setminus N_j^1} c_{ij_k} z_i / c_{pj_k}$ and $C = \sum_{i \in N_{j_k}^\alpha \setminus N_j^1} z_i$. We want to prove concavity of the function $(A + \varepsilon)(B + \varepsilon)/(C + \varepsilon)$. Note that $C + \varepsilon > 0$ since $\varepsilon \in [-\min\{z_p, 1 - z_q\}, \min\{1 - z_p, z_q\}]$ and there are at least two fractional facilities in the set $N_{j_k}^\alpha \setminus N_j^1$. The first derivative of this function is

$$\frac{A + \varepsilon}{C + \varepsilon} + \frac{B + \varepsilon}{C + \varepsilon} - \frac{(A + \varepsilon)(B + \varepsilon)}{(C + \varepsilon)^2}$$

and the second derivative is equal to

$$\frac{2}{C + \varepsilon} + \frac{2(A + \varepsilon)(B + \varepsilon)}{(C + \varepsilon)^3} - \frac{2(2\varepsilon + A + B)}{(C + \varepsilon)^2}$$

We claim that

$$(C + \varepsilon)^2 + (A + \varepsilon)(B + \varepsilon) \leq (2\varepsilon + A + B)(C + \varepsilon) .$$

And therefore the second derivative is nonpositive when $\varepsilon \in [-\min\{z_p, 1 - z_q\}, \min\{1 - z_p, z_q\}]$. Indeed, the above inequality is equivalent to the inequality $C^2 + AB \leq AC + BC$, which is equivalent to $B(A - C) \leq C(A - C)$. We prove it by proving two inequalities $A \geq C$ and $B \leq C$.

The inequality $A \geq C$ is equivalent to $1 - z_q \geq \sum_{i \in N_{j_k}^\alpha \setminus N_j^1} z_i$ which holds since $q \in N_{j_k}^\alpha \cap N_j^1$ and $1 = \sum_{i \in N_{j_k}^\alpha} z_i$.

The inequality $B \leq C$ is equivalent to $\sum_{i \in N_{j_k}^\alpha \setminus N_j^1} c_{ij_k} z_i \leq c_{pj_k} (\sum_{i \in N_{j_k}^\alpha \setminus N_j^1} z_i)$ which holds since we chose p, q to be furthest facility points from the cluster center j_k. Therefore the second derivative of $\varphi_j(\varepsilon, p, q)$ is negative when $\varepsilon \in [-\min\{z_p, 1 - z_q\}, \min\{1 - z_p, z_q\}]$, this implies the statement of the lemma.

7   Property 1: Technical Lemmas

We now prove few technical lemmas on the auxiliary properties of F_j. Let $r_j(t) = R_j(1/t)$ and recall that $R_j(1/t)$ is a smallest number such that a ball of this radius around demand point j contains facilities with total sum of facility assignment variables of least t. The first technical lemma establishes the connection between $r_j(t), R_j(\alpha)$ and $C_j(\alpha)$.

Lemma 2.

$$C_j(\alpha) = \alpha \int_0^{1/\alpha} r_j(t) dt = \alpha \int_\alpha^{+\infty} \frac{R_j(\gamma)}{\gamma^2} d\gamma .$$

Proof. Shmoys, Tardos and Aardal (Lemma 10, [32]) proved this statement for $\alpha = 1$, our proof is very similar. Recall that $c_{\pi_1 j} \leq \ldots \leq c_{\pi_m j}$. The function $r_j(t)$ is a step function and it is equal to $c_{\pi_k j}$ for every $t \in (\sum_{s=1}^{k-1} y_{\pi_s}^*, \sum_{s=1}^k y_{\pi_s}^*]$ (we use the fact that $y_i^* = x_{ij}^*$). Let q be an index such that

$$1/\alpha \in (\sum_{s=1}^{q-1} y_{\pi_s}^*, \sum_{s=1}^q y_{\pi_s}^*] .$$

Hence,

$$C_j(\alpha) = c_{\pi_q j} \left(1 - \sum_{s=1}^{q-1} \bar{x}_{\pi_s j}\right) + \sum_{s=1}^{q-1} c_{\pi_s j} \bar{x}_{\pi_s j} =$$

$$\alpha \left(c_{\pi_q j} \left(\frac{1}{\alpha} - \sum_{s=1}^{q-1} x_{\pi_s j}^*\right) + \sum_{s=1}^{q-1} c_{\pi_s j} x_{\pi_s j}^*\right) = \alpha \int_0^{1/\alpha} r_j(t) dt .$$

The second equality is trivial since we just should change variable t into variable $\gamma = 1/t$.

The next lemma is basically a reformulation of some property of expectation used in [12,14].

Lemma 3. *In the case when $N_{j_k}^\alpha \setminus N_j^1 \neq \emptyset$ for some cluster center j_k and some demand point j assigned to it, the following inequality holds*

$$c_{jj_k} + \frac{\sum_{i \in N_{j_k}^\alpha \setminus N_j^1} c_{ij_k} z_i}{\sum_{i \in N_{j_k}^\alpha \setminus N_j^1} z_i} \leq R_j(1) + R_j(\alpha) + C_j(\alpha) \ .$$

Proof. If there is a facility $q \in N_j^1 \cap N_{j_k}^\alpha$ such that $c_{qj_k} \leq C_{j_k}(\alpha)$ then by triangle inequality $c_{jj_k} \leq c_{jq} + c_{qj_k}$ and since $c_{jq} \leq R_j(1)$ and $c_{ij_k} \leq R_{j_k}(\alpha)$ we obtain

$$c_{jj_k} + \frac{\sum_{i \in N_{j_k}^\alpha \setminus N_j^1} c_{ij_k} z_i}{\sum_{i \in N_{j_k}^\alpha \setminus N_j^1} z_i} \leq R_j(1) + C_{j_k}(\alpha) + R_{j_k}(\alpha) \leq R_j(1) + C_j(\alpha) + R_j(\alpha)$$

where the last inequality follows from the definition of cluster centers.

If $c_{qj_k} > C_{j_k}(\alpha)$ for all facilities $q \in N_j^1 \cap N_{j_k}^\alpha$ then

$$\frac{\sum_{i \in N_{j_k}^\alpha \setminus N_j^1} c_{ij_k} z_i}{\sum_{i \in N_{j_k}^\alpha \setminus N_j^1} z_i} \leq \frac{\sum_{i \in N_{j_k}^\alpha} c_{ij_k} z_i}{\sum_{i \in N_{j_k}^\alpha} z_i} = C_{j_k}(\alpha) \ .$$

Using the fact that by triangle inequality $c_{jj_k} \leq R_j(1) + R_{j_k}(\alpha)$ we obtain the statement of the lemma.

Lemma 3 implies that

$$F_j(\bar{y}_1, \ldots, \bar{y}_n, \alpha) \leq c_{\pi_1 j} \bar{y}_{\pi_1} + c_{\pi_2 j}(1 - \bar{y}_{\pi_1})\bar{y}_{\pi_2} + \ldots +$$

$$c_{\pi_m j}\left(\prod_{k=1}^{m-1}(1 - \bar{y}_{\pi_k})\right)\bar{y}_{\pi_m} + \prod_{k=1}^{m}(1 - \bar{y}_{\pi_k})\left(R_j(1) + R_j(\alpha) + C_j(\alpha)\right).$$

Lemma 4.

$$c_{\pi_1 j}\bar{y}_{\pi_1} + c_{\pi_2 j}(1 - \bar{y}_{\pi_1})\bar{y}_{\pi_2} + \ldots + c_{\pi_m j}\left(\prod_{k=1}^{m-1}(1 - \bar{y}_{\pi_k})\right)\bar{y}_{\pi_m} +$$

$$R_j(1)\prod_{k=1}^{m}(1 - \bar{y}_{\pi_k}) \leq \int_0^1 \alpha e^{-\alpha t} r_j(t)dt + R_j(1)e^{-\alpha} \ .$$

Proof. Recall that the way we modified the solution (x^*, y^*) in Sect. 2 guarantees that $\bar{y}_i = \alpha y_i^* = \alpha x_{ij}^*$. Let $Y_1 = \bar{y}_{\pi_1}, Y_2 = (1 - \bar{y}_{\pi_1})\bar{y}_{\pi_2}, \ldots, Y_m = \left(\prod_{k=1}^{m-1}(1 - \bar{y}_{\pi_k})\right)\bar{y}_{\pi_m}$ and $Y_{m+1} = \prod_{k=1}^{m}(1 - \bar{y}_{\pi_k})$. Let $Z_1 = 1 - e^{-\alpha y_1^*}, Z_2 = e^{-\alpha y_1^*} - e^{-\alpha(y_1^* + y_2^*)}, \ldots, Z_m = e^{-\alpha \sum_{k=1}^{m-1} y_k^*} - e^{-\alpha \sum_{k=1}^{m} y_k^*}$ and $Z_{m+1} = e^{-\alpha}$. We claim that these two sequences of numbers satisfy the following properties:

P1. $\sum_{i=1}^{m+1} Y_i = \sum_{i=1}^{m+1} Z_i = 1$,

P2. $\sum_{i=1}^{k} Y_i \geq \sum_{i=1}^{k} Z_i, k = 1, \ldots, m$.

The first property follows directly from the definition of sequences and the fact that $\sum_{k=1}^{m} y_{\pi_k}^* = 1$. We now derive the second property:

$$\sum_{i=1}^{k} Y_i = 1 - \prod_{i=1}^{k}(1 - \bar{y}_{\pi_i}) \geq 1 - e^{-\alpha \sum_{i=1}^{k} y_{\pi_i}^*} = \sum_{i=1}^{k} Z_i$$

where we applied the well-known inequality $\prod_{i=1}^{k}(1 - p_i) \leq e^{-\sum_{i=1}^{k} p_i}$ for $0 \leq p_i \leq 1$.

Since $r_j(t)$ is a step function and $\int_a^b \alpha e^{-\alpha t} dt = -e^{-\alpha t}|_a^b$ we obtain

$$\int_0^1 \alpha e^{-\alpha t} r_j(t) dt = c_{\pi_1 j} Z_1 + c_{\pi_2 j} Z_2 + \ldots + c_{\pi_m j} Z_m .$$

Therefore, what we really need to prove is the inequality

$$c_{\pi_1 j} Y_1 + c_{\pi_2 j} Y_2 + \ldots + c_{\pi_m j} Y_m + R_j(1) Y_{m+1} \leq$$

$$c_{\pi_1 j} Z_1 + c_{\pi_2 j} Z_2 + \ldots + c_{\pi_m j} Z_m + R_j(1) Z_{m+1} . \tag{9}$$

We claim that the inequality (9) holds for any sequences of positive numbers Y and Z satisfying properties $P1, P2$ and any sequence $0 \leq c_{\pi_1 j} \leq c_{\pi_2 j} \leq \ldots \leq c_{\pi_m j} \leq R_j(1)$, the similar inequality was proved by Hardy, Littlewood and Polya [20].

Lemma 5. *For any real number $B \geq 1$ the following inequality holds*

$$\int_0^1 \alpha e^{-\alpha t} r_j(t) dt \leq B(1 - e^{-\alpha/B}) \int_0^{1/B} r_j(t) dt + \frac{B}{B-1}(e^{-\alpha/B} - e^{-\alpha}) \int_{1/B}^1 r_j(t) dt.$$

Proof. The proof is a direct application of the following integral inequality. We are given two monotone functions $g(x)$ and $r(x)$ on the interval (a, b). If function $g(x)$ is nonincreasing and function $r(x)$ is nondecreasing then

$$\int_a^b g(x) r(x) dx \leq \frac{\left(\int_a^b g(x) dx \right) \left(\int_a^b r(x) dx \right)}{b - a} . \tag{10}$$

The above inequality is a special case of the Chebychev Inequality (see inequality 236 for an integral version or Theorem 43 for the finite version of this inequality in [20]). Note also that Chebychev Inequality we mentioned above is different from the well-known probabilistic inequality with the same name.

To prove Lemma 5 using the inequality (10), we just split the integral $\int_0^1 \alpha e^{-\alpha t} r_j(t) dt$ into two integrals, one on the interval $[0, 1/B]$ and another on the interval $[1/B, 1]$ and apply the inequality (10) to each of these integrals.

Lemma 6 (Chudak & Shmoys [14]).

$$\sum_{j \in \mathcal{D}} R_j(1) \leq LP^* \ .$$

We reformulated this lemma using our notations. The proof uses the connection between $R_j(1)$ and dual variable corresponding to the demand point j in the dual LP to the linear relaxation (1)-(4),(6). We omit the proof from this version of the paper. The following lemma can be checked by a straightforward computation.

Lemma 7. *The function* $f(x) = \frac{e^x}{x^3}$ *is a solution of the following integral equation*

$$x^2 e^{-x} f(x) + \int_1^x t e^{-t} f(t) dt = 1 \ .$$

From above lemmas we derive the following upper bound on $F_j(\bar{y}_1, \ldots, \bar{y}_n, \alpha)$ for any $j \in \mathcal{D}$.

Theorem 3.

$$F_j(\bar{y}_1, \ldots, \bar{y}_n, \alpha) \leq B(1 - e^{-\alpha/B}) \int_B^{+\infty} \frac{R_j(\gamma)}{\gamma^2} d\gamma +$$

$$\frac{B}{B-1}(e^{-\alpha/B} - e^{-\alpha}) \int_1^B \frac{R_j(\gamma)}{\gamma^2} d\gamma + e^{-\alpha} \left(R_j(1) + R_j(\alpha) + \alpha \int_\alpha^{+\infty} \frac{R_j(\gamma)}{\gamma^2} d\gamma \right) \ .$$

8   Property 1: Proof of the Theorem 2

We want to estimate above the following quantity

$$\int_1^{+\infty} \left(\sum_{j \in \mathcal{D}} F_j(\bar{y}_1, \ldots, \bar{y}_n, \alpha) + \sum_{i \in \mathcal{F}} f_i \bar{y}_i \right) g(\alpha) d\alpha$$

using optimal value of linear relaxation LP^*.

By Theorem 3,

$$\int_1^{+\infty} F_j(\bar{y}_1, \ldots, \bar{y}_n, \alpha) g(\alpha) d\alpha \leq$$

$$\int_1^{+\infty} \left(B(1 - e^{-\alpha/B}) \int_B^{+\infty} \frac{R_j(\gamma)}{\gamma^2} d\gamma + \frac{B}{B-1}(e^{-\alpha/B} - e^{-\alpha}) \int_1^B \frac{R_j(\gamma)}{\gamma^2} d\gamma \right.$$

$$\left. + e^{-\alpha} \left(R_j(\alpha) + \alpha \int_\alpha^{+\infty} \frac{R_j(\gamma)}{\gamma^2} d\gamma + R_j(1) \right) \right) g(\alpha) d\alpha =$$

We will simplify this expression using Fubini's Theorem (switching the order of integration).

$$= \int_B^{+\infty} \frac{R_j(\gamma)}{\gamma^2} \left(\int_1^{+\infty} B(1 - e^{-\alpha/B}) g(\alpha) d\alpha \right) d\gamma +$$

$$\int_1^B \frac{R_j(\gamma)}{\gamma^2} \left(\int_1^{+\infty} \frac{B}{B-1}(e^{-\alpha/B} - e^{-\alpha})g(\alpha)d\alpha \right) d\gamma + \int_1^{+\infty} e^{-\alpha} R_j(\alpha)g(\alpha)d\alpha +$$

$$\int_1^{+\infty} \frac{R_j(\gamma)}{\gamma^2} \left(\int_1^\gamma \alpha e^{-\alpha}g(\alpha)d\alpha \right) d\gamma + R_j(1) \int_1^{+\infty} e^{-\alpha}g(\alpha)d\alpha =$$

Renaming variable α to γ in the third integral we continue.

$$= \int_1^{+\infty} \frac{R_j(\gamma)}{\gamma^2} F(\gamma)d\gamma + R_j(1) \int_1^{+\infty} e^{-\alpha}g(\alpha)d\alpha$$

where

$$F(\gamma) = \int_1^{+\infty} \frac{B}{B-1}(e^{-\alpha/B} - e^{-\alpha})g(\alpha)d\alpha + \gamma^2 e^{-\gamma}g(\gamma) + \int_1^\gamma \alpha e^{-\alpha}g(\alpha)d\alpha,$$

if $1 \leq \gamma \leq B$ and

$$F(\gamma) = \int_1^{+\infty} B(1 - e^{-\alpha/B})g(\alpha)d\alpha + \gamma^2 e^{-\gamma}g(\gamma) + \int_1^\gamma \alpha e^{-\alpha}g(\alpha)d\alpha,$$

if $\gamma > B$. Using Lemma 7 and the fact that $g(\gamma) = 0$ if $\gamma > B$ we simplify the expressions for $F(\gamma)$:

$$F(\gamma) = \begin{cases} \int_1^B \frac{B}{B-1}(e^{-\alpha/B} - e^{-\alpha})g(\alpha)d\alpha + K, & \text{if } 1 \leq \gamma \leq B, \\ \int_1^B B(1 - e^{-\alpha/B})g(\alpha)d\alpha + \int_1^B \alpha e^{-\alpha}g(\alpha)d\alpha, & \text{if } \gamma > B. \end{cases}$$

Therefore, $F(\gamma)$ is a step function and

$$F(\gamma) \leq \max \left\{ \int_1^B B(1 - e^{-\alpha/B})g(\alpha)d\alpha + \int_1^B \alpha e^{-\alpha}g(\alpha)d\alpha, \right.$$

$$\left. \int_1^B \frac{B}{B-1}(e^{-\alpha/B} - e^{-\alpha})g(\alpha)d\alpha + K \right\}.$$

We are ready now to prove Theorem 2

$$\int_1^{+\infty} \left(\sum_{j \in \mathcal{D}} F_j(\bar{y}_1, \ldots, \bar{y}_n, \alpha) + \sum_{i \in \mathcal{F}} f_i \bar{y}_i \right) g(\alpha)d\alpha \leq$$

$$\sum_{j \in \mathcal{D}} \left(\int_1^{+\infty} \frac{R_j(\gamma)}{\gamma^2} F(\gamma)d\gamma + R_j(1) \int_1^{+\infty} e^{-\alpha}g(\alpha)d\alpha \right) +$$

$$\int_1^{+\infty} \alpha g(\alpha)d\alpha \sum_{i \in \mathcal{F}} f_i y_i^* \leq$$

We continue applying Lemmas 2 and 6.

$$\max_{\gamma \geq 1} F(\gamma) \sum_{j \in \mathcal{D}} C_j(1) + \left(\int_1^{+\infty} e^{-\alpha}g(\alpha)d\alpha \right) LP^* + \int_1^{+\infty} \alpha g(\alpha)d\alpha \sum_{i \in \mathcal{F}} f_i y_i^* \leq$$

$$\left(\max \left\{ \max_{\gamma \geq 1} F(\gamma), \int_1^{+\infty} \alpha g(\alpha)d\alpha \right\} + \int_1^{+\infty} e^{-\alpha}g(\alpha)d\alpha \right) LP^*.$$

9   Conclusion

In this paper we obtained an approximation algorithm for the metric uncapacitated facility location problem using new rounding technique and many new
technical lemmas. The most interesting new technical lemma is Lemma 4 which
is a refinement of Lemma 17 from [14]. The rounding technique we used in this
paper is rather general and can be implemented in the similar way for other
more general facility location problems like fault tolerant facility location [19] or
facility location with outliers [11].

The performance guarantee in this paper can be easily improved by using
more numerical methods. Instead of using Lemma 5 we could choose the density
function $g(x)$ as scaled solution of the following integral equation

$$\int_1^B te^{-t/x}f(t)dt + x^2e^{-x}f(x) + \int_1^x te^{-t}f(t)dt = 1 .$$

Unfortunately, this equation does not seem to have an analytic solution, it can
be solved by discretizing an interval $[1, B]$ and solving a corresponding system
of linear algebraic equations. Notice that numerical computations show that for
big values of constant B the function $g(x)$ becomes negative and therefore we
should be careful in choosing a value for B. Since all our numerical experiments
gave a constant worse then recently announced 1.52 [28], we decided to omit
the description of more exact estimation of the performance guarantee of our
algorithm.

References

1. K. Aardal, F. Chudak and D. Shmoys, A 3-approximation algorithm for the k-level
   uncapacitated facility location problem, Inform. Process. Lett. **72** (1999), 161–167.
2. A. Ageev, An approximation scheme for the uncapacitated facility location problem
   on planar graphs, manuscript.
3. A. Ageev and V. Beresnev, Polynomially solvable special cases of the simple plant
   location problem, in: R. Kannan and W. Pulleyblank (eds.), Proceedings of the
   First IPCO Conference, Waterloo University Press, Waterloo, 1990, 1–6.
4. A. Ageev and M. Sviridenko, An 0.828-approximation algorithm for the uncapacitated facility location problem, Discrete Appl. Math. **93** (1999), 149–156.
5. A. Ageev and M. Sviridenko, Approximation algorithms for maximum coverage
   and max cut with given sizes of parts, Integer programming and combinatorial
   optimization (Graz, 1999), 17–30, Lecture Notes in Comput. Sci., 1610, Springer,
   Berlin, 1999.
6. A. Ageev, R Hassin and M. Sviridenko, An 0.5-Approximation Algorithm for the
   MAX DICUT with given sizes of parts, SIAM Journal of Discrete Mathematics
   v.14 (2001), pp. 246–255
7. A. Ageev and M. Sviridenko, An approximation algorithm for Hypergraph Max
   Cut with given sizes of parts, in Proceedings of European Simposion on Algorithms
   (ESA00).

8. S. Arora, P. Raghavan and S. Rao, Approximation schemes for Euclidean k-medians and related problems, STOC '98 (Dallas, TX), 106–113, ACM, New York, 1999.

9. V. Arya, N. Garg, R. Khandekar, V. Pandit, A. Meyerson and K. Munagala, Local Search Heuristics for k-median and Facility Location Problems, to appear in STOC01.

10. M. Charikar and S. Guha, Improved Combinatorial Algorithms for Facility Location and K-Median Problems, In Proceedings of IEEE Foundations of Computer Science, 1999.

11. M. Charikar, S. Khuller, G. Narasimhan and D. Mount, Facility Location with Outliers, in Proceedings of Symposium on Discrete Algorithms (SODA), (Jan 2001).

12. F. Chudak, Improved approximation algorithms for uncapacitated facility location, Integer programming and combinatorial optimization (Houston, TX, 1998), 180–194, Lecture Notes in Comput. Sci. **1412**, Springer, Berlin, 1998.

13. F. Chudak and D. Shmoys, Improved approximation algorithms for the capacitated facility location problem, In the Proceedings of the 10th Annual ACM-SIAM Symposium on Discrete Algorithms (1999), 875–876.

14. F. Chudak and D. Shmoys, Improved approximation algorithms for uncapacitated facility location, manuscript.

15. F. Chudak nad D. Williamson, Improved approximation algorithms for capacitated facility location problems, Integer programming and combinatorial optimization (Graz, 1999), 99–113, Lecture Notes in Comput. Sci., 1610, Springer, Berlin, 1999.

16. G. Cornuejols, M. L. Fisher and G. L. Nemhauser, Location of Bank Accounts to Optimize Float: An Analytic Study Exact and Approximate Algorithms, Management Science **23** (1977), 789–810.

17. U. Feige, A Threshold of $\ln n$ for Approximating Set Cover, Journal of ACM **45** (1998), 634–652.

18. S. Guha and S. Khuller, Greedy strikes back: improved facility location algorithms, Ninth Annual ACM-SIAM Symposium on Discrete Algorithms (San Francisco, CA, 1998), J. Algorithms **31** (1999), 228–248.

19. S. Guha, A. Meyerson and K. Munagala, Improved Algorithms for Fault Tolerant Facility Location, Proceedings of 12th ACM-SIAM Symposium on Discrete Algorithms, 2001.

20. G. Hardy, J. Littlewood and G. Pólya, Inequalities, Reprint of the 1952 edition. Cambridge Mathematical Library. Cambridge University Press, Cambridge, 1988.

21. D. Hochbaum, Heuristics for the fixed cost median problem, Math. Programming 22 (1982), 148–162.

22. K. Jain, M. Mahdian and A. Saberi, A new greedy approach for facility location problems, to appear in STOC01.

23. K. Jain and V. Vazirani, Approximation Algorithms for Metric Facility Location and k-Median Problems Using the Primal-Dual Schema and Lagrangian Relaxation, Proc. 1999 FOCS, to appear in JACM.

24. S. Kolliopoulos and S. Rao, A nearly linear-time approximation scheme for the Euclidean k-median problem, Algorithms—ESA '99 (Prague), 378–389, Lecture Notes in Comput. Sci., 1643, Springer, Berlin, 1999.

25. M. Korupolu, G. Plaxton, and R. Rajaraman, Analysis of a local search heuristic for facility location problems, Ninth Annual ACM-SIAM Symposium on Discrete Algorithms (San Francisco, CA, 1998), J. Algorithms 37 (2000), 146–188.

26. J. Lin and J. Vitter, Approximation algorithms for geometric median problems, Inform. Process. Lett. **44** (1992), 245–249.

27. M. Mahdian, E. Markakis, A. Saberi and V. Vazirani, A greedy facility location algorithm ananlyzed using dual fitting, APPROX 2001, LNCS 2129, pp.127-137.

28. M. Mahdian, Y. Ye, and J. Zhang, A 1.52-Approximation Algorithm for the Uncapacitated Facility Location Problem, manuscript, 2002.

29. R. Mettu and G. Plaxton, The online median problem, In Proceedings of FOCS00, 339–348.

30. A. Meyerson, K. Munagala and S. Plotkin, Web Caching using Access Statistics, in Proceedings of 12th ACM-SIAM Symposium on Discrete Algorithms, 2001.

31. P. Mirchandani and R. Francis, eds. Discrete location theory, Wiley-Interscience Series in Discrete Mathematics and Optimization. A Wiley-Interscience Publication. John Wiley & Sons, Inc., New York, 1990.

32. D. Shmoys, E. Tardos and K. Aardal, Approximation algorithms for facility location problems, In 29th ACM Symposium on Theory of Computing (1997), 265–274.

33. N. Young, k-medians, facility location, and the Chernoff-Wald bound, Proceedings of the Eleventh Annual ACM-SIAM Symposium on Discrete Algorithms (San Francisco, CA, 2000), 86–95, ACM, New York, 2000.

A Polyhedral Approach to Surface Reconstruction from Planar Contours[*]

Ernst Althaus[1] and Christian Fink[2]

[1] International Computer Science Institute,
1947 Center St., Suite 600, Berkeley, CA 94704-1198, USA
[2] Max-Planck-Institute für Informatik, Im Stadtwald, 66123 Saarbrücken

Abstract. We investigate the problem of reconstruction a surface given its contours on parallel slices. We present a branch-and-cut algorithm which computes the surface with the minimal area. This surface is assumed to be the best reconstruction since a long time. Nevertheless there were no algorithms to compute this surface. Our experiments show that the running time of our algorithm is very reasonable and that the computed surfaces are highly similar to the original surfaces.

1 Introduction

We investigate the problem of reconstruction a surface given its contours on parallel slices. This problem has its applications mainly in medical imaging such as computer tomography (CT) and magnetic resonance imaging (MRI). The result of those scans is a series of slices of the object. These slices are converted by edge-detection methods to sets of simple, non-intersecting polygons, called the contours. These are the input of our algorithm. The goal is to visualize the original surface in a computer. Typical visualization software requires triangulated surfaces as input. Thus, we have to solve the surface reconstruction from planar contours problem, i.e. we have to convert the polygonal slices to a triangulated surface that looks like the surface of the original object. The problem is illustrated in Fig. 1.

It is assumed since a long time [7,10] that the triangulated surface with minimal area is the best reconstruction. As we will see, the problem to compute this surface is NP-complete. Up to now there where only algorithms that solve the area minimization problem either for special cases [10,8] or only approximatively [7,8]. We present a branch-and-cut algorithm which computes the surface with the minimal area exactly. Our experiments show that the running time of the algorithm is very reasonable and that the computed surfaces are highly similar to the original surfaces.

We assume that the reader is familiar with basic polyhedral theory and branch-and-cut algorithms. For an introduction, we refer to [14]. In Sect. 2, we relate our work to extant work. In Sect. 3, we introduce some basic notations

---

[*] Research partially supported by the IST Programme of the EU under contract number IST-1999-14186 (ALCOM-FT).

W.J. Cook and A.S. Schulz (Eds.): IPCO 2002, LNCS 2337, pp. 258–272, 2002.

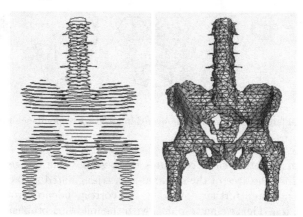

Fig. 1. The input is a series of parallel slices. We compute a triangulated surface

and definitions. In particular, we give two possible characterizations of triangulated surfaces. These characterizations lead to integer linear programming (ILP) formulations formulations, which are given in Sect. 4. Our investigations on the corresponding polyhedra follow in Sect. 5 and 6, respectively. In Sect. 7, we show the NP-completeness of both problems. In Sect. 8 and 9 we describe the resulting branch-and-cut algorithms and the experiments we made with the algorithms. Finally we give some conclusion.

2 Comparison to Extant Work

Our algorithm as well as most previous algorithms compute the triangulated surface by iteratively computing the part of the surface that lies between two neighbored slices. Thus we assume in the following, that we want to compute a surface between two parallel slices. Most previous algorithms split the surface reconstruction into three subproblems.

1. the correspondence problem, i.e. the problem to decide which contours on the one slice are connected to which contours on the other slice.
2. the branching problem, i.e. if there is no one-to-one correspondence between the contours on the two slices, the contours must be split (see Fig. 2d) or connected (see Fig. 2c) by adding one or two additional segments to the contour. These segments are called the *bridge*.
3. the tiling problem, i.e. given one contour in each slice which triangles should be chosen to get an appropriate surface

Keppel [10] proposes to solve the tiling problem by area minimization, i.e. given two contours, the triangulated surface with the minimal surface area should be chosen as reconstruction. Sederberg, Hong, Kaneda, and Klimaszewski [8] go one step beyond and propose to solve the branching problem by area minimization, too. They describe an algorithm that, given two contours in one slice and

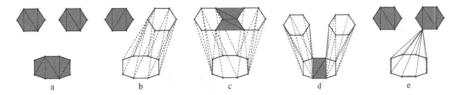

Fig. 2. Different solutions for the branching and correspondence problem

one contour in the other slice, computes the best possible bridge, i.e. the two segments, so that the area of the minimal-area triangulated surface is minimal.

We go even a step further and solve the correspondence problem by area minimization, too. Thus we have to deal with the following problem: Given a set of contours in two parallel slices, we compute the triangulated surface with the minimal area.

3   Definitions and Notation

We introduce some notations. We assume that we are given the set of contours on the two slices as a graph $G = (V, S)$, whose nodes V correspond to the corners of the contours and where two points are connected by an edge in S, if the points are connected by a segment in a contour. Let $n = |V|$ and k be the number of polygons in G. Let furthermore E be the set of edges of the complete graph over V and T all possible triangles over V. A triangle is *adjacent* to a point p or an edge pq if it contains the point or the edge, respectively. We call edges in S segment edges and edges in $E \setminus S$ non-segment edges. Let F be a set of triangles over V. For two points $p, q \in V$ let $F_p = \{t \in F \mid p \in t\}$, $F_{pq} = \{t \in F_p \mid q \in t\}$ and $\mathcal{E}^p(F) = \{rs \mid prs \in F\}$. Furthermore we assume that the contours are directed in some way. For a point p let p^- its predecessor on the contour and p^+ its successor. We also assume that the contours are numbered from 0 to $k - 1$ and that $\oplus$ is the addition modulo k. For a point p let c_p the index of the contour of p. For an set of edges $E' \subseteq E$ let $E'^i = \{pq \in E' \mid c_p = c_q = i\}$. For a subset $V' \subseteq V$, let $V'^i = \{p \in V' \mid c_p = i\}$.

First we have to characterize those sets of triangles that correspond to triangulated surfaces for the input. We define a triangulated surface between two slices as a set of triangles with the following two properties.

1. For each segment-edge of an input contour, exactly one triangle adjacent to it is in the surface.
2. Each non-segment-edge is either in the surface or not. In the first case, exactly two adjacent triangles are in the surface, in the second case no adjacent triangle is in the surface.

One could further restrict the set of surfaces by insisting to the following property.

3. Let F be the set of the triangles of the surface. Then $\mathcal{E}^p(F)$ is a path betweeen p^- and p^+ for every $p \in V$.

The last property forbids configurations as shown in Fig. 2e to be surfaces. We call surfaces with property 3 *normal* surfaces. If we do not restrict the surfaces to have property 3 we talk about *general* surfaces. We present ILP-formulations and derive branch-and-cut algorithms for both cases. In all our experiments, the general surface with the minimal area was a normal surface.

Let $C = (p_0, \ldots, p_l, p_0)$ be a cycle. A set of triangles F *closes* the cycle, if $|F_{pq}| = 1$ for all edges pq of the cycle and $|F_{pq}|$ is 0 or 2 for other pairs of points in $\{p_0, \ldots, p_l\}$. Furthermore $\mathcal{E}^{p_i}(F)$ is a path between the two adjacent points of p_i. A cycle can be closed e.g. by the triangles $p_i p_{i+1} p_l$, $0 \leq i \leq l-2$. One possible normal surface is obtained by closing all contours. It is an easy observation that all segment and non-segment conditions are satisfied for this surface.

4   ILP-Formulations and Cutting Planes

Both ILP-formulations use binary decision variables for the triangles, indicating whether a triangle belongs to a surface or not. For all triangles $t \in T$, let a_t be the surface area of the triangle t. The first two conditions above are easily formulated as linear constraints.

$$\min \sum_{t \in T} a_t x_t$$

$$s.t. \sum_{t \in T_{pq}} x_t = 1 \qquad \text{for all } pq \in S, \tag{1}$$

$$\sum_{t \in T_{pq}} x_t \leq 2 \qquad \text{for all } pq \in E \setminus S, \tag{2}$$

$$x_{pqr} \leq \sum_{s \in V^p \setminus \{q\}} x_{pqs} \quad \text{for all } pq \in E \setminus S, r \in V \setminus \{p, q\}, \tag{3}$$

$$x_t \in \{0, 1\} \qquad \text{for all } t \in T .$$

The Equalities (1) enforce for each segment that exactly one adjacent triangle in the surface, thus property 1 is satisfied. To guarantee property 2, we need two classes of inequalities. The Inequalities (2) enforce that at most two triangles adjacent to a non-segment-edge are selected and Inequalities (3) enforce that either no triangle or at least two triangles adjacent to a non-segment are selected.

The following lemma is crucial for the formulation for normal surfaces and for the validity of the cutting planes we use.

Lemma 1. *Let F be a surface and $p \in V$ be fixed. The edges $\mathcal{E}^p(F)$ form a unique path P between p^- and p^+ and a, possibly empty, set of cycles $C_1, \ldots, C_l$. Furthermore $P, C_1, \ldots, C_l$ are pairwise node-disjoint.*

Proof. From property 1 we conclude that there is exactly one edge adjacent to p^- and p^+. From property 2 we conclude that for all other points there are either 0 or 2 adjacent edges. □

We now introduce some additional notation. For a point p let $G^p = (V^p, E^p)$ be the complete graph whose nodes correspond to $V \setminus \{p\}$. We call a set of edges

$E' \subseteq E^p$ a p-path configuration, if E' is the union of a path P between p^- and p^+ and cycles $C_1, \ldots C_l$, so that $P, C_1, \ldots, C_l$ are pairwise node-disjoint. This name is justified in Lemma 2, where we show that for every p-path configuration P there is a surface F, so that $\mathcal{E}^p(F) = P$.

Observation 1. *If $\sum_{q,r \in V^p} a_{qr} x_{qr} \leq a_0$ is a valid inequality of the p-path configuration problem in G^p, the inequality $\sum_{pqr \in T_p} a_{qr} x_{pqr} \leq a_0$ is valid for the general surface problem. Analogously, if $\sum_{q,r \in V^p} a_{qr} x_{qr} \leq a_0$ is a valid inequality of the (p^-, p^+)-path problem in G^p, the inequality $\sum_{pqr \in T_p} a_{qr} x_{pqr} \leq a_0$ is valid for the normal surface problem.*

This transformation of an inequality for G^p to an inequality for the surface problem is called *lifting*. The reverse transformation is called *projection*.

Look at any (p^-, p^+)-cut in G^p. At least one edge of any path between p^- and p^+ crosses this cut. Lifting the corresponding inequalities leads to the following (exponential) class of valid inequalities for the general and normal surface polytope.

$$\sum_{q \in V', r \in V^p \setminus V'} x_{pqr} \geq 1 \text{ for all } V', \{p^-\} \subseteq V' \subseteq V^p \setminus \{p^+\} . \tag{4}$$

We use this inequalities as cutting planes. We show a separation algorithm in Sect. 8.

To restrict the set of feasible solutions of the ILP above to normal surfaces, we have to eliminate those surfaces, where at least one set $\mathcal{E}^p(F)$ is not only a path from p^- to p^+ but contains also a cycle C. The following (exponential) set of inequalities can be used.

$$\sum_{r \in V^p \setminus \{q\}} x_{pqr} \leq \sum_{r \in V', s \in V^p \setminus V'} x_{prs} \text{ for all } p, q \in V, \{q\} \subseteq V' \subseteq V^p \setminus \{p^-, p^+\} . \tag{5}$$

We first argue the validity of the inequality. Look at a node p and the path between p^- and p^+ of the edges $\mathcal{E}^p(F)$. Let q be any node. If q is not on the path, the left hand side of the inequality is 0 and thus the inequality is satisfied for all V'. If q is on the path, the left hand side of the inequality is 2. The right hand side of the inequality is at least two, since there is a path from q to p^- and a path from q to p^+.

We now argue that the inequalities restrict the set of feasible solutions as desired. Assume that for one point p the edges $\mathcal{E}^p(F)$ form a path P and a cycle C. Choosing q as any point in C and V' as the set of nodes in C, the left hand side of the inequality is 2 whereas the right hand side is 0. Thus this inequality is violated.

We handle these inequalities also by separation. The separation algorithm is given in Sect. 8.

5   Polyhedral Investigations for General Surfaces

We want to show that all described inequalities of the general surface ILP are facet defining.

If one takes a closer look to the inequalities, one makes the following observation: for every described inequality $ax \leq b$ for the general surface polytope, there is at least one point p, so that $a_t = 0$ for all $t \in T \setminus T_p$. Thus we can project these inequalities to an inequality in G^p. Looking at the resulting inequalities one observes that they are valid for the p-path configuration problem.

We first proof, that all p-path configurations can indeed be extended to general surfaces (Lemma 2). Furthermore we will see that for all facets of the p-path configuration polytope with a special property, the lifted inequality defines a facet of the general surface polytope. We finally show that the projected inequalities of all described inequalities for the general surface polytope are facets of the p-path configuration polytope and that they have the special property.

We will use the following observation several times.

Observation 2. *Let $P = (q_i, \ldots, q_j)$ be a path of segment edges, so that the length of the path is smaller than the size of the contour. For any (normal) surface F of the modified problem, where the path P is replaced by an edge (q_i, q_j), the set of triangles $F \cup \{q_l q_{l+1} q_j \mid i \leq l \leq j - 2\}$ is a (normal) surface for the original input (see Fig. 3).*

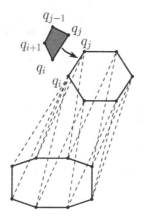

Fig. 3. The nodes $q_{i+1}, \ldots q_{j-1}$ were removed and a surface is constructed. Then the triangles $\{q_l q_{l+1} q_j \mid i \leq l \leq j - 2\}$ are added to the surface

We now show the crucial lemma of our proof. Every p-path configuration can be extended to a general surface.

Lemma 2. *Let $p \in V$ and P be a set of edges forming a p-path configuration in G^p. There is a general surface with $T_p = \{pqr \mid qr \in P\}$.*

Proof. Let F be the set of triangles induced by P. First notice that no segment edge has more than one adjacent triangle in F and no non-segment edge has more than two adjacent triangles in F. Furthermore only non-segment edges adjacent to p can have two adjacent triangles. Let E' be the set of edges that need a further triangle, i.e. the set of segment edges without adjacent triangle in F and the set of non-segment edges which have exactly one adjacent triangle in F. We claim that E' is a union of cycles, not containing the point p. Thus closing all those cycles will lead to a correct surface.

We show that every point q has an even number of adjacent edges in E'. If q is not in P, the two adjacent segment edges are in E', namely qq^- and qq^+. If q is on the path and q^- and q^+ are on P, too, E' contains no edge adjacent to q, since both segment edges have an adjacent triangle. If exactly one of q^-, q^+ is also on the path, say q^-, the segment q^-q^+ and qr are in E', where r is the adjacent point of q in P different from q^-. Finally, if neither q^- nor q^+ is on the path, there are 4 adjacent edges in E', namely q^-q, q^+q, rq and sq, where r and s are the two adjacent points of q in P. The point p is not on a cycle, since there are triangles adjacent to pp^- and to pp^+ in F.   □

Note that, if the path is different from (p^-, p^+), the surface can be chosen, so that it does not use the segment p^-p^+. This is obvious, if p^- and p^+ are on different cycles. If p^- and p^+ are on the same cycle, we have to chose the triangles that close the cycle carefully (Note that the edge p^-p^+ is not on a cycle). If the path contains no further point on c_p, p^- and p^+ lie on the same cycle, since all edges on c_p different from p^-p and pp^+ are in this cycle (all those edges are in E' and the nodes have degree 2 in E'). In this case there is also an surface that contains the edge p^-p^+, since we can close the cycle so that the edge p^-p^+ is chosen.

Before we can make a facet proof, we need to know the dimension of the polytope.

Lemma 3. *The dimension of the general surface polytope is $\binom{n}{3} - n$, where n is the number of nodes in G.*

Proof. Let $r_i \in V$ be any point with $c_{r_i} = i$. We have n equalities of type 1 in our description, one for each segment. Let $pq \in S^i$. The coefficient of the triangle $pqr_{i\oplus 1}$ is 1 for the equality for pq and 0 for all other equalities. Thus all equalities are linearly independent. Hence the dimension of the surface polytope is at most $\binom{n}{3} - n$.

We now look at the special case of k contours each containing exactly 3 points. For $k = 2$ and $k = 3$ the lemma can be easily checked by enumerating all surfaces. Let $k \geq 4$ and assume that we have proven the statement for all $l \leq k - 1$. It is known that the dimension of the polytope is $\binom{n}{3} - n$, if and only if all valid equalities are linearly dependent to the known equalities.

We can use the fact that $a_{pqr_{i\oplus 1}}$ is 1 for one equality and 0 for all other equalities to define a normal form of equalities. An equality is in *normal form*, if $a_{pqr_{i\oplus 1}} = 0$ for all segments pq. Every equality that is linearly independent to the equalities in our description, has an equivalent equality in normal form.

There exists a valid equality linearly independent to the known equalities if and only if there exists a equality in normal form different from $0 = 0$.

Assume there is a valid inequality $\sum_{t \in T} a_t x_t = a_0$ in normal form and $a_{pqr} \neq 0$ for three points p, q, r. Let i be any contour with $i \neq c_p, i \neq c_q$ and $i \neq c_r$. We look at all surfaces closing polygon i and choosing no other triangle adjacent to any of the three points. The equality is valid for all those surfaces, which is equivalent to $\sum_{t \in T \setminus (T_p \cup T_q \cup T_r)} a_t x_t = a_0 - a_{pqr}$. This inequality must be a valid inequality for $k - 1$ contours, thus we conclude that $a_t = 0$ for all $t \in T \setminus (T_p \cup T_q \cup T_r)$. This contradicts the assumption that $a_{pqr} \neq 0$.

Let's turn to the general case. In the facet proof, we will show that there are $\binom{n}{3} - n$ affinely independent points that are active for a specific facet, e.g. for the facet $x_{p^- p p^+} \geq 0$. This proof is by induction and uses as basis the case above. Observing that a facet is not active for all surfaces (in the example, notice that there is a surface that uses the triangle $p^- p p^+$), we conclude that there are at least $\binom{n}{3} - n + 1$ affinely independent surfaces. This proves that the dimension is $\binom{n}{3} - n$. $\qquad \square$

We resign to proof the following lemma.

Lemma 4. *The dimension of the p-path configuration polytope is $\binom{n-1}{2} - 2$.*

We are now turning to our main theorem.

Theorem 1. *Let $p \in V$ be fixed so that the size of the contour of p is at least 4. Let $\sum_{q,r \in V^p} a_{qr} x_{qr} \leq a_0$ be a facet of the p-path configuration polytope. If (C1) $a_{p^- p^+} = a_0$ and (C2) there is a $p^- p^+$-path P with $P \cap V^{c_p} = \{p^-, p^+\}$ and $\sum_{e \in P} a_e = a_0$ then $\sum_{pqr \in T_p} a_{qr} x_{pqr} \leq a_0$ defines a facet for the general surface polytope.*

Proof. Let k be the number of contours. For every k we make an inductive proof over the total number of points in the contours to show that the inequalities that have the properties of the lemma are active for at least $\binom{n}{3} - n$ affinely independent surfaces. For $n = 3k$ there is nothing to show (since there is no contour with size at least 4). Thus let $n > 3k$.

We define the *p-reduced surface polytope* as the surface polytope where the path $p^- p p^+$ is replaced by the edge $p^- p^+$. If $n - 1 > 3k$, there is at least one inequality of the p-reduced surface polytope that has the properties of the lemma. Since this inequality is not active for all surfaces, we can conclude that the dimension of the p-reduced surface polytope is $\binom{n-1}{3} - n + 1$. If $n - 1 = 3k$, we have already shown that the dimension of the p-reduced surface polytope is $\binom{n-1}{3} - n + 1$.

There are $\binom{n-1}{3} - n + 2$ affinely independent surfaces which have the triangle $pp^- p^+$, since fixing the triangle $p^- p p^+$ is equivalent to replacing the path $p^- p p^+$ by the edge $p^- p^+$ and thus we have the same number of affinely independent surfaces as in the p-reduced surface polytope. From condition (C1) we conclude that all of them satisfy the facet with equality. Furthermore there are $\binom{n-1}{2} - 3$ affinely independent p-path configurations different from the path (p^-, p^+) in

G^p which satisfy the projected facet with equality. These path can be extended to surfaces which do not use the edge p^-p^+. These surfaces also satisfy the facet with equality. We resign to give the rather simple proof that all those surfaces are affinely independent. Thus we have $\binom{n-1}{3} - n + 2 + \binom{n-1}{2} - 3 = \binom{n}{3} - n - 1$ affinely independent surfaces. Let $\mathcal{F}$ be the set of these surfaces.

We need one further surface to finish the proof. All surfaces in $\mathcal{F}$ satisfy the equality

$$x_{pp^-p^+} = \sum_{r \in V^p \setminus \{p^-, p^+\}} x_{p^-p^+r} \; . \tag{6}$$

Due to (C2), there is a path P with $P \cap V^{c_p} = \{p^-, p^+\}$ and $\sum_{e \in P} a_e = a_0$. A surface that does not satisfy this equality can be obtained by extending the path P so that the non-segment edge p^-p^+ is realized. The resulting surface is active for the facet. Since it does not satisfy the Equality (6) it is affinely independent from the surfaces $\mathcal{F}$. □

We resign to proof the following lemma which summarizes our polyhedral investigations on the p-path configuration polytope.

Lemma 5. *The following valid inequalities for the p-path configuration polytope are facet defining.*

1. $x_{qr} \geq 0$ *for all* $qr \in E^p$.
2. $x_{qr} \leq \sum_{s \in V^p, s \neq r} x_{qs}$ *for all* $q, r \in V^p$, $q \neq r, p^-, p^+$.
3. $\sum_{q \in V', r \in V^p \setminus V'} x_{qr} \geq 1$ *for all* $\{p^-\} \subseteq V' \subseteq V^p \setminus \{p^+\}$ *with* $2 \leq |S| \leq |V^p| - 2$.

Now we can formulate the following lemma.

Lemma 6. *Let* $p \in V$ *so that the size of the contour of p is at least 4. The following inequalities are facet defining for the general surface polytope.*

1. $x_{pqr} \geq 0$ *for all* $q, r \in V$, p, q, r *pairwise disjoint.*
2. $x_{pqr} \leq \sum_{s \in V \setminus \{p,q,r\}} x_{pqs}$ *for all* $q, r \in V^p$, $q \neq r$.
3. $\sum_{q \in V', r \in V^p \setminus V'} x_{pqr} \geq 1$ *for all* $\{p^-\} \subseteq V' \subseteq V^p \setminus \{p^+\}$ *with* $2 \leq |S| \leq |V^p| - 2$.

Proof. We start with the first statement. If $qr \neq p^-p^+$, $x_{qr} \geq 0$ is facet defining for the p-path configuration polytope. Furthermore $a_{p^-p^+} = 0 = a_0$ and $a_{p^-r} + a_{rp^+} = 0 = a_0$ for any point $r \neq q$. If $qr = p^-p^+$, the inequality $x_{pr} \geq 0$ is facet defining for the q-path configuration polytope and the conditions (C1) and (C2) hold.

For the other two items notice that the corresponding path inequality is facet defining for the p-path configuration polytope of G^p and that the condition (C1) holds. It is easy to find a path P so that (C2) holds. □

6 Polyhedral Investigations for Normal Surfaces

The polyhedral investigations for normal surfaces follow almost the same path as for general surfaces. There is one additional complication. We were not able to show that all (p^-, p^+)-paths are extensible to normal surfaces. If the degree of a node q in E' is 4, the adjacent edges of q in the surface we described needn't form only a path, but can have additional cycles. If every node has degree at most 2, we can show that the constructed surface is normal.

From the following set of path we can proof that we can extend them to normal surfaces.

Definition 1. *A p-ordered path is a path P in G^p between p^- and p^+ with the following properties.*

- *For any point q in P different from p^- and p^+, at least one of its adjacent points in P lies in the same contour as q.*
- *For every contour i let s be the first point of P in contour i and t be the last point. The subpath of P consisting of all nodes in contour i is a subpath of one of the two s, t paths in the contour i.*

We were able to show that the dimension of the p-ordered path polytope is $\binom{n-1}{2} - 2$, where $n - 1$ is the number of nodes of G^p. Thus we can exactly argue as in the case of general surfaces and obtain the following theorem.

Theorem 2. *Let p be fixed so that the size of the contour of p is at least 4. Let $\sum_{q,r \in V^p} a_{qr} x_{qr} \leq a_0$ be a facet of the p-ordered path polytope. Let furthermore $\sum_{pqr \in T_p} a_{qr} x_{pqr} \leq a_0$ be a valid inequality for the surface polytope. If (C1) $a_{p^- p^+} = a_0$ and (C2') there is a p-ordered path P with $P \cap V^{c_p} = \{p^-, p^+\}$ and $\sum_{e \in P} a_e = a_0$, then $\sum_{pqr \in T_p} a_{qr} x_{pqr}$ defines a facet for the normal surface polytope.*

Note that in the theorem there is the additional precondition that the facet of the p-ordered path polytope is valid for the normal surface problem.

Lemma 7. *Let $p \in V$ and P be a set of edges forming an p-ordered path in G^p. There is a normal surface with $T_p = \{pqr \mid qr \in P\}$.*

Proof. Assume first that every edge e in P that lies completely in one contour is a segment edge. We conclude that for every node of the path, at least one adjacent edge is a segment edge.

We show that the surface we constructed in Lemma 2 is a normal surface. Let q be any point. Assume first that q lies on the path P. If q^- and q^+ are on the path P, the triangles qq^-p and qq^+p are the triangles that are adjacent to q. Assume now that exactly one of q^- and q^+ are on the path, say q^-. Let r be the other adjacent edge of q on the path. The path contains the triangles qq^-p and qrp that are adjacent to q. The cycle containing q contains the path rqq^+ and thus the triangles that close the cycle form a path between q^+ and r. Thus all triangles adjacent to q form a path between q^- and q^+. If q is not on

P, the cycle containing q contains the path $q^- q q^+$. Thus closing this cycle will guarantee that the triangles adjacent to q form a path between q^- and q^+.

Assume now that there is a non-segment edge of P that lies in one contour, say i. Let q_0 be the first node of P on contour i and $q_1, \ldots q_l$ be the remaining points of the contour going along the segment edges, so that the indices of the nodes in contour i in P are increasing. Let $q_a q_b$, $b > a$ be a non-segment edge of P in this contour. We can close the cycle $q_a, q_{a+1}, \ldots, q_b$ and handle $q_a q_b$ as a segment edge. This is possible since no point between q_a and q_b is on the path P.   □

We now summarize our investigations on the p-ordered path polytope.

Lemma 8. *The following inequalities define facets for the p-ordered path polytope.*

1. $x_{qr} \geq 0$, *for all* $qr \in E^p$ *with* $qr \neq p^- p^{--}$ *and* $qr \neq p^+ p^{++}$. *If q and r lie in the same contour different from c_p, we assume furthermore that the size of the contour is at least 4.*

2. $x_{qr} \leq \sum_{s \in V^p, s \neq r} x_{qs}$ *for all* $q, r \in V^p$ *with* $q, r \neq p^-, p^+$ *and* $c_q = c_r \neq c_p$.

3. $\sum_{q \in V', r \in V^p \setminus V'} x_{qr} \geq 1$ *for all* $\{p^-\} \subseteq V' \subseteq V^p \setminus \{p^+\}$ *with* $|V' \cap V^i| \neq 1$ *and* $|(V^p \setminus V') \cap V^i| \neq 1$ *for all* $i \neq c_p$. *The nodes of the path* $(p^-, p^{--}, \ldots, p^{++}, p^+)$ *have an initial part of nodes in V' followed by a path of nodes that are not in V'.*

4. $\sum_{r \in V^p \setminus \{q\}} x_{qr} \leq \sum_{r \in V', s \in V^p \setminus V'} x_{rs}$ *for all* $\{q\} \subseteq V' \subseteq V^p \setminus \{p^-, p^+\}$ *if either*
   (a) $c_q \neq c_p$, $|V' \cap V^i| \neq 1$, $|(V^p \setminus V') \cap V^i| \neq 1$ *for all i*, $|V' \cap V^{c_q}| \geq 3$, *and* $|V^{c_q}| \geq 4$, *or*
   (b) $c_q = c_p$, $|V' \cap V^i| \neq 1$, $|(V^p \setminus V') \cap V^i| \neq 1$ *for all i*, $V' \cap V^{c_p} = \{q\}$.

We finally state the following lemma.

Lemma 9. *Let $p \in V$ be a point so that the size of the contour of p be at least 4. The following inequalities define facets for the normal surface polytope.*

1. $x_{pqr} \geq 0$, *for all* $p, q, r \in V$, p, q, r *pairwise disjoint. If q and r lie in the same contour, we assume furthermore that the size of the contour is at least 4. If p, q, r lie in the same contour we assume that at most one of the three edges pq, qr, rp is a segment edge.*

2. $x_{pqr} \leq \sum_{s \in V^p, s \neq r} x_{pqs}$ *for all* $p, q, r \in V^p$ *with* $c_q = c_r \neq c_p$.

3. $\sum_{q \in V', r \in V^p \setminus V'} x_{pqr} \geq 1$ *for all* $\{p^-\} \subseteq V' \subseteq V^p \setminus \{p^+\}$ *with* $|V' \cap V^i| \neq 1$ *and* $|(V^p \setminus V') \cap V^i| \neq 1$ *for all* $i \neq c_p$. *The nodes of the path* $(p^-, p^{--}, \ldots, p^{++}, p^+)$ *have an initial part of nodes in V' followed by a path of nodes that are not in V'.*

4. $\sum_{r \in V^p \setminus \{q\}} x_{pqr} \leq \sum_{r \in V', s \in V^p \setminus V'} x_{prs}$ *for all* $\{q\} \subseteq V' \subseteq V^p \setminus \{p^-, p^+\}$ *if either*
   (a) $c_p \neq c_q$, $|V' \cap V^i| \neq 1$, $|(V^p \setminus V') \cap V^i| \neq 1$ *for all i.* $|V_q^c \cap V'| \geq 3$ *and* $|V^{c_q}| \geq 4$, *or*
   (b) $c_p = c_q$, $|V' \cap V^i| \neq 1$, $|(V^p \setminus V') \cap V^i| \neq 1$ *for all i and* $V' \cap V^{c_p} = \{q\}$.

Note that we now have to show the existence of a p-ordered path P which satisfies condition (C2').

7 NP-Completeness

We will now show the NP-completeness of the minimal general and normal surface problem for general cost functions. We do not analyze the Euclidean case, where any point of a contour has a coordinate in the Euclidean Space and the cost of a triangle is defined as its surface area.

Theorem 3. *Given a set of contours, a function c that maps triples of points of the contours to integers, and a constant k, the question whether there exists a general surface of cost at most k is NP-complete. The same holds for normal surfaces.*

Proof. It is easy to see that the problem is in NP.

For the completeness, we show a reduction from maximal 3-dimensional matching. The maximal 3-dimensional matching problem is defined as follows. Given three disjoint sets X, Y, and Z and a subset $D \subseteq X \times Y \times Z$, find a matching of maximal cardinality. A matching is a subset $M \subseteq D$ so that no elements in M agree in any coordinate. The maximal 3-dimensional matching problem is NP-complete (see [9]).

The reduction will be so that for any matching M there will be a normal surface of cost $|D| - |M|$ and no (general) surface of smaller cost.

For every $v \in X \cup Y \cup Z$, we introduce a contour v_1, v_2, v_3. For every element $d \in D$, we introduce a contour d_1, d_2, d_3. The cost function c is defined as follows. For all $d = (x, y, z) \in D$ (where $x \in X$, $y \in Y$, and $z \in Z$), we define $c(d_1 d_2 d_3) = 1$, $c(d_1 d_2 x_1) = 0$, $c(d_1 x_1 x_2) = 0$, $c(d_2 d_3 y_1) = 0$, $c(d_2 y_1 y_2) = 0$, $c(d_3 d_1 z_1) = 0$, $c(d_3 z_1 z_2) = 0$, $c(d_1 z_1 x_2) = 0$, $c(z_1 z_3 x_2) = 0$, $c(z_3 x_2 x_3) = 0$, $c(d_2 x_1 y_2) = 0$, $c(x_1 x_3 y_2) = 0$, $c(x_3 y_2 y_3) = 0$, $c(d_3 y_1 z_2) = 0$, $c(y_1 y_3 z_2) = 0$, $c(y_3 z_2 z_3) = 0$, and $c(x_3 y_3 z_3) = 0$. We call these triangles the triangles associated with d. For all $v \in X \cup Y \cup Z$, we define $c(v_1 v_2 v_3) = 0$. For the remaining triangles t, we define $c(t) = \infty$. (It is easy to avoid ∞, by defining $c(t) = |D| + 1$.) See Fig. 4 for an example.

Let $M \subseteq D$ be a matching. We can construct a surface of cost $|D| - |M|$ as follows. For all $d \in M$, choose the triangles associated with d which have cost 0. For all $d \notin M$, choose the triangle $d_1 d_2 d_3$. For all $v \in X \cup Y \cup Z$ which have no adjacent triangle in M, choose the triangle v_1, v_2, v_3. This constructs a normal surface of cost $|D| - |M|$.

Assume now there is a general surface F of cost less than $|D| - |M|$, where M is a maximal matching. Let $M' = \{d \in D \mid d_1 d_2 d_3 \notin F\}$. Obviously $|M'| > |M|$ and thus M' is not a matching. Let d and d' be two elements in M', which share a common node, say x. Without loss of generality, let $x \in X$. There must be a triangle in F adjacent to the edge $d_1 d_2$. The only such triangle with finite cost different from $d_1 d_2 d_3$ is $d_1 d_2 x_1$ and hence it must be in F. Hence there is an other triangle adjacent to $d_1 x_1$ in F. The only such triangle with finite cost is $d_1 x_1 x_2$ and thus it must be in F. In the same way, we argue that the triangle $d_1' x_1 x_2$ must be in F. Thus F contains at least two triangles adjacent to the segment edge $x_1 x_2$. This contradicts the property 1 of general surfaces.   □

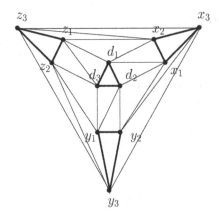

Fig. 4. The figure shows a tiny example: $X = \{x\}$, $Y = \{y\}$, $Z = \{z\}$ and $D = \{(x, y, z)\}$. The segment-edges are shown as bold lines. The faces of the graph (including the outer face) correspond to the triangles with finite cost. The triangle $d_1 d_2 d_3$ has cost 1, the remaining triangles have cost 0. There are exactly two surfaces consisting only of triangles of finite cost. We can either close all four contours, or we chose all other faces of the planar graph (including the outer face)

8   Branch-and-Cut Algorithm

The input size of our problem instances, i.e. the total number of points in the two slices, was up to 267 points in our experiments. This would lead to over $3,000,000$ triangles. Thus we cannot afford to compute the minimal area surface with respect to all triangles. Therefore we restrict the possible triangles to the Delaunay triangles. Only considering Delaunay Triangles is wide-spread for surface reconstruction problems.

Assuming that the contours in one slice are non-intersecting, Cheng and Dey [4] show how to modify the input so that there is a surface consisting of Delaunay triangles. They showed that it is sufficient that all segments are edges of the 2D Delaunay Triangulation of the individual slices. If a segment is not a Delaunay edge, they propose to add an additional point in the middle of the segment and recompute the Delaunay triangulation. This process is iterated until all segments are Delaunay edges. We also use this approach in our experiments.

We now turn to the separation algorithms. Both classes of inequalities can be separated by computing a number of minimal cuts.

Fix a $p \in V$ and look at the inequalities $\sum_{q \in V', r \in V^p \setminus V'} x_{pqr} \geq 1$ for all $\{p^-\} \subseteq V' \subseteq V^p \setminus \{p^+\}$. The most violated inequality is achieved for the set $V', \{p^-\} \subseteq V' \subseteq V^p \setminus \{p^+\}$ that minimizes $\sum_{q \in V', r \in V^p \setminus V'} x_{pqr}^*$ where x^* is the actual LP solution. Thus we have to solve a minimal (p^-, p^+)-cut problem in G^p with edge weights x^*. To find the most violated inequality, we have to iterate over all $p \in V$. Thus the total running time is $O(n^4)$.

For the inequalities (5) we fix p and q. Than we can find the most violated inequality by finding the subset $V', \{q\} \subseteq V' \subseteq V^p \setminus \{p^-, p^+\}$ that minimizes

$\sum_{r \in V', s \in V^p \setminus V'} x^*_{prs}$. This set can be found by computing a minimal (q, p^+)-cut in G^p and by setting the edge capacity of an edge rs to x^*_{prs} if $rs \neq p^- p^+$ and to infinity for the edge $p^- p^+$. The total running time is $O(n^5)$.

9 Experiments

We implemented a branch-and-cut algorithm for both ILP-formulations. We used the branch-and-cut system SCIL [12] which is based on ABACUS [1] and LEDA [11]. The LP solver can be chosen among CPLEX [5], XPRESS [15] and SOPLEX [13]. All our experiments were performed with CPLEX. SCIL provides an easy handling interface to implement branch-and-cut algorithms and LEDA provides the functionality to implement the separation algorithms. Furthermore we use CGAL [3] to compute the Delaunay triangulations.

As input for our experiments, we use a dataset given by Barequest [2]. It contains several sets of polygonal slices which originally came from CT-scans of human organs.

In all our experiments the general surface with minimal area was a normal surface.

Our experiments show that too many polygons were closed in the optimal surface. The reason is that the slices of the input are rather far away from each other compared to the size of the polygons. To reduce the number of closed polygons, we compute the area of the triangles according to slices whose distance is decreased by a constant factor. This results in surfaces that were very close to the original surfaces. All correspondence and branching problems were solved correctly.

The running time for our algorithm was very reasonable. For example the reconstruction of the 50 slices of pelvis data-set takes about 9 minutes if we compute the minimal area general surface and about 3.5 minutes if we restrict the surface to be normal. The running time to compute all Delaunay triangulations was about 25 seconds.

More experimental results can be found in [6].

10 Conclusion

We approached a problem arising from practical application with integer programming techniques. We studied the corresponding polyhedra and experimented with the resulting branch-and-cut algorithm. The results we obtained are promising.

It would be interesting to find further facets and corresponding separation algorithms for the polytopes. These could either be further lifted inequalities from the p-path configuration or the p-ordered path polytope, or inequalities that can not be obtained by lifting.

For the surface reconstruction problem for general point sets, there are algorithms that provably find an surface that is within a certain distance and homeomorphic to the original surface, if the point set satisfies some conditions. It would be interesting to obtain a similar result for our algorithm.

The next step seems to attack the surface reconstruction problem from points in general position with integer programming methods. We formulated this problem as ILP, but we were not able to solve practical instances with our formulation.

References

1. ABACUS–A Branch And CUt System. www.informatik.uni-koeln.de/ls_juenger/projects/abacus.html.
2. Dr. Gill Barequet's home page. http://www.cs.technion.ac.il/ barequet/.
3. CGAL–Computational Geometry Algorithms Library. www.cgal.org.
4. Siu-Wing Cheng and Tamal K. Dey. Improved constructions of delaunay based contour surfaces. In *Proceedings of the ACM Symposium on Solid Modeling and Applications*, pages 322–323, 1999.
5. CPLEX. www.cplex.com.
6. C. Fink. Oberflächenrekonstruktion von planaren Konturen. Master's thesis, Universität des Saarlandes, 2001.
7. H. Fuchs, Z. M. Kedem, and S. P. Uselton. Optimal surface reconstruction from planar contours. *Graphics and Image Processing*, 20(10):693–702, 1977.
8. M. Hong, T.W. Sederberg, K.S. Klimaszewski, and K. Kaneda. Triangulation of branching contours using area minimization. *International Journal of Computational Geometry & Applications*, 8(4):389–406, 1998.
9. V. Kann. Maximum bounded 3-dimensional matching is MAXSNP-complete. *Information Processing Letters*, 37:27–35, 1991.
10. E. Keppel. Approximating complex surfaces by triangulation of contour lines. *IBM J. of Research and Development*, 19:2–11, 1975.
11. LEDA (Library of Efficient Data Types and Algorithms). www.mpi-sb.mpg.de/LEDA/leda.html.
12. SCIL–Symbolic Constraints for Integer Linear programming. www.mpi-sb.mpg.de/SCIL.
13. SoPlex. www.zib.de/Optimization/Software/Soplex.
14. Laurence A. Wolsey. *Integer programming*. Wiley-interscience series in discrete mathematics and optimization. Wiley & Sons, New York, 1998.
15. XPRESS. www.dash.co.uk/.

The Semidefinite Relaxation
of the k-Partition Polytope Is Strong[*]

Andreas Eisenblätter

Konrad-Zuse-Zentrum für Informationstechnik Berlin (ZIB)
Takustr. 7, D-14195 Berlin, Germany
eisenblaetter@zib.de

Abstract Radio frequency bandwidth has become a very scarce resource. This holds true in particular for the popular mobile communication system GSM. Carefully planning the use of the available frequencies is thus of great importance to GSM network operators. Heuristic optimization methods for this task are known, which produce frequency plans causing only moderate amounts of disturbing interference in many typical situations. In order to thoroughly assess the quality of the plans, however, lower bounds on the unavoidable interference are in demand. The results obtained so far using linear programming and graph theoretic arguments do not suffice. By far the best lower bounds are currently obtained from semidefinite programming. The link between semidefinite programming and the bound on unavoidable interference in frequency planning is the semidefinite relaxation of the graph minimum k-partition problem.

Here, we take first steps to explain the surprising strength of the semidefinite relaxation. This bases on a study of the solution set of the semidefinite relaxation in relation to the circumscribed k-partition polytope. Our focus is on the huge class of hypermetric inequalities, which are valid and in many cases facet-defining for the k-partition polytope. We show that a "slightly shifted version" of the hypermetric inequalities is implicit to the semidefinite relaxation. In particular, no feasible point for the semidefinite relaxation violates any of the facet-defining triangle inequalities for the k-partition polytope by more than $\sqrt{2} - 1$ or any of the (exponentially many) facet-defining clique constraints by $\frac{1}{2}$ or more.

1 Introduction

Frequency planning is the key for GSM network operators to fully exploit the radio spectrum available to them. This spectrum is slotted into channels of 200 kHz bandwidth, characterized by their central frequency. Frequency planning has a significant impact on the quantity as well as on the quality of the radio communication services. Roughly speaking, radio communication requires a radio signal of sufficient strength which is not suffering too severely from interference by other signals. In a cellular system like GSM, these two properties, strong signals and little interference, are in conflict.

---

[*] This article is based on results contained in the Ph.D. thesis of the author.

W.J. Cook and A.S. Schulz (Eds.): IPCO 2002, LNCS 2337, pp. 273–290, 2002.
© Springer-Verlag Berlin Heidelberg 2002

1.1   Frequency Planning

The task of finding a "good" frequency plan can be described as follows:

> Given are the transmitters, the set of generally available frequencies to-
> gether with the local unavailabilities, and three upper-diagonal square
> matrices. The first matrix specifies for every pair of transmitters how
> far apart their frequencies have to be at least. These are called separa-
> tion constraints. The second matrix specifies for every transmitter pair
> how much co-channel interference is incurred in case they use the same
> frequency (where permissible), and the third matrix specifies how much
> adjacent-channel interference is incurred in case two transmitters use
> neighboring frequencies (where permissible). One frequency has to be as-
> signed to every transmitter such that the following holds. All separation
> requirements are met, and all assigned frequencies are locally available.
> The optimization goal is to find a frequency assignment resulting in the
> least possible interference.

A precise mathematical model of our GSM frequency planning problem is
given in [2,12], for example. This model is widely accepted in practice, compare
with [7], but there are alternative models as well.

Finding a feasible assignment is $\mathcal{NP}$-hard in general. This is easily derived by
exploiting the connection to list coloring [13] or to T-coloring of graphs [20]. Neither can an approximation algorithms with reasonable performance exist unless
$\mathcal{P} = \mathcal{NP}$, see [12].

Numerous heuristic and a few exact methods for solving the frequency assignment problem are known, see [2,25] for surveys. Although we are nowadays in the
situation that a number of heuristic methods performs well in practice, there is
an uncomfortable lack of quality guarantees on the per-instance-level—not just
in theory. Substantial efforts have been made to remedy this shortcoming over
the last decade, see [1,22,23,25,31] but the results are not satisfactory. In particular, no exact method is presently known that is capable of solving large realistic
planning problems in practice.

1.2   Quality Guarantees

A novel approach was recently proposed to derive a lower bound on the unavoidable co-channel interference: The frequency planning problem is simplified
to a minimum graph k-partition problem, where the edge-weights represent co-
channel interference or separation constraints and k is the number of overall
available frequencies. A lower bound for the optimal k-partition partition (and,
thus, for the unavoidable co-channel interference) is computed from a semidefinite relaxation of the resulting minimum graph k-partition problem.

Table 1 lists computational results for five realistic planning problems, which
are publicly available over the Internet [14]. The first column gives the instance's
name. The next five columns describe several characteristics of the instance,
mostly in terms of underlying graph $G = (V, E)$. This graph contains one vertex

Table 1. Realistic frequency planning problems with quality guarantees

instance	characteristics					best assignment		lower bound	gap [%]
	$\|V(G)\|$	$\rho(G)$	$\Delta(G)$	$\omega(G)$	freqs. (k)	co-ch.	ad.-ch.		
K	267	0.56	238	69	50	0.37	0.00	**0.1836**	102
B[4]	2775	0.13	1133	120	75	17.29	0.44	**4.0342**	339
B[10]	4145	0.13	1704	174	75	142.09	4.11	**54.0989**	170
SIE2	977	0.49	877	182	$4 + 72$	12.57	2.18	**6.9463**	112
SIE4	2785	0.11	752	100	39	71.09	9.87	**27.6320**	193

for each transmitter, and two vertices are joined by an edge whenever at least one of the corresponding three matrix entries is non-zero. We list the number $|V(G)|$ of vertices in G, the edge density $\rho(G)$ of the graph, the maximum degree $\Delta(G)$ of a vertex, the clique number $\omega(G)$, and the number of globally available frequencies. (In the case of instance SIE2, the available spectrum consists of two contiguous bands of size 4 and 72. In all other cases there is just one band.) Columns 7 and 8 show the amount of co- and adjacent-channel interference incurred by the best presently known frequency assignment for each instance, see [12] for details. The last but one column contains the lower bound on the unavoidable co-channel interference. Finally, the last column gives the relative quality of the best known solution with respect to the bound on the unavoidable (co-channel) interference.

The lower bound figures are computed using the semidefinite programming solvers of Burer, Monteiro, and Zhang [4] and Helmberg [21]. Both are dual solvers, that is, their results are dual approximations of the optimal value without quality guarantee. The corresponding semidefinite programs involve a $|V| \times |V|$ square matrix variable and $\binom{|V|+1}{2}$ many linear constraints. Looking again at Tab. 1 reveals that these semidefinite programs are (very) large.

From the applications points of view, gaps of a few hundred percent are by far not satisfying. Nevertheless, for large realistic planning problems these are the first significant lower bounds to the best of our knowledge. We want to indicate why the (approximate) solution of the semidefinite relaxation yields stronger bounds than previous approaches.

1.3   Sketch of Results

In the remainder of this paper, we give first reasons for the surprisingly strong bounds on the unavoidable interference obtained from semidefinite programming. To this end, we study the relation between a class of polytopes associated to the minimum graph k-partition problem (via an integer linear programming formulation) and their circumscribing semidefinite relaxations. This is done in terms of the hypermetric inequalities, which form a huge class of valid and in many cases facet-defining inequalities for the polytopes. We show in Prop. 3 that a "slightly shifted version" of the hypermetric inequalities is implicit in the semidefinite relaxation. Propositions 5, 6, 7, and 8 state to which extent the "shift" may by reduced under specific conditions. We also show, see Prop. 9, that neither the semidefinite relaxation nor the linear relaxation, which is obtained from the standard integer linear programming formulation of minimum k-partition by dropping the integrality conditions, is generally stronger than the other. This is particularly noteworthy since the semidefinite relaxation contains only polynomially many linear constraints (in the size of the graph and k), whereas the linear relaxation is of exponential size.

1.4 Preliminaries

The definitions and notations used here are mostly taken from [32] for graph theory, from [30] for (integer) linear programming, and from [21] for semidefinite programming. A thorough exposition of the notions used is contained in [12].

2   The Minimum k-Partition Problem

The minimum graph k-partition problem or MINIMUM K-PARTITION problem we want to solve is formally defined as follows.

Definition 1. *An instance of the* MINIMUM K-PARTITION *problem consists of an undirected graph* $G = (V, E)$*, a weighting* $w\colon E \to \mathbb{Q}$ *of the edges, and a positive integer* k*. The objective is to find a partition of* V *into at most* k *disjoint sets* $V_1, \ldots, V_p$ *such that the value* $\sum_{l=1}^{p} \sum_{vw \in E(G[V_l])} w(vw)$ *is minimized.*

The MINIMUM K-PARTITION problem is studied in [5,6]. The complementing MAXIMUM K-CUT problem is investigated in [8,9,15,17,24]. Recall that both problems have quite different characteristics in terms of their approximability. Whereas the MAXIMUM K-CUT can be solved $(1 - k^{-1})$–approximately in polynomial time, the MINIMUM K-PARTITION problem is not $\mathcal{O}(|E|)$–approximable in polynomial time unless $\mathcal{P}= \mathcal{NP}$. Results on the approximation of the MAXIMUM K-CUT problem are obtained in [15,24]. These results are, however, of no use for the bounding of unavoidable interference in GSM frequency planning. The optimal cut value is underestimated so that the value of the MINIMUM K-PARTITION is overestimated. No lower bound is supplied that way. Due to the opposite sign restriction, the results from [17] do not apply either.

The class of polytopes associated with the MINIMUM K-PARTITION is defined through the convex hull of the feasible solutions to the integer linear programming formulation (1)–(4) of the problem. This formulation is only valid for complete graphs K_n with $n \geq k \geq 2$. (The graph $G = (V, E)$ with edge weights w is completed to $K_{|V|}$, and the edge weighting is extended to all new edges by assigning a weight of zero.) We assume for notational convenience that the vertex set of K_n is $\{1, \ldots, n\}$ and that the edge set is $\{ij \mid 1 \leq i < j \leq n\}$. One binary variable is used for every edge of the complete graph. The intended meaning is that $z_{ij} = 1$ if and only if the vertices i and j are in the same partite set of the partition.

$$\min \sum_{i,j \in V} w_{ij} z_{ij} \tag{1}$$

$$z_{ih} + z_{hj} - z_{ij} \leq 1 \qquad \forall\, h, i, j \in V \tag{2}$$

$$\sum_{i,j \in Q} z_{ij} \geq 1 \qquad \forall\, Q \subseteq V \text{ with } |Q| = k+1 \tag{3}$$

$$z_{ij} \in \{0,1\} \tag{4}$$

The triangle inequalities (2) require the setting of the variables to be consistent, that is, transitive. The clique constraints (3) impose that at least two from a set of $k+1$ vertices have to be in the same partite set. Together with the constraints (2) this implies that there are at most k partite sets. In total, there are $3 \binom{|V|}{3}$ many triangle and $\binom{|V|}{k+1}$ many clique inequalities. The number of clique inequalities grows roughly as fast as $|V|^k$ as long as $2k \leq |V|$. We come back to this point at the end of this section.

The well-known integer linear programming formulation of MAXIMUM K-CUT is obtained by complementing the variables.

For the sake of simplicity, we assume $k \geq 3$ in the following. Clearly, if $k = 1$, then there is only one "partition;" in the case of $k = 2$, the MINIMUM 2-PARTITION problem is equivalent to the well-known MAXIMUM CUT problem, see [11,28] for surveys.

For all $3 \leq k \leq n$, we define the polytope

$$\mathcal{P}_{\leq k}(K_n) = \mathrm{conv}\{z \in \{0,1\}^{E(K_n)} \mid z_{hi} + z_{ij} - z_{hj} \leq 1 \quad \forall\, h, i, j \in V;$$
$$\sum_{i,j \in Q} z_{ij} \geq 1 \quad \forall\, Q \subseteq V, |Q| = k+1\} \ .$$

$\mathcal{P}_{\leq k}(K_n)$ is the set of convex combinations of all feasible solutions to the integer linear program (1)–(4). The resulting class of polytopes is studied in [6,8,9,12] and the special case $k = n$ is also investigated in [29]. Here, we merely highlight a few points.

The polytopes are full-dimensional in the space spanned by the edge variables, and none contains the origin. All valid inequalities for $\mathcal{P}_{\leq k}(K_n)$ that are violated by the origin have large support. To be more precise, given some inequality $a^T z \geq a_0$, the *support graph* of $a^T z \geq a_0$ is the subgraph of K_n induced by all edges ij with $a_{ij} \neq 0$. The following holds.

Proposition 1 ([12]). *Let $a^T z \leq a_0$ be a valid inequality for $\mathcal{P}_{\leq k}(K_n)$ and H its support graph. If H is k-partite, then $a_0 \leq 0$, and if $a_0 > 0$, then $|\{ij \in E \mid a_{ij} \neq 0\}| \geq \binom{k+1}{2}$.*

The hypermetric inequalities form a large and important class of valid and in many cases facet-defining inequalities for $\mathcal{P}_{\leq k}(K_n)$ [6]. They sometimes do separate the origin from $\mathcal{P}_{\leq k}(K_n)$, and they generalize a number of previously known inequalities, such as the triangle inequalities [19], the 2-partition inequalities [6,19], the (general) clique inequalities [5], and the claw inequalities [29]. The right-hand sides of the hypermetric inequalities involve the function

$f_{hm}(\eta, k) = \max\left\{\sum_{1 \leq i < j \leq k} x_i x_j \mid \sum_{i=1}^{k} x_i = \eta, x_i \in \mathbb{Z}_+\right\}$, which depends on two integral parameters η and k, $\eta \geq 0$, $k \geq 1$. For given η and k, the value $f_{hm}(\eta, k)$ is the maximum number of edges in a k-partite graph with η many vertices. For our purposes, the following equivalent definition is more convenient:

$$f_{hm}(\eta, k) = \binom{\eta \bmod k}{2}\left\lceil\frac{\eta}{k}\right\rceil^2 + \binom{k - \eta \bmod k}{2}\left\lfloor\frac{\eta}{k}\right\rfloor^2$$
$$+ (\eta \bmod k)(k - \eta \bmod k)\left\lceil\frac{\eta}{k}\right\rceil\left\lfloor\frac{\eta}{k}\right\rfloor .$$

The hypermetric inequalities for $\mathcal{P}_{\leq k}(K_n)$ are defined as follows.

Definition 2. *Given $k \geq 3$ and a complete graph K_n and vertex weights $b_v \in \mathbb{Z}$ with $\eta = \sum_{v \in V(K_n)} b_v \geq 0$. The associated hypermetric inequality is*

$$\sum_{vw \in E(K_n)} b_v b_w z_{vw} \geq \sum_{vw \in E(K_n)} b_v b_w - f_{hm}(\eta, k) . \tag{5}$$

The hypermetric inequalities are shown to be valid for $\mathcal{P}_{\leq k}(K_n)$ in [6,10]. Let us consider the simple case in which all vertex weights are equal to one. Then (5) bounds the number of edges that run within partite sets from below. The corresponding inequality for the MAXIMUM K-CUT polytopes bounds the number of edges in the k-cut from above.

Before concluding this section, let us return to the integer linear program (1)–(4). A linear programming relaxation is obtained by replacing the feasible variable values $\{0, 1\}$ by their convex hull $[0, 1]$. We call the resulting polytope $\mathcal{P}_{\leq k}^{LP}(K_n)$. Recall that all facets separating the origin from the polytope have a support of at least $\binom{k+1}{2}$. The smallest such examples are the exponentially many clique inequalities (3). Among others, these inequalities with large support make linear programming relaxation of (1)–(4) hard to deal with. Not surprisingly, no major successes in solving MINIMUM K-PARTITION instances with a few hundred vertices and $k \gg 2$ using the classical branch-and-cut approach are reported in literature.

3   A Semidefinite Relaxation of Min k-Partition

Semidefinite programming is the task of minimizing (or maximizing) a linear objective function over the convex cone of positive semidefinite matrices subject to linear constraints. Here, a square matrix X with real-valued entries is positive semidefinite ($X \succeq 0$) if it is symmetric and its eigenvalues are non-negative. We assume that the reader is familiar with semidefinite programming, see, for example, the introductions/surveys given in [3,21,33].

An alternative formulation of the MINIMUM K-PARTITION problem can be seen as a semidefinite program with "integrality" constraints, compare with [15,24]. The next lemma is the basis for this formulation.

Lemma 1 ([12,15,24]). *For all integers n and k satisfying $2 \leq k \leq n + 1$ the following holds:*

1. k unit vectors $\bar{u}_1, \ldots, \bar{u}_k \in \mathbb{R}^n$ exist such that $\langle \bar{u}_i, \bar{u}_j \rangle = \frac{-1}{k-1}$ for all $i \neq j$;

2. any given k unit vectors $u_1, \ldots, u_k \in \mathbb{R}^n$ satisfy $\sum_{i<j} \langle u_i, u_j \rangle \geq -\frac{k}{2}$ and if $\langle u_i, u_j \rangle \leq \delta$ for all $i \neq j$, then $\delta \geq \frac{-1}{k-1}$.

According to the lemma, we may fix a set $U = \{u_1, \ldots, u_k\} \subseteq \mathbb{R}^n$ of unit vectors with $\langle u_i, u_j \rangle = \frac{-1}{k-1}$ for $i \neq j$ (and $\langle u_i, u_i \rangle = 1$ for $i = 1, \ldots, k$). These k vectors are used as labels (or representations) for the k partite sets. The MINIMUM K-PARTITION problem is then the task of finding an assignment $\phi : V \mapsto U$ that minimizes the expression $\sum_{i,j \in V(K_n)} c_{ij} \frac{(k-1)\langle \phi(i), \phi(j) \rangle + 1}{k}$. The quotient in the summands evaluates to either 1 or 0, depending on whether the same vector or distinct vectors are assigned to the respective two vertices.

We assemble the scalar products $\langle \phi(i), \phi(j) \rangle$ into a square matrix X, being indexed row- and column-wise by V. The matrix X has the following properties: all entries on the principal diagonal are ones, all off-diagonal elements are either $\frac{-1}{k-1}$ or 1, and X is positive semidefinite. Conversely, every matrix X satisfying the above properties defines a k-partition of V in the same way as ϕ does [12]. Hence,

$$\min \sum_{ij \in E(K_n)} c_{ij} \frac{(k-1)X_{ij} + 1}{k} \tag{6}$$

$$\text{s. t.}$$

$$X_{ii} = 1 \qquad \forall i \in V \tag{7}$$

$$X_{ij} \in \left\{ \frac{-1}{k-1}, 1 \right\} \qquad \forall i, j \in V \tag{8}$$

$$X \succeq 0 \tag{9}$$

is an alternative formulation of the MINIMUM K-PARTITION problem.

Replacing the constraints (8) by $\frac{-1}{k-1} \leq X_{ij} \leq 1$ yields a semidefinite program. Notice that $X_{ij} \leq 1$ can be dropped as constraint since it is enforced implicitly by X being positive semidefinite and $X_{ii} = 1$. The semidefinite programming relaxation obtained from (6)–(9) is ϵ-approximately solvable in polynomial time [18].

This type of relaxation of a combinatorial optimization problem was first introduced to compute the Shannon capacity of a graph [27] and later formed the basis for the famous 0.878-approximation algorithm for the MAXIMUM CUT problem [16]. The approximation algorithm was generalized to the MAXIMUM K-CUT problem with positive edge-weights in [15] and [24] independently. Notice that the semidefinite relaxation of MAXIMUM K-CUT differs from that of (6)–(9) only in the objective function. Due the previously cited inapproximability of MINIMUM K-PARTITION, however, a comparable approximation result as that for MAXIMUM K-CUT is not to be expected for MINIMUM K-PARTITION.

As stated before, we want to investigate the strength of the semidefinite relaxation. In order to do so, we relate the solution set of the semidefinite relaxation to the polytope $\mathcal{P}_{\leq k}(K_n)$. This is done by considering a projection of an affine image of the solution set into $\mathbb{R}^{\binom{n}{2}}$ in such a way that the objective function values are preserved. The image of the projection is called $\Theta_{k,n}$ and contains $\mathcal{P}_{\leq k}(K_n)$.

Let $\Psi_{k,n}$ denote the set of feasible solutions of the semidefinite relaxation of (6)–(9), that is, $\Psi_{k,n} = \{ X \in \mathbb{R}^{n \times n} \mid X \succeq 0, X_{ii} = 1, X_{ij} \geq \frac{-1}{k-1}, i,j \in \{1,\ldots,n\} \}$. In the case of $k = 2$, $\Psi_{k,n}$ is the *elliptope* $\mathcal{E}_n = \{ X \in \mathbb{R}^{n \times n} \mid X \succeq 0, X_{ii} = 1, i \in \{1,\ldots,n\} \}$. For $k > 2$, $\Psi_{k,n}$ is obtained by intersecting the elliptope $\mathcal{E}_n$ with the half-spaces defined by $X_{ij} \geq \frac{-1}{k-1}$ for all $i,j \in \{1,\ldots,n\}$. The elliptope is studied extensively in the literature, see [11] for a survey. It will later be exploited that the elliptope can also be characterized as

$$\mathcal{E}_n = \Big\{ X \in S_n \mid X_{ii} = 1 \text{ for } i = 1,\ldots,n;$$

$$2 \sum_{1 \leq i < j \leq n} b_i b_j X_{ij} \geq - \sum_{i=1}^{n} b_i^2 \quad \text{for all } b \in \mathbb{Z}^n \Big\} . \tag{10}$$

Next, we define an affine mapping that projects $\Psi_{k,n}$ into $\mathbb{R}^{\binom{n}{2}}$. Let $T_k \colon \mathbb{R} \to \mathbb{R}$ be the affine transformation $x \mapsto \frac{k-1}{k} x + \frac{1}{k}$, mapping 1 onto 1 and $\frac{-1}{k-1}$ onto 0. The affine transformation T_k is extended to the set S_n of $n \times n$ symmetric real-valued matrices (which is isomorphic to $\mathbb{R}^{\binom{n+1}{2}}$) by letting $T_k \colon S_n \to S_n$, $S \mapsto \frac{k-1}{k} S + \frac{1}{k} E(n,n)$. Here, $E(n,n)$ is the $n \times n$ matrix with all entries being equal to one. Let $\zeta_{k,n} \colon S_n \to \mathbb{R}^{\binom{n}{2}}$, $X \mapsto \zeta_{k,n}(X) = z$ with $z_{ij} = (T_k(X))_{ij}$ for $i < j$, and consider

$$\Theta_{k,n} = \zeta_{k,n}(\Psi_{k,n}) = \{ \zeta_{k,n}(X) \mid X \in \Psi_{k,n} \} .$$

The restriction of $\zeta_{k,n}$ onto $\Psi_{k,n}$ is one-to-one, and $\zeta_{k,n}|_{\Psi_{k,n}} \colon \Psi_{k,n} \to \Theta_{k,n}$ is an affine bijection. Moreover, for any given $X \in \Psi_{k,n}$ and any given $w \in \mathbb{R}^{\binom{n}{2}}$ the identity $\frac{1}{2} \langle W, T_k(X) \rangle = \langle w, \zeta_{k,n}(X) \rangle$ holds. Here, W is the symmetric matrix obtained from w by letting $W_{ii} = 0$ and $W_{ij} = w_{ij}$ for all $1 \leq i < j \leq n$. $\langle \cdot, \cdot \rangle$ denotes the ordinary scalar products on the vector spaces S_n and $\mathbb{R}^{\binom{n+1}{2}}$, respectively.

A direct consequence of our definitions is that the optimization problems

$$\min \frac{1}{2} \langle W, T_k(X) \rangle \text{ s.t. } X \in \Psi_{k,n} \quad \text{and} \quad \min \langle w, z \rangle \text{ s.t. } z \in \Theta_{k,n} \tag{11}$$

are equivalent.

The affine image $\Theta_{k,n}$ of the truncated elliptope $\Psi_{k,n}$ contains the polytope $\mathcal{P}_{\leq k}(K_n)$ and is itself contained in the hypercube $[0,1]^{\binom{n}{2}}$, see [12]. A related connection between the MAXIMUM CUT polytope and the elliptope is studied in [26]. We call $\Theta_{k,n}$ a *semidefinite relaxation* of $\mathcal{P}_{\leq k}(K_n)$. $\Theta_{k,n}$ and $\mathcal{P}_{\leq k}(K_n)$ contain the same integral points.

Proposition 2. *Given integers k, n with $2 \leq k < n$, then $\Theta_{k,n}$ and $\mathcal{P}_{\leq k}(K_n)$ contain the same integral points.*

Proof. Let $\bar{z}$ be an integral (binary) vector in $\Theta_{k,n}$. Let $\bar{X}$ denote the pre-image of $\bar{z}$ under the mapping $\zeta_{k,n}$. All entries of the positive semidefinite matrix $\bar{X}$ are either $\frac{-1}{k-1}$ or $+1$.

No triangle constraint (2) is violated, because such a violation would imply that $\bar{X}$ has one of the matrices

$$\begin{bmatrix} 1 & \frac{-1}{k-1} & 1 \\ \frac{-1}{k-1} & 1 & 1 \\ 1 & 1 & 1 \end{bmatrix}, \quad \begin{bmatrix} 1 & 1 & \frac{-1}{k-1} \\ 1 & 1 & 1 \\ \frac{-1}{k-1} & 1 & 1 \end{bmatrix}, \quad \begin{bmatrix} 1 & 1 & 1 \\ 1 & 1 & \frac{-1}{k-1} \\ 1 & \frac{-1}{k-1} & 1 \end{bmatrix}$$

as a principal sub-matrix. In all cases, the determinant is $-\left(\frac{k}{k-1}\right)^2 < 0$. Hence, none of these matrices may appear as a principal sub-matrix of $\bar{X}$.

According to Lemma 1, no subset Q of size larger than k can induce a sub-matrix $\bar{X}_{QQ}$ with all its off-diagonal elements equal to $\frac{-1}{k-1}$. Thus, at least one off-diagonal element in $\bar{X}_{QQ}$ equals 1 for each set Q of size $k+1$, and, consequently, no clique constraint (3) is violated by $\bar{z}$.   □

4   The Semidefinite Relaxation $\Theta_{k,n}$ and $\mathcal{P}_{\leq k}(K_n)$

In this section, we investigate the relation between the semidefinite relaxation $\Theta_{k,n}$ and the polytope $\mathcal{P}_{\leq k}(K_n)$ in more detail. The hypermetric inequalities (5) play the central role. Our first result shows that a "slightly shifted version" of the hypermetric inequality is valid for $\Theta_{k,n}$. We also show, that neither $\mathcal{P}_{\leq k}^{LP}(K_n)$ nor $\Theta_{k,n}$ is generally contained in the other.

Proposition 3. *Given an integer $k \geq 3$ and an integral weight b_i for every vertex $i \in V(K_n)$. The following inequality is valid for $\Theta_{k,n}$:*

$$\sum_{ij \in E(K_n)} b_i\, b_j\, z_{ij} \geq \frac{1}{2k}\left(\left(\sum_{i \in V(K_n)} b_i\right)^2 - k \sum_{i \in V(K_n)} b_i^2\right). \qquad (12)$$

Proof. The function $\zeta_{k,n}$ maps any matrix $X \in \Psi_{k,n}$ to a vector $z \in \Theta_{k,n}$ such that $X_{ij} = \frac{1}{k-1}(k\,z_{ij} - 1)$ holds for every $1 \leq i < j \leq n$. Plugging this into (10) yields

$$2 \sum_{1 \leq i < j \leq n} b_i b_j \frac{k\,z_{ij} - 1}{k-1} = \frac{2k}{k-1} \sum_{1 \leq i < j \leq n} b_i b_j z_{ij} - \frac{2}{k-1} \sum_{1 \leq i < j \leq n} b_i b_j \geq -\sum_{i=1}^{n} b_i^2 \ .$$

This is equivalent to (12).   □

The difference between the right-hand side of hypermetric inequalities (5) and the right-hand side of (12) is bounded by a term depending on the relation between k and the sum of the vertex weights b_i.

Proposition 4. *Given an integer $k \geq 3$ and an integral weight b_i for every vertex $i \in V(K_n)$. Then the difference between the right-hand side of the hypermetric inequalities (5) for the polytope $\mathcal{P}_{\leq k}(K_n)$ and the right-hand side of the hypermetric inequalities (12) for $\Theta_{k,n}$ is*

$$\left(\left(\sum_{i\in V(K_n)} b_i\right) \bmod k\right) \frac{k - (\sum_{i\in V(K_n)} b_i) \bmod k}{2k} \ .$$

The upper bounded $\frac{k}{8}$ for this expression is attained if $(\sum_{i\in V(K_n)} b_i) \bmod k = \frac{k}{2}$.

Proof. Let $p = \left\lfloor (\sum_{i\in V(K_n)} b_i)/k \right\rfloor$ and $r = (\sum_{i\in V(K_n)} b_i) \bmod k$, then simply plugging these parameters into the right-hand side of (5) yields

$$f_{hm}\left(\sum_{i\in V(K_n)} b_i, k\right) + \left(\left(\sum_{i\in V(K_n)} b_i\right) \bmod k\right) \frac{k - (\sum_{i\in V(K_n)} b_i) \bmod k}{2k}$$

$$= f_{hm}(pk + r, k) + r\frac{k - r}{2k} \ .$$

We expand this expression:

$$f_{hm}(pk + r, k) + r\frac{k - r}{2k}$$

$$= \left(-\frac{k\,p^2}{2} + \frac{k^2\,p^2}{2} - \frac{r}{2} - pr + kpr + \frac{r^2}{2}\right) + \left(\frac{r}{2} - \frac{r^2}{2k}\right)$$

$$= \frac{1}{2k}\left(-k^2\,p^2 + k^3\,p^2 - kr - 2kpr + 2k^2\,pr + kr^2 + kr - r^2\right)$$

$$= \frac{1}{2k}\left(k^3\,p^2 + 2k^2\,pr + kr^2 - k^2\,p^2 - 2kpr - r^2\right)$$

$$= \frac{(k-1)(pk + r)^2}{2k} \ . \tag{13}$$

The right-hand side of (12) is

$$\frac{1}{2k}\left(\left(\sum_{i\in V(K_n)} b_i\right)^2 - k\sum_{i\in V(K_n)} b_i^2\right)$$

$$= \frac{1}{2k}\left(2k\sum_{ij\in E(K_n)} b_ib_j - k\left(\sum_{i\in V(K_n)} b_i\right)^2 + \left(\sum_{i\in V(K_n)} b_i\right)^2\right)$$

$$= \sum_{ij\in E(K_n)} b_ib_j - \frac{(k-1)\left(\sum_{i\in V(K_n)} b_i\right)^2}{2k}$$

$$\stackrel{(13)}{=} \sum_{ij\in E(K_n)} b_ib_j - f_{hm}\left(\sum_{i\in V(K_n)} b_i, k\right)$$

$$- \left(\left(\sum_{i\in V(K_n)} b_i\right) \bmod k\right) \frac{k - (\sum_{i\in V(K_n)} b_i) \bmod k}{2k} \ .$$

The first part of the claim follows from this. As far as the second part is concerned, we observe that $r\frac{k-r}{2k}$ is a quadratic polynomial in r. Its maximum of $\frac{k}{8}$ is attained for $r = \frac{k}{2}$. This completes the proof.   □

In addition to the binary restrictions on the variables, the integer linear programming formulation (1)–(4) linked to $\mathcal{P}_{\leq k}(K_n)$ contains only constraints on triangles and on cliques of size $k+1$. Both classes of constraints are facet-defining hypermetric inequalities for $\mathcal{P}_{\leq k}(K_n)$ [5]. For these two types of inequalities, the following holds with respect to $\Theta_{k,n}$. Notice that (14) improves on the general hypermetric inequality (12) for $\Theta_{k,n}$.

Proposition 5. *Given the complete graph K_n and an integer k with $4 \leq k \leq n$, then for every $z \in \Theta_{k,n}$*

$$z_{ij} + z_{jl} - z_{il} \leq 1 + \frac{\sqrt{2(k-2)(k-1)} - (k-2)}{k} \quad \left[< \sqrt{2}\right] \qquad (14)$$

holds for every triangle and

$$\sum_{ij \in E(Q)} z_{ij} \geq 1 - \frac{k-1}{2k} \quad \left[> \frac{1}{2}\right] \qquad (15)$$

holds for every clique Q of size $k+1$ in K_n. Both bounds are tight.

All triangle inequalities (2) and all clique constraints (3) are "more than half satisfied" by every point in $\Theta_{k,n}$ in the sense that the violation is bounded by $\frac{1}{2}$ rather than by the worst possible 1.

In both cases, tightness follows from more general statements in Props. 6 and 8. These results, together with that from Prop. 7, are proven by means of semidefinite programming duality in Sec. 5.

Proposition 6. *Given the complete graph K_n and an integer k with $3 \leq k < n$. Let Q be a clique in K_n of size larger than k. Then*

$$\sum_{ij \in E(Q)} z_{ij} \geq \frac{|Q|}{2k}(|Q| - k) \qquad (16)$$

is valid for $\Theta_{k,n}$, and there is a point $\bar{z} \in \Theta_{k,n}$ satisfying (16) at equality.

Under certain conditions on the relation among k, $|S|$, and $|T|$, a "shifted 2-partition inequality" is tight for $\Theta_{k,n}$.

Proposition 7. *Given the complete graph K_n, $n \geq 4$, and an integer k with $4 \leq k \leq n$. Let S and T be non-empty, disjoint subsets of $V(K_n)$ with $|S| \leq |T|$. Then*

$$z(E(S)) + z(E(T)) - z([S,T]) \geq \frac{1}{2k}\left((|T| - |S|)^2 - k(|T| + |S|)\right) \qquad (17)$$

is valid for $\Theta_{k,n}$. Furthermore, there is a point $\bar{z} \in \Theta_{k,n}$ satisfying (17) at equality if one of the following conditions holds:

1. $|S| = 1$ and $|T| \geq k - 1$;
2. $|S| \geq 2$, $|S| + |T| \leq k$ and either $|T| \leq |S|^2$, or $|T| > |S|^2$ together with $k \leq \frac{|T|^2 - |S|^2}{|T| - |S|^2}$.

The treatment of the case $|S| = 1$ in Prop. 7 is not entirely satisfying, because the most prominent representative of the 2-partition inequalities, namely, the triangle inequalities, is not covered. The case of $|S| = 1$ and $2 \leq |T| \leq k - 2$ is therefore considered separately.

Proposition 8. *Given the complete graph K_n and an integer k with $4 \leq k \leq n$. Let S and T be disjoint subsets of $V(K_n)$ with $1 = |S| < |T| \leq k - 2$. Then*

$$z(E(S)) + z(E(T)) - z([S,T])$$
$$\geq -1 - \frac{\sqrt{t(k-t)(k-1)} - (k-t)}{k} \quad \left[> -\sqrt{t} \right] \tag{18}$$

is valid for $\Theta_{k,n}$, and a point $\bar{z} \in \Theta_{k,n}$ fulfills (18) at equality.

Finally, we show that $\Theta_{k,n}$ and the polytope $\mathcal{P}_{\leq k}^{LP}(K_n)$ are incomparable in general. Recall that $\mathcal{P}_{\leq k}^{LP}(K_n)$ is associated to the linear program, which is obtained from (1)–(4) by replacing $[0,1]$ for $\{0,1\}$ in (4). It follows from Prop. 5 that $\Theta_{k,n}$ contains points which are not contained in $\mathcal{P}_{\leq k}^{LP}(K_n)$, that is, $\Theta_{k,n} \not\subset \mathcal{P}_{\leq k}^{LP}(K_n)$. In general, the reverse inclusion does not hold either. In order to see this, we fix integers k and n such that $4 \leq k < \sqrt{n}$. Let $\tilde{z} \in \mathbb{R}^{\binom{n}{2}}$ be the vector with all coordinates equal to $\frac{1}{k+1}$. Then $\tilde{z} \in \mathcal{P}_{\leq k}^{LP}(K_n)$ because $0 < \tilde{z}_{ij} < 1$ for all ij and $\tilde{z}$ satisfies all triangle inequalities (2) as well as all clique inequalities (3). The vector $\tilde{z}$ is, however, not contained in $\Theta_{k,n}$, because the valid inequality (12) with $b_i = 1$ for every vertex i is violated by $\tilde{z}$:

$$\sum_{ij \in E(K_n)} \tilde{z}_{ij} = \binom{n}{2} \frac{1}{k+1} = \frac{n(n-1)}{2(k+1)} \not\geq \frac{n(n-k)}{2k} = \frac{1}{2k}(n^2 - kn) \ .$$

This follows from $k(n-1) < (k+1)(n-k) \iff 0 < n - k^2$ and our assumption $\sqrt{n} > k$. In summary, the following holds.

Proposition 9. *Given two integers k and n with $4 \leq k < \sqrt{n}$, then neither $\Theta_{k,n}$ is contained in $\mathcal{P}_{\leq k}^{LP}(K_n)$ nor is the converse true.*

5   Proving Tightness Results Using Duality Theory

Two types of matrices are of importance in our proofs of Prop. 6, 7, and 8. Let $E(m,n) \in \mathbb{R}^{m \times n}$ be the matrix with all entries equal to 1. Let $D^{\alpha,\beta}(n)$ denote the symmetric square matrix of order $n \geq 1$ with all entries on the principal diagonal equal to α and all other entries equal to β. Some basic properties of $D^{\alpha,\beta}(n)$ are easily observed.

Proposition 10. *For $n \geq 1$, the determinant of $D^{\alpha,\beta}(n)$ is given by*

$$\det(D^{\alpha,\beta}(n)) = (\alpha - \beta)^{n-1} (\alpha + (n-1)\beta) .$$

For $\beta \notin \{\frac{-\alpha}{n-1}, \alpha\}$ the matrix $D^{\alpha,\beta}(n)$ is regular and its inverse is

$$D^{\alpha,\beta}(n)^{-1} = \frac{1}{(\alpha - \beta)(\alpha + (n-1)\beta)} D^{\alpha+(n-2)\beta,-\beta}(n) .$$

$D^{\alpha,\beta}(n)$ *is positive semidefinite if and only if $\alpha \geq \beta \geq \frac{-\alpha}{n-1}$; it is positive definite if and only if strict inequality holds in both cases. (In case $n = 1$, $D^{\alpha,\beta}(n) = [\alpha]$ and $\beta = 0$ is assumed. The condition "$\beta \geq \frac{-\alpha}{n-1}$" becomes void.)*

We need to know the conditions on α, β, γ, δ, ε, and s, t under which the matrix

$$A = \begin{bmatrix} D^{\alpha,\beta}(s) & \gamma E(s,t) \\ \gamma E(t,s) & D^{\delta,\varepsilon}(t) \end{bmatrix}$$

is positive semidefinite. These conditions can be derived by means of the Schur Complement Theorem. (This theorem states that the composite matrix with matrices X and Z as diagonal blocks and Y and Y^T as off-diagonal blocks is positive semidefinite if and only if $Z \succeq Y^T X^{-1} Y$ when X is a positive definite $(n \times n)$–matrix, Z is a symmetric $(m \times m)$–matrix, and Y is a $(m \times n)$–matrix.)

Proposition 11. *Given integers $s, t \geq 1$, the matrix*

$$A = \begin{bmatrix} D^{\alpha,\beta}(s) & \gamma E(s,t) \\ \gamma E(t,s) & D^{\delta,\varepsilon}(t) \end{bmatrix}$$

is positive semidefinite if and only if $D^{\alpha,\beta}(s)$, $D^{\delta,\varepsilon}(t)$ are both positive semidefinite and $(\alpha + (s-1)\beta)(\delta + (t-1)\varepsilon) \geq st\gamma^2$ holds.

Our tightness proofs are based on a pair of dual semidefinite programs. We denote with $E^{ij}(n) \in S_n$ the symmetric $(n \times n)$–matrix with entries 1 at positions (i, j) and (j, i), and zeros elsewhere. We simply write E^{ij} if the dimension is clear from the context.

Remark 1. For every matrix $C \in S_n$, the semidefinite programs

$$\min \sum_{1 \leq i,j \leq n} C_{ij} X_{ij} \quad \text{s.t.}$$

$$\langle E^{ii}, X_{ii} \rangle = 1, \quad \langle E^{ij}, X_{ij} \rangle \geq \frac{-1}{k-1}, \quad \forall\, i, j \in \{1, \ldots, n\}, i < j \qquad (19)$$

$$X \in S_n^+$$

and

$$\max \sum_{i=1}^{n} y_{ii} - \sum_{1 \leq i < j \leq n} \frac{y_{ij}}{k-1} \quad \text{s.t.}$$

$$C - \sum_{1 \leq i \leq j \leq n} y_{ij} E^{ij} \in S_n^+, \quad y_{ii} \in \mathbb{R}, \quad y_{ij} \in \mathbb{R}_+ \qquad (20)$$

are dual to each other. They are both strictly feasible.

The dual variable associated to the primal constraint $\langle E^{ii}, X \rangle = 1$ is y_{ii} and that associated to the primal constraint $\langle E^{ij}, X \rangle \geq \frac{-1}{k-1}$ is y_{ij}.

Proof. Duality is easily checked. We merely prove the strict feasibility here.

The identity matrix I_n is positive definite and strictly meets all inequality constraints of (19). Hence, I_n is in the relative interior of the solution space, and the first program is strictly feasible.

The vector $y \in \mathbb{R}^{\binom{n}{2}}$ with $y_{ii} = -\sum_{j=1}^{n} |C_{ij}| - n$ for all i and $y_{ij} = 1$ for all $i < j$ is a feasible dual solution. All sign restrictions on y are strictly met, and the matrix $C - \sum_{1 \leq i \leq j \leq n} y_{ij} E^{ij}$ is positive definite, because it is strictly diagonally dominant. Therefore, the program (20) is strictly feasible, too.   □

The Prop. 6, 7, and 8 can all be proven by exhibiting primal and dual feasible solutions of matching objective function values for an appropriate semidefinite program of the form given in Remark 1.

We give a complete proof only for Prop. 8, because in this case more than just a tightness result for (12) for a special case is needed. The right-hand side of (18) is strictly larger that of (12), and a new validity proof is required. In the other two cases, we merely give the matrix C for the objective function as well as suitable primal and dual feasible solutions. In fact, since the objective function values of the primal solutions attain the bound from (12), (new) dual solutions are strictly speaking not necessary.

Proof (of Prop. 8). We have to prove that (18) is valid and tight for $\Theta_{k,n}$. This is done by considering the optimization problem (19) with the left-hand side of (18) as objective function and by showing that the right-hand side of (18) is the optimal solution value.

Let $\tilde{b}$ be the edge weights obtained by setting $\tilde{b}_{ij} = b_i b_j$ with $b_i = 1$ if $i \in T$, $b_i = -1$ if $i \in S$, and $b_i = 0$ otherwise. Moreover, let $\tilde{B}$ denote the symmetric matrix with $\tilde{b}$ on its off-diagonal positions and zeros on the principal diagonal. Recall from (11) that solving $\min \sum_{ij \in E(K_n)} b_i b_j z_{ij}$ s.t. $z \in \Theta_{k,n}$ is equivalent to solving $\min \frac{1}{2} \langle \tilde{B}, T_k(X) \rangle$ s.t. $X \in \Psi_{k,n}$. It suffices to consider the subproblem on $S \cup T$, because every pair of dual solutions for the reduced problem can be extended to the full problem without changing the objective function value.

We give solutions X and y to the dual programs (19) and (20), respectively, with matching objective function values for the primal cost matrix $\tilde{B}_{S \cup T, S \cup T} = C^{1,t}$. We then compute $\frac{k-1}{2k} \langle C^{1,t}, X \rangle + \frac{1}{2k} \langle C^{1,t}, E(1+t, 1+t) \rangle$ and show that this is the desired value.

Let us first consider the maximization problem (20). We fix $a = \sqrt{\frac{k-1}{t(k-t)}}$. A short computation reveals that $0 < a \leq 1$ provided $1 \leq t \leq k-2$. Let $y_{11} = -\frac{1}{a}$, $y_{ii} = -a$ for $i = 2, \ldots, 1+t$, $y_{1j} = y_{j1} = 0$ for all $j = 2, \ldots 1+t$, and $y_{ij} = 1-a$ for all $i, j \in \{2, \ldots, 1+t\}, i < j$. Then, the vector y is a feasible solution because $y_{ij} \geq 0$ for all $i < j$ and

$$C^{1,t} - \sum_{1 \leq i < j \leq n} y_{ij} E^{ij} = \begin{bmatrix} 1/a & -E(1,t) \\ -E(t,1) & D^{a,a}(t) \end{bmatrix} \succeq 0 .$$

The latter is a direct consequence of Prop. 11.

The objective function evaluates to

$$\sum_{i=1}^{n} y_{ii} - \frac{1}{k-1} \sum_{i \neq j} y_{ij}$$

$$= 1 \left(-\sqrt{\frac{k-1}{t(k-t)}} \right)^{-1} + t \left(-\sqrt{\frac{k-1}{t(k-t)}} \right) - \frac{t(t-1)}{k-1} \left(1 - \sqrt{\frac{k-1}{t(k-t)}} \right)$$

$$= -\sqrt{\frac{t(k-t)}{k-1}} - \underbrace{\frac{t(k-1) - t(t-1)}{k-1} \sqrt{\frac{k-1}{t(k-t)}}}_{= \sqrt{\frac{t(k-t)}{k-1}}} - \frac{t(t-1)}{k-1}$$

$$= -2 \sqrt{\frac{t(k-t)}{k-1}} - \binom{t}{2} \frac{2}{k-1} .$$

Next, we argue that the matrix

$$X = \begin{bmatrix} 1 & \sqrt{\frac{k-t}{t(k-1)}} \, E(1,t) \\ \sqrt{\frac{k-t}{t(k-1)}} \, E(t,1) & D^{1,-1/(k-1)}(t) \end{bmatrix}$$

is a primal feasible solution. Given that $1 < t \leq k - 2$, all off-diagonal entries are at least as large as $\frac{-1}{k-1}$. By Prop. 11, X is positive semidefinite. We check the only condition that is not trivially fulfilled, namely,

$$1 \left(1 + (t-1) \frac{-1}{k-1} \right) = \frac{k-t}{k-1} = 1 \, t \left(\sqrt{\frac{k-t}{t(k-1)}} \right)^2 .$$

The corresponding objective function value is

$$\langle C^{1,t}, X \rangle = -2t \sqrt{\frac{k-t}{t(k-1)}} - \binom{t}{2} \frac{2}{k-1} = -2 \sqrt{\frac{t(k-t)}{k-1}} - \binom{t}{2} \frac{2}{k-1} .$$

For the dual transformed objective function value we obtain

$$\frac{k-1}{2k} \left(-2 \sqrt{\frac{t(k-t)}{k-1}} - \binom{t}{2} \frac{2}{k-1} \right) + \frac{t(t-1) - 2t}{2k}$$

$$= -\frac{k-1}{k} \sqrt{\frac{t(k-t)}{k-1}} - \binom{t}{2} \frac{1}{k} - \frac{t}{k} + \binom{t}{2} \frac{1}{k}$$

$$= -1 - \frac{\sqrt{t(k-t)(k-1)} - (k-t)}{k} .$$

This proves the claims concerning the validity and tightness of (18).

Finally, we show that $-\sqrt{t}$ bounds the above term from below. An application of l'Hôspital's rule yields that the expression $-1 - \frac{\sqrt{t(k-t)(k-1)}-(k-t)}{k}$ converges to $-\sqrt{t}$ as k goes to infinity. It remains to check that the value of the expression is bounded from below by $-\sqrt{t}$:

$$- \sqrt{t} < -1 - \frac{\sqrt{t(k-t)(k-1)} - (k-t)}{k}$$

$$\iff k\sqrt{t} - t > \sqrt{t(k-t)(k-1)}$$

$$\overset{k \geq t \geq 1}{\iff} tk^2 - 2tk\sqrt{t} + t^2 > t(k-t)(k-1) \quad [= tk^2 - t(t+1)k + t^2]$$

$$\overset{t \geq 0}{\iff} -2k\sqrt{t} > -t(t+1)k$$

$$\overset{k > 0}{\iff} \sqrt{t} < \frac{t+1}{2} .$$

The last inequality holds for all $t \geq 2$. This completes the proof. $\qquad\square$

Finally, we give hints for proving Prop. 6 and Prop. 7 analogously to the previous proof.

Proof (of Prop. 6, Sketch). Use $C = D^{0,1}(q)$ as objective and $X = D^{1,\frac{-1}{q-1}}(q)$ as primal solution. The resulting optimal value is $\frac{q}{2k}(q-k)$. This value is also attained for the dual solution y with $y_{ii} = 1$ for all i and $y_{ij} = 0$ for all $i \neq j$. $\quad\square$

Proof (of Prop. 7, Sketch). Use the objective $C^{s,t} = \begin{bmatrix} D^{0,1}(s) & -E(s,t) \\ -E(t,s) & D^{0,1}(t) \end{bmatrix}$. In case of $s = 1$, use $X = \begin{bmatrix} 1 & \frac{1}{t}E(1,t) \\ \frac{1}{t}E(t,1) & D^{1,-1/t}(t) \end{bmatrix}$ as primal solution, and $X = \begin{bmatrix} D^{1,\alpha}(s) & \gamma E(s,t) \\ \gamma E(t,s) & D^{1,\beta}(t) \end{bmatrix}$ in case of $s \geq 2$, $k \geq s+t$ and either $t \leq s^2$, or $t > s$ together with $k \leq \frac{t^2-s^2}{t-s^2}$. The resulting optimal value is $\frac{(t-s)^2-k(t+s)}{2k}$. This value is also attained for the dual solution y with $y_{ii} = 1$ for all i and $y_{ij} = 0$ for all $i \neq j$. $\quad\square$

6   Conclusion

The semidefinite relaxation from Sect. 3 of the combinatorial MINIMUM K-PARTITION problem has been known for several years. The fact that such a semidefinite program is (ϵ-approximately) solvable in polynomial time is known even longer. But just within the last one or two years SDP solvers have matured to the point, where the semidefinite programs associated to graphs of sizes in the order of a few hundred vertices are computationally tractable in practice. Recall, however, that the semidefinite relaxation of MINIMUM K-PARTITION cannot serve as the basis for a constant-factor approximation for MINIMUM K-PARTITION unless $\mathcal{P} = \mathcal{NP}$.

Our motivation has been to derive lower bounds on the unavoidable interference in GSM frequency planning. In that sense, MINIMUM K-PARTITION is already a relaxations of our original problem. Nevertheless, the numerical bounds shown in Tab. 1 are surprisingly strong. Clearly, similar investigations should be performed for MINIMUM K-PARTITION directly.

We attribute the strength of the numerical bounds to a large extent to the "shifted hypermetric inequalities" (12), which are implicit in the semidefinite relaxation. Hence, with respect to the integer linear programming formulation (1)–(4) of the MINIMUM K-PARTITION problem, all triangle constraints (2) are violated by at most $\sqrt{2} - 1$ and all clique constraints (3) by less than $\frac{1}{2}$. This may serve as a preliminary explanation, but the relation between the MINIMUM K-PARTITION and its semidefinite relaxation is certainly not yet settled.

Moreover, the substantial progress in the development of SDP solvers and our promising computational results call for fathoming the following two questions: Which semidefinite-programming-based k-partitioning heuristics perform well in practice? (Simply applying randomized rounding may not be sufficient.) How successful is a branch-and-cut approach to the MINIMUM K-PARTITION problem based on semidefinite programming?

Acknowledgement

I am indebted to Christoph Helmberg for several discussions and his stimulating comments.

References

1. Aardal, K., Hurkens, C., Lenstra, J., Tiourine, S. Algorithms for Frequency Assignment Problems. *CWI Quaterly*, 9(1 & 2):1–8 (1996).
2. Aardal, K. I., Hoesel, C. P. M. v., Koster, A. M. C. A., Mannino, C., Sassano, A. Models and solution techniques for the frequency assignment problem. ZIB Rep. 01–40, Konrad-Zuse-Zentrum für Informationstechnik Berlin, Berlin, Germany (2001). Available at `http://www.zib.de/PaperWeb/abstracts/ZR-01-40/`.
3. Alizadeh, F. Interior point methods in semidefinite programming with applications to combinatorial optimization. *SIAM J. Optim.*, 5(1):12–51 (1995).
4. Burer, S., Monteiro, R. D., Zhang, Y. Interior-point algorithms for semidefinite programming based on a nonlinear programming formulation. Tech. Rep. TR 99–27, Department of Computational and Applied Mathematics, Rice University (1999).
5. Chopra, S., Rao, M. R. The partition problem. *Math. Program.*, 59:87–115 (1993).
6. Chopra, S., Rao, M. R. Facets of the k-partition polytope. *Discrete Appl. Math.*, 61:27–48 (1995).
7. Correia, L. M. (ed.). *COST 259: Wireless Flexible Personalized Communications.* J. Wiley & Sons (2001).
8. Deza, M. M., Grötschel, M., Laurent, M. Complete Descriptions of Small Multicut Polytopes. *Applied Geometry and Discrete Mathematics—The Victor Klee Festschrift*, 4:221–252 (1991).
9. Deza, M. M., Grötschel, M., Laurent, M. Clique-web facets for multicut polytopes. *Math. Oper. Res.*, 17:981–1000 (1992).
10. Deza, M. M., Laurent, M. Facets for the cut cone I. *Math. Program.*, 56:121–160 (1992).
11. Deza, M. M., Laurent, M. *Geometry of Cuts and Metrics*, vol. 15 of *Algorithms and Combinatorics*. Springer-Verlag (1997).

12. Eisenblätter, A. *Frequency Assignment in GSM Networks: Models, Heuristics, and Lower Bounds*. Ph.D. thesis, TU Berlin, Germany (2001). Cuvillier-Verlag, ISBN 3-8987-3213-4, also available at ftp://ftp.zib.de/pub/zib-publications/books/PhD_eisenblaetter.ps.Z.
13. Erdős, P., Rubin, A. L., Taylor, H. Choosability in graphs. *Congr. Numer.*, 26:125–157 (1979).
14. Eisenblätter, A., Koster, A. FAP web—a website about frequency assignment problems (2000). URL http://fap.zib.de/.
15. Frieze, A., Jerrum, M. Improved Approximation Algorithms for MAX k-CUT and MAX BISECTION. *Algorithmica*, 18:67–81 (1997).
16. Goemans, M. X., Williamson, D. P. Improved approximation algorithms for maximum cut and satisfiability problems using semidefinite programming. *J. ACM*, 42(6):1115–1145 (1995).
17. Goldschmidt, O., Hochbaum, D. S. A polynomial algorithm for the k-cut problem for fixed k. *Math. Oper. Res.*, 19:24–37 (1994).
18. Grötschel, M., Lovász, L., Schrijver, A. *Geometric Algorithms and Combinatorial Optimization*. Springer-Verlag, 2nd ed. (1994).
19. Grötschel, M., Wakabayashi, Y. Facets of the clique partitioning polytope. *Math. Program.*, 47:367–387 (1990).
20. Hale, W. K. Frequency Assignment: Theory and Applications. In *Proceedings of the IEEE*, vol. 68, pp. 1497–1514. IEEE (1980).
21. Helmberg, C. Semidefinite programming for combinatorial optimization. Habilitationsschrift, TU Berlin, Berlin, Germany (2000).
22. Hurkens, C., Tiourine, S. Upper and lower bounding techniques for frequency assignment problems. Tech. Rep., Eindhoven University of Technology, The Netherlands (1995).
23. Jaumard, B., Marcotte, O., Meyer, C. Mathematical models and exact methods for channel assignment in cellular networks. In Sansò, B., Soriano, P. (eds.), *Telecommunications Network Planning*, chap. 13, pp. 239–255. Kluwer Academic Publishers (1999).
24. Karger, D., Motwani, R., Sudan, M. Approximate graph coloring by semidefinite programming. *J. ACM*, 45(2):246–265 (1998).
25. Koster, A. M. C. A. *Frequency Assignment - Models and Algorithms*. Ph.D. thesis, Universiteit Maastricht, Maastricht, The Netherlands (1999).
26. Laurent, M., Poljak, S. On a positive semidefinite relaxation of the cut polytope. *Linear Algebra Appl.*, 223/224:439–461 (1995).
27. Lovász, L. On the Shannon capacity of a graph. *IEEE Transactions on Information Theory*, IT-25:1–7 (1979).
28. Poljak, S., Tuza, Z. Maximum cuts and large bipartite subgraphs. In Cook, W., Lovász, L., Seymour, P. (eds.), *Combinatorial Optimization*, vol. 20 of *DIMACS Ser. in Discr Math. and Theoretical Comput. Sci.*, pp. 188–244. American Mathematical Society (1995).
29. Rutten, J. *Polyhedral Clustering*. Ph.D. thesis, Universiteit Maastricht, Maastricht, The Netherlands (1998).
30. Schrijver, A. *Theory of Linear and Integer Programming*. J. Wiley & Sons (1986).
31. Verfaillie, G., Lemaître, M., Schiex, T. Russian doll search for solving constraint optimization problems. In *Proceedings of the 13th National Conference on Artificial Intelligence (AAAI-96)*, pp. 181–187. Portland, OR, USA (1996).
32. West, D. B. *Introduction to Graph Theory*. Prentice Hall (1996).
33. Wolkowicz, H., Saigal, R., Vandenberghe, L. (eds.). *Handbook on Semidefinite Programming*, vol. 27. Kluwer Academic Publishers (2000).

A Polyhedral Study of
the Cardinality Constrained Knapsack Problem*

Ismael R. de Farias, Jr.[1] and George L. Nemhauser[2]

[1] CORE, 34 Voie du Roman Pays, 1348 Louvain-la-Neuve, Belgium
[2] School of Industrial and Systems Engineering, Georgia Institute of Technology,
Atlanta, GA 30332, USA

Abstract. A cardinality constrained knapsack problem is a continuous knapsack problem in which no more than a specified number of non-negative variables are allowed to be positive. This structure occurs, for example, in areas such as finance, location, and scheduling. Traditionally, cardinality constraints are modeled by introducing auxiliary 0-1 variables and additional constraints that relate the continuous and the 0-1 variables. We use an alternative approach, in which we keep in the model only the continuous variables, and we enforce the cardinality constraint through a specialized branching scheme and the use of strong inequalities valid for the convex hull of the feasible set in the space of the continuous variables. To derive the valid inequalities, we extend the concepts of cover and cover inequality, commonly used in 0-1 programming, to this class of problems, and we show how cover inequalities can be lifted to derive facet-defining inequalities. We present three families of non-trivial facet-defining inequalities that are lifted cover inequalities. Finally, we report computational results that demonstrate the effectiveness of lifted cover inequalities and the superiority of the approach of not introducing auxiliary 0-1 variables over the traditional MIP approach for this class of problems.

1   Introduction

Let n and K be two positive integers, and $N = \{1, \ldots, n\}$. For each $j \in N$, let u_j be a positive number. The *cardinality constrained knapsack problem* (CCKP) is

$$\max \quad \sum_{j \in N} c_j x_j$$
$$\sum_{j \in N} a_j x_j \leq b, \tag{1}$$
$$\text{at most } K \text{ variables can be positive,} \tag{2}$$
$$x_j \leq u_j, \quad j \in N, \tag{3}$$
$$x_j \geq 0, \quad j \in N . \tag{4}$$

* This research was partially supported by NSF grants DMI-0100020 and DMI-0121495. An expanded version of this report can be found in [15].

W.J. Cook and A.S. Schulz (Eds.): IPCO 2002, LNCS 2337, pp. 291–303, 2002.

We denote by S the set of feasible solutions of CCKP, i.e. $S = \{x \in \mathbb{R}^n : x$ satisfies (1)-(4)\}, $PS =$conv(S), and LPS is the set of feasible solutions of the LP relaxation, i.e. $LPS = \{x \in \mathbb{R}^n : x$ satisfies (1), (3), and (4)\}. We assume that: $a_1 \geq \cdots \geq a_n \geq 0$ and $b > 0$; a_j is scaled such that $u_j = 1$ and $a_j \leq b$ $\forall j \in N$; and $\sum_{j=1}^{K} a_j > b$ and $2 \leq K \leq n-1$ (otherwise the problem is trivial).

Theorem 1. *CCKP is NP-hard.*   □

Constraint (2) is present in a large number of applications, such as portfolio optimization [6,27], p-median [10], synthesis of process networks [5,28], etc. It is usually modeled by introducing 0-1 variables $y_j, j \in N$, and the constraints

$$x_j \leq u_j y_j, j \in N, \tag{5}$$
$$\sum_{j \in N} y_j \leq K, \tag{6}$$

see [8,22].

Rather than introducing auxiliary 0-1 variables and the inequalities (5) and (6) to model (2), we keep only the continuous variables, and we enforce (6) algorithmically. This is done in the branch-and-cut algorithm by using a specialized branching scheme and by using strong inequalities that are valid for PS (which is defined in the space of the continuous variables). The idea of dispensing with auxiliary 0-1 variables to model certain combinatorial constraints on continuous variables and enforcing the combinatorial constraints directly in the branch-and-bound algorithm through a specialized branching scheme was pioneered by Beale and Tomlin [3,4] in the context of special ordered sets (SOS) of types I and II.

For several NP-hard combinatorial optimization problems, branch-and-cut has proven to be more effective than a branch-and-bound algorithm that does not account for the polyhedral structure of the convex hull of the set of feasible solutions of the problem, see [7,16,17,19,21,24,26,31,34]. In this paper we study the facetial structure of PS, with the purpose of using strong inequalities valid for PS as cuts in a branch-and-cut scheme without auxiliary 0-1 variables for the *cardinality constrained optimization problem* (CCOP):

$$\begin{aligned}
\max \quad & \sum_{j \in N} c_j x_j \\
& \sum_{j \in N} a_{ij} x_j \leq b_i, i \in M, \\
& x \text{ satisfies (2)-(4)},
\end{aligned} \tag{7}$$

where $M = \{1, \ldots, m\}$ and m is a positive integer.

Some potential benefits of this approach are [12]:

- Faster LP relaxations. Adding the auxiliary 0-1 variables and the new constraints substantially increases the size of the model. Also, the inclusion of variable upper bound constraints, such as (5), may turn the LP relaxation into a highly degenerate problem.
- Less enumeration. It is easy to show that a relaxation of the 0-1 mixed integer problem may have fractional basic solutions that satisfy (2). In this case, additional branching may be required, even though the solution satisfies the cardinality constraint.

This approach has been studied by de Farias [9], and recently it has been used by Bienstock in the context of portfolio optimization [6], and by de Farias, Johnson, and Nemhauser in the context of complementarity problems [13], and the generalized assignment problem [11,14]. It has also been explored in the context of logical programming, see for example [12,20].

In Sect. 2 we present polyhedral results. These results are used in Sect. 3, where we give a branch-and-cut algorithm to solve the model with continuous variables only, and we compare its performance to solving a MIP formulation with CPLEX. We conclude with some directions for further research.

2   Polyhedral Results

First we present some families of facet-defining inequalities for *PS* that are easily implied by the problem, and which we call trivial, and a necessary and sufficient condition for them to completely characterize *PS*.

Proposition 1. *PS is full-dimensional. The vertices of PS have at most one fractional component. If x is a vertex of PS that has a fractional component, then x must satisfy (1) at equality.* □

Proposition 2. *Inequality (1) is facet-defining iff $\sum_{j=1}^{K-1} a_j + a_n \geq b$. Inequality (3) is facet-defining iff $a_j < b, j \in N$. Inequality (4) is facet-defining $\forall j \in N$.* □

Example 1. Let $n = 4$, $K = 2$, and (1) be

$$6x_1 + 4x_2 + 3x_3 + x_4 \leq 6 \ . \tag{8}$$

Then, (3) for $j \in \{1, 2, 3\}$, (4) $\forall j \in N$, and (8) are facet-defining. Note that (8) is stronger than $x_1 \leq 1$, which, therefore, is not facet-defining. □

We now give a necessary and sufficient condition for *PS = LPS*.

Proposition 3. *PS = LPS iff $\sum_{j=n-K+1}^{n} a_j \geq b$.* □

When *PS ≠ LPS*, all vertices of *LPS* that do not satisfy (2) can be cut off by a single inequality. This inequality is presented in the next proposition.

Proposition 4. *The inequality*

$$\sum_{j \in N} x_j \leq K \tag{9}$$

is facet-defining iff $a_1 + \sum_{j=n-K+2}^{n} a_j \leq b$ and $\sum_{j=n-K}^{n-1} a_j \leq b$. □

Example 2. Let $n = 5$, $K = 2$, and (1) be

$$4x_1 + 3x_2 + 2x_3 + x_4 + x_5 \leq 6 \ . \tag{10}$$

Then, $\sum_{j=1}^{5} x_j \leq 2$ is facet-defining. On the other hand, (9) does not define a facet for *PS* in Example 1. □

Proposition 5. *Inequality (9) cuts off all vertices of LPS that do not satisfy (2).*   □

We now give a necessary and sufficient condition for the system defined by (1), (3), (4), and (9) to define *PS*.

Proposition 6. *$PS = LPS \cap \{x : x \text{ satisfies } (9)\}$ iff $a_1 + \sum_{j \in T - \{t\}} a_j \leq b$ $\forall T \subseteq N - \{1\}$ such that $|T| = K$, $\sum_{j \in T} a_j < b$, and $a_t = min\{a_j : j \in T\}$.*   □

Now we study facet-defining inequalities for *PS* that are derived by applying the sequential lifting procedure [25,33] to a family of inequalities that define facets for lower dimensional projections of *PS*. We call these inequalities *cover inequalities*. They are defined by sets of indices that we call *covers*. Our definition of a cover is based on the similar concept used in 0-1 programming [1,18,32]. A major difference is that in our case the cover inequalities are valid for *LPS*, whereas in 0-1 programming they are not valid for the feasible set of the LP relaxation. However, by lifting our cover inequalities, we obtain facet-defining inequalities for *PS* that are not valid for *LPS*, and therefore can be used as cuts in a branch-and-cut scheme to solve CCOP. We present three families of facet-defining inequalities for *PS* obtained by lifting cover inequalities in a specific order. In the remainder of this section we denote $PS(N_0, N_1) = PS \cap \{x \in \Re^n : x_j = 0 \ \forall j \in N_0 \text{ and } x_j = 1 \ \forall j \in N_1\}$, where N_0 and N_1 are two disjoint subsets of N, and similarly for $LPS(N_0, N_1)$. Given a polyhedron P, we denote by $V(P)$ the set of all vertices of P.

Definition 1. *Let C, N_0, and N_1 be three disjoint subsets of N with $N = C \cup N_0 \cup N_1$ and $|C| = K - |N_1|$. If $\sum_{j \in C} a_j > b - \sum_{j \in N_1} a_j$, we say that C is a cover for $PS(N_0, N_1)$, and that*

$$\sum_{j \in C} a_j x_j \leq b - \sum_{j \in N_1} a_j \qquad (11)$$

is a cover inequality for $PS(N_0, N_1)$.

It is easy to show that

Proposition 7. *Inequality (11) is valid and facet-defining for $PS(N_0, N_1)$.*   □

The lifting procedure is based on a theorem of Wolsey [33], see [9] for a proof. Now we lift (11) with respect to a variable x_l, $l \in N_0$, to give a facet-defining inequality for $PS(N_0 - \{l\}, N_1)$ that is not valid for $LPS(N_0 - \{l\}, N_1)$.

Proposition 8. *Let C, N_0, and N_1 be three disjoint subsets of N with $N = C \cup N_0 \cup N_1$ and $|C| = K - |N_1|$. Suppose $N_0 \neq \emptyset$ and C is a cover for $PS(N_0, N_1)$. Let $i \in C$ and $l \in N_0$ be such that $a_i = min\{a_j : j \in C\}$ and $\sum_{j \in C - \{i\}} a_j + a_l < b - \sum_{j \in N_1} a_j$. Then,*

$$\sum_{j \in C} a_j x_j + \left(b - \sum_{j \in N_1} a_j - \sum_{j \in C - \{i\}} a_j \right) x_l \leq b - \sum_{j \in N_1} a_j \qquad (12)$$

defines a facet of $PS(N_0 - \{l\}, N_1)$.   □

Example 1 (Continued). The set $\{1\}$ is a cover for $PS(\{2,3\},\{4\})$, and the cover inequality is

$$6x_1 \leq 5 \ . \tag{13}$$

Lifting (13) with respect to x_2, we obtain $6x_1 + 5x_2 \leq 5$, which defines a facet of $PS(\{3\},\{4\})$. Lifting now with respect to x_4, the lifting coefficient is

$$\alpha_4 = \max\left\{\frac{6x_1 + 5x_2 - 5}{1 - x_4} : x \in V(PS(\{3\},\emptyset)) \text{ and } x_4 < 1\right\} \ .$$

If $\hat{x} \in V(PS(\{3\},\emptyset))$ is such that $\hat{x}_4 < 1$ and

$$\alpha_4 = \frac{6\hat{x}_1 + 5\hat{x}_2 - 5}{1 - \hat{x}_4} \ ,$$

then clearly $\hat{x}_1 > 0$. Because of Proposition 1, $\hat{x}_1 = 1$ if $\hat{x}_4$ is fractional. However, $\hat{x}_1 = 1 \Rightarrow \hat{x}_4 = 0$. Therefore, $\hat{x}_4 = 0$. So,

$$\alpha_4 = 6\hat{x}_1 + 5\hat{x}_2 - 5 \ .$$

Solving the (trivial) continuous knapsack problem that results from the lifting problem when the value of $\hat{x}_4$ is fixed, we obtain, $\hat{x}_1 = \frac{1}{3}$ and $\hat{x}_2 = 1$. Therefore, $\alpha_4 = 2$, and $6x_1 + 5x_2 + 2x_4 \leq 7$ defines a facet of $PS(\{3\},\emptyset)$. The lifting coefficient of x_3 is

$$\alpha_3 = \min\left\{\frac{7 - 6x_1 - 5x_2 - 2x_4}{x_3} : x \in V(PS) \text{ and } x_3 > 0\right\} \ .$$

If $\tilde{x} \in V(PS)$ is such that $\tilde{x}_3 > 0$ and

$$\alpha_3 = \frac{7 - 6\tilde{x}_1 - 5\tilde{x}_2 - 2\tilde{x}_4}{\tilde{x}_3} \ ,$$

then, since $K = 2$, at most one of $\tilde{x}_1$, $\tilde{x}_2$, and $\tilde{x}_4$ can be positive. If $\tilde{x}_3 = 1$, either $\tilde{x}_4 = 1$ and $\alpha_3 = 5$, or, by solving the (trivial) continuous knapsack problem that results from the lifting problem when $\tilde{x}_3 = 1$ and $\tilde{x}_4 = 0$, $\tilde{x}_2 = \frac{3}{4}$, and $\alpha_3 = 3.25$. If $\tilde{x}_3$ is fractional, Proposition 1 implies that $\tilde{x}_2 = 1$, and $\alpha_3 = 3$. Therefore, $\alpha_3 = 3$, and

$$6x_1 + 5x_2 + 3x_3 + 2x_4 \leq 7$$

defines a facet of PS.   □

From Proposition 8 it follows that

Theorem 2. *The inequality*

$$\sum_{j=1}^{n} max\left\{a_j, b - \sum_{j=1}^{K-1} a_j\right\} x_j \leq b \tag{14}$$

defines a facet of PS.   □

Example 2 (Continued). Inequality (10) is not facet-defining, and it can be replaced with

$$4x_1 + 3x_2 + 2x_3 + 2x_4 + 2x_5 \leq 6 \ .$$

□

We proved in Proposition 4 that (9) is facet-defining only if $a_1 + \sum_{j=n-K+2}^{n} a_j \leq b$. We next present a family of facet-defining inequalities that are stronger than (9) when $a_1 + \sum_{j=n-K+2}^{n} a_j > b$.

Theorem 3. *Suppose that* $a_1 + \sum_{j=n-K+2}^{n} a_j > b$, $\sum_{j=n-K+2}^{n} a_j < b$, *and* $\sum_{j=n-K}^{n-1} a_j \leq b$. *Then,*

$$a_1 x_1 + \sum_{j=2}^{n-K-1} max\left\{ a_j, b - \sum_{i=n-K+2}^{n} a_i \right\} x_j + \left(b - \sum_{i=n-K+2}^{n} a_i \right) \sum_{j=n-K}^{n} x_j$$

$$\leq K \left(b - \sum_{j=n-K+2}^{n} a_j \right) \quad (15)$$

defines a facet of PS.

□

Note that when $a_1 + \sum_{j=n-K+2}^{n} a_j > b$, (15) is stronger than (9).

Example 3. Let $n = 4$, $K = 2$, and (1) be

$$6x_1 + 3x_2 + 2x_3 + x_4 \leq 6 \ .$$

Then,

$$6x_1 + 5x_2 + 5x_3 + 5x_4 \leq 10 \quad (16)$$

defines a facet of *PS*. Inequality (16) is stronger than $\sum_{j=1}^{4} x_j \leq 2$. □

Next, we present a family of valid inequalities that under certain conditions is facet-defining for *PS*. The family of inequalities is particularly useful when the constraint matrix is not dense, a situation that is common in applications.

Proposition 9. *Let* C, N_0, *and* N_1 *be three disjoint subsets of* N *with* $N = C \cup N_0 \cup N_1$ *and* $|C| = K - |N_1|$. *Assume that* C *is a cover for* $PS(N_0, N_1)$, *and* $a_p = min\{a_j : j \in C\}$. *Suppose that* $\sum_{j \in C - \{p\}} a_j < b - \sum_{j \in N_1} a_j$, *and* $a_l = 0$ *for some* $l \in N_0$. *Then,*

$$\sum_{j \in C} a_j x_j + \Delta \sum_{j \in N_0} x_j + \sum_{j \in N_1} \alpha_j x_j \leq b + \sum_{j \in N_1} (\alpha_j - a_j) \quad (17)$$

is valid for PS, where $\Delta = b - \sum_{j \in C - \{p\}} a_j - \sum_{j \in N_1} a_j$, *and*

$$\alpha_j = \begin{cases} \Delta + a_j & if\ a_p > \Delta + a_j \\ max\{a_p, a_j\} & otherwise. \end{cases}$$

□

Theorem 4. *If $a_j \leq \Delta \; \forall j \in N_0$, (17) is facet-defining for PS.*   □

Example 4. Let $n = 5$, $K = 3$, and (2) be given by

$$5x_1 + 5x_2 + 3x_3 + 0x_4 + 0x_5 \leq 9 \;.$$

Then,

$$5x_1 + 5x_2 + 4x_3 + 4x_4 + 4x_5 \leq 13$$

is facet-defining for *PS*, where $C = \{1, 2\}$, $N_0 = \{3, 4\}$, and $N_1 = \{5\}$.   □

3 Solving CCOP by Branch-and-Cut

We tested the performance on difficult instances of CCOP of the

- MIP formulation, in which one introduces in the model auxiliary 0-1 variables, and models Constraint (2) with Constraints (5) and (6)
- continuous formulation, in which one keeps in the model only the continuous variables, and enforces Constraint (2) algorithmically through a branch-and-bound algorithm by using a specialized branching scheme
- continuous formulation through a branch-and-cut algorithm by using a specialized branching scheme and the lifted cover inequalities introduced in Sect. 2 as cuts.

The number of knapsack constraints, m, and the number of variables, n, in the instances of CCOP tested are given in the first column of Table 1. We tested 3 different instances for each pair $m \times n$. The values of K for the instances varied from 150 to 3,000, and the densities of the constraint matrix from 13% to 50%. These values were chosen by selecting the hardest ones determined by performing preliminary computational tests. The instances with the same m and n had the same cardinality and density.

The instances were randomly generated as follows. The profit coefficients $c_j, j \in N$, were integers uniformly generated between 10 and 25. The nonzero knapsack coefficients, $a_{ij}, i \in M, j \in N$, were integers uniformly generated between 5 and 20. The $m \times$ density indices of the nonzero knapsack coefficients were uniformly generated between 1 and n. The right-hand-sides of the knapsack constraints were given by $b_i = \max\{\lfloor .3 \sum_{j \in N} a_{ij} \rfloor$, greatest coefficient of the ith knapsack+1$\}$, $i \in M$.

The continuous formulation was tested using MINTO 3.0 [23,29] with CPLEX 6.6 as LP solver. Our motivation for using MINTO was the flexibility that it offers to code alternative branching schemes, feasibility tests, and separation routines. Initially we implemented the MIP formulation with MINTO. However, MINTO proved to be too slow when compared to CPLEX 6.6 in solving the MIPs. Also, CPLEX 6.6 has Gomory cuts, which we thought could be helpful in reducing the effort required to complete the enumeration [2]. Note that by using CPLEX to solve the MIPs and MINTO to solve the continuous formulation, the results are biased in favor of the MIP formulation, since CPLEX is much faster than MINTO.

We adopted the specialized branching scheme proposed by Bienstock in [6], which we briefly describe. Suppose that more than K variables are positive in a solution $\tilde{x}$ of the LP relaxation of CCOP, and that $\tilde{x}_l$ is one of them. Then, we may divide the solution space by requiring in one branch, which we call *down*, that $x_l = 0$, and in the other branch, which we call *up*, that

$$\sum_{j \in N - \{l\}} x_j \leq K - 1 \ . \tag{18}$$

Let $S_{\text{down}} = S \cap \{x \in \Re^n : x_l = 0\}$ and $S_{\text{up}} = S \cap \{x \in \Re^n : x \text{ satisfies } (18)\}$. Clearly $S = S_{\text{down}} \cup S_{\text{up}}$, although it may happen that $\tilde{x} \in S_{\text{up}}$.

In general, suppose that at the current node of the enumeration tree the variables $x_j, j \in F$, are free, i.e. have not been branched on, and that t variables have been branched up. We branch at the current node on variable x_p by imposing in the down branch that $x_p = 0$, and in the up branch that

$$\sum_{j \in F - \{p\}} x_j \leq K - t - 1 \ .$$

Even though the current LP relaxation solution may reoccur in the up branch, this branching scheme ends within a finite number of nodes, since we can fathom any node that corresponds to K up branches.

Since each knapsack constraint of CCOP has a large number of 0 coefficients in the data we used to test CCOP, the conditions of Proposition 4 are satisfied in general, and (9) is facet-defining for the convex hull of the feasible sets of the instances tested. Because of this, we included (9) in the initial formulation (LP relaxation) of the instances of CCOP we tested, which considerably tightened the model.

Initially, we wanted to use (14) of Theorem 2 in a preprocessing phase to tighten any inequality in the initial formulation for which $a_{ij} < b - \sum_{l=1}^{K-1} a_{il}$ for some $i \in M, j \in N$. However, we did not detect this condition in any of the instances we tested. Note that the specific way we defined $b_i, i \in M$, in the instances was such that the knapsack inequalities were already tight, in the sense that $a_{ij} \geq b - \sum_{l=1}^{K-1} a_{il} \ \forall i \in M, j \in N$. On the other hand, (14) might be useful in real world instances where some defined models are not tight.

We also wanted initially to use (15) to define the up branches every time it was stronger than (9), as would happen if the conditions of Theorem 3 were present in at least one of the knapsack constraints. However, because of the number of 0 coefficients in the data we tested, in general $a_{i1} + \sum_{j=n-K+2}^{n} a_{ij} \leq b_i, i \in M$.

We used (17) as a cut. Unlike (14) and (15), (17) proved to be very useful in the instances we tested, and therefore we used it in our branch-and-cut algorithm.

Given a point $\tilde{x}$ that belongs to the LP relaxation of CCOP and that does not satisfy (2), to find a violated inequality (17) we have to select disjoint subsets $C, N_1 \subset N$, such that $C \neq \emptyset$, $|C| + |N_1| = K$, $\sum_{j \in C} a_{ij} + \sum_{j \in N_1} a_{ij} > b_i$, $\sum_{j \in C - \{p\}} a_{ij} + \sum_{j \in N_1} a_{ij} < b_i$, with $a_{ip} = \min\{a_{ij} : j \in C\}$, $a_{ij} = 0$ for some $j \in N - (C \cup N_1)$ with $\tilde{x}_{ij} > 0$, and

$$\sum_{j \in C} a_{ij}\tilde{x}_j + \Delta \sum_{j \in N_0} \tilde{x}_j + \sum_{j \in N_1} \alpha_j \tilde{x}_j > b_i + \sum_{j \in N_1} (\alpha_j - a_{ij}), \qquad (19)$$

where $N_0 = N - (C \cup N_1)$. It appears then the separation problem for (17) is difficult. Thus we used a heuristic, which we now describe, to solve the separation problem.

Let $i \in M$ be such that $\sum_{j \in N} a_{ij}\tilde{x}_i = b_i$. Because the terms corresponding to $j \in C$ in (17) are $a_{ij}x_j$, we require that $a_{ij} > 0$ and $\tilde{x}_j > 0 \; \forall j \in C$. Because $j \in N_1$ contributes $\alpha_j x_j$ to the left-hand-side and α_j to the right-hand-side of (17), we require that $\tilde{x}_j = 1 \; \forall j \in N_1$. Since C cannot be empty, we first select its elements. We include in C as many $j \in N$ with $a_{ij} > 0$ and $\tilde{x}_j \in (0,1)$ as possible, u p to K elements. If $|C| = K$, we include in N_0 every $j \in N - C$ (in this case $N_1 = \emptyset$). If $|C| < K$, we include in N_1 all $j \in N - C$ with $\tilde{x}_j = 1$, in nonincreasing order of a_{ij} until $|C| + |N_1| = K$ or there are no more components of $\tilde{x}$ with $\tilde{x}_j = 1$. If $|C| + |N_1| < K$ or if there is no $j \in N - (C \cup N_1)$ with $a_{ij} = 0$ and $\tilde{x}_j > 0$, we fail to generate a cut for $\tilde{x}$ out of row i. If $|C| + |N_1| = K$ and $\exists j \in N - (C \cup N_1)$ with $a_{ij} = 0$ and $\tilde{x}_j > 0$, we make $N_0 = N - (C \cup N_1)$. Finally, if (19) holds, we succeed in generating a cut for $\tilde{x}$ out of row i. Otherwise, we fail.

The following example comes from our preliminary computational experience.

Example 5. Let $m = 2$, $n = 100$, and $K = 20$. The solution of the LP relaxation is $\tilde{x}$ given by $\tilde{x}_1 = \tilde{x}_{11} = \tilde{x}_{17} = \tilde{x}_{22} = \tilde{x}_{43} = \tilde{x}_{45} = \tilde{x}_{56} = \tilde{x}_{57} = \tilde{x}_{60} = \tilde{x}_{61} = \tilde{x}_{62} = \tilde{x}_{64} = \tilde{x}_{68} = \tilde{x}_{70} = \tilde{x}_{79} = \tilde{x}_{80} = \tilde{x}_{86} = \tilde{x}_{95} = 1$, $\tilde{x}_{19} = 0.434659$, $\tilde{x}_{75} = 0.909091$, $\tilde{x}_{78} = 0.65625$, and $\tilde{x}_j = 0$ otherwise. (Note that 3 variables are fractional even though $m = 2$. This is because (9) was included in the original formulation.) Let $N_0' = \{j \in N : \tilde{x}_j = 0\}$. Below we give the terms of one of the knapsack constraints for which $\tilde{x}_j > 0$ and $a_{ij} > 0$:

$$14x_1 + 30x_{11} + 32x_{19} + 20x_{22} + 12x_{45} + 14x_{56} + 12x_{61} + 24x_{70} + 32x_{75}$$
$$+24x_{79} \leq 193 \; .$$

The 18 components of $\tilde{x}$ equal to 1 are all included in N_1, i.e. $N_1 = \{1, 11, 17, 22, 43, 45, 56, 57, 60, 61, 62, 64, 68, 70, 79, 80, 86, 95\}$. We choose the cover $C = \{19, 75\}$ and we let $N_0 = \{78\} \cup N_0'$, $\Delta = 11$. Therefore the inequality (17) is

$$32x_{19} + 32x_{75} + 11x_{78} + 11 \sum_{j \in N_0'} x_j + 25x_1 + 32x_{11} + 11x_{17} + 31x_{22} + 11x_{43}$$

$$+ 23x_{45} + 25x_{56} + 11x_{57} + 11x_{60} + 23x_{61} + 11x_{62} + 11x_{64} + 11x_{68} + 32x_{70}+$$
$$32x_{79} + 11x_{80} + 11x_{86} + 11x_{95} \leq 376. \quad (20)$$

Since this inequality is violated by $\tilde{x}$, it is included in the formulation at the root node. Because we fail to generate a cut for $\tilde{x}$ out of the other knapsack inequality, this is the only cut included at this time, and we re-solve the LP relaxation with (20) included in the formulation. $\qquad \square$

We search for a cut (19) for $\tilde{x}$ from every knapsack inequality (7) and we include in the formulation in the current node as many cuts as we find. In case we find cuts, we re-solve the LP relaxation with the cuts added. Otherwise, we branch. We search for cuts in every node of the enumeration tree.

We used a Sun Ultra 2 with two UltraSPARC 300 MHz CPUs and 256 MB memory to perform the computational tests. The results are summarized in Table 1, which gives for each pair $m \times n$ the average number of nodes, CPU seconds, and number of cuts for the MIP and continuous formulations. The second column gives the average number of nodes over the 3 instances generated by CPLEX to solve the MIP formulation to proven optimality. The next column, Cont. B&B, gives the average number of nodes generated by MINTO to solve the continuous formulation exactly with Bienstock's branching scheme without the use of cuts, i.e. through a pure branch-and-bound approach. The following column, Cont. B&C, gives the number of nodes generated by MINTO to solve the continuous formulation exactly with Bienstock's branching scheme and (17) as cuts, i.e. through a branch-and-cut approach. The column "% Red." gives the percentage reduction in number of nodes by using the continuous formulation and a branch-and-cut algorithm with (17) as cuts over the MIP formulation. Note that overall, the percentage reduction in number of nodes by using branch-and-cut over branch-and-bound for the continuous formulation was 70%. Thus the use of a branch-and-cut approach to solve the continuous formulation can be considerably more effective than pure branch-and-bound. The great overall reduction of 97% in the average number of nodes by using the continuous formulation with a branch-and-cut approach over the MIP approach indicates that by adding auxiliary 0-1 variables and enforcing (2) through their integrality does not take as much advantage of the combinatorial structure of the problem as in the continuous approach, where we fathom a node when (2) is satisfied, and where we branch by using Bienstock's scheme. As mentioned earlier, (9) is usually facet-defining, and thus using it to define the up branches may considerably help to reduce the upper bound on the up branches.

The four columns under "Time" in Table 1 have similar meanings to the four columns under "Nodes". The overall time reduction by using (17) as cuts in a branch-and-cut scheme to solve the continuous formulation over pure branch-and-bound was 62%. Because CPLEX is much faster than MINTO, the overall time reduction of 78% of branch-and-cut on the continuous formulation over the MIP formulation is very significant. We believe that such a great time reduction is not just the result of the reduction in the number of nodes, but also due to the fact that the size of the MIP formulation is 2 times greater than the continuous formulation, which becomes significant in the larger instances. Also, the degeneracy introduced by the variable upper bound constraints (5) may be harmful.

The only cuts generated by CPLEX with default options were the Gomory cuts. The column "Cuts Gom." at the end of Table 1 gives the average number of Gomory cuts generated by CPLEX. As we mentioned in Sect. 3, initially we tested the MIP approach with MINTO. MINTO generated a very large number

Table 1. Average number of nodes, time, and number of cuts for the MIP and the continuous formulations

$m \times n$	Nodes				Time				Cuts	
		Cont.		%		Cont.		%	Gom.	(17)
	MIP	B&B	B&C	Red.	MIP	B&B	B&C	Red.		
20×500	36,364	681	531	98	527	17	94	82	12	10
$20 \times 1,000$	109,587	1,360	315	99	2,141	792	208	90	75	23
$20 \times 1,500$	33,761	1,423	229	99	1,746	203	109	93	67	18
$20 \times 2,000$	12,738	2,753	729	94	802	839	318	60	153	72
$20 \times 2,500$	46,873	3,479	1,157	97	9,959	1,558	770	92	148	69
$30 \times 3,000$	196,010	3,927	1,092	99	36,288	3,570	720	98	205	162
$30 \times 3,500$	20,746	161	3	99	14,507	44	12	99	54	6
$50 \times 4,000$	18,529	289	112	99	14,798	57	49	99	97	11
$50 \times 4,500$	26,811	4,230	1,358	94	35,601	19,758	9,505	73	79	91
$50 \times 5,000$	39,776	5,553	1,749	95	53,388	23,570	11,320	78	161	211
$50 \times 5,500$	43,829	6,763	2,129	95	65,423	27,140	12,639	80	134	287
$50 \times 6,000$	49,574	7,981	2,727	94	72,751	28,923	13,547	81	231	320
$50 \times 6,500$	54,251	8,975	3,152	94	85,721	34,188	15,754	81	166	289
$50 \times 7,000$	17,524	163	158	99	24,439	577	715	97	61	15
$70 \times 7,500$	45,279	8,572	1,359	96	96,714	98,576	32,711	66	244	412
$70 \times 8,000$	32,497	9,085	2,168	93	89,238	106,684	31,730	64	315	601
TOTAL	784,149	65,395	18,968	97	604,043	346,496	130,201	78	2,202	2,597

of lifted flow cover inequalities (LFCIs), typically tens of thousands, even for the smallest instances. The reason is because the violation tolerance for LFCIs in MINTO is much smaller than in CPLEX. To verify the effectiveness of LFCIs in CPLEX we increased their violation tolerance. Our preliminary tests indicated that even though LFCIs help reduce the integrality gap faster, they are not effective in closing the gap, and we then kept CPLEX's default. The average number of inequalities (17) generated by MINTO is given in the last column of Table 1.

Given the encouraging computational results, it is important to study the following questions on branch-and-cut for CCOP:

– how can lifted cover inequalities be separated efficiently?
– how can cover inequalities be lifted efficiently in any order, either exactly or approximately, to obtain strong cuts valid for *PS*?
– in which order should cover inequalities be lifted?
– are there branching strategies more effective than Bienstock's [6]? (See [12] for an alternative branching strategy.)

Note that it is not possible to complement a variable x_j with $a_j < 0$, as is usually done for the 0-1 knapsack problem, and keep the cardinality constraint (2) intact. This means that the assumptions that $b > 0$ and $a_{ij} \geq 0 \; \forall j \in N$ of Sect. 1 imply in loss of generality. Thus, it is important to investigate the cardinality knapsack polytope when $a_j < 0$ for some of the knapsack coefficients.

Besides cardinality, there exists a small number of other combinatorial constraints, such as semi-continuous and SOS [12], that are pervasive in practical applications. We suggest investigating their polyhedral structure in the space of the continuous variables, and comparing the performance of branch-and-cut without auxiliary 0-1 variables for these problems against the usual MIP approach.

Recently, there has been much interest in unifying the tools of integer programming (IP) and constraint programming (CP) [30]. Traditionally in IP, combinatorial constraints on continuous variables, such as semi-continuous, cardinality or SOS are modeled as mixed-integer programs (MIPs) by introducing auxiliary 0-1 variables and additional constraints. Because the number of variables and constraints becomes larger and the combinatorial structure is not used to advantage, these MIP models may not be solved satisfactorily, except for small instances. Traditionally, CP approaches to such problems keep and use the combinatorial structure, but do not use linear programming (LP) bounds. The advantage of our approach is exploiting both the combinatorial structure and LP bounds simultaneously.

References

1. E. Balas, "Facets of the Knapsack Polytope," *Mathematical Programming* 8, 146-164 (1975).
2. E. Balas, S. Ceria, G. Cornuéjols, and N. Natraj, "Gomory Cuts Revisited," *Operations Research Letters* 19, 1-9 (1996).
3. E.L.M. Beale, "Integer Programming," in: K. Schittkowski (Ed.), *Computational Mathematical Programming*, NATO ASI Series, Vol. F15, Springer-Verlag, 1985, pp. 1-24.
4. E.L.M. Beale and J.A. Tomlin, "Special Facilities in a General Mathematical Programming System for Nonconvex Problems Using Ordered Sets of Variables," in: J. Lawrence (Ed.), *Proceedings of the fifth Int. Conf. on O.R.*, Tavistock Publications, 1970, pp. 447-454.
5. L.T. Biegler, I.E. Grossmann, and A.W. Westerberg, *Systematic Methods of Chemical Process Design*, Prentice Hall, 1997.
6. D. Bienstock, "Computational Study of a Family of Mixed-Integer Quadratic Programming Problems," *Mathematical Programming* 74, 121-140 (1996).
7. H. Crowder, E.L. Johnson, and M.W. Padberg, "Solving Large Scale Zero-One Problems," *Operations Research* 31, 803-834 (1983).
8. G.B. Dantzig, "On the significance of Solving Linear Programming Problems with some Integer Variables," *Econometrica* 28, 30-44 (1960).
9. I.R. de Farias, Jr., "A Polyhedral Approach to Combinatorial Complementarity Programming Problems," *Ph.D. Thesis*, School of Industrial and Systems Engineering, Georgia Institute of Technology, Atlanta, GA (1995).
10. I.R. de Farias, Jr., "A Family of Facets for the Uncapacitated p-Median Polytope," *Operations Research Letters* 28, 161-167 (2001).
11. I.R. de Farias, Jr., E.L. Johnson, and G.L. Nemhauser "A Generalized Assignment Problem with Special Ordered Sets: A Polyhedral Approach," *Mathematical Programming* 89, 187-203 (2000).

12. I.R. de Farias, Jr., E.L. Johnson, and G.L. Nemhauser "Branch-and-Cut for Combinatorial Optimization Problems without Auxiliary Binary Variables," *Knowledge Engineering Review* 16, 25-39 (2001).

13. I.R. de Farias, Jr., E.L. Johnson, and G.L. Nemhauser, "Facets of the Complementarity Knapsack Polytope," to appear in *Mathematics of Operations Research*.

14. I.R. de Farias, Jr. and G.L. Nemhauser, "A Family of Inequalities for the Generalized Assignment Polytope," *Operations Research Letters* 29, 49-51 (2001).

15. I.R. de Farias, Jr. and G.L. Nemhauser, "A Polyhedral Study of the Cardinality Constrained Knapsack Problem," CORE Discussion Paper (2001).

16. Z. Gu, G.L. Nemhauser, and M.W.P. Savelsbergh, "Lifted Flow Cover Inequalities for Mixed 0-1 Integer Programs," *Mathematical Programming* 85, 439-467 (1999).

17. M. Grötschel, M. Jünger, and G. Reinelt, "A Cutting Plane Algorithm for the Linear Ordering Problem," *Operations Research* 32, 1195-1220 (1984).

18. P.L. Hammer, E.L. Johnson, and U.N. Peled, "Facets of Regular 0-1 Polytopes," *Mathematical Programming* 8, 179-206 (1975).

19. K. Hoffman and M.W. Padberg, "LP-Based Combinatorial Problem Solving," *Annals of Operations Research* 4, 145-194 (1985).

20. J.N. Hooker, G. Ottosson, E.S. Thornsteinsson, and H.-J. Kim, "A Scheme for Unifying Optimization and Constraint Satisfaction Methods," *Knowledge Engineering Review* 15, 11-30 (2000).

21. H. Marchand, A. Martin, R. Weismantel, and L.A. Wolsey, "Cutting Planes in Integer and Mixed-Integer Programming," CORE Discussion Paper (1999).

22. H.M. Markowitz and A.S. Manne, "On the Solution of Discrete Programming Problems," *Econometrica* 25, 84-110 (1957).

23. G.L. Nemhauser, G.C. Sigismondi, and M.W.P. Savelsbergh, "MINTO, a Mixed-INTeger Optimizer," *Operations Research Letters* 15, 47-58 (1994).

24. G.L. Nemhauser and L.A. Wolsey, *Integer Programming and Combinatorial Optimization*, John Wiley and Sons, 1988.

25. M.W. Padberg, "A Note on Zero-One Programming," *Operations Research* 23, 883-837 (1975).

26. M.W. Padberg and G. Rinaldi, "Optimization of a 532-City Symmetric Traveling Salesman Problem by Branch-and-Cut," *Operations Research Letters* 6, 1-7 (1987).

27. A.F. Perold, "Large-Scale Portfolio Optimization," *Management Science* 30, 1143-1160 (1984).

28. R. Raman and I.E. Grossmann, "Symbolic Integration of Logic in MILP Branch-and-Bound Methods for the Synthesis of Process Networks," *Annals of Operations Research* 42, 169-191 (1993).

29. M.W.P. Savelsbergh, "Functional Description of MINTO, a Mixed INTeger Optimizer (version 3.0)," http://udaloy.isye.gatech.edu/ mwps/projects/minto.html.

30. P. van Hentenryck, *Constraint Satisfaction in Logic Programming*, MIT Press, 1989.

31. T.J. van Roy and L.A. Wolsey, "Solving Mixed-Integer Programming Problems Using Automatic Reformulation," *Operations Research* 35, 45-57 (1987).

32. L.A. Wolsey, "Faces for a Linear Inequality in 0-1 Variables," *Mathematical Programming*, 8 165-178 (1975).

33. L.A. Wolsey, "Facets and Strong Valid Inequalities for Integer Programs," *Operations Research* 24, 367-372 (1976).

34. L.A. Wolsey, *Integer Programming*, John Wiley and Sons, 1998.

A PTAS for Minimizing Total Completion Time of Bounded Batch Scheduling*

Mao-Cheng Cai[1], Xiaotie Deng[2], Haodi Feng[2], Guojun Li[3], and Guizhen Liu[3]

[1] Institute of Systems Science, Academy of Siences, Beijing, China
cscaimc@cityu.edu.hk
[2] Department of Computer Science, City University of Hong Kong, Hong Kong
{csdeng,fenghd}@cs.cityu.edu.hk
[3] School of Mathematics and System Science, Shandong University, Jinan, 250100
gjli@math.sdu.edu.cn

Abstract. We consider a batch processing system $\{p_i : i = 1, 2, \cdots, n\}$ where p_i is the processing time of job i, and up to B jobs can be processed together such that the handling time of a batch is the longest processing time among jobs in the batch. The number of job types m is not fixed and all the jobs are released at the same time. Jobs are executed non-preemptively. Our objective is to assign jobs to batches and sequence the batches so as to minimize the total completion time. The best previously known result is a $2-$approximation algorithm due to D. S. Hochbaum and D. Landy [6]. In this paper, we establish the first *polynomial time approximation scheme* (PTAS) for the problem.

1 Introduction

The non-preemptive batch processing system arises as a model of jobs processing for the burn-in stage of manufacturing integrated circuits [2,3]. At that stage, circuits are loaded into ovens and exposed to high temperature in order to test the quality of chips and to weed out the chips that might endure an early failure under normal operation conditions. Usually, each chip has a pre-specified minimum burn-in time, dependent on its type or customer's requirement. Since a chip might stay in the oven for a longer time, different types of chips can be assigned into one batch and the batch will stay in the oven for the longest time of the burn-in times of the chips. Once a batch is being processed, no preemption is permitted. The scheduling problem is to assign chips into batches and sequence the batches. We are interested in minimizing the total completion time.

There are two major classes of the batch processing problems: the unbounded case where there is no limit on the number of jobs that can be assigned to a batch and the bounded case. For the unbounded case with one release date, P. Brucker et al. [2] designed a polynomial time algorithm via a dynamic programming approach and they also made a thorough discussion of the scheduling problem on

---

* Research is partially supported by a grant from the Research Grants Council of Hong Kong SAR (CityU 1056/01E) and a grant from CityU of Hong Kong (Project No.7001215).

W.J. Cook and A.S. Schulz (Eds.): IPCO 2002, LNCS 2337, pp. 304–314, 2002.

the batch machine with various constraints and objective functions. And for the unbounded weighted case with different release dates, X. Deng and Y. Zhang [4] proved that it is NP-hard to find the optimal schedule and they also presented polynomial time algorithms for some special cases. A PTAS was obtained recently by X. Deng, et al., [5] for unbounded case with different release dates. There has not been much work done for the bounded case, however. For a fixed number m of job types and one release date, D.S. Hochbaum and D. Landy [6] presented an optimal algorithm MTB with a running time of $O(m^2 3^m)$ and for variable m they gave a highly non-trivial $2-$approximation algorithm when all the jobs arrive at the same time.

The above 2-approximation algorithm is based on several important properties of a specially structured optimal schedule.

A major break-through in methodology of designing PTAS for scheduling problems came very recently in the seminal work of Foto Afrati et al. [1]. They made a novel combination of time stretching, geometric rounding and dynamic programming approaches, to obtain a general framework for obtaining PTAS for several classical scheduling problems. Our work is motivated by their success. In Sect. 2, we introduce notations and simplify the bounded batch processing problem by applying the geometric rounding technique as well as the structural properties of the work of D.S. Hochbaum and D. Landy [6].

Our main contribution is an optimal solution for jobs with processing time in geometric series: $(1 + \epsilon)^i$, where ϵ is a given positive number and i is an integer. The solution takes a quite non-trivial dynamic programming approach. Even though one may argue that the approach of Foto Afrati et al. may render the task of PTAS design simple for scheduling problems and we are inspired by their work, our solution is not a consequence of theirs. Our result builds upon a concise mathematical structural lemma of D.S. Hochbaum and D. Landy [6]. It requires further development of structural analysis of the optimal solution and a sophisticated dynamic programming. In Sect. 3 we state several structural properties of the optimal schedule for this class of jobs. These properties form the basis of our algorithm that is presented in Sect. 4. We end this paper with remarks and discussion in Sect. 5.

2 Preliminaries

We consider one non-preemptive batch processing machine and m job types numbered $1, \ldots, m$, each type $t \in \{1, \ldots, m\}$ has n_t jobs and a processing time, i.e., the minimum burn-in time, p_t, where $p_1 < p_2 < \ldots < p_m$. The batch machine can handle up to $B < n = \sum_{t=1}^{m} n_t$ jobs simultaneously and we call such a machine a bounded batch processing machine. For a batch of jobs, its handling time is defined to be the longest processing time among the jobs in the batch. Jobs processed in the same batch have the same start time and the same handling time(i.e., the start time and the handling time of the batch). For each job $J_{t,k}$ of type t, its completion time, denoted by $C_{t,k}$, is the sum of the start time and the handling time of the batch that includes it, where the subscript t $(1 \leq t \leq m)$ is the index of the job type and k is the index of the job in type t under an arbitrary index method and $1 \leq k \leq n_t$. Therefore, all the jobs in the same batch have the same completion time. Our target is to schedule all jobs so that the total completion time $\sum\limits_{t=1}^{m} \sum\limits_{k=1}^{n_t} C_{t,k}$ is minimized.

We need several important structural properties due to Chandru V. et al., [3], D.S. Hochbaum and D. Landy [6], as well as some lemmas given by Foto Afrati et al., [1]. We present them in subsequent subsections.

2.1   Geometric Rounding

To establish a PTAS, for any given positive number ϵ, we need to find a $1 + \epsilon-$ optimal solution by our algorithm. Our framework is that for a given ϵ, we first apply geometric rounding to reduce the problem to a well–structured set of processing times, with a loss of no more than $1 + \epsilon$. Then we design a polynomial time algorithm for the rounded problem. The following crucial lemma is due to Foto Afrati et al. [1]

Lemma 1. *[1] With $1 + \epsilon$ loss, we can assume that all processing times are integer powers of $1 + \epsilon$.*

For simplicity, we still use previous symbols for the problem produced by rounding the processing times, i.e., m for the number of job types, n_t and p_t for the number and the processing time of jobs in type t, respectively, $t \in \{1, \ldots, m\}$, and $p_1 < p_2 < \ldots < p_m$. Notice that the processing times now are integer powers of $1 + \epsilon$, and we call the new problem *the rounded problem*. When there is no ambiguity, we always refer to the rounded problem in the following discussion.

2.2   Reduction to Leftover Batches

Given two types s and t, we say that type s is smaller than type t if $s < t$ (i.e., $p_s < p_t$). For a batch B_i, we define its handling time $p(B_i)$ as the longest processing time among jobs in B_i. Using notations given by Chandru V. et al. [3], a batch is called *full* if it contains B jobs, and *homogeneous* if it contains jobs of a single type, where B is the maximum batch size that the batch machine can handle. D.S. Hochbaum and D. Landy [6] extended those terminologies as follows: a batch is called $t-pure$ if it is homogeneous with jobs of type t. A batch which is not full is called a *partial* batch. If $p_t = p(B_i)$ is the longest processing time in a batch B_i, then t is called the *dominant type* of B_i and alternatively t *dominates* batch B_i.

The following three lemmas are due to Chandru V. et al. [3].

Lemma 2. *[3] Consider k batches $\{B_1, ..., B_k\}$ of jobs. Let $p(B_i)$ and $|B_i|$ be the handling time and the number of jobs in batch B_i, respectively. Then the* *sequence* $S = \{B_1, ..., B_k\}$ *of batches in the increasing order of the indices is optimal if*

$$\frac{p(B_1)}{|B_1|} \le \frac{p(B_2)}{|B_2|} \le ... \le \frac{p(B_k)}{|B_k|} . \tag{1}$$

Following definitions of Chandru V. et al. [3], we say that a sequence of batches satisfying (1) is in batch weighted shortest handling time (BWSHT) order.

Lemma 3. *[3] There exists an optimal solution containing $[n_t/B]$ homogeneous and full batches of type t, where $[r]$ denotes the lower integer part of a real number r. Furthermore, these batches will be sequenced consecutively.*

By Lemma 3, we can possibly first assign some jobs into homogeneous full batches for each type $t \in \{1, \ldots, m\}$. Then there are still $n_t - B[n_t/B]$ remaining jobs for each type $t \in \{1, \ldots, m\}$, we call these jobs *leftover jobs*, and the batches containing those jobs *leftover batches*. We have the following lemma for leftover jobs and leftover batches.

Lemma 4. *[3] There exists an optimal solution in which the jobs in each leftover batch are consecutive with respect to their processing times, that is, for any three leftover jobs with processing times $p_i \le p_j \le p_k$, if a batch contains the jobs with processing times p_i and p_k, it must also contains the job with processing time p_j.*

2.3   Reduction to Leftmost Batches

The following three lemmas are given by D.S. Hochbaum and D. Landy [6].

Lemma 5. *[6] In every optimal schedule that satisfies the properties of Lemmas 2 – 4, each job type $t \in \{1, \ldots, m\}$ dominates at most one leftover batch.*

Lemma 6. *[6] In any optimal schedule, if there are two leftover jobs i and j with processing times $p_i < p_j$ such that the batch containing job j precedes the batch containing job i, then job j must be in a full batch.*

Define a *basic* schedule as one that satisfies the properties described in Lemmas 2 – 6. To reduce the number of possible schedulings of the leftover jobs, D.S. Hochbaum and D. Landy [6] introduced the concept of *leftmost batching* as follows.

They first defined *a batching* as a list of leftover batches and a batching is called *basic* when it satisfies the properties described in Lemmas 4 and 5, namely:

- jobs are consecutive with respect to processing time in each batch, and
- each job type dominates at most one leftover batch.

Definition 1. *[6] Given two sets of job types, $F, P \subseteq \{1, \ldots, m\}$, with $F \cap P = \emptyset$, we say that a basic batching $\mathcal{A}$ agrees with the pair (F, P) if for each job type $t \in \{1, \ldots, m\}$,*

- $t \in F$ if and only if t dominates a full batch in $\mathcal{A}$;
- $t \in P$ if and only if t dominates a partial batch in $\mathcal{A}$;
- $t \notin F \cup P$ otherwise.

For a given pair (F, P), there may be more than one basic batchings that agree with it, or there may be no basic batching that agrees with it, in which case the pair (F, P) is called *infeasible*. If there is at least one basic batching that agrees with a pair (F, P), we define a new batching, denoted by $L(F, P)$, which is called the *leftmost batching* with respect to (F, P) as the *unique* basic batching that agrees with (F, P) and in which all jobs are pushed to the "left" as much as possible. This means that if the batches are arranged from left to right in order of increasing handling times (i.e., processing times of dominating types), then it is impossible to move a job from its current batch into a batch to the left and the resulting batching still agrees with (F, P).

The following essential lemma is due to D.S. Hochbaum and D. Landy [6].

Lemma 7. *[6] There exists an optimal schedule S for the burn-in problem such that S is basic and the leftover batches of S form a leftmost batching.*

3   More Structural Properties

Given a batch B_i dominated by type i and a batch B_j dominated by type j, if $i < j$ and $\dfrac{p(B_j)}{|B_j|} < \dfrac{p(B_i)}{|B_i|}$, then we call B_i an *out–of–order small* batch of B_j, and correspondingly B_j is called an *out–of–order large* batch of B_i. According to Lemmas 6 and 2, in an optimal schedule, we must have $|B_j| = B$ and B_j must be ahead of B_i if B_i is an *out–of–order small* batch of B_j. Since all the processing times are integer powers of $1 + \epsilon$, we have the following lemma.

Lemma 8. *In an optimal schedule, if B_j is an out–of–order large batch of B_i, then we have*
$$j - i \leq \left\lceil \log_{1+\epsilon} B \right\rceil.$$

Proof. Suppose $p(B_i) = (1 + \epsilon)^x$ and $p(B_j) = (1 + \epsilon)^y$, where i and j are the types dominating batches B_i and B_j and $1 \leq i < j \leq m$, x and y are integers and $x < y$. Recall that different types have different processing times, i.e., different powers of $1 + \epsilon$ and that the job types are indexed in the order of increasing processing times, we have that the processing time of each job type between i and j must be a number of the power of $1 + \epsilon$ between $(1 + \epsilon)^x$ and $(1 + \epsilon)^y$, i.e., $j - i \leq y - x$. Recall that we consider leftover batches only and that each type dominates at most one leftover batch. According to the previous discussion, since B_j is an *out–of–order large* batch of B_i, we must have
$$1 \leq |B_i| < B, \ |B_j| = B \text{ and } \frac{p(B_i)}{|B_i|} > \frac{p(B_j)}{|B_j|}$$
implying
$$\frac{(1 + \epsilon)^x}{1} \geq \frac{p(B_i)}{|B_i|} > \frac{p(B_j)}{|B_j|} = \frac{(1 + \epsilon)^y}{B}$$

and subsequently $y - x < \log_{1+\epsilon} B \implies y - x \leq \left\lceil \log_{1+\epsilon} B \right\rceil$ by the integrality of x and y. Finally, we have

$$j - i \leq y - x \leq \left\lceil \log_{1+\epsilon} B \right\rceil .$$

The proof is complete. □

For any given type t, let

$$t_s = t - \left\lceil \log_{1+\epsilon} B \right\rceil , \quad t_l = t + \left\lceil \log_{1+\epsilon} B \right\rceil .$$

For a given type t and a batch B_t dominated by t, we denote by OS_t a set of some *out-of-order small* batches with respect to B_t, and call it an *out-of-order small batch set* of B_t. Similarly, we denote by OL_t a set of some *out-of-order large* batches of B_t and call it an *out-of-order large batch set* of B_t. When there is no ambiguity, we call OS_t an *out-of-order small batch set* without specifying the possessor B_t, and similarly call OL_t an *out-of-order large batch set*. Notice that for any given batch B_t, the number of $OS_t's$ and the number of $OL_t's$ can be exponentially large in general. However, the following analysis shows that Lemma 8 allows us to focus on a polynomial number of them.

According to Lemma 8, all batches in OS_t must be dominated by types between t_s and $t - 1$.

Given a type t and a batch B_t dominated by t, let $(F, P)_{SLB_t}$ denote any possible (F, P)-pair of job types between t_s and $t - 1$, and $L(F, P)_{SLB_t}$ denote the leftmost batching of jobs in $\bigcup_{x=t_s}^{t-1} LJ(x) \backslash B_t$ with respect to $(F, P)_{SLB_t}$, where $LJ(x)$ is the leftover–job set of type x.

Given a type t and a batch B_t dominated by t, let $(F, P, B_t)_s$ be any possible batching of the following types:

type 1_s. $L(F, P)_{SLB_t}$ for some $(F, P)_{SLB_t}$–pairs,

type 2_s. the leftmost batching of jobs in $\bigcup_{x=t_s}^{t-1} LJ(x) \backslash B_t$ and j largest smaller jobs (jobs smaller than type t_s, refer to Lemma 4) with respect to some $(F, P)_{SLB_t}$–pairs(in this case, all the batches are still dominated by types between t_s and $t - 1$), where j is any possible number between 1 and $B - 1$.

Note that for a given batch B_t, the numbers of $(F, P, B_t)_s's$ of type 1_s and type 2_s are $O(B^{(\log_{1+\epsilon} 3)})$ and $O(B^{(1+\log_{1+\epsilon} 3)})$, respectively, due to the fact that the number of possible $(F, P)_{SLB_t}$–pairs is $O(B^{(\log_{1+\epsilon} 3)})$ (there are at most $\log_{1+\epsilon} B$ different types between type $t - 1$ and t_s, and each type t can belong to one the three categories, i.e., t dominates a full batch; t dominates a partial batch; t does not dominate any batch. Therefore there are totally $3^{(\log_{1+\epsilon} B)} = B^{(\log_{1+\epsilon} 3)}$ possible ways for deciding the $(F, P)_{SLB_t}$–pairs).

For each $(F, P, B_t)_s$ described above, let

$$OS(F, P, B_t)_s = \{\text{out-of-order small batches of } B_t \text{ in } (F, P, B_t)_s\}.$$

The following lemma ensures the reduction of the number of $OS_t's$ to polynomially small.

Lemma 9. *In an optimal schedule as described in Lemma 7, for each batch B_t, its out–of–order small batch set OS_t must be the same as one $OS(F, P, B_t)_s$ for some $(F, P, B_t)_s$ and the number of $OS_t's$ is no more than $O(B^{(1+\log_{1+\epsilon} 3)})$.*

Proof. Otherwise, in the optimal schedule, the batches dominated by types between t_s and $t-1$ is not a locally leftmost batching. Suppose the corresponding (F, P)–pair of these batches is $(F, P)_s$ and the (F, P)–pair of the all the batches in the schedule is $(F, P)_a$. Then, we can move some jobs to the left and the resulting local batching still agrees with $(F, P)_s$, and subsequently the resulting whole batching still agrees with $(F, P)_a$, i.e., the handling time of each batch does not change. However, since some jobs are pushed to the left which means their completion times become smaller, the completion time of the new schedule is smaller than the original optimal schedule, which leads to contradiction (the reason we consider type2$_s$ is that a batch dominated by a type between t_s and $t-1$ may also include jobs of types smaller than t_s). We can learn the number of $OS_t's$ from the above discussion.

Now we discuss the case for OL_t. Given a type t and a batch B_t dominated by t, according to Lemma 8, all batches in OL_t must be dominated by types between $t+1$ and t_l.

Let $(F, P)_{LLB_t}$ be any possible (F, P)–pair of types between $t+1$ and t_l and $L(F, P)_{LLB_t}$ the leftmost batching of jobs in $\bigcup_{x=t}^{t_l} LJ(x) \setminus B_t$ with respect to $(F, P)_{LLB_t}$ (notice that type t does not dominate any batch in $L(F, P)_{LLB_t}$).

Given a type t and a batch B_t dominated by type t, let $(F, P, B_t)_{LL}$ be any possible batching of the following types:

type 1$_l$. $L(F, P)_{LLB_t}$ for some $(F, P)_{LLB_t}$–pairs, and we denote $(F, P, B_t)_{LL}$ of this type as $T_1(F, P, B_t)_{LL}$;

type 2$_l$. the leftmost batching of jobs in $\bigcup_{x=t}^{t_l} LJ(x) \setminus B_t \setminus LJ_t$ with respect to some $(F, P)_{LLB_t}$–pairs, where LJ_t is the set of j consecutively largest jobs of types between $t+1$ and t_l (in this case, some largest types between $t+1$ and t_l may not dominate any batch, refer to Lemma 4) and j is any possible number between 1 and $B-1$. We denote $(F, P, B_t)_{LL}$ of this type as $T_2(F, P, B_t)_{LL}$.

Notice that, for a given type t and a given batch B_t, the number of possible $(T_1(F, P, B_t)_{LL})'s$ is $O(B^{(\log_{1+\epsilon} 3)})$ and the number of possible $(T_2(F, P, B_t)_{LL})'s$ is $O(B^{(1+\log_{1+\epsilon} 3)})$ as the number of $(F, P)_{LLB_t}$–pairs is $O(B^{(\log_{1+\epsilon} 3)})$ (the reason is like that in the previous discussion).

For a given $T_1(F, P, B_t)_{LL}$ (resp. a $T_2(F, P, B_t)_{LL}$), let $T_1(F, P, B_t)_L$ (resp. $T_2(F, P, B_t)_L$) denote the batching which contains all the batches in $T_1(F, P, B_t)_{LL}$ (resp. $T_2(F, P, B_t)_{LL}$) and all the full homogeneous batches of types between $t+1$ and t_l.

Given a $T_1(F, P, B_t)_L$ and $T_2(F, P, B_t)_L$, set

$$OBT_1(F, P, B_t)_L = \{\text{out-of-order large batches of } B_t \text{ in } T_1(F, P, B_t)_L\},$$
$$OBT_2(F, P, B_t)_L = \{\text{out-of-order large batches of } B_t \text{ in } T_2(F, P, B_t)_L\}.$$

The following lemma restricts the number of $OL'_t s$ to polynomially small.

Lemma 10. *In an optimal schedule as described in Lemma 7, for each batch B_t, its out–of–order large batch set OL_t can only be the same as either one $OBT_1(F, P, B_t)_L$ (now we say that OL_t is of type 1) or one $OBT_2(F, P, B_t)_L$ (now we say OL_t is of type 2) and consequently the number of $OL'_t s$ is no more than $O(B^{(1+\log_{1+\epsilon} 3)})$.*

Proof. The proof is similar to that of Lemma 9 (the reason we consider type 2_l is that some largest jobs of type between $t + 1$ and t_l may be processed in a batch which is dominated by a type larger than t_l).

Let $\overline{OL_t}$ denote the remaining batching of $T_1(F, P, B_t)_L$ by omitting batches in OL_t if $OL_t = OBT_1(F, P, B_t)_L$, and the remaining batching of $T_2(F, P, B_t)_L$ if $OL_t = OBT_2(F, P, B_t)_L$.

4   The Dynamic Programming Solution

The state variables in our dynamic programming solution is indexed in the form of (t, B_t, OS_t, OL_t), which means the "present" type is t, the "present" batch in the scheduling is a leftover batch B_t dominated by type t, the "present" *out–of–order small batch set* of B_t is OS_t and the "present" *out–of–order large batch set* of B_t is OL_t. Our dynamic programming entry $O(t, B_t, OS_t, OL_t)$ denotes both the optimal schedule and its value of all jobs in $\bigcup\limits_{x=t}^{m} J(x) \cup B_t \cup OS_t \setminus OL_t$ with the assumption that batch B_t and batches in OS_t keep unchanged, where $J(x)$ is the job set of type x.

What we should point out here is that in $O(t, B_t, OS_t, OL_t)$, the full homogeneous batches of type t are always scheduled first. Let $O(t, B_t, OS_t, OL_t)'$ denote the schedule and its value obtained from $O(t, B_t, OS_t, OL_t)$ by omitting the full homogeneous batches of type t.

When computing $O(t, B_t, OS_t, OL_t)$, the most important thing is to enumerate all the possible next "present" batches. Notice that according to Lemma 8, all the batches in OS_t can't be scheduled later than batches dominated by types larger than t_l, and that $t_l + 1$ is the smallest type by which a batch is dominated can't be scheduled earlier than B_t. Hence, we enumerate all possible next "present" batches from batches dominated by type $t_l + 1$ on.

Given (t, B_t, OS_t, OL_t), next "present" batch $B_{t'}$ and next "present" type t' can only be as follows:

- if OL_t is of type 1, then $B_{t'}$ includes j consecutive leftover jobs of types larger than t_l for some $1 \le j \le B$, and t' is the largest type in $B_{t'}$;
- if OL_t is of type 2, then $B_{t'}$ includes all the jobs in LJ_t and j consecutive larger jobs(smallest jobs larger than type t_l, refer to Lemma 4) for some $1 \le j \le B - |LJ_t|$, and t' is the largest type in $B_{t'}$.

Given t, B_t, OS_t, OL_t, t' and $B_{t'}$, next "present" *out-of-order small batch set* $OS_{t'}$ of $B_{t'}$ is a subset of *out-of-order small* batches of $B_{t'}$ in $\overline{OL_t}$, and the next "present" *out-of-order large batch set* $OL_{t'}$ of $B_{t'}$ is a possible subset of *out-of-order large* batches of $B_{t'}$ that can be similarly defined as the definition of OL_t of B_t by substituting t' for t and $B_{t'}$ for B_t.

Let $M(B_t, B_{t'})$ denote the batching that contains the batches in $OS_t \cup \overline{OL_t} \cup OL_{t'} \setminus OS_{t'}$ (notice that $M(B_t, B_{t'})$ maybe include full homogeneous batches). It is easy to see that in an optimal schedule all the batches in $M(B_t, B_{t'})$ must be scheduled between B_t and $B_{t'}$ once t' and $B_{t'}$ are given. Therefore, the cost caused by the batches in $M(B_t, B_{t'})$ is only dependent on the local schedule, i.e., sequence of these batches (because the delay caused by these batches to the later batches is equal to the sum of the handling times of all these batches). Let $CM(B_t, B_{t'})$ denote the optimal local schedule and its value of batches in $M(B_t, B_{t'})$.

For each type t, let $f_t = [n_t/B]$ denote the number of full homogeneous batches of type t, where $n_t = |J(t)|$ denotes the number of jobs of type t.

Now, we are ready to present our dynamic programming equations.

4.1   Dynamic Programming Equations

Base case. For each t between $m - [\log_{1+\epsilon} B]$ and m, for each possible leftover batch B_t that is dominated by t, for all possible *out-of-order small batch set* OS_t and *out-of-order large batch set* OL_t of B_t as described in Lemma 9 and Lemma 10 (in this case, OL_t can be only of type 1),

$$O(t, B_t, OS_t, OL_t)' = p_t \cdot (|B_t| + |\overline{OL_t} \cup OS_t|) + opt(\overline{OL_t} \cup OS_t),$$

where $|\overline{OL_t} \cup OS_t|$ is the number of jobs in $\overline{OL_t} \cup OS_t$ and $opt(\overline{OL_t} \cup OS_t)$ denotes both the optimal sequence and its value of batches in $\overline{OL_t} \cup OS_t$ (notice that we can compute it with a time of no more than $(\log_{1+\epsilon} B)^2$), and

$$O(t, B_t, OS_t, OL_t) = O(t, B_t, OS_t, OL_t)' + p_t \cdot \left(\frac{f_t(f_t + 1)}{2} B \right.$$
$$\left. + f_t \cdot (|B_t| + |\overline{OL_t} \cup OS_t|) \right) .$$

Recursion. For each t between 1 and $m - [\log_{1+\epsilon} B] - 1$, for each possible leftover batch B_t that is dominated by type t, for all possible *out-of-order small batch set* OS_t and *out-of-order large batch set* OL_t of B_t as described in Lemma 9 and Lemma 10, let

$$Y = \left(\bigcup_{x=t+1}^{m} J(x) \right) \cup LJ(t) \cup OS_t \setminus OL_t,$$

then

$$O(t, B_t, OS_t, OL_t)'$$
$$= \min_{(t', B_{t'}, OS_{t'}, OL_{t'})} \{ p_t |Y| + CM(B_t, B_{t'}) + O(t', B_{t'}, OS_{t'}, OL_{t'})' \},$$

where the minimum is taken over all possible $(t', B_{t'}, OS_{t'}, OL_{t'})$ as described before, and

$$O(t, B_t, OS_t, OL_t) = O(t, B_t, OS_t, OL_t)' + p_t \cdot \left(\frac{f_t(f_t + 1)}{2} B + f_t \cdot |Y| \right) \ .$$

Conclusion.

$$\min_{(t, B_t, OS_t, OL_t)} \{ O(t, B_t, OS_t, OL_t) \},$$

where the minimum is take over all possible dynamic programming states (t, B_t, OS_t, OL_t).

4.2   The Theorem

After the above analysis, we finally obtain the following results.

Theorem 1. *Within a time of no more than $O(m \cdot B^{(7 + \log_{1+\epsilon} 27)} \cdot (\log_{1+\epsilon} B)^2)$, our algorithm gives a solution to the rounded problem.*

Proof. It is easy to verify that our algorithm produces a schedule no worse than the optimal schedule described in Lemma 7. Suppose $\mathcal{S}$ is an optimal schedule described in Lemma 7. Suppose the first leftover batch of $\mathcal{S}$ is dominated by type t_0 and we denote this batch as B_{t_0}, the first leftover batch dominated by a type larger than $t_0 - 1 + \log_{1+\epsilon} B$ is dominated by type t_1 and we denote this batch as B_{t_1} and the first leftover batch dominated by a type larger than $t_1 - 1 + \log_{1+\epsilon} B$ is dominated by type t_2 and we denote this batch as B_{t_2} and so on. Suppose the last batch we get in this way is dominated by type t_r and we denote this batch by B_{t_r} (Since full homogeneous batches can be processed easily, we don't discuss them). For short, we denote by A_{ij} as the sequence of leftover batches between B_{t_i} and B_{t_j}, by A_{00} the sequence before B_{t_0} and by A_{rr} the sequence after B_{t_r}. It is easy to find the corresponding OS_{t_i} (batches in $A_{i(i+1)}$ dominated by types smaller than t_i) and OL_{t_i} (batches in $A_{(i-1)i}$ dominated by types larger than t_i). Then the optimal schedule can be viewed as the concatenation of all the described A_{ij}'s and B_{t_i}'s, which can not be better than the schedule given by our algorithm which is the best among all this kind of concatenations.

Regarding the computing complexity, we need consider at most $O(m \cdot B^{(4 + \log_{1+\epsilon} 9)})$ states where there are at most m types, for each type t, there are at most B^2 batches B_t which may include 1 to $B - 1$ *type-t* jobs and some other consecutively smaller jobs and for the fixed B_t, there are at most $B^{(1 + \log_{1+\epsilon} 3)} OS_t's$ and $B^{(1 + \log_{1+\epsilon} 3)} OL_t's$ according to Lemma 9 and Lemma 10. For each state, there are at most $O(B^{(3 + \log_{1+\epsilon} 3)} \cdot (\log_{1+\epsilon} B)^2)$ cases, where there are at most B^2 next "present" batches $B_{t'}$, and there are at most $O(B^{(1 + \log_{1+\epsilon} 3)})$ *out-of-order large batch set's* $OL_{t'}$, and the computation of $CM(B_t, B_{t'})$ spends a time of no more than $(\log_{1+\epsilon} B)^2$. □

Corollary 1. *We give a PTAS for the original burn-in problem of minimizing total completion time of bounded batch scheduling with all jobs released at the same time.*

Proof. We can complete the proof immediately by Lemma 1 and Theorem 1. □

5   Remarks and Discussion

In this work, we obtain the first PTAS for the batch processing problem when the batch size B is finite, number of job types is variable and all jobs arrive at time zero. The techniques we apply to achieve this include geometric rounding [1] and several structural properties of an optimal solution described in [3,6] and Sect. 3, as well as a quite complicated dynamic programming solution.

References

1. Foto Afrati, Evripidis Bampis, Chandra Chekuri, David Karger, Claire Kenyon, Sanjeev Khanna, Ioannis Milis, Maurice Queyranne, Martin Skutella, Cliff Stein, Maxim Sviridenko: Approximation Schemes for Minimizing Average Weighted Completion Time with Release Dates. the proceeding of the 40th Annual Symposium on Foundations of Computer Science, New York (Oct. 1999) 32–43.
2. P. Brucker, A. Gladky, H. Hoogeveen, M. Kovalyov, C. Potts, T. Tautenhahn, S. van de Velde: Scheduling a batching machine. Journal of Scheduling 1 (1998) 31–54.
3. Chandru V., C. Y. Lee, R. Uzsoy: Minimizing Total Completion Time on a Batch Processing Machine with Job Families. O. R. Lett. 13 (1993b) 61–65.
4. X. Deng, Y. Zhang: Minimizing Mean Response Time in Batch Processing System. Lecture Notes in Computer Science, Springer-Verlag 1627 (COCOON'99, Tokyo, Japan, July 1999), 231–240.
5. X. Deng and H. Feng, P. Zhang and H. Zhu: A Polynomial Time Approximation Scheme for Minimizing Total Completion Time of Unbounded Batch Scheduling. In proceeding of ISAAC'01, LNCS 2223, 26–35.
6. D. S. Hochbaum, D. Landy: Scheduling Semiconductor Burn-in Operations to Minimize Total Flowtime. Operations Research 45 (6)(1997) 874–885.

An Approximation Scheme for the Two-Stage, Two-Dimensional Bin Packing Problem

Alberto Caprara, Andrea Lodi, and Michele Monaci

D.E.I.S., Università di Bologna,
Viale Risorgimento 2, 40136 Bologna, Italy
{acaprara,alodi,mmonaci}@deis.unibo.it

Abstract. We present an asymptotic PTAS for Two-Dimensional Bin Packing, which requires packing (or cutting) a given set of rectangles from the minimum number of square bins, with the further restriction that cutting the rectangles from the bins can be done in two stages, as is frequently the case in real-world applications. To the best of our knowledge, this is the first approximation scheme for a nontrivial two-dimensional (and real-world) generalization of classical one-dimensional Bin Packing in which rectangles have to be packed in (finite) squares. A simplification of our method yields an asymptotic PTAS for the two-stage packing of rectangles in a bin of unit width and infinite height. Moreover, we point out that our method may lead to a better approximation guarantee for Two-Dimensional Bin Packing without stage restrictions, provided some structural property holds.

1 Introduction

The classical one-dimensional *Bin Packing* (BP) is one of the most widely studied problems in Combinatorial Optimization. In particular, a huge amount of work has been done to analyze the worst-case performance of heuristic algorithms [4]. The approximability status of BP, namely the existence of *Asymptotic (Fully) Polynomial Time Approximation Schemes* (A(F)PTASs), was essentially settled in the early 80s by the work of Fernandez de la Vega and Lueker [8] and Karmarkar and Karp [10]. In contrast, many open questions exist about the approximability of important two- (and more-)dimensional generalizations of BP, many of which arise from real-world applications.

In this paper, we will consider two-dimensional geometric generalizations of BP, in which rectangles (called also *items* in the following) of specified size have to be packed in other (larger) rectangles (called *bins*). The most studied version of the problem is the one in which a specified edge for each item has to be packed parallel to a specified edge of a bin. Such a requirement is called *orthogonal packing without rotation*. Different results were obtained for the two cases in which, respectively, there is a unique bin (called also *strip*) where one edge is infinitely long, called *Two-Dimensional Strip Packing* (2SP), and there are "enough" many identical bins of finite width and height, called *Two-Dimensional (Finite) Bin Packing* (2BP). For 2SP, a long series of results, starting with the

W.J. Cook and A.S. Schulz (Eds.): IPCO 2002, LNCS 2337, pp. 315–328, 2002.
© Springer-Verlag Berlin Heidelberg 2002

work of Baker, Coffman and Rivest [1] and Coffman, Garey, and Johnson [5] and ending with the elegant AFPTAS proposed by Kenyon and Rémila [11], settled the approximability of the problem. On the other hand, for 2BP the best known asymptotic approximation guarantee is $2 + \varepsilon$ (for any $\varepsilon > 0$) that can be achieved by applying the method of [11], packing all the items in an infinite strip and then cutting this strip into slices so as to get bins of the required size, packing the items that are split between two bins in additional bins. Apparently, there is no way to extend the method of [11] to 2BP so as to get an approximation guarantee strictly better than 2. On the other hand, to the best of our knowledge, there is no result that rules out the existence of an APTAS for 2BP. For the special case in which *squares* have to be packed, Seiden and van Stee [13] presented an approximation algorithm with asymptotic guarantee $14/9 + \varepsilon$ (for any $\varepsilon > 0$). It has to be remarked that, in the case of squares, it is trivial to round-up the sizes of the items so as to get a constant number of distinct sizes without significantly increasing the optimal value. On the other hand, the same is apparently impossible for general rectangles.

The above situation for the case of orthogonal packing without rotation is common to all the two-dimensional variants of bin packing that we are aware of: when the items have to be packed in *finite* bins, there is a huge gap between the approximation guarantee that can be achieved by known methods and the one that is excluded by negative results. The only problem that one may consider as an exception is the classical BP itself, that coincides with the *one-stage*, two-dimensional orthogonal bin packing without rotation. In this paper, we make one step further, deriving an APTAS for the *two-stage*, two-dimensional orthogonal bin packing without rotation, also called *Two-Dimensional Shelf Bin Packing* (2SBP), which arises in real-world applications. Roughly speaking, the problem calls for packing the items in the bins in rows so as to form a sequence of *shelves* (or *levels*), see Fig. 1. To the best of our knowledge, this is the first approximation scheme for a nontrivial two-dimensional (and real-world) generalization of BP in which rectangles have to be packed in other finite rectangles. As a byproduct, we also obtain an APTAS for the two-stage, two-dimensional orthogonal strip packing without rotation, also called *Two-Dimensional Shelf Strip Packing* (2SSP). (Actually, as one may expect, the derivation of such a result for this latter problem is substantially simpler than for 2SBP.)

Our approximation scheme is subdivided into several steps: although basically all the techniques used in each single step are analogous to those in other approximation schemes (not necessarily for BP), their combination into the overall picture is certainly far from trivial. In particular, many known techniques are applied with an unusual purpose. For instance, we apply the *geometric grouping* technique of [10] in order to round the heights of "small" (thin) items, whereas grouping was used so far only to round the size of "large" items. For "large" items, we first round the heights and then the widths, achieving a constant number of "large" rectangle sizes. This typically does not work for other geometric problems. Moreover, we combine enumeration of the "large" item packings with the solution of *Integer Linear Programs* (ILPs) in order to pack optimally the

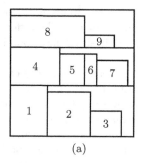 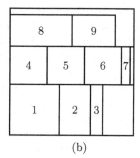

Fig. 1. Examples of two-staged patterns: non-exact (a) and exact (b) cases

"small" items, whereas for the other schemes that we know the solution of an LP is typically sufficient for this purpose. Finally, the structure of the ILPs in our method appears to be more complex than the one of the ILPs used in other approximation schemes.

The degree of sophistication of our method, which may be unavoidable, perhaps reflects the difficulty in designing approximation schemes (if any exist) for the more complex 2BP or, anyway, for k-stage versions of the problem with $k \geq 3$.

Finally, it has to be remarked that our algorithm has an astronomical running time for any reasonable value of the required accuracy. Such a situation is common to all the other (A)PTASs that we are aware of, although ours is much worse than most others since we must perform several times some complete enumeration combined with the solution of an ILP with a huge (constant) number of variables. On the other hand, from a practical viewpoint, the structure of an APTAS may inspire the design of (parts of) heuristic algorithms that are effective in practice. Moreover, the existence itself of an APTAS is often a good theoretical indicator that practical instances of the problem will not be "too hard" to solve.

Two-Stage Packing

The *two-stage* name is originated from cutting applications in which the items have to be *cut* from the bins. Formally, by following the definition introduced by Gilmore and Gomory [9], two-stage cutting/packing requires that the final pattern is such that each item can be obtained from the associated bin by at most two stages of cutting. Namely, after a first stage in which the bins are cut into *shelves*, the second stage produces *slices* which contain a single item. If an additional stage is allowed in order to separate an item from a waste area, as in Fig. 1 (a), the problem is called two-stage *with trimming* (also known as *non-exact case*); otherwise, the problem is called two-stage *without trimming* (*exact case*), as in Fig. 1 (b). The case with trimming is the one considered in this paper, being more frequently found in practical applications. Gilmore and

Gomory [9] also introduced the case in which the items have to be obtained by at most *three* stages of cutting (*three-stage cutting*). The difference between two-stage with trimming and three-stage is that slices may contain more than one item in the three-stage case.

Staged cutting problems have practical relevance in all the cases where the raw material to be cut has a low cost with respect to the industrial costs involved by the cutting process, such as, e.g., in many wood, paper or glass cutting applications. Real-world applications involving two-stage cutting problems in the steel industry can be found in de Carvalho and Rodrigues [6,7], while an industrial three-stage cutting problem was recently considered by Vanderbeck [14].

Note that two-stage packing with trimming is equivalent to packing the items in the bins in shelves, where a *shelf* is a row of items having their bases on a line which is either the base of the bin or the line drawn at the top of the tallest item packed in the shelf preceding the current one in the bin (see again Fig. 1 (a)). The fact that many classical heuristics for 2SP [5,2] and 2BP [3] pack the items in this way gives additional relevance to two-stage cutting (with trimming).

Preliminaries and Outline

In all the problems that we consider we are given a set $N = \{1, \ldots, n\}$ of items, the j-th having *width* w_j and *height* h_j; we also say that item j has *size* (w_j, h_j). A *shelf* is a set $S \subseteq N$ whose *width*, defined by $\sum_{j \in S} w_j$, does not exceed 1. The *height* of the shelf is defined by $\max_{j \in S} h_j$.

In 2SSP the items have to be packed in shelves within a strip of width 1 and infinite height, so as to minimize the overall height of the shelves. In 2SBP the items have to be packed in shelves and the resulting shelves in identical bins of width and height 1 (i.e. the overall height of the shelves in each bin must not exceed 1) so as to minimize the number of nonempty bins. Clearly, both 2SSP and 2SBP are generalizations of BP, which arises when $h_j = 1$ for all $j \in N$.

Consider a minimization problem P, letting $\mathcal{I}_P$ denote the set of all its instances. Given an instance $I \in \mathcal{I}_P$, let $opt_P(I)$ denote the value of the optimal solution for I. If no confusion arises, we will write $\mathcal{I}$ for $\mathcal{I}_P$ and $opt(I)$ for $opt_P(I)$. For a given constant $\rho \geq 1$, an *asymptotic ρ-approximation algorithm* is an algorithm that produces a solution of value at most $\rho \cdot opt(I) + \delta$ for every $I \in \mathcal{I}$, where δ is a constant independent of I. An *APTAS* is an algorithm that receives on input also a required *accuracy* $\varepsilon > 0$, runs in time polynomial in the size of the instance, and produces a solution of value at most $(1 + \varepsilon) \cdot opt(I) + \delta$ for every $I \in \mathcal{I}$, where δ is a constant independent of I. (Actually we may replace the constant δ by $o(opt(I))$ above, but this simpler definitions are sufficient for our purposes.)

The main result of this paper is an APTAS for 2SBP, which is presented in Sect. 2 and whose approximation guarantee is shown in Sect. 3. A simplification of this algorithm leads to an APTAS for 2SSP, mentioned in Sect. 4. Finally, in that section we also discuss the possibility for our algorithm to be a $(\gamma + \varepsilon)$-approximation algorithm for the famous 2BP for any $\varepsilon > 0$ and for some $\gamma < 2$, that would improve on the current best approximation achievable, equal to $2 + \varepsilon$.

Algorithm APTAS$_{2SBP}$:

Let $\bar{\varepsilon}$ be the required accuracy and define $\varepsilon := \bar{\varepsilon}/574$.

(a) **Decomposition:**
  - (a.1) Guess the parameter f $(1 \le f \le t := \lceil 1/\varepsilon^2 \rceil)$ and for $i \in \mathbb{N}$ round-up all heights in $(\varepsilon^{f+it}, \varepsilon^{f+it-1})$ to ε^{f+it-1}.
  - (a.2) Form the shelves in the final solution by considering *separately* each set A_i, $i \in \mathbb{N}$, containing the items with heights in $[\varepsilon^{f+it-1}, \varepsilon^{f+(i-1)t}]$, (i.e. pack the items belonging to different sets in different shelves).

(b) **Item Subdivision, Vertical Melting and Grouping:**
  For each set of items A_i, $i \in \mathbb{N}$, distinguish between *wide* items (having width $\ge \varepsilon$) and *thin* items (having width $< \varepsilon$), and:
  - (b.1) Allow thin items to be cut *vertically* in the packing.
  - (b.2) Round the *heights* of the wide items by *linear grouping*.
  - (b.3) Round the *widths* of the wide items with the *same height* after Step (b.2) by *linear grouping*.
  - (b.4) Round the *heights* of the thin items by *geometric grouping*.

(c) **Definition of Low Shelves:**
  Solve the 2SSP instance defined by the items in each set A_i, $i \in \mathbb{N} \setminus \{0\}$, by enumeration of the shelves with wide items combined with ILP (1)–(4), determining the packing in (low) shelves for the items in each set.

(d) **Horizontal Melting and Definition of Bins:**
  - (d.1) Allow the low shelves found in Step (c) to be cut *horizontally* in the packing.
  - (d.2) Solve the 2SBP instance defined by the items in A_0 and by the shelves found in Step (c) by enumeration of the shelves with wide items in A_0 combined with ILP (5)–(9), determining the packing in the bins for the items in A_0 and the low shelves found in Step (c).

(e) **Horizontal Solidification:**
  Pack the low shelves that are cut horizontally in the solution in separate bins by *next fit*.

(f) **Vertical Solidification:**
  Pack the thin items that are cut horizontally in the solution in separate bins by *next fit decreasing height*.

Fig. 2. General structure of the algorithm

2   Description of the Algorithm

The general structure of our APTAS for 2SBP is shown in Fig. 2, where $\mathbb{N}$ denotes the set of integer nonnegative numbers (including 0). Below, we describe in detail each step, showing that the running time is polynomial for every fixed value of the required accuracy $\bar{\varepsilon}$. In the next section, we will prove the approximation guarantee of the method. Generally, since we modify the widths and heights of the items *several times*, when describing a step we will denote by w_j and h_j, respectively, the width and height of each item j *before* performing the step.

Decomposition. In Step (a.1), we suppose that the guess of f is *correct*, i.e. the one that leads to the best final solution. This can be achieved by trying all possible $O(1/\varepsilon^2)$ values of f. For each item $j \in N$ such that $h_j \in (\varepsilon^{f+it}, \varepsilon^{f+it-1})$ for some $i \in \mathbb{N}$ we redefine $h_j := \varepsilon^{f+it-1}$.

In Step (a.2), let $A_i := \{j \in N : h_j \in [\varepsilon^{f+it-1}, \varepsilon^{f+(i-1)t}]\}$ for $i \in \mathbb{N}$, noting that $A_0 = \{j \in N : h_j \in [\varepsilon^{f-1}, 1]\}$. Of course, we can avoid considering all sets A_i such that $A_i = \emptyset$. Accordingly, there are $O(n)$ nonempty sets A_i that can be defined in $O(n)$ time.

Grouping. In Step (b), let A_i $(i \in \mathbb{N})$ be the instance considered and define the set of wide and thin items in A_i, respectively, by $C_i := \{j \in A_i : w_j \geq \varepsilon\}$ and $B_i := \{j \in A_i : w_j < \varepsilon\}$.

In Steps (b.2) and (b.3), *linear grouping* is the classical grouping proposed in [8]. Namely, in Step (b.2), let $h_1 \geq h_2 \geq \ldots \geq h_{\ell_i}$ be the heights of the items in C_i after sorting. If $\ell_i < 2/\varepsilon^{t+1}$, we leave these heights unchanged, defining for convenience $g_i := \ell_i$. Otherwise, letting $k_i := \lfloor \ell_i \varepsilon^{t+1} \rfloor$, we form $g_i := \lceil \ell_i/k_i \rceil$ groups of k_i consecutive items, starting from the k_i tallest ones, and letting the last group contain $\ell_i - (g_i - 1)k_i$ items. We redefine the heights of the items in each group as the height of the first (tallest) item in the group. After this step, it is easy to check that the number g_i of different heights in C_i is upper bounded by $3/\varepsilon^{t+1} + 1 = O(1/\varepsilon^{t+1})$ (details are left to the full paper).

In Step (b.3), if $\ell_i < (g_i + 1)/\varepsilon^{t+1}$, we leave the widths of all items in C_i unchanged. Otherwise, we consider *separately* the sets of items in C_i with the same height h_j after Step (b.2), for $j = 1, \ldots, g_i$, each denoted by D_{ij}. For each set D_{ij}, let $w_1 \geq w_2 \geq \ldots \geq w_{m_{ij}}$ be the widths of the items after sorting. We form $d_{ij} := \lceil m_{ij}/p_i \rceil$ groups of $p_i := \lfloor \ell_i \varepsilon^{t+1}/g_i \rfloor$ consecutive items starting from the p_i widest ones and letting the last group contain $m_{ij} - (d_{ij} - 1)p_i$ items (possibly forming only one group). We redefine the width of the items in each group as the width of the first (widest) item in the group. After this step, it is easy to verify that the number of different widths in C_i is upper bounded by $g_i(g_i + 2)/\varepsilon^{t+1} + g_i = O(1/\varepsilon^{3(t+1)})$ (again, details are in the full paper).

According to the above discussion, for each i, the number s_i of distinct item sizes in each set C_i after Steps (b.2) and (b.3) is bounded by a constant, namely $s_i \leq g_i \cdot (g_i(g_i + 2)/\varepsilon^{t+1} + g_i) = O(1/\varepsilon^{4(t+1)})$.

In Step (b.4) we group the heights of the thin items (whose widths are unchanged afterwards) according to the geometric scheme of [10]. To this aim, let B_i be the set of thin items considered, $h_1 \geq h_2 \geq \ldots \geq h_{s_i}$ be the heights of the items in B_i after sorting, and $W_i := \sum_{j \in T_i} w_j$ be the overall width of these items. We form groups with consecutive items, starting from the first (tallest) one, so that the overall *width* of the items in each group adds up to *exactly* $\varepsilon^t W_i$, with the possible exception of the last group, for which the overall width of the items is $W_i - \lfloor 1/\varepsilon^t \rfloor \varepsilon^t W_i$ (if this value is not 0). Note that such a subdivision is always possible since we allow thin items to be cut vertically, and therefore we may insert only a vertical fraction of an item in a group. As for linear grouping, we redefine the heights of the items in each group as the height of the first (tallest) item in the group. The overall number of heights for the items in B_i after this step is $\lceil 1/\varepsilon^t \rceil = O(1/\varepsilon^t)$, i.e. it is bounded by a constant. We can represent each group by a single *equivalent item* that may be cut vertically.

Enumeration Plus ILP. In Steps (c) and (d) we separately pack the items in each set A_ℓ ($\ell \in \mathbb{N}$) by complete enumeration of the possible ways of packing the wide items in shelves, and, for each possibility, solution of an ILP with a constant number of variables to pack the thin items. (The final solution for each set is the one associated with the ILP with the best value among all the possibilities.) In particular, in Step (c) we consider each set A_ℓ, $\ell \in \mathbb{N} \setminus \{0\}$, and solve the associated 2SSP instance, optimally packing the items in so-called *low* shelves so as to minimize the overall height of the shelves. In Step (d), we consider set A_0 along with the shelves produced in Step (c) and optimally solve the associated 2SBP instance, i.e. we pack the items in A_0 in shelves and the shelves in bins, together with the low shelves (that may be cut horizontally) so as to minimize the number of bins used.

For Step (c), consider a generic set A_ℓ ($\ell \in \mathbb{N} \setminus \{0\}$). Since the number s_ℓ of different sizes for the wide items in C_ℓ is bounded by a constant, and each of the associated widths is at least ε, i.e. at most $\lfloor 1/\varepsilon \rfloor$ such items fit in a shelf, the number of different ways of packing the items in C_ℓ in shelves is polynomially bounded. More precisely, each shelf can be represented by a vector with s_ℓ non-negative integer components, where the j-th component represents the number of items of size j in the shelf, each component being at most $\lfloor 1/\varepsilon \rfloor$. So the number of *shelf types* is $O(\lfloor 1/\varepsilon \rfloor^{s_\ell})$, and each solution is defined by the number n_r of shelves of each type r used. Noting that $n_r \leq n$ for each r, the number of possibilities to be considered in a complete enumeration is $O(|C_\ell|^{\lfloor 1/\varepsilon \rfloor^{s_\ell}}) = O(n^{(1/\varepsilon)^{1/\varepsilon^{4(t+1)}}})$, i.e. polynomial in n. Consider a possible way of packing the wide items in shelves. Since we are considering all the possibilities, we can assume that, in the optimal solution of 2SSP for A_ℓ, the wide items are partitioned into shelves exactly in this way. Clearly, the number of different sizes of the resulting shelves is bounded by a constant. Specifically, let p be the number of these sizes and, for each size $(\overline{w}_i, \overline{h}_i)$ ($i = 1, \ldots, p$), let $\overline{n}_p$ be the number of shelves with this size. In particular, recall that if a shelf contains the items in set $S \subseteq C_k$, its width is defined by $\sum_{j \in S} w_j$ and its height by $\max_{j \in S} h_j$. Moreover, let q be the number of groups of thin items, letting (w_j, h_j) ($j = 1, \ldots, q$) denote the size of the equivalent item. For convenience, we assume $h_1 > h_2 > \ldots > h_q$ (*strict* inequality). This is allowed since we can replace possible groups with the same height by a unique group of items (hence not all w_j are necessarily equal).

We construct an ILP in which the variables represent the possible shelves that can be obtained by combining a shelf with wide items with some thin items. First of all, we consider shelves with thin items only, the tallest of which has height h_j, called *shelves of type j*. Such a shelf has height h_j and can contain thin items (of height h_j or smaller) for a total width of up to 1. Moreover, we consider *shelves of type (i, j)*, obtained by combining a shelf with wide items of size $(\overline{w}_i, \overline{h}_i)$ with some thin items, the tallest of which has height h_j. The height of a shelf of type (i, j) is given by $h_{ij} := \max\{\overline{h}_i, h_j\}$. We also define $w_{ij} := 1 - \overline{w}_i$ to be the horizontal space for thin items (of height h_{ij} or smaller) in the shelf. As a special case, we let a shelf with wide items only be a shelf of type $(i, q + 1)$, letting $h_{i,q+1} := \overline{h}_i$.

In the ILP, we have variables x_{ij} and y_j to denote, respectively, the number of shelves of type (i, j) and j in the optimal solution. The corresponding ILP reads as follows:

$$\text{minimize} \sum_{i=1}^{p} \sum_{j=1}^{q+1} h_{ij} x_{ij} + \sum_{j=1}^{q} h_j y_j \tag{1}$$

$$\text{subject to} \sum_{j=1}^{q+1} x_{ij} = \overline{n}_i, \quad i = 1, \ldots, p, \tag{2}$$

$$\sum_{i=1}^{p} \sum_{k=1}^{j} w_{ik} x_{ik} + \sum_{k=1}^{j} 1 \cdot y_k \geq \sum_{k=1}^{j} w_k, \quad j = 1, \ldots, q, \tag{3}$$

$$x_{ij}, y_j \geq 0 \text{ integer}, \quad i = 2, \ldots, p; j = 1, \ldots, q + 1. \tag{4}$$

Objective function (1) requires the minimization of the overall height of the shelves. Constraints (2) ensure that all the shelves of wide items formed in the enumeration step are packed. Finally, constraints (3) guarantee that there is enough space for the thin items in the shelves, noting that these items can be cut vertically and that thin items of height h_j can be packed either in a shelf of type k or in a shelf of type (i, k) with $k \leq j$. Since the number of variables is constant, this ILP can be solved in polynomial time by the algorithm of [12]. In particular, the number of variables is $O(pq) = O((1/\varepsilon)^{t+1/\varepsilon^{4(t+1)}})$.

Step (d) is analogous to Step (c) in many respects. Shelves of type j or (i, j) are defined exactly in the same way. Note that each self of type j or (i, j) has height at least ε^{f-1}, i.e. at most $\lfloor 1/\varepsilon^{f-1} \rfloor$ such shelves can be packed in a bin. Along with the fact that the number of shelf types is constant, and that the low shelves produced in Step (c) can be cut horizontally in this step, this yields a constant number of bin types, say d, where, for $s = 1, \ldots, d$, we let u_{ijs} and v_{js} be, respectively, the number of shelves of type (i, j) and type j packed in bin of type s. Moreover, we let r_s be the residual vertical space for the low shelves in bin of type s, i.e. $r_s := 1 - \sum_{j=1}^{q} h_j v_{js} - \sum_{i=1}^{p} \sum_{j=1}^{q+1} \overline{h}_{ij} u_{ijs}$. In order to pack all low shelves, we must simply be sure that the overall residual vertical space in the bins is not smaller than the overall height of the low shelves, say H. Therefore, letting z_s denote the number of bins of type s used, the optimal bin packing solution can be found by solving the following ILP:

$$\text{minimize} \sum_{s=1}^{d} z_s \tag{5}$$

$$\text{subject to} \sum_{s=1}^{d} \left(\sum_{j=1}^{q+1} u_{ijs} \right) z_s = \overline{n}_i, \quad i = 0, \ldots, p, \tag{6}$$

$$\sum_{s=1}^{d} \left(\sum_{i=1}^{p} \sum_{k=1}^{j} w_{ik} u_{ijs} + \sum_{k=1}^{j} 1 \cdot v_{ks} \right) z_s \geq \sum_{k=1}^{j} w_k, \quad j = 1, \ldots, q, \tag{7}$$

$$\sum_{s=1}^{d} r_s z_s \geq H, \tag{8}$$

$$z_s \geq 0 \text{ integer}, \quad s = 1, \ldots, d. \tag{9}$$

The overall number of variables is $d = O((pq + p)^{1/\varepsilon^{f-1}}) = O((1/\varepsilon^{1/\varepsilon^{4(t+1)}}(1 + 1/\varepsilon^t))^{1/\varepsilon^{t-1}})$.

Solidification. Clearly, given a solution to ILP (5)-(9), we can pack all the low shelves in the bins since they can be cut horizontally. In particular, we can find a packing in which at most one shelf per bin is cut. These shelves, whose height does not exceed ε^f, are packed in Step (e) in additional bins by *next fit* [4], i.e. they are considered in an arbitrary order and packed in the current bin as long as they fit. When the current shelf does not fit, the current bin is closed and a new bin is considered, ending the procedure when all shelves have been packed. The corresponding running time is $O(n)$.

Finally, in Step (f), we consider thin items in each set A_i ($i \in \mathbb{N}$) that are cut vertically in the solution. Note that there is at most one such item for each shelf in the solution. We consider these items in *decreasing order of height* and pack them in new shelves by a next fit policy, closing the current shelf and starting a new one when the current item does not fit in the shelf. The shelves formed in this way (for all sets A_i) are packed in new bins by next fit. The corresponding algorithm is known as *next fit decreasing height* [5] and runs in $O(n \log n)$ time.

Overall, the running time of APTAS_{2SBP} is polynomial for fixed $\bar{\varepsilon}$, the bottleneck of the algorithm being by far Steps (c) and (d). For future reference, we state this result formally.

Lemma 1. *$APTAS_{2SBP}$ runs in polynomial time for every fixed $\bar{\varepsilon}$.*

3 Proof of Approximation Guarantee

In order to prove that our algorithm is an APTAS we analyze each step in Fig. 2. Note that Steps (b.1) and (d.1) correspond to a *relaxation* of the problem, therefore the optimal solution value cannot increase after performing these steps. Moreover, Step (d.2) finds an optimal solution for the problem, after rounding and melting, hence it cannot be responsible for an increase in the optimal solution value. On the other hand, Steps (a.1), (a.2), (b.2), (b.3) and (b.4) define a new instance for which the solution value is worse than before. Moreover, Step (c) defines the shelves for the items in $\bigcup_{i \in \mathbb{N} \setminus \{0\}} A_i$ so as to guarantee that the overall height of the shelves is minimized – note that this is not necessarily optimal for 2SBP. Finally, Steps (e) and (f) perform an adjustment of the (infeasible) solution, increasing the number of bins.

Generally, in this section, for a generic step we will denote by z the value of the optimal solution before performing the step and by z' the value of the optimal solution afterwards. The approximation guarantee follows from a series of simple lemmas, each showing that $z' \leq (1 + \gamma\varepsilon) + \delta$ for (small) constants γ and δ for some step in the algorithm. We will repeatedly use the following well-known fact for BP (stated here in terms of 2SBP):

Fact 1 *Given a set of shelves whose overall height is H (and each having height at most 1), these shelves are packed in at most $2H + 1$ bins by next fit.*

We analyze the steps that possibly increase the optimal solution value in order of appearance in the algorithm. In the analysis, we will often implicitly assume $\varepsilon \leq 1/2$, which is certainly true for any reasonable value of the required accuracy $\bar{\varepsilon}$.

Lemma 2. *There exists a value $f \in \{1, \ldots, t\}$ such that, for Step (a.1), $z' \leq (1 + 2\varepsilon)z + 1$.*

Proof. Consider an optimal solution for the 2SBP instance before rounding-up heights and let H denote the overall height of the shelves in the solution. Clearly $z \geq H$. We say that a shelf is *rounded-up* if its height increases after Step (a.1). Note that the this may happen only if the height of the tallest item(s) in the shelf is rounded-up, and therefore the sets of rounded-up shelves in the solution are disjoint for distinct values of f. Hence, by an obvious average argument, there exists a value of f such that the overall height of the rounded-up shelves is not larger than $1/t \cdot H$. The increase in the height of these shelves after rounding is by a factor of at most $1/\varepsilon$. Accordingly, a feasible solution after round-up is obtained by removing from their bins the shelves whose height is increased and packing them in at most $2(1/\varepsilon)(1/t)H + 1$ additional bins by next fit, since their overall height is at most $(1/\varepsilon)(1/t)H$. Noting that $(1/\varepsilon)(1/t) \leq \varepsilon$ by the definition of t completes the proof.

Lemma 3. *For Step (a.2), $z' \leq (1 + 4\varepsilon)z + 1$.*

Proof. Recall that Step (a.2) corresponds to imposing that the items in each set A_i be packed in distinct shelves. Consider an optimal solution for the instance before decomposing, letting H denote the overall height of the corresponding shelves and noting that $z \geq H$. Consider a shelf containing item set $S \subseteq N$ and let $P := \{i \in \mathbb{N} : S \cap A_i \neq \emptyset\}$. If $|P| = 1$, this shelf is feasible also after decomposing. Otherwise, letting h_j be the height of the shelf and k the minimum index in P, we may leave the items in $S \cap A_k$ in the shelf, as well as define $|P| - 1$ new shelves, one for each set $S \cap A_i$, $i \in P \setminus \{k\}$. Since the height of each item in A_{k+i+1} is not larger than ε^{1+it} times the height of each item in A_k ($i \in \mathbb{N}$), the overall height of these new shelves does not exceed

$$\sum_{i \in \mathbb{N}:k+i+1 \in P} \varepsilon^{1+it} h_j \leq \sum_{i \in \mathbb{N}} \varepsilon^{1+it} h_j = \varepsilon \sum_{i \in \mathbb{N}} (\varepsilon^t)^i h_j = \frac{\varepsilon}{1 - \varepsilon^t} h_j \leq 2\varepsilon h_j.$$

Therefore, the overall height of the new shelves after performing the above for all shelves in the original solution does not exceed $2\varepsilon H$, i.e. the new shelves can be packed in at most $4\varepsilon H + 1$ additional bins by next fit.

Lemma 4. *For Step (b.2), $z' \leq (1 + 2\varepsilon)z + 1$.*

Proof. Consider an optimal solution for the instance before grouping heights, and let $h_1^i \geq h_2^i \ldots \geq h_{s_i}^i$ be the heights of the shelves with items in A_i in this solution, i.e. the overall height of the shelves in the solution is given by $H := \sum_{i \in \mathbb{N}} H_i$ where $H_i := \sum_{j=1}^{s_i} h_j^i$. Clearly, $z \geq H$. We will use ℓ_i and k_i as defined in the previous section for Step (b.2). Note that $H_i \geq \ell_i \varepsilon^{f+it}$ since at most $1/\varepsilon$ wide items can be in a shelf, and the height of each item is at least ε^{f+it-1}.

Let $\overline{h}_1^i \geq \overline{h}_2^i \ldots \geq \overline{h}_{s_i}^i$ be the heights of the shelves with items in A_i after Step (b.2). We have $\overline{h}_{j+k_i}^i \leq h_j^i$ for $j = 1, \ldots, s_i - k_i$. This means that the vertical space for shelf of height h_j^i in the original solution can be used for the shelf of height $\overline{h}_{j+k_i}^i$ after grouping. The height of the remaining shelves is $\sum_{j=1}^{k_i} \overline{h}_j^i \leq k_i \varepsilon^{f+(i-1)t} \leq \varepsilon H_i$ by the definition of k_i. The overall height of the remaining shelves (considering all sets A_i) is therefore at most $\sum_{i \in \mathbb{N}} \varepsilon H_i = \varepsilon H$, and they can be packed by next fit in at most $2\varepsilon H + 1$ additional bins.

Lemma 5. *For Step (b.3), $z' \leq (1 + 2\varepsilon)z + 1$.*

Proof. Consider an optimal solution for the instance before grouping widths. Letting H_i denote the overall height of the shelves with items in A_i, we have $H_i \geq \ell_i \varepsilon^{f+it}$ as already observed in the proof of Lemma 4. Moreover, let $H := \sum_{i \in \mathbb{N}} H_i$ and note $z \geq H$. Let ℓ_i, g_i, m_{ij}, p_i be defined as in the previous section for Step (b.3), and consider the items in each set D_{ij}, assuming they are sorted according to nondecreasing values of the widths before grouping and letting $\overline{w}_j$ denote the width of item j after grouping. If $m_{ij} > p_i$, we have $\overline{w}_{k+p_i} \leq w_k$ for $k = 1, \ldots, m_{ij} - p_i$. Hence, the horizontal space for item k in the original solution can be used for item $k + p_i$ after grouping. The remaining items in D_{ij} are $1, \ldots, p_i$. Analogously, if $m_{ij} \leq p_i$, we have items $1, \ldots, m_{ij}$ remaining. The height of the remaining items in A_i is therefore at most $g_i p_i \varepsilon^{f+(i-1)t} \leq \varepsilon H_i$ by the definition of p_i. Hence, the overall height of the remaining items (considering all sets A_i) is at most $\sum_{i \in \mathbb{N}} \varepsilon H_i = \varepsilon H$, and these items can be packed by next fit in no more than $2\varepsilon H + 1$ bins (in the worst case, each item goes in a separate shelf).

Lemma 6. *For Step (b.4), $z' \leq (1 + 2\varepsilon)z + 3$.*

Proof. Consider an optimal solution for the instance before grouping heights of the thin items. Letting W_i denote the overall width of the thin items in A_i and H_i the overall height of shelves with items in A_i, we have $H_i \geq W_i \varepsilon^{f+it-1}$ (each shelf can contain thin items for a total width of at most 1). As usual, let $H := \sum_{i \in \mathbb{N}} H_i$ and note $z \geq H$. Considering groups $1, \ldots, q_i$ formed in Step (b.4), the space for the items in group j in the original solution can be used for the items in group $j+1$ after grouping ($j = 1, \ldots, q_i - 1$). As to the items in group 1, their overall width is $\varepsilon^t W_i$. Recalling that thin items can be cut vertically, these items fit in $\lceil \varepsilon^t W_i \rceil \leq \varepsilon^t W_i + 1$ shelves of height at most $\varepsilon^{f+(i-1)t}$ (actually, if $i = 0$ this height is at most 1). The overall height of these shelves (considering all sets A_i) is at most

$$\sum_{i\in\mathbb{N}} (\varepsilon^t W_i + 1)\varepsilon^{f+(i-1)t} = \sum_{i\in\mathbb{N}} W_i \varepsilon^{f+it} + \sum_{i\in\mathbb{N}} \varepsilon^{f+(i-1)t} \leq$$

$$\varepsilon H + 1 + \varepsilon^f \sum_{i\in\mathbb{N}} (\varepsilon^t)^i \leq \varepsilon H + 2,$$

(see also the proof of Lemma 3), where on the left of the first "$\leq$" we wrote ε^{f-t} instead of 1 for convenience. Hence, these shelves can be packed in no more than $2\varepsilon H + 3$ bins by next fit.

Lemma 7. *If the low shelves can be cut horizontally (Step (d.1)) then it is optimal to define these shelves so that their overall height is minimized (Step(c)).*

Proof. Obvious.

Lemma 8. *For Step (e), $z' \leq (1 + 2\varepsilon)z + 1$.*

Proof. Let B be the number of bins in the solution after Step (d). Clearly, one can assume that at most B low shelves are cut horizontally in this solution. Their overall height is at most $\varepsilon^f B$ since the height of each low shelf is at most ε^f. Hence, these shelves are packed in at most $2\varepsilon^f B + 1 \leq 2\varepsilon B + 1$ bins by next fit.

Lemma 9. *For Step (f), $z' \leq (1 + 4\varepsilon)z + 3$.*

Proof. The main argument in the proof is from [5]. Let H_i be the height of the shelves with items in A_i in the solution, letting $h_1^i \leq \ldots \leq h_{s_i}^i$ be the height of these shelves in nonincreasing order. As usual, let $H := \sum_{i\in\mathbb{N}} H_i$ and note $z \geq H$. For each shelf ℓ, at most one thin item j of width $w_j \leq \varepsilon$ and height $h_j \leq h_\ell^i$ is vertically cut. If we pack these items in new shelves in decreasing order of height by next fit, at least $\lfloor 1/\varepsilon \rfloor$ of them will fit in each shelf. Hence, letting $\overline{h}_p^i$ denote the height of the p-th new shelf constructed this way, we have $\overline{h}_p^i \leq h_{p\lfloor 1/\varepsilon \rfloor + 1}^i$ (i.e. in the worst case the first new shelf has height h_1^i, the second height $h_{\lfloor 1/\varepsilon \rfloor + 1}^i$, ...). We have that

$$h_{p\lfloor 1/\varepsilon \rfloor + 1}^i \leq \frac{\varepsilon}{1-\varepsilon} \sum_{\ell=(p-1)\lfloor 1/\varepsilon \rfloor + 1}^{p\lfloor 1/\varepsilon \rfloor} h_\ell^i,$$

since $h_{p\lfloor 1/\varepsilon \rfloor + 1}^i$ is smaller than all the $\lfloor 1/\varepsilon \rfloor$ values h_ℓ^i for $\ell = (p-1)\lfloor 1/\varepsilon \rfloor + 1, \ldots, p\lfloor 1/\varepsilon \rfloor$ and $1/\lfloor 1/\varepsilon \rfloor \leq \varepsilon/(1-\varepsilon)$. Accordingly, the total height of the new shelves is at most

$$h_1^i + \sum_{p\geq 0 : p\lfloor 1/\varepsilon \rfloor + 1 \leq s_i} h_{p\lfloor 1/\varepsilon \rfloor + 1}^i \leq \varepsilon^{f+(i-1)t} + \frac{\varepsilon}{1-\varepsilon} H_i .$$

This implies that the overall height of the new shelves (considering all sets H_i) is at most

$$\sum_{i \in \mathbb{N}} \varepsilon^{f+(i-1)t} + \frac{\varepsilon}{1-\varepsilon} H_i = \frac{\varepsilon}{1-\varepsilon} H + 2 \leq 2\varepsilon H + 2,$$

(see also the proof of Lemma 6), and these shelves are packed in at most $4\varepsilon H + 3$ bins by next fit.

According to Lemmas 2-9, the final solution produced by APTAS$_{2\mathrm{SBP}}$ for an instance I has value at most

$$(1+2\varepsilon)^5 (1+4\varepsilon)^2 \mathrm{opt}(I) + 11 \leq (1+574\varepsilon)\mathrm{opt}(I) + 11 = (1+\bar{\varepsilon})\mathrm{opt}(I) + 11,$$

by the definition of ε (the inequality is tight for $\varepsilon = 1/2$). Hence, the following theorem follows from Lemmas 1-9.

Theorem 1. *APTAS$_{2SBP}$ is an APTAS for 2SBP.*

4   Further Implications: Better Approximation for 2BP?

The first thing to observe is that our algorithm can easily be modified to get an APTAS for 2SSP, the case in which the items must be packed in shelves within a unique bin of infinite height. In this case, it is sufficient to scale all item heights so that they are all $\leq \varepsilon^t$, i.e. $A_0 = \emptyset$ for every possible f, skipping steps (d) and (e). The overall height of the shelves produced, i.e. the value of the solution found, will be near-optimal for at least one value of f. We call the resulting algorithm APTAS$_{2\mathrm{SSP}}$. The proof of the following theorem is analogous to (actually slightly simpler than) the proof of Theorem 1.

Theorem 2. *APTAS$_{2SSP}$ is an APTAS for 2SSP.*

Of course, algorithm APTAS$_{2\mathrm{SBP}}$ is also an approximation algorithm for the classical 2BP without level (i.e. two-stage) restrictions. Since the asymptotic 2.125-approximation algorithm for 2BP described in [3] actually constructs a solution which is feasible for 2SBP, we have that the asymptotic approximation guarantee of our algorithm is not worse than $2.125 + \varepsilon$ for any $\varepsilon > 0$. Actually, we suspect that the situation may be much better, i.e. our method may improve on the asymptotic $2 + \varepsilon$ approximation guarantee of [11] which is the best known so far. Clearly, letting γ be the smallest constant for which there exists another constant δ such that

$$\mathrm{opt}_{2\mathrm{SBP}}(I) \leq \gamma \, \mathrm{opt}_{2\mathrm{BP}}(I) + \delta, \text{ for all } I \in \mathcal{I}_{2\mathrm{BP}},$$

the following is implied by Theorem 1

Corollary 1. *Algorithm APTAS$_{2SBP}$ is a polynomial-time asymptotic $(\gamma + \varepsilon)$-approximation algorithm for 2BP for any $\varepsilon > 0$.*

If the correct value of γ is < 2, we would get a better approximation for 2BP. On the other hand, proving $\gamma < 2$ (if true) may be quite difficult. All we know so far is $1.6910\ldots \leq \gamma \leq 2.125$, where the upper bound of course is from [3]. Considering the counterpart of γ for the strip case, i.e. the smallest constant α for which there exists another constant β such that

$$\mathrm{opt}_{2\mathrm{SSP}}(I) \leq \alpha \,\mathrm{opt}_{2\mathrm{SP}}(I) + \beta, \text{ for all } I \in \mathcal{I}_{2\mathrm{SP}},$$

we can show that α is arbitrarily close to $1.6910\ldots$ This part is deferred to the full paper.

Acknowledgments

This work was partially supported by CNR and MIUR, Italy. We are grateful to Gerhard Woeginger for helpful discussions on the subject.

References

1. B.S. Baker, E.G. Coffman, Jr., and R.L. Rivest. Orthogonal packing in two dimensions. *SIAM Journal on Computing*, 9:846–855, 1980.
2. B.S. Baker and J.S. Schwartz. Shelf algorithms for two-dimensional packing problems. *SIAM Journal on Computing*, 12:508–525, 1983.
3. F.R.K. Chung, M.R. Garey, and D.S. Johnson. On packing two-dimensional bins. *SIAM Journal on Algebraic and Discrete Methods*, 3:66–76, 1982.
4. E.G. Coffman, Jr., M.R. Garey, and D.S. Johnson. Approximation algorithms for bin packing: A survey. In D.S. Hochbaum, editor, *Approximation Algorithms for NP-Hard Problems*. PWS Publishing Company, Boston, 1997.
5. E.G. Coffman, Jr., M.R. Garey, D.S. Johnson, and R.E. Tarjan. Performance bounds for level-oriented two-dimensional packing algorithms. *SIAM Journal on Computing*, 9:801–826, 1980.
6. J.M.V. De Carvalho and A.J.G. Rodrigues. A computer based interactive approach to a two-stage cutting stock problem. *INFOR*, 32:243–252, 1994.
7. J.M.V. De Carvalho and A.J.G. Rodrigues. An LP-based approach to a two-stage cutting stock problem. *European Journal of Operational Research*, 84:580–589, 1995.
8. W. Fernandez de la Vega and G.S. Lueker. Bin packing can be solved within $1 + \epsilon$ in linear time. *Combinatorica*, 1:349–355, 1981.
9. P.C. Gilmore and R.E. Gomory. Multistage cutting problems of two and more dimensions. *Operations Research*, 13:94–119, 1965.
10. N. Karmarkar and R.M. Karp. An efficient approximation scheme for the one-dimensional bin-packing problem. In *Proc. 23rd Annual IEEE Symp. Found. Comput. Sci.*, pages 312–320, 1982.
11. C. Kenyon and E. Rémila. A near-optimal solution to a two-dimensional cutting stock problem. *Mathematics of Operations Research*, 25:645–656, 2000.
12. H.W. Lenstra. Integer programming with a fixed number of variables. *Mathematics of Operations Research* 8:538–548, 1983.
13. S.S. Seiden and R. van Stee. New bounds for multi-dimensional packing. In *Proc. 13th Annual ACM-SIAM Symp. Discr. Alg. (SODA'02)*, 485–495, 2002.
14. F. Vanderbeck. A nested decomposition approach to a 3-stage 2-dimensional cutting stock problem. *Management Science*, 47:864–879, 2001.

On Preemptive Resource Constrained Scheduling: Polynomial-Time Approximation Schemes[*]

Klaus Jansen[1] and Lorant Porkolab[2]

[1] Institut für Informatik und praktische Mathematik,
Christian-Albrechts-Universität zu Kiel, 24 098 Kiel, Germany
kj@informatik.uni-kiel.de

[2] Applied Decision Analysis, PricewaterhouseCoopers London, United Kingdom
lorant.porkolab@uk.pwcglobal.com

Abstract. We study resource constrained scheduling problems where the objective is to compute feasible preemptive schedules minimizing the makespan and using no more resources than what are available. We present approximation schemes along with some inapproximibility results showing how the approximability of the problem changes in terms of the number of resources. The results are based on linear programming formulations (though with exponentially many variables) and some interesting connections between resource constrained scheduling and (multidimensional, multiple-choice, and cardinality constrained) variants of the classical knapsack problem. In order to prove the results we generalize a method by Grigoriadis et al. for the max-min resource sharing problem to the case with weak approximate block solvers (i.e. with only constant, logarithmic, or even worse approximation ratios). Finally we present applications of the above results in fractional graph coloring and multiprocessor task scheduling.

1 Introduction

In this paper we consider the general preemptive resource constrained scheduling problem denoted by $P|res..., pmtn|C_{\max}$: There are given n tasks $\mathcal{T} = \{T_1, \ldots, T_n\}$, m identical machines and s resources such that each task $T_j \in \mathcal{T}$, is processed by one machine requiring p_j units of time and r_{ij} units of resource i, $i = 1, \ldots, s$, from which only c_i units are available at each time. One may assume w.l.o.g. that $r_{ij} \in [0, 1]$ and $c_i \geq 1$. The objective is to compute a preemptive schedule of the tasks minimizing the maximum completion time $C_{\max}$. The three dots in the notation indicate that there are no restrictions on the number of resources s, the largest possible capacity o and resource requirement r values, respectively. If any of these is limited, the corresponding fixed limit replaces the corresponding dot in the notation (e.g. if $s \leq 1$, then $P|res1.., pmtn|C_{\max}$ is used, or if $r_{ij} \leq r$, then $P|res..r, pmtn|C_{\max}$). We will study different variants of the

[*] Supported in part by EU Thematic Network APPOL I + II, Approximation and Online Algorithms, IST-1999-14084 and IST-2001-30012 and by the EU Research Training Network ARACNE, Approximation and Randomized Algorithms in Communication Networks, HPRN-CT-1999-00112.

W.J. Cook and A.S. Schulz (Eds.): IPCO 2002, LNCS 2337, pp. 329–349, 2002.

problem and their applications in multiprocessor task scheduling and fractional graph coloring.

1.1   Related Results

Resource constrained scheduling is one of the classical scheduling problems. Garey and Graham [12] proposed approximation algorithms for the non-preemptive variant $P|res...|C_{\max}$ with approximation ratios $s + 1$ (when the number of machines is unbounded, $m \geq n$) and $\min(\frac{m+1}{2}, s + 2 - \frac{2s+1}{m})$ (when $m \geq 2$). Further results are known for some special cases: Garey et al. [13] proved that when $m \geq n$ and each task T_j has unit-execution time, i.e. $p_j = 1$, the problem (denoted by $P|res..., p_j = 1|C_{\max}$) can be solved by First Fit and First Fit Decreasing heuristics providing asymptotic approximation ratio $s + \frac{7}{10}$ and a ratio between $s + ((s - 1)/s(s + 1))$ and $s + \frac{1}{3}$, respectively. De la Vega and Lueker [6] gave a linear-time algorithm with asymptotic approximation ratio $s + \epsilon$ for each fixed $\epsilon > 0$. Further results and improvements for non-preemptive variant are given in [5,38,39].

For the preemptive variant substantially less results are known: Blazewicz et al. [2] proved that when m is fixed, the problem $Pm|pmtn, res...|C_{\max}$ (with identical machines) and even the variant $Rm|pmtn, res...|C_{\max}$ (with unrelated machines) can be solved in polynomial time. Krause et al. [25] studied $P|pmtn, res1..|C_{\max}$, i.e. where there is only one resource ($s = 1$) and proved that both First Fit and First Fit Decreasing heuristics can guarantee $3 - 3/n$ asymptotic approximation ratio.

A related problem is multiprocessor task scheduling, where a set $\mathcal{T}$ of n tasks has to be executed by m processors such that each processor can execute at most one task at a time and each task must be processed by several processors in parallel. In the parallel (non-malleable) model $P|size_j|C_{\max}$, there is a value $size_j \in M = \{1, ..., m\}$ given for each task T_j indicating that T_j can be processed on any subset of processors of cardinality $size_j$ [1,7,8,18,23]. In the malleable variant $P|fctn|C_{\max}$, each task can be executed on an arbitrary subset of processors, and the execution time $p_j(\ell)$ depends on the number ℓ of processors assigned to it [27,31,40]. Regarding the complexity, it is known [7,8] that the preemptive variant $P|size_j, pmtn|C_{\max}$ is NP-hard. In [20], focusing on computing optimal solutions, we presented an algorithm for solving the problem $P|size_j, pmtn|C_{\max}$ and showed that this algorithm runs in $O(n) + poly(m)$ time, where $poly(.)$ is a univariate polynomial. Furthermore, we extended this algorithm also to malleable tasks with running time polynomial in m and n. These results are based on methods by Grötschel et al. [17] and use the ellipsoid method.

Another related problem is fractional graph coloring, see e.g. [10,16,28,30,32,35,36]. Grötschel et al. [16] proved that the weighted fractional coloring problem is NP-hard for general graphs, but can be solved in polynomial time for perfect graphs. They have proved the following interesting result: For any graph class $\mathcal{G}$, if the problem of computing $\alpha(G, w)$ (the weight of the largest weighted independent set in G) for graphs $G \in \mathcal{G}$ is NP-hard, then the problem of determining the weighted fractional chromatic number $\chi_f(G, w)$ is also NP-hard. This gives a negative result of the weighted fractional coloring problem even for planar cubic graphs. Furthermore, if the weighted independent set problem for graphs in $\mathcal{G}$ is polynomial-time solvable, then the weighted fractional coloring problem for $\mathcal{G}$ can also be solved in polynomial time. The first inapproximability result for the unweighted version of the problem (i.e. where $w_v = 1$ for each vertex $v \in V$) was obtained by Lund and Yannakakis [28] who proved that there exists a $\delta > 0$ such that there is no polynomial-time approximation algorithm for the problem with approximation ratio n^δ, unless P=NP. Feige and Kilian [10] showed that the fractional chromatic number $\chi_f(G)$ cannot be approximated within $\Omega(|V|^{1-\epsilon})$ for any $\epsilon > 0$, unless ZPP=NP. Recently, the authors [19] proved that fractional coloring is NP-hard even for graphs with $\chi_f(G) = 3$ and constant degree 4. Similarly, as it was shown by Gerke and McDiarmid [14], the problem remains NP-hard even for triangle-free graphs. Regarding the approximability of the fractional chromatic number, Matsui [30] gave a polynomial-time 2-approximation algorithm for unit disk graphs.

1.2   New Results

The results presented in this paper are based on linear programming formulations. They are typically of the following form:

$$
\begin{aligned}
\min \ & \sum_{h \in H} x_h \\
\text{s.t.} \ & \sum_{h \in H: s \in h} x_h \geq w_s, \qquad \forall s \in S, \\
& x_h \geq 0, \qquad\qquad \forall h \in H,
\end{aligned}
\tag{1}
$$

where S is a finite set (usually the set of all tasks), and $H \subseteq 2^S$ is a set consisting of subsets of S satisfying some combinatorial property (usually each contains tasks that can be scheduled together at the same time). These linear programs will have – in general – exponentially many variables, but special underlying structures allowing efficient approximations. A linear program of form (1) can be solved (approximately) by using binary search on its optimum and computing at each stage an approximate solution of a special max-min resource sharing problem of the following type:

$$
\begin{aligned}
\lambda^* = \max \ & \lambda \\
\text{s.t.} \ & f_i(x_1, \ldots, x_N) \geq \lambda, \qquad i = 1, \ldots, M, \\
& (x_1, \ldots, x_N) \in P,
\end{aligned}
\tag{2}
$$

where $f_i : P \to \mathbb{R}^+$, $i = 1, \ldots, M$, are – in general – nonnegative concave functions defined on a nonempty convex set $P \subseteq \mathbb{R}^N$. Furthermore, approximate solutions for Problem (2) can be computed by an iterative procedure that requires in each iteration for a given M-vector $(p_1, \ldots, p_M)$ the approximate maximization of $\sum_{i=1}^M p_i f_i(x)$ over all $x = (x_1, \ldots, x_N) \in P$. Interestingly, these subproblems for different variants of resource constrained scheduling turn out to be knapsack type problems (multiple-choice, multi-dimensional, and cardinality constrained knapsack) with efficient approximation algorithms. For fractional graph coloring the subproblem is the well-known maximum weighted independent set problem.

In Sect. 2 we describe the methodology used for solving the max-min resource sharing problem approximately. Let $f(x) = (f_1(x), \ldots, f_M(x))$ and $\Lambda(p) = \max_{x \in P} p^T f(x)$. Based on the paper of Grigoriadis et al. [15], we derive the following result extending some of the previous works on computing approximate solutions for fractional covering problems [15,33,41]: If there exists a polynomial-time approximation algorithm with approximation ratio c for the subproblem, i.e. for finding a vector $\hat{x}(p) \in P$ such that $p^T f(\hat{x}) \geq \frac{1}{c}\Lambda(p))$, then there is also a polynomial time approximation algorithm for the max-min resource sharing problem that computes a feasible solution with objective function value $\frac{1-\epsilon}{c}\lambda^*$. Interestingly, the number of iterations (hence also the number of calls to the solver for the subproblem) is bounded by $O(M(\ln M + \ln c\epsilon^{-3} + \epsilon^{-2}))$, independently of the width [33] of P and the number of variables. If – in particular – there is a (fully) polynomial time approximation scheme ((F)PTAS) for the subproblem, one also gets a (F)PTAS for the original problem and the number of iterations is at most $O(M(\ln M + \epsilon^{-2}))$ [15]. This fact can be particularly useful for models with exponentially many variables.

In Sect. 3 we describe a linear programming approach for the preemptive resource constrained scheduling problem, where there are no assumptions on s or $m \geq n$, they are arbitrary numbers and parts of the input. We show by using the linear programming formulation that there is an approximation algorithm for our scheduling problem with approximation ratio $O(s^{\frac{1}{c_{\min}}})$, where the minimum resource capacity $c_{\min} = \min_i c_i \geq 1$. Furthermore, we argue that if for each resource i, capacity $c_i \geq \frac{12}{\epsilon^2}\log(2s)$, then there is a polynomial time $(1 + \epsilon)$ approximation algorithm.

Then with the aim of obtaining stronger approximation results we study restricted variants, where s is fixed. In particular, we show that for any constant $s \geq 2$, there exists a PTAS computing an ϵ-approximate preemptive schedule satisfying the s resource constraints. In fact this is the best one can expect regarding approximation, since as we show it, this variant even with $s = 2$ cannot have a FPTAS unless P=NP. However, if we assume that $s = 1$ (i.e. the number of resources is pushed to its lower extreme), the problem posses a FPTAS. Next, we apply our approach to the case where there are only a limited number of processors. We give a FPTAS for the variant of the problem with one resource improving the previously known best $(3 - \frac{1}{n})$-approximation algorithm by Krause et al. [25]. The method can be used to obtain a PTAS for a more general variant with fixed number of resources, where the input also contains release and delivery times for each task. In Sect. 5 we study the preemptive multiprocessor task scheduling problem $P|size_j, pmtn|C_{\max}$ and its generalization $P|fctn_j, pmtn|C_{\max}$ to malleable tasks. We show the existence of FPTASs for both problems.

Finally, we apply our linear programming based approach – that was initially introduced for preemptive resource constrained scheduling – to the problem of computing the fractional weighted chromatic number. We prove an approximation analogue of the above mentioned classical result of Grötschel, Lovász and Schrijver [16] on the equivalence between polynomial-time (exact) computations of $\alpha(G, w)$ and $\chi_f(G, w)$: If for a graph class $\mathcal{G}$ there exists a polynomial-time $\frac{1}{c}$-approximation algorithm for computing $\alpha(G, w)$, then there is also a polynomial-time $c(1 + \epsilon)$-approximation algorithm for computing $\chi_f(G, w)$ for graphs in $\mathcal{G}$. By applying this general result for intersection graphs of disks in the plane, we also obtain a PTAS for the fractional coloring problem providing a substantial improvement on Matsui's result [30].

2 Approximate Max-Min Resource Sharing

In this section we will follow the presentation of [15] and use the notation introduced there. Let $f : B \rightarrow \mathbb{R}_+^M$ be a vector with M non-negative, continuous, concave functions f_m, block B a non-empty, convex, compact set and $e^T = (1, \ldots, 1) \in \mathbb{R}_+^M$. Consider the optimization problem

$$(P) \qquad \lambda^* \;=\; \max \left\{ \, \lambda \, : \, f(x) \, \geq \, \lambda e, \; x \in B \, \right\},$$

and assume without loss of generality that $\lambda^* > 0$. Let $\lambda(f) = \min_{1 \leq m \leq M} f_m$ for any given function f. Here we are interested in computing a (c, ϵ)-approximate solution for (P), i.e. for an approximation guarantee $c = c(M) > 1$ and an additional error tolerance $\epsilon \in (0, 1)$ we want to solve the following problem:

$$(P_{c,\epsilon}) \qquad \text{compute } x \in B \text{ such that } f(x) \, \geq \, [\tfrac{1}{c}(1 - \epsilon)\lambda^*]e \ .$$

In order to solve this resource sharing problem we study the subproblem

$$\Lambda(p) = \max \left\{ \, p^T f(x) \, : \, x \in B \, \right\}$$

for $p \in P = \{p \in \mathbb{R}^M \, : \, \sum_{i=1}^M p_i = 1, \; p_i \geq 0\}$. Here we use an approximate block solver (ABS) that solves the following subproblem:

$$\text{ABS}(p, c) \qquad \text{compute } \hat{x} = x(p) \in B \text{ such that } p^T f(\hat{x}) \, \geq \, \tfrac{1}{c}\Lambda(p) \ .$$

By duality we have $\lambda^* \;=\; \max_{x \in B} \min_{p \in P} p^T f(x) \;=\; \min_{p \in P} \max_{x \in B} p^T f(x)$. This implies that $\lambda^* = \min\{\Lambda(p) \, : \, p \in P\}$. Based on this equality, one can naturally define the problem of finding a (c, ϵ)-approximate dual solution:

$$(D_{c,\epsilon}) \qquad \text{compute } p \in P \text{ such that } \Lambda(p) \, \leq \, c(1 + \epsilon)\lambda^* \ .$$

Then the following result holds:

Theorem 1. *If there exists a polynomial time block solver $ABS(p, c)$ for some $c \geq 1$ and any $p \in P$, then there is an approximation algorithm for the resource sharing problem that computes a solution whose objective function value is at least $\frac{1}{c}(1 - \epsilon)\lambda^*$.*

The running time of the approximation algorithm depends only on c, M and $\frac{1}{\epsilon}$. In particular, if there is a (F)PTAS for the block problem computing an $\hat{x} \in B$

such that $p^T f(\hat{x}) \geq (1-\epsilon)\Lambda(p)$ for any constant $\epsilon > 0$, then there is a (F)PTAS for the resource sharing problem [15].

The algorithm uses the logarithmic potential function

$$\Phi_t(\theta, f) = \ln\theta + \frac{t}{M}\sum_{m=1}^{M}\ln(f_m - \theta),$$

where $\theta \in \mathbb{R}$, $f = (f_1, \ldots, f_M)$ are variables associated with the coupling constraints $f_m \geq \lambda$, $1 \leq m \leq M$ and $t > 0$ is a tolerance (that depends on ϵ). For $\theta \in (0, \lambda(f))$, the function Φ_t is well defined. The maximizer $\theta(f)$ of function $\Phi_t(\theta, f)$ is given by the first order optimality condition

$$\frac{t\theta}{M}\sum_{m=1}^{M}\frac{1}{f_m - \theta} = 1 . \tag{3}$$

This has a unique root since $g(\theta) = \frac{t\theta}{M}\sum_{m=1}^{M}\frac{1}{f_m-\theta}$ is a strictly increasing function of θ. The logarithmic dual vector $p = p(f)$ for a fixed f is defined by

$$p_m(f) = \frac{t}{M}\frac{\theta(f)}{f_m - \theta(f)} . \tag{4}$$

By (3), we have $p(f) \in P$. We will also use the following properties [15]:

Proposition 1.

(a) $p(f)^T f = (1+t)\theta(f)$.
(b) $\frac{\lambda(f)}{1+t} \leq \theta(f) \leq \frac{\lambda(f)}{1+t/M}$.

Now define parameter $v = v(x, \hat{x})$ by

$$v(x, \hat{x}) = \frac{p^T\hat{f} - p^T f}{p^T\hat{f} + p^T f}, \tag{5}$$

where $p \in P$, $f = f(x)$, $\hat{f} = f(\hat{x})$ and $\hat{x} \in B$ is an approximate block solution produced by ABS(p, c). The following lemma provides a generalization of a useful result in [15]:

Lemma 1. *Suppose $\epsilon \in (0, 1)$ and $t = \epsilon/5$. For a given $x \in B$, let $p \in P$ computed from (4) and $\hat{x}$ computed by ABS(p, c). If $v(x, \hat{x}) \leq t$, then the pair (x, p) solves $(P_{c,\epsilon})$ and $(D_{c,\epsilon})$, respectively.*

Proof. First rewrite condition $v \leq t$ by using (5): $p^T\hat{f}(1-t) \leq p^T f(1+t)$. Then use that $p^T\hat{f} \geq \frac{1}{c}\Lambda(p)$, $p^T f = (1+t)\theta$ and $\theta(f) < \lambda(f)$ by Proposition 1. This gives

$$\Lambda(p) \leq cp^T\hat{f} \leq c\frac{(1+t)}{(1-t)}p^T f = c\frac{(1+t)^2}{(1-t)}\theta(f) < c\frac{(1+t)^2}{(1-t)}\lambda(f) \leq c(1+\epsilon)\lambda(f).$$

Using $\lambda^* \leq \Lambda(p) \leq c(1+\epsilon)\lambda(f)$, one has $\lambda(f) \geq \frac{1}{c}\frac{1}{1+\epsilon}\lambda^* > \frac{1}{c}(1-\epsilon)\lambda^*$ for any $\epsilon > 0$, which gives $(P_{c,\epsilon})$. Using $\lambda(f) \leq \lambda^*$, one gets $\Lambda(p) \leq c(1+\epsilon)\lambda(f) \leq c(1+\epsilon)\lambda^*$ which is $(D_{c,\epsilon})$. $\square$

The main algorithm works as follows:

Algorithm $Improve(f, B, \epsilon, x)$
(1) set $t := \epsilon/5$; $v := t + 1$;
(2) while $v > t$ **do**
    (2.1) compute $\theta(f)$ and $p \in P$;
    (2.2) set $\hat{x} := \text{ABS}(p, c)$;
    (2.3) compute $v(x, \hat{x})$;
    (2.4) if $v > t$ **then** set $x = (1 - \tau)x + \tau\hat{x}$, where $\tau \in (0, 1)$ is an appropriate
        step length
    end
(3) return(x, p).

The step length can be defined by $\tau = \frac{t\theta v}{2M(p^T \hat{f} + p^T f)}$. Notice that $\theta < p^T \hat{f} + p^T f$ (by Proposition 1), and therefore $\tau \in (0, 1)$. Furthermore, $v > t > 0$ implies that $\tau > 0$. For the initial solution, let $x^0 = \frac{1}{M} \sum_{m=1}^{M} \hat{x}^{(m)}$, where $\hat{x}^{(m)}$ is the solution given by $\text{ABS}(e_m, c)$ obtained for unit vector e_m with all zero coordinates except for its m-th component which is 1. The next lemma provides a bound on $f(x^0)$.

Lemma 2. *For each $p \in P$, $\lambda^* \leq \Lambda(p) \leq cMp^T f(x^0)$. Furthermore, $f_m(x^0) \geq \frac{1}{M}\frac{1}{c}\lambda^*$ for each $m = 1, \ldots, M$.*

Proof. The first inequality follows from duality. For the second inequality,

$$\Lambda(p) = \max\{p^T f(x) : x \in B\} = \max\{\textstyle\sum_{m=1}^{M} p_m f_m(x) : x \in B\}$$
$$\leq \textstyle\sum_{m=1}^{M} p_m \max\{f_m(x) : x \in B\},$$

where $\max\{f_m(x) : x \in B\} = \Lambda(e_m)$. Since $\hat{x}^{(m)}$ is the solution computed by $\text{ABS}(e_m, c)$, $f_m(\hat{x}^{(m)}) \geq \frac{1}{c}\Lambda(e_m)$ implying that $\Lambda(e_m) \leq cf_m(x^{(m)})$. Therefore, $\Lambda(p) \leq c\sum_{m=1}^{M} p_m f_m(\hat{x}^{(m)})$. Using the concavity of f_m we get

$$f_m(\hat{x}^{(m)}) \leq \sum_{\ell=1}^{M} f_m(\hat{x}^{(\ell)}) \leq M f_m \left(1/M \sum_{\ell=1}^{M} \hat{x}^{(\ell)}\right) = M f_m(x^0) .$$

Combining the two inequalities, we obtain

$$\Lambda(p) \leq c \sum_{m=1}^{M} p_m f_m(\hat{x}^{(m)}) \leq cM \sum_{m=1}^{M} p_m f_m(x^0) = cMp^T f(x^0) .$$

Finally, $f_m(x^0) \geq \frac{1}{M} f_m(\hat{x}^{(m)}) \geq \frac{1}{M}\frac{1}{c}\Lambda(e_m) \geq \frac{1}{M}\frac{1}{c}\lambda^*$. $\square$

Let $\phi_t(f) = \Phi_t(\theta(f), f)$, which is called the reduced potential function. The following two lemmas proved in [15] are used here to bound the number of iterations.

Lemma 3. *For any two consecutive iterates x and x' of Algorithm Improve, it holds that $\phi_t(f') - \phi_t(f) \geq tv^2/4M$.*

Lemma 4. *For any two points* $x' \in B$ *and* $x \in B$ *with* $\lambda(f) > 0$, $\phi_t(f') - \phi_t(f) \leq (1+t)\ln\frac{\Lambda(p)}{p^T f}$, *where* p *is the vector defined by (4).*

Theorem 2. *Algorithm Improve solves* $(P_{c,\epsilon})$ *and* $(D_{c,\epsilon})$ *in* $O(\frac{M \ln M}{\epsilon} + \frac{M}{\epsilon^2} + \frac{M \ln c}{\epsilon^3})$ *iterations.*

Proof. Let N_0 be the number of iterations to reach an iterate x^1 with corresponding error $v \leq 1/2$ starting from our initial solution x^0. For all iterations with $v \geq 1/2$, each iteration increases the potential by at least $tv^2/4M \geq t/16M$ (see Lemma 3). By Lemma 4, the total increase is bounded by $\phi_t(f^1) - \phi_t(f^0) \leq (1+t)\ln\frac{\Lambda(p^0)}{p^{0T}f^0}$. Since $t = \epsilon/5$ and $\Lambda(p^0) \leq cMp^{0T}f^0$ (by Lemma 2), we obtain that

$$N_0 \leq \frac{(1+\epsilon/5)16M\ln(cM)}{\epsilon/5} = O\left(\frac{M\ln(cM)}{\epsilon}\right).$$

Now suppose that the error $v_\ell \leq 1/2^\ell$ for iterate $x^\ell \in B$, and let N_ℓ be the number of iterations to halve this error. We get $\phi_t(f^{\ell+1}) - \phi_t(f^\ell) \geq \frac{N_\ell t v_{\ell+1}^2}{4M} = \frac{N_\ell t v_\ell^2}{16M}$. On the other hand, the definition of v_ℓ implies $p^{\ell T}\hat{f}^\ell(1 - v_\ell) = p^{\ell T}f^\ell(1 + v_\ell)$. Using ABS$(p^\ell, c)$, we get a solution $\hat{x}^\ell$ with $p^{\ell T}\hat{f}^\ell \geq \frac{1}{c}\Lambda(p^\ell)$. Combining the two inequalities,

$$\frac{\Lambda(p^\ell)}{p^{\ell T}f^\ell} \leq \frac{c(1+v_\ell)}{(1-v_\ell)} \leq c(1 + 4v_\ell).$$

The last inequality holds since $v_\ell \leq 1/2$. Since $d \geq 0$, Lemma 4 implies that

$$\phi_t(f^{\ell+1}) - \phi_t(f^\ell) \leq (1+t)(\ln c + \ln(1 + 4v_\ell)) \leq (1+t)(\ln c + 4v_\ell).$$

This gives now an upper bound

$$N_\ell \leq \frac{16M(1+t)(\ln c + 4v_\ell)}{tv_\ell^2} = O\left(\frac{M(\ln c + v_\ell)}{\epsilon v_\ell^2}\right).$$

One gets the total number of iterations by summing N_ℓ over all $\ell = 0, 1, \ldots, \lceil\ln(\frac{1}{t})\rceil$. Therefore, the total number of iterations is bounded by

$$N_0 + O\left(\frac{M\ln c}{\epsilon}\sum_{\ell=1}^{\lceil\ln(\frac{1}{t})\rceil}2^{2\ell} + \frac{M}{\epsilon}\sum_{\ell=1}^{\lceil\ln(\frac{1}{t})\rceil}2^\ell\right) \leq O\left(\frac{M\ln(cM)}{\epsilon} + \frac{M\ln c}{\epsilon^3} + \frac{M}{\epsilon^2}\right).$$

$\square$

The total number of iterations can be improved by the scaling method used in [33,15]. The idea is to reduce the parameter t step by step to the desired accuracy. In the s-th scaling phase we set $\epsilon_s = \epsilon_{s-1}/2$ and $t_s = \epsilon_s/5$ and use the current approximate point x^{s-1} as its initial solution. For phase $s = 0$, use the initial point $x^0 \in B$. For this point we have $p^T f(x^0) \geq \frac{1}{cM}\Lambda(p)$. We set $\epsilon_0 = (1 - 1/M)$. Using Lemma 1, $f_m(x^0) \geq \frac{1}{M}\frac{1}{c}\lambda^*$. This implies $f_m(x^0) \geq \frac{1}{cM}\lambda^* = \frac{1}{c}(1 - 1 + \frac{1}{M})\lambda^* = \frac{1}{c}(1 - \epsilon_0)\lambda^*$, for each $m = 1, \ldots, M$.

Theorem 3. *For any accuracy $\epsilon > 0$, the error scaling implementation computes solutions x and p of $(P_{c,\epsilon})$ and $(D_{c,\epsilon})$, respectively, in*

$$N \;=\; O(M \ln M + M \ln c/\epsilon^3 + M/\epsilon^2)$$

iterations.

Proof. To reach the first $\epsilon_0 \in (1/2, 1)$ in the primal and dual problem we need $O(M(\ln c + \ln M))$ iterations (by Theorem 2). Let N_s be the number of iterations in phase s to reach ϵ_s for $s \geq 1$. By Lemma 3, each iteration of phase s increases the potential function by at least $t_s^3/4M = \theta(\epsilon_s^3/M)$. Lemma 4 implies that for $x = x^s$ and $x' = x^{s+1}$,

$$\phi_{t_s}(f^{s+1}) - \phi_{t_s}(f^s) \;\leq\; (1 + t_s) \ln \frac{\Lambda(p^s)}{p^{sT} f^s} \;.$$

Note that x^s is a $\epsilon_{s-1} = 2\epsilon_s$ solution of $(P_{c,\epsilon_{s-1}})$, and therefore $f(x^s) \geq (1 - 2\epsilon_s)\frac{1}{c}\lambda^* e$. Furthermore, since $\Lambda(p^s) \leq c(1 + 2\epsilon_s)\lambda^*$, $\Lambda(p^s) \leq c^2 \frac{1+2\epsilon_s}{1-2\epsilon_s}\lambda(f^s) \leq c^2 \frac{1+2\epsilon_s}{1-2\epsilon_s} p^{sT} f^s$, implying that $\frac{\Lambda(p^s)}{p^{sT} f^s} \leq c^2(1 + 8\epsilon_s)$. Then one can bound N_s by $O(M(\ln c + \epsilon_s)/\epsilon_s^3)$, and as before, the overall number of iterations is bounded by

$$N_0 + \sum_{s \geq 1} N_s \;\leq\; O\left(M \ln M + \frac{M \ln c}{\epsilon^3} + \frac{M}{\epsilon^2} \right) \;.$$

$\square$

Remark: The root $\theta(f)$ can often be computed only approximately, but an accuracy of $O(\epsilon^2/M)$ for $\theta(f)$ is sufficient such that the iteration bounds remain valid. With this required accuracy, the number of evaluations of the sum $\sum_{m=1}^{M} \frac{1}{f_m - \theta}$ is bounded by $O(\ln(M/\epsilon))$. This gives $O(M(\ln M/\epsilon))$ arithmetic operations to determine $\theta(f)$ approximately. The overhead can be further improved by using the Newton's method to $O(M(\ln \ln(M/\epsilon)))$ [15].

3   General Linear Programming Approach

In this section we study the preemptive resource constrained scheduling problem. First we consider the case with unlimited number of machines $m \geq n$. In fact, if $m \leq n$, the machines can be handled as the $(s+1)$st resource with requirement $r_{s+1,j} = 1$ and capacity $c_{s+1} = m$. For our scheduling problem, a *configuration* is a compatible (or feasible) subset of tasks that can be scheduled simultaneously. Let F be the set of all configurations, and for every $f \in F$, let x_f denote the length (in time) of configuration f in the schedule. Clearly, $f \in F$ iff $\sum_{j \in f} r_{ij} \leq c_i$, for $i = 1, \ldots, s$.

By using these variables, the problem of finding a preemptive schedule of the tasks with smallest makespan value (subject to the resource constraints) can be formulated as the following linear program [20]:

$$\begin{aligned}
\min \; &\sum_{f \in F} x_f \\
\text{s.t.} \; &\sum_{f \in F: j \in f} x_f \geq p_j, \quad j = 1, \ldots, n, \\
&x_f \geq 0, \quad \forall f \in F \;.
\end{aligned} \tag{6}$$

One can solve (6) by using binary search on the optimum value and testing at each stage the feasibility of the following linear system for a given $r \in [p_{max}, np_{max}]$:

$$\sum_{f \in F : j \in f} x_f \geq p_j, \quad j = 1, \ldots, n, \quad (x_f)_{f \in F} \in P,$$

where

$$P = \{ (x_f)_{f \in F} : \sum_{f \in F} x_f = r, \quad x_f \geq 0, \ f \in F \} .$$

This can be done approximately (hence leading to an approximate decision procedure) by computing an approximate solution for the following max-min resource sharing problem:

$$\lambda^* = \max \left\{ \lambda : \sum_{f \in F : j \in f} \frac{1}{p_j} \cdot x_f \geq \lambda, \ j = 1, \ldots, n, \ (x_f)_{f \in F} \in P \right\} . \quad (7)$$

The latter problem can also be viewed as a fractional covering problem with one block P, and n coupling constraints. Let the coupling (covering) constraints be represented by $Ax \geq \lambda$. By using the approach presented in Sect. 2, problem (7) can be solved approximately in $O(n(\delta^{-2} + \delta^{-3} \ln c + \ln n))$ iterations (coordination steps), each requiring for a given n-vector $y = (y_1, \ldots, y_n)$ a $\frac{1}{c}$-approximate solution of the problem,

$$\Lambda(y) = \max\{ y^T Ax : x \in P \} . \quad (8)$$

Since P in (8) is just a simplex, the optimum of this linear program is also attained at a vertex $\tilde{x}$ of P corresponding to a (single) configuration $\tilde{f}$. A similar argument was used for the bin packing problem by Plotkin et al. [33]. At this vertex $\tilde{x}_{\tilde{f}} = r$ and $\tilde{x}_f = 0$ for $f \neq \tilde{f}$. Therefore, it suffices to find a subset $\tilde{f}$ of tasks that can be executed in parallel and has the largest associated profit value $c_{\tilde{f}}$ in the profit vector $c^T = y^T A$. But for given multipliers $y_1, \ldots, y_n$, this problem can also be formulated as follows,

$$\max \left\{ \sum_{j \in f} \frac{y_j}{p_j} : f \in F \right\} ,$$

or equivalently as a general s-dimensional Knapsack Problem (sD-KP) or Packing Integer Program (PIP),

$$\max \sum_{j=1}^{n} \frac{y_j}{p_j} x_j$$
$$\text{s.t.} \sum_{j=1}^{n} r_{ij} x_j \leq c_i, \quad i = 1, \ldots, s, \quad (9)$$
$$x_j \in \{0, 1\}, \quad j = 1, \ldots, n .$$

Let $K(n, s, c)$ denote the time required (in the worst case) to compute a $\frac{1}{c}$-approximate solution for (9). At each iteration, in addition to solving (9) (approximately), we also need to compute the new y vector based on Ax for the current x. Though the dimension of x is exponential, the computation requires only updating the previous Ax value, since the current x is $(1 - \tau)x + \tau\hat{x}$ (for an appropriate step length $\tau \in (0, 1]$), where $\hat{x}$ is the vertex of P corresponding to the solution of (9) at the current iteration. Thus the number of non-zero components of x can increase by at most one at each iteration, and each update of Ax takes $O(n)$ operations.

Initially, x^0 has at most n non-zero components obtained from solving n sub-problems (one for each n-dimensional unit vector as y) requiring $O(nK(n, s, c))$ time, and computing the initial y^0 in $O(n^2)$ time. Approximating the root and determining the next price vector p can be done in $O(n \ln \ln \frac{n}{\delta}) = O(n^2)$ time (for e.g. $\delta \geq 1/n$). Later each update of Ax^k can be done in $O(n)$ time. For any fixed r, the algorithm requires $O(n(\delta^{-2} + \delta^{-3} \ln c + \ln n)(K(n, s, c) + n \ln \ln(n\delta^{-1}))$ time.

By binary search on r one can obtain a solution $(x_f)_{f \in F}$ with $\sum_{f \in F} x_f = (1 + \epsilon/4)r^*$ and $\sum_{f \in F: j \in f} x_f \geq \frac{1}{c}(1 - \delta)p_j$, where r^* is the length of an optimal schedule. Now one can define $\tilde{x}_f = x_f c(1 + 4\delta)$ and obtain $\sum_{f \in F: j \in f} \tilde{x}_f \geq (1 - \delta)(1 + 4\delta)p_j = (1 + 3\delta - 4\delta^2)p_j \geq p_j$ for $\delta \leq 3/4$. In this case the length of the generated schedule is at most $cr^*(1 + 4\delta)(1 + \epsilon/4) = cr^*(1 + 4\delta + \epsilon/4 + \delta\epsilon) \leq cr^*(1 + \epsilon)$ by choosing $\epsilon \leq 1$ and $\delta \leq 3\epsilon/20$. Since the optimum of (6) lies within interval $[p_{max}, np_{max}]$, the overall complexity of the algorithm can be bounded by $O(n \ln \frac{n}{\epsilon}(\epsilon^{-2} + \epsilon^{-3} \ln c + \ln n)(K(n, s, c) + n \ln \ln(n\epsilon^{-1})))$ time. For $K(n, s, c) \geq O(n \ln \ln \frac{n}{\epsilon})$ we obtain $O(n \ln(n\epsilon^{-1})(\epsilon^{-2} + \epsilon^{-3} \ln c + \ln n)K(n, s, c))$ time.

The number of iterations can be improved by computing an approximate non-preemptive schedule with a greedy algorithm. The main idea is to use a modified list scheduling algorithm. The classical list scheduling algorithm is defined as follows. First consider the tasks in any fixed order $L = (T_{i_1}, \ldots, T_{i_n})$. At any time if there are positive quantities available from all resources, the algorithm scans L from the beginning and selects the first task T_k (if there is any) which may validly be executed and which has not been already (or is not currently) executed. If a task is finished, it will be removed from the list. Garey and Graham [12] showed that this list scheduling algorithm for non-preemptive tasks gives a $(s + 1)$-approximation ratio (comparing the length of the produced schedule and the optimum non-preemptive schedule). To compare with the optimal preemptive makespan C^*_{max}, we allow to overpack the resources with one task at each time. Let $C_{max}(H)$ be the length of this (infeasible) pseudo-schedule and consider a task T_k that is finished at time $C_{max}(H)$. Then for each time $t \in [0, C_{max}(H) - p_k)$ at least one resource is completely used by the tasks. Let $l(i)$ be the total length of intervals where resource i is overpacked. Clearly, we have $l(i) \leq C^*_{max}$ and $p_k \leq p_{max} \leq C^*_{max}$. The length of the pseudo schedule is at most $\sum_{i=1}^{s} l(i) + p_k \leq (s + 1)C^*_{max}$. By replacing the overpacked tasks to the end we obtain a feasible schedule of length $C^{(MLS)}_{max} \leq (2s + 1)C^*_{max}$. This implies that $C^*_{max} \leq C^{(MLS)}_{max} \leq (2s + 1)C^*_{max}$, i.e. $1/(2s + 1)C^{(MLS)}_{max} \leq C^*_{max} \leq C^{(MLS)}_{max}$. Hence the binary search for the optimum of (6) requires only $O(\ln \frac{s}{\epsilon})$ steps (instead of $O(\ln \frac{n}{\epsilon})$) improving the previous running time to

$$O\left(n\left(K(n,s,c)+n\ln\ln\frac{n}{\epsilon}\right)\min(\ln(s\epsilon^{-1}),\ln(n\epsilon^{-1}))(\epsilon^{-2}+\epsilon^{-3}\ln c+\ln n)\right)\ .$$

If the block problem posses an approximation scheme, then the factor $\epsilon^{-3}\ln c$ can be removed. As main result we obtain:

Theorem 4. *Let $\mathcal{I}$ be a set of instances of the preemptive resource constrained scheduling problem. If there is a polynomial time approximation algorithm for the corresponding s - dimensional knapsack instance with ratio c, then for any $\epsilon > 0$ there is a polynomial time algorithm for preemptive resource constrained scheduling restricted to $\mathcal{I}$ with approximation ratio $c(1+\epsilon)$.*

The number of configurations in the final solution can be reduced from $O(n(\epsilon^{-2}+\epsilon^{-3}\ln c+\ln n))$ to $O(n)$ within $O(n(\epsilon^{-2}+\epsilon^{-3}\ln c+\ln n)\mathcal{M}(n))$ time where $\mathcal{M}(n)$ is the time to invert a $(n\times n)$ matrix. The maximum number of tasks per configuration is bounded by $t=\min(n,\min_{i=1}^{s}\frac{c_i}{min_j r_{ij}}))$. Therefore, the number of preemptions can be bounded by $O(nt)$.

4   Approximability as a Function of the Number of Resources

As we have seen in the previous section, the s-dimensional Knapsack Problem is a key subproblem in our approach whose solution (for various inputs) is required repeatedly. It is well known that the approximability of this problem varies with the dimension s. Therefore in this section we will specialize the above general result by making different assumptions on s and using different approximation algorithms for the sD-KP. In particular, we will obtain a sequence of approximation results for our scheduling problem where the approximation will improve (constant, PTAS, then FPTAS) as move from arbitrary to fixed number of resources and eventually to the case with a single resource. To contrast these approximation algorithms, we will also present some inapproximability results for the first two variants.

4.1   Arbitrary Number of Resources

In this section we consider the case when s is arbitrary, i.e. it is part of the input. First we give the presentation of our approximation algorithms, then we briefly discuss some simple inapproximability results.

Approximation Algorithms. It is known [34,37] that for general s, sD-KP or equivalently PIP has a $\Omega(1/s^{1/c_{min}})$ approximation algorithm when all $r_{ij}\in[0,1]$, $c_{min}=\min_i c_i\geq 1$ and $\frac{y_j}{p_j}\geq 0$. This implies the following result:

Theorem 5. *For any number s of resources, there is a polynomial-time approximation algorithm with performance ratio $O(s^{\frac{1}{c_{min}}})$ for the preemptive resource constrained scheduling problem.*

This result can be further improved by using the algorithm by Srinivasan [37]. Furthermore, Srivastav and Stangier [38,39] showed that if $c_{min}\geq\frac{16}{\epsilon^2}\log(2s)$

and $OPT \geq 12/\epsilon^2$ (where OPT is the optimum value of the linear relaxation of (9)), an ϵ-approximate solution for sD-KP can be computed in polynomial-time. The running time of the algorithm is bounded by $O(K_r(n, s, 1) + sn^2 \ln(sn))$, where $K_r(n, s, 1)$ is the time required to solve (exactly) the linear programming relaxation of (9). Combining these with our approach presented in the previous section and extending the algorithm to arbitrary OPT, we obtain the following.

Theorem 6. *For any number s of resources, if $c_i \geq \frac{12}{\epsilon^2} \log(2s)$ and $r_{ij} \in [0, 1]$, for each i and j, there is a polynomial time approximation algorithm that computes a $(1 + \epsilon)$-approximate solution for the preemptive resource constrained scheduling problem.*

The last two results can also be generalized to $P|res..., pmtn|C_{\max}$ with limited number of machines, i.e. when $m \leq n$. Note that Theorems 5 and 6 hold only under some special conditions on resource capacities and requirements. Therefore it is natural to ask whether they can be eliminated at least when s is fixed. After presenting some inapproximability results, we will show in Sect. 4.2 that if s is fixed, though the problem remains NP-hard, approximating its optimum becomes much easier. In particular we prove that for any fixed s the general problem has a PTAS.

Inapproximability. For any graph $G = (V, E)$ one can construct a resource constrained scheduling problem with $n = |V|$ tasks and $s = |E|$ resources, where vertices correspond to tasks and edges to resources in the following way: The resource capacities are all 1, i.e. $c_e = 1$ for each $e \in E$, while resource requirement $r_{ev} = 1$, if $v \in e$, and 0 otherwise. Independent sets of vertices correspond to sets of tasks that can be executed together at the same time, therefore the (fractional) coloring problem for graphs can be viewed as a special case of (preemptive) resource constrained scheduling. Hence the inapproximability results in [28,10] imply the following:

Theorem 7. *For any $\delta > 0$, the preemptive resource constrained scheduling problem with n tasks and s resources has no polynomial-time approximation algorithm with approximation ratio $n^{1-\delta}$, neither for some $\delta > 0$, unless $P = NP$; nor for any $\delta > 0$, unless $ZPP = NP$.*

Note that this negative result holds even for the restricted case when each processing time is of unit length, and all capacities and resource requirements are either 0 or 1. This shows that for arbitrary s the problem is not only hard, but even approximating its optimum is difficult. Using that $s \leq n^2$ in the special case above we get:

Corollary 1. *The preemptive resource constrained scheduling problem with n tasks and s resources has no polynomial-time approximation algorithm with approximation ratio $s^{1/2-\delta}$, neither for some $\delta > 0$, unless $P = NP$; nor for any $\delta > 0$, unless $ZPP = NP$.*

4.2   Fixed Number of Resources – PTAS

In this section we study how the approximability of the problem changes under a restricting assumption on the number of resources. We consider here the case, when $s \geq 1$ is a fixed constant larger than one. As we argue below this restriction allows us to prove substantially better approximation result than the one above. Namely, we will show that under the discussed assumption, the problem posses a PTAS, and then we will also prove that - in fact - this is the best one can expect (unless P=NP).

Approximation Algorithms. It is known [29,11] that for any fixed s, sD-KP has a PTAS. Let $K(n, s, \delta)$ denote the time required (in the worst case) to compute a δ-approximate solution for the sD-KP. Using that s is constant, the running time of our scheduling algorithm is bounded by $O((K(n, s, c) + n \ln \ln(n\epsilon^{-1}))n \ln(\epsilon^{-1})(\epsilon^{-2} + \ln n))$. The currently known best bound [3] for $K(n, s, \Theta(\epsilon))$ is $O(n^{\lfloor \frac{s}{\epsilon} \rfloor - s}) = n^{O(\frac{s}{\epsilon})}$. By using this bound and the above argument, we obtain the following result.

Theorem 8. *For any fixed number s of resources, there is a PTAS for the preemptive resource constrained scheduling problem with running time $n^{O(\frac{s}{\epsilon})}$.*

Notice that for fixed s, there is also a $O(n)$ time $(s + 1)$-approximation algorithm for the sD-KP [3], which implies the following:

Corollary 2. *There is a $(s+1)(1+\epsilon)$-approximation algorithm for the preemptive resource constrained scheduling problem with running time $O((n^2 \ln \ln(n\epsilon^{-1})) \ln(\epsilon^{-1})(\epsilon^{-2} + \ln n))$.*

Inapproximability. The running time of the previously described algorithm depends exponentially on the accuracy, and as the next result shows this dependence cannot be improved to polynomial, unless P=NP.

Theorem 9. *For any $s \geq 2$, there is no FPTAS for the preemptive resource constrained scheduling problem with s resources, unless $P = NP$.*

Proof. We use a reduction from the NP-complete problem Partition: Given a set A and a size $s(a) \in \mathbb{N}$ for each $a \in A$, where $n = |A|$ is assumed to be even, decide whether there is a subset I of A such that $|I| = n/2$ and $\sum_{a \in I} s(a) = \frac{1}{2} \sum_{a \in A} s(a)$.

Let $s_{\max} = \max_{a \in A} s(a)$. Now construct n tasks and two resources with capacities $\frac{1}{2} \sum_{a \in A} s(a)$ and $\frac{1}{2} \sum_{a \in A}(s_{\max} - s(a))$, where each task $a \in A$ requires $(s(a), s_{\max} - s(a))$ of the two resources and has processing time $p_a = 1$. If there is a solution I of the partition problem, then $|I| = n/2$, $\sum_{a \in I} s(a) = \frac{1}{2} \sum_{a \in A} s(a)$ and $\sum_{a \in I}(s_{\max} - s(a)) = \frac{1}{2} \sum_{a \in A}(s_{\max} - s(a))$. This means that set I can be executed in parallel on both resources. Furthermore, the set $A \backslash I$ is also a solution for the partition problem and can be executed also parallel on both resources. Therefore, one can schedule all tasks in two phases in a non-preemptive way:

in one phase all tasks in I (of length 1), and in the other phase all tasks in $A \setminus I$ (also of length 1). This gives a schedule with makespan $C_{\max} = 2$ and the minimum makespan is $C^*_{\max} = 2$ (by using an argument based on the required minimum area for all tasks). If there is no solution of the partition problem, then we can still split the set in three parts $I_1, \ldots, I_3$ according to resource 1 such that $\sum_{a \in I_j} s(a) \leq \frac{1}{2} \sum_{a \in A} s(a)$. Now only one of these parts can have $\sum_{a \in I_j} (s_{\max} - s(a)) > \frac{1}{2} \sum_{a \in A} (s_{\max} - s(a))$. By splitting this set (according to resource 2) in three parts, we obtain a feasible non-preemptive schedule of length $C_{\max} \leq 5$. This implies that $C^*_{\max} \leq 5$.

Assume now that there is a FPTAS for the preemptive 2-resource constrained scheduling problem and then show that this leads to a contradiction. The FPTAS gives for each $\epsilon > 0$ a $poly(n, 1/\epsilon)$ time algorithm to obtain a schedule with length $\leq C^*_{\max}(1 + \epsilon) \leq C^*_{\max} + 5\epsilon$. If we choose $\epsilon = 1/5n$ then we obtain in $poly(n)$ time a preemptive schedule with length $\leq C^*_{\max} + 1/n$. The length of the preemptive schedule (given by the FPTAS with $\epsilon = 1/5n$) is larger than $2 + 1/n$ if and only if the partition problem has no solution. If the length of the schedule is larger than $2 + 1/n$, then $C^*_{\max} > 2$ implying that we have a no-instance of the partition problem. Consider the other direction: For each time step t there are at least $n/2 + 1$ tasks which are not executed at step t; otherwise we had a solution of the partition problem. To see this consider a set I of $n/2$ tasks executed at one time step. This implies that $\sum_{a \in I} s(a) \leq 1/2 \sum_{a \in A} s(a)$ and $\sum_{a \in I} (s_{\max} - s(a)) \leq \frac{1}{2} \sum_{a \in A} (s_{\max} - s(a))$. Both of these inequalities can be transformed into:

$$(n/2)s_{\max} - \sum_{a \in I} s(a) \leq \sum_{a \in I} (s_{\max} - s(a))$$
$$\leq \tfrac{1}{2} \sum_{a \in A} (s_{\max} - s(a)) = (n/2)s_{\max} - \sum_{a \in I} s(a) \ .$$

Then, $\sum_{a \in I} s(a) = \frac{1}{2} \sum_{a \in A} s(a)$ and $\sum_{a \in I} (s_{\max} - s(a)) = \frac{1}{2} \sum_{a \in A} (s_{\max} - s(a))$. Therefore, I is a solution of the partition problem.

Let $ne(i)$ be the total length in interval $[0, 2]$ where task T_i is not executed. Using the property above, $\sum_{j=1}^{n} ne(i) \geq 2(n/2 + 1) = n + 2$. This implies that at least one task T_k has $ne(k) \geq 1 + 2/n$. Therefore, this task is executed at most $1 - 2/n$ in interval $[0, 2]$ and the schedule length is at least $2 + 2/n$. This argument implies that we can test (using the FPTAS) the existence of a solution for the partition problem in polynomial time, which is impossible, unless P=NP. $\square$

In the proof above we have used an idea of Korte and Schrader [24]. They proved that there is no FPTAS for the sD-KP with $s = 2$, unless $P = NP$. Since it was essential in the proof of Theorem 9 that s is (at least) two, it is natural to ask again, what happens when $s = 1$. In this case - as it will be demonstrated in the next section - there is a FPTAS for the problem, and hence the negative result of Theorem 9 does not hold any longer.

4.3   Single Resource – FPTAS

Clearly, the general approach presented in Sect. 4.2 can also be used for the special case when there is only one resource. Note that the number of iterations in computing an approximate solution for (6) is independent of s, so it remains $O(n(\delta^{-2} + \ln n))$, as above. The only difference is that the subproblem one has to solve (approximately) at each iteration becomes the classical (1-dimensional) Knapsack Problem (instead of the s-dimensional variant). This can be solved approximately with any $\Theta(\delta)$ accuracy in $O(n \min(\ln n, \ln(1/\delta)) + 1/\delta^2 \min(n, 1/\delta \ln(1/\delta))) = O(n\delta^{-2})$ time [22]. In addition, we have to count the overhead of $O(n \ln \ln(n\epsilon^{-1}))$ operations in each iteration (i.e. the computation of the root and the new price vector). Hence the previous bound can be substituted for $K(n, s, \Theta(\delta))$ in the analysis above in Sect. 4.2, and therefore for any fixed r the procedure requires $O(n^2 \max(\delta^{-2}, \ln \ln(n\delta^{-1}))(\delta^{-2} + \ln n))$ time (including also the overheads arising from computing the initial solution).

Similarly to the discussion in Sect. 4.2, one can use binary search on r to find a good approximation for the optimum of (6). The initial interval for r can be determined by a strip packing algorithm (called longest task first) [4,40] that computes a non-preemptive schedule of length at most three times the length of the an optimal preemptive schedule. These all imply the following:

Theorem 10. *If there is only one resource, the resource constrained scheduling problem has a FPTAS which runs in $O(n^2 \ln(\epsilon^{-1}) \max(\epsilon^{-2}, \ln \ln(n\epsilon^{-1}))(\epsilon^{-2} + \ln n))$ time.*

So far we have assumed that m is sufficiently large (e.g. $m \geq n$), or otherwise processors can be treated as an extra resource. But having seen above the the dividing line (regarding approximability) between instances with 1 and 2 resources, one may naturally ask how easy or difficult it is to compute approximate solutions for the problem when there is one resource and a limited number of machines. Krause, Shen and Schwetman [25] gave a polynomial-time $(3 - \frac{1}{n})$-approximation algorithm for the problem. This can be substantially improved by following our approach and extending Theorem 10 to this variant. First formulate it as a restricted preemptive 2-resource constrained scheduling problem, where the m identical machines correspond to the 2nd resource with $r_{2j} = 1$ for each task j and capacity $c_2 = m$. It is easy to check, that the subproblem in this case is the cardinality constrained $(\sum_{j=1}^{n} x_j \leq m)$ knapsack problem, which has a FPTAS with running time $O(nm^2\epsilon^{-1})$ [3]. In addition, the initial interval for the binary search on r can be bounded as for $s = 2$ resources. Hence the following holds:

Theorem 11. *There is a FPTAS for $P|res1..,pmtn|C_{\max}$ with running time $O(n^2 \ln(\epsilon^{-1}) \max(m^2\epsilon^{-1}, \ln \ln(n\epsilon^{-1})(\epsilon^{-2} + \ln n)))$.*

This result can also be extended to the variant $P|res\,s..,r_j,pmtn|L_{\max}$, where the input contains release r_j and delivery q_j dates for each task T_j, and the objective is to find a schedule minimizing the maximum delivery completion time $L_{\max} = \max_j C_j + q_j$.

Theorem 12. *There is a FPTAS for $P|res\,1..,r_j,pmtn|L_{\max}$ that runs in time $poly(n, 1/\epsilon)$.*

5   Multiprocessor Task Scheduling

In this section we address preemptive multiprocessor task scheduling problems, where a set $\mathcal{T} = \{T_1, \ldots, T_n\}$ of n tasks has to be executed by m processors such that each processor can execute at most one task at a time and a task must be processed simultaneously by several processors.

Since we consider here the preemptive model, each task can be interrupted any time at no cost and restarted later possibly on a different set of processors. We will focus on those preemptive schedules where migration is allowed, that is where each task may be assigned to different processor sets during different execution phases [1,7,8]. The *malleable* variant of multiprocessor task scheduling, $P|fctn_j, pmtn|C_{\max}$ can be formulated as the following linear program [20], where M_j denotes the set of different cardinalities that processor sets executing task T_j can have.

$$
\begin{aligned}
\min \ & \textstyle\sum_{f \in F} x_f \\
\text{s.t.} \ & \textstyle\sum_{\ell \in M_j} \frac{1}{p_j(\ell)} \sum_{f \in F : |f^{-1}(j)| = \ell} x_f \ \geq \ 1, \qquad j = 1, \ldots, n, \\
& x_f \ \geq \ 0, \qquad \forall f \in F .
\end{aligned}
\tag{10}
$$

Here the goal is to find for a given r a vector $x \in P = \{(x_f) \ : \ \sum_{f \in F} x_f = r, x_f \geq 0, f \in F\}$ that satisfies all the other constraints in (10). This corresponds to a vector $x \in P$ such that $Ax \geq 1 - \delta$. Again, we get a subroutine to find a vertex $\tilde{x}$ in P such that $c^T \tilde{x} \geq c^T x$ for all $x \in P$, where $c = y^T A$. For each task T_j we have now different values in M_j and hence in the corresponding Knapsack Problem the profit $y_j / p_j(\ell)$ depends on the cardinality $\ell \in M_j$, while the capacity of the knapsack remains m, as before. The subroutine corresponds now to a generalized Knapsack Problem with different choices for tasks (items). The problem we have to solve (approximately) for a given n-vector $(y_1, \ldots, y_n)$ can be formulated as follows:

$$
\begin{aligned}
\max \ & \textstyle\sum_{j=1}^{n} \sum_{\ell \in M_j} \frac{y_j}{p_j(\ell)} \cdot x_{j\ell} \\
\text{s.t.} \ & \textstyle\sum_{j=1}^{n} \sum_{\ell \in M_j} \ell \cdot x_{j\ell} \ \leq \ m, \\
& \textstyle\sum_{\ell \in M_j} x_{j\ell} \ \leq \ 1, \qquad j = 1, \ldots, n, \\
& x_{j\ell} \in \{0,1\}, \qquad \ell \in M_j, \ j = 1, \ldots, n .
\end{aligned}
\tag{11}
$$

In fact, this is the Multiple-choice Knapsack Problem.

Lawler [26] showed for this problem that an ϵ-approximate solution can be computed in $O(\sum_j |M_j| \ln |M_j| + \sum_j |M_j| n / \epsilon) = O(nm \ln m + n^2 m / \epsilon)$ time. In order to obtain a lower bound, one can compute $d_j = \min_{1 \leq \ell \leq m} p_j(\ell)$ and $d_{\max} = \max_j d_j$. Then $d_{\max} \leq OPT \leq n d_{\max}$. In this case, the overhead $O(n \ln \ln(n/\epsilon)) = O(n \ln \ln \ln n + n \ln \ln(1/\epsilon)) = O(n^2 + n\epsilon^{-1})$ is less than the running time required by the knapsack subroutine. Hence by using an argument similar to the one in the previous section, one can obtain the following result.

Theorem 13. *There exists a FPTAS for $P|fctn_j, pmtn|C_{\max}$ whose running time is bounded by $O(n(\epsilon^{-2} + \ln n) \ln(n\epsilon^{-1})(nm \ln m + n^2 m\epsilon^{-1}))$.*

Other variants of $P|fctn_j, pmtn|C_{\max}$ concern preemptive scheduling on parallel processors, where the underlying interconnection network is not completely disregarded (note that in the original formulation, we assumed nothing about the network architecture). Based on the above results the following can be shown:

Theorem 14. *If the processors are arranged in a line or hypercube network, $P|fctn_j, pmtn|C_{\max}$ has a FPTAS that runs in $O(n(\epsilon^{-2}+\ln n)\ln(n\epsilon^{-1})(nm\ln m+ n^2m\epsilon^{-1}))$ time.*

6 Weighted Fractional Coloring

Let $G = (V, E)$ be a graph with a positive weight w_v for each vertex $v \in V$. Let $\mathcal{I}$ be the set of all independent sets of G. The weighted fractional coloring problem consists of assigning a non-negative real value x_I to each independent set I of G such that each vertex $v \in V$ is completely covered by independent sets containing v (i.e. the sum of their values is at least w_v) and the total value $\sum_I x_I$ is minimized. This problem can also be formulated as a linear program of form (6). Similarly to Sect. 3, this linear program can be solved approximately by using binary search on the optimum value r^* and computing at each stage for the current r an approximate solution for a fractional covering problem of form (7). Let $w_{\max} = \max_{v \in V} w_v$ be the maximum weight of a vertex. By binary search, one can obtain a solution $(x_I)_{I \in \mathcal{I}}$ with $\sum_{I \in \mathcal{I}} x_I = (1 + \epsilon/4)r^*$ and $\sum_{I \in \mathcal{I}:v \in I} x_I \geq 1/c(1 - \delta)w_v$. Now one can define $\tilde{x}_I = x_I c(1 + 4\delta)$ and obtain $\sum_{I \in \mathcal{I}:v \in I} \tilde{x}_I \geq (1-\delta)(1+4\delta)w_v = (1+3\delta-4\delta^2)p_j \geq p_j$ for $\delta \leq 3/4$. In this case, the length of the generated fractional coloring is at most $cr^*(1 + 4\delta)(1 + \epsilon/4) = cr^*(1 + 4\delta + \epsilon/4 + \delta\epsilon) \leq cr^*(1 + \epsilon)$ by choosing $\epsilon \leq 1$ and $\delta \leq 3\epsilon/20$. Since the optimum lies within the interval $[w_{\max}, nw_{\max}]$, the overall complexity of the algorithm can be bounded by $O((n\ln n+n\ln c/\epsilon^3 +n/\epsilon^2)(WIS(G,n,c,d)+ n\ln\ln(n/\epsilon))\ln(n\epsilon^{-1}))$, where $WIS(n,c,d)$ is the time required to compute an approximate weighted independent set for a weighted graph (G, w). The above arguments imply the following result.

Theorem 15. *Let $\mathcal{G}$ be a graph class. If there is a polynomial time algorithm for the weighted independent set problem restricted to graphs $G \in \mathcal{G}$ with approximation ratio $1/c$ for $c \geq 1$, then for any $\epsilon > 0$ there is a polynomial time algorithm for the fractional weighted coloring problem restricted to $\mathcal{G}$ with approximation ratio $c(1 + \epsilon)$.*

Corollary 3. *Let $\mathcal{G}$ be a graph class. If there is a (F)PTAS for the computation of the weighted independent set in a graph $G \in \mathcal{G}$ and weights w, then we obtain a (F)PTAS for the fractional weighted coloring problem for graphs $G \in \mathcal{G}$.*

Using a recent result [9] for computing the maximum weighted independent set in intersection graphs of disks in the plane, we obtain the following:

Corollary 4. *There is a PTAS for the computation of the fractional weighted chromatic number for intersection graphs of disks in the plane.*

Since this graph class contains planar graphs and unit disk graphs, Corollary 3 implies the following result which also provides a substantial improvement on Matsui's polynomial-time 2-approximation algorithm [30] for unit disk graphs:

Corollary 5. *There is a PTAS for the computation of the fractional weighted chromatic number for planar and unit disk graphs.*

7 Conclusion

In this paper we have studied preemptive variants of resource constrained scheduling and the closely related fractional coloring problem. The approach we presented is based on linear programming formulations with exponentially many variables but with special structures allowing efficient approximations. The linear programs are solved (approximately) in an iterative way as covering problems, where at each iteration subproblems of the same type have to be solved. Interestingly, for resource constrained scheduling these subproblems turned out to be knapsack type problems (multiple-choice, multi-dimensional, and cardinality constrained knapsack) with efficient approximation algorithms. For fractional coloring, it is the well known maximum weighted independent set problem.

For some of the subproblems we have encountered, there are only relatively weak polynomial-time approximation results (i.e. with constant, logarithmic, or with even worse approximation ratios). To handle these cases too, we have extended some of the methods in [15,33,41] to the case where the subproblem can be solved only approximatively. The underlying algorithm is independent from the width [33] and the number of variables. We note that by using other techniques [21] (via the ellipsoid method and approximate separation) with higher running time the approximation ratio $c(1 + \epsilon)$ in Theorems 4 and 15 can be improved to ratio c.

We mention in closing, that by using the same approach, similar approximation results can be expected for various other preemptive scheduling and fractional graph problems, e.g. for fractional path coloring, call scheduling, bandwidth allocation, scheduling multiprocessor tasks on dedicated processors, as well as open, flow and job shop scheduling.

Acknowledgement

The authors thank R. Schrader and A. Srivastav for helpful comments on the complexity of the 2-dimensional knapsack problem and the approximation of sD-KP, respectively.

References

1. J. Blazewicz, M. Drabowski and J. Weglarz, Scheduling multiprocessor tasks to minimize schedule length, *IEEE Transactions on Computers*, C-35-5 (1986), 389-393.

2. J. Blazewicz, J.K. Lenstra and A.H.G. Rinnooy Kan, Scheduling subject to resource constraints: Classification and Complexity, *Discrete Applied Mathematics* 5 (1983), 11-24.
3. A. Caprara, H. Kellerer, U. Pferschy and D. Pisinger, Approximation algorithms for knapsack problems with cardinaliy constraints, *European Journal of Operational Research* 123 (2000), 333-345.
4. E. Coffman, M. Garey, D. Johnson and R. Tarjan, Performance bounds for level-oriented two dimensional packing algorithms, *SIAM Journal on Computing*, 9 (1980), 808-826.
5. C. Chekuri and S. Khanna, On multi-dimensional packing problems, *Proceedings 10th ACM-SIAM Symposium on Discrete Algorithms* (1999), 185-194.
6. W.F. de la Vega and C.S. Lueker, Bin packing can be solved within $1 + \epsilon$ in linear time, *Combinatorica*, 1 (1981), 349-355.
7. M. Drozdowski, Scheduling multiprocessor tasks - an overview, *European Journal on Operations Research*, 94 (1996), 215-230.
8. J. Du and J. Leung, Complexity of scheduling parallel task systems, *SIAM Journal on Discrete Mathematics*, 2 (1989), 473-487.
9. T. Erlebach, K. Jansen and E. Seidel, Polynomial-time approximation schemes for geometric graphs, *Proceedings of the 12th ACM-SIAM Symposium on Discrete Algorithms* (2001), 671-679.
10. U. Feige and J. Kilian, Zero knowledge and the chromatic number, *Journal of Computer and System Sciences*, 57 (1998), 187-199.
11. A.M. Frieze and M.R.B. Clarke, Approximation algorithms for the m-dimensional $0 - 1$ knapsack problem, *European Journal of Operational Research*, 15 (1984), 100-109.
12. M.R. Garey and R.L. Graham, Bounds for multiprocessor scheduling with resource constraints, *SIAM Journal on Computing*, 4 (1975), 187-200.
13. M.R. Garey, R.L. Graham, D.S. Johnson and A.C.-C. Yao, Resource constrained scheduling as generalized bin packing, *Journal Combinatorial Theory A*, 21 (1976), 251-298.
14. S. Gerke and C. McDiarmid, Graph imperfection, unpublished manuscript, 2000.
15. M.D. Grigoriadis, L.G. Khachiyan, L. Porkolab and J. Villavicencio, Approximate max-min resource sharing for structured concave optimization, *SIAM Journal on Optimization*, 41 (2001), 1081-1091.
16. M. Grötschel, L. Lovász and A. Schrijver, The ellipsoid method and its consequences in combinatorial optimization, *Combinatorica*, 1 (1981), 169-197.
17. M. Grötschel, L. Lovász and A. Schrijver, Geometric Algorithms and Combinatorial Optimization, Springer Verlag, Berlin, 1988.
18. K. Jansen and L. Porkolab, Linear-time approximation schemes for scheduling malleable parallel tasks, *Algorithmica*, 32 (2002), 507-520.
19. K. Jansen and L. Porkolab, Preemptive scheduling on dedicated processors: applications of fractional graph coloring *Proceedings 25th International Symposium on Mathematical Foundations of Computer Science* (2000), LNCS 1893, Springer Verlag, 446-455.
20. K. Jansen and L. Porkolab, Computing optimal preemptive schedules for parallel tasks: Linear Programming Approaches, *Proceedings 11th Annual International Symposium on Algorithms and Computation* (2000), LNCS 1969, Springer Verlag, 398-409, and *Mathematical Programming*, to appear.
21. K. Jansen, Approximate separation with application in fractional coloring and preemptive scheduling, *Proceedings 9th International Symposium on Theoretical Aspects of Computer Science*, (2002), LNCS, Springer Verlag, to appear.

22. H. Kellerer and U. Pferschy, A new fully polynomial approximation scheme for the knapsack problem, *Proceedings 1st International Workshop on Approximation Algorithms for Combinatorial Optimization* (1998), 123-134.
23. C. Kenyon and E. Remila, Approximate strip packing, *Proceedings 37th IEEE Symposium on Foundations of Computer Science* (1996), 31-36.
24. B. Korte and R. Schrader, On the existence of fast approximation schemes, *Nonlinear Programming*, 4 (1981), 415-437.
25. K.L. Krause, V.Y. Shen and H.D. Schwetman, Analysis of several task scheduling algorithms for a model of multiprogramming computer systems, *Journal of the ACM*, 22 (1975), 522-550 and Errata, *Journal of the ACM*, 24 (1977), 527.
26. E. Lawler, Fast approximation algorithms for knapsack problems, *Mathematics of Operations Research*, 4 (1979), 339-356.
27. W. Ludwig and P. Tiwari, Scheduling malleable and nonmalleable parallel tasks, *Proceedings 5th ACM-SIAM Symposium on Discrete Algorithms* (1994), 167-176.
28. C. Lund and M. Yannakakis, On the hardness of approximating minimization problems, *Journal of the ACM*, 41 (1994), 960-981.
29. O. Oguz and M.J. Magazine, A polynomial time approximation algorithm for the multidimensional $0-1$ knapsack problem, *working paper, University of Waterloo*, 1980.
30. T. Matsui, Approximation algorithms for maximum independent set problems and fractional coloring problems on unit disk graphs, *Proceedings Symposium on Discrete and Compuational Geometry*, LNCS 1763 (2000), 194-200.
31. G. Mounie, C. Rapine and D. Trystram, Efficient approximation algorithms for scheduling malleable tasks, *Proceedings ACM Symposium on Parallel Algorithms* (1999), 23-32.
32. T. Niessen and J. Kind, The round-up property of the fractional chromatic number for proper circular arc graphs, *Journal of Graph Theory*, 33 (2000), 256-267.
33. S.A. Plotkin, D.B. Shmoys, and E. Tardos, Fast approximation algorithms for fractional packing and covering problems, *Mathematics of Operations Research*, 20 (1995), 257-301.
34. P. Raghavan, C.D. Thompson, Randomized rounding: a technique for provably good algorithms and algorithmic proofs, *Combinatorica* 7 (1987), 365-374.
35. E.R. Schreinerman and D.H. Ullman, Fractional graph theory: A rational approach to the theory of graphs, Wiley Interscience Series in Discrete Mathematics, 1997.
36. P.D. Seymour, Colouring series-parallel graphs, *Combinatorica*, 10 (1990), 379-392.
37. A. Srinivasan, Improved approximation guarantees for packing and covering integer programs, *Proceedings 27th ACM Symposium on the Theory of Computing* (1995), 268-276.
38. A. Srivastav and P. Stangier, Algorithmic Chernoff-Hoeffding inequalities in integer programming, *Random Structures and Algorithms*, 8(1) (1996), 27-58.
39. A. Srivastav and P. Stangier, Tight approximations for resource constrained scheduling and bin packing, *Discrete Applied Mathematics*, 79 (1997), 223-245.
40. J. Turek, J. Wolf and P. Yu, Approximate algorithms for scheduling parallelizable tasks, *Proceedings 4th ACM Symposium on Parallel Algorithms and Architectures* (1992), 323-332.
41. N.E. Young, Randomized rounding without solving the linear program, *Proceedings 6th ACM-SIAM Symposium on Discrete Algorithms* (1995), 170-178.

Hard Equality Constrained Integer Knapsacks

Karen Aardal[1] and Arjen K. Lenstra[2]

[1] Mathematisch Instituut, Universiteit Utrecht, Budapestlaan 6,
3584 CD Utrecht, The Netherlands
aardal@math.uu.nl
[2] Emerging Technology, Citibank N.A., 1 North Gate Road,
Mendham, NJ 07945-3104, USA,
and Faculteit Wiskunde en Informatica, Technische Universiteit Eindhoven,
Postbus 513, 5600 MB Eindhoven, The Netherlands
arjen.lenstra@citibank.com

Abstract. We consider the following integer feasibility problem: "Given positive integer numbers a_0, $a_1, \ldots, a_n$, with $\gcd(a_1, \ldots, a_n) = 1$ and $\boldsymbol{a} = (a_1, \ldots, a_n)$, does there exist a nonnegative integer vector $\boldsymbol{x}$ satisfying $\boldsymbol{ax} = a_0$?" Some instances of this type have been found to be extremely hard to solve by standard methods such as branch-and-bound, even if the number of variables is as small as ten. We observe that not only the sizes of the numbers a_0, $a_1, \ldots, a_n$, but also their structure, have a large impact on the difficulty of the instances. Moreover, we demonstrate that the characteristics that make the instances so difficult to solve by branch-and-bound make the solution of a certain reformulation of the problem almost trivial. We accompany our results by a small computational study.

1 Introduction

1.1 Problem Statement and Summary of Results

In the past decade there has been a substantial progress in computational integer programming. Many large and complex instances can now be solved. There are, however, still many small instances that seem extremely hard to tackle by standard methods such as branch-and-bound or branch-and-cut, and it is still quite unclear what makes these instances so hard. Examples are the so-called market share problems [6,1], some feasibility problems reported on by Aardal et al. [2], and certain portfolio planning problems [16]. All of these are generalizations of the following simple problem. Let $a_0, a_1, \ldots, a_n$ be positive integer numbers with $\gcd(a_1, \ldots, a_n) = 1$ and $a_i \leq a_0$, $1 \leq i \leq n$, and let

$$P = \{\boldsymbol{x} \in \mathbb{R}^n : \boldsymbol{ax} = a_0, \boldsymbol{x} \geq \boldsymbol{0}\} . \tag{1}$$

The problem is:

$$\text{Is } X = P \cap \mathbb{Z}^n \neq \emptyset \text{ ?} \tag{2}$$

If the components of $\boldsymbol{x}$ may take any integer value, then the problem is easy. There exists a vector $\boldsymbol{x} \in \mathbb{Z}^n$ satisfying $\boldsymbol{ax} = a_0$ if and only if a_0 is an integer multiple of $\gcd(a_1, \ldots, a_n)$. The nonnegativity requirement on $\boldsymbol{x}$ makes the

W.J. Cook and A.S. Schulz (Eds.): IPCO 2002, LNCS 2337, pp. 350–366, 2002.

problem NP-complete. In this study we focus on infeasible instances to rule out that a search algorithm terminates fast because we found a feasible solution by luck.

In our study we demonstrate that if it is possible to decompose the $\boldsymbol{a}$-coefficients as $a_i = p_i M + r_i$ with $p_i, M \in \mathbb{Z}_{>0}$, $r_i \in \mathbb{Z}$, and with M large compared to p_i and $|r_i|$, then the maximum size of the right-hand side coefficient a_0 for which $\boldsymbol{a}\boldsymbol{x} = a_0$ has no nonnegative integer solution becomes large. This is proved in Theorem 1 in Sect. 3.1. An infeasible instance with a large value of a_0 is particularly difficult for branch-and-bound applied to formulation P. Moreover, we show that if the $\boldsymbol{a}$-coefficients can be decomposed as described above, then the reformulation obtained by using the projection suggested by Aardal, Hurkens, and Lenstra [2] is computationally very easy to solve by a search algorithm similar to the algorithm suggested by H.W. Lenstra, Jr. [14], since the projected polytope will be thin in the direction of the last coordinate. This is demonstrated in Sect. 3.2. The projection, based on lattice basis reduction, is briefly described in Sect. 2.

To illustrate our observations we report on a modest computational study on infeasible instances from the literature, and infeasible instances that we generated ourselves. About half of the instances have $\boldsymbol{a}$-coefficients that decompose as described in Sect. 3, and the rest of the instances have random $\boldsymbol{a}$-coefficients of the same size as the first group of instances. The smallest of the instances has five variables, and the largest has ten. The computational results, presented in Sect. 4, clearly confirm our theoretical observations.

1.2 Integer Programming and Branching on Hyperplanes

The polytope P as defined by (1) has dimension $n - 1$, i.e., it is not full-dimensional. In the full-dimensional case the following is known. Let S be a full-dimensional polytope in $\mathbb{R}^n$ given by integer input. The *width* of S along the nonzero vector $\boldsymbol{d}$ is defined as $W(S, \boldsymbol{d}) = \max\{\boldsymbol{d}^T\boldsymbol{x} : \boldsymbol{x} \in S\} - \min\{\boldsymbol{d}^T\boldsymbol{x} : \boldsymbol{x} \in S\}$. Notice that this is different from the definition of the geometric width of a polytope. Consider the problem: "Does the polytope S contain a vector $\boldsymbol{x} \in \mathbb{Z}^n$?" Khinchine [12] proved that if S does not contain a lattice point, then there exists a nonzero integer vector $\boldsymbol{d}$ such that $W(S, \boldsymbol{d})$ is bounded from above by a constant depending only on the dimension. H. W. Lenstra, Jr., [14] developed an algorithm that runs in polynomial time for fixed dimension n, and that either finds an integer vector in S, or a lattice hyperplane H such that at most $c(n)$ lattice hyperplanes parallel to H intersect S, where $c(n)$ is a constant depending only on the dimension n. This is the same as to say that an integer nonzero direction is found such that the width of S in that direction is bounded by a constant depending only on n. The intersection of each lattice hyperplane with S gives rise to a problem of dimension at most $n - 1$, and each of these lower-dimensional problems is solved recursively to determine whether or not S contains an integer vector. One can illustrate the algorithm by a search tree having at most n levels. The number of nodes created at each level is bounded from above by a constant depending only on the dimension at that level. A search node is pruned if, in the given direction, no lattice hyperplane is intersecting the polytope defined by the search node. We are not aware of any implementation of Lenstra's algorithm. Cook et al. [5] implemented the integer programming algorithm by Lovász and Scarf [15], which is similar in structure to Lenstra's algorithm, and they observed that, for their instances, the number of search nodes created by the Lovász-Scarf algorithm was much less than the number of nodes of a branch-and-bound tree. To compute a good search direction in each node was, however, more time consuming than computing an LP-relaxation. This raises the question of understanding if there are situations in which good search directions can be determined fast.

Here we will consider a full-dimensional projection of P as suggested by Aardal, Hurkens, and Lenstra [2]. An important conclusion of our study is that for the most difficult infeasible instances, i.e., the instances that have decomposable a-coefficients as outlined in the previous subsection and described in more detail in Sect. 3, this projection itself yields an integer direction in which the projected polytope is provably thin. This direction is the last coordinate direction. So, if we apply a tree search algorithm as that of Lenstra to the projected polytope, but branch only in coordinate directions in the order of decreasing variable indices, then we observe that the instances become very easy. If the a-coefficients do not possess the decomposition property, then the instances are relatively easy to solve using branch-and-bound on formulation P.

1.3   Notation

We conclude this section by introducing some definitions and notation. $|\boldsymbol{x}|$ denotes the Euclidean length in $\mathbb{R}^n$ of the vector $\boldsymbol{x}$. The *identity matrix* is denoted by $\boldsymbol{I}$, and the matrix consisting of zeros only is denoted by $\boldsymbol{0}$. If we want to make the dimension of a matrix clear, then it is indicated as a superscript, for example, $\boldsymbol{I}^{(n)}$ is the $n \times n$ identity matrix, and $\boldsymbol{0}^{(p \times q)}$ is the $p \times q$ matrix consisting of zeros. The matrix $\boldsymbol{A}$ is said to be in *Hermite normal form* if it has the form $[\boldsymbol{C}, \boldsymbol{0}]$, where $\boldsymbol{C}$ is a lower triangular, nonnegative matrix in which each diagonal element is the unique maximum row entry. The Hermite normal form of the matrix $\boldsymbol{A}$ is denoted by HNF $(\boldsymbol{A})$. A set of the form $L = L(\boldsymbol{b}_1, \ldots, \boldsymbol{b}_l) = \{\sum_{i=1}^{l} \lambda_i \boldsymbol{b}_i, \ \lambda_i \in \mathbb{Z}, 1 \leq i \leq l\}$, where $\boldsymbol{b}_1, \ldots, \boldsymbol{b}_l$ are linear independent vectors in $\mathbb{R}^n$, $l \leq n$, is called a *lattice*. The set of vectors $\{\boldsymbol{b}_1, \ldots, \boldsymbol{b}_l\}$ is called a *lattice basis*. Notice that a lattice may have several different bases. The *determinant* of the lattice L, $d(L)$, is defined as $d(L) = \sqrt{\boldsymbol{B}^T \boldsymbol{B}}$, where $\boldsymbol{B}$ is a basis for L. If the lattice L is full-dimensional we have $d(L) = |\det \boldsymbol{B}|$. The *rank* of the lattice L, rk L, is the dimension of the Euclidean vector space spanned by L. The *integer width* of a polytope S in the non-zero integer direction $\boldsymbol{d}$ is defined as $W_I(S, \boldsymbol{d}) = \lfloor \max\{\boldsymbol{d}^T x : x \in S\} \rfloor - \lceil \min\{\boldsymbol{d}^T x : x \in S\} \rceil$. The number of lattice hyperplanes in the direction $\boldsymbol{d}$ that intersect S is equal to $W_I(S, \boldsymbol{d}) + 1$, so if $W_I(S, \boldsymbol{d}) = -1$, then S does not contain an integer vector.

2   The Reformulation and the Search Algorithm

The starting point of the reformulation of (2) suggested by Aardal, Hurkens, and Lenstra [2] is the integer relaxation $X_{\mathbb{R}} = \{\boldsymbol{x} \in \mathbb{Z}^n : \boldsymbol{a}\boldsymbol{x} = a_0\}$ of X. The relaxation $X_{\mathbb{R}}$ can be rewritten as $X_{\mathbb{R}} = \{x \in \mathbb{Z}^n : x = x_f + B_0 y, \ y \in \mathbb{Z}^{n-1}\}$, where x_f is an integer vector satisfying $ax_f = a_0$, and where B_0 is a basis for the lattice $L_0 = \{x \in \mathbb{Z}^n : ax = 0\}$. That is, any integer vector x, not necessarily nonnegative, satisfying $ax = a_0$ can be written as a sum of an integer vector x_f satisfying $ax = a_0$ and an integer linear combination of vectors x_0 satisfying $ax_0 = 0$. Since $\gcd(a_1, \ldots, a_n) = 1$ and a_0 is integer, we know that a vector x_f exists, and determining x_f can be done in polynomial time. Aardal et al. determine x_f and B_0 as follows. Consider the lattice

$$\left\{ z \in \mathbb{Z}^{n+2} : z = B \begin{pmatrix} x \\ y \end{pmatrix}, \ x \in \mathbb{Z}^n, \ y \in \mathbb{Z} \right\}, \tag{3}$$

where

$$B = \begin{pmatrix} I^{(n)} & 0^{(n \times 1)} \\ 0^{(1 \times n)} & N_1 \\ N_2 a & -N_2 a_0 \end{pmatrix}. \tag{4}$$

Aardal et al. showed that if N_1 and N_2 are chosen appropriately, then the basis B' obtained by applying Lovász' basis reduction algorithm [13] to B (4) is of the following form:

$$B' = \begin{pmatrix} B_0^{(n \times (n-1))} & x_f & u \\ 0^{(1 \times (n-1))} & N_1 & v \\ 0^{(1 \times (n-1))} & 0 & N_2 \end{pmatrix}, \tag{5}$$

where B_0 is a basis for the lattice L_0, and where x_f is an integer vector satisfying $ax_f = a_0$. Let

$$Q = \{y \in \mathbb{R}^{n-1} : B_0 y \geq -x_f\}. \tag{6}$$

Problem (2) can now be restated as:

$$\text{Is } Q \cap \mathbb{Z}^{n-1} \neq \emptyset ? \tag{7}$$

The polytope Q is a full-dimensional formulation, and as mentioned in the previous section we can apply Lenstra's [14] algorithm, or any other integer programming algorithm, to Q. Here we will consider a tree search algorithm inspired by Lenstra's algorithm, but using only unit directions in the search.

At the root node of the search tree, choose an index $i : 1 \leq i \leq n - 1$. Determine the integer width in the direction e_i, i.e., determine $u = \lfloor \max\{e_i^T y : y \in Q\} \rfloor$, and $l = \lceil \min\{e_i^T y : y \in Q\} \rceil$. If $l \leq u$, then create a search node for each value $k = l, \ldots, u$, and if $l > u$, then stop. Suppose that $l \leq u$. Choose a new index i', $1 \leq i' \leq n - 1$, $i' \neq i$, and any of the nodes just created, say the node corresponding with value $k = k'$. Fix $y_i = k'$, and determine $u = \lfloor \max\{e_{i'} y : y \in Q, y_i = k'\} \rfloor$, and $l = \lceil \min\{e_{i'} y : y \in Q, y_i = k'\} \rceil$. Repeat the procedure described above until all nodes have been investigated. For an example of a search tree, see Fig. 1. Notice that the search tree created in this way is similar to the search tree of Lenstra's algorithm in that the number of levels of the tree is no more than the number of variables in the problem

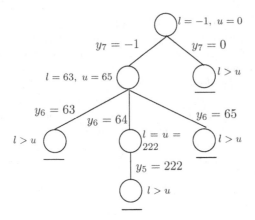

Fig. 1. The search tree for instance prob2 (cf. Sect. 4)

instance, and that the number of nodes created at a certain level corresponds to the width of the polytope in the chosen search direction.

Here we will investigate a class of instances that are exceptionally hard to solve by branch-and-bound when using the original formulation in $\boldsymbol{x}$-variables, but that become easy to solve when applying the branching scheme described above to the reformulated problem in y-variables (6) because the width in the unit direction $\boldsymbol{e}_{n-1}$ is small. Below we give an example of such an instance.

Example 1. Let

$$P = \{\boldsymbol{x} \in \mathbb{R}^3 : 12223x_1 + 12224x_2 + 36672x_3 = 149389505, \boldsymbol{x} \ge 0\} .$$

A vector $\boldsymbol{x}_f$ and a basis $\boldsymbol{B}_0$ for this instance is:

$$\boldsymbol{x}_f = \begin{pmatrix} -1 \\ 1221 \\ 3667 \end{pmatrix} \qquad \boldsymbol{B}_0 = \begin{pmatrix} 0 & 12224 \\ -3 & -1222 \\ 1 & -3667 \end{pmatrix} .$$

The polytope Q is:

$$Q = \{\boldsymbol{y} \in \mathbb{R}^2 : 12224y_2, \ge 1, \ -3y_1 - 1222y_2 \ge -1221, \ y_1 - 3667y_2 \ge -3667\} .$$

Moreover, we have $W_I(Q, \boldsymbol{e}_1) = 4074$ and $W_I(Q, \boldsymbol{e}_2) = -1$, so if consider the search direction $\boldsymbol{e}_2$ first, we can immediately conclude that $Q \cap \mathbb{Z}^2 = \emptyset$.

If we solve the formulation in $\boldsymbol{x}$-variables by branch-and-bound with objective function 0 using the default settings of CPLEX 6.5, it takes 1,262,532 search nodes to verify infeasibility. ☐

An instance such as the one given in Example 1 may seem quite artificial, but many of the instances reported on in [6,1,2,16] show a similar behavior, and some of these instances stem from applications. To try to explain this behavior therefore seems relevant also from the practical viewpoint.

3   The Class of Instances

3.1   The Coefficient a_0

The linear relaxation of Problem (2) is an n-simplex $\{x \in \mathbb{R}^n_{\geq 0} : ax = a_0\}$. An instance of problem (2) is particularly hard to solve by branch-and-bound if it is infeasible and if the intersection points of the n-simplex with the coordinate axes have large values. Branch-and-bound will then be forced to enumerate many of the possible combinations of $x_1, \ldots, x_n$ with $0 \leq x_i \leq a_0/a_i$. Since the instance is infeasible we cannot "get lucky" in our search, which may happen if the instance is feasible, and if we by chance have chosen an objective function that takes us to a feasible solution quickly. Example 1 of the previous section illustrates such a hard infeasible instance. Similar, but larger, instances are virtually impossible to solve using a state-of-the-art branch-and-bound algorithm such as implemented in CPLEX.

To create infeasible instances with maximum values of a_0/a_i we choose a_0 as the largest number such that $ax = a_0$ has no nonnegative integer solution. This number is called the *Frobenius number* for $a_1, \ldots, a_n$, denoted by $F(a_1, \ldots, a_n)$. Computing the Frobenius number for given natural numbers $a_1, \ldots, a_n$ with $\gcd(a_1, \ldots, a_n) = 1$ is NP-hard. In Appendix 1 we discuss the algorithm that we used in our computational study. For $n = 2$ it is known that $F(a_1, a_2) = a_1 a_2 - a_1 - a_2$. (In "Mathematics from the Educational Times, with Additional Papers and Solutions", Sylvester published the problem of proving that if a_1 and a_2 are relatively prime integers, then there are exactly $1/2(a_1 - 1)(a_2 - 1)$ nonnegative integers α less than $a_1 a_2 - a_1 - a_2$ for which $a_1 x_1 + a_2 x_2 = \alpha$ does not have a nonnegative integer solution. The solution to this problem was provided by Curran Sharp in volume 41 (1884) of the journal. The precise reference is [21]. See also Schrijver [18] p. 376.) For $n = 3$ the Frobenius number can be computed in polynomial time, see Selmer and Beyer [20], Rödseth [17], and Greenberg [9]. Kannan [11] developed a polynomial time algorithm for computing the Frobenius number for every *fixed* n. His algorithm is based on the relation between the Frobenius number and the covering radius of a certain polytope. Some upper bounds on the Frobenius number are also known. If $a_1 < a_2 < \cdots < a_n$, Brauer [3] showed that $F(a_1, \ldots, a_n) \leq a_1 a_n - a_1 - a_n$. Other upper bounds were provided by Erdős and Graham [8] and Selmer [19].

Below we determine a lower and an upper bound on $F(a_1, \ldots, a_n)$ that are of the same order. In particular, we observe that if a_i is written as $a_i = p_i M + r_i$, with $p_i, M \in \mathbb{Z}_{>0}$, $r_i \in \mathbb{Z}$, and with M large compared to p_i and $|r_i|$, then the lower bound on $F(a_1, \ldots, a_n)$ becomes large. This implies that instances in which the a-coefficients decompose like this have relatively large Frobenius numbers.

Theorem 1. *Let $a_i = p_i M + r_i$, for $i = 1, \ldots, n$, let $(r_j/p_j) = \max_{i=1,\ldots,n}\{r_i/p_i\}$, and let $(r_k/p_k) = \min_{i=1,\ldots,n}\{r_i/p_i\}$. Assume that:*

1. $a_1 < a_2 < \cdots < a_n$,
2. $p_i, M \in \mathbb{Z}_{>0}$, $r_i \in \mathbb{Z}$ for all $i = 1, \ldots n$,

3. $\sum_{i=1}^{n} |r_i| < 2M$,
4. $M > 2 - (r_j/p_j)$,
5. $M > (r_j/p_j) - 2(r_k/p_k)$.

Then, we obtain $f(\boldsymbol{p}, \boldsymbol{r}, M) \leq F(a_1, \ldots, a_n) \leq g(\boldsymbol{p}, \boldsymbol{r}, M)$, where

$$f(\boldsymbol{p}, \boldsymbol{r}, M) = \frac{(M^2 p_j p_k + M(p_j r_k + p_k r_j) + r_j r_k)(1 - \frac{2}{M+(r_j/p_j)})}{p_k r_j - p_j r_k} - 1,$$

and

$$g(\boldsymbol{p}, \boldsymbol{r}, M) = M^2 p_1 p_n + M(p_1 r_n + p_n r_1 - p_1 - p_n) + r_1 r_n - r_1 - r_n .$$

Proof. The upper bound $g(\boldsymbol{p}, \boldsymbol{r}, M)$ is derived from the result by Brauer [3] that $F(a_1, \ldots, a_n) \leq a_1 a_n - a_1 - a_n$.

In our proof of the lower bound we introduce the following notation:

$$B = \{\boldsymbol{x} \in \mathbb{R}^n : \boldsymbol{ax} = 0\},$$
$$\Delta_n = \{\boldsymbol{x} \in \mathbb{R}^n : \boldsymbol{ax} = n, \ \boldsymbol{x} \geq \boldsymbol{0}\},$$
$$C = \{\boldsymbol{x} \in \mathbb{R}^n : \boldsymbol{px} = 0, \ \boldsymbol{rx} = 0\} .$$

The lattice L_0 is as defined in Sect. 2.

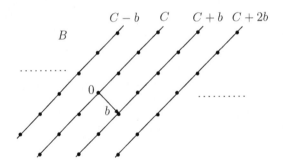

Lattice L_0 is contained in parallel hyperplanes generated by C

Fig. 2. B, C, and L_0

The idea behind the proof is as follows. We define a homomorphism from Δ_n to $\mathbb{R}/\mathbb{Z}$ such that $\boldsymbol{x} \in \mathbb{Z}^n \cap \Delta_n$ maps to 0, if such a vector $\boldsymbol{x}$ exists. An integer number n for which 0 is not contained in the image of Δ_n under this map then provides a lower bound on the Frobenius number. We define such a homomorphism by first defining a projection π_b along the vector $\boldsymbol{b}$ of Δ_1 onto B, where $\boldsymbol{b}$ is in the same plane as Δ_1. Then we consider a homomorphism $f : B \to \mathbb{R}/\mathbb{Z}$. We show that the kernel of f is $L_0 + C$. Due to the First Isomorphism Theorem (see e.g. Hungerford[10], p. 44) we know that B divided out by (ker f), i.e., $B/(L_0 + C)$, is isomorphic to $\mathbb{R}/\mathbb{Z}$. The image of $\pi_b(\Delta_1)$ under the isomorphism $B/(L_0+C) \to \mathbb{R}/\mathbb{Z}$ is an interval $[l, u]$ in $\mathbb{R}/\mathbb{Z}$. Finally we determine an integer number n such that $[nl, nu]$ does not contain an integer point. The integer n then yields a lower bound on the Frobenius number under the conditions given in the theorem.

We first define a linear mapping $\pi : \mathbb{R}^n \to B$ given by $\pi_b(x) = x - (ax)b$, where b satisfies $pb = 0$ and $rb = 1$, and hence $ab = 1$. When $M \to \infty$, then $\pi_b(\Delta_1) \to -b$. Notice that $-b \notin C$ as $rb = 1$.

Next we define the homomorphism $f : B \to \mathbb{R}/\mathbb{Z}$ given by $x \mapsto (px \mod 1)$.

Claim: The kernel of f is $L_0 + C$.

First we show that $(L_0 + C) \subseteq (\ker f)$. If $x \in L_0$ then $x \in \mathbb{Z}$, which implies $px \in \mathbb{Z}$, and hence $(px \mod 1) = 0$. If $x \in C$, then $px = 0$.

Next, show that $(\ker f) \subseteq (L_0 + C)$. Notice that each element in B can be written as $c + y_1 + y_2$ with $c \in C$, $y_1 \in L_0$, and $y_2 \notin C$ such that the absolute value of each element of y_2 is in the interval $(-1/2, 1/2)$ and such that $y_2 \notin C$. Since $f(c + y_1 + y_2) = f(c) + f(y_1) + f(y_2) = 0$ and since $f(c) = f(y_1) = 0$, as $c \in C$ and $y_1 \in L_0$, we obtain $f(y_2) = 0$, and hence $(py_2 \mod 1) = 0$. If $py_2 = 0$, then, since $y_2 \in B$, $ry_2 = 0$, but this contradicts $y_2 \notin C$. So, $0 = ay_2 = Mpy_2 + ry_2$, and since py_2 is integral we have that ry_2 is an integer multiple of M. Now, observe that since the absolute value of each element of y_2 is less than $1/2$, then, due to Assumption 3 of the theorem, $ry_2 < 1/2 \sum_{i=1}^n |r_i| < M$, and therefore $y_2 = 0$ is the only possible solution. This concludes the proof of our claim.

Due to the First Isomorphism Theorem the homomorphism f induces an isomorphism $f' : B/(L_0 + C) \to \mathbb{R}/\mathbb{Z}$. Below we determine the image of Δ_1 under the composition of the mappings $\pi : \mathbb{R}^n \to B$ and $B \to B/(L_0 + C) \to \mathbb{R}/\mathbb{Z}$. This composition of mappings is a homomorfism.

We use v_i to denote vertex i of Δ_1. Vertex v_i, is the vector $(0, \ldots, 0, 1/a_i, 0, \ldots, 0)^T = (0, \ldots, 0, 1/(p_i M + r_i), 0, \ldots, 0)$, where $1/a_i = 1/(p_i M + r_i)$ is the ith component of v_i. Applying the linear mapping π to v_i yields $\pi_b(v_i) = v_i - b$. Next, by the isomorphism $x \mapsto (px \mod 1)$, $\pi_b(v_i)$ becomes

$$\frac{p_i}{Mp_i + r_i} = \frac{1}{M + r_i/p_i} .$$

Let d_i denote $1/(M + r_i/p_i)$, and recall that $(r_j/p_j) = \max_{i=1,\ldots,n}\{r_i/p_i\}$, and $(r_k/p_k) = \min_{i=1,\ldots,n}\{r_i/p_i\}$. Then, the image of Δ_1 is an interval $[d_j, d_k]$ of length

$$L = \frac{p_k r_j - p_j r_k}{M^2 p_j p_k + M(p_j r_k + p_k r_j) + r_j r_k} .$$

Now we will demonstrate that there exists an integer $n \geq \lfloor \frac{1-2d_j}{L} \rfloor$ such that the interval $[nd_j, nd_k]$ does not contain an integer point. This implies that

$\lfloor \frac{1-2d_j}{L} \rfloor$ is a lower bound on the Frobenius number. Notice that $1 - 2d_j > 0$ due to Assumption 4.

Let $k = \lfloor \frac{1-2d_j}{L} \rfloor$. The interval $[I_1,\ I_2] = [kd_j,\ kd_k]$ has length less than or equal to $1 - 2d_j$. Let $\ell = \lfloor kd_j \rfloor$. Notice that $\ell \leq I_1$. Now define $k' = \ell/d_j$. The number $k' \leq k$ yields an interval $[I_1',\ I_2'] = [k'd_j,\ k'd_k]$, such that I_1' is integral. The interval $[I_2',\ I_2'+2d_j)$ does not contain an integer since the length of $[I_1',\ I_2']$ is less than or equal to the length of the interval $[I_1,\ I_2]$, and since I_1' is integral. Now define $k^* = \lfloor k' \rfloor + 1$. We claim that the interval $[I_1^*,\ I_2^*] = [k^*d_j,\ k^*d_k]$ does not contain an integer point. To prove the claim, first assume that k' is integer. Then $k^*d_j = I_1' + d_j$ and $k^*d_k < I_1' + 1$ due to Assumption 5 that implies that $d_k < 2d_j$. Next, assume that k' is fractional. In this case we obtain $I_1' < k^*d_j < I_1' + d_j$, and, due to the same reasoning as for k' integer, we obtain $k^*d_k < I_1' + 1$. This finishes the proof of our claim. We finally notice that $k^* \geq \lfloor \frac{1-2d_j}{L} \rfloor$, so we can conclude that $\lfloor \frac{1-2d_j}{L} \rfloor$ yields a lower bound on the Frobenius number. We obtain

$$\left\lfloor \frac{1 - 2d_j}{L} \right\rfloor \geq \frac{1 - 2d_j}{L} - 1 =$$

$$\frac{(M^2 p_j p_k + M(p_j r_k + p_k r_j) + r_j r_k)(1 - \frac{2}{M+(r_j/p_j)})}{p_k r_j - p_j r_k} - 1 \ .$$

$\square$

Example 2. The a-coefficients in Example 1 decompose as follows. Let $M = 12223$.
$$a_1 = M + 0,$$
$$a_2 = M + 1,$$
$$a_3 = 3M + 3 \ .$$
Theorem 1 yields a lower bound on the Frobenius number equal to 149381362. The Frobenius number for this instance is 149389505. $\square$

For all our instances that decompose with short vectors p and r relative to M, the Frobenius number is large, see the computational study in Sect. 4. We have computed the lower bound on the Frobenius number for these instances and in all cases it was close to the actual value. It would be interesting to investigate whether it is possible to use similar techniques to tighten the upper bound on the Frobenius number for instances of this sort.

In the following subsection we demonstrate that instances with a-coefficients that decompose with large M and relatively short p and r are trivial to solve using the reformulation outlined in Sect. 2. These are the instances that are extremely hard to solve by branch-and-bound due to the large Frobenius numbers.

3.2 The Coefficients $a_1, \ldots, a_n$

For the further analysis of our class of instances we wish to express the determinant of the lattice $L_0 = \{x \in \mathbb{Z}^n : ax = 0\}$ in terms of the input a. We will begin by stating a theorem that is more general than what we actually need here. The following notation is used. Let A be an integer $m \times n$ matrix with $m \leq n$. Moreover, let

$$L_0 = \{x \in \mathbb{Z}^n : Ax = 0\},$$
$$L_A = \{y \in \mathbb{Z}^m : Ax = y, \ x \in \mathbb{Z}^n\},$$
$$L_{A^T} = \{z \in \mathbb{Z}^z : A^T x = z, \ x \in \mathbb{Z}^m\},$$
$$L_{0^T} = \{x \in \mathbb{Z}^m : A^T x = 0\} .$$

Remark 1. Suppose that A has full row rank. Then rk $L_0 = m - n$, and $L_{0^T} = \{0\}$, which implies that $d(L_{0^T}) = 1$.

Remark 2. Suppose that HNF $(A) = (I, 0)$. Then $L_A = \mathbb{Z}^m$, and $d(L_A) = 1$.

Theorem 2.

$$d(L_0) = \frac{d(L_{A^T})}{d(L_A)} \cdot d(L_{0^T}) . \tag{8}$$

Proof. The proof will be presented in the complete version of the paper.   □

Corollary 1. *If A has full row rank, then*

$$d(L_0) = \frac{d(L_{A^T})}{d(L_A)} . \tag{9}$$

Proof. See Remark 1 regarding $d(L_{0^T})$.   □

Corollary 2. *If $m = 1$, that is, $A = a$, and if $\gcd(a_1, \ldots, a_n) = 1$, then*

$$d(L_0) = d(L_{A^T}) = |a| . \tag{10}$$

Proof. We have $d(L_{0^T}) = 1$, see Remark 1. We also have HNF $(a) = (1, 0, \ldots, 0)$, since $\gcd(a_1, \ldots, a_n) = 1$, which implies that $L_A = \mathbb{Z}$ and, hence, $d(L_A) = 1$, see Remark 2. The vector a forms a basis for L_{A^T}, so $d(L_{A^T}) = \sqrt{a^T a} = |a|$.   □

Remark 3. Notice that $d(L_0)$ can also be computed as $d(L_0) = \sqrt{B_0^T B_0}$.

Write $a_i = p_i M + r_i$ with $p_i, M \in \mathbb{Z}_{>0}$ and $r_i \in \mathbb{Z}$. Recall that C denotes the orthogonal complement of the hyperplane spanned by p and r, cf. the proof of Theorem 1.

Proposition 1. *The lattice L_0 contains the lattice $C \cap \mathbb{Z}^n$. The rank of the lattice $C \cap \mathbb{Z}^n$ is equal to $n - 2$.*

Observe that the lattice $C \cap \mathbb{Z}^n$ is generated by elements independent of M, and if p and r are short, then $C \cap \mathbb{Z}^n$ will consist of relatively short vectors. The determinant of L_0 is equal to the length of the vector a, Corollary 3, and this value does depend on the value of M. Since L_0 contains $C \cap \mathbb{Z}^n$, and since the rank of L_0 is just one higher than the rank of $C \cap \mathbb{Z}^n$, the size of $d(L_0)$ has to be realized mainly by one generating vector.

Suppose p and r are short relative to M. Lovász' [13] basis reduction algorithm yields a basis in which the basis vectors are ordered according to increasing length, up to a certain factor. In a basis B_0 for L_0, such as we generate it, the first $n-2$ vectors form a basis for the lattice $C \cap \mathbb{Z}^n$. These vectors are short, while the last, $(n-1)^{\text{st}}$, vector is long since it has to contribute to a large extent to $d(L_0)$.

Example 3. Recall the decomposition of the a-coefficients from Examples 1 and 2. Let $M = 12223$.
$$a_1 = M + 0,$$
$$a_2 = M + 1,$$
$$a_3 = 3M + 3,$$
so $p = (1,1,3)^T$ and $r = (0,1,3)^T$. The first column of B_0 is short. This vector, $(0,-3,1)^T$, is orthogonal to a, p, and r. The second, and last, column of B_0, $(12224, -1222, -3667)^T$, is long. $\square$

The question is now what the implications are for the polytopes P as defined by (1) and Q as defined by (6). For P it means that one step in the direction of the last basis vector b_{n-1} is long, so the width of P is small in direction b_{n-1}. For the reformulation Q it means that the width of Q in the direction e_{n-1} is small. In Example 1 we observed that $W_I(Q, e_2) = -1$, which immediately gave us a certificate for infeasibility. The argument regarding the length of the columns of B_0 presented above also holds in the more general case that the a-coefficients decompose as follows:

$$a_i = p_i M + r_i N, \quad \text{for } i = 1, \dots, n,$$

where $p_i, M, N \in \mathbb{Z}_{\geq 0}$, $r_i \in \mathbb{Z}$, and where M and N are assumed to be large compared to p_i and $|r_i|$.

4   Computational Results

To illustrate our results we have solved various instances of type (2). The instances are given in Table 1. In the first column the instance name is given. Next, in column "a", the a_i-coefficients are given, and in the last column the Frobenius number can be found. For all the instances we computed the Frobenius number using the algorithm described in Appendix 1. The instances can be divided into two groups. The first group contains instances cuww1-cuww5 and prob1-prob10, and the second group consists of instances prob11-prob20. Instances cuww1-cuww5 were generated by Cornuéjols, Urbaniak, Weismantel, and Wolsey [7], and the remaining instances were generated for this study. For each of the instances cuww1-cuww5 there is a decomposition $a_i = p_i M + r_i$ with short vectors p and r. In Table 2 we give values of M that yield short vectors p and r for these instances. Instances prob1-prob10 were generated such that

Table 1. The instances

Instance	a										Frobenius number
cuww1	12223	12224	36674	61119	85569						89643481
cuww2	12228	36679	36682	48908	61139	73365					89716838
cuww3	12137	24269	36405	36407	48545	60683					58925134
cuww4	13211	13212	39638	52844	66060	79268	92482				104723595
cuww5	13429	26850	26855	40280	40281	53711	53714	67141			45094583
prob1	25067	49300	49717	62124	87608	88025	113673	119169			33367335
prob2	11948	23330	30635	44197	92754	123389	136951	140745			14215206
prob3	39559	61679	79625	99658	133404	137071	159757	173977			58424799
prob4	48709	55893	62177	65919	86271	87692	102881	109765			60575665
prob5	28637	48198	80330	91980	102221	135518	165564	176049			62442884
prob6	20601	40429	40429	45415	53725	61919	64470	69340	78539	95043	22382774
prob7	18902	26720	34538	34868	49201	49531	65167	66800	84069	137179	27267751
prob8	17035	45529	48317	48506	86120	100178	112464	115819	125128	129688	21733990
prob9	3719	20289	29067	60517	64354	65633	76969	102024	106036	119930	13385099
prob10	45276	70778	86911	92634	97839	125941	134269	141033	147279	153525	106925261
prob11	11615	27638	32124	48384	53542	56230	73104	73884	112951	130204	577134
prob12	14770	32480	75923	86053	85747	91772	101240	115403	137390	147371	944183
prob13	15167	28569	36170	55419	70945	74926	95821	109046	121581	137695	765260
prob14	1828	14253	46209	52042	55987	72649	119704	129334	135589	138360	680230
prob15	13128	37469	39391	41928	53433	59283	81669	95339	110593	131989	663281
prob16	35113	36869	46647	53560	81518	85287	102780	115459	146791	147097	1109710
prob17	14054	22184	29952	64696	92752	97364	118723	119355	122370	140050	752109
prob18	20303	26239	33733	47223	55486	93776	119372	136158	136989	148851	783879
prob19	20212	30662	31420	49259	49701	62688	74254	77244	139477	142101	677347
prob20	32663	41286	44549	45674	95772	111887	117611	117763	141840	149740	1037608

Table 2. A value of M for instances cuww1–5 yielding short p and r

	cuww1	cuww2	cuww3	cuww4	cuww5
M	12223	12228	12137	13211	13429

the a-coefficients have a decomposition $a_i = p_i M + r_i N$ with short p and r. We randomly generate M from the uniform distribution $U[10000, 20000]$, N from $U[1000, 2000]$, p_i from $U[1, 10]$, and r_i from $U[-10, 10]$. In contrast, the second group of instances prob11-prob20 were randomly generated such that the a-coefficients are of the same size as in prob1-prob10, but they do not necessarily decompose with short vectors p and r. We chose the same size of the a-coefficients since this yields values of $d(L_0)$ of approximately the same size as for the instances prob1-prob10. For instances prob11-prob20 coefficient a_i is randomly generated from $U[10000, 150000]$.

The computational results of verifying infeasibility for the instances is reported on in Table 3. For each of the instances we computed $d(L_0)$, the length of each of the basis vectors of the basis B_0, and the number of lattice hyperplanes intersecting Q in the coordinate directions e_1 and e_{n-1}. We then applied the integer branching algorithm described in Sect. 2 to Q. The number of nodes that

Table 3. Verification of infeasibility

Instance	$d(L_0)$	$\lvert b_i\rvert$	$W_I(Q,e_1)+1$	$W_I(Q,e_{n-1})+1$	# Search tree nodes	Time	# B&B nodes	B&B time
cuww1	112700.5	2.0 3.5 3.5 4823.1	1862	0	1	.001	$> 50\times10^6$	> 8139.3
cuww2	119803.3	2.0 2.2 2.6 3.9 2922.9	1291	1	3	.001	0^*	0.0
cuww3	97088.2	2.0 2.4 2.8 4.0 2218.0	1155	2	3	.001	$> 50\times10^6$	> 8079.9
cuww4	154638.3	1.7 2.4 2.4 4.0 3.0 2726.8	2429	1	2	.001	$> 50\times10^6$	> 7797.5
cuww5	123066.9	2.0 2.2 2.6 2.6 2.8 1711.4	1279	1	3	.001	$> 50\times10^6$	> 6080.6
prob1	227895.5	2.0 2.0 2.6 2.8 3.2 4.7 678.4	347	2	7	.001	$> 50\times10^6$	> 7912.6
prob2	256849.8	1.7 2.0 2.6 3.0 3.2 4.4 1016.0	274	2	7	.001	$> 50\times10^6$	> 6529.2
prob3	337663.2	2.2 2.4 3.0 3.3 3.6 988.4	466	2	11	.002	$> 50\times10^6$	> 6872.1
prob4	226877.3	2.4 2.6 2.6 3.2 3.6 3.5 1058.4	468	2	8	.001	$> 50\times10^6$	> 8432.2
prob5	324461.5	2.0 2.4 2.8 3.0 3.0 3.7 937.6	964	2	10	.002	$> 50\times10^6$	> 8368.4
prob6	191805.0	2.0 2.2 2.2 2.4 2.8 2.6 646.6	502	2	8	.001	$> 50\times10^6$	> 5550.1
prob7	207240.4	1.7 1.7 2.2 2.4 2.4 2.4 888.6	588	2	9	.002	$> 50\times10^6$	> 5411.5
prob8	288168.2	2.2 2.2 2.6 2.4 2.4 2.4 773.4	455	2	7	.001	$> 50\times10^6$	> 5565.4
prob9	235618.6	1.7 2.8 2.6 2.4 2.4 2.8 788.6	430	2	18	.003	$> 50\times10^6$	> 6944.7
prob10	363052.5	2.0 2.2 2.4 2.2 2.4 2.4 1165.2	880	2	10	.002	0^*	0.0
prob11	225420.4	3.6 4.0 4.5 4.4 4.6 4.4 4.7 6.1	4	5	37	.005	88858	9.3
prob12	307211.3	4.4 4.5 4.6 4.6 4.5 4.4 6.0 5.4	2	4	86	.012	445282	51.0
prob13	266246.9	4.6 4.2 4.6 4.0 4.8 5.3 5.1 5.8	6	6	41	.006	580565	62.6
prob14	286676.3	4.4 4.1 4.0 4.4 4.7 5.1 5.1 5.6	9	7	112	.012	371424	43.4
prob15	238047.7	3.6 4.5 4.1 3.9 5.1 4.8 5.5 6.0	3	3	66	.080	426692	49.4
prob16	297717.2	4.0 3.7 3.7 4.2 4.2 4.6 4.7 9.7	3	2	67	.080	549483	61.4
prob17	294591.6	4.6 4.4 4.2 4.6 5.1 4.0 4.2 5.7	2	4	126	.150	218374	24.1
prob18	300087.6	3.5 4.6 4.6 4.5 5.1 5.2 5.5 5.0 5.8	4	5	90	.120	425727	46.9
prob19	249577.9	3.9 4.6 3.7 4.1 5.1 5.2 5.6 4.8 5.5 4.6	11	6	78	.100	255112	27.7
prob20	314283.7	3.7 4.7 4.7 4.5 4.6 4.7 5.1 5.5 6.2	5	3	39	.005	423608	46.1

*) CPLEX Presolve determines problem is infeasible or unbounded.

were generated, and the computing time in seconds are given in the columns "#
Search tree nodes" and "Time". Finally, we attempted to solve the instances,
using the original formulation P, by standard linear programming based branch-
and-bound using CPLEX version 6.5.3 . The number of nodes needed by branch-
and-bound, and the computing time in seconds are reported on in the columns
"# B&B nodes" and "B&B time". For the branch-and-bound algorithm we set
the node limit to 50 million nodes. If an instance was not solved within this
node limit, this is indicated by "$> 50 \times 10^6$" in the column "# B&B nodes".
The time t needed to evaluate the 50 million nodes is then indicated as "$> t$" in
the column "B&B time". All the computations were carried out on a Sun Ultra
60 Model 2360 workstation with two UltraSPARC-II 359 MHz processors (our
implementation is sequential) and 512 MB of memory.

We make the following observations. First, the Frobenius number of the instances cuww1-cuww5 and prob1-prob10 is about two orders of magnitude larger
than the Frobenius number of instances prob11-prob20 (see Table 1). Infeasible instances are typically harder to solve than feasible ones, and the larger the
intersection points a_0/a_i between the n-simplex P and the coordinate axes, the
harder the instance becomes for branch-and-bound. So, as a class, the first group
of instances is harder for branch-and-bound, even if we would use a smaller
value of the right hand side coefficient than the Frobenius number, since the
large Frobenius number indicates that there are other large integer values α for
which $ax = \alpha$ has no nonnegative integer solution. In Table 3 we can see that
instances cuww1-cuww5 and prob1-prob10 are considerably harder to solve by
branch-and-bound than instances prob11-prob20. The presolver of CPLEX was
able to verify infeasibility for instances cuww2 and prob10, but none of the other
instances in the first group was solved within the node limit of 50 million nodes.
All of the instances prob11-prob20 were solved by branch-and-bound within
half a million search nodes and one minute of computing time.

We also observe that the shape of the polytope Q is very much influenced by
the decomposition of the a-coefficients. If the coefficients decompose with short
vectors p and r relative to M, then the width of the corresponding polytope
Q in the unit direction e_{n-1} is very small. This made the instances trivial for
our tree search algorithm applied to Q. All instances were solved using less than
twenty search nodes and a fraction of a second computing time. For instances
prob11-prob20 where the a-coefficients are generated randomly from a certain
interval we observe that the width of Q is of the same size in all unit directions,
and in general greater than two. Our tree search algorithm applied to Q therefore needed more nodes and longer computing times than for the first group
of instances. Still, none of the instances prob11-prob20 needed more than 126
nodes and about a tenth of a second computing time.

Acknowledgments

We want to thank Hendrik Lenstra for his valuable suggestions and in particular
for the outline of the proof of Theorem 1. We also wish to thank Bram Verweij

for providing a framework code, based on his general enumeration library, for our integer branching algorithm.

The research of the first author was partially financed by the project TMR-DONET nr. ERB FMRX-CT98-0202 of the European Community.

References

1. Aardal K., Bixby R.E., Hurkens C.A.J., Lenstra A.K., Smeltink J.W. (1999) Market Split and Basis Reduction: Towards a Solution of the Cornuéjols-Dawande Instances. In: Cornuéjols G., Burkard R.E., Woeginger G.J. (eds.) *Integer Programming and Combinatorial Optimization, 7th International IPCO Conference.* Lecture Notes in Computer Science 1610. Springer-Verlag, Berlin, Heidelberg, 1–16
2. Aardal K., Hurkens, C.A.J., Lenstra A.K. (2000) Solving a system of diophantine equations with lower and upper bounds on the variables. *Mathematics of Operations Research* 25:427–442
3. Brauer A. (1942) On a problem of partitions. *American Journal of Mathematics* 64:299–312
4. Brauer A., Shockley J.E. (1962) On a problem of Frobenius. *Journal für reine und angewandte Mathematik* 211:399–408
5. Cook W., Rutherford T., Scarf H. E., Shallcross D. (1993) An implementation of the generalized basis reduction algorithm for integer programming. *ORSA Journal on Computing* 5:206–212
6. Cornuéjols G., Dawande M. (1998) A class of hard small 0-1 programs. In: Bixby R.E., Boyd E.A., Ríos-Mercado R.Z. (eds.) *Integer Programming and Combinatorial Optimization, 6th International IPCO Conference.* Lecture Notes in Computer Science 1412. Springer-Verlag, Berlin Heidelberg, 284–293
7. Cornuéjols G., Urbaniak R., Weismantel R., Wolsey L.A. (1997) Decomposition of integer programs and of generating sets. In: Burkard R.E., Woeginger G.J. (eds.) *Algorithms – ESA '97.* Lecture Notes in Computer Science **1284**. Springer-Verlag, Berlin, Heidelberg, 92–103
8. Erdős P., Graham R.L. (1972) On a linear diophantine problem of Frobenius. *Acta Arithmetica* 21:399–408
9. Greenberg H. (1988) Solution to a linear Diophantine equation for nonnegative integers. *Journal of Algorithms* 9:343–353
10. Hungerford T.W. (1996) *Algebra*; corrected eighth printing. Springer-Verlag, New York
11. Kannan R. (1991) Lattice translates of a polytope and the Frobenius Problem. *Combinatorica* 12:161–177
12. Khinchine A. (1948) A quantitative formulation of Kronecker's theory of approximation (In Russian). *Izvestiya Akademii Nauk SSR Seriya Matematika* 12:113–122
13. Lenstra A.K., Lenstra H.W., Jr., Lovász L. (1982) Factoring polynomials with rational coefficients. *Mathematische Annalen* 261:515–534
14. Lenstra H.W., Jr., (1983) Integer programming with a fixed number of variables. *Mathematics of Operations Research* 8:538–548
15. Lovász L., Scarf H.E. (1992) The generalized basis reduction algorithm, *Mathematics of Operations Research* 17:751–764
16. Louveaux Q., Wolsey L.A. (2000) Combining problem structure with basis reduction to solve a class of hard integer programs. CORE Discussion Paper 2000/51, CORE, Université Catholique de Louvain, Louvain-la-Neuve, Belgium (to appear in *Mathematics of Operations research*)

17. Rödseth Ö.J. (1978) On a linear diophantine problem of Frobenius. *Journal für reine und angewandte Mathematik* 301:171–178
18. Schrijver A. (1986) *Theory of linear and integer programming*, Wiley, Chichester
19. Selmer E.S. (1977) On the linear diophantine problem of Frobenius. *Journal für reine und angewandte Mathematik* 93/294:1–17
20. Selmer E.S., Beyer Ö. (1978) On the linear diophantine problem of Frobenius in three variables. *Journal für reine und angewandte Mathematik* 301:161–170
21. Sylvester J.J., Curran Sharp W.J. (1884). [Problem] 7382. *Mathematics from the Educational Times, with Additional Papers and Solutions* 41:21

Appendix 1: Computing the Frobenius Number

Since the main aim of this paper is not to compute the Frobenius number – we use the Frobenius number to create infeasible instances – our approach is quite simple and based on the following theorem by Brauer and Shockley [4].

Theorem 3. [4] *Let $a_1, \ldots, a_n$ be positive integer numbers with $\gcd(a_1, \ldots, a_n) = 1$. Given $i \in \{1, \ldots, n\}$, let t_l be the smallest positive integer congruent to l mod a_i that can be expressed as a nonnegative integer combination of a_j, $1 \leq j \leq n$, $j \neq i$. Then*

$$F(a_1, \ldots, a_n) = \max_{l \in \{1, 2, \ldots, a_i - 1\}} t_l - a_i . \tag{11}$$

One can compute the Frobenius number using a sequence of integer optimization problems as follows. To compute as few optimization problems as possible we assume that a_i in Theorem 3 is the smallest of the $\boldsymbol{a}$-coefficients. Here we assume that a_1 is the smallest $\boldsymbol{a}$-coefficient. The value t_l for $l = 1, \ldots, a_1 - 1$ can be computed as:

$$t_l = \min\{\sum_{i=2}^{n} a_i x_i : \sum_{i=2}^{n} a_i x_i = l + a_1 x_1, \ \boldsymbol{x} \in \mathbb{Z}_{\geq 0}^n\} . \tag{12}$$

Since the instances of type (12) that we tackled are hard to solve by branch-and-bound we again applied the reformulation described in Sect. 2 to each subproblem and solved the reformulated subproblems by branch-and-bound. Notice that the reformulation only has to be determined for $l = 1$. The basis for $L = \{x \in \mathbb{Z}^n : -a_1 x_1 + \sum_{i=2}^{n} a_i x_i = 0\}$ is independent of l, and if we have computed $\boldsymbol{x}_f$ for $l = 1$, then $l\boldsymbol{x}_f$ can be used in the subsequent computations of subproblems $l = 2, \ldots, a_1 - 1$. Cornuéjols et al. [7] used a formulation similar to (12) for computing the Frobenius number, but instead of using the reformulation described in Sect. 2 combined with branch-and-bound, they used test sets after having decomposed the $\boldsymbol{a}$-coefficients.

In Table 4 we give the computational results for the Frobenius number computations. In the two first columns the instance name and number of variables are given. Then, the computing time and the total number of branch-and-bound nodes needed for all $a_1 - 1$ subproblems are given. Since a_1 can vary quite a lot, we report on the average number of branch-and-bound nodes per subproblem in the last column.

Table 4. Results for the Frobenius number computations

Instance	# Vars	Time	Total # B&B nodes	Ave. # nodes per subprob.
cuww1	5	50.0	11652	1.0
cuww2	6	62.3	25739	2.1
cuww3	6	64.6	39208	3.2
cuww4	7	76.3	28980	2.2
cuww5	8	130.2	210987	15.7
prob1	8	891.3	3782264	150.9
prob2	8	90.2	53910	4.5
prob3	8	396.2	571199	14.4
prob4	8	371.1	204191	4.2
prob5	8	257.6	349320	12.2
prob6	10	9057.3	39164012	1901.1
prob7	10	200.7	93987	5.0
prob8	10	304.8	577948	33.9
prob9	10	162.6	91223	24.5
prob10	10	586.8	445777	9.8
prob11	10	241.3	577134	49.7
prob12	10	515.8	1518531	102.8
prob13	10	391.8	998415	65.8
prob14	10	476.7	1551241	848.6
prob15	10	418.0	1178543	89.8
prob16	10	821.7	2063690	58.8
prob17	10	385.4	1027115	73.1
prob18	10	567.3	1494456	73.6
prob19	10	499.0	1289971	63.8
prob20	10	799.2	2070667	63.4

The Distribution of Values
in the Quadratic Assignment Problem

Alexander Barvinok and Tamon Stephen*

Department of Mathematics
University of Michigan
Ann Arbor, MI, 48109-1109
{barvinok,tamon}@umich.edu

Abstract. We obtain a number of results regarding the distribution of values of a quadratic function f on the set of $n \times n$ permutation matrices (identified with the symmetric group S_n) around its optimum (minimum or maximum). We estimate the fraction of permutations σ such that $f(\sigma)$ lies within a given neighborhood of the optimal value of f and relate the optimal value with the average value of f over a neighborhood of the optimal permutation. We describe a natural class of functions (which includes, for example, the objective function in the Traveling Salesman Problem) with a relative abundance of near-optimal permutations. Also, we identify a large class of functions f with the property that permutations close to the optimal permutation in the Hamming metric of S_n tend to produce near optimal values of f, and show that for general f just the opposite behavior may take place.

1 Introduction

The Quadratic Assignment Problem (QAP for short) is an optimization problem on the symmetric group S_n of $n!$ permutations of an n-element set. The QAP is one of the hardest problems of combinatorial optimization, whose special cases include the Traveling Salesman Problem (TSP) among other interesting problems.

Recently the QAP has been of interest to many people. An excellent survey is found in [5]. Despite this work, it is still extremely difficult to solve QAP's of size $n = 20$ to optimality, and the solution to a QAP of size $n = 30$ is considered noteworthy, see, for example, [1] and [4]. Moreover, it appears that essentially no positive approximability results for the general QAP are known, although some special cases have been established, see [2], and it is known that it is NP-complete to get an approximation with even exponential fall-off for MIN QAP with positive coefficients [3].

The goal of this paper is to study the distribution of values of the objective function of the QAP. We hope that our results would allow one on one hand to understand the behavior of various heuristics, and, on the other hand, to

---

* This research was partially supported by NSF Grant DMS 9734138.

W.J. Cook and A.S. Schulz (Eds.): IPCO 2002, LNCS 2337, pp. 367–383, 2002.
© Springer-Verlag Berlin Heidelberg 2002

estimate the optimum using some simple algorithms based on random or partial enumeration with guaranteed complexity bounds. In particular, we estimate how well the sample optimum from a random sample of a given size approximates the global optimum. This gives an approximation guarantee with respect to the average.

1.1   The Quadratic Assignment Problem

Let Mat_n be the vector space of all real $n \times n$ matrices $A = (a_{ij})$, $1 \leq i, j \leq n$ and let S_n be the group of all permutations σ of the set $\{1, \ldots, n\}$. There is an action of S_n on the space Mat_n by simultaneous permutations of rows and columns: we let $\sigma(A) = B$, where $A = (a_{ij})$ and $B = (b_{ij})$, provided $b_{\sigma(i)\sigma(j)} = a_{ij}$ for all $i, j = 1, \ldots, n$. One can check that $(\sigma\tau)A = \sigma(\tau A)$ for any two permutations σ and τ. There is a standard scalar product on Mat_n :

$$\langle A, B \rangle = \sum_{i,j=1}^{n} a_{ij}b_{ij} \ , \quad \text{where} \quad A = (a_{ij}) \quad \text{and} \quad B = (b_{ij}) \ .$$

Let us fix two matrices $A = (a_{ij})$ and $B = (b_{ij})$ and let us consider a real-valued function $f : S_n \longrightarrow \mathbb{R}$ defined by

$$f(\sigma) = \langle B, \sigma(A) \rangle = \sum_{i,j=1}^{n} b_{\sigma(i)\sigma(j)}a_{ij} = \sum_{i,j=1}^{n} b_{ij}a_{\sigma^{-1}(i)\sigma^{-1}(j)} \ . \tag{1}$$

The problem of finding a permutation σ where the maximum or minimum value of f is attained is known as the *Quadratic Assignment Problem*. It is one of the hardest problems of Combinatorial Optimization. From now on we assume that $n \geq 4$.

Our approach produces essentially identical results for a more general problem. Suppose we are given a 4-dimensional array (tensor) $C = \left\{ c_{kl}^{ij} : 1 \leq i, j, k, l \leq n \right\}$ of n^4 real numbers and the function f is defined by

$$f(\sigma) = \sum_{i,j=1}^{n} c_{\sigma(i)\sigma(j)}^{ij} \ . \tag{2}$$

If $c_{kl}^{ij} = a_{ij}b_{kl}$ for some matrices $A = (a_{ij})$ and $B = (b_{kl})$, we get the special case (1) we started with. The convenience of working with the generalized problem is that the set of objective functions (2) is a vector space.

We remark in Sect. 2.1 that the Traveling Salesman Problem (TSP) can be considered as a special case of the QAP.

1.2   Definitions

For two permutations $\tau, \sigma \in S_n$, let the distance $\text{dist}(\sigma, \tau)$ be the standard Hamming distance, that is the number of indices $1 \leq i \leq n$ where σ and τ disagree:

$$\text{dist}(\tau, \sigma) = |i : \sigma(i) \neq \tau(i)| \ .$$

One can observe that the distance is invariant under the left and right actions of S_n :

$$\text{dist}(\sigma\sigma_1, \sigma\sigma_2) = \text{dist}(\sigma_1, \sigma_2) = \text{dist}(\sigma_1\sigma, \sigma_2\sigma)$$

for all $\sigma_1, \sigma_2, \sigma \in S_n$. For a permutation τ and an integer $k \geq 0$, we consider the "k-th ring" around τ :

$$U(\tau, k) = \{\sigma \in S_n : \text{dist}(\sigma, \tau) = n - k\} \ .$$

Hence for any permutation τ the group S_n splits into the disjoint union of n rings $U(\tau, k)$ for $k = 0, 1, \ldots, n - 2, n$. The ring $U(\tau, k)$ consists of the permutations that agree with τ on precisely k symbols.

Let $f : S_n \longrightarrow \mathbb{R}$ be a function of type (1) or (2). Let

$$\overline{f} = \frac{1}{n!} \sum_{\sigma \in S_n} f(\sigma)$$

be the average value of f on the symmetric group and let

$$f_0 = f - \overline{f}$$

be the "shifted" function. Hence the average value of f_0 is 0. Let τ be a permutation where the maximum value of f_0 is attained, so $f_0(\tau) \geq f_0(\sigma)$ for all $\sigma \in S_n$ and $f_0(\tau) > 0$ unless $f_0 \equiv 0$ (the problem with minimum instead of maximum is completely similar). We maintain the definitions of $\overline{f}$, f_0 and τ in future sections.

We remark that it is easy to compute the average value $\overline{f}$, see Lemma 10.

We are interested in the following questions:

• Given a constant $0 < \gamma < 1$, how many permutations $\sigma \in S_n$ satisfy $f_0(\sigma) \geq \gamma f_0(\tau)$? In particular, how well does the sample optimum of a set of randomly chosen permutations approximates the true optimum?

• How does the average value of f_0 over the k-th ring $U(\tau, k)$ compare with the optimal value $f_0(\tau)$? In particular, is a random permutation from the vicinity of the optimal permutation better than a random permutation from the whole group S_n?

We define the function $\nu(m)$ for $m \geq 2$:

$$\nu(m) = \frac{d_{m-2}}{d_m} \qquad \text{where} \qquad d_m = \sum_{j=0}^{m}(-1)^j \frac{1}{j!} \ .$$

For $m = 0$ we let $\nu(m) = 0$. From the definition $0 \leq \nu(m) \leq 2$, and $\nu(m)$ approaches 1 rapidly as m increases.

The following functions p and t on the symmetric group S_n play a special role in our approach. For a permutation $\sigma \in S_n$, let

$$p(\sigma) = |i : \sigma(i) = i|$$

be the number of fixed points of σ and let

$$t(\sigma) = \big| i < j : \sigma(i) = j \quad \text{and} \quad \sigma(j) = i \big|$$

be the number of 2-cycles in σ.

One can show that $p(\sigma)$, $p^2(\sigma)$ and $t(\sigma)$ are functions of type (2) for some particular tensors $\{c^{ij}_{kl}\}$. We denote by ε the identity permutation in S_n and by $|X|$ the cardinality of a set X.

2   Results

Our results are divided into four cases, which arise from considering the representations of Mat_n under the action of S_n (see Sect. 3.1). We begin with the most special case, which we call the "bullseye," and which already includes the symmetric TSP. We then consider two different generalizations of the bullseye – the "pure" QAP and the symmetric QAP. Finally, we briefly mention the general case, which is technically more complicated than the symmetric case, but has essentially similar bounds.

The proofs are discussed in Sect. 3.

2.1   Bullseye Distribution

Suppose that the matrix $A = (a_{ij})$ in (1) is symmetric and has constant row and column sums and a constant diagonal. For example,

$$A = \begin{pmatrix} 0 & 1 & 0 & \dots & 0 & 1 \\ 1 & 0 & 1 & \dots & 0 & 0 \\ 0 & 1 & 0 & \dots & 0 & 0 \\ \dots & \dots & \dots & \dots & \dots & \dots \\ 0 & 0 & 0 & \dots & 0 & 1 \\ 1 & 0 & 0 & \dots & 1 & 0 \end{pmatrix} , \quad a_{ij} = \begin{cases} 1 & \text{if } |i - j| = 1 \bmod n \\ 0 & \text{otherwise} \end{cases}$$

satisfies these properties and the corresponding optimization problem is the *Symmetric Traveling Salesman Problem*.

Similarly, for the generalized problem (2) we assume that for any k and l the matrix $A = (a_{ij})$, where $a_{ij} = c^{ij}_{kl}$, is symmetric with constant row and column sums and has a constant diagonal.

It turns out that the optimum has a characteristic "bullseye" feature with respect to the averages over the rings $U(\tau, k)$.

Theorem 1 (Bullseye Distribution). *Let*

$$\alpha(n, k) = \frac{k^2 - 3k + \nu(n - k)}{n^2 - 3n} ,$$

where $k = 0, 1, \dots, n - 2, n$ and ν is as defined in Sect. 1.2.

Then we have

$$\frac{1}{|U(\tau,k)|} \sum_{\sigma \in U(\tau,k)} f_0(\sigma) = \alpha(n,k)f_0(\tau)$$

for $k = 0, 1, \ldots, n-2, n$.

We observe that as the ring $U(\tau, k)$ contracts to the optimal permutation τ (hence k increases), the *average* value of f on the ring steadily improves (as long as $k \geq 3$). It is easy to construct examples where *some* values of f in a very

Distribution of values of the objective function with respect to the Hamming distance

from the maximum point

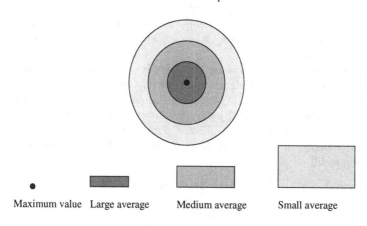

Maximum value   Large average   Medium average   Small average

Fig. 1. Bullseye distribution

small neighborhood of the optimum are particularly bad, but as follows from Theorem 1, such values are relatively rare. In our opinion, this suggests that this special case may be more amenable to various heuristics (see, for example, [5]) than the general symmetric case. Incidentally, one can observe the same type of the "bullseye" behavior for the Linear Assignment Problem and some other polynomially solvable problems, such as the weighted Matching Problem.

Estimating the size of the ring $U(\tau, k)$, we get the following result.

Corollary 2. *Let us choose an integer* $3 \leq k \leq n - 3$ *and a number* $0 < \gamma < 1$ *and let*

$$\beta(n,k) = \frac{k^2 - 3k}{n^2 - 3n} .$$

The probability that a random permutation $\sigma \in S_n$ *satisfies the inequality*

$$f_0(\sigma) \geq \gamma\beta(n,k)f_0(\tau)$$

is at least

$$\frac{(1-\gamma)\beta(n,k)}{3k!} .$$

2.2 Pure Distribution

We can consider a slightly more general problem where we relax the condition that A is symmetric, but keep the constant row and sum columns. For example, the matrix

$$A = \begin{pmatrix} 0 & 1 & 0 & \ldots & \ldots & 0 \\ 0 & 0 & 1 & 0 & \ldots & 0 \\ \ldots & \ldots & \ldots & \ldots & \ldots & \ldots \\ 0 & 0 & \ldots & \ldots & 0 & 1 \\ 1 & 0 & \ldots & \ldots & \ldots & 0 \end{pmatrix}, \qquad a_{ij} = \begin{cases} 1 & \text{if } j = i+1 \bmod n \\ 0 & \text{otherwise} \end{cases}$$

satisfies these properties and the corresponding optimization problem is the *Asymmetric Traveling Salesman Problem.*

Similarly, for generalized problems (2) we assume that for any k and l the matrix $A = (a_{ij})$, where $a_{ij} = c_{kl}^{ij}$ has constant row and column sums and has a constant diagonal.

We call this case pure, because the objective function f lacks the component attributed to the Linear Assignment Problem. Generally, an arbitrary objective function f in the Quadratic Assignment Problem can be represented as a sum $f = f_1 + f_2$, where f_1 is the objective function in a Linear Assignment Problem and f_2 is the objective function in some pure case.

The behavior of averages of f_0 over the rings $U(\tau, k)$ is described by the following result.

Theorem 3. *Let us define three functions of n and k:*

$$\alpha_1(n, k) = 1 - \nu(n - k)$$
$$\alpha_{2e}(n, k) = \frac{k^2 - 3k - n - 3\nu(n - k) + \nu(n - k)n + 4}{n^2 - 4n + 4} \quad and$$
$$\alpha_{2o}(n, k) = \frac{k^2 - 3k - n - 2\nu(n - k) + \nu(n - k)n + 3}{n^2 - 4n + 3},$$

where $k = 0, 1, \ldots, n - 2, n$ and ν is the function of Sect. 1.2.

If n is even, then for some non-negative γ_1 and γ_2 such that $\gamma_1 + \gamma_2 = 1$ we have

$$\frac{1}{|U(\tau, k)|} \sum_{\sigma \in U(\tau, k)} f_0(\sigma) = \Big(\gamma_1 \alpha_1(n, k) + \gamma_2 \alpha_{2e}(n, k) \Big) f_0(\tau)$$

for $k = 0, 1, \ldots, n - 2, n$.

If n is odd, then for some non-negative γ_1 and γ_2 such that $\gamma_1 + \gamma_2 = 1$ we have

$$\frac{1}{|U(\tau, k)|} \sum_{\sigma \in U(\tau, k)} f_0(\sigma) = \Big(\gamma_1 \alpha_1(n, k) + \gamma_2 \alpha_{2o}(n, k) \Big) f_0(\tau)$$

for $k = 0, 1, \ldots, n - 2, n$.

We observe that there are two extreme cases. If $\gamma_1 = 0$ and $\gamma_2 = 1$ then f exhibits a bullseye type distribution of Sect. 2.1. If $\gamma_1 = 1$ and $\gamma_2 = 0$ then f

exhibits a "damped oscillator" type of distribution: the average value of f_0 over $U(\tau, k)$ changes its sign with the parity of k and approaches 0 fast as k grows. In short, if f has a damped oscillator distribution, there is no particular advantage in choosing a permutation in the vicinity of the optimal permutation τ. For a

Distribution of values of the objective function with respect

to the Hamming distance from the maximum point

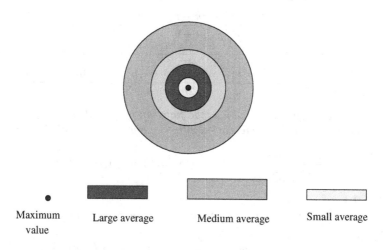

Maximum value   Large average   Medium average   Small average

Fig. 2. Damped oscillator distribution

typical function f one can expect both γ_1 and γ_2 positive, so f would show a "weak" bullseye distribution: the average value of f_0 over $U(\tau, k)$ improves moderately as k gets smaller, but not as dramatically as in the bullseye case of Theorem 1.

Still, it turns out that we can sufficiently many reasonably good permutations to get a version of Corollary 2 with only slightly worse constants.

Corollary 4. *Let us choose an integer $3 \leq k \leq n - 5$ and a number $0 < \gamma < 1$ and let*

$$\beta(n, k) = \frac{k^2 - 3k + 1}{n^2 - 3n + 1} \ .$$

The probability that a random permutation $\sigma \in S_n$ satisfies the inequality

$$f_0(\sigma) \geq \gamma \beta(n, k) f_0(\tau)$$

is at least

$$\frac{(1 - \gamma)\beta(n, k)}{10k!} \ .$$

2.3 Symmetric Distribution

We now consider what happens when we relax the condition of constant row and column sums. We will focus on the case of a symmetric matrix A, but the situation for general A is similar. Overall, the distribution of values of f turns out to be much more complicated than in the bullseye special case.

Theorem 5 (Symmetric Distribution). *Let us define three functions of n and k:*

$$\alpha_1(n, k) = \frac{2nk - 2n - k^2 - 3k - \nu(n - k) + 6}{n^2 - 5n + 6}$$

$$\alpha_{2e}(n, k) = \frac{-nk + n + k^2 + k + \nu(n - k) - 4}{2n - 4} \quad and$$

$$\alpha_{2o}(n, k) = \frac{-n^2k + nk^2 + n^2 + nk + n\nu(n - k) - 4n - 3k + 3}{2n^2 - 7n + 3} \quad,$$

where $k = 0, 1, \ldots, n - 2, n$ and ν is the function of Sect. 1.2.

If n is even, then for some non-negative γ_1 and γ_2 such that $\gamma_1 + \gamma_2 = 1$ we have

$$\frac{1}{|U(\tau, k)|} \sum_{\sigma \in U(\tau,k)} f_0(\sigma) = \Big(\gamma_1 \alpha_1(n, k) + \gamma_2 \alpha_{2e}(n, k)\Big) f_0(\tau)$$

for $k = 0, 1, \ldots, n - 2, n$.

If n is odd, then for some non-negative γ_1 and γ_2 such that $\gamma_1 + \gamma_2 = 1$ we have

$$\frac{1}{|U(\tau, k)|} \sum_{\sigma \in U(\tau,k)} f_0(\sigma) = \Big(\gamma_1 \alpha_1(n, k) + \gamma_2 \alpha_{2o}(n, k)\Big) f_0(\tau)$$

for $k = 0, 1, \ldots, n - 2, n$.

For any choice of $\gamma_1, \gamma_2 \geq 0$ such that $\gamma_1 + \gamma_2 = 1$ there is a function f of type (2) for which the averages of f_0 over $U(\tau, k)$ are given by the formulas of Theorem 5. Moreover, at least in the case where n is even, one can choose f to be a function of type (1). We plan to include these proofs in [9].

Remark 6 (Spike Distribution). In Theorem 5, we see that there are two extreme cases. If $\gamma_1 = 1$ and $\gamma_2 = 0$ then f has a bullseye type distribution described in Sect. 2.1. If $\gamma_1 = 0$ and $\gamma_2 = 1$ then f has what we call a "spike" distribution. In this case, for $2 \leq k \leq n - 3$ the average value of f_0 over $U(\tau, k)$ is negative. Thus an average permutation $\sigma \in U(\tau, n - 3)$ presents us with a worse choice than the average permutation in S_n. However, the average value of f_0 over $U(\tau, 0)$ is about one half of the maximum value $f_0(\tau)$. Thus there are plenty of reasonably good permutations very far from τ and we can easily get such a permutation by random sampling.

The distribution of a typical function f around its optimal permutation in the symmetric QAP is a certain mixture of the bullseye and spike distributions. This interference of the bullseye and spike distributions (which, in some sense, are "pulling in the opposite directions") provides, in our opinion, a plausible

Distribution of values of the objective function with respect to the Hamming distance

from the maximum point

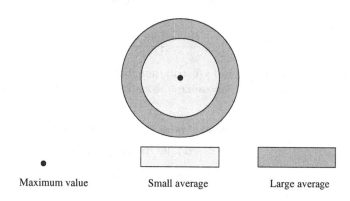

Maximum value   Small average   Large average

Fig. 3. Spike distribution

explanation of the computational hardness of the general symmetric QAP even in comparison with other NP-hard problems such as TSP.

We obtain the following estimate for the number of near-optimal permutations.

Corollary 7. *Let us choose an integer* $3 \leq k \leq n - 3$ *and a number* $0 < \gamma < 1$ *and let*

$$\beta(n, k) = \frac{3k - 5}{n^2 - kn + k + 2n - 5} .$$

The probability that a random permutation $\sigma \in S_n$ *satisfies the inequality*

$$f_0(\sigma) \geq \gamma\beta(n, k)f_0(\tau)$$

is at least

$$\frac{(1 - \gamma)\beta(n, k)}{5k!2^k} .$$

One can notice that the obtained bound is essentially weaker than the bound of Corollary 2.

2.4   Notes on General Distribution and Algorithms

If we consider the general (possibly asymmetric) case, the distribution is more complicated. An arbitrary general distribution can be expressed as a convex combination of 4 extreme distributions if n is even, and 5 if n is odd. Using this, we can derive a bound not much worse than that of Corollary 7.

Corollary 8. *Let us choose an integer* $3 \leq k \leq n - 5$ *and a number* $0 < \gamma < 1$ *and let*

$$\beta(n, k) = \frac{k - 2}{n^2 - nk + k - 2} .$$

The probability that a random permutation $\sigma \in S_n$ satisfies

$$f_0(\sigma) \geq \gamma\beta(k,n)f_0(\tau)$$

is at least

$$\frac{(1-\gamma)\beta(k,n)}{5k!}.$$

We do not get any new type of a distribution, but we can find a sharper spike than in the symmetric case. It is possible to construct a function f of type (2) where

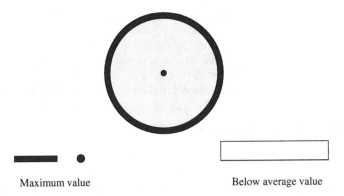

Distribution of values of the objective function

with respect to the Hamming distance

from the maximum point

Maximum value   Below average value

Fig. 4. Sharp spike distribution

the maximum value of 1 is attained at the identity and at every permutations without fixed points. The value of f is negative on any permutation with at least 1 and at most $n-2$ fixed points.

By choosing the appropriate k in Corollaries 4 and 8, we get the following algorithmic interpretation of our results.

Corollary 9. *1. Let us fix any $\alpha > 1$. Then there exists a $\delta = \delta(\alpha) > 0$ such that for all sufficiently large $n \geq N(\alpha)$ the probability that a random permutation σ in S_n satisfies the inequality*

$$f_0(\sigma) \geq \frac{\alpha}{n^2} f_0(\tau)$$

is at least δn^{-2}. In particular, one can choose $\delta = \exp\{-c\alpha \ln \alpha\}$ for some absolute constant $c > 0$. If, in problem (1), matrix A has constant row and column sums, then one can choose $\delta = \exp\{-c\sqrt{\alpha} \ln \alpha\}$.

2. *Let us fix any $\epsilon > 0$. Then there exists a $\delta = \delta(\epsilon) < 1$ such that for all sufficiently large $n \geq N(\epsilon)$ the probability that a random permutation σ in S_n satisfies the inequality*

$$f_0(\sigma) \geq n^{-1-\epsilon} f_0(\tau)$$

is at least $\exp\{-n^\delta\}$. In particular, one can choose any $\delta > 1 - \epsilon$.
If, in problem (1), matrix A has constant row and column sums, this bound improves to

$$f_0(\sigma) \geq n^{-\epsilon} f_0(\tau)$$

for any $\delta > 1 - \epsilon/2$.

We conclude that for any fixed $\alpha > 1$ there is a randomized algorithm of $O(n^2)$ complexity which produces a permutation σ satisfying part 1 of Corollary 9. This simple randomized algorithm produces a better bound in a more general situation than some known deterministic algorithms based on semidefinite relaxations of the QAP (cf. [10] and references therein).

If we are willing to settle for an algorithm of mildly exponential complexity (that is, $\exp\{n^\beta\}$ for some $\beta < 1$), we can achieve the bound of Corollary 9, part 2.

3   Methods

First, we comment that it is easy to compute the average value $\overline{f}$ of a function f defined by (1) or (2).

Lemma 10. *Let $f : S_n \longrightarrow \mathbb{R}$ be a function defined by (1) for some $n \times n$ matrices $A = (a_{ij})$ and $B = (b_{ij})$. Let us define*

$$\alpha_1 = \sum_{1 \leq i \neq j \leq n} a_{ij}, \quad \alpha_2 = \sum_{i=1}^{n} a_{ii} \quad and$$

$$\beta_1 = \sum_{1 \leq i \neq j \leq n} b_{ij}, \quad \beta_2 = \sum_{i=1}^{n} b_{ii} \ .$$

Then

$$\overline{f} = \frac{\alpha_1 \beta_1}{n(n-1)} + \frac{\alpha_2 \beta_2}{n} \ .$$

Similarly, if f is defined by (2) for some tensor $C = \{c_{kl}^{ij}\}$, $1 \leq i, j, k, l \leq n$ then

$$\overline{f} = \frac{1}{n(n-1)} \sum_{1 \leq i \neq j \leq n} \sum_{1 \leq k \neq l \leq n} c_{kl}^{ij} + \frac{1}{n} \sum_{1 \leq i, l \leq n} c_{ll}^{ii} \ .$$

The proof is found in [8].

Remark 11. Suppose that $f(\sigma) = \langle B, \sigma(A) \rangle$ for some matrices A and B and all $\sigma \in S_n$ is the objective function in the QAP (1) and suppose that the maximum value of f is attained at a permutation τ. Let $A_1 = \tau(A)$ and let $f_1(\sigma) = \langle B, \sigma(A_1) \rangle$. Then $f_1(\sigma) = f(\sigma\tau)$, hence the maximum value of f_1 is attained at the identity permutation ε and the distribution of values of f and f_1 is the same. We observe that if A is symmetric then A_1 is also symmetric, and if A has constant row and column sums and a constant diagonal then so does A_1. Hence, as far as the distribution of values of f is concerned, without loss of generality we may assume that the maximum of f is attained at the identity permutation ε.

Next, we introduce our main tool.

Definition 12. *Let* $f : S_n \longrightarrow \mathbb{R}$ *be a function. Let us define function* $g : S_n \longrightarrow \mathbb{R}$ *by*

$$g(\sigma) = \frac{1}{n!} \sum_{\omega \in S_n} f(\omega^{-1}\sigma\omega) \ . \tag{3}$$

We call g *the* central projection *of* f.

It turns out that the central projection captures some important information regarding the distribution of values of a function.

Lemma 13. *Let* $f : S_n \longrightarrow \mathbb{R}$ *be a function and let* g *be the central projection of* f. *Then*

- *The averages of* f *and* g *over the* k-*th ring* $U(\varepsilon, k)$ *around the identity permutation coincide:*

$$\frac{1}{|U(\varepsilon, k)|} \sum_{\sigma \in U(\varepsilon, k)} f(\sigma) = \frac{1}{|U(\varepsilon, k)|} \sum_{\sigma \in U(\varepsilon, k)} g(\sigma) \ ;$$

- *The average values of* f *and* g *on the symmetric group coincide:* $\overline{f} = \overline{g}$;
- *Suppose that* $f(\varepsilon) \geq f(\sigma)$ *for all* $\sigma \in S_n$. *Then* $g(\varepsilon) \geq g(\sigma)$ *for all* $\sigma \in S_n$.

We study the distribution of values in the central projection and, using Lemma 13, deduce various facts about the distribution of values of objective functions of types (1) and (2).

3.1   Action of the Symmetric Group in the Space of Matrices

The crucial observation for our approach is that the vector space of all central projections g of functions f defined by (1) or (2) is 4-, 3-, or 2- dimensional depending on whether we consider the general case, the symmetric case, or the bullseye case of Sect. 2. If we require, additionally, that $\overline{f} = 0$ then the dimensions drop by 1 to 3, 2 and 1, respectively. This fact is explained by the representation theory of the symmetric group (see, for example, [6]).

In particular, Mat_n decomposes as the sum of four *isotypical components* of the irreducible representations of the symmetric group in the space of matrices. We write $\mathrm{Mat}_n = L_n + L_{n-1,1} + L_{n-2,2} + L_{n-2,1,1}$ where each of the four subspaces is invariant under conjugation by S_n. We describe these subspace, recalling that $n \geq 4$.

Subspace L_n: Let L_n^1 be the space of constant matrices A:

$$a_{ij} = \alpha \quad \text{for some} \quad \alpha \quad \text{and all} \quad 1 \le i, j \le n \ .$$

Let L_n^2 be the subspace of scalar matrices A:

$$a_{ij} = \begin{cases} \alpha & \text{if } i = j \\ 0 & \text{if } i \ne j \end{cases} \quad \text{for some} \quad \alpha \ .$$

Finally, Let $L_n = L_n^1 + L_n^2$. One can observe that $\dim L_n = 2$ and that L_n is the subspace of all matrices that remain fixed under the action of S_n.

Subspace $L_{n-1,1}$: Let $L_{n-1,1}^1$ be the subspace of matrices with identical rows and such that the sum of entries in each row is 0. Similarly, let $L_{n-1,1}^2$ be the subspace of matrices with identical columns and such that the sum of entries in each column is 0. Finally, let $L_{n-1,1}^3$ be the subspace of diagonal matrices whose diagonal entries sum to zero. Let $L_{n-1,1} = L_{n-1,1}^1 + L_{n-1,1}^2 + L_{n-1,1}^3$. One can check that the dimension of each of $L_{n-1,1}^1$, $L_{n-1,1}^2$ and $L_{n-1,1}^3$ is $n-1$ and that $\dim L_{n-1,1} = 3n - 3$. Moreover, the subspaces $L_{n-1,1}^1$, $L_{n-1,1}^2$ and $L_{n-1,1}^3$ do not contain non-trivial invariant subspaces. The action of S_n in $L_{n-1,1}$, although non-trivial, is not very complicated. One can show that if $A \in L_{n-1,1} + L_n$, then the problem (1) of optimizing $f(\sigma)$ reduces to the Linear Assignment Problem.

Subspace $L_{n-2,2}$: Let us define $L_{n-2,2}$ as the subspace of all *symmetric* matrices A with row and column sums equal to 0 and zero diagonal. One can check that $L_{n-2,2}$ is an invariant subspace and that $\dim L_{n-2,2} = (n^2 - 3n)/2$. Also, $L_{n-2,2}$ contains no non-trivial invariant subspaces.

Subspace $L_{n-2,1,1}$: Let us define $L_{n-2,1,1}$ as the subset of all *skew symmetric* matrices A with row and column sums equal to 0. One can check that $L_{n-2,1,1}$ is an invariant subspace and that $\dim L_{n-2,1,1} = (n^2 - 3n)/2 + 1$. Similarly, $L_{n-2,1,1}$ contains no non-trivial invariant subspaces.

The following proposition from representation theory describes the central projection g obtained from the quadratic function (1). We recall the definitions of the central projection (Definition 12) and the functions $p(\sigma)$ and $t(\sigma)$ from Sect. 1.2.

Proposition 14. *For $n \times n$ matrices A and B, where $n \ge 4$, let $f : S_n \longrightarrow \mathbb{R}$ be the function defined by (1) and let $g : S_n \longrightarrow \mathbb{R}$ be the central projection of f.*

1. If $A \in L_n$ then g is a scalar multiple of the constant function

$$\chi_n(\sigma) = 1 \quad \text{for all} \quad \sigma \in S_n \ ;$$

2. If $A \in L_{n-1,1}$ then g is a scalar multiple of the function

$$\chi_{n-1,1}(\sigma) = p(\sigma) - 1 \quad \text{for all} \quad \sigma \in S_n \ ;$$

3. If $A \in L_{n-2,2}$ then g is a scalar multiple of the function

$$\chi_{n-2,2}(\sigma) = t(\sigma) + \frac{1}{2}p^2(\sigma) - \frac{3}{2}p(\sigma) \quad \text{for all} \quad \sigma \in S_n \ ;$$

4. If $A \in L_{n-2,1,1}$ then g is a scalar multiple of the function

$$\chi_{n-2,1,1}(\sigma) = \frac{1}{2}p^2(\sigma) - \frac{3}{2}p(\sigma) - t(\sigma) + 1 \quad \text{for all} \quad \sigma \in S_n \ .$$

Proposition 14 follows from the representation theory of the symmetric group (see, for example, Part 1 of [6]). The functions $\chi_n, \chi_{n-1,1}, \chi_{n-2,2}$ and $\chi_{n-2,1,1}$ are the characters of corresponding irreducible representations of S_n for $n \geq 4$ (see Lecture 4 of [6]). The irreducible characters are linearly independent, and, moreover orthonormal.

3.2   Proof Techniques

Theorem 1: Consider the situation of Theorem 1. The conditions on A are exactly that $A \in L_n + L_{n-2,2}$. Without loss of generality, we may assume that the maximum of $f_0(\sigma)$ is attained at the identity permutation ε. Excluding the non-interesting case of $f_0 \equiv 0$, by scaling f, if necessary, we can assume that $f_0(\varepsilon) = 1$. Let g be the central projection of f_0. Then we have $\bar{g} = 0$ and $1 = g(\varepsilon) \geq g(\sigma)$ for all $\sigma \in S_n$. Moreover, since $A \in L_n + L_{n-2,2}$, by Parts 1 and 3 of Proposition 14, g must be a linear combination of the constant function χ_n and $\chi_{n-2,2}$. Since $\bar{g} = 0$, g should be proportional to $\chi_{n-2,2}$ and since $g(\varepsilon) = 1$, we have

$$g(\sigma) = \frac{2}{n^2 - 3n}\chi_{n-2,2} = \frac{2t(\sigma) + p^2(\sigma) - 3p(\sigma)}{n^2 - 3n} \ .$$

On a given ring $U(\varepsilon, k)$, the value of $p(\sigma)$ is constant. A technical calculation gives us the sum of t over the permutations in $U(\varepsilon, k)$.

Lemma 15. *We have*

$$\frac{1}{|U(\varepsilon, k)|} \sum_{\sigma \in U(\varepsilon, k)} t(\sigma) = \frac{1}{2}\nu(k),$$

where $\nu(k)$ is the function of 1.2 and ε is the identity permutation.

The proof is a combinatorial exercise, and will be included in [9].

Theorem 3: Now consider the situation of Theorem 3, where A has constant row and column sums, but may not be symmetric. We have $A \in L_n + L_{n-2,2} + L_{n-2,1,1}$. Using the normalizations we applied to the bullseye case, we get the central projection g of f_0 to be a linear combination of $\chi_{n-2,2}$ and $\chi_{n-2,1,1}$. The condition that g is maximized at the identity can be interpreted as a finite set of linear inequalities $g(\varepsilon) \geq g(\sigma)$ for $\varepsilon \neq \sigma \in S_n$. These define a cone in $\mathbb{R}^2$. After eliminating redundant inequalities, we find the cone is generated by two functions of (p, t) with one depending on the parity of n. With the added restriction that $g(\varepsilon) = 1$, we find that g is a convex combination of two extreme functions.

Summing these two extreme functions over the ring $U(\varepsilon, k)$ gives Theorem 3.

Theorem 5: If we take A to be symmetric, as in Theorem 5 but don't require it to have constant row and column sums, then we have $A \in L_n + L_{n-1,1} + L_{n-2,2}$. Proceeding as in the pure case, we can express the central projection g of f_0 as a convex combination of two extreme functions, and sum over the ring $U(\varepsilon, k)$.

For general A, we can extend our methods to the three-dimensional linear space spanned by $\chi_{n-1,1}$, $\chi_{n-2,2}$ and $\chi_{n-2,1,1}$. In this case we will arrive at four or five extreme distributions depending on the parity of n. Further details will be included in [9].

Notes and Lemmas: To get the frequency estimate of Corollary 2, we need to estimate the number of permutations with a given number fixed points and 2-cycles. We recall the conjugacy class structure of the symmetric group.

Let us fix a permutation $\rho \in S_n$. As ω ranges over the symmetric group S_n, the permutation $\omega^{-1}\rho\omega$ ranges over the conjugacy class $X(\rho)$ of ρ, that is the set of permutations that have the same cycle structure as ρ. The central projection (3) is exactly the function g whose value at each permutation σ is the average of f on the equivalence class of that permutation.

To get estimates on f from our estimates on g, we will rely on a Markov type lemma, which asserts, roughly, that a function with a sufficiently large average takes sufficiently large values sufficiently often.

Lemma 16. *Let X be a finite set and let $f : X \longrightarrow \mathbb{R}$ be a function. Suppose that $f(x) \leq 1$ for all $x \in X$ and that*

$$\frac{1}{|X|} \sum_{x \in X} f(x) \geq \beta \quad \text{for some} \quad \beta > 0 .$$

Then for any $0 < \gamma < 1$ we have

$$\left|\{x \in X : f(x) \geq \beta\gamma\}\right| \geq \beta(1-\gamma)|X| .$$

Proof. We have

$$\beta \leq \frac{1}{|X|} \sum_{x \in X} f(x) = \frac{1}{|X|} \sum_{x : f(x) < \beta\gamma} f(x) + \frac{1}{|X|} \sum_{x : f(x) \geq \beta\gamma} f(x)$$

$$\leq \beta\gamma + \frac{\left|\{x : f(x) \geq \beta\gamma\}\right|}{|X|} .$$

Hence

$$\left|\{x : f(x) \geq \beta\gamma\}\right| \geq \beta(1-\gamma)|X| .$$

We use generating series to calculate the number of permutations with given cycle structures. Let us fix some positive integers $c_i : i = 1, \ldots, m$ and let a_n be the number of permutations in S_n that have no cycles of length c_i for $1 \leq i \leq m$. The exponential generating function for a_n is given by

$$\sum_{n=0}^{\infty} \frac{a_n}{n!} x^n = \frac{1}{1-x} \exp\left\{ -\sum_{i=1}^{m} \frac{x^{c_i}}{c_i} \right\} ,$$

where we agree that $a_0 = 1$, see, for example, pp. 170–173 of [7]. It follows that the number of permutations $\sigma \in S_n$ without fixed points is asymptotically $e^{-1}n!$. More precisely, it is equal to $d_n n!$, where d_n is as defined in Sect. 1.2. Similarly, the number of permutations without fixed points and 2-cycles is asymptotically $e^{-3/2}n!$. We will use that the first number exceeds $n!/3$ and the second number exceeds $n!/5$ for $n \geq 5$.

Corollary 2: We can now prove Corollary 2. We again scale f_0 so that its maximum is 1.

Let us estimate the cardinality $|U(\tau, n-k)| = |U(\varepsilon, n-k)|$. Since $\sigma \in U(\varepsilon, n-k)$ if and only if σ has k fixed points, to choose a $\sigma \in U(\varepsilon, n-k)$ one has to choose k points in $\binom{n}{k}$ ways and then choose a permutation of the remaining $n-k$ points without fixed points. Estimating the number of permutations with no fixed points, we get

$$|U(\tau, n-k)| \geq \binom{n}{k}(n-k)!/3 = \frac{n!}{3k!} .$$

Applying Lemma 16 with $\beta = \beta(n, k)$ and $X = U(\tau, n-k)$, from Theorem 1 we conclude that

$$P\left\{\sigma \in S_n : f_0(\sigma) \geq \gamma\beta(n, k)\right\} \geq \frac{(1-\gamma)\beta(n, k)|U(\tau, n-k)|}{n!}$$
$$\geq \frac{(1-\gamma)\beta(n, k)}{3k!} .$$

Corollaries 4, 7 and 8: To prove Corollaries 4, 7 and 8, we look at the extreme distributions found in the corresponding theorems. For the pure case, both extreme distributions (the bullseye and the damped oscillator) have enough good values near the optimum to produce the estimate of Corollary 4. Then it is not hard to show that the estimates also hold on convex combinations of the two extreme distributions.

The strategy of picking permutations close to the optimum fails in the general symmetric case of Theorem 5, because there are relatively few good permutations close to the optimum in the spike distribution. However, in the case of the spike distribution there are many permutations with no fixed points that yield good objective values. It is possible to show that for any convex combination of the spike and bullseye distributions from the symmetric case that either the strategy of picking points close to the optimum or the strategy of picking points far from the optimum yields enough permutations to give the bound of Corollary 7. Note that this bound is weaker than the bound of Corollary 4. Similarly, the bound of Corollary 8 is obtained by showing that a suitable strategy works for any convex combination of the four or five extreme distributions of the general case.

These computations are technical, and will be included in [9].

4   Concluding Remarks and Open Questions

The estimates of Theorems 1 and 5 on the number of near-optimal permutations can be used to bound the optimal value by a sample optimum in branch-and-bound algorithms. Those estimates are (nearly) best possible for the generalized problem (2). However, it is not clear whether they can be improved in the case of standard QAP (1), or how to improve them in interesting special cases. In particular, we ask the following

- Question: Let $f : S_n \longrightarrow \mathbb{R}$ be the objective function in the Traveling Salesman Problem (cf. Sect. 2.1), let $\overline{f}$ be the average value of f and let $f_0 = f - \overline{f}$. Let τ be an optimal permutation, so that $f_0(\tau) \geq f_0(\sigma)$ for all $\sigma \in S_n$. Is it true that for any fixed $\gamma > 0$ there is a number $\delta = \delta(\gamma) > 0$ such that the probability that a random permutation $\sigma \in S_n$ satisfies the inequality $f_0(\sigma) \geq \dfrac{\gamma}{n} f_0(\tau)$ is at least $n^{-\delta}$ for all sufficiently large n?

Another question is whether approximations can be obtained *deterministically*. The random sampling algorithm in the "bullseye" case can be relatively easily derandomized. Whether the same is true for the sampling algorithms in the non-bullseye cases is not clear at the moment.

Our methods can be applied to study the distribution of values in the Assignment Problems of higher order and their special cases, such as the Weighted Hypergraph Matching Problem.

References

1. Anstreicher, K., Brixius, N., Goux, J.-P., Linderoth, J.: Solving large quadratic assignment problems on computational grids. Math Programming B (to appear)
2. Arkin, E., Hassin, R., Sviridenko, M.: Approximating the maximum quadratic assignment problem. Inform. Process. Lett. **77** (2000) no. 1. 13–16
3. Ausiello, G., Crescenzi, P., Gambosi, G., Kann, V., Marchetti-Spaccamela, A., Protasi, M.: Complexity and approximation. Combinatorial optimization problems and their approximability properties. Springer-Verlag, Berlin (1999)
4. Brüngger, A., Marzetta, A., Clausen, J., Perregaard M.: Solving large scale quadratic assignment problems in parallel with the search library ZRAM. Journal of Parallel and Distributed Computing **50** (1998) 157-66
5. Burkard, R., Çela, E., Pardalos, P., Pitsoulis, L.: The quadratic assignment problem. In: Du, D.-Z., and Pardalos, P. M. (eds.): Handbook of Combinatorial Optimization. Kluwer Academic Publishers (1999) 75-149
6. Fulton, W., Harris, J.: Representation theory. Springer-Verlag, New York (1991)
7. Goulden, I. P., Jackson, D. M.: Combinatorial enumeration. John Wiley & Sons, Inc., New York (1983)
8. Graves, G. W., Whinston, A. B.: An algorithm for the quadratic assignment problem. Management Science **17** (1970) no. 7. 452-71
9. Stephen, T.: The distribution of values in combinatorial optimization problems. Ph.D. Dissertation, University of Michigan (in preparation)
10. Ye, Y.: Approximating quadratic programming with bound and quadratic constraints. Math. Programming **84** (1999) no. 2. 219-226

A New Subadditive Approach
to Integer Programming

Diego Klabjan

University of Illinois at Urbana-Champaign, Urbana, IL
klabjan@uiuc.edu

Abstract. The linear programming duality is well understood and the reduced cost of a column is frequently used in various algorithms. On the other hand, for integer programs it is not clear how to define a dual function even though the subadditive dual theory was developed a long time ago. In this work we propose a family of computationally tractable subadditive dual functions for integer programs. We develop a solution methodology that computes an optimal primal solution and an optimal subadditive dual function. We report computational results for set partitioning instances. To the best of our knowledge these are the first computational experiments on computing the optimal subadditive dual function.

1 Introduction

The LP duality was established a long time ago. With each primal problem there is an associated dual problem with the same optimal objective value. Many algorithms compute both a primal and a dual solution, e.g. simplex and primal-dual algorithms. Given a dual vector we define the reduced cost of a column, which estimates how much would the addition of the column change the objective value, and the sensitivity analysis can be carried out using the dual information. Large-scale LPs can be efficiently solved with SPRINT, see e.g. Anbil *et al.* (1992). The idea of SPRINT is to solve many small LP subproblems and gradually add columns to the subproblem based on the reduced cost values. Columns with small reduced cost are more likely to improve the incumbent solution and therefore they are appended to the subproblem.

This paper addresses the questions of a dual function, reduced cost, and sensitivity analysis for IPs. The original motivation for this study was in designing an algorithm for large-scale IPs that mimics SPRINT. Such an approach solves at each iteration an IP consisting of a small subset of columns. Columns are then appended to the current IP and the problem is reoptimized. The key question related to the success of this approach is what columns to append, i.e. what is the equivalent notion to the LP reduced cost. Dual vectors are also used in the Benders decomposition algorithms, see e.g. Nemhauser and Wolsey (1988). Currently these approaches can be applied only to mixed integer linear programs since they require a dual vector. But if applied to IPs, they raise the question of a dual function for IPs. In LP it is known that all the alternative optimal

W.J. Cook and A.S. Schulz (Eds.): IPCO 2002, LNCS 2337, pp. 384–400, 2002.

solutions can be found among the columns with zero reduced cost. It would be very useful if we can produce alternative IP solutions, e.g. among several optimal solutions we can select the most robust one. Again, as shown in this work, an available dual function for IPs makes such a task much easier.

For integer programs the subadditive duality developed first by Johnson (1973) gives us a partial answer to these questions.

Definition 1. *A function* $F : \mathbb{R}^m \to \mathbb{R}$ *is subadditive on* $Z \subseteq \mathbb{R}^m$ *if* $F(x+y) \leq F(x) + F(y)$ *for all* $x \in Z, y \in Z$ *such that* $x + y \in Z$.

If Z is not specified, we assume $Z = \mathbb{R}^m$. Johnson showed that for a feasible IP

$$
\begin{array}{llll}
\min & cx & \max & F(b) \\
& Ax = b & = & F(a_i) \leq c_i \quad i = 1, \ldots, n \quad (1) \\
& x \in \mathbb{Z}_+^n & & F \text{ subadditive},
\end{array}
$$

where $A = (a_1, \ldots, a_n) \in \mathbb{Z}^{m \times n}, b \in \mathbb{Z}^m, c \in \mathbb{Z}^n$. We refer to the second problem as the *subadditive dual problem*. At least theoretically the answer to all of the raised questions are in the *optimal subadditive function* (OSF) F. In other words, the analog to the optimal dual vector in LP is the OSF. The reduced cost of a column i can be defined as $c_i - F(a_i)$ and most of the other properties from LP carry over to IP, e.g. complementary slackness, $F(b)$ provides a lower bound on the optimal IP value, and the alternative optimal solutions can be found only among the columns i with $c_i = F(a_i)$. However there are still two fundamental issues that need to be addressed; how to encode F and how to compute F. Theory tells us that an *OSF* can always be obtained as a composition of C-G inequalities, see e.g. Nemhauser and Wolsey (1988), but such a function would be hard to encode and hard to evaluate. Very little is known about how to compute an OSF. Llewellyn and Ryan (1993) show how an OSF can be constructed from Gomory cuts. Our work originates from the work done by Burdet and Johnson (1977), where an algorithm for solving an IP based on subadditivity is presented. Both of these two works do not present any computational experiments.

We give a new family of subadditive functions that is easy to encode and often easy to evaluate. We present an algorithm that computes an OSF. As part of the algorithm we give several new theorems that further shed light on OSFs. The contribution of this research goes beyond a novel methodology for computing an OSF. There are many implications of having an OSF: the reduced cost can be defined, the sensitivity analysis can be performed, we can obtain alternative optimal solutions, and new approaches for large-scale IPs can be developed ('integer' SPRINT, Benders decomposition for IPs).

In Sect. 2 we present a new subadditive function that is computationally tractable. We give several interesting properties of this function. In addition we generalize the concept of reduced cost fixing. Sect. 3 first outlines the algorithm that computes an optimal primal solution and an OSF. We show how to obtain an OSF given an OSF of the preprocessed problem and we show how to obtain an initial subadditive function from the LP formulation with clique inequalities. In this section we also present the entire algorithm that computes an optimal primal solution and an OSF. In the last section we report the computational experiments. We conclude the introduction with a brief description of the Burdet-Johnson algorithm.

The Burdet-Johnson Algorithm

Burdet and Johnson show that there is a subadditive function $\pi : \mathbb{R}^n \to \mathbb{R}$ such that $\pi(e_i) \leq c_i$ for every column i, and the optimal value z^{IP} to the IP equals to $\min_x$ feasible to IP $\sum_{i \in N} \pi_i x_i$. Here e_i is the ith unit vector and $N = \{1, 2, \ldots, n\}$ are the column indices. Note that such a function does not serve our purpose since it is defined on the set of all the columns and not rows. Based on π we do not see an appropriate way to define the reduced cost. However we do use such a π to obtain a good starting point for our desired F.

Their subadditive function is based on the concept of generator subsets. Given a subinclusive set $E \in \mathbb{Z}_+^n$, i.e. if $y \in E$ then for every $0 \leq x \leq y$ it follows that $x \in E$, and a vector $\delta \in \mathbb{R}^n$ they define a function $\pi(x) = \max_{y \in E \cap \underline{S}(x)} cy + \delta(x - y)$, where $\underline{S}(x) = \{y \in \mathbb{Z}_+^n : y \leq x\}$. A candidate set H is defined as $H = \{x \in \mathbb{Z}_+^n : x \notin E, \underline{S}(x) \setminus \{x\} \subseteq E\}$. They showed that if

$$\pi(x_1 + x_2) \leq cx_1 + cx_2 \tag{2}$$

for all $x_1 \in E, x_2 \in E$ and $x_1 + x_2 \in H$, then π is subadditive.

Their algorithm starts with $E = \emptyset$ and δ equal to the optimal dual vector of the LP relaxation. In every iteration first E is expanded, which is called the enumeration step, and then δ is adjusted. The enumeration step selects an element from H and appends it to E. This operation is followed by an update of H and δ has to be adjusted to satisfy (2), i.e. subadditivity. Given fixed E and H, the maximum dual objective value is obtained by solving the LP

$$\max \ \pi_0 \tag{3a}$$

$$\pi_0 - \delta(x - y) \leq cy \qquad \text{for all } y \in E, x \text{ feasible to IP, and } y \leq x \tag{3b}$$

$$\delta x \leq cx \qquad x \in H \tag{3c}$$

$$\delta, \pi_0 \text{ unrestricted} . \tag{3d}$$

Burdet and Johnson show that (3b) give the dual objective value and that (3c) are equivalent to (2). In other words, we maximize the dual objective value by (3b) while maintaining subadditivity with (3c). The enumeration step and the adjustment of δ are then iterated until an optimal solution is found. The algorithm is given in Algorithm 1, where we denote by w^D the best lower bound on z^{IP}.

We tried to implement this algorithm but we were not able to solve even very small problems. We took this framework and we substantially enhanced the algorithm. We present the enhanced algorithm in Sect. 3.3.

1: $E = \emptyset, H = \{e_i : i \in N\}, \delta =$ optimal dual vector of the LP relaxation, $w^D = -\infty$.

2: **loop**

3:     $\bar{x} = \mathrm{argmin}\{\frac{cx}{\delta x} : x \in H, \delta x > 0\}$

4:     $E = E \cup \{\bar{x}\}, H = H \cup \{\bar{x} + e_i : i \in N, y \le \bar{x} + e_i, y \in \mathbb{Z}_+^n\}$

5:     Update δ by solving (3). Let π_0^* be the optimal value.

6:     $w^D = \max\{w^D, \pi_0^*\}$

7:     If $w^D = \min\{cx : x \in H, Ax = b\}$, then we have solved the IP and exit.

8: **end loop**

Algorithm 1. The Burdet-Johnson algorithm

---

2   The Generator Subadditive Functions

Here we present the approach only for the set partitioning problems $\min\{cx : Ax = \mathbf{1}, x \text{ binary}\}$, where $\mathbf{1}$ is a vector with all components equal to 1. We denote by z^{IP} the optimal value. In addition, we assume that the problem is feasible. The extension to general IPs, the theory for infeasible IPs, and additional results are given in Klabjan (2001b) and Klabjan (2001a).

Given a vector $\alpha \in \mathbb{R}^m$, we define a generator subadditive function $F_\alpha : \mathbb{R}^m \to \mathbb{R}$ as

$$F_\alpha(d) = \alpha d - \max \sum_{i \in E} (\alpha a_i - c_i) x_i$$

$$A^E x \le d$$

$$x \text{ binary},$$

where $E = \{i \in N : \alpha a_i > c_i\}$ is a generator set and A^E is the submatrix of A corresponding to the columns in E. The generator set E depends on α but for simplicity of notation we do not show this dependence in our notation. Whenever an ambiguity can occur, we write $E(\alpha)$. In addition, for simplicity of notation we write $H = N \setminus E$.

It is easy to see that F_α is a subadditive function. It is also easy to see that $F_\alpha(a_i) \le \alpha a_i \le c_i$ for all $i \in H$ by taking $x = 0$ in $\max\{(\alpha A^E - c^E)x : A^E x \le a_i, x \in \mathbb{Z}_+^{|E|}\}$ and that $F_\alpha(a_i) \le c_i$ for all $i \in E$ by considering $x = e_i$ in $\max\{(\alpha A^E - c^E)x : A^E x \le a_i, x \in \mathbb{Z}_+^{|E|}\}$. Therefore $F_\alpha(a_i) \le c_i$ for all $i \in N$. This shows that F_α is a feasible subadditive function and therefore $F_\alpha(\mathbf{1})$ provides a lower bound on z^{IP}. The vector α is a generalization of dual vectors of the LP relaxation. Every dual feasible vector α to the LP relaxation has to satisfy $\alpha a_i \le c_i$ for all $i \in N$, however α in the definition of F_α can violate some of these constraints. Indeed, if y^* is an optimal solution to the dual of the LP relaxation of the IP, then $E = \emptyset$ and F_{y^*} gives the value of the LP relaxation.

The equivalence (1) states that among all the subadditive dual functions there is one that attains the equality however it does not say anything for specially structured subadditive functions like F_α.

Theorem 1. *There exists an α such that F_α is an OSF, i.e. $F_\alpha(1) = z^{IP}$.*

Since potentially we want to use F_α to compute reduced costs $c_i - F(a_i)$ for many columns i, e.g. the idea of an integral SPRINT, it is desirable that the cardinality of E is small. Our computational experiments show that this is indeed the case in practice. Even problems with 100,000 columns have only up to 300 columns in E.

Theorem 1 can be proved directly but here we give a constructive proof based on valid inequalities.

Proposition 1. *Let $\pi^j x \le \pi_0^j, j \in J$ be valid inequalities for $\{Ax \le 1, x \in \mathbb{Z}_+^n\}$. Let*

$$z^* = \min \; cx$$
$$Ax = 1 \tag{4}$$
$$\pi^j x \le \pi_0^j \qquad j \in J$$
$$x \ge 0$$

and let (α, γ) be an optimal dual vector to this LP, where α corresponds to constraints (4). Then $F_\alpha(1) \ge z^*$.

Proof. The dual of the LP stated in the proposition reads

$$\max \; 1\alpha - \sum_{j \in J} \pi_0^j \gamma_j$$
$$a_i \alpha - \sum_{j \in J} \pi_i^j \gamma_j \le c_i \qquad i \in N \tag{5}$$
$$\alpha \text{ unrestricted}, \gamma \ge 0 \,.$$

The optimal value of this LP is z^* and let (α, γ) be an optimal vector. The statement $F_\alpha(1) \ge z^*$ is equivalent to

$$\max\{(\alpha A^E - c^E)x : A^E x \le 1, x \text{ binary}\} \le 1\alpha - z^* \,. \tag{6}$$

Let x be a binary vector such that $A^E x \le 1$. We have

$$(\alpha A^E - c^E)x \le \sum_{i \in E} \sum_{j \in J} x_i \pi_i^j \gamma_j \tag{7}$$
$$= \sum_{j \in J} \gamma_j \sum_{i \in E} \pi_i^j x_i \le \sum_{j \in J} \gamma_j \pi_0^j = 1\alpha - z^* \,, \tag{8}$$

where (7) follows from (5), and the inequality in (8) holds since $\pi^j x \le \pi_0^j, j \in J$ are valid inequalities for $\{A^E x \le 1, x \text{ binary}\}$ and $\gamma \ge 0$. This shows (6) and it proves the claim. $\square$

Theorem 1 now easily follows from Proposition 1 since the convex hull of $\{Ax = 1, x \in \mathbb{Z}_+^n\}$ can be obtained from valid inequalities to $\{Ax \le 1, x \in \mathbb{Z}_+^n\}$. Next we give two theorems that have a counterpart in LP and are used in our algorithm.

Theorem 2 (Complementary slackness). *Let x^* be an optimal primal solution. If $x_i^* = 1$, then $\alpha a_i \geq c_i$ in a generator OSF F_α.*

This theorem easily follows from the general complementary slackness condition $x_i^*(c_i - F(a_i)) = 0$, where x^* is an optimal solution and F an OSF.

Since F_α is always a valid subadditive function, $F_\alpha(\mathbf{1})$ is a lower bound on z^{IP}. In IP the reduced cost fixing based on an LP solution is a commonly used technique for fixing the variables to 0. The next theorem establishes an equivalent condition based on a subadditive dual function.

Theorem 3 (Extended reduced cost fixing). *Let F be a feasible subadditive dual function, i.e. F is subadditive and $F(a_i) \leq c_i$ for all $i \in N$, and let $\bar{z}^{IP}$ be an upper bound on z^{IP}. If $c_k - F(a_k) \geq \bar{z}^{IP} - F(\mathbf{1})$ for a column $k \in N$, then there is an optimal solution with $x_k = 0$.*

Proof. Let k be a variable index such that $c_k - F(a_k) \geq \bar{z}^{IP} - F(\mathbf{1})$. Consider the IP $\min\{cx : Ax = \mathbf{1}, x_k = 1, x \in \mathbb{Z}_+^n\}$. We show that the optimal value of this IP is greater or equal to $\bar{z}^{IP}$.

The subadditive dual problem of this IP reads

$$
\begin{aligned}
\max \quad & G(\mathbf{1}, 1) \\
& G(a_i, 0) \leq c_i & i \in N - \{k\} \\
& G(a_k, 1) \leq c_k \\
& G \text{ subadditive}\,,
\end{aligned}
\tag{9}
$$

where the extra coordinate in columns corresponds to the constraint $x_k = 1$. Consider the feasible subadditive function $\bar{G}(d, s) = F(d) + (c_k - F(a_k))s$ to (9). The objective value of this function is $\bar{G}(\mathbf{1}, 1) = F(\mathbf{1}) + (c_k - F(a_k))1 \geq \bar{z}^{IP}$ and therefore the objective value of the subadditive dual problem (9) is at least $\bar{z}^{IP}$. This in turn implies that the objective value of the IP with $x_k = 1$ is at least $\bar{z}^{IP}$, which concludes the proof. □

If F is a subadditive valid function, then $\sum_{i \in N} F(a_i)x_i \geq F(\mathbf{1})$ is a valid subadditive function, Nemhauser and Wolsey (1988). Therefore for $F = F_\alpha$ we get that

$$
\sum_{i \in E} c_i x_i + \sum_{i \in H} (\alpha a_i)x_i \geq F_\alpha(\mathbf{1})
\tag{10}
$$

is a valid inequality. These inequalities are used in our computational experiments.

2.1   Basic and Minimal Generator Subadditive Functions

Note that the generator subadditive functions form an infinite family of functions since α is arbitrary. In linear programming extreme points suffice to solve the dual problem and there are only a finite number of them. A similar result holds for the generator subadditive functions. Namely, it suffices to consider only those generator subadditive functions, called *basic generator subadditive functions*, with α an extreme point of a certain polyhedra. By using these functions we can extend the traditional Benders decomposition algorithm for mixed integer programs to integer programs. The details are given in Klabjan (2001b). In addition, if (10) is a facet, then F_α is a basic generator subadditive function.

A valid inequality is minimal if is not dominated by any other valid inequality. A generator subadditive function is *minimal* if (10) is a minimal valid inequality. Minimal generator subadditive functions have two interesting properties. Consider the set packing problem $\max\{(\alpha A^E - c^E)x : A^E x \leq \mathbf{1}, x \text{ binary}\}$. F_α is a minimal generator subadditive function if and only if this set packing problem has an optimal solution x^* such that there is a binary vector x' with $A(x' + x^*) = \mathbf{1}$. The following more interesting property is shown in Klabjan (2001b). If F_α is minimal generator subadditive function, than for every $i \in E$ there is an optimal solution x^* to the set packing problem with $x_i^* = 1$. This statement shows that this set packing problem has a wide variety of optimal solutions. Since a generator OSF is minimal, these properties can facilitate us in designing algorithms for obtaining all optimal primal solutions.

3   Solution Methodology

We first briefly describe the main ideas of our algorithm that finds a primal optimal solution and it simultaneously computes an OSF F_α.

We first preprocess the problem. We use all of the 9 preprocessing rules described in Borndorfer (1998) except that we do not check for dominated columns. In addition, we have detected 2 more preprocessing rules. In the next step we solve the LP relaxation of the preprocessed problem and we add clique inequalities, Nemhauser and Wolsey (1988). They are separated with a standard heuristic. Next we form an initial E by considering the dual prices of the LP relaxation with cliques.

Computational experiments have shown that finding in one 'attempt' an OSF is not computationally tractable. Instead we gradually remove columns that do not change the optimal value from the problem. This has an implication that the resulting OSF is not necessarily subadditive in $\mathbb{R}^m$ but it is only subadditive on a subset of still active columns. The algorithm proceeds in 3 stages. In stage 1 an optimal primal solution and an optimal subadditive function is found. However the subadditive function is only subadditive on a small subset of columns. In this stage we use the Burdet-Johnson framework. In stage 2 we improve subadditivity by obtaining a subadditive function that satisfies the complementary slackness conditions. The resulting function is still not necessarily subadditive on all the columns but it turns out that only few columns violate subadditivity. In stage 3 we do the last correction to make the function subadditive on $\mathbb{R}^m$.

3.1   Preprocessing and the Generator Subadditive Functions

The problem is first preprocessed and then a generator OSF is found. Since we want to obtain an OSF to the original problem, we have to show how to get such a function given an OSF to the preprocessed problem. In preprocessing the preprocessing rules are iteratively applied. All of the 10 preprocessing rules are applied in several passes until in the last pass we do not further reduce the problem. To be able to construct an OSF of the original problem, we have to show how each preprocessing rule affects the subadditive function. If we want to easily construct the function, then we have to show how to obtain an OSF $F_{\bar{\alpha}}$ from an OSF F_α of the preprocessed problem, after one step of a given preprocessing rule. It is crucial here that the new function is again a generator subadditive function since otherwise we loose the structure and it would be hard to successfully unwind the preprocessing steps.

Fortunately for most of the preprocessing rules we can apply the following proposition. For an $i \in N$ let $A_i = \{j \in M : a_{ji} = 1\}$, where $M = \{1, \ldots, m\}$ is the set of all the rows, and similarly for $j \in M$ let $A^j = \{i \in N : a_{ji} = 1\}$.

Proposition 2. *Let F_α be an OSF for $\min\{cx : Ax = 1, x$ binary$\}$ and let r be a row of A. Let a_{n+1} be a new column with the cost c_{n+1}. Furthermore, let $r \notin A_{n+1}$, let $\sum_{i \in A^r} x_i + x_{n+1} \le 1$ be a valid inequality for $\{(x, x_{n+1}) : Ax + a_{n+1}x_{n+1} \le 1, x$ binary$, x_{n+1}$ binary$\}$, and let us define $F_\alpha(a_{n+1}) = \alpha a_{n+1}$. Then the LP*

$$\min \mathbf{1}y$$
$$a_i y \le c_i - F_\alpha(a_i) \qquad i \in N - A^r$$
$$(a_i - 1)y \le c_i - F_\alpha(a_i) \qquad i \in A^r \text{ and } i = n+1$$
$$\mathbf{1}y \ge 0$$

has an optimal solution y^* and $F_{\alpha+y^*}$ is an OSF for $\min\{cx + c_{n+1}x_{n+1} : Ax + a_{n+1}x_{n+1} = 1, x$ binary$, x_{n+1}$ binary$\}$.

Let us show how to use the theorem for dominated rows. If $r, n+1$ are two rows such that $A^r \subseteq A^{n+1}$, then we can fix to 0 all the variables in $A^{n+1} \setminus A^r$ and we can remove row $n+1$ from the problem. Suppose now that $F_{\bar{\alpha}}$ is an OSF for the problem without row $n+1$ and the columns in $A^{n+1} \setminus A^r$. First we can append back row $n+1$ without the columns in $A^{n+1} \setminus A^r$ to the matrix and define $\alpha = (\bar{\alpha}, 0) - (\bar{\alpha}_r/2)e_r + (\bar{\alpha}_r/2)e_{n+1}$. It is easy to see that F_α is an OSF for the new problem. Now we can handle the removed columns from $A^{n+1} \setminus A^r$ by Proposition 2 to get an OSF for the original problem.

3.2 Clique Inequalities and the Generator Subadditive Functions

Consider the LP relaxation together with some clique inequalities. The LP value gives a lower bound on z^{IP}. In our algorithm we use the following corollary to Proposition 1.

Corollary 1. *Let*

$$z^{clq} = \min\{cx : Ax = 1, \sum_{i \in C_j} x_i \le 1 \quad j \in J, x \ge 0\},$$

where $C_j, j \in J$, are cliques in the conflict graph. Let α be the optimal dual vector corresponding to constraints $Ax = \mathbf{1}$. Then F_α has the value at least z^{clq}, i.e. $F_\alpha(\mathbf{1}) \geq z^{clq}$.

If we can solve a set partitioning problem to optimality by adding clique inequalities, then by Corollary 1 the generator OSF is readily available.

In our implementation we first run a primal heuristic to obtain $\bar{z}^{IP}$, which is the best known upper bound on z^{IP}. Then we solve the LP relaxation with cliques and we run the primal heuristic again but this time with the strengthen formulation with cliques. At the end we apply reduced cost fixing.

3.3   Stage 1: Obtaining an Optimal Primal Solution with the Use of Subadditive Functions

In this stage we find an optimal primal solution and an approximate subadditive dual function by using the Burdet-Johnson framework. Since we are interested in obtaining a subadditive function in $\mathbb{R}^m$, we choose $\delta = \alpha A$, where α is an unknown vector. This mapping also substantially simplifies (3b) since now they read $\pi_0 - \mathbf{1}\alpha + \alpha(Ay) \leq cy$ for all $y \in E$.

We first enhance their algorithm with a concept that is the equivalent to pruning in branch-and-bound algorithms. Instead of searching for a subadditive function in all $\mathbb{R}^n$ we require subadditivity only on $\{x : Ax = \mathbf{1}, cx < \bar{z}^{IP}, x \text{ binary}\}$. It means that we can remove from H all the elements that yield an objective function value larger than $\bar{z}^{IP}$. Namely, if $h \in H$ and

$$\min\{cx : Ax = \mathbf{1}, x_i = 1 \text{ for all } i \text{ with } h_i = 1, x \geq 0\} \tag{11}$$

is greater or equal to $\bar{z}^{IP}$, then h can be removed from H. However computing this LP for every element in H is too expensive and therefore we compute the exact value only for the element that is selected to be added to E. Note that the cardinality of H can double in each iteration. For the candidate element h, if the objective value of (11) is greater or equal to $\bar{z}^{IP}$, then h is permanently removed from consideration and the selection process is repeated.

The second major enhancement we employ is the selection of an element from H that is added to E. In view of the above discussion, we would like to append to E such an element that introduces the fewer number of new elements in H. Note that the elements in H yield constraints (3c) and therefore it is desirable to have only a small number of them. It can be seen that this is equivalent to maximizing (11) and therefore selecting an element from H that has the highest objective value of (11) is beneficial for both goals, i.e. pruning and having a small H. We avoid computing (11) for every element in H by using pseudo costs, Nemhauser and Wolsey (1988). On the other hand, when an element is moved from H to E, the corresponding constraint (3c) is relaxed into a constraint (3b). Therefore we would like to relax the most binding constraint, i.e. a constraint with $\alpha Ah = ch$.

The overall selection can be summarized as follows. Among all the elements h in H that satisfy $\alpha Ah = ch$, we select the element with the largest pseudo cost since such an element is most likely to be pruned and it keeps H small. Next we solve (11) and if the element can be pruned, we permanently delete it and we repeat the selection procedure. Otherwise the element is added to E. If the solution to (11) is integral and better than the best known primal solution so far, we update $\bar{z}^{IP}$.

At every iteration we apply the extended reduced cost fixing and we add to the problem (11) the subadditivity cut resulting from the current subadditive function.

Note that at the end we assert that there are no feasible solution, i.e. the current solution is optimal. We are optimal when $\{x : Ax = \mathbf{1}, cx < \bar{z}^{IP}, x \text{ binary}\}$ is empty, which implies that there is a subadditive function with infinity objective value. Therefore the stopping criteria is when the dual objective π_0 becomes greater or equal to $\bar{z}^{IP}$.

3.4 Stage 2: Improving the Feasibility of the Subadditive Dual Function

The goal of this stage is to obtain a subadditive function that is subadditive on the set $\{x : Ax = \mathbf{1}, cx \le z^{IP}, x \text{ binary}\}$. Such a function has to satisfy the complementary slackness conditions.

This stage is very similar to stage 1. In stage 1 we record E and H every time we obtain a better primal solution. The last recorded E and H are used as a warm start for stage 2. The algorithm follows closely the algorithm in stage 1 with a few changes. The selection of an element from H that is added to E is now more targeted toward improving the dual objective.

3.5 Stage 3: Computing a Generator Optimal Subadditive Function

In this stage we obtain a generator OSF. We take α from stage 2 as a starting point. F_α is an OSF on all the columns that have not been pruned in stage 2. This F_α is not necessarily an OSF since the columns i that have been pruned may have $\alpha a_i > c_i$. In this case we would have to add i to E, which can potentially decrease $F_\alpha(\mathbf{1})$. We call the columns that have been pruned in stage 2 and have $\alpha a_i > c_i$ *infeasible* columns. Typically we have several thousand infeasible columns. Stage 3 makes these columns feasible by modifying E and α.

First we reduce the number of infeasible columns by slightly adjusting α. The procedure is based on the following theorem.

Theorem 4. *Let F_α be a generator OSF for the set partitioning problem with the input data c and A, let c_{n+1}, a_{n+1} be a column with $\alpha a_{n+1} > c_{n+1}$, and assume that*

$$\min \sum_{i \in N} (c_i - F_\alpha(a_i)) x_i + (c_{n+1} - \alpha a_{n+1}) x_{n+1}$$

$$Ax + a_{n+1} x_{n+1} = \mathbf{1} \tag{12}$$

$$x \ge 0, x_{n+1} \ge 0$$

is greater or equal to 0. If γ is an optimal dual vector to this LP, then $F_{\alpha+\gamma}$ is a generator OSF for the set partitioning problem with the input data (c, c_{n+1}) and (A, a_{n+1}).

Proof. First note that the objective value of (12) is 0 since the optimal IP solution gives the objective value 0. Therefore $\gamma \mathbf{1} = 0$. For all $i \in H \cup \{n + 1\}$ clearly $(\alpha+\gamma)a_i \leq c_i$ since γ is dual feasible to (12). This implies that $E(\alpha+\gamma) \subseteq E$.

It suffices to show that

$$\max\{((\alpha+\gamma)A^E - c^E)x : A^E x \leq \mathbf{1}, x \in \mathbb{Z}_+^{|E|}\} \leq (\alpha+\gamma)\mathbf{1} - F_\alpha(\mathbf{1}) = \alpha\mathbf{1} - F_\alpha(\mathbf{1}). \tag{13}$$

Let x be a binary vector with $A^E x \leq \mathbf{1}$. For every $i \in E$ let $F_\alpha(a_i) = \alpha a_i - (\alpha A^E - c^E)z^i$, where $A^E z^i \leq a_i, z^i \in \mathbb{Z}_+^{|E|}$. Then we have

$$\sum_{i \in E}(\gamma a_i + \alpha a_i - c_i)x_i \leq \sum_{i \in E}(\alpha A^E - c^E)z^i x_i = (\alpha A^E - c^E)\sum_{i \in E} z^i x_i$$

$$\leq \max\{(\alpha A^E - c^E)x : A^E x \leq \mathbf{1}, x \in \mathbb{Z}_+^{|E|}\} = \alpha\mathbf{1} - F_\alpha(\mathbf{1}),$$

where the first inequality follows from dual feasibility of γ to (12) and the second one from

$$A^E(\sum_{i \in E} z^i x_i) = \sum_{i \in E}(A^E z^i)x_i \leq \sum_{i \in E} a_i x_i \leq \mathbf{1}.$$

The last equality holds by optimality of F_α. This shows (13) and therefore the claim.   □

Note that if $x_{n+1} = 0$ in the above LP, then the objective value is 0 since all of the coefficients are nonnegative and the optimal IP solution gives a solution with 0 objective value. Therefore the condition in the theorem is likely to hold. We apply this theorem iteratively for all the infeasible columns. If the condition is not met, then we leave the column infeasible. This procedure reduces the number of infeasible columns from several thousand to only a dozen.

Obtaining feasibility for remaining infeasible columns is the most computationally intensive part of the entire algorithm and we call the next algorithm the *enlarge generator algorithm*. For each infeasible column i we proceed as follows. We add i to E to obtain a generator subadditive function that satisfies the complementary slackness but it is not necessarily optimal since the addition of a new column to E can reduce the value of $\max\{\sum_{i \in E}(\alpha a_i - c_i)x_i : A_E x \leq \mathbf{1}, x \text{ binary}\}$. Given E we maximize the dual value by solving the LP

$$\max \eta \tag{14a}$$

$$\eta + \alpha(A^E x - \mathbf{1}) \leq c^E x \qquad\qquad A^E x \leq \mathbf{1}, x \text{ binary} \tag{14b}$$

$$\alpha a_i \leq c_i \qquad\qquad\qquad\qquad i \in H \tag{14c}$$

$$\alpha a_i \geq c_i \qquad\qquad\qquad\qquad x_i^* = 1 \tag{14d}$$

$$(\eta, \alpha) \in \mathbb{R} \times \mathbb{R}^m. \tag{14e}$$

Here we assume that x^* is an optimal solution to the set partitioning problem, which has already been obtained in phase 1. Constraints (14d) express the complementary slackness conditions (see Theorem 2). Given a solution to this LP, we choose a column from H with the largest dual value and we move it from H to E. The process is repeated until the objective value π becomes $\bar{z}^{\text{IP}}$.

Next we describe how we solve (14). This LP has a large number of rows due to the large number of feasible solutions to $\{A^E x \leq \mathbf{1}, x \text{ binary}\}$ and therefore it is solved by row generation. Given an optimal solution (η^*, α^*) to (14) with only a subset of rows (14b), we have to find the most violated row by solving the set packing problem

$$z^* = \max\{(\alpha^* A^E - c^E)x : A^E x \leq \mathbf{1}, x \text{ binary}\} . \tag{15}$$

If $z^* \leq \mathbf{1}\alpha^* - \eta^*$, then (η^*, α^*) is an optimal solution to (14). Otherwise, we add the constraint $\eta + \alpha(A^E x^* - \mathbf{1}) \leq c^E x^*$ to (14), where x^* is the optimal solution to (15), and we repeat the procedure.

Since in general, (15) is NP-hard, solving this problem at every iteration with a commercial branch-and-cut solver to optimality is too time consuming and therefore we solve (15) approximately. It has been observed in the past that if separation is solved approximately, then it is beneficial to add several rows at the same time. Therefore we have developed a heuristic that generates several 'good' solutions. In the context of set covering, Balas and Carrera (1996) assign to each column several greedy estimates and they generate several set covers by randomly selecting an estimate type. We use their idea to generate the initial set packing solutions. For simplicity of notation let $d = \alpha^* A^E - c^E$, let t_i be the number of nonzero elements in column i, and let $E = \{i_1, i_2, \ldots, i_{|E|}\}$. For every column $i \in E$ we define the greedy estimates

$$g_i^1 = d_i \qquad\qquad g_i^2 = \frac{d_i}{t_i} \qquad\qquad g_i^3 = \frac{d_i}{1 + \log t_i}$$

$$g_i^4 = \frac{d_i}{1 + t_i \log t_i} \qquad\qquad g_i^5 = \frac{d_i}{t_i^2} \qquad\qquad g_i^6 = \frac{\sqrt{d_i}}{t_i^2} .$$

Several random set packing solutions are obtained by repeating 4 times the following procedure. We first choose a random estimate type k, i.e. k is a random number between 1 and 6. Next we greedily find a set packing $\bar{x}$ based on the greedy estimates g^k. We find additional set packing solutions by either randomly selecting an element to add to or to remove from the set packing $\bar{x}$. To obtain high cost set packing solutions, we want to remove elements with low cost and add elements with high cost. Let X be the random variable that selects an element from E. We define the probability distribution of X as

$$P[X = i_j] = \begin{cases} \dfrac{d_{i_j}}{w} & \bar{x}_{i_j} = 0 \\ \dfrac{M - d_{i_j}}{w} & \bar{x}_{i_j} = 1 , \end{cases}$$

where $M = \max\{d_{i_j} : \bar{x}_{i_j} = 1\}$ and $w = \sum_{i_j : \bar{x}_{i_j} = 0} d_{i_j} + \sum_{i_j : \bar{x}_{i_j} = 1}(M - d_{i_j})$. Note that the probability distribution depends on $\bar{x}$.

Table 1. Computational Results for the Instances from Eso (1999)

name	size rows	cols	preprocessing rows	cols	time	time stage 1	stage 2	stage 3	| E |	inf	CPLEX time
sp11	104	2775	63	519	47	438					240
sp2	173	3686	150	3528	59	39	0	0	223	0	5
sp3	111	1668	73	967	26	3	0	0	97	0	18

To generate alternative set packing solutions resulting from $\bar{x}$, we iterate the following K times, where K is a parameter. Generate a random number X. If $\bar{x}_X = 1$, then we set $\bar{x}_X = 0$. If $\bar{x}_X = 0$ and $\bar{x} + e_X$ is a feasible solution to (15), then we set $\bar{x}_X = 1$. Every time we find a different $\bar{x}$, we add the corresponding constraint (14b) to the current LP if it is violated.

If this row generation procedure does not find a single violated constraint, then we solve (15) to optimality by a branch-and-cut algorithm.

To further reduce the computational time for solving (14) we use complementary slackness. For any vector x we use the notation $\text{supp}(x) = \{i : x_i \neq 0\}$. By complementary slackness it follows that in a generator OSF F_α we have $\alpha a_i \geq c_i$ for every column $i \in \text{supp}(x^*)$, which is guaranteed by (14d). Therefore we can assume that the columns in $\text{supp}(x^*)$ are in E. If we include them explicitly in E, then (15) becomes hard to solve, leading to high execution times. Instead we consider some of them only implicitly in E. We denote by $E' \subseteq E$ the set of columns that have to be included explicitly. In each iteration of the enlarge generator algorithm in stage 3, let $R = \{i \in \text{supp}(x^*) : \text{supp}(a_i) \cap \text{supp}(a_j) = \emptyset$ for every $j \in E'\}$. We maintain the property that $E = E' \cup R$. By definition of R, and since x^* is a set packing vector with $d_i \geq 0$ for every $i \in \text{supp}(x^*)$, it follows that z^* from (15) is equal to $\sum_{i \in R} d_i + \max\{dx : A^{E'} x \leq \mathbf{1}, x \text{ binary}\}$. Thus in separation it suffices to solve $\max\{dx : A^{E'} x \leq \mathbf{1}, x \text{ binary}\}$. Every time we add a new column $j \notin \text{supp}(x^*)$ to E, we need to reduce R and expand E'.

This strategy reduced the computational time for solving (14) on average by 50%. In early iterations, when E' is small, the reduction is larger. As E' grows, more columns have to be moved from R to E' and the benefit gradually disappears.

4   Computational Experiments

The computational experiments were carried out on the set partitioning instances used by Hoffman and Padberg (1993) and Eso (1999). They were performed on an IBM Thinkpad 570 with a 333 MHz Pentium processor and 196 MBytes of main memory. We used Visual Studio C++, version 6.0, and CPLEX, CPLEX Optimization (1999), as a linear programming solver.

The computational results for the instances from Hoffman and Padberg (1993) are presented in Table 2 and 3 instances from Eso (1999) are given in Table 1. Instances with an integral solution to the LP relaxation are omitted since F_{y^*}

Table 2. Computational Results for the Instances from Hoffman and Padberg (1993)

name	size rows	cols	preprocessing rows	cols	time	time stage 1	stage 2	stage 3	$\|E\|$	inf	CPLEX time
aa01	823	8904	605	7531	32	3803	84	?	?	95	723
aa03	825	8627	537	6695	30	25	7	164	142	0	88
aa04	426	7195	342	6123	17	?	?	?	?	?	908
aa05	801	8308	521	6236	21	106	19	?	?	1	97
aa06	646	7292	486	5858	23	52	23	?	?	5	39
kl01	55	7479	47	5915	49	12	63	273	100	13	14
kl02	71	36699	69	16542	8	240	271	?	?	60	118
nw03	59	43749	53	38958	6	29	0	0	19	0	48
nw04	36	87482	35	46189	16	5757	445	?	?	998	378
nw06	50	6774	37	5964	45	10	0	0	58	0	8
nw11	39	8820	28	6436	28	4	0	0	3	0	6
nw13	51	16043	48	10900	4	9	0	0	4	0	18
nw18	124	10757	81	7861	90	100	0	0	292	0	13
nw20	22	685	22	536	0	0	0	0	13	0	0
nw21	25	577	25	421	0	0	0	0	4	0	0
nw22	23	619	23	521	0	0	0	0	7	0	0
nw23	19	711	15	416	0	4	0	0	52	0	2
nw24	19	1366	19	926	2	1	0	0	52	0	1
nw25	20	1217	20	844	0	0	0	0	9	0	0
nw26	23	771	21	468	1	0	0	0	7	0	1
nw27	22	1355	22	817	3	0	0	0	4	0	0
nw28	18	1210	18	582	3	0	0	0	10	0	0
nw29	18	2540	18	2034	6	3	5	14	50	4	1
nw30	26	2653	26	1877	22	1	0	0	27	0	1
nw31	26	2662	26	1728	14	1	0	0	13	0	1
nw32	19	294	17	250	0	2	0	0	28	0	1
nw33	23	3068	23	2308	26	2	2	3	16	1	1
nw34	20	899	20	718	2	0	0	0	3	0	1
nw35	23	1709	23	1191	7	0	0	0	3	0	1
nw36	20	1783	20	1244	15	5	0	14	67	12	0
nw37	19	770	19	639	0	0	0	0	3	0	0
nw38	23	1220	20	722	9	0	0	0	6	0	1
nw39	25	677	25	565	1	0	0	0	9	0	0
nw40	19	404	19	336	0	0	0	0	11	0	0
nw41	17	197	17	177	0	0	0	0	4	0	0
nw42	23	1079	20	791	4	1	0	0	30	0	0
nw43	18	1072	17	982	0	0	0	0	3	0	0
us02	100	13635	44	5197	289	247	0	0	287	0	17
us04	163	28016	95	4080	53	3	0	0	25	0	68

is a generator OSF for these instances, where y^* is an optimal dual solution to the LP relaxation. In the Padberg-Hoffman instances, due to the low physical memory on the notebook, we were not able to solve instances nw05, nw17, us01, and nw16 is omitted since it is solved by preprocessing. Only 3 instances from Eso are solved by CPLEX and therefore all remaining sp instances are not presented. The sp11 instance is infeasible and our algorithm finds an unbounded LP (14). All the times are CPU times in seconds. The column 'inf' shows the number of infeasible columns before we apply the enlarge generator algorithm and the last column gives the CPU time for solving the problem by the branch-and-cut solver CPLEX. The '?' denotes that we have exceeded the time limit of 2 hours.

The problem aa04 is hard and we were not able to finish stage 1 in the given time limit. For 5 other problems we were able to solve stages 1 and 2, but the enlarge generator algorithm exceeded the time limit. Note that only for the problems with positive number in the 'inf' column we have to apply the enlarge generator algorithm. It takes a substantially amount of time to apply the enlarge generator algorithm (for example kl01). The problem aa05 is interesting since we only have a single infeasible column and yet it takes more than 2 hours to find the final generator OSF.

An important fact from these two tables is the observation that $|E|$ is relatively low for all of the instances. Even instances with more than 10,000 columns have the cardinality of E less than 300.

The overall computational times, which are the sum of the preprocessing times and the times for solving the 3 stages, are acceptable for a methodology that reveals much more information about an IP instance than just an optimal IP solution. It is unreasonable to expect that the computational times would be lower than branch-and-cut computational times since the latter algorithm finds only a primal optimal solution. Nevertheless, these computational results show that it is doable to obtain an optimal subadditive dual function.

4.1   Obtaining All Optimal Solutions

Given an optimal generator OSF F_α, by complementary slackness, all optimal solutions to the IP are found only among the columns i with $\alpha a_i \geq c_i$. These columns have zero reduced cost, which for a column i is defined as $c_i - F_\alpha(a_i)$. Given all such columns, we found all optimal solutions to the set partitioning problem by a naive methodology of enumerating all of them, i.e. solving several IPs. For selected instances we present the computational results in Table 3. From the second column we observe that we do not have many columns with zero reduced cost and therefore enumerating all optimal solution is acceptable, which is confirmed by the execution times given in the last column. The fourth column shows the number of columns that are at 1 in all optimal solutions. We observe that larger instances have alternative optimal solutions, which is important for robustness. A decision maker can select the best optimal solution based on alternative objective criteria.

4.2   Sensitivity Analysis

Given a generator OSF, we can perform sensitivity analysis. We have performed two computational experiments.

Table 3. All Optimal Solutions

name	no. 0 reduced cost cols	no. cols at 1 in x^*	no. cols at 1 in all	no. optimal solutions	time in seconds
nw36	8	4	4	1	0
nw33	13	5	5	1	0
nw29	12	4	4	1	0
nw04	17	9	9	1	0
kl01	42	13	32	6	2
aa06	312	93	95	2	17
aa05	449	101	103	2	30
aa03	403	102	102	1	28
aa01	255	101	101	1	15

Table 4. Addition of Negative Reduced Cost Columns

name	[0, 25%]		[25%, 50%]	
	obj. improvement	no. improvements	obj. improvement	no. improvements
nw18	4.70%	5.5%	15.0%	99%
nw06	5.10%	1.0%	3.50%	2.0%
nw24	11.5%	2.5%	13.1%	3.5%
us02	3.30%	1.5%	7.80%	1.5%
kl01	12.5%	3.0%	24.0%	0.5%

In the first one we have generated random columns with negative reduced cost. The generated columns reflect the structure of the instances, e.g. the cost is of the same order and the number of nonzeros per column is also of the same order as those of the existing columns in the constraint matrix. On average, among 200,000 randomly generated columns only 10% had negative reduced cost and therefore we know beforehand that the remaining columns do not decrease the IP objective value. We added, one at the time, each generated column with negative reduced cost to the IP and solve it to find the new optimal value, which is clearly at most z^{IP}. The computational results are presented in Table 4. The objective improvement shows the average ratio $(z^{IP} - z^{new})/z^{IP}$, where z^{new} is the IP objective value to the set partitioning problem after adding a column with the negative reduced cost. The number of improvements represents the percentage of the negative reduced cost columns that actually improve the objective value. Note that because of degeneracy after adding such a column we might not decrease the objective value. The last two columns show the improvements if the absolute value of the reduced cost is between 25% and 50% of z^{IP} and the remaining two columns if it is between 0 and 25%. With the exception of nw18, the problems are degenerate since only approximately 2% of the columns with negative reduced cost actually decrease the objective value. On the other hand, the objective improvements are solid. With the exception of nw06, the objective value improvements are larger when the reduced cost is smaller.

Table 5. Changing Right Hand Sides

name	F_α	LP relaxation
nw18	2.7%	17.2%
nw13	0.7%	4.20%
nw24	5.8%	4.80%
us04	2.5%	4.80%

In the second experiment we have considered changing right hand sides. If F_α is a generator OSF to the set partitioning problem, then for any row i the value $F_\alpha(\mathbf{1} - e_i)$ gives a lower bound on $\bar{z}_i = \min\{cx : Ax = \mathbf{1} - e_i, x \text{ binary}\}$. This holds since F_α is a subadditive dual feasible function to this IP. We have compared this bound with the lower bound obtained by the optimal dual vector of the LP relaxation to the set partitioning problem. Table 5 shows the average over all rows i of values $(\bar{z}_i - lb)/\bar{z}_i$, where lb denotes the corresponding lower bound. Except for the nw24 instance, the bound obtained by the generator OSF is substantially better. Note that $F_\alpha(\mathbf{1} - e_i)$ does not necessarily produce a better lower bound, which is shown by instance nw24.

References

Anbil, R., Johnson, E., and Tanga, R. (1992). A global approach to crew pairing optimization. *IBM Systems Journal*, **31**, 71–78.

Balas, E. and Carrera, M. (1996). A dynamic subgradient-based branch-and-bound procedure for set covering. *Operations Research*, **44**, 875–890.

Borndorfer, R. (1998). *Aspects of Set Packing, Partitioning, and Covering*. Ph.D. thesis, Technical University of Berlin.

Burdet, C. and Johnson, E. (1977). A subadditive approach to solve integer programs. *Annals of Discrete Mathematics*, **1**, 117–144.

CPLEX Optimization (1999). *Using the CPLEX Callable Library*. ILOG Inc., 6.5 edition.

Eso, M. (1999). *Parallel Branch and Cut for Set Partitioning*. Ph.D. thesis, Cornell University.

Hoffman, K. and Padberg, M. (1993). Solving airline crew scheduling problems by branch-and-cut. *Management Science*, **39**, 657–682.

Johnson, E. (1973). Cyclic groups, cutting planes and shortest path. In T. Hu and S. Robinson, editors, *Mathematical Programming*, pages 185–211. Academic Press.

Klabjan, D. (2001a). A new subadditive approach to integer programming: Implementation and computational results. Technical report, University of Illinois at Urbana-Champaign. Available from
http://www.staff .uiuc.edu/~klabjan/professional.html.

Klabjan, D. (2001b). A new subadditive approach to integer programming: Theory and algorithms. Technical report, University of Illinois at Urbana-Champaign. Available from
http://www.staff .uiuc.edu/~klabjan/professional.html.

Llewellyn, D. and Ryan, J. (1993). A primal dual integer programming algorithm. *Discrete Applied Mathematics*, **45**, 261–275.

Nemhauser, G. and Wolsey, L. (1988). *Integer and combinatorial optimization*. John Wiley & Sons.

Improved Approximation Algorithms
for Resource Allocation

Gruia Calinescu[1], Amit Chakrabarti[2,*], Howard Karloff[3], and Yuval Rabani[4,**]

[1] Department of Computer Science, Illinois Institute of Technology,
Stuart Building, 10 West 31st Street, Chicago, IL 60616
calinesc@cs.iit.edu
[2] Department of Computer Science, Princeton University, Princeton, NJ 08544
amitc@cs.princeton.edu
[3] AT&T Labs – Research, 180 Park Ave., Florham Park, NJ 07932
howard@research.att.com
[4] Computer Science Department, Technion – IIT, Haifa 32000, Israel
rabani@cs.technion.ac.il

Abstract. We study the problem of finding a most profitable subset of
n given tasks, each with a given start and finish time as well as profit and
resource requirement, that at no time exceeds the quantity B of available
resource. We show that this NP-hard RESOURCE ALLOCATION problem
can be $(1/2 - \varepsilon)$-approximated in polynomial time, which improves upon
earlier approximation results for this problem, the best previously published result being a 1/4-approximation. We also give a simpler and faster
1/3-approximation algorithm.

1 Introduction

We consider the following optimization problem. Suppose we have a limited
supply of one reusable resource and are given a set of n tasks each of which
occupies a fixed interval of time and requires a given amount of the resource.
Further, each task has an associated profit which we obtain if we schedule that
task. Our goal is to select a subset of the n tasks to schedule, so that the resource
available is sufficient at all times for all simultaneously scheduled tasks, and so
that the profit obtained is maximized. Let us call this problem the RESOURCE
ALLOCATION PROBLEM or RAP for short.

* Part of this work was done while visiting AT&T Labs – Research. Work at Princeton
University supported by NSF Grant CCR-99817 and ARO Grant DAAH04-96-1-
0181.
** Part of this work was done while visiting AT&T Labs – Research. Work at the
Technion supported by Israel Science Foundation grant number 386/99, by BSF
grants 96-00402 and 99-00217, by Ministry of Science contract number 9480198, by
EU contract number 14084 (APPOL), by the CONSIST consortium (through the
MAGNET program of the Ministry of Trade and Industry), and by the Fund for the
Promotion of Research at the Technion.

W.J. Cook and A.S. Schulz (Eds.): IPCO 2002, LNCS 2337, pp. 401–414, 2002.
© Springer-Verlag Berlin Heidelberg 2002

This abstractly specified problem actually occurs in several concrete guises. For instance, we may be given a set of network sessions, with fixed start and finish times, that compete for a limited amount of bandwidth between two fixed endpoints in the network. Leonardi et al. [6] observe that a research project on scheduling requests for remote medical consulting on a satellite channel requires solving a slight variant of this very problem. Hall and Magazine [5] were interested in maximizing the value of a space mission by deciding what set of projects to schedule during the mission, where an individual project typically occupies only a part of the mission. Due to its various guises, RAP is known by several other names such as BANDWIDTH ALLOCATION PROBLEM, RESOURCE CONSTRAINED SCHEDULING and CALL ADMISSION CONTROL.

To formulate the problem precisely, we are given an integer B (the total available amount of the shared resource) and a collection of n tasks, with starting times s_i, finishing times t_i, resource requirements b_i, where $b_i \leq B$, and profits p_i. All these numbers are nonnegative integers and $s_i < t_i$ for all i. We want to identify a subset $S \subseteq [n]$ of tasks to schedule which maximizes $\sum_{i \in S} p_i$ among those S satisfying the following constraint:

$$\forall t \quad \sum_{i:\, [s_i, t_i) \ni t} b_i \leq B \,.$$

RAP is a natural generalization of the (polynomial-time solvable) problem of finding a maximum weight independent set in a weighted interval graph.

1.1   Prior Work

There has been a flurry of research activity focusing either on RAP itself or on problems with a similar flavor. All tend to be NP-hard; in RAP, setting each $s_i = 0$ and $t_i = 1$ gives KNAPSACK as a special case of RAP. Accordingly, research on RAP strives to obtain polynomial-time approximation algorithms. Since it is not known whether RAP is MaxSNP-hard, the possibility of a polynomial-time approximation scheme (PTAS) remains open.

Using LP rounding techniques, Phillips et al. [8] obtain a 1/6-approximation algorithm[1] for RAP. Actually they solve a more general problem in which each task occupies not a fixed interval of time but has a fixed length and a window of time in which it is allowed to slide. Using the local ratio technique, Bar-Noy et al. [2] obtain results which imply a 1/4-approximation algorithm for RAP. Bar-Noy [3] has since informed us that the techniques in [2] can in fact yield a 1/3-approximation algorithm. Using different ideas, Chen et al. [4] have recently obtained a 1/3-approximation for RAP in the special case where the profit of each task i equals its "area" $(t_i - s_i)b_i$.

The STORAGE ALLOCATION PROBLEM is a related problem in which, in addition, the resource must be allocated to the scheduled tasks in contiguous

---

[1] Since RAP is a maximization problem, approximation factors of algorithms for it will be at most 1 – the larger the factor, the better.

blocks, which must not change during the lifetime of the task. Both Bar-Noy et al. [2] and Leonardi et al. [6] study this problem and the latter paper obtains a 1/12-approximation algorithm, which is the current best. The former paper obtains results for several related problems.

1.2   Our Results

We present three algorithms for restricted versions of RAP. Suitably combining these algorithms leads to approximation algorithms for the general case. The best approximation ratio that we can obtain improves upon previous results mentioned above.

Theorem 1.1. *Consider* RAP *with the restriction that every task satisfies* $b_i > \delta B$, *for a constant* $0 < \delta \leq 1$. *There is an algorithm that solves the problem exactly and runs in time* $O(n^{2/\delta + O(1)})$.

Theorem 1.2. *For every* δ, $0 < \delta \leq 0.976$, *there is a randomized* $O(n^2 \log^2 n)$-*time algorithm for the special case of* RAP *in which all* $b_i \leq \delta B$, *which achieves a* $(1 - 4\varepsilon)$-*approximation with high probability, where* $\varepsilon = \sqrt{(8/3)\delta \ln(1/\delta)}$ [2].

Since $\lim_{\delta \to 0} \delta \ln(1/\delta) = 0$, we conclude that for any $\varepsilon > 0$ there is a $\delta > 0$ such that there is a randomized, polynomial-time, $(1 - 4\varepsilon)$-approximation algorithm if all $b_i \leq \delta B$.

Theorem 1.3. *The restriction of* RAP *to inputs having all* $b_i \leq B/2$ *has a deterministic polynomial-time* 1/2-*approximation algorithm.*

In what follows, we shall refer to our algorithms which prove the above theorems as the *Large Tasks Algorithm*, the *Small Tasks Algorithm* and the *List Algorithm*, respectively.

To obtain results for unrestricted RAP, we can "combine" the algorithms for special cases in a certain manner. The combined algorithm's running time will be the sum of the running times of the constituents. Combining the Large Tasks and Small Tasks algorithms gives the following result.

Theorem 1.4. *Given any constant* $\varepsilon > 0$, *there is a randomized polynomial-time algorithm that approximates* RAP *within a factor of at least* $1/2 - \varepsilon$, *with high probability. The exponent in the running time is* poly$(1/\varepsilon)$. (*More precisely, the exponent is a constant times the smaller of the two positive* δ's *that satisfy* $\varepsilon = \sqrt{\delta \log(1/\delta)}$.)

One can trade approximation guarantee for running time, and simplicity, by combining the List Algorithm with a well-known interval graph algorithm (details in Sect. 4):

Theorem 1.5. *There is a deterministic* 1/3-*approximation algorithm for* RAP *with running time* $O(n^2 \log^2 n)$.

The technique of the proof of Theorem 1.5 is very simple and quite different from that used by Bar-Noy [3] to obtain the same approximation ratio.

---

[2] This statement is vacuous unless $\delta < 0.0044$.

1.3 Organization of the Rest of the Paper

The remainder of the paper proves the above theorems. In Sect. 2 we make some basic definitions and explain how to combine algorithms. Section 3 describes the Large Tasks Algorithm, the Small Tasks Algorithm and the recipe for combining them to obtain Theorem 1.4. Finally, Sect. 4 describes the List Algorithm and proves Theorem 1.5.

2 Preliminaries

Assumption 2.1 *Throughout the paper we shall assume that the tasks are numbered 1 to n in order of increasing starting time, ties broken arbitrarily, i.e.,* $i < j \Rightarrow s_i \leq s_j$.

Definition 2.2. *We say that task i is* active *at an instant t of time (where t is any real number) if $t \in [s_i, t_i)$. Also, for each i, we define S_i to be the set of tasks in $\{1, 2, \ldots, i\}$ that are active at time s_i: $S_i = \{j \leq i \,:\, t_j > s_i\}$. Note that $i \in S_i$.*

Notice that a task is considered active at its start time s_i but not at its finish time t_i; this prevents task i from competing for resources with a task which ends at time s_i.

We observe that RAP can be easily expressed as an integer linear program as follows:

$$\text{Maximize } \sum_{i=1}^{n} p_i x_i \tag{1}$$

$$\text{s.t. } \sum_{j \in S_i} b_j x_j \leq B, \qquad 1 \leq i < n , \tag{2}$$

$$\text{and } x_i \in \{0, 1\}, \qquad 1 \leq i \leq n . \tag{3}$$

This integer program has a natural LP relaxation, obtained by replacing (3) by the constraint

$$0 \leq x_i \leq 1, \qquad 1 \leq i \leq n . \tag{4}$$

We shall refer to the LP given by (1), (2) and (4) as LPMAIN.

Definition 2.3. *A set $U \subseteq \{1, \ldots, n\}$ of tasks is called a* packing *if its characteristic vector satisfies the constraints (2)–(3).*

2.1 Solving the LP

We note that LPMAIN can be solved by a min-cost flow algorithm, based on the following previously-known construction similar to the one described in [1]. Construct the network $\mathcal{N}$, with vertex set $V(\mathcal{N}) = \{1, 2, \ldots, n, n+1\}$. For every $1 \leq i \leq n$, add an arc $(i, i+1)$ of capacity B and cost 0. For every task i, define $r_i = \min\{j : 1 \leq j \leq n \text{ and } s_j \geq t_i\}$, with the convention that the minimum over an empty set is $n+1$. Then add an arc (r_i, i) of capacity b_i and cost $-p_i/b_i$.

These are all the arcs of $\mathcal{N}$. A valid flow in $\mathcal{N}$ with flow on the arc (r_i, i) equal to f_i corresponds to a feasible solution to LPMAIN with $x_i = f_i/b_i$ and has cost $\sum_{i=1}^n p_i x_i$. Therefore a min-cost flow in $\mathcal{N}$ corresponds to an optimum solution to LPMAIN.

The best known strongly polynomial-time min-cost flow algorithm, due to Orlin [7], solves LPMAIN in $O(n^2 \log^2 n)$ time.

2.2   Combining Two Algorithms for Restricted RAP

Suppose algorithm $\mathcal{A}_L$ works when each $b_i > \tau$ and yields an exact (optimal) solution, and algorithm $\mathcal{A}_S$ works when each $b_i \leq \tau$ and yields an α-approximation for some constant $\alpha \leq 1$. The dividing point τ may depend on B.

1. Divide the tasks in the input instance into *small tasks* (those with $b_i \leq \tau$), and *large tasks* (those with $b_i > \tau$).
2. Using $\mathcal{A}_S$, compute an α-approximation to the optimal packing that uses small tasks alone.
3. Using $\mathcal{A}_L$, compute an optimal packing that uses large tasks alone.
4. Of the two packings obtained in Steps 2 and 3, choose the one with greater profit.

Note that in any problem instance, if OPT denotes the profit of the optimal packing and OPT_L (respectively, OPT_S) denotes the profit of the optimal packing using only large (respectively, small) tasks, then

$$\text{either}\ \ \mathsf{OPT}_S \geq \frac{1}{1+\alpha} \cdot \mathsf{OPT} \ \ \text{or}\ \ \mathsf{OPT}_L \geq \frac{\alpha}{1+\alpha} \cdot \mathsf{OPT} \ .$$

Therefore the above combined algorithm achieves an approximation ratio of at least $\alpha/(1+\alpha)$.

3   The Large and Small Tasks Algorithms

In this section, we prove Theorems 1.1 and 1.2. For the former, we give a dynamic programming algorithm that does not consider LPMAIN at all. For the latter, we solve LPMAIN and use a randomized rounding scheme to obtain a good approximation. We then derive Theorem 1.4 from these two results as indicated above.

3.1   Dynamic Programming for the Large Tasks

We describe how to solve RAP exactly, provided each $b_i > \delta B$. We begin with some definitions.

Let U be a packing. We shall denote the profit of U, i.e., the sum of profits of all tasks in U, by $p(U)$. If $U \neq \emptyset$, we denote the highest numbered task in U by $\text{top}(U)$; this is a task in U which starts last.

Definition 3.1. *For a packing $U \neq \emptyset$, we define its* kernel $\ker(U)$ *to be the set of all tasks in U which are active at the time when* $\text{top}(U)$ *starts:* $\ker(U) = \{i \in U : [s_i, t_i) \ni s_{\text{top}(U)}\} = \{i \in U : t_i > s_{\text{top}(U)}\}$. *We also define* $\ker(\emptyset) = \emptyset$.

Definition 3.2. *A packing U is called a* pile *if $\ker(U) = U$.*

A pile is a clique in the underlying interval graph and it is a feasible solution to the RAP instance. Any kernel is a pile. Now define the real function f on piles V as follows:

$$f(V) = \max \{p(U) : U \text{ is a packing with } \ker(U) = V\} .$$

Note that $f(\emptyset) = 0$. Clearly, if we can compute $f(V)$ for all V, we are done. We show below how to do this efficiently.

For any packing $U \neq \emptyset$, let $U' = U \setminus \{\text{top}(U)\}$. For a pile $V \neq \emptyset$, let $\mathcal{T}(V)$ denote the following set of piles:

$$\mathcal{T}(V) = \{W \supseteq V' : W \text{ is a pile and } t_j \leq s_{\text{top}(V)} \text{ for all } j \in W \setminus V'\} . \quad (5)$$

Clearly $\mathcal{T}(V) \neq \emptyset$, since $V' \in \mathcal{T}(V)$. In words, $\mathcal{T}(V)$ consists of all piles obtained from V by removing its highest numbered ("rightmost") task and adding in zero or more tasks that end exactly when or before this removed task starts. We shall need some facts about the functions defined so far:

Lemma 3.3. *In any instance of RAP, the following hold:*
(i) *For any packing $U \neq \emptyset$, $\text{top}(U) = \text{top}(\ker(U))$.*
(ii) *For any packing $U \neq \emptyset$, $\ker(U') \supseteq (\ker(U))'$.*
(iii) *Let $V \neq \emptyset$ be a pile and $W \in \mathcal{T}(V)$. If $W \neq \emptyset$ then $\text{top}(W) < \text{top}(V)$.*
(iv) *Let $V \neq \emptyset$ be a pile, $W \in \mathcal{T}(V)$, and U be a packing with $\ker(U) = W$. Then $U \cup \{\text{top}(V)\}$ is a packing.*
(v) *Let $V \neq \emptyset$ be a pile, $W \in \mathcal{T}(V)$, and U be a packing with $\ker(U) = W$. Then $\ker(U \cup \{\text{top}(V)\}) = V$.*
(vi) *Let $V \neq \emptyset$ be a pile and X be a packing with $\ker(X) = V$. Then $\ker(X') \in \mathcal{T}(V)$.*

Proof. The proofs are straightforward.
(i) This is trivial from the definitions.
(ii) When U is a singleton this is trivial, so we assume that U has at least two elements; so $U' \neq \emptyset$.

Let $k = \text{top}(U)$ and $k' = \text{top}(U') < k$. Let $j \in (\ker(U))'$; by part (i), this means $j \in \ker(U) \setminus \{k\}$, i.e., $j \in \ker(U) \subseteq U$ and $j < k$; therefore $j \in U'$ and in particular $j \leq k'$. By Assumption 2.1 we have $s_j \leq s_{k'}$ and since $j \in \ker(U)$, we have $t_j > s_k \geq s_{k'}$; thus j is active when $k' = \text{top}(U')$ starts and so $j \in \ker(U')$.
(iii) For any $j \in W$, by (5), either $j \in V'$ whence $j < \text{top}(V)$, or else $j \in W \setminus V'$ and hence $s_j < t_j \leq s_{\text{top}(V)}$ whence $j < \text{top}(V)$. Thus, in particular, $\text{top}(W) < \text{top}(V)$.
(iv) First of all, note that if $\ker(U) = W = \emptyset$, then $U = \emptyset$ and $\{\text{top}(V)\}$ is trivially a packing, since every singleton is. So we may assume $W \neq \emptyset$ and $U \neq \emptyset$.

Let $k = \text{top}(V)$ and $\ell = \text{top}(U)$. By parts (i) and (iii), $\ell = \text{top}(U) = \text{top}(\ker(U)) = \text{top}(W) < \text{top}(V) = k$. Thus $\text{top}(U \cup \{k\}) = \max\{\ell, k\} = k$. This means that at any time $\tau < s_k$, the set $U \cup \{k\}$ of tasks satisfies the feasibility constraint (2) at time τ (because U is given to be a packing). It remains to show that the feasibility constraint is satisfied at time s_k as well. To this end, we shall show that the set of tasks in $U \cup \{k\}$ which are active at time s_k is a subset of V; since V is a packing, this will complete the proof.

Let $j \in U \cup \{k\}$ be active at time s_k. If $j = k$ we immediately have $j \in V$. Otherwise, since j is active at time s_k, and $k > \ell$ as shown above, $t_j > s_k \geq s_\ell$. Also, $j \leq \text{top}(U) = \ell$, so $s_j \leq s_\ell$. Thus j is active at time s_ℓ, so $j \in \ker(U) = W$.

Suppose $j \notin V$. Then $j \notin V'$ and by (5) we see that $t_j \leq s_k$, a contradiction. Therefore $j \in V$ and we are done.

(v) First, note that $W = \emptyset$ implies $U = \emptyset$ and, by (5), $V' = \emptyset$, which means V is a singleton: $V = \{\text{top}(V)\}$. The statement is now trivial. So we may assume $W \neq \emptyset$ and $U \neq \emptyset$.

Let $k = \text{top}(V)$. As shown above, $\text{top}(U \cup \{k\}) = k$. Suppose $j \in \ker(U \cup \{k\})$. Then j is active at time s_k. In the proof of part (iv) we have already seen that this implies $j \in V$. Thus $\ker(U \cup \{k\}) \subseteq V$.

Now suppose $j \in V$. Since V is a pile, j is active at time $s_{\text{top}(V)} = s_k$. As shown above, $k = \text{top}(U \cup \{k\})$, so j is active at the start time of job $\text{top}(U \cup \{k\})$. We claim that $j \in U \cup \{k\}$. Indeed, if $j \neq k$, then $j \in V \setminus \{k\} = V'$. Since $W \in \mathcal{T}(V)$, by definition of $\mathcal{T}(V)$ (see (5)) we have $W \supseteq V'$. Thus $j \in V' \subseteq W \subseteq U$, which proves the claim. This in turn shows that $j \in \ker(U \cup \{k\})$. Thus $V \subseteq \ker(U \cup \{k\})$.

(vi) Since V is nonempty, so is X. Let $k = \text{top}(V) = \text{top}(X)$. By part (ii), $\ker(X') \supseteq (\ker(X))' = V'$. Let $j \in \ker(X') \setminus V'$. Looking at (5), in order to prove that $\ker(X') \in \mathcal{T}(V)$, we must show that $t_j \leq s_k$. Suppose not. Since $j \in \ker(X') \subseteq X' \subseteq X$, we have $j \leq k$. This means j is active at time s_k, since we've assumed that $t_j > s_k$; therefore $j \in \ker(X) = V$. But $j \notin V'$; so $j = k$. However, $j \in X' = X \setminus \{k\}$, which is a contradiction. $\qquad\square$

The key to the dynamic programming algorithm is the following lemma:

Lemma 3.4. *Let $V \neq \emptyset$ be a pile. Then*

$$f(V) = p_{\text{top}(V)} + \max_{W \in \mathcal{T}(V)} f(W) . \tag{6}$$

Proof. Let R denote the right-hand side of (6).

We first establish that $f(V) \geq R$. Let $W \in \mathcal{T}(V)$ be a pile that maximizes $f(W)$ among all such W, and let U be a packing with $\ker(U) = W$ that maximizes $p(U)$ among all such U. If $W = \emptyset$ we have $R = p_{\text{top}(V)}$. Since V is a pile, it is a packing with kernel V, so $f(V) \geq p(V) \geq p_{\text{top}(V)} = R$.

So assume $W \neq \emptyset$. Then $U \neq \emptyset$. Now $R = p_{\text{top}(V)} + f(W) = p_{\text{top}(V)} + p(U)$. By parts (i) and (iii) of Lemma 3.3, $\text{top}(U) = \text{top}(\ker(U)) = \text{top}(W) < \text{top}(V)$. Therefore $\text{top}(V) \notin U$. By parts (iv) and (v) of Lemma 3.3, $U \cup \{\text{top}(V)\}$ is a packing with kernel V. By definition of f this means

$$f(V) \geq p(U \cup \{\text{top}(V)\}) = p(U) + p_{\text{top}(V)} = R .$$

Now we establish that $f(V) \leq R$. Let X be a packing with $\ker(X) = V$ that maximizes $p(X)$ among all such X; then $f(V) = p(X)$. Also, X is nonempty since V is. By part (i) of Lemma 3.3, $\mathrm{top}(X) = \mathrm{top}(V) = k$, say. Since $k \in X$, $p(X) = p_k + p(X \setminus \{k\}) = p_k + p(X') \leq p_k + f(\ker(X'))$, where the inequality follows from the definition of f. By part (vi) of Lemma 3.3, $\ker(X') \in \mathcal{T}(V)$. Therefore

$$f(V) \ = \ p(X) \ \leq \ p_k + f(\ker(X')) \ \leq \ p_k + \max_{W \in \mathcal{T}(V)} f(W) \ = \ R \, ,$$

which completes the proof of the lemma. □

We now describe the dynamic programming algorithm.

Proof (of Theorem 1.1). Compute $f(V)$ for all piles V in increasing order of $\mathrm{top}(V)$, using formula (6). By part (iii) of Lemma 3.3, every pile $W \in \mathcal{T}(V)$, $W \neq \emptyset$, satisfies $\mathrm{top}(W) < \mathrm{top}(V)$ so that when computing $f(V)$ we will have already computed $f(W)$ for every W involved in (6). Then compute the profit of the optimal packing as the maximum of $f(V)$ over all piles V.

The above algorithm computes only the profit of an optimal packing, but it is clear how to modify it to find an optimal packing as well.

The running time is clearly at most $\mathrm{poly}(n)$ times a quantity quadratic in the number of piles. Since each task has $b_i > \delta B$, the size of a pile is at most $1/\delta$ and so the number of distinct nonempty piles is at most $\sum_{k=1}^{\lfloor 1/\delta \rfloor} \binom{n}{k} \leq (1+n)^{1/\delta}$. This gives a running time of $O(n^{2/\delta + O(1)})$, as claimed. □

3.2   Randomized Rounding for the Small Tasks

In this subsection we prove Theorem 1.2. Accordingly, we assume that the tasks in the input instance satisfy $b_i \leq \delta B$ and $\delta \leq 0.976$. We set

$$\varepsilon = \sqrt{\frac{8}{3} \delta \ln(1/\delta)} \, . \tag{7}$$

We shall describe an algorithm, based on randomized rounding of an LP solution, which returns a solution to RAP whose expected performance ratio is at least $1 - 2\varepsilon$. Repeating this algorithm several times one can get a $(1 - 4\varepsilon)$-approximation with high probability, thereby proving the theorem.

If $1 - 4\varepsilon \leq 0$, there is nothing to prove, so we may assume that $\varepsilon < 1/4$. Since $f(\delta) = \sqrt{(8/3)\delta \ln(1/\delta)}$ is increasing on $(0, 1/e)$, decreasing on $(1/e, 1]$, $f(0.976) > 1/4$, $f(0.0044) > 1/4$, and $\delta \leq 0.976$, we infer that

$$\delta < 0.0044 \, . \tag{8}$$

We solve LPMAIN using the algorithm indicated in Sect. 2.1. Suppose this gives us an optimal fractional solution $(x_1, x_2, \dots, x_n)$ with profit OPT^*. We then choose independent random variables $Y_i \in \{0, 1\}$, $i = 1, 2, \dots, n$, with

$\Pr[Y_i = 1] = (1 - \varepsilon)x_i$. Now, if we define the (dependent) random variables $Z_1, Z_2, ..., Z_n$, in that order, as follows:

$$Z_i = \begin{cases} 1, & \text{if } Y_i = 1 \text{ and } \sum_{j \in S_i \setminus \{i\}} b_j Z_j \leq B - b_i \ , \\ 0, & \text{otherwise} \ , \end{cases}$$

then clearly $\{i : Z_i = 1\}$ is a packing, and its expected profit is $\sum_{i=1}^{n} p_i \cdot \Pr[Z_i = 1]$. We shall lower bound the probability that $Z_i = 1$. For this purpose we need a simple probabilistic fact and a tail estimate, collected together in the following lemma.

Lemma 3.5. *Let $X_1, X_2, \ldots, X_m$ be independent random variables and let $0 \leq \beta_1, \beta_2, ..., \beta_m \leq 1$ be reals, where for $i \in \{1, 2, \ldots, m\}$, $X_i = \beta_i$ with probability p_i, and $X_i = 0$ otherwise. Let $X = \sum_i X_i$ and $\mu = E[X]$. Then*
(i) $\sigma(X) \leq \sqrt{\mu}$.
(ii) *For any λ with $0 < \lambda < \sqrt{\mu}$, $\Pr[X > \mu + \lambda\sqrt{\mu}] < \exp\left(-\frac{\lambda^2}{2}\left(1 - \lambda/\sqrt{\mu}\right)\right)$.*

Proof. For part (i):

$$\sigma^2(X) = \sum_{i=1}^{m} \left(E[X_i^2] - (E[X_i])^2\right)$$
$$\leq \sum_{i=1}^{m} E[X_i^2]$$
$$\leq \sum_{i=1}^{m} E[X_i]$$
$$= \mu \ ,$$

where the second inequality follows from $\beta_i \leq 1$.

To prove part (ii), put $t = \ln\left(1 + \lambda/\sqrt{\mu}\right) \geq 0$. Trivially, $t \geq \lambda/\sqrt{\mu} - \lambda^2/(2\mu)$ since $0 \leq \lambda/\sqrt{\mu} < 1$. By Markov's inequality, $\Pr\left[X > \mu + \lambda\sqrt{\mu}\right] = \Pr\left[e^{tX} > e^{t\mu + t\lambda\sqrt{\mu}}\right] < E[e^{tX}]/\exp(t\mu + t\lambda\sqrt{\mu})$. Now

$$E[e^{tX}] = E\left[\prod_{i=1}^{m} e^{tX_i}\right]$$
$$= \prod_{i=1}^{m} E\left[e^{tX_i}\right]$$
$$= \prod_{i=1}^{m} \left(1 - p_i + p_i e^{t\beta_i}\right)$$
$$\leq \prod_{i=1}^{m} \exp\left(p_i(e^{t\beta_i} - 1)\right)$$
$$= \exp\left(\sum_{i=1}^{m} p_i\left(e^{t\beta_i} - 1\right)\right) \ .$$

Further,

$$\sum_{i=1}^{m} p_i \left(e^{t\beta_i} - 1 \right) = \sum_{i=1}^{m} p_i \left(t\beta_i + \frac{1}{2!}(t\beta_i)^2 + \frac{1}{3!}(t\beta_i)^3 + \cdots \right)$$

$$\leq \sum_{i=1}^{m} p_i \beta_i \left(e^t - 1 \right)$$

$$= \mu \left(e^t - 1 \right),$$

where the inequality follows from the fact that each $\beta_i \leq 1$. This gives

$$\Pr\left[X > \mu + \lambda\sqrt{\mu} \right] < \exp\left(\mu(e^t - 1) - t\mu - t\lambda\sqrt{\mu} \right)$$

$$\leq \exp\left(\lambda\sqrt{\mu} - \lambda\sqrt{\mu} + \lambda^2/2 - \lambda^2 + \lambda^3/(2\sqrt{\mu}) \right)$$

$$= \exp(-\lambda^2/2 + \lambda^3/(2\sqrt{\mu})),$$

where the last inequality follows from the lower bound on t. □

Lemma 3.6. *Under conditions* (7) *and* (8), $\Pr[Z_i = 1] \geq (1 - 2\varepsilon)x_i$.

Proof. Conditions (7) and (8) imply $\varepsilon > 57\delta$. This fact will be used twice below.

Fix an i. We shall estimate the probability $\pi_i = \Pr[Z_i = 0 \mid Y_i = 1]$. Notice that Z_j does not depend on Y_i when $j < i$. Since $Z_j \leq Y_j$ for all j and $b_i \leq \delta B$, we now have

$$\pi_i = \Pr\left[\sum_{j \in S_i \setminus \{i\}} b_j Z_j > B - b_i \right] \leq \Pr\left[\sum_{j \in S_i \setminus \{i\}} \frac{b_j Y_j}{\delta B} > \frac{1 - \delta}{\delta} \right]. \quad (9)$$

Now the random variables $\{b_j Y_j/(\delta B)\}_{j \in S_i \setminus \{i\}}$, and $\beta_j = b_j/(\delta B)$, $p_j = (1-\varepsilon)x_j$ satisfy the conditions of Lemma 3.5. Let Y be the sum of the new random variables and let $\mu = E[Y]$. We have

$$\mu = \sum_{j \in S_i \setminus \{i\}} \frac{b_j}{\delta B} \cdot (1 - \varepsilon)x_j = \frac{1 - \varepsilon}{\delta} \sum_{j \in S_i \setminus \{i\}} \frac{b_j x_j}{B} \leq \frac{1 - \varepsilon}{\delta}, \quad (10)$$

by constraint (2) of LPMAIN. We now consider two cases.

Case 1: $\mu < (7/8)(1 - \delta)/\delta$. Part (i) of Lemma 3.5 gives $\sigma(Y) \leq \sqrt{\mu}$. Using (9) and Chebyshev's inequality we get

$$\pi_i \leq \Pr\left[Y > \frac{1 - \delta}{\delta} \right]$$

$$\leq \Pr\left[|Y - \mu| > \frac{1}{8} \cdot \frac{1 - \delta}{\delta} \cdot \frac{\sigma(Y)}{\sqrt{\mu}} \right]$$

$$\leq \frac{64\delta^2 \mu}{(1 - \delta)^2}.$$

By our assumption about μ, this is less than $56\delta/(1-\delta)$ and now by (8), $\pi_i < 57\delta$.

Case 2: $\mu \geq (7/8)(1-\delta)/\delta$. Set λ such that $\mu + \lambda\sqrt{\mu} = (1-\delta)/\delta$. Then λ, considered as a function of μ, is decreasing. From this and (10) we have

$$\lambda = \frac{\frac{1-\delta}{\delta} - \mu}{\sqrt{\mu}} \geq \frac{\frac{1-\delta}{\delta} - \frac{1-\varepsilon}{\delta}}{\sqrt{\frac{1-\varepsilon}{\delta}}} = \frac{\varepsilon - \delta}{\sqrt{\delta(1-\varepsilon)}} \geq \frac{\varepsilon - \delta}{\sqrt{\delta}} \geq \frac{56}{57}\frac{\varepsilon}{\sqrt{\delta}} .$$

Also, $1 - \lambda/\sqrt{\mu} = 2 - (1-\delta)/\delta\mu \geq 6/7$, by the assumption about μ, and further, we trivially get $\lambda < \sqrt{\mu}$. By (9) and part (ii) of Lemma 3.5, applied to the variables $\{b_j Y_j/(\delta B)\}_{j \in S_i \setminus \{i\}}$ and their sum Y, we obtain

$$\begin{aligned}
\pi_i &\leq \Pr[Y > \mu + \lambda\sqrt{\mu}] \\
&< \exp\left(-\frac{\lambda^2}{2}(1 - \lambda/\sqrt{\mu})\right) \\
&< \exp\left(-\frac{1}{2}\left(\frac{56}{57}\right)^2\frac{\varepsilon^2}{\delta}\frac{6}{7}\right) \\
&< \delta .
\end{aligned}$$

In either case, $\pi_i < 57\delta$. Hence,

$$\Pr[Z_i = 1] = (1 - \pi_i) \cdot \Pr[Y_i = 1] \geq (1 - 57\delta)(1-\varepsilon)x_i \geq (1-\varepsilon)^2 x_i \geq (1 - 2\varepsilon)x_i ,$$

which completes the proof of the lemma. $\qquad\square$

Proof (of Theorem 1.2). The randomized rounding procedure computes a packing; let the random variable P denote its profit. As noted in the comments preceding Lemma 3.5, $E[P] = \sum_{i=1}^{n} p_i \cdot \Pr[Z_i = 1]$ which, by Lemma 3.6, is at least $(1 - 2\varepsilon)\mathsf{OPT}^* \geq (1 - 2\varepsilon)\mathsf{OPT}$. However, P never exceeds OPT. Markov's inequality now implies $\Pr[P \geq (1 - 4\varepsilon)\mathsf{OPT}] \geq 1/2$. By repetition, we can now suitably amplify the probability of obtaining at least $1 - 4\varepsilon$ of the profit. The bound on the running time follows easily from Subsect. 2.1. $\qquad\square$

3.3   Proof of Theorem 1.4

We have already described in Subsect. 2.2 how to combine the two algorithms described above. In the terminology of that section, we would have $\alpha = 1 - 4\varepsilon$ and $\tau = \delta B$ where ε and δ are related according to (7). As argued at the end of that section, this would lead to an approximation ratio of $\alpha/(1 + \alpha) = (1 - 4\varepsilon)/(2 - 4\varepsilon) \geq 1/2 - 2\varepsilon$ for general RAP.

Given an ε, in order to achieve this $1/2 - 2\varepsilon$ approximation, we solve (7) for δ (we use the smaller δ obtained) and use this δ in the Large and Small Tasks Algorithms.

4   The List Algorithm

In this section we give a fast and simple 1/2-approximation algorithm in the case when each $b_i \leq B/2$; this will prove Theorem 1.3. It is inspired by a similar "coloring" algorithm of Phillips et al [8]. Our algorithm has the advantage that it does not have to round to a large common denominator; therefore it is faster and obtains a 1/2-approximation, rather than a $(1/2 - \varepsilon)$-approximation.

Our algorithm generates a list of packings with the property that the "average" of the packings in the list has large profit. To be precise, we first solve LPMAIN and obtaining an optimal fractional solution $(x_1, x_2, \ldots, x_n)$ whose profit is $\sum_{i=1}^{n} p_i x_i$. Our rounding algorithm will produce a list $U_1, U_2, \ldots, U_m$ of sets of tasks, together with non-negative real weights $x(U_1), \ldots, x(U_m)$ for these sets. These sets and weights will satisfy the following properties:

1. Each set U_k is a packing.
2. $0 \leq x(U_k) \leq 1$ for each set U_k.
3. For each i we have $\sum_{k:\, U_k \ni i} x(U_k) = x_i$.
4. $\sum_{k=1}^{m} x(U_k) \leq 2$.

Define the profit of a set to be the sum of the profits of the constituent tasks. Our algorithm will select a maximum profit set from the list it constructs. Let P be the profit of the set picked by the algorithm. Then, assuming that the above properties hold, $\frac{1}{2} \sum_{i=1}^{n} p_i x_i = \frac{1}{2} \sum_{i=1}^{n} \sum_{k:\, U_k \ni i} p_i x(U_k) = \sum_{k=1}^{m} \frac{1}{2} x(U_k) \sum_{i:\, i \in U_k} p_i \leq \sum_{k=1}^{m} \frac{1}{2} x(U_k) \cdot P \leq P$, where the first equality follows from Property 3, the first inequality from Property 1 and the final inequality from Property 4. Therefore we indeed have a 1/2-approximation algorithm.

We now describe the procedure for generating the sets U_k and weights $x(U_k)$. Initialize a list $\mathcal{L}$ of sets to an empty list. Then consider the tasks in the order $1, 2, \ldots, n$. For task i:

---

1. If $x_i = 0$ proceed to the next task. (If there are no more tasks, stop.)
2. Search $\mathcal{L}$ for a set not containing i to which i can be added without violating Property 1.
3. If no such set exists, create a new set $V = \{i\}$ with weight $x(V) = x_i$ and add V to $\mathcal{L}$. Set $x_i = 0$ and return to step 1.
4. Otherwise, suppose $U \in \mathcal{L}$ is such a set.
   4a. If $x_i < x(U)$ then decrease $x(U)$ to $x(U) - x_i$, create a new set $V = U \cup \{i\}$ with weight $x(V) = x_i$ and add V to $\mathcal{L}$. Set $x_i = 0$ and return to step 1.
   4b. If $x_i \geq x(U)$, add i to U and decrease x_i to $x_i - x(U)$. Return to step 1.

Lemma 4.1. *After processing all the tasks as above, $\mathcal{L}$ will hold a list of tasks satisfying all four properties.*

Proof. Let $\hat{x}_i$ denote the *original* values of x_i that were input to the above procedure. It is easy to verify that the procedure maintains the following invariant:

$$x_i + \sum_{j:U_j \ni i} x(U_j) = \hat{x}_i, \qquad 1 \le i \le n . \tag{11}$$

Properties 1 and 2 are clearly satisfied. Property 3 follows from (11) and the fact that after all tasks have been processed, each $x_i = 0$.

It remains to prove that Property 4 holds; we shall show that it holds as an invariant of the procedure. A new set may be added to the list $\mathcal{L}$ either from step 3 or step 4a. In the latter case, the weight of a set is split amongst itself and the newly created set, leaving the sum of all weights unaffected. In the former case, the newly created set is a singleton consisting of task i, say. Consider the list $\mathcal{L}$ immediately after this singleton is added. Note that every task j in a set in $\mathcal{L}$ satisfies $j \le i$. For each set $U \in \mathcal{L}$ let $b(U)$ be the sum of b_j over all $j \in U$ such that task j is active at the time when task i starts (i.e., $s_j \le s_i < t_j$, since $j \le i$). Then

$$\sum_{U \in \mathcal{L}} b(U)x(U) = \sum_{U \in \mathcal{L}} \sum_{\substack{j \in U \\ s_j \le s_i < t_j}} b_j x(U) \tag{12}$$

$$= \sum_{\substack{j:j \le i, \\ s_j \le s_i < t_j}} b_j \sum_{U \in \mathcal{L}:U \ni j} x(U) \tag{13}$$

$$= \sum_{\substack{j:j \le i, \\ s_j \le s_i < t_j}} b_j \hat{x}_j \tag{14}$$

$$\le B . \tag{15}$$

Equation (13) holds because we process tasks in the order $1, \ldots, n$, (14) holds because of (11) and the fact that when task i is being processed we have $x_j = 0$ for all $j < i$, and (15) holds because of constraint (2) of LPMAIN.

Let $\mathcal{L}'$ be the sublist of $\mathcal{L}$ consisting of those sets which contain i. For $U \in \mathcal{L}'$ we clearly have $b(U) \ge b_i$. For $U \in \mathcal{L} \setminus \mathcal{L}'$, since we're in the case when step 3 is executed, it follows that task i did not fit into any set in $\mathcal{L} \setminus \mathcal{L}'$ and so $b(U) > B - b_i$. Therefore we get

$$B \ge \sum_{U \in \mathcal{L}'} b_i x(U) + \sum_{U \in \mathcal{L} \setminus \mathcal{L}'} (B - b_i)x(U) = b_i \hat{x}_i + (B - b_i) \sum_{U \in \mathcal{L} \setminus \mathcal{L}'} x(U)$$

where the equality follows from (11). Consider the last expression; if we have $\sum_{U \in \mathcal{L} \setminus \mathcal{L}'} x(U) > \hat{x}_i$, then the expression is a decreasing function of b_i. Since $b_i \le B/2$, we get

$$B \ge \frac{B}{2}\hat{x}_i + \frac{B}{2} \sum_{U \in \mathcal{L} \setminus \mathcal{L}'} x(U) . \tag{16}$$

If, on the other hand, we have $\sum_{U \in \mathcal{L} \setminus \mathcal{L}'} x(U) \le \hat{x}_i \le 1$, then (16) clearly holds. Thus (16) always holds, and applying (11) to it gives $B \ge \frac{B}{2}\sum_{U \in \mathcal{L}} x(U)$, whence Property 4 follows. $\qquad\square$

Proof (of Theorem 1.5). We combine two RAP algorithms as described in Sect. 2.2, with $\alpha = 1/2$ and $\tau = B/2$ in the terminology of that section. For one half we use the above List Algorithm.

For the other half, observe that if each task in a RAP instance has $b_i > B/2$, then two tasks which are active together (at some instant of time) cannot both be in a packing. Therefore, a packing in this case is simply an independent set in the underlying interval graph of the RAP instance; so the problem reduces to (weighted) MAX-INDEPENDENT-SET for interval graphs. This problem is well-known to be solvable (exactly) in $O(n \log n)$ time (see, e.g., [2]).

The combination gives us an approximation guarantee of $\alpha/(1 + \alpha) = 1/3$.

Finally, consider the running time of the List Algorithm. Note that the size m of list $\mathcal{L}$ is at most n, since each task creates at most one new set to be added to $\mathcal{L}$. This means that the List Algorithm does at most $O(n^2)$ work after solving the LP. From Subsect. 2.1 we know that the LP can be solved in $O(n^2 \log^2 n)$ time. This completes the proof. □

References

1. E. M. Arkin, E. B. Silverberg. Scheduling jobs with fixed start and end times. Discrete Applied Mathematics, **18** (1987), 1–8.
2. A. Bar-Noy, R. Bar-Yehuda, A. Freund, J. Naor, B. Schieber. A unified approach to approximating resource allocation and scheduling. In Proceedings of the 32nd Annual ACM Symposium on Theory of Computing (2000), 735–744.
3. A. Bar-Noy. Private communication (2001).
4. B. Chen, R. Hassin, M. Tzur. Allocation of bandwidth and storage. IIE Transactions, **34** (2002), 501–507.
5. N. G. Hall, M. J. Magazine. Maximizing the value of a space mission. European Journal of Operational Research, **78** (1994), 224–241.
6. S. Leonardi, A. Marchetti-Spaccamela, A. Vitaletti. Approximation algorithms for bandwidth and storage allocation problems under real time constraints. In Proceedings of the 20th Conference on Foundations of Software Technology and Theoretical Computer Science (2000), 409–420.
7. J. B. Orlin. A faster strongly polynomial minimum cost flow algorithm. Operations Research, **41** (1993), 338–350.
8. C. Phillips, R. N. Uma, J. Wein. Off-line admission control for general scheduling problems. In Proceedings of the 11th Annual ACM-SIAM Symposium on Discrete Algorithms (2000), 879–888.

Approximating the Advertisement Placement Problem

Ari Freund and Joseph (Seffi) Naor

Department of Computer Science, Technion,
Haifa 32000, Israel*
{arief,naor}@cs.technion.ac.il

Abstract. The *advertisement placement problem* deals with space and time sharing by advertisements on the Internet. Consider a WWW page containing a rectangular display area (e.g., a banner) in which advertisements may appear. The display area can be utilized efficiently by allowing several small advertisements to appear simultaneously side by side, as well as by cycling through a schedule of ads, displaying different ads at different times. A customer wishing to purchase advertising space specifies an ad size and a *display count*, which is the number of times his advertisement should appear during each cycle. The scheduler may accept or reject any given advertisement, but must be able to schedule all accepted ones within the given time and space constraints, while honoring the display count of each. The objective is to schedule a maximum-profit subset of ads. We present a $(3 + \epsilon)$-approximation algorithm for the general problem, as well as $(2 + \epsilon)$-approximation algorithms for two special cases.

1   Introduction

The Advertisement Placement Problem. The last decade has witnessed the advent of the World Wide Web and graphical browsers, and the consequent explosion in the use of the Internet. From a network serving mainly academia, the Internet has grown into a global communications network providing a diverse range of services such as search engines, email, real-time video conferencing, software archives, etc. Virtually all of these services are conveniently accessible through the World Wide Web, and many of them are provided free of charge.

The evolution of the Internet has tracked a positive feedback loop, in which easy accessibility and low (or no) cost create an ever-expanding user base, which, in turn, attracts investors and developers, who bring about a further increase in the quality, diversity, and accessibility of services. While many of the services remain free of charge, investors still expect to make money. It has always been assumed that this would be achieved, at least in part, through advertising on the Web (as well as more sinister activities, such as collecting and selling information on users). Although the quick-profit promise of the Internet has yet

---

* Part of this work was done while the second author was at Bell Labs, Lucent Technologies, 700 Mountain Ave., Murray Hill, NJ 07974 and the first author was visiting there.

to be realized, the fact remains that nearly every commercially operated web site providing some free service contains advertising, most often in the form of a banner stretching across the top of the page. Such banners are often utilized to better advantage by periodically changing their contents, and it is also common practice to place several side-by-side advertisements in a single banner. Thus, advertising on the web combines time and space sharing between advertisers.

In the *advertisement placement* problem we are given a schedule length of L time slots and a collection of ads which we must schedule within this time frame. (Presumably the schedule is cyclic, repeating every L time units.) The advertisements must be placed in a rectangular display area whose contents can change every time unit. The advertisements all share the same height, which is the height of the display area, but may have different widths. Several ads may be displayed simultaneously (side by side), as long as their combined width does not exceed the width of the display area, which we normalize to 1 for simplicity. In addition, each advertisement specifies a *display count* (in the range $1, \ldots, L$), which is the number of time slots during which the ad must be displayed. The actual time slots in which the advertisement will be displayed may be chosen arbitrarily by the scheduler, and, in particular, need not be consecutive. There are two problems to consider. In the *resource minimization* problem *all* ads must be scheduled, and the objective is to minimize the width of the display area required to make this possible. In the *profit maximization* problem each advertisement has a non-negative profit associated with it, and the scheduler may accept or reject any given ad. The objective is to schedule a maximum-profit subset of ads within a display area of given width. Both problems are NP-hard [1].

Although the prime motivation for these problem has to do with advertising, we can also consider continuous time and arbitrary display *durations* (rather than discrete time slots and integral display *counts*). This generalized setting models any scheduling situation in which jobs require the use of a certain amount of resource (or a certain number of machines) for varying amounts of time, all jobs have a common release-date and a common due-date, and preemption and migration are allowed at no cost. This model has been studied quite extensively in the scheduling community [2], particularly, the profit maximization version, in both the single machine and the multiple machine settings [3,4,5]. In the models studied, a job was assumed to require the use of only one machine at any given moment. Therefore, the results obtained for these models do not carry over to our setting, where each job may require a different number of machines (corresponding to different advertisement widths).

Previous Work. The advertisement placement problem was introduced and studied by Adler et al. [1]. They considered both resource minimization and profit maximization. Nearly all of their results pertain to the special case of *divisible* ad widths, i.e., when the advertisement widths are $h_1, \ldots, h_n$ such that h_i divides h_{i+1} for all i, and in addition, for the profit maximization case, $h_1 = 1/k$ for some positive integer k. For the resource minimization problem, they gave an efficient algorithm that finds an optimal schedule in the case of divisible ad widths. This algorithm implies an efficient 2-approximation algorithm for general ad widths (by rounding up the widths to the nearest (non-positive) power of 2). For the profit maximization problem, Adler et al. only considered the special case where the ad widths are divisible and the profit of each ad is proportional to its "volume" (width times display count). They devised a 2-approximation algorithm for this case. They also obtained several results for the on-line version of the profit maximization problem. Their work was implemented, and a Java demonstration is available at http://www.bell-labs.com/project/collager.

Independently of our work, Dawande et al. [6] also considered the advertisement placement problem. They suggested a simple $\max\{2 - 1/(L - l_{\min} + 1), 2 - 2/(L+1)\}$-approximation algorithm for resource minimization (where $l_{\min}$ is the minimum ad display count) and a $\min\{10/3, 4L/(L + 1)\}$-approximation algorithm for profit maximization. They also obtained better results for several special cases.

Our Results. Our main result is a $(3+\epsilon)$-approximation algorithm for the profit maximization problem with *arbitrary* ad widths and profits. The complexity of our algorithm is polynomial in the problem parameters and $1/\epsilon$. Our approach is opposite to that of Adler et al. [1] in that whereas they consider ads for assignment in non-increasing order of width, we consider them in non-decreasing order. Although non-increasing order has a better chance of yielding optimal solutions, (indeed, as pointed out by Adler et al., it results in an assignment of all ads if such an assignment exists), the opposite order maintains a more even distribution of ads in time slots, making for a better approximation guarantee. This is also true for the resource minimization problem; our approach yields an approximation factor of 2 without having to round the ad widths.

In addition to the general profit maximization problem, we also consider two special cases, for which we show $(2 + \epsilon)$-approximation algorithms. The special cases are: (1) ad widths do not exceed one half of the display area width, and (2) each advertisement must either occupy the entire display area, or else have width at most one half of the display area width (i.e., no ad width is in the real interval $(0.5, 1)$). Note that the second special case contains the case of divisible widths. It also contains the first special case, but we consider them separately because our solution for the first case is much simpler than for the second. In fact, the $(2+\epsilon)$ approximation factor for the first case is achieved by a slight modification of our algorithm for the general problem, and this modified algorithm is then used as a stepping stone to obtain the $(2 + \epsilon)$ performance guarantee for the second special case.

Finally, we demonstrate that the integrality gap of the straightforward linear programming relaxation of the ad placement problem is 3. Actually, we show a stronger result: we observe that the familiar *knapsack* problem can be obtained from the ad placement problem by relaxing some, but not all, of the integrality constraints, and show that this relaxation incurs a gap of 3.

Organization of This Paper. In Sect. 2 we define the profit maximization problem precisely and introduce some notation and terminology. In Sect. 3 we

present the main ideas of our approach and the 2-approximation algorithm for
the resource minimization problem. In Sect. 4 we describe the algorithm for
the profit maximization problem and our treatment of the two special cases. In
Sect. 5 we show that the integrality gap is 3.

2   Problem Statement, Notation, and Terms Used

For simplicity, we develop the algorithm in terms of the following *discrete bins*
formulation, which is essentially the formulation given by Adler et al. [1]. This
formulation is equivalent to the informal description of the ad placement problem
in the previous section. In Sect. 4.2 we discuss briefly how our algorithm may
be adapted to the more general scheduling problem that results when time is
continuous and ad widths are arbitrary rationals.

The *discrete bins* formulation of the problem follows. The input consists of
$L > 0$ unit size *bins* numbered $\{0, \ldots, L-1\}$, and n *jobs* labeled $j_1, \ldots, j_n$.
Each job j_i is defined by a triplet (h_i, l_i, w_i), where $0 < h_i \leq 1$ is the job's
height, $l_i \in \{1, \ldots, L\}$ is its *length*, and $w_i > 0$ is its *weight*. In terms of the
advertisement scheduling problem discussed earlier, job heights correspond to
ad widths and job lengths correspond to display counts. A *feasible solution* is
a subset of jobs $S \subseteq \{j_1, \ldots, j_n\}$ together with a *feasible assignment* for S. A
feasible assignment for S is defined as an assignment of l_i bins to each job $j_i \in S$,
such that the total height of jobs to which any given bin is assigned does not
exceed 1. More formally, a feasible assignment of S is a mapping $\alpha : S \to \mathcal{P}(L)$
(where $\mathcal{P}(L)$ is the power set of $\{0, \ldots, L-1\}$) such that:

1. For all $j_i \in S$, $|\alpha(j_i)| = l_i$.
2. For all $b \in \{0, \ldots, L-1\}$, $\displaystyle\sum_{i \mid j_i \in S \wedge b \in \alpha(j_i)} h_i \leq 1.$

The weight of a feasible solution (S, α) is the total weight of jobs in S, denoted
$w(S)$. The goal is to find a maximum-weight feasible solution.

We will find it convenient to think of feasible assignments as placing *copies* of
jobs in bins rather than assigning subsets of bins to jobs. In other words, rather
than assigning bins 0, 3, 4, and 7 to job j_9 we speak of placing the four copies
of j_9 in bins 0, 3, 4, and 7. (Here we have assumed $l_9 = 4$.) Indeed, we formulate
the algorithm in terms of progressively placing job copies in bins. Taking this
metaphor one step further, we introduce the notions of *volume* and *capacity*.
The *volume* of a single copy of job j_i is simply its height h_i and the volume of
the entire job is $h_i \cdot l_i$. The *capacity* of each bin is one volume unit. Also, since
our algorithm places job copies one at a time, it is useful to define the *occupancy*
of a bin at a given moment as the total volume of job copies present in the bin
at that moment.

3   Elements of the Algorithms

Our algorithm combines two key elements: *Knapsack Relaxation* and *Smallest
Size Least Full (SSLF) heuristic*.

Knapsack Relaxation. Consider a feasible solution. It consists of selecting a subset of jobs and distributing all of their copies among the available bins. Clearly, the solution must obey the condition that the total volume of jobs selected does not exceed the total capacity of all bins, namely L. Thus, the familiar *knapsack* problem may be viewed as a relaxation of our problem: for a given input instance I we define an instance of *knapsack* in which the knapsack capacity is L and the objects are the jobs in I. The size and weight of each object are equal to the volume and weight of the corresponding job, as defined by I. Every feasible solution to our problem defines a feasible solution with the same weight to the *knapsack* instance. Thus, the optimum weight for the *knapsack* relaxation is an upper bound on the optimum weight for our problem. It is well known that a fully polynomial time approximation scheme (FPTAS) exists for *knapsack* (see, e.g. [7, Chapter 2, pp. 69–74]). We use this fact in our algorithm.

Smallest Size Least Full (SSLF) Heuristic. The SSLF heuristic actually consists of two separate heuristics governing the order in which jobs are considered and the order in which bins are considered.

Smallest Size[1] First (SS). Sort the jobs by height in non-decreasing order. The jobs will be considered for placement in the bins in this order.

Least Full First (LF). When a job is up for placement, place its copies in the following manner. Let l be the length of the job. Repeat the following l times: *place one copy of the job in the bin with minimum occupancy among the bins that do not currently contain a copy of the job (breaking ties arbitrarily).*

The salient property of the LF heuristic is expressed in the next lemma, which is implied by the proof of Claim A.3 in [1]. We provide a proof here for completeness.

Lemma 1. *Suppose jobs are placed in bins according to the LF heuristic (regardless of the order in which the jobs are considered, and allowing bins to overflow). At any given moment during the assignment let $h_{\max}$ be the maximum height of a job considered so far, and denote by $f(b)$ the occupancy of bin b. Then $f(b') - f(b'') \leq h_{\max}$ for all bins b' and b''.*

Proof. The proof is by induction. Consider a pair of bins b' and b''. Initially, both bins are empty and the claim holds. Suppose the claim is true after the first $k-1$ jobs are processed, and consider the placement of the kth job, whose height we denote h. The claim can only be violated if $f(b') - f(b'')$ increases (since $h_{\max}$ can only increase), and this can only happen if the algorithm places a copy of the kth job in b' but not in b''. However, in this case, the LF heuristic implies that $f(b') \leq f(b'')$ before the assignment, and thus $f(b') - f(b'') \leq h \leq h_{\max}$ afterwards. A simple extension of this argument shows that the claim holds at all times, not just between jobs.

An immediate consequence of Lemma 1 is a 2-approximation algorithm for the resource minimization problem.

---

[1] Following [1] we use the term *size* rather than *height*.

Theorem 1. *Assigning jobs by the SSLF heuristic yields a 2-approximate solution for the resource minimization problem.*

Proof. The LF heuristic does not necessarily place every job copy in a bin of minimum occupancy, but rather in a bin of minimum occupancy *among the bins not containing a copy of the same job*. The SSLF heuristic, on the other hand, does place every job copy in a bin of minimum occupancy overall, and this is implied by Lemma 1. Thus the SSLF heuristic can be viewed as an instance of Graham's algorithm for scheduling on identical machines [8], which achieves an approximation guarantee of 2.

4   Algorithms for the Profit Maximization Problem

Our $(3 + \epsilon)$-approximation algorithm for the general problem proceeds by three steps.

1. Use an FPTAS to solve the *knapsack* relaxation, and obtain a set of jobs S such that $w(S)$ is at least $1/(1 + \epsilon/3)$ times the optimum value for *knapsack*.
2. Place the copies of the jobs in S in the bins according to the SSLF heuristic. Do so until some bin overflows. Let j_i be the offending job. Partition S into three subsets $S_1 = \{$all jobs placed successfully$\}$, $S_2 = \{j_i\}$, and $S_3 = \{$all remaining jobs in $S\}$. If no overflow occurs, then define $S_1 = S$ and $S_2 = S_3 = \emptyset$.
3. Return the maximum-weight set among S_1, S_2, and S_3, together with an assignment. If S_1 is returned, the assignment is the one found in Step 2; if S_2 is returned, the assignment is trivial; if S_3 is returned, the assignment is constructed as described below.

To complete the description of the algorithm, it remains to show how to construct an assignment for S_3. Let us examine the state of the bins just before the algorithm attempts to place job j_i (the offending job). Let k be the number of bins whose occupancy is $1 - h_i$ or less. We know that $k < l_i$, for otherwise job j_i would not cause an overflow. Furthermore, since the placement of j_i *does* cause an overflow, we know that the occupancy of at least one bin is strictly greater than $1 - h_i$. Thus, by Lemma 1, the occupancy of every bin is greater than $1 - 2h_i$ (because the SSLF heuristic ensures that jobs are considered in non-decreasing order of height and thus $h_{\max} = h_i$). Thus, we have k bins with occupancy greater than $1 - 2h_i$ and $L - k$ bins with occupancy greater than $1 - h_i$. Hence the total volume of the jobs in S_1 is more than $(L - k)(1 - h_i) + k(1 - 2h_i) = L - Lh_i - kh_i > L - Lh_i - l_ih_i$. Now, the total volume of jobs in S is at most L (by the *knapsack* relaxation) and the volume of j_i is l_ih_i, so the total volume of jobs in S_3 is less than Lh_i. Finally, by the SSLF heuristic, the height of every job in S_3 is at least h_i, and thus the total length of these jobs is less than L. Placing a set of jobs with total length less than L in L bins is easy: simply place each job copy in a bin of its own.

Theorem 2. *The algorithm finds a $(3 + \epsilon)$-approximate solution.*

Proof. Let OPT and OPT_K be the optimum value for our problem and for the knapsack relaxation, respectively, and let S_i be the solution returned by the algorithm. Then

$$w(S_i) = \max\{w(S_1), w(S_2), w(S_3)\},$$
$$\geq \frac{1}{3}(w(S_1) + w(S_2) + w(S_3)),$$
$$= \frac{1}{3}w(S),$$
$$\geq \frac{1}{3} \cdot \frac{OPT_K}{1 + \epsilon/3},$$
$$\geq \frac{OPT}{3 + \epsilon}.$$

4.1   Two Special Cases

As we show in Sect. 5, the analysis of our algorithm is tight (up to ϵ) for general input. There are, however, special cases in which our algorithm can be used as a basis to achieving an approximation factor of $2 + \epsilon$. We present two such cases next.

The first special case we consider is the case in which $h_i \leq 1/2$ for all i. Suppose we run the algorithm and obtain S_1, S_2, and S_3. Observe that the total length of jobs in $S_2 \cup S_3$ is less than $2L$, since the contribution of S_3 is less than L (as we have seen in the previous section) and S_2 consists of a single job, whose length is at most L (by definition). Since no job height is greater than $1/2$, we can construct an assignment for $S_2 \cup S_3$ as follows. Create a list of the job copies such that the copies of each job appear consecutively and place the ith copy on the list in bin number $i \bmod L$. We obtain a $(2 + \epsilon)$-approximate solution by choosing the better solution between S_1 and $S_2 \cup S_3$.

The second special case is when every job of height greater than $1/2$ has height 1. We combine our algorithm with the dynamic programming approach of the standard *knapsack* FPTAS. The idea is to find a combination of *full-height* jobs (i.e., jobs of height 1) and *small* jobs (jobs of height at most $1/2$) that can be scheduled successfully. Recall that the *knapsack* FPTAS scales the job weights and rounds them to integers. It then calculates an upper bound $\omega_{\max}$ on the optimal rounded weight (i.e., the optimum with respect to the new job weights), and uses dynamic programming to find $S(\omega)$ for $\omega = 0, 1, \ldots, \omega_{\max}$, where $S(\omega)$ is a minimum total-size subset of jobs with total weight precisely ω. (Note that $S(\omega)$ may be undefined for some values of ω.) Let $L(S(\omega))$ be the total size of the jobs in $S(\omega)$. The FPTAS returns $S(\omega^*)$, where ω^* is maximum such that $L(S(\omega^*))$ does not exceed the knapsack's capacity.

We are not interested in the solution returned by the FPTAS, but rather in the sets $S(\omega)$ that it generates. The key property of these sets is the following. Let $\omega_1, \omega_2, \ldots, \omega_k$ be the sequence of ωs (in ascending order) for which $S(\omega)$ is defined. Let $OPT_K(c)$ be the optimum value for the knapsack problem with a knapsack of capacity c. If the FPTAS is run with a knapsack of capacity L, then for all i, $w(S(\omega_i)) \geq OPT_K(c) - OPT_K(L) \cdot \epsilon/(1+\epsilon)$ for all $c \in [L(S(\omega_i)), L(S(\omega_{i+1})))$. (We define $L(S(\omega_{k+1})) = \infty$.) We make use of this as follows. Let $\epsilon' > 0$ be sufficiently small relative to ϵ. (The exact choice of ϵ' is discussed later.) We run the *knapsack* FPTAS with accuracy parameter ϵ' on the full-height jobs, using a knapsack capacity of L, and obtain the sets $S(\omega_i)$. For every ω_i we first place the copies of the jobs belonging to $S(\omega_i)$ in $L(S(\omega_i))$ bins, and then run our previously described $(2+\epsilon')$-approximation algorithm on the small jobs to find a solution using the remaining $L - L(S(\omega_i))$ bins. Thus, we obtain for each ω_i a feasible solution that is a combination of full-height jobs and small jobs. As we show next, an appropriate choice of ϵ' ensures that the best of these solutions is $(2+\epsilon)$-approximate.

Consider an optimal solution $S_{\mathrm{opt}} = X \cup Y$ to the *knapsack* relaxation (with the original weights, not the scaled and rounded ones) using a knapsack of capacity L, where X consists of full-height jobs and Y consists of small jobs. Let $x = \max\{i \mid L(S(\omega_i)) \leq |X|\}$. Then, $L(S(\omega_x)) \leq |X| < L(S(\omega_{x+1}))$, and therefore $w(S(\omega_x)) \geq OPT_K(|X|) - OPT_K(L) \cdot \epsilon'/(1+\epsilon') \geq w(X) - w(S_{\mathrm{opt}}) \cdot \epsilon'/(1+\epsilon')$. (Recall that $w(S_{\mathrm{opt}}) = OPT_K(L)$ by definition.) In addition, because the number of bins used for the jobs in Y is at most $L - |X| \leq L - L(S(\omega_x))$, running our $(2+\epsilon')$-approximation algorithm on the small jobs with $L - L(S(\omega_x))$ bins will yield a solution whose weight is at least $w(Y)/(2+\epsilon')$. Thus, the feasible solution we construct for ω_x has weight at least

$$w(X) - \frac{\epsilon'}{1+\epsilon'} w(S_{\mathrm{opt}}) + \frac{w(Y)}{2+\epsilon'} \geq w(S_{\mathrm{opt}}) \left(\frac{1}{2+\epsilon'} - \frac{\epsilon'}{1+\epsilon'} \right) \geq \frac{w(S_{\mathrm{opt}})}{2+\epsilon},$$

where the last inequality holds if ϵ' is chosen small enough relative to ϵ. One can easily verify that $\epsilon' = \epsilon/8$ is sufficient (assuming $\epsilon \leq 1$). The solution we find is therefore $(2+\epsilon)$-approximate, since the *knapsack* optimum value is an upper bound on the optimum value for the advertisement placement problem.

4.2   Polynomial-Time Implementation

A straightforward implementation of our algorithm runs in super-polynomial time, since the placement of jobs in bins according to the SSLF heuristic requires $\Omega(L)$ time. This cannot be avoided if the contents of each bin must be specified explicitly in the output. However, a more compact representation is possible, which results in a polynomial-time implementation of the algorithm.

Recall that when the algorithm selects a bin according to the LF heuristic, it chooses the currently least full bin, breaking ties arbitrarily. Let us not break ties arbitrarily, but rather always select the lowest numbered bin among the contestants. Doing so leads to each job being placed in "consecutive runs" of bins. Thus, after processing $i - 1$ jobs the set of bins can be partitioned into at most i *super-bins*, which are simply sets of consecutive bins, such that the contents of all bins belonging to the same super-bin are identical. Let us define the *occupancy* of a super-bin to be the occupancy of any one of its bins. Then, to assign the ith job we repeatedly select a super-bin of minimum current occupancy and place a copy of the job in each of its constituent bins. If necessary, we split the last super-bin in order to maintain the invariant that all the bins constituting a given super-bin have identical contents. Thus, we can formulate the algorithm entirely in terms of super-bins and dispense with the original bins completely. The running time and output size in a straightforward implementation of this algorithm depend polynomially on n and $1/\epsilon$ (the output size depends only on n), and are independent of L.

It is now clear that the requirement that L and the l_i's be integers serves no real purpose. Our algorithm can deal just as easily with the generalized problem in which L and the l_i's are arbitrary rationals, and jobs may be split into "pieces" at arbitrary points along their l_i axis and the resulting pieces may be placed anywhere in the interval $[0, L]$. This generalized problem models any scheduling situation in which jobs require the use of a certain amount of resource for varying amounts of time, all jobs have a common release-date and a common due-date, and preemption and migration are allowed at no cost.

5   The Integrality Gap of the Knapsack Relaxation

For simplicity, we return to the *discrete bins* version of the problem, although everything we say in this section applies to the general problem (defined at the end of the previous section) as well.

The feasibility constraints in our problem contain two types of integrality constraints. (In fact, the problem can be formulated as an integer programming problem in which these two sets of constraints translate into integrality constraints on two different sets of 0-1 variables.) The integrality constraints are: (1) every job must either be accepted in its entirety or rejected completely, and (2) every job copy must be either present in its entirety in a given bin or absent from it completely (e.g., it is impossible to place one half of a job copy in one bin and the other half in another bin). Relaxing the constraints of the second type yields the *knapsack* problem. Hence, the term *integrality gap* is appropriate in connection with the *knapsack* relaxation.

As we have shown in the previous section, the integrality gap is at most 3. We now demonstrate that it is at least 3 as well. Consider the following problem instance. The input consists of three identical jobs, each with length $L/2 + 1$, height $1/2 + \delta$, for sufficiently small δ, and weight w. Clearly, the total volume of all three jobs is roughly $3L/4$ (for sufficiently large L) and thus the optimum for the *knapsack* relaxation is $3w$. On the other hand, no feasible assignment exists for more than one job, since every such assignment would have to place at least one pair of job copies in the same bin, and that is not allowed because their combined height exceeds 1. Thus, the optimum value for our problem is w, and hence the integrality gap is 3.

The integrality gap of 3 does not necessarily imply that the worst case performance ratio of our algorithm is at least 3, only that no analysis comparing it against the *knapsack* optimum can yield a ratio better than 3. To see that our algorithm's worst case ratio is indeed greater than or equal to 3, we can extend the above example by adding a fourth job of height 1, length L, and weight $3w$. In the first step of the algorithm, the FPTAS might return the first three jobs, resulting eventually in a solution of weight w, while the optimal solution consists of the fourth job and has weight $3w$.

Acknowledgments

We thank the anonymous reviewer for pointing out Reference [6]. He (or she) also mentioned a writeup titled *Algorithmic improvements for scheduling advertisements on web* (sic), which we were unable to track down. It is apparently another working paper by the authors of [6].

References

1. Adler, M., Gibbons, P., Matias, Y.: Scheduling space-sharing for internet advertising. Journal of Scheduling (to appear). Also available electronically at http://www.cs.umass.edu/~micah/pubs/pubs.html.
2. Lawler, E., Lenstra, J., Rinnooy Kan, A., Shmoys., D.: Sequencing and scheduling: algorithms and complexity. In Graves, S., Rinnooy Kan, A., Zipkin, P., eds.: Handbooks in Operations Research and Management Science, Vol.4: Logistics for Production and Inventory. North Holland (1993) 445–522
3. Lawler, E.: Sequencing to minimize the weighted number of tardy jobs. Recherche Operationnel **10** (1976) 27–33
4. Kise, H., Ibaraki, T., Mine, H.: A solvable case of one machine scheduling problem with ready and due dates. Operations Research **26** (1978) 121–126
5. Lawler, E.: A dynamic programming algorithm for preemptive scheduling of a single machine to minimize the number of late jobs. Annals of Operations Research **26** (1990) 125–133
6. Dawadne, M., Kumar, S., Srishkandarajah, C.: Performance bounds of algorithms for scheduling advertisements on a web page. Working paper (2001). Available at http://www.utdallas.edu/~chelliah/wpapers.html.
7. Ausiello, G., Crescenzi, P., Gambosi, G., Kann, V., Marchetti-Spaccamela, A., Protasi, M.: Complexity and Approximation. Springer-Verlag (1999)
8. Graham, R.: Bounds for certain multiprocessing anomalies. Bell System Technical Journal **45** (1966) 1563–1581

Algorithms for Minimizing Response Time in Broadcast Scheduling*

Rajiv Gandhi[1], Samir Khuller[2], Yoo-Ah Kim[1], and Yung-Chun (Justin) Wan[1]

[1] Department of Computer Science, University of Maryland, College Park, MD 20742
{gandhi,ykim,ycwan}@cs.umd.edu
[2] Department of Computer Science and Institute for Advanced Computer Studies,
University of Maryland, College Park, MD 20742
samir@cs.umd.edu

Abstract. In this paper we study the following problem. There are n pages which clients can request at any time. The arrival times of requests for pages are known. Several requests for the same page may arrive at different times. There is a server that needs to compute a good broadcast schedule. Outputting a page satisfies all outstanding requests for the page. The goal is to minimize the average waiting time of a client. This problem has recently been shown to be NP-hard. For any fixed α, $0 < \alpha \leq \frac{1}{2}$, we give a $\frac{1}{\alpha}$- speed, polynomial time algorithm with an approximation ratio of $\frac{1}{1-\alpha}$. For example, setting $\alpha = \frac{1}{2}$ gives a 2-speed, 2-approximation algorithm. In addition, we give a 4-speed, 1-approximation algorithm improving the previous bound of 6-speed, 1-approximation algorithm.

1 Introduction

There has been a lot of interest lately in data dissemination services, where clients request information from a source. Advances in networking and the need to provide data to mobile and wired devices have led to the development of large-scale data dissemination applications (election results, stock market information etc). While the WWW provides a platform for developing these applications, it is hard to provide a completely scalable solution. Hence researchers have been focusing their attention on Data Broadcasting methods.

Broadcasting is an appropriate mechanism to disseminate data since multiple clients can have their requests satisfied simultaneously. A large amount of work in the database and algorithms literature has focused on scheduling problems based on a broadcasting model (including several PhD theses from Maryland and Brown) [7,9,4,1,5,2,8,23,6]. Broadcasting is used in commercial systems, including the Intel Intercast System [15], and the Hughes DirecPC [13]. There are two primary kinds of models that have been studied – the first kind is a *push*-based scheme, where some assumptions are made on the access probability for a certain data item and a broadcast schedule is generated [3,9,7,17,6]. We focus our attention on the second kind, namely *pull-based* schemes, where clients request

---

* Research supported by NSF Awards CCR-9820965 and NSF CCR-0113192.

W.J. Cook and A.S. Schulz (Eds.): IPCO 2002, LNCS 2337, pp. 425–438, 2002.

the data that they need (for example via phone lines) and the data is delivered on a fast broadcast medium (often using satellites) [5]. This model is motivated by wireless web applications. This work deals entirely with the *pull-based* model, where requests for data arrive over time and a good broadcast schedule needs to be created.

A key consideration is the design of a good broadcast schedule. The challenge is in designing an algorithm that generates a schedule that provides good average response time. Several different scheduling policies have been proposed for on-line data broadcast. Aksoy and Franklin [5] proposed one such online algorithm. Their algorithm, called RxW, takes the product of the number of outstanding requests for page with the longest waiting time to compute the "demand" for a page. The page with the maximum demand is then broadcast. Another reasonable heuristic is to take the sum of waiting times of all outstanding requests for a page to compute its demand. Tarjan et al [22] have recently shown how to implement this algorithm efficiently. Not surprisingly, the worst case competitive ratio of these algorithms is unbounded, as shown in [18].

While the practical problem is clearly online, it is interesting to study the complexity of the offline problem as well. In trying to evaluate the performance of online algorithms, it is useful to compare them to an optimal offline solution. In addition, when the demands are known for a small window of time into the future (also called the look-ahead model in online algorithms) being able to quickly compute an optimal offline solution can be extremely useful. Many kinds of demands for data (e.g., web traffic) exhibit good predictability over the short term, and thus knowledge of requests in the immediate future leads to a situation where one is trying to compute a good offline solution.

One could also view the requests in the offline problem as release times of jobs, and one is interested in minimizing average (weighted) flow time. While this problem has been very well studied (see [11] and references therein), the crucial difference between our problem and the problem that has been studied is the fact that scheduling a job satisfies *many* requests simultaneously. (The term "overlapping jobs" has also been used to describe such scheduling problems in the past.)

The informal description of the problem is as follows. There are n data items, $1, \ldots, n$, called pages. Time is broken into "slots". A time slot is defined as the unit of time to transmit one page on the wireless channel. A request for a page j arrives at time t and then waits. When page j has been transmitted, this request has been satisfied. Arrival times of requests for pages is known, and we wish to find a broadcast schedule that *minimizes the average waiting time*. There has been a lot of work on this problem when we assume some knowledge of a probability distribution for the demand of each page [6,7,17].

In the same model, the paper by Bartal and Muthukrishnan [8] studies the problem of minimizing the maximum response time. The offline version has a PTAS, and there is a 2-competitive algorithm for the online version [8]. (They also credit Charikar, Khanna, Motwani and Naor for some of these results that were obtained independently.)

The recent paper by Kalyanasundaram et al. [18] studies this problem as well. They showed that for any fixed $\epsilon, 0 < \epsilon \le \frac{1}{3}$, it is possible to obtain a $\frac{1}{\epsilon}$-speed $\frac{1}{1-2\epsilon}$-approximation algorithm for minimizing the average response time, where a k-speed algorithm is one where the server is allowed to broadcast k pages in each time slot. For example by putting $\epsilon = \frac{1}{3}$ they obtain a 3-speed, 3-approximation. The approximation factor bounds the cost of the k-speed solution compared to the cost of an optimal 1-speed solution. (This kind of approximation guarantee is also referred to as a "bicriteria" bound in many papers.) Note that we cannot set $\epsilon = \frac{1}{2}$ to get a 2-speed, constant approximation. Their algorithm is based on rounding a fractional solution for a "network-flow" like problem that is obtained from an integer programming formulation. This problem has recently shown to be NP-hard by Erlebach and Hall [14].

Our Results: We consider a different Integer Program(IP) for this problem, and show that by relaxing this IP, for any fixed $\alpha \in (0, \frac{1}{2}]$, we can obtain a $\frac{1}{\alpha}$-speed solution that is a $\frac{1}{1-\alpha}$-approximation. For example, by setting $\alpha = \frac{1}{2}$ we obtain a 2-speed, 2-approximation. By setting $\alpha = \frac{1}{3}$ we obtain a 3-speed, 1.5-approximation. Note that our algorithm improves both the speed and approximation guarantee of their algorithm [18]. This can be viewed as a step towards ultimately obtaining a 1-speed algorithm. The rounding method is of independent interest and quite different from the rounding method proposed by Kalyanasundaram et al. [18]. Moreover, our formulation draws on rounding methods developed for the k-medians problem [10]. This connection with scheduling is perhaps a surprising aspect of our work. Erlebach and Hall [14] showed that one can get a 1-approximation via a 6-speed algorithm. Here we show how to use the method in this paper to get a 1-approximation via a 4-speed algorithm.

2   Problem

The problem is formally stated in [18], but for the sake of completeness we will describe it here. There are n possible pages, $P = \{1, 2, \ldots, n\}$. We assume that time is discrete and at time t, any subset of pages can be requested. Let (p, t) represent a request for page p at time t. Let r_t^p denote number of requests (p, t). Let T be the time of last request for a page. Without loss of generality, we can assume that T is polynomially bounded. A time slot t is the window of time between time $t - 1$ and time t. We also have a k-speed server that can broadcast up to k pages at any time t. We say that a request (p, t) is satisfied at time S_t^p, if S_t^p is the first time instance *after* t when page p is broadcast. In this paper, we work in the offline setting in which the server is aware of all the future requests. Our goal is to schedule the broadcast of pages in a way so as to minimize the total response time of all requests. The total response time is given by

$$\sum_p \sum_t r_t^p (S_t^p - t) \ .$$

Consider the example shown in Fig. 1. The table on the left shows requests for the three pages $A, B,$ and C at different times. An optimal schedule for this

Input:r_t^p					
	t=0	t=1	t=2	t=3	t=4
page A	3	2	2	0	0
page B	2	0	2	0	0
page C	0	2	0	0	2

Response time:$r_t^p(S_t^p - t)$					
	t=0	t=1	t=2	t=3	t=4
page A	9	4	2	0	0
page B	2	0	4	0	0
page C	0	2	0	0	2

Fig. 1. The table on the left is an example input and the table on the right shows the response time for each request in an optimal schedule of broadcasting pages B, C, A, B, C at times $1, 2, 3, 4, 5$ respectively

instance broadcasts pages B, C, A, B, C at times $1, 2, 3, 4, 5$ respectively. The table on the right of the figure shows the response time for each request in the optimal schedule. Adding up the response time of each request gives us the total response time of 25.

3   Integer Programming Formulation

The Broadcast Scheduling Problem can be formulated as an integer program as follows. The binary variable $y_{t'}^p = 1$ iff page p is broadcast at time t'. The binary variable $x_{tt'}^p = 1$ iff a request (p, t) is satisfied at time $t' > t$ i.e. $y_{t'}^p = 1$ and $y_{t''}^p = 0, t < t'' < t'$. The constraints (2) ensure that whenever a request (p, t) is satisfied at time t', page p is broadcast at t'. Constraints (3) ensure that every request (p, t) is satisfied at some time $t' > t$. Constraints (4) ensure that at most one page is broadcast at any given time.

$$\min \sum_p \sum_t \sum_{t'=t+1}^{T+n} (t' - t) \cdot r_t^p \cdot x_{tt'}^p \tag{1}$$

subject to

$$y_{t'}^p - x_{tt'}^p \geq 0, \quad \forall p, t, t' > t \tag{2}$$

$$\sum_{t'=t+1}^{T+n} x_{tt'}^p \geq 1, \quad \forall p, t \tag{3}$$

$$\sum_p y_{t'}^p \leq 1, \quad \forall t' \tag{4}$$

$$x_{tt'}^p \in \{0, 1\}, \quad \forall p, t, t' \tag{5}$$

$$y_{t'}^p \in \{0, 1\}, \quad \forall p, t' \tag{6}$$

The corresponding linear programming (LP) relaxation can be obtained by letting the domain of $x_{tt'}^p$ and $y_{t'}^p$ be $0 \leq x_{tt'}^p, y_{t'}^p \leq 1$. For the example in Fig. 1, running an LP solver produces a fractional schedule that broadcasts pages $\{A, B\}, \{A, C\}, \{A, B\}, \{A, B\}, \{C\}$ at times $1, 2, 3, 4, 5$ respectively, where broadcasting $\{P_1, P_2\}$ at any time t means that exactly half of page P_1 and half of page P_2 are broadcast at time t. The cost of this fractional solution is 24.5.

```
REQUEST CONSOLIDATION(page p, α)
1      N'_p ← {t_{f_p}}
2      l ← f_p
3      r'^p_{t_l} ← r^p_{t_l}
4      for k ← l − 1 down to 1 do
5          ft(α, p, t_k) ← min_{t'}{∑_{t=t_k+1}^{t'} x^p_{t_k t} ≥ α}
6          if (ft(α, p, t_k) ≤ t_l) then
7              N'_p ← N'_p ∪ {t_k}
8              l ← k
9              r'^p_{t_l} ← r^p_{t_l}
10         else
11             g(p, t_k) ← t_l
12             r'^p_{t_l} ← r'^p_{t_l} + r^p_{t_k}
13             r'^p_{t_k} ← 0
14     return N'_p
```

Fig. 2. Algorithm for Consolidating Requests

4   Outline of the Algorithm

Let I be the given instance of the problem. The algorithm solves the LP for I to obtain an optimal (fractional) solution. It uses the LP solution to create a simplified instance I'. A $\frac{1}{\alpha}$-speed ($\frac{1}{\alpha}$ is an integer) fractional solution is constructed for instance I', which is converted into a $\frac{1}{\alpha}$-speed integral solution for I' using a min-cost flow computation. The integral solution for I' is then converted into a schedule for I.

5   Algorithm

Step I: Let I be the given instance of the problem. For any page p, let $N_p = \{t_1, t_2, \ldots, t_{f_p}\}$ denote the times at which requests for page p are made in instance I. We first solve the LP for I to obtain an optimal fractional solution (x, y). Let $ft(\alpha, p, t)$ be the first time instance when α-fraction of request (p, t) get satisfied in the LP solution, i.e. $ft(\alpha, p, t) = \min\{t'' | \sum_{t'=t+1}^{t''} x^p_{tt'} \geq \alpha\}$. We consolidate the requests in I, transforming the instance I into a simplified instance I' which has the following property. If N'_p represents the times of positive requests for page p in I' then for any times $\{t'_u, t'_v\} \subseteq N'_p$, such that $t'_u < t'_v$, we have $ft(\alpha, p, t'_u) \leq t'_v$, where α is any fixed fraction in $(0, \frac{1}{2}]$. For every request (p, t) that is grouped with a request $(p, g(p, t)), g(p, t) \geq t$, we have $ft(\alpha, p, t) > g(p, t)$. Let $r'^p_{g(p,t)}$ denote the number of requests $(p, g(p, t))$ in I'. This transformation is done separately for each page. The pseudo-code for this transformation is given in Fig. 2.

Step II: Now we find a $\frac{1}{\alpha}$-speed fractional solution to instance I'. Let $N'_p = \{t'_1, t'_2, \ldots, t'_{f_p}\}$ be the times of positive requests for page p in I'. Note that at

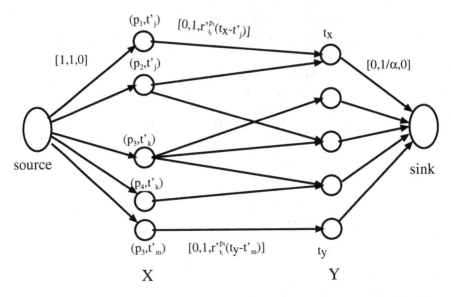

Fig. 3. The minimum cost flow network N. Each edge $e \in N$ has a label of the form $[l, u, c]$, where l and u denote the lower and the upper bounds respectively on the amount of flow through e and c denotes the cost per unit flow

least α-fraction of each request $(p, t'_i) \in I'$ gets satisfied before t'_{i+1} in the LP solution (x, y), i.e. $ft(\alpha, p, t'_i) \le t'_{i+1}$. Consider the solution (x_α, y), where

$$
x^p_{\alpha t t'} = \begin{cases} x^p_{tt'} & \text{if } t' < ft(\alpha, p, t) \\ \alpha - \sum_{t''=t+1}^{t'-1} x^p_{tt''} & \text{if } t' = ft(\alpha, p, t) \\ 0 & \text{otherwise} \end{cases}
$$

By scaling all the y values and the x_α values by $\frac{1}{\alpha}$ we obtain a feasible $\frac{1}{\alpha}$-speed fractional solution, $(\frac{1}{\alpha} x_\alpha, \frac{1}{\alpha} y)$.

Step III: To obtain a $\frac{1}{\alpha}$-speed integer solution to instance I', we construct a minimum cost flow network N. N is the flow network that consists of a bipartite graph $G = (X, Y, E)$. Each node in X corresponds to a request $(p, t') \in I'$. Each node in Y corresponds to a time instance at which a page can be broadcast. There is an edge between $(p, t) \in X$ and $t' \in Y$ if $x_{\alpha t t'} > 0$. Note that for any two nodes in X, say (p_1, t'_1) and (p_2, t'_2), that share a neighbor in Y, $p_1 \ne p_2$. This edge has a lower capacity of 0 and an upper capacity of 1 and a cost of $r'^p_t(t' - t)$. In addition, we have a source and a sink in N. There is an edge of 0 cost from the source to a node $(p, t) \in X$ with a lower and upper capacity of 1. Each edge from a node $t \in Y$ to the sink is of 0 cost and has a lower capacity of 0 and an upper capacity of $\frac{1}{\alpha}$. Figure 3 illustrates the min-cost flow network N. A $\frac{1}{\alpha}$-speed integral schedule can be obtained by setting $y^p_{t'} = f((p, t), t')$, where $f(a, b)$ denotes the flow along edge $(a, b) \in N$.

6 Analysis

Lemma 1. *In Step I, recall that each request* $(p, t) \in I$ *is grouped with a request* $(p, g(p, t))$ *to form instance* I', *where* $g(p, t) \geq t$. *If* S *and* S' *are integral solutions to instance* I *and* I' *respectively, then*

$$cost(S) \leq cost(S') + \sum_{(p,t) \in I} r_t^p(g(p,t) - t) .$$

Proof. The inequality follows easily because the additional cost of converting a solution for request $(p, g(p, t)) \in I'$ to a solution for request $(p, t) \in I$ is exactly $r_t^p(g(p, t) - t)$. Summing this over all requests will give us the required inequality.

Recall that (x, y) is an optimal LP solution for I. Define $C_{(p,t)}$ to be the cost of satisfying one request $(p, t) \in I$ in the LP solution (x, y), i.e. $C_{(p,t)} = \sum_{t'=t+1}^{T+n}(t' - t)x_{tt'}^p$. For any fixed fraction $\alpha \in (0, \frac{1}{2}]$, let $\beta = 1 - \alpha$. For any request $(p, g(p, t)) \in I'$, define $C_{(p,g(p,t))}^{'\theta}$ as the cost of satisfying exactly θ-fraction of request $(p, g(p, t)) \in I'$, i.e. $C_{(p,g(p,t))}^{'\theta} = \sum_{t'=g(p,t)+1}^{T+n}(t' - g(p,t))x_{\theta g(p,t)t'}^p$. Note that the cost of the $\frac{1}{\alpha}$-speed fractional solution for instance I' is given by $\frac{1}{\alpha} \sum_{(p,g(p,t)) \in I'} r_{g(p,t)}^{'p} C_{(p,g(p,t))}^{'\alpha} = \frac{1}{\alpha} \sum_{(p,t) \in I} r_t^p C_{(p,g(p,t))}^{'\alpha}$ and the cost of the optimal fractional solution for instance I is given by $\sum_{(p,t) \in I} r_t^p C_{(p,t)}$. Lemma 2 relates the two costs.

Lemma 2. *The cost of* $\frac{1}{\alpha}$-*speed fractional solution in* I' *and the cost of an optimal fractional solution for instance* I *are related as follows.*

$$\frac{1}{\alpha} \sum_{(p,t) \in I} r_t^p C_{(p,g(p,t))}^{'\alpha} \leq \frac{1}{\beta} \sum_{(p,t) \in I} r_t^p C_{(p,t)} - \sum_{(p,t) \in I} r_t^p(g(p,t) - t) .$$

Proof. For the case when $t = g(p, t)$, the inequality follows because $(g(p, t) - t) = 0$ and $C_{(p,t)}$ is the cost of satisfying a complete request whereas $C_{(p,g(p,t))}^{'\alpha}$ is the cost of satisfying α-fraction of the request. For the case when $t < g(p, t)$, we know that $ft(\alpha, p, t) > g(p, t)$. This means that at least $\beta = (1 - \alpha)$-fraction of the request $(p, t) \in I$ gets satisfied after $g(p, t)$. In other words, broadcast of page p between $g(p, t)$ and $ft(\beta, p, g(p, t))$ partially satisfies request (p, t) and $(p, g(p, t))$ in I. Thus we have

$$C_{(p,g(p,t))}^{'\beta} + \beta(g(p,t) - t) \leq C_{(p,t)} . \tag{7}$$

Multiplying (7) by $\frac{1}{\beta}$ and rearranging the terms gives us

$$\frac{1}{\beta} C_{(p,g(p,t))}^{'\beta} \leq \frac{1}{\beta} C_{(p,t)} - (g(p,t) - t) . \tag{8}$$

The left side of (8) represents the average cost of satisfying β-fraction of request $(p, g(p, t)) \in I'$. The cost of satisfying exactly α-fraction of the request, $C_{(p,g(p,t))}^{'\alpha}$, is upper bounded by $\frac{\alpha}{\beta}C'^{\beta}_{(p,g(p,t))}$, which is obtained by multiplying (8) by α, thus giving us

$$C'^{\alpha}_{(p,g(p,t))} \leq \frac{\alpha}{\beta}C'^{\beta}_{(p,g(p,t))} \leq \frac{\alpha}{\beta}C_{(p,t)} - \alpha(g(p,t) - t) \ . \tag{9}$$

Multiplying (9) by $\frac{1}{\alpha}r^p_t$ and summing over all requests gives us the cost of $\frac{1}{\alpha}$-speed fractional solution as required.

Lemma 3. *For any feasible flow in the minimum cost flow network, N, there is a $\frac{1}{\alpha}$-speed feasible fractional solution for instance I' of the same cost.*

Proof. Let $f(a,b)$ denote the flow along edge $(a,b) \in N$. Set $x^p_{\alpha tt'} = \alpha f((p,t),t')$ and $y^p_{t'} = x^p_{\alpha tt'}$. Since we know that for any node $(p,t) \in X$, $\sum_{t'} f((p,t),t') = 1$, we have $\frac{1}{\alpha}\sum_{t'} x^p_{\alpha tt'} = 1$. We can show that we have a $\frac{1}{\alpha}$-speed solution as follows. We know that $\sum_{(p,t)} f((p,t),t') \leq \frac{1}{\alpha}$, which implies that $\frac{1}{\alpha}\sum_{(p,t)} x^p_{\alpha tt'} \leq \frac{1}{\alpha}$ and hence $\frac{1}{\alpha}\sum_p y^p_{t'} \leq \frac{1}{\alpha}$. Thus feasibility is ensured. The cost of the $\frac{1}{\alpha}$-speed solution equals

$$\frac{1}{\alpha}\sum_{(p,t)\in I'}\sum_{t'>t} r^p_t(t'-t)x^p_{\alpha tt'} = \sum_{(p,t)\in X}\sum_{t'\in Y} r^p_t(t'-t)f((p,t),t') \ .$$

This proves that the $\frac{1}{\alpha}$-speed solution has the same cost as the flow in N.

Lemma 4. *There is a flow in N of the same cost as the $\frac{1}{\alpha}$-speed feasible fractional solution.*

Proof. Let the flow along an edge $((p,t),t')$ equal $\frac{1}{\alpha}x^p_{\alpha tt'}$. By definition of x_α, the capacity constraint of each edge is satisfied. Since we have a $\frac{1}{\alpha}$-speed *feasible* solution, $\frac{1}{\alpha}\sum_{t'} x^p_{\alpha tt'} = 1$ and hence the flow is conserved at every node in X which has an incoming flow of 1 unit. Since our solution is $\frac{1}{\alpha}$-speed, $\frac{1}{\alpha}\sum_p y^p_{t'} \leq \frac{1}{\alpha}$. Set flow along any edge from node t' to the sink equal to $\frac{1}{\alpha}\sum_p y^p_{t'}$. Note that the flow is also conserved at a node $t' \in Y$ as we know that $y^p_{t'} \geq x^p_{\alpha tt'}$. The cost of the solution equals

$$\sum_{(p,t)\in X}\sum_{t'\in Y} r^p_t(t'-t)f((p,t),t') = \frac{1}{\alpha}\sum_{(p,t)\in I'}\sum_{t'} r^p_t(t'-t)x^p_{\alpha tt'} \ .$$

Thus the cost of the flow in N is the same as the cost of the $\frac{1}{\alpha}$-speed solution.

Lemma 5. *There exists a $\frac{1}{\alpha}$-speed integral solution for I' of the same cost as the $\frac{1}{\alpha}$-speed fractional solution, $(\frac{1}{\alpha}x_\alpha, \frac{1}{\alpha}y)$.*

Proof. From Lemma 3 and Lemma 4 we know that the minimum cost flow in N equals the cost of the $\frac{1}{\alpha}$-speed fractional solution $(\frac{1}{\alpha}x_\alpha, \frac{1}{\alpha}y)$. By the integrality theorem, we can determine in polynomial time a minimum cost integral flow f^* in N that satisfies the capacity constraints. Using this integral flow f^* we can derive a $\frac{1}{\alpha}$-speed integral solution for instance I' using Lemma 3.

Theorem 6. *There is a $\frac{1}{\alpha}$-speed, $\frac{1}{1-\alpha}$-approximation solution for the Broadcast Scheduling Problem.*

Proof. We will prove that the algorithm in Sect. 5 gives a $\frac{1}{\alpha}$-speed, $\frac{1}{1-\alpha}$-approximation solution. From Lemma 2 and Lemma 5, we know that the cost of the $\frac{1}{\alpha}$-speed integral solution equals to

$$\frac{1}{\alpha} \sum_{(p,t)\in I} r_t^p C_{(p,g(p,t))}^{\prime\alpha} \le \frac{1}{\beta} \sum_{(p,t)\in I} r_t^p C_{(p,t)} - \sum_{(p,t)\in I} r_t^p (g(p,t) - t) \ .$$

Substituting this expression for $cost(S')$ in Lemma 1, we get

$$cost(S) \le \frac{1}{\beta} \sum_{(p,t)\in I} r_t^p C_{(p,t)} \le \frac{1}{\beta}OPT = \frac{1}{1-\alpha}OPT \ .$$

Corollary 7. *There is a 2-speed, 2-approximation solution and a 3-speed, 1.5-approximation solution for the Broadcast Scheduling Problem.*

Proof. The proof follows easily from Theorem 6 by setting $\alpha = \frac{1}{2}$ and $\alpha = \frac{1}{3}$ respectively for a 2-speed and a 3-speed solution.

7   4-Speed 1-Approximation Algorithm

Erlebach and Hall [14] showed that one can use a 6-speed algorithm to get a 1-approximation. This is done by taking a fractional solution obtained by solving a linear program, and then applying randomized rounding to generate schedules for 2 channels. On the remaining 4 channels they use the 4-speed algorithm by [18]. They are able to prove that the expected cost of each request is upper bounded by its fractional cost.

In this section, we show that one can obtain a random *fractional* schedule using the approach outlined in our paper. We can establish an upper bound on the cost of this fractional schedule. Finally, we convert this fractional schedule to an integral schedule using the network flow approach.

We obtain a 4-speed fractional solution as follows. On two channels, we select two pages (independently) with probability y_t^p at each time t. For the remaining two channels, we take our 2-speed *fractional* solution constructed in step II of Sect. 5. Note that in this 2-speed fractional solution, a request (p,t) is satisfied by the same broadcasting as $(p, g(p,t))$ because we construct a 2-speed fractional solution after merging (p,t) to $(p, g(p,t))$. Therefore the cost of satisfying one request (p,t) in the 2-speed fractional solution is $2C_{(p,g(p,t))}^{\prime 1/2} + (g(p,t) - t)$.

Lemma 8. *The expected cost of satisfying a request (p,t) by the 4-speed fractional schedule is at most the optimal LP cost for a request (p,t).*

Proof. In our 4-speed fractional schedule, each request (p,t) could be satisfied by either of the two random channels and *if* it is not satisfied by time $g(p,t)$, the

2-speed fractional schedule will satisfy it fractionally. Thus the expected cost of satisfying a request (p, t) by the 4-speed fractional schedule is at most

$$A = \sum_{t'=t+1}^{g(p,t)} ((\prod_{t''=t+1}^{t'-1} (1 - y_{t''}^p)^2)(1 - (1 - y_{t'}^p)^2)(t' - t)) + (\prod_{t'=t+1}^{g(p,t)} (1 - y_{t'}^p)^2)(t^* - t),$$

where $(t^* - t) = 2C_{(p,g(p,t))}^{'1/2} + (g(p,t) - t)$. The optimal LP cost for a request (p, t) is at least

$$B = \sum_{t'=t+1}^{g(p,t)} y_{t'}^p \cdot (t' - t) + (1 - \sum_{t'=t+1}^{g(p,t)} y_{t'}^p) \cdot (t^* - t) .$$

Because $\sum_{t'=t+1}^{g(p,t)} y_{t'}^p < 1/2$ and $t^* > g(p, t)$, we can get $A \le B$ using the same proof as [14].

Lemma 9. *We can find a 4-speed fractional schedule of the cost at most OPT in polynomial time.*

Proof. By Lemma 8 and the linearity of expectation, the expected cost of this 4-speed fractional schedule is at most OPT. Using standard derandomization techniques, we can obtain a deterministic polynomial-time algorithm that yields a 4-speed fractional schedule of the cost at most OPT.

Now we convert the 4-speed fractional schedule to a 4-speed integral schedule. Because the schedule on two random channels is already integral, we only need to convert the 2-speed fractional schedule to an integral schedule without losing any cost.

Lemma 10. *We can convert the 4-speed fractional schedule to a 4-speed integral schedule of the same cost.*

Proof. By constructing the minimum cost flow network only for the requests satisfied by the fractional schedule, we can convert the 2-speed fractional solution to 2-speed integral solution of the same cost. Combining with two random channels, we have a 4-speed integral solution.

Theorem 11. *There is a 4-speed, 1-approximation solution for the Broadcast Scheduling Problem.*

8   Experiments

Although we cannot obtain 1-speed algorithm which has an $O(1)$ approximation guarantee, we devised several simple heuristics and compared their performance with LP and IP solutions. The results showed that we can get very close to the optimal solution for randomly generated instances.

We have two types of heuristics. The first heuristics use the number of outstanding requests and the time of next request. In the second heuristics, we first solve LP and use the solution to guide the construction of our schedule.

8.1   Heuristic 1

Let N_t^p be the number of requests for page p which are not yet satisfied at the end of timeslot t. We may want to broadcast a page with the largest N_t^p. Another useful information to consider is the next request time. Let C_t^p be the next request time for page p after time t. If C_t^p is small, it may be better to delay broadcasting the page till the next request time, by which a single broadcast satisfies them with a small delay. We tried several heuristics using combinations of these two values.

One simple heuristic is to compute N_t^p (C_t^p) for all pages p at each time t and broadcast a page j which satisfies $N_t^j \geq N_t^k$ ($C_t^j \geq C_t^k$) for any page k. These heuristics which consider one metric (N_t^p or C_t^p) did not perform very well because the other value can be too small.

To reflect both values at the same time, a reasonable metric is $N_t^p \cdot C_t^p$. In addition, if N_t^p is very small, then we may ignore the page even if its C_t^p value is large enough to make $N_t^p \cdot C_t^p$ the largest. At each time t, we compute both N_t^p and C_t^p and broadcast a page j which satisfies $N_t^j \cdot C_t^j \geq N_t^k \cdot C_t^k$ for any page k of which N_t^k is within top $\alpha\%$. We varied α from 20% to 100% in the experiments.

8.2   Heuristic 2

In this type of heuristic, we solve LP to get an optimal fractional solution (x, y) and compute z_t^p as follows.

$$z_t^p = \begin{cases} z_{t-1}^p + y_t^p & \text{if } N_t^p > 0 \\ 0 & \text{otherwise} \end{cases}$$

z_t^p can be interpreted as preference to broadcasting page p at time t. Thus we choose a page with the largest z_t^p value at each time. We call this heuristic as *Deterministic LP Rounding*.

A variation of this heuristic is to use randomization. We normalize z_t^p by dividing it by $\sum_p z_t^p$ so that z_t^p represents the probability to broadcast page p at time t. Then we choose a page randomly based on z_t^p distribution. We call this heuristic as *Randomized LP Rounding*.

8.3   Settings

We experimented these heuristics with two types of inputs.

- Uniformly random input generator: Over all possible time and possible pages, there are approximately $(density) \cdot n \cdot T$ uniformly distributed nonzero demands. If a demand is nonzero, it is a random number from 1 to $(maximum\ demand)$.
- Zipf Distribution input generator: At every time, the total number of requests, summing over all pages, is a random number from 1 to $(density) \cdot n \cdot$

Table 1. Percentage change relative to OPT for uniformly distributed input

Different Heuristics	Mean	Median	Min	Max	S.D.	% same as OPT
LP	-0.02	0.00	-0.37	0.00	0.06	68.00
N_i	25.76	25.27	10.01	41.40	6.13	0.00
C_i	60.13	57.86	34.26	116.70	16.32	0.00
$N_i \cdot C_i$	9.57	9.08	3.15	19.61	3.38	0.00
$N_i \cdot C_i$ (top 20%)	12.96	12.60	4.50	27.33	3.99	0.00
$N_i \cdot C_i$ (top 50%)	9.31	8.83	3.15	19.21	3.38	0.00
$N_i \cdot C_i$ (top 80%)	9.57	9.08	3.15	19.61	3.38	0.00
Deterministic LP Rounding	0.90	0.00	0.00	11.47	2.21	68.67
Randomized LP Rounding	7.17	0.00	0.00	39.95	11.61	67.33
Ditto (best over 100 runs)	2.15	0.00	0.00	18.45	4.67	67.33

(*maximum demand*). Each request has probability of $\frac{1}{i \cdot H(n)}$ of requesting page i, where $H(n) = \sum_{i=1}^{n} 1/i$. In other words, the choice of pages follows Zipf distribution [24] with $\theta = 0$. Thus, page 1 has most requests and page n has least. This corresponds to the *measurements* performed in [12] (for a movies-on-demand application).

In our experiment, the uniformly random input consists of 10 pages ($n = 10$). The time of last request is 50 ($T = 50$). The *density* of demand distribution in the uniformly random input generator is 0.4. The *maximum demand* of a request is 20.

Similar to the uniformly random input generator, the Zipf distribution input consists of 10 pages, and the time of last request is 50. The maximum total number of requests, summing over all pages, is $0.4 \cdot 10 \cdot 20 = 80$, so that the expected total demand is the same as that in the uniformly random input.

To get the expected result of the Randomized LP Rounding heuristic, the result of this heuristic is the averaged cost over 100 runs. *Randomized LP Rounding (best over 100 run)* is the lowest cost over 100 randomized LP rounding.

We ran our experiments on a Sun Ultra-5 workstation, using ILOG CPLEX 6.5.2 to solve LP and IP, and Matlab 5.3 to implement all heuristics.

8.4   Results

Table 1 and 2 shows the ratio of our heuristic solutions to IP optimal solution. We repeated the experiment 150 times for uniformly random input, and another 150 times for Zipf distribution input and took averages. The Zipf input seems harder to solve. It takes 1000 to 4000 seconds to find an optimal solution, while it takes less than 100 seconds to find a LP solution. The uniformly random input problems are easier to solve. It takes less than 20 seconds to find an optimal solution for most problems. To find a LP solution, it takes a few seconds.

To find a solution using first set of heuristics, it takes around 0.1 seconds. LP rounding takes around 0.1 seconds too.

Table 2. Percentage change relative to OPT for Zipf distribution input

Different Heuristics	Mean	Median	Min	Max	S.D.	% same as OPT
LP	-0.08	-0.07	-0.31	0.00	0.08	20.00
N_i	17.82	17.35	13.21	24.54	2.48	0.00
C_i	120.80	109.20	62.86	195.20	34.82	0.00
$N_i \cdot C_i$	12.81	12.71	8.12	18.18	1.64	0.00
$N_i \cdot C_i$ (top 20%)	15.25	14.97	9.59	19.79	2.49	0.00
$N_i \cdot C_i$ (top 50%)	12.65	12.65	6.25	16.68	1.98	0.00
$N_i \cdot C_i$ (top 80%)	12.61	12.36	8.12	16.55	1.72	0.00
Deterministic LP Rounding	1.54	1.47	0.00	5.29	1.34	22.00
Randomized LP Rounding	19.18	24.63	0.00	30.21	10.41	20.00
Ditto (best over 100 runs)	10.33	12.94	0.00	19.08	6.19	20.00

Deterministic LP rounding method performed best for both uniform and Zipf inputs. The difference was only about 0.9% for uniform input and 1.6% for Zipf input. Moreover, in a few cases, deterministic LP rounding can round a non optimal solution to an optimal solution, as shown from its "% same as OPT" is higher than that of LP.

As for the first type of heuristics, $N_i \cdot C_i$ metric performed best. Although it does not perform as good as deterministic LP rounding, it runs much faster since it needs not to solve a LP. In this method, we only considered the pages which have large N_i and varied the percentage as 20%, 50%, 80%, 100%. 50% performed best for the uniform input and 80% for the Zipf input. If the heuristic considers only the requests with N_i larger than the highest 20-percentile among all N_i, it cuts too many good candidates. If the heuristic simply ranks all requests by $N_i \cdot C_i$, it may broadcast a page with small N_i but very large C_i. However, it is better delay this page and broadcast a page with larger N_i.

References

1. S. Acharya. "Broadcast Disks": Dissemination-based data management for asymmetric communication environments. Ph.D. Thesis, Brown University, 1998.
2. S. Acharya, M. Franklin, and S. Zdonik. Dissemination-based data delivery using broadcast disks. In *IEEE Personal Communications*, 2(6), 1995.
3. S. Acharya, R. Alonso, M. Franklin, and S. Zdonik. Broadcast Disks: Data management for asymmetric communications Environments. In *Proc. of ACM SIGMOD International Conference on Management of Data* (SIGMOD), 199-210, 1995.
4. D. Aksoy. On-demand data broadcast for large-scale and dynamic applications. Ph.D. Thesis, University of Maryland at College Park, 2000.
5. D. Aksoy, and M. Franklin. RxW: A scheduling approach for large-scale on-demand data broadcast. In *IEEE/ACM Transactions On Networking*, Volume 7, Number 6, 486-860, 1999.
6. M. H. Ammar and J. W. Wong. The design of teletext broadcast cycles. In *Performance Evaluation*, Vol. 5(4), 235-242, 1985.

7. A. Bar-Noy, R. Bhatia, J. Naor, and B. Schieber. Minimizing service and operation costs of periodic scheduling. In *Proceedings of 9th Annual ACM-SIAM Symposium on Discrete Algorithms*, 11-20, 1998.
8. Y. Bartal and S. Muthukrishnan. Minimizing maximum response time in scheduling broadcasts. In *Proceedings of 11th Annual ACM-SIAM Symposium on Discrete Algorithms*, 558-559, 2000.
9. R. Bhatia. Approximation algorithms for scheduling problems. Ph.D. Thesis, University of Maryland at College Park, 1998.
10. M. Charikar, S. Guha, É. Tardos, and D. Shmoys. A constant factor approximation for the k-median problem. In *Proc. of 31st Annual ACM Symposium on Theory of Computing*, 1-10, 1999.
11. C. Chekuri, S. Khanna and A. Zhu. Algorithms for minimizing weighted flow time. In *Proc. of 33rd Annual ACM Symp. on Theory of Computing*, 84-93, 2001.
12. A. L. Chervenak. Tertiary Storage: An Evaluation of New Applications. *Ph.D. Thesis, UC Berkeley*, 1994.
13. DirecPC website, http://www.direcpc.com
14. T. Erlebach, A. Hall. NP-Hardness of broadcast scheduling and inapproximability of single-source unsplittable min-cost flow. In *Proc. of 13th Annual ACM-SIAM Symposium on Discrete Algorithms*, 194-202, 2002.
15. Intel intercast website, http://www.intercast.com
16. T. S. Jayram, T. Kimbrel, R. Krauthgamer, B. Schieber, and M. Sviridenko. Online server allocation in a server farm via benefit task systems. In *Proc. of 33rd Annual ACM Symp. on Theory of Computing*, 540-549, 2001.
17. C. Kenyon, N. Schabanel, and N. Young. Polynomial-time approximation scheme for data broadcast. In *Proc. of 32nd Annual ACM Symposium on Theory of Computing* 659-666, 2000.
18. B. Kalyanasundaram, K. Pruhs, and M. Velauthapillai. Scheduling broadcasts in wireless networks. In *European Symposium of Algorithms*, LNCS 1879, Springer-Verlag, 290-301, 2000.
19. B. Kalyanasundaram and K. Pruhs. Speed is as powerful as clairvoyance. In *IEEE Symposium on Foundations of Computation*, 214-221, 1995.
20. C. Phillips, C. Stein, E. Torng, and J. Wein. Optimal time-critical scheduling via resource augmentation. In *Proc. of 29th Annual ACM Symposium on Theory of Computing*, 140-149, 1997.
21. M. S. Squillante, D. D. Yao and L. Zhang. Web traffic modelling and web server performance analysis. In *Proc. of 38th IEEE Conf. on Decision and Control*, 4432-4437, 1999.
22. H. Kaplan, R. E. Tarjan, and K. Tsioutsiouliklis. Faster kinetic heaps and their use in broadcast scheduling. In *Proc. of 12th Annual ACM-SIAM Symposium on Discrete Algorithms*, 836-844, 2001.
23. J. Wong. Broadcast Delivery. In *Proc. of the IEEE*, 76(12), 1988.
24. D. E. Knuth. The Art of Computer Programming, Volume 3. Addison-Wesley, 1973.

Building Edge-Failure Resilient Networks

Chandra Chekuri[1], Anupam Gupta[1], Amit Kumar[2],
Joseph (Seffi) Naor[3], and Danny Raz[3],[*]

[1] Lucent Bell Labs, 600 Mountain Ave., Murray Hill, NJ 07974
{chekuri,anupamg}@research.bell-labs.com
[2] Computer Science Dept., Cornell University, Ithaca, NY 14853
amitk@cs.cornell.edu
[3] Computer Science Department, Technion, Israel Institute of Technology,
Haifa 32000, Israel
{naor,danny}@cs.technion.ac.il

Abstract. We consider the design of resilient networks that are fault tolerant against single link failures. Resilience against single link failures can be built into the network by providing backup paths, which are used when an edge failure occurs on a primary path. We consider several network design problems in this context: the goal is to provision primary and backup bandwidth while minimizing cost. Our models are motivated by current high speed optical networks and we provide approximation algorithms for the problems below.

The main problem we consider is that of *backup allocation*. In this problem, we are given an already provisioned (primary) network, and we want to reserve backup capacity on the edges of the underlying network so that all the demand can be routed even in the case of a single edge failure. We also consider a variant where the primary network has a tree topology and it is required that the restored network retains the tree topology. We then address the problem of simultaneous *primary and backup allocation*, in which we are given specifications of the traffic to be handled, and we want to both provision the primary as well as the backup network. We also investigate the online model where the primary network is not known in advance. Demands between source-sink pairs arrive online and the goal is to provide a primary path and set of backup edges that can be used for restoring a failure of any of the primary edges.

1 Introduction

Fault tolerance in networks is an important and well studied topic with many applications. Telephone networks and other proprietary networks adopt a variety of techniques to provide reliability and resilience to network failures and have been in use for many years now. On the other hand data networks such as the Internet have very little centralized fault tolerance. Instead, the network relies on the routing protocols that adapt to failures by sending traffic on alternate

---

[*] Part of this work was done while the last three authors were visiting Lucent Bell Labs.

W.J. Cook and A.S. Schulz (Eds.): IPCO 2002, LNCS 2337, pp. 439–456, 2002.

paths. This has been acceptable since there are no guarantees on the quality of service on the Internet. With maturity of the Internet, many applications now require quality of service guarantees. The emergence of very high capacity optical networks has enabled the move towards providing users with their own virtual private networks (VPNs). Several networks are accommodated on the underlying high capacity optical network by splitting the available bandwidth among them. Although this approach helps in providing QoS to applications and users, fault tolerance becomes a very critical issue: failure of a single high capacity link can disrupt many VPNs that use the link.

One way to provide VPN over optical networks is using MPLS [9]. Survivability issues of IP over optical networks are discussed in [24] and [12], and restoration in MPLS tunnels are discussed in [19] and [20]. In many cases the speed and the capacity of the links do not allow, unlike the Internet, to rely on the routing protocol to successfully reroute traffic on alternate routes *after* the failure. Thus, one needs to provision the network in advance to handle failures. This places two constraints on these networks: 1) resources for re-routing traffic should be reserved at the same time the sub-networks are provisioned, and 2) the routing protocol should be simple both for the regular routing and when a fault occurs.

For the reasons mentioned above, there has been much recent interest in obtaining algorithmic solutions for problems of guaranteeing resilience against failures. A variety of failure and recovery models have been proposed. It is not feasible to give even an overview of all the models and their intricacies, hence we mention only a few high level assumptions that we make in this paper. The precise model is given later in this section. One central assumption we make is the *single* link failure assumption, that is we assume that at most one link can fail at any particular time. This is a common assumption that seems to work reasonably well in practice. Further, the resulting optimization problems are already hard and it is useful to obtain insight into this case.

Clearly, resilience against single edge failures can be built into the network by providing backup paths, which are used when an edge failure occurs on the primary path. However, note that these backup paths could intersect each other and *share* the same amount of bandwidth, provided they are used for the failures of *different* edges in the network. This multiplexing is one of the factors that makes this problem especially difficult; we shall spell out some of the others as we explain the models and our results.

Finally, most of our results are for an *uncapacitated* network. In other words, we assume that capacities of the underlying network are unlimited and there is a cost on each edge per unit of bandwidth. Although this assumption is not true for any practical network, we make it for two reasons. First, we believe it is a reasonable approximation since the capacities of the underlying network are usually much larger than the capacity of any *single* VPN. Second, the capacitated versions are much harder in theory and we believe that the domain in which they are hard does not apply to real settings. For example, the disjoint paths problem is notoriously hard for small capacities, but it is much easier if the capacities of the edges are sufficiently large compared to the individual demands. See [6,11,17] for similar assumptions.

We consider several network design problems with the above assumptions. Though our problems have similarities to traditional network design problems, they also differ in some respects. Our contributions include providing models and building upon existing techniques to give algorithms for these new problems. We hope that our techniques and ideas will be useful in related contexts.

The first problem we consider is that of *Backup Allocation*. In this problem, we are given an already *provisioned (primary) network*, and we want to reserve *backup* capacity on the edges of the underlying network so that all the demand can be routed even in the case of an edge failure. At this point, we point a requirement of the network: the restoration has to be handled *locally*; i.e., if edge $e = (i, j)$ carrying $u(e)$ bandwidth fails, there must be a *single* path $P(e)$ in the backup network between i and j with capacity at least $u(e)$, which stands in for the edge e.

Local restoration is important for timing guarantees, otherwise it could take too much time before other portions of the network are aware of a failure at e; it is also useful since it does not require any of the other paths being currently used in the network to be changed or rerouted. It is imperative that there is a *single* path between u and v that routes all of $u(e)$; having a backup network that is able to push the right amount of "flow" would not suffice. This is necessary in optical networks, where splitting the traffic is not feasible. This is also the reason that the traffic between two hosts in the originally provisioned network is routed on a single path. (As an aside, the reader curious about the actual mechanisms of efficient local restoration of paths is pointed to the literature on MPLS [9].)

The second problem we consider is that of *Primary and Backup Allocation*. In this paper we consider two cases of the problem. In the first case, we are given specifications of the traffic to be handled, and we want to provision both the primary as well as the backup network. The second case is related to an online version of the problem, where demands arrive one by one. Here, we must find both a primary path and a backup path between a pair of terminals $\{s, t\}$, but where some of the edges may have already been chosen as part of a backup path for previous demands. Since we are allowed to share those edges between backups of different pairs of terminals, we model this by allowing different costs for an edge depending on whether it is a part of a primary or a backup path.

1.1   Models and Results

We now give detailed and precise formulations of the problems studied and results obtained in this paper.

Backup Allocation: In this paper, we look at (undirected) *base networks* $G = (V, E)$ with edge costs c_e. In backup allocation, we are given a provisioned network $G^p = (V^p, E^p)$, with each edge $e \in E^p$ having provisioned capacity $u^p(e)$. The objective is to find an edge set $E^b \subseteq E$ (which could intersect with E^p) and backup capacities u^b for these edges, so that for each $e = (u, v) \in E^p$, there is a path $P(e) \in E^b \backslash \{e\}$ between its endpoints u and v on edges of capacity at least $u^p(e)$. This path $P(e)$ can be used to locally restore e lest it fail.

In Sect. 3, we describe an $O(1)$ approximation algorithm for this problem. We first examine the uniform capacity case, that is when $u^p(e) = 1$ for all $e \in E^p$. This special case is similar to the Steiner network problem [23,15,18], in that it prescribes connectivity requirements for vertex pairs, except that now there is a forbidden edge for each pair. We give an algorithm to handle this in Sect. 2, and then use scaling to handle the general case with non-uniform primary capacities.

Primary and Backup Allocation: The most common model for specifying traffic characteristics is the *point-to-point demand model*, where a *demand* matrix $D = (d_{ij})$ gives demands between each pair of terminals, and the objective is to find the cheapest network capable of sustaining traffic specified by D. In the uncapacitated case which is of interest here, allocating the primary can be done trivially by routing flow on shortest-paths between the terminals.

Considering that good estimates are often not known for the pairwise demands in real networks, Duffield et al. [10] proposed an alternate way to specify traffic patterns, the so-called *VPN hose model*. In its simplest form, each terminal i is given a *threshold* $b(i)$, and a symmetric demand matrix $D = (d_{ij})$ is called *valid* if it respects all thresholds, i.e., if $\sum_j d_{ij} \le b(i)$ for all i. The primary network is specified by a vector u^p indicating the bandwidth allocated on the various edges, and also paths P_{ij} on which all the flow between terminals (i, j) takes place unsplittably. Feasibility of a solution implies that for each valid demand matrix (d_{ij}),

$$\sum_{i<j} d_{ij}\, \chi(P_{ij}) \le u^p,$$

where $\chi(P)$ is the characteristic vector of P, and the sum is a vector sum. Provisioning the primary network in the hose model was studied in [16], where among other results, an optimal algorithm was given when the provisioned network was required to be a tree; it was also shown that this tree is a 2-approximation for general networks (without the tree restriction).

These are just some ways to specify the traffic requirements; given a primary network, the backup network is defined just as before. In this paper, we show that if there is a α-approximation algorithm for allocating the primary, there is an $O(\alpha \log n)$ approximation for both primary and backup allocation. The simple two-stage algorithm that achieves this first allocates the primary primary network G^p using the α-approximation algorithm, and then uses the algorithm of Sect. 3 to find a near-optimal backup network for G^p.

Tree Networks: Simplicity, along with good routing schemes and error recovery, make trees particularly attractive. This prompted [16] to give an algorithm for primary allocation in the VPN hose model which outputs the optimal tree (which is within a factor 2 of the optimal network). However, when some edge e in this tree fails and is locally restored by $P(e)$, the new network may no longer be a tree. For some applications, and also for simplicity of routing schemes, it is convenient that the network remains a tree at all times, even after restoration. In Sect. 5, we study the problem of allocating backup while retaining the tree topology of a given primary network. We show that this problem is hard to approximate within $\Omega(\log n)$. We also give a backup allocation algorithm whose cost is $O(\log^2 n)$ times the optimal cost of primary and backup allocation.

The Online Problem: In practical applications, the demands often appear in an online manner, i.e., new demands for both primary and backup paths between pairs of nodes arrive one by one. Here we need to solve the primary and backup allocation problem in the point-to-point demand case, i.e., when there is a single source-sink pair in the network with a given demand. Concretely, the goal is to construct both a primary path and a set of backup edges that can be used for restoring a failure of any of the primary edge. As explained before, a backup edge can be used to back up more than one primary edge, and hence some edges may have already been paid for by being on a previous backup path. We model this by allowing different *primary costs* and *backup costs* for an edge, depending on the purpose for which we will use this edge. Clearly, the primary cost of an edge should be at least as high as the backup cost. We present a simple 2-approximation algorithm for this case. We then present two natural linear programming formulations of the problem and show that one formulation dominates the other for all instances. Note, that we are considering the local optimization problem that needs to be solved each time a new demand arrives, and do not aim at performing the usual competitive analysis where the online algorithm is compared to the best offline solution.

Related Work: There have been several papers on (splittable) flow networks resilient to edge-failures; see, e.g., [6,7,11]. The papers [19,20] formulate the on-line restoration problem as an integer program, and give some empirical evidence in favor of their methods. The paper of [17] considers backup allocation in the VPN hose model and gives a constant-factor approximation when accounting *only* for the cost of edges not used in the primary network. The paper [1] looks at the problem of limited-delay restoration; however, it does not consider the question of bandwidth allocation.

The problem of survivable network design has also been investigated extensively (see, e.g., [2] and the references therein). Most of this work has been focused on obtaining strong relaxations to be used in cutting plane methods. In fact, the linear programs we use have been studied in these contexts, and have been found to give good empirical performance. For more details on these, and on polyhedral results related to them, see [3,4,5,8]. In contrast to most of these papers, we focus on worst-case approximation guarantees, and our results perhaps explain the good empirical performance of relaxations considered in the literature. Our models and assumptions also differ in some ways from those in the literature. We are interested in local restoration, and not necessarily in end-to-end restoration. This allows our results to be applicable to the VPN hose model as well, in contrast to the earlier literature, which is concerned with the point-to-point model. We also focus on path restoration as opposed to flow restoration. On the other hand, we do consider a simpler model and limit ourselves only to the case of uncapacitated networks.

2   Constrained Steiner Network Problem

Recall that our model assumes the following: when link $e = (u, v)$ fails, the backup path for (u, v) *locally* replaces e, i.e, any path that used e now uses the backup path in place of e without altering the rest of the path. Given provisioned network G^p, the Backup problem seeks to find a set of edges such that for each $(u, v) \in G^p$ there is a backup path that does not use (u, v) itself. Note that we can share edges in the backup paths for different edges. If all the capacities are the same, this is similar in spirit to traditional network design problems. Motivated by this we study a variant of the Steiner network problem that is described below.

In the Steiner network problem we are given an undirected graph $G = (V, E)$ and cost function $c : E \to \mathbb{R}^+$. We are given a *requirement* $r_{ij} \in \mathbb{Z}^+$ for pairs of vertices $(i, j) \in V$. (We can assume that $r_{ij} = 0$ for pairs (i, j) for which there is no requirement.) The goal is to select a minimum cost set of edges $E' \subseteq E$ such that there are r_{ij} edge-disjoint paths between i and j in E'. In a seminal paper, Jain [18] gave a 2-approximation for this problem, improving upon the earlier $2H_{r_{max}}$ approximation [23] where r_{max} is the largest requirement.

For our application, we add the constraint that for pairs of vertices (i, j), $E' - \{(i, j)\}$ must support r_{ij} edge-disjoint paths between i and j. Note that though we are not allowing the edge (i, j) to be used in connecting i and j, (i, j) could be used in E' to connect some other pair (i', j'). We will refer to the Steiner network problem as the SN problem, and our modified problem as the CSN (constrained SN) problem.

We show that any α-approximation algorithm for SN can be used to solve the CSN problem with an additional loss of a factor of 2 in the approximation ratio. The algorithm is simple and is given below.

- Let I_1 be the instance of SN with requirement r on G. Solve I_1 approximately, and let E' be the set of edges chosen.
- Define a new requirement function r' as follows. For $(i, j) \in E'$ such that $r_{ij} > 0$, set $r'_{ij} = r_{ij} + 1$, else set $r'_{ij} = r_{ij}$.
- Let I_2 be the instance of SN on G with requirement function r' and with the cost of edges in E' reduced to zero. Let E'' be an approximate solution to I_2. Output $E'' \cup E'$.

It is easy to see that the above algorithm produces a feasible solution. Indeed, if $(i, j) \notin E'$ then $E' - \{(i, j)\}$ contains r_{ij} edge-disjoint paths between i and j. If $(i, j) \in E'$ then E'' contains $r_{ij} + 1$ edge-disjoint paths between (i, j), and hence $E'' - \{(i, j)\}$ contains r_{ij} edge-disjoint paths.

Lemma 1. *The cost of the solution produced is at most $2\alpha\,\mathrm{OPT}$ where α is the approximation ratio of the algorithm used to solve SN.*

Proof. It is easy to see that $\mathrm{OPT}(I_1) \leq \mathrm{OPT}$, and hence $c(E') \leq \alpha\,\mathrm{OPT}$. We claim that $\mathrm{OPT}(I_2) \leq \mathrm{OPT}$. Indeed, if $A \subseteq E$ is an optimal solution to I, then $A \cup E'$ is feasible for requirements r'. Therefore, $c(E'' - E') \leq \alpha\,\mathrm{OPT}(I_2) \leq \alpha\,\mathrm{OPT}$, and $c(E'' \cup E') \leq 2\alpha\,\mathrm{OPT}$.

2.1   Integrality Gap of LP Relaxation for CSN

We used the algorithm for the Steiner network problem as a black box in obtaining the above approximation ratios. Consider the following integer linear programming formulation for CSN, where x_e is the indicator variable for picking edge e in the solution. For compactness we use the following notation to describe the constraints. We say that a function $\bar{x}$ on the edges *supports* a flow of f between s and t if the maximum flow between s and t in the graph with capacities on the edges given by $\bar{x}$ is at least f.

$$\min \quad \sum_e c_e x_e \tag{IP1}$$
$$\text{s.t.} \quad \bar{x} \in \{0,1\}^{|E|} \text{ supports } r_{ij} \text{ flow between } (i,j) \text{ in } E - \{(i,j)\} \quad \text{for all } i,j$$

We relax the integrality constrains to obtain a linear program (LP1), and claim that the integrality gap is at most 4, the same as the approximation guarantee for the algorithm above. Jain [18] showed that the integrality gap of the natural cut formulation for SN is 2. The following linear programming relaxation (LP2) for SN is flow based and is equivalent to the cut formulation, and hence its integrality gap is also 2.

$$\min \quad \sum_e c_e x_e \tag{LP2}$$
$$\text{s.t.} \quad \bar{x} \in [0,1]^{|E|} \text{ supports } r_{ij} \text{ flow between } (i,j) \text{ in } E \quad \text{for all } i,j$$

Note that the optimal solutions to (LP2) for the instances I_1 and I_2 cost no more than an optimal solution to (LP1) for I. This, combined with the fact that (LP2) has an integrality gap of at most 2, gives the following result.

Lemma 2. *The integrality gap of (LP1), the LP for the CSN problem is upper bounded by 4.*

3   Backup Allocation

In this section we show an $O(1)$ approximation for the problem of computing the cheapest backup network for a given network. Let $G = (V, E)$ be the underlying base network and let $G^p = (V^p, E^p)$, a subgraph of G, denote the primary network. We are also given a real valued function u^p on the edge set E that gives the primary bandwidth provisioned on the edges. Our goal is to find an edge set $E^b \subseteq E$ and a function $u^b : E^b \to \mathbb{R}^+$ such that E^b backs up the network G^p for single link failures. Note that we are working in the uncapacitated case which implies that we can buy as much bandwidth as we want on any edge of E and the cost for buying b units of bandwidth on edge e is $b \cdot c_e$, where c_e is the cost of e.

Let $u^p_{\max} = \max_{e \in E^p} u^p(e)$. Our algorithm for backup allocation given below is based on scaling the capacities and solving the resulting uniform capacity problems separately.

- Let $E_i^p = \{e \in E^p \mid u^p(e) \in [2^i, 2^{i+1})\}$. For all $e \in E_i^p$, round up $u^p(e)$ to 2^{i+1}.
- For $1 \leq i \leq \lceil \log u_{\max}^p \rceil$, *independently* backup E_i^p.

Let E_i^b be the edges for backing up E_i^p and u_i^b be the backup bandwidth on E_i^b. Note that rounding up the bandwidths of E_i^p causes the the backup allocation problem on E_i^p to be a uniform problem. The lemma below states that solving the problems separately does not cost much in the approximation ratio.

Lemma 3. *Let α be the approximation ratio for the uniform capacity backup allocation problem. Then there is an approximation algorithm for the backup allocation problem with ratio 4α.*

Proof. Let E^{r*} be an optimal solution for backup allocation, with u^{r*} being the bandwidth on E^{r*}. For $1 \leq i \leq \lceil \log u_{\max}^p \rceil$ construct solutions E_i^{r*}, where $e \in E_i^{r*}$ with capacity $u_i^{r*}(e) = 2^{i+1}$ if $u^{r*}(e) \geq 2^i$. Clearly $\sum_i u_i^{r*}(e) \leq 4u^{r*}(e)$, and so $\sum_i c(E_i^{r*}) \leq 4c(E^{r*})$. However, note that E_i^{r*} is a feasible backup for E_i^p, since every edge in E^{r*} of bandwidth at least 2^i lies E_i^{r*} with bandwidth 2^{i+1}. Hence, for each i, using the approximation algorithm for the uniform case for E_i^p would give us a solution with cost at most $\alpha c(E_i^{r*})$. This completes the proof.

However, the backup allocation problem for the uniform bandwidth case can be approximated to within $\alpha = 4$. Given a set of edges E^p with uniform bandwidth $u^p(e) = U$ that need to be backed up, the problem can be scaled so that $u^p(e) = 1$ for $e \in E^p$. This is just a problem of finding, for $(i, j) \in E^p$, a path between i and j that does not use the edge (i, j) itself, which in turn is the problem described in Sect. 2 with $r_{ij} = 1$ for $(i, j) \in E^p$. Combining this with Lemma 3, we get a 16-approximation algorithm for the backup allocation problem.

The ratio of 4α can be improved to $e\alpha$ by randomness: instead of grouping by powers of 2, grouping can be done by powers of e (with a randomly chosen starting point). This technique is fairly standard by now (e.g., [21,13]) and the details are deferred to the final version.

Theorem 1. *Given any G^p, there is a $4e \simeq 10.87$-approximation for the backup allocation problem for G^p.*

3.1 Integrality Gap of an LP Relaxation

We showed an $O(1)$ approximation for the backup allocation problem. We now analyze the integrality gap of a natural LP relaxation for the problem and show that it is $\Theta(\log n)$. This will allow us to analyze an algorithm for simultaneous allocation of primary and backup networks in the next section. The LP formulation uses variables y_e which indicate the backup bandwidth bought on edge e.

We relax the requirement that the flow uses a *single* path.

$$\min \quad \sum_e c_e y_e \qquad\qquad\qquad \text{(LP3)}$$

s.t. $\bar{y}$ supports u_e^p flow between (i, j) in $E - \{e\}$ $\qquad$ for all $e \in E^p$

$\qquad y_e \geq 0$.

We now analyze the integrality gap. Recall the definition of E_i^p as the set of edges in E^p such that $u^p(e) \in [2^i, 2^{i+1})$. As before we round up the bandwidth of these edges to 2^{i+1}. Let $x_e(i) = \min\{1, y_e/2^i\}$. Note that $x_e(i) \in [0, 1]$. We claim the following.

Proposition 1. *The variables $x_e(i)$ are feasible for the uniform bandwidth backup allocation problem induced by E_i^p where the bandwidths are scaled to 1.*

From the analysis in Sect. 2.1 it follows that the integrality gap of (LP1), the LP for the uniform bandwidth problem is at most 4. Hence we can find a solution that backs up the edges in E_i^p with cost at most $4 \sum_e c_e y_e$. Since we have to only look at $\lceil \log u_{max}^p \rceil$ values of i, there is a solution that backs up all edges in E^p with cost at most $4 \log u_{max}^p \sum_e c_e y_e$. We can make the upper bound on the integrality gap $O(\log n)$ via a simple argument. We set $x_e(i) = 0$ if $y_e/2^i \leq 1/n^3$, otherwise we set $x_e(i) = \min\{1, (1 + 1/n)y_e/2^i\}$. It is straightforward to argue that Proposition 1 still holds for the variables $x_e(i)$ defined in this modified fashion. The cost goes up by a $(1 + 1/n)$ factor. Each edge e participates in the backup of at most $O(\log n)$ groups E_i^p, hence the overall cost is at most $O(\log n)$ times the LP cost. This gives us the following theorem.

Theorem 2. *The integrality gap of (LP3) is $O(\min\{\log n, \log u_{max}^p\})$.*

The following theorem shows that our analysis is tight.

Theorem 3. *The integrality gap of (LP3) is $\Omega(\log n)$.*

Proof. We construct a graph G with the required gap as follows. The graph consists of a binary tree T rooted at r with some additional edges. The cost of each edge in T is 1. The additional edges go from leaves to the root and each of them is of cost d, where d is the depth of T. Primary bandwidth is provisioned only on the edges of T and is given by $u^p(e)$: for an edge e at depth $d(e)$, $u^p(e) = 2^d/2^{d(e)}$. Backup bandwidth allocation defined by the following function $u^b(e)$ is feasible for (LP3): $u^b(e) = 1$ for each edge e that goes from a leaf to the root and $u^b(e) = u^p(e)$ for each edge of T. It is easy to check that the cost of this solution is $O(d2^d)$. We claim that any path solution to the backup of T in G has a cost of $\Omega(d^2 2^d)$. To prove this claim we observe that in any path backup solution the number of edges from the leaves to the root of backup capacity 2^{d-i} is at least 2^i. This follows since there are 2^i edges of capacity 2^{d-i} in T each of which could fail and each of them requires a backup edge from a leaf in its subtree to the root. The subtrees are disjoint and hence these back up edges cannot be shared. We set d to be $\log n$ to obtain the desired bound.

We note that the primary network in the above proof is valid for both the point to point demand model and the VPN hose model. That the former is true is clear: every edge implicitly defines a point to point demand between its end points of value equal to the primary bandwidth allocated to the edge. To see that the above primary network is valid in the VPN hose model consider the leaves of T as demand points, each with a bandwidth bound of 1.

4   Simultaneous Primary & Backup Allocation

In this section we examine the problem of simultaneously building a primary network and the backup network of minimum overall cost. We have already seen an $O(1)$ approximation algorithm to provide backup given the primary network. We adopt a natural two-phase strategy where we build the primary network first and then build a backup network for it. If α is the approximation guarantee for the problem of building the primary network then we obtain an $O(\alpha \log n)$ approximation for building the primary and backup networks. For the two primary networks of interest, namely the pairwise demand model and the VPN hose model, we have $O(1)$ approximation algorithms for building the primary network, hence we obtain an $O(\log n)$ approximation for the combined problem.

An $O(\log n)$ approximation: We analyze the two-stage approach for primary and backup allocation. Let G^p be the subgraph of G that is chosen in this first step. We provide backup for this network using the algorithm described in Sect. 3. To analyze this algorithm we use the LP relaxation (LP3) for the backup allocation problem. In the following lemma we will be using extra capacity on the edges of provisioned network itself. Note that this is allowed.

Lemma 4. *Let u^p be any solution to the primary problem. Let u^{p*} and u^{r*} be the primary and backup in some optimal solution. Then, $u^p + u^{p*} + u^{r*}$ is a feasible solution for (LP3), the LP relaxation for the backup of u^p.*

Proof. We assume that the solution u^p is minimal. Let $e = (i, j)$ be such that $u^p(e) > 0$. Since we have a minimal solution it implies that there exists some traffic between terminals that requires at least $u^p(e)$ flow on e. Let the flow paths that use e in that traffic be $P_1, P_2, \ldots, P_k$ and let f_i be the flow on P_i. In the graph G^p, let X be the set of terminals that are connected to i if the edge (i, j) is removed from the paths, and let Y be those terminals that are connected to j. Let P_h connect terminal x_h to terminal y_h where $x_h \in X$ and $y_h \in Y$. We need to argue that in the graph with out the edge (i, j), we can a send a flow of value $u^p(e)$ from i to j with capacities defined by $u^p + u^{p*} + u^{r*}$. We do this as follows. We send flow f_i from i to each x_i using capacities $u^p(e)$. Since the optimum solution is resilient against single edge failures, for $1 \le i \le k$, there must exist flow paths that can route a flow of f_i units from x_i to y_i, none of which use the edge (i, j). Since $\sum_i f_i = u^p(e)$, it follows that we can route a flow of $u^p(e)$ from i to j without using (i, j) in the capacitated graph defined by $u^p + u^{p*} + u^{r*}$.

Theorem 4. *The two-stage approach yields an $O(\alpha \log n)$ approximation to the combined primary and backup allocation problem where α is the approximation ratio for finding the primary allocation.*

Proof. Let P be the cost of the primary allocation and B the cost of backup allocation in the two stage approach. From the approximation guarantee on finding P, we have $P \leq \alpha$ OPT. From Lemma 4, it follows that there is a feasible (LP3) relaxation for the backup allocation problem of value at most $P +$ OPT, hence at most $(\alpha + 1)$ OPT. From Theorem 2, the backup solution we obtain is at most $O(\log n)$ times the LP value. Hence, $B = O(\alpha \log n)$ OPT and the theorem follows.

Corollary 1. *There is an $O(\log n)$ approximation for the combined primary and backup allocation problems for the pairwise demand model and the VPN hose model.*

It turns out that the two-stage approach loses an $\Omega(\log n)$ factor even if the first step obtains a primary network of optimum cost; the example in the proof of Theorem 3 demonstrates this.

5   Backup for Tree Networks

In this section, we consider the case when the provisioned network T is a tree. Furthermore, we require that when an edge e fails, the network $T - \{e\} + P(e)$ also be a tree. We show that finding a minimum cost backup network in this case is at least as hard as the group Steiner problem [14] on trees, which in turn is $\Omega(\log n)$-hard. We also give an algorithm whose approximation ratio is within constant factors of the approximation ratio for the group Steiner problem on trees, which at most $O(\log^2 n)$ due to Garg et al. [14].

There is a slight difference in the manner in which we define the approximation ratio in this section. Let $T = G^p = (V^p, E^p)$ be the provisioned network as usual, and let E^b be the backup edges, with u^p and u^b be the primary and backup bandwidths as usual. We shall consider the cost of the solution to be $\sum_e (u^p(e) + u^b(e))$, and our approximation will be with respect to this measure. The problem turns out to be hard, even when the measure of goodness is taken to be $\sum_e u^b(e)$. We omit the proof of the following theorem.

Theorem 5. *The tree backup problem is at least as hard as the group Steiner problem on trees.*

5.1   Approximation Algorithm

Let $T = (V, E)$ be the already provisioned tree. When an edge e fails, it splits T into two components, and $P(e)$ must be a path between these two components which is internally node disjoint from the tree T. We must reserve enough bandwidth on the edges in the graph such that the tree formed thus can support traffic between the demand nodes.

Our basic strategy is the same as in Sect. 3: Let E_i be the set of edges in T on which the bandwidth u^p lies in the interval $[2^i, 2^{i+1})$. Let $u^p_{\max}$ lie in the interval $[2^s, 2^{s+1})$. Our algorithm will proceed in stages — in the ith stage, we will "protect" the edges in E_{s+1-i}. When we have already protected the edges in $E_{i+1}, \ldots, E_s$ by reserving bandwidth on some edges, we contract edges in $E_{i+1} \cup \ldots \cup E_s$. This will not affect our performance by more than a constant, since the bandwidth we may later reserve on some edge e in this set will be at most $\sum_{j \leq i} 2^{i+1} \leq 4u^p(e)$. Let T_i and G_i be the resulting tree and base graph after the contraction. We shall now consider protecting the edges in E_i, using the edges of G_i.

The procedure has a few conceptual steps, so let us describe these here, with the intent of convincing the reader. The proofs closely follow the ideas mentioned here, but are omitted due to lack of space.

- It can be seen that the edges of E_i form a "spider"; i.e., there is a root r, and a collection of paths $\{P_i\}_{i=1}^k$ which meet at r but are otherwise node-disjoint. This is because of the structure of the VPN trees as given in [16], where the allocated demand on the edges from the root to the leaves is non-increasing.
- The graph G_i can be transformed into a graph G'_i of which T_i is a spanning tree. All non-T_i edges go between vertices of T_i. This can be done so that the backup solutions for T_i in G_i and in G'_i can be translated between each other with only a constant factor difference in cost.
  This transformation is based on the following idea: we consider the backup edges for E_i which do not belong to T_i; these must form a tree (else we could drop one of the edges). Take an Eulerian tour of this tree and consider the subpaths between consecutive nodes in T_i; these can be replaced by new edges with the appropriate weight, and all other vertices can be disposed off. Note that the optimal solution in this new instance will be at most twice the optimal solution before, since the Eulerian tour counted every edge twice. One has to take care that there may be backup edges in T_i itself, and accounting for these requires a slightly more complicated argument, which we omit here.
- We now have a simpler problem: a graph G'_i, with a spanning tree T_i in which we need to find a tree backup for the edges of E_i, which form a spider. Let r_i be the root of T_i, and let P_j's be the paths of the spider. Also let $T_{i,j}$ be the subtree of T_i hanging off P_j. (See Fig. 1 for a picture.) Call a non-tree edge a *back* edge if both its end points belong to the same tree $T_{i,j}$, and a *cross edge* otherwise. For example, the edge e_b is a back edge, and e_c a cross edge in the figure. Now each edge e of E_i has a *savior edge* $sav(e)$ which is used to connect the two components formed if e fails. A crucial fact is that if a cross edge from $T_{i,j}$ to $T_{i,j'}$ is a savior for some edge e in P_j, then it is a savior for all edges on P_j which are above e. Hence, fixing the lowest edge e in P_j whose savior is a cross edge implies that all edges above it are also saved by that same savior edge, and all edges below it on P_j must be saved

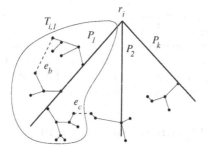

Fig. 1. The tree T_i, with tree edges shown in solid lines and non-tree edges in dotted ones

by back edges. The cost $Q(e)$ of saving the rest by back edges depends on the portion of T_i attached to these edges and the back edges between this portion; note that this is entirely independent of the rest of the problem.

Suppose we know, for each edge $e \in \cup_j P_j$, the cost $Q(e)$. (We shall discharge this assumption later.) Then the cost of backing up all the edges in E_i consists of the following: for each P_j, picking the single cross edge (say going to $T_{i,j'}$) which is going to be savior (and reserving 2^{i+1} capacity on it), and reserving 2^{i+1} capacity on the edges in $T_{i,j'}$ from the other end of this edge to the root r_i. Of course, we have to add the cost of saving the edges that were not saved by these cross edges to the solution.

- We now claim that this can be modeled as a minor variant of the group Steiner tree problem with vertex costs. Each vertex $v \in T_i$ which is the endpoint of some cross edge e from $T_{i,j}$ is belongs to the group S_j and has a "cost" $2^{i+1}c_e + Q(e')$, where e' is the lowest edge in P_j that can be saved by e. (Note that S_j must be a multi-set, with the vertex v occurring several times in S_j with various costs, if there are several such cross edges.) As a pedantic aside, there may be no such cross edge, and so r_i also belongs to S_j with cost equal to saving P_j entirely with back edges. This is done for every vertex and every value of j. Now the objective is to find the minimum cost subtree of T_i that contains the root and hits every group S_j at least once, where we also have to pay for the vertex of S_j picked in the tree. It is fairly easy to see that this can be transformed into a regular group Steiner tree problem, and the algorithm of Garg et al. [14] then gives us a $O(\log^2 n)$ approximation.

- There is one more assumption to be discharged: we have to show how to compute all the $Q(e)$. We will not be able to do this optimally, but we give a constant factor approximation for this as well. Since these are independent problems, let us consider the case when we want to find the cost of backing up P_1 using only back edges. We would like to just use the Eulerian trick done above to reduce the problem to edges only between vertices of P_1, and then find the least cost augmentation. The technical problem that arises is that that the optimal solution could be using edges in $T_{i,j} - P_j$, and doing this trick naively could result in our paying several times for this reservation, when paying once would have sufficed. We can however show that we pay for no such edge more than twice, and hence the cost of the min-cost augmentation is off by a factor of at most 2. Of course, we can only compute the augmentation within a factor of 2 of optimal, and hence we can compute $Q(e)$ within a factor 4 of optimal.

6   The Online Optimization Problem

In this section we consider a unit-capacity MPLS primary and backup allocation problem which is motivated by the online problem of choosing the best primary and backup paths for demands arriving one by one. Suppose that we are given source and destination vertices, denoted by s and t, respectively. The goal is to simultaneously provision a primary path p from s to t and a backup set of edges q of minimum overall cost. Since we are dealing with a single source-sink pair we can scale the bandwidth requirement to 1, hence all edges have unit capacity, i.e., the primary and backup edge sets are disjoint. We require that for any failure of an edge $e \in p$, $q \cup p - \{e\}$ contains a path from s to t. We call this problem SSSPR (Single Source Sink Provisioning and Restoration). Note that this requirement is slightly different from the backup model discussed earlier in the paper; here, we do not insist on local restoration. The backup edges together with the primary edges are required to provide connectivity from s to t. This problem is in the spirit of the work of Kodialam and Lakshman [19,20]. As explained before, we model the online nature of the problem by using two different costs. Formally, there are two non-negative cost functions associated with the edges: cost function c_1 denotes the cost of provisioning a primary edge the primary, and cost function c_2 denotes the cost of an edge when used as a backup edge. We assume that $c_1(e) \geq c_2(e)$ for all edges $e \in E$.

Let p be a primary path from the source s to the destination t. The following procedure due to Suurballe [22] computes a minimum cost backup set of edges for a given primary path p. The idea is to direct the edges on the path p in the "backward" direction, i.e., from t to s and set their cost to be zero. All other edges are replaced by two anti-symmetric arcs. For each edge e which is replaced by arcs a and a^-, the cost of both a and a^- is set to $c_2(e)$. We now compute a shortest path q from s to t. It can be shown that the edges of q that do not belong to p define a minimum cost local backup [22].

A 2-approximation algorithm for the SSSPR problem can be obtained as follows. First, find a shortest path p from s to t with respect to the c_1-cost function. Then, use Suurballe's [22] procedure to compute an optimal backup q to the path p with respect to the c_2-cost function. We show below that p and q together induce a 2-approximate solution.

Theorem 6. *The two stage approach yields a 2-approximation to SSSPR.*

Proof Sketch. Let OPT be the cost of an optimal primary and backup solution and let $P = \sum_{e \in p} c_1(e)$ be the cost of p and $Q = \sum_{e \in q} c_2(e)$ be the cost of q. It is clear that $P \leq$ OPT since we find the cheapest primary path. We next argue that $Q \leq$ OPT. Consider Suurballes [22] algorithm to find the optimum backup path for p. As described earlier, the algorithm finds a shortest path in a directed graph obtained from G and p. We can express the computation of this shortest path as a linear program L in a standard way – essentially as a minimum cost flow computation of sending one unit from s to t. The main observation is that any primary path p' and a q' that backs up p' yield a feasible solution to linear program L. We omit the formal proof of this observation but it is not difficult to see. In particular, this holds for the set of edges of p^* and q^*, where p^* is an optimal primary path and q^* is a set of edges that back up p^*. Therefore, it follows that $Q \leq \sum_{e \in p^* \cup q^*} c_2(e) \leq \sum_{e \in p^*} c_1(e) + \sum_{e \in q^*} c_2(e)$. Here is where we crucially use the assumption that $c_2(e) \leq c_1(e)$ for all e. Hence, $Q \leq$ OPT and the theorem follows.

Although we provide an approximation algorithm, we note that it is not known whether the problem is NP-hard or not.

6.1   Linear Programming Formulations for SSSPR

We provide two linear programming relaxations of SSSPR. The first formulation is based on cuts and the second formulation is based on flows. We show that the second formulation dominates the first one on all instances.

A *cut* in a graph G is a partition of V into two disjoint sets V_1 and V_2. The edges of the cut are those edges that have precisely one endpoint in both V_1 and V_2. Let T be a subgraph of G which is a tree. A cut (V_1, V_2) of G is a *canonical cut* of G with respect to T if there exists an edge $e \in T$, decomposing T into T_1 and T_2, such that $T_1 \subseteq V_1$ and $T_2 \subseteq V_2$.

Let p be a primary path from the source s to the destination t. It follows from Suurballe's [22] procedure that a set of edges q is a backup to a path p if it covers all the canonical cuts of p. This leads us to the following linear programming formulation which is based on covering cuts. For an edge e, let $x(e)$ denote the primary indicator variable and let $y(e)$ denote the backup indicator variable.

$$
\begin{aligned}
\min \quad & \sum_{e \in E} c_1(e) \cdot x(e) + c_2(e) \cdot y(e) & & \text{(Cut-LP)} \\
\text{s.t.} \quad & \sum_{e \in C}(x(e) + y(e)) \geq 2 & & \text{for all } \{s, t\}\text{-cuts } C \\
& \sum_{e \in C} x(e) \geq 1 & & \text{for all } \{s, t\}\text{-cuts } C \\
& x(e) + y(e) \leq 1 & & \text{for all } e \in E \\
& x(e), y(e) \geq 0 & & \text{for all } e \in E
\end{aligned}
$$

It is not hard to see that the value of an optimal (fractional) solution to Cut-LP is a lower bound on the value of an optimal integral solution to SSSPR. We now present a second linear programming formulation of SSSPR which is based on flows. Our formulation relies on the following lemma.

Lemma 5. *Let p be a primary path from s to t and let q be a set of backup edges. Replace each edge from p and q by two parallel anti-symmetric unit capacity arcs. Then, two units of flow can be sent from s to t.*

This leads us to the following bidirected flow relaxation. We replace each edge by two parallel anti-symmetric unit capacity arcs. Denote by $D = (V, A)$ the directed graph obtained. The goal is to send two units of flow in D from s to t, one from each commodity, while minimizing the cost. Denote the two commodities by blue and red, corresponding to primary and backup edges, respectively. The cost of the blue commodity on an arc a (obtained from edge e) is equal to $c_1(e)$. The cost of the red commodity on an arc a (obtained from edge e) is defined as follows. Suppose there is blue flow on arc a^- of value f. Then, red flow on a up to value of f is *free*. Beyond f, the cost of the red flow is $c_2(e)$.

$$\min \quad \sum_{e \in E} c_1(e) \cdot f_1(e) + c_2(e) \cdot f_2(e) \qquad \text{(Flow-LP)}$$

$$\text{s.t.} \quad \bar{x} \quad \text{supports a unit flow } (f_1) \text{ between } s \text{ and } t$$

$$\bar{y} \quad \text{supports a unit flow } (f_2) \text{ between } s \text{ and } t$$

$$f_1(e) \geq \max(f_1(a), f_1(a^-)) \qquad \text{for all } e = (a, a^-)$$

$$f_2(e) \geq \max((f_2(a) - f_1(a^-)), 0) + \max((f_2(a^-) - f_1(a)), 0)$$

$$\text{for all } e = (a, a^-)$$

$$x(a) + y(a) \leq 1 \qquad \text{for all } a \in A$$

$$x(a), y(a) \geq 0 \qquad \text{for all } a \in A \ .$$

Given a solution to the SSSPR problem, Lemma 5 tells us how to obtain a two-commodity flow solution from it. We claim that the cost of the two-commodity flow solution is equal to the cost of the solution to the SSSPR problem. Notice that the blue flow costs the same as the blue edges in the SSSPR solution. The cost of the red flow is zero on arcs which are obtained from blue edges. On other edges, the cost of the red flow and the cost of the SSSPR solution are the same. Therefore, the value of an optimal (fractional) solution to the Flow-LP is a lower bound on the value of an optimal integral solution. We now prove that Flow-LP dominates Cut-LP.

Theorem 7. *For any instance of the SSSPR problem, the cost of the optimal solution produced by Flow-LP is at least as high as the cost of the optimal solution produced by Cut-LP.*

Proof. We show that given a feasible solution to Flow-LP, we can generate a feasible solution to Cut-LP without increasing the cost. Consider edge $e \in E$ which is replaced by two anti parallel arcs a and a^- in Flow-LP. Without loss of generality, we can assume that at most one of $\{f_1(a), f_1(a^-)\}$ is non-zero and at most one of $\{f_2(a), f_2(a^-)\}$ is non-zero. Define $x(e) = f_1(e)$ (or $x(e) = f_1(a) + f_1(a^-)$) and $y(e) = \min(f_2(e), 1 - f_1(e))$. We show that $\{x(e), y(e)\}_{e \in E}$ defines a feasible solution for Cut-LP. Let the x-capacity (y-capacity) of a cut be the sum of the variables $x(e)$ ($y(e)$) taken over the edges e belonging to the cut. Clearly, the x-capacity of all $\{s, t\}$-cuts is at least one, since flow f_1 in D sends one unit of flow from s to t. It remains to show that the x-capacity together with the y-capacity of all $\{s, t\}$-cuts is at least two.

Consider a particular $\{s, t\}$-cut C. Decompose flow function f_1 in D into flow paths, each of flow value ε. Let $n(k)$ denote the number of flow paths in the decomposition that use precisely $2k+1$ edges from C. Clearly, $\sum_{k=0}^{\infty} n(k) \cdot \varepsilon = 1$, and so the contribution of flow f_1 in D to the x-capacity of C is

$$\sum_{k=0}^{\infty} (2k+1) \cdot n(k) \cdot \varepsilon = 2 \sum_{k=0}^{\infty} k \cdot n(k) \cdot \varepsilon + \sum_{k=0}^{\infty} n(k) \cdot \varepsilon$$
$$= 2 \sum_{k=0}^{\infty} k \cdot n(k) \cdot \varepsilon + 1 .$$

Suppose $\sum_{k=0}^{\infty} k \cdot n(k) \cdot \varepsilon < 1/2$, otherwise we are done. The red flow in D, f_2, can send for "free" flow of value at most $\sum_{k=0}^{\infty} k \cdot n(k) \cdot \varepsilon$ using arcs belonging to cut C. (For each arc a carrying blue flow of value ε, red flow of value ε can be sent on a^- for free.) Therefore, the red flow must send flow of value at least $1 - \sum_{k=0}^{\infty} k \cdot n(k) \cdot \varepsilon$ using capacity "paid" for by f_2. (Note that for this flow we have $y(e) = f_2(e)$ for all edges e.) Hence, the y-capacity of C is at least $1 - \sum_{k=0}^{\infty} k \cdot n(k) \cdot \varepsilon$, yielding that the capacity of cut C (x-capacity and y-capacity) is $1 + 2 \sum_{k=0}^{\infty} k \cdot n(k) \cdot \varepsilon + 1 - \sum_{k=0}^{\infty} k \cdot n(k) \cdot \varepsilon \geq 2$, thus completing the proof.

Integrality Gap: It is not hard to show that a fractional solution to both formulations can be rounded to an integral solution while increasing the cost by at most a factor of 2. The proof is along the lines of the proof for the combinatorial 2-approximation algorithm presented earlier.

We also have an example of an instance where there is a (multiplicative) gap of at least $5/4$ between the optimal solution to Flow-LP and any integral solution.

Acknowledgments

We would like to thank Rajeev Rastogi for providing us a copy of [17], and for useful discussions.

References

1. A. Bremler-Barr, Y. Afek, E. Cohen, H. Kaplan and M. Merritt. Restoration by Path Concatenation: Fast Recovery of MPLS Paths. In *Proceedings of the ACM PODC*, pages 43–52, 2001.
2. A. Balakrishnan, T. Magnanti, and P. Mirchandani. Network Design. *Annotated Bibliographies in Combinatorial Optimization*, M. Dell'Amico, F. Maffioli, and S. Martello (eds.), John Wiley and Sons, New York, 311-334, 1997.
3. A. Balakrishnan, T. Magnanti, J. Sokol, and Y. Wang. Modeling and Solving the Single Facility Line Restoration Problem. Working Paper OR 327-98, Operations Research Center, MIT, 1998. To appear in *Operations Research*.
4. A. Balakrishnan, T. Magnanti, J. Sokol, and Y. Wang. Telecommunication Link Restoration Planning with Multiple Facility Types. To appear in *Annals of Operations Research*, volume "Topological Network Design in Telecommunications" edited by P. Kubat and J. M. Smith.

5. D. Bienstock and G. Muratore. Strong Inequalities for Capacitated Survivable Network Design Problems. *Math. Programming*, 89:127–147, 2001.
6. G. Brightwell, G. Oriolo and F. B. Shepherd. Reserving resilient capacity in a network. In SIAM J. Disc. Math., **14**(4), 524–539, 2001.
7. G. Brightwell, G. Oriolo and F. B. Shepherd. Reserving Resilient Capacity with Upper Bound Constraints. Manuscript.
8. G. Dahl and M. Stoer. A Cutting Plane Algorithm for Multicommodity Survivable Network Design Problems. *INFORMS Journal on Computing*, 10, 1-11, 1998.
9. B. Davie and Y. Rekhter. *MPLS: Technology and Applications*. Morgan Kaufmann Publishers, 2000.
10. N. G. Duffield, P. Goyal, A. G. Greenberg, P. P. Mishra, K.K. Ramakrishnan, and J. E. van der Merwe. A flexible model for resource management in virtual private networks. In *Proceedings of the ACM SIGCOMM, Computer Communication Review*, volume 29, pages 95–108, 1999.
11. L. Fleischer, A. Meyerson, I. Saniee, F. B. Shepherd and A. Srinivasan. Near-optimal design of MPλS tunnels with shared recovery. DIMACS Mini-Workshop on Quality of Service Issues in the Internet, 2001.
12. A. Fumagalli and L. Valcarenghi. IP restoration vs. WDM Protection: Is there an Optimal Choice? *IEEE Network*, 14(6):34-41, November/December 2000.
13. M. Goemans and J. Kleinberg. *An improved approximation ratio for the minimum latency problem*. In *Proceedings of 7th ACM-SIAM SODA*, pages 152–157, 1996.
14. N. Garg, G. Konjevod, and R. Ravi. A polylogarithmic approximation algorithm for the group Steiner tree problem. *Journal of Algorithms*, 37(1):66–84, 2000. (Preliminary version in: *9th Annual ACM-SIAM Symposium on Discrete Algorithms*, pages 253–259, 1998).
15. M. X. Goemans, A. V. Goldberg, S. Plotkin, D. B. Shmoys, É. Tardos, and D. P. Williamson. Improved approximation algorithms for network design problems. In *Proceedings of the 5th Annual ACM-SIAM Symposium on Discrete Algorithms*, pages 223–232, 1994.
16. A. Gupta, A. Kumar, J. Kleinberg, R. Rastogi, and B. Yener. Provisioning a Virtual Private Network: A network design problem for multicommodity flow. In *Proceedings of the 33rd Annual ACM Symposium on Theory of Computing*, pages 389–398, 2001.
17. G. F. Italiano, R. Rastogi and B. Yener. Restoration Algorithms for Virutal Private Networks in the Hose Model. In *Proceedings of Infocom 2002*, to appear.
18. K. Jain. A factor 2 approximation algorithm for the generalized Steiner network problem. *Combinatorica*, 21(1):39–60, 2001. (Preliminary version in: *39th Annual Symposium on Foundations of Computer Science*, pages 448–457, 1998).
19. M. Kodialam and T. V. Lakshman. Minimum Interference Routing with Applications to MPLS Traffic Engineering. *Infocom 2000*, pages 884–893, 2000.
20. M. Kodialam and T. V. Lakshman. Dynamic Routing of Bandwidth Guaranteed Tunnels with Restoration. *Infocom 2000*, pages 902–911, 2000.
21. R. Motwani, S. Phillips and E. Torng. *Non-clairvoyant scheduling*. *Theoretical Computer Science*, 130:17–47, 1994.
22. J. W. Suurballe. Disjoint paths in a network. *Networks*, 4: 125–145, 1974.
23. D. P. Williamson, M. X. Goemans, M. Mihail, and V. V. Vazirani. A primal-dual approximation algorithm for generalized Steiner network problems. *Combinatorica*, 15(3):435–454, 1995. (Preliminary version in: *25th Annual ACM Symposium on Theory of Computing*, pages 708–717, 1993).
24. D. Zhou and S. Subramaniam. Survivability in Optical Network. *IEEE Network*, 14(6):16–23, November/December 2000.

The Demand Matching Problem

Bruce Shepherd[1] and Adrian Vetta[2]

[1] Bell Laboratories
bshep@research.bell-labs.com
[2] Massachusetts Institute of Technology
avetta@math.mit.edu

Abstract. We examine formulations for the well-known b-matching problem in the presence of integer demands on the edges. A subset M of edges is feasible if for each node v, the total demand of edges in M incident to it is at most b_v. We examine the system of star inequalities for this problem. This system yields an exact linear description for b-matchings in bipartite graphs. For the demand version, we show that the integrality gap for this system is at least $2\frac{1}{2}$ and at most $2\frac{13}{16}$. For general graphs, the gap lies between 3 and $3\frac{5}{16}$. A fully polynomial approximation scheme is also presented for the problem on a tree, thus generalizing a well-known result for the knapsack problem.

1 Introduction

A combinatorial maximum packing problem is determined by a ground set V, each element v of which has an associated *profit*, denoted p_v, and a collection $\mathcal{F}$ of *feasible* subsets of V. The objective is to find a feasible set $F \in \mathcal{F}$ inducing a profit $p(F) \equiv \sum_{v \in F} p_v$ of maximum value. Feasibility for such problems is often determined by some set of resource capacities into which at most some bounded number of ground elements can be packed. For instance, in the b-matching problem on a graph, a subset M of its edges is feasible if each node v is incident to at most b_v edges of M. We are interested in understanding the relationship of such *base problems*, where ground elements in effect have a unit size, to their *demand version* where the elements each come with an integer demand value. For instance, in the case of b-matchings, each edge e could be additionally supplied with a demand d_e. A set M of edges is then feasible if for each node v, the total demand of edges in M incident to v does not exceed the capacity b_v. Naturally, this gives rise to a completely different collection of feasible sets.

A traditional attack in solving such base packing problems is to find a linear description for the convex hull of incidence vectors of feasible sets. This reduces the original problem to a linear program of the form $\max\{\mathbf{p} \cdot \mathbf{x} : A\mathbf{x} \leq \mathbf{b}, \mathbf{x} \geq 0\}$ since any basic solution is then a $0-1$ vector which identifies an optimal feasible set. For example, if G is bipartite then its node-edge incidence matrix is totally unimodular. Hence this matrix immediately yields such a good formulation for the b-matching problem [9]. The demand version of the problem can then be expressed as an optimization problem: $\max\{\mathbf{p} \cdot \mathbf{x} : A\mathbf{x} \leq \mathbf{b}, x_i \in \{0, d_i\}\}$, where

W.J. Cook and A.S. Schulz (Eds.): IPCO 2002, LNCS 2337, pp. 457–474, 2002.

$d_i \in \mathbb{Z}$ is the size, or *demand* of the element i. However, analysis of the linear relaxations of these problems seems to require different methods from those used for the base problem itself. This is largely because the whole demand must be satisfied before its profit may be reaped. Such "all or nothing" constraints arise quite naturally in practice; bandwidth trading in communication networks, for instance, is one such example.

We may equivalently view the demand version of the original packing problem as optimizing over the set $\{\mathbf{x} : A^d \mathbf{x} \leq \mathbf{b}, x_i \in \{0, 1\}\}$, where A^d is obtained from A by multiplying each column i by the value d_i. One may ask whether applying this process to well-behaved matrices A leads to linear relaxations whose integrality gaps are also well bounded in some way. The examples in the present article, for instance, always have at worst a constant integrality gap. This is not always the case. Chekuri [5] has shown instances of multicommodity flow in a path graph which lead to logarithmic integrality gaps. The base problem in this case is finding stable sets in an interval graph, and hence the original constraint matrix is totally unimodular.

Theoretical work on "all or nothing" demand versions of combinatorial packing problems is rather limited. The obvious starting point for problems of this genre is the knapsack problem; this is the case where A consists of a single row. Multiple knapsack problems have been studied from the perspective of providing good approximation algorithms, see [4]. This is the case where all rows of A are equal. As stated, however, our focus is upon such multiple knapsack problems that arise from some base combinatorial packing problem.

A demand version of network flows was introduced by Dinitz et el. [6]. Specifically, they develop techniques for the maximum single-source unsplittable flow problem. There, the base combinatorial problem is that of packing paths into a capacitated network. In the unsplittable flow setting, one is also given a source s and a collection of terminals $t_1, t_2, \ldots, t_k$ with demands $d_1, d_2, \ldots, d_k$, respectively. The packing problem is to satisfy a maximum number of the demands subject to the edge capacity constraints. This problem may be viewed in our present framework as follows. Let each $s - t_i$ path have the demand d_i. If we add a sink node t and edges $t_i t$ of capacity d_i, then the goal is to find a maximum packing of the weighted $s - t$ flow paths.

In this paper, we focus on the "all-or-nothing" profit model as imposed on one of the best understood combinatorial optimization problems: b-matchings [7]. Note that the maximization form of the assignment problem has this form. Namely, the base problem can be viewed as bipartite b-matching where every node on one side of the bipartition has the same b_v-value as a common demand value on each of its incident edges. Indeed, our analysis of basic solutions follows along similar lines to that of [1] for the generalized assignment problem. Results of Shmoys and Tardos [12] on the generalized assignment problem, although focused on congestion minimization, also bear resemblance to certain findings in the present paper. In general, the relationship between the congestion minimization problem and the maximization problem is unclear. For multicommodity flow where the supply graph is a tree, for instance, the congestion problem is trivial.

Finding a maximum routable (obeying capacities) subset of the commodities, however, is not. Even the case of unit demands requires an interesting analysis [8]. We know nothing about the demand version of this problem. Recently, however, a constant-approximation has been found in the case that the supply graph is a path [3] (assuming the maximum demand is bounded by the minimum edge capacity).

1.1   The Demand Matching Problem

Take a graph $G = (V, E)$ and let each node $v \in V$ have an integral *capacity*, denoted by b_v. Let each edge $e = uv \in E$ have an integral *demand*, denoted by d_e. In addition, associated with each edge $e \in E$ is a *profit*, denoted by p_e. A *demand matching* is a subset $M \subseteq E$ such that $\sum_{e \in \delta(v) \cap M} d_e \leq b_v$ for each node v. Here $\delta(v)$ denotes the set of edges of G incident to v. We assume, throughout, that the demand of any edge is less than the capacities of both its end-vertices; otherwise such an edge is not contained in any demand matching and may be discarded. The demand matching problem is to find a demand matching which maximizes $\sum_{e \in M} p_e$. We also associate with an edge e a value, $\pi_e = p_e/d_e$, which we call the *marginal profit* of that edge. Marginal profits play an important role in the understanding of demand matchings. We use them now in a formulation of the problem.

$$\max \sum_{e \in E} \pi_e x_e$$

$$\text{s. t.} \qquad \sum_{e \in \delta(v)} x_e \quad \leq \quad b_v \qquad \forall v \in V, \qquad (1)$$

$$x_e \quad \in \quad \{0, d_e\} \qquad \forall e \in E .$$

There are several special cases of the demand matching problem that are interesting in their own right. We begin by considering specific demand and profit functions.

(i) Unit Demand Function: the demand of each edge is one. Observe that this problem is just the familiar b-matching problem. Hence, the unit demand version can be solved in polynomial time.

(ii) Maximum Cardinality Demand Matching Problem: the profit associated with each edge is one, and hence the objective is to find a feasible demand matching containing as many edges as possible.

(iii) Constant Marginal Profit Function: the profit associated with each edge is proportional to its demand. As a result the marginal profit of each edge is the same.

Particular underlying graphs also give rise to interesting subproblems. For example, suppose that the underlying graph is a *star*, i.e., a tree in which every node is a leaf except for the root node r. In this case, the demand matching problem is equivalent to the *knapsack problem*, where the knapsack capacity is b_r and there is an item of weight d_e for each edge e. In addition, if we have a unit marginal profit function, then the demand matching problem on a star includes the familiar *bin-packing problem*. As a consequence, we note that the demand matching problem is NP-hard even where G is a star.

1.2   An Overview of the Paper

The paper is organized as follows. In Sect. 2 we show that the demand matching problem, even in the cardinality case, is MAXSNP complete. In Sect. 3 the natural linear programming relaxation is presented and studied. Here it is seen that the integrality gap for the formulation is at least $2\frac{1}{2}$ for bipartite graphs (whereas, it is 1 for the unit-demand case) and at least 3 for general graphs. We are discuss an extension of Berge's Augmenting Path Theorem which gives an optimality certificate for fractional demand matchings. Most of the remainder of the paper seeks upper bounds on this gap by way of approximation algorithms that turn fractional solutions into integral solutions. The basic scheme uses augmenting paths and is introduced in Sect. 4. It yields a 3-approximation for bipartite graphs and a $3\frac{1}{2}$-approximation for general graphs. In Sect. 4.3 it is seen that understanding the integrality gap is related to determining the fractional chromatic number of bipartite graphs where some of the edges have been subdivided. A randomized algorithm is then devised for this problem which improves the general bound for bipartite graphs to $2\frac{13}{16}$ (and $3\frac{5}{16}$ for general graphs). In Sect. 5 algorithms for the cardinality problem are presented with approximation guarantees that coincide with the lower bounds on the gap of the linear program. Finally, in Sect. 6 we present a fully polynomial time approximation scheme for the problem in which the underlying graph is a tree; this generalizes the well known result for the knapsack problem.

2   Hardness Results

We have already seen that the demand matching problem is hard for instances with unit marginal profit functions. However, we have also seen that instances with a unit demand function are polynomially solvable. In this section we examine the hardness of the demand matching problem in more detail. In particular, we show that the maximum cardinality demand matching problem (and hence the general demand matching problem) is MAXSNP-hard. Thus, there exists a constant $\epsilon > 0$ such the problem admits no $(1 - \epsilon)$-approximation algorithm unless $P = NP$. We say that $\frac{1}{1-\epsilon}$ is the *inapproximability constant* for the problem.

Theorem 1. *The cardinality demand matching problem is MAXSNP-complete (even if the demands are restricted to be 1 or 3).*

Proof. We first give a reduction from the the stable set problem to the cardinality demand matching problem. Let (G, k) be an instance of the stable set problem. We construct an instance G' of the (decision) cardinality demand matching problem such that G has a stable set of size k if and only if G' has a demand

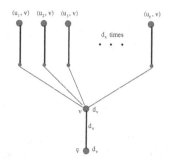

Fig. 1. A Gadget

matching of size $2m + k$, where m is the number of edges in G. For each node v in G we have two nodes v and $\bar{v}$ connected by an edge in G'. In addition, for each edge uv in G there is a node (u, v) in G'. Between nodes v and (u, v) there is a path consisting of two edges. Similarly there is a a path consisting of two edges between the nodes u and (u, v). Thus, for each node v in G we have an associated gadget in G'; such a gadget is shown in Fig. 1. Here v is adjacent to $u_1, u_2, \ldots, u_r$ in G, where r equals the degree, d_v, of v in G. We say that node v is *selected* by a demand matching if the matching contains all the $d_v + 1$ "outer" edges from v's gadget (these outer edges are shown in bold in Fig. 1. If the demand matching chooses the d_v "inner" edges from the gadget, then v is said to be *deselected*. If each node is either selected or deselected, then the size of the demand matching is $\sum_{v \notin S} d(v) + \sum_{v \in S} (d(v) + 1) = \sum_v d(v) + |S|$, where S is the set of selected nodes.

To enforce these selection rules, we set the capacity of each node in a gadget to be 1 with the exception of v and $\bar{v}$ which have capacity d_v. In addition, every edge has demand 1, except for the edges $v\bar{v}$ which have demand d_v. This ensures that if $v\bar{v}$ is chosen, then none of the inner edges may be, and that only one of the two edges on the path from the node v to the node (u, v) may be chosen. One sees that any demand matching which contains $d(v) + 1$ edges from a node v's gadget must use the edge $v\bar{v}$. Moreover, take a demand matching of size $2m + k$, for some k. Such a matching can easily be transformed into a demand matching, of the same or greater magnitude, in which each node is either selected or deselected. By our choice of capacities, we then have that the selected nodes form a stable set in G. Conversely, suppose S is a stable set of size k in G. Then the union of the outer edges from gadgets corresponding to nodes in S with the inner edges from gadgets corresponding to nodes not in S gives a matching of cardinality $2m + k$. Thus, we have the desired reduction.

We now complete the proof of max-SNP completeness. Consider the maximum stable set problem in graphs with bounded maximum degree. Inapproximability results are given in [2] for such graphs; for graphs of maximum degree 5 the inapproximability constant is at least 1.003. So, suppose we have an $(1 - \epsilon)$ approximation algorithm for the cardinality demand matching problem. Now take a graph G of maximum degree 5 with a maximum stable set of

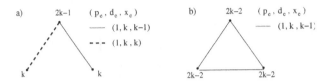

Fig. 2. Lower bound examples for trees and non-bipartite graphs

size k. Observe that $k \geq \frac{n}{6} \geq \frac{m}{15}$. In addition, our reduction produces a graph G' with maximum demand matching of cardinality OPT $= 2m + k$. Applying our approximation algorithm we obtain a demand matching of cardinality at least $(1 - \epsilon)(2m + k)$. This corresponds to a stable set in G of size at least $(1 - \epsilon)(2m + k) - 2m = (1 - \epsilon)k - \epsilon \cdot 2m \geq (1 - \epsilon)k - \epsilon \cdot 30k = k(1 - 31\epsilon)$. So $1 - 31\epsilon < 0.997$ and hence the inapproximability constant for the demand matching problem is at least 1.000097.   □

We remark that we do not know the problem complexity if the demands are restricted to be either 1 or 2.

3   A Linear Programming Relaxation

We now consider a linear program relaxation of the formulation (1), i.e. for each edge we now have the linear constraints $0 \leq x_e \leq d_e$. We call the solution space of the resultant linear program the *fractional demand matching polytope*. We say the a point **x** in the polytope is a *fractional demand matching*. In this section we investigate the structure of this polytope and its extreme points; we apply the results later when we return to the integral problem.

3.1   Lower Bounds on the Integrality Gap

We first describe some lower bounds on the integrality gap of the linear programming relaxation. For trees the integrality gap is exactly two. This is well-known as there is a class of knapsack problems for which the fractional optimum is twice the integral optimum. An example is shown in Fig. 2a), where the vertices are labeled by their capacities. For non-bipartite graphs, we now show that the integrality gap of the linear program is at least three. Consider the simple example shown in Fig. 2b). Note that no node can satisfy both of its incident edge demands. Thus, the optimal integral solution contains exactly one edge from the triangle. Since each edge has value $p_e = 1$ the optimal integral solution has value 1. However, consider setting each edge weight to be $x_e = k - 1$. This gives a feasible fractional demand matching with profit $3 - \frac{3}{k}$. Thus the integrality gap is at least three. Finally, for bipartite graphs, Fig. 3 shows that the integrality gap is at least $2\frac{1}{2}$. The optimal fractional solution, shown in Fig. 3a), gives a profit of value $20 - \frac{12(d+1)}{k}$, for some constants d and k. The optimal integral solutions, shown in shown in Fig. 3i) and 3ii), give a profit of value 8. We obtain the claimed gap as k becomes large relative to d.

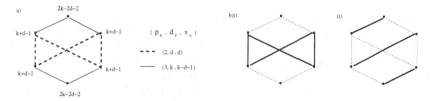

Fig. 3. A lower bound example for bipartite graphs

3.2 Basic Solutions of the Fractional Demand Matching Polytope

Given a fractional demand matching $\mathbf{x}$ we say that a node v is *tight* if

$$\sum_{e \in \delta(v)} d_e x_e = b_v \; ;$$

otherwise the node is termed *slack*. We say that an edge e is *tight* if $x_e = d_e$; we say that e is *fractional* if $0 < x_e < d_e$. Let $F(\mathbf{x}) \subseteq E$ be the set of fractional edges induced by the fractional demand matching $\mathbf{x}$. Let $G(\mathbf{x})$ denote the graph induced by $F(\mathbf{x})$.

Lemma 1. *Let $\mathbf{x}$ be an extreme point of the demand matching polytope. Then each component of $G(\mathbf{x})$ consists of a tree plus (possibly) one edge. In addition, any cycle in $G(\mathbf{x})$ has odd length.*

Proof. First, suppose that $G(\mathbf{x})$ contains an even cycle $C = \{e_1, e_2, \ldots, e_{2k}\}$. For any ϵ, we define $\mathbf{y}$ by setting $y_i = x_{e_i} + (-1)^i \epsilon$ for each $i = 1, 2 \ldots, 2k$ and $y_e = x_e$ for any other edge e. Similarly we define $\mathbf{z} := \mathbf{x} - (\mathbf{y} - \mathbf{x})$. Observe that, since the edges on the cycle are fractional, there exists a positive ϵ such that both $\mathbf{y}$ and $\mathbf{z}$ are feasible. We then obtain the contradiction that $\mathbf{x} = \frac{1}{2}(\mathbf{y} + \mathbf{z})$. Next suppose that some component contains two odd cycles C_1 and C_2. The cycles do not share an edge; otherwise they induce an even cycle. Thus C_1 and C_2 are edge disjoint and connected by a (possibly empty) path P. Observe that the edge set $C_1 \cup P \cup C_2 \cup P$ is an even length circuit. We then obtain a contradiction by similar reasoning to that given above except that increments along P are by $\pm 2\epsilon$. The result follows.   □

3.3 Berge Conditions for Fractional Demand Matchings

Berge's classic result on matchings asserts that a matching is of maximum cardinality if and only if it has no augmenting path. This result does not have a simple generalization for demand matchings. We can, however, extend this result to give optimality conditions for the fractional demand matching problem. We begin with some definitions. Suppose we partition some subset of the edges into two sets $\mathcal{P} \cup \mathcal{N}$. We call the resultant edge sets the *positive* edge set and the *negative* edge set, respectively. The marginal value $\pi(F)$ of a set of edges $F \subseteq \mathcal{P} \cup \mathcal{N}$ is just the sum of the marginal values of its positive edges minus the sum of the marginal values of its negative edges i.e., $\pi(C) = \sum_{e \in F \cap \mathcal{P}} \pi_e - \sum_{e \in F \cap \mathcal{N}} \pi_e$. A *balloon* is formed by an odd length cycle attached to a (possibly empty) path known as the *string*. A *dumbbell* is formed by two edge-disjoint odd length cycles connected by a (possibly empty) path known as the *rod*. A *structure* is either a path, an even length cycle, a balloon or a dumbbell.

A path P is called *augmenting* (with respect to $\mathbf{x}$) if there is a partition $\mathcal{P} \cup \mathcal{N}$ of E such that (i) Edges on the path are alternately positive and negative (or vice versa), (ii) $\pi(P) > 0$, (iii) If an endpoint v of the path is tight, then its incident edge is negative, (iv) None of the positive edges in P is tight. The motivation behind this definition is as follows. The first two conditions state that it is beneficial to *augment* around the structure i.e., add ϵ weight to the positive edges, remove ϵ weight from the negative edges. Such an augmentation maintains feasibility with respect to the node constraints at internal nodes in the path. The third condition ensures that, for small ϵ, the node constraints remain satisfied at the end nodes as well. Finally the fourth constraint ensures that, for small ϵ, the edge constraints also remain satisfied. Thus if we find an augmenting path we may improve our current demand matching.

Other structures may also be augmenting. An even length cycle is just a closed path of even length, and a dumbbell is an even length circuit in which the edges in the rod are traversed twice. Thus, augmenting cycles (and dumbbells) can be defined analogously to augmenting paths (and balloons). Here, however, we may omit condition (iii). This is because in an even circuit all the nodes may be considered internal to the path. Observe that a balloon is just a closed odd length path in which edges on the string are traversed twice. Thus, augmenting balloons can be defined analogously to augmenting paths. Note that as the circuit is of odd length, some node, say v, is incident only to positive edges (or only to negative edges). We may think of v as the "endpoint" of the balloon and, thus, in defining augmenting balloons we do require condition (iii). It is also easy to see that it is beneficial to augment along any of these other augmenting structures. We omit a proof of the following result due to space constraints.

Theorem 2. *A fractional demand matching is optimal if and only if it induces no augmenting structure.*   □

4   Linear Programming Based Algorithms

4.1   Transforming a Fractional Demand Matching

We now present a procedure which takes a fractional demand matching and transforms it so as to reveal a two coloring of the edges which can be used later. The algorithm for trees lies at the heart of the main algorithms so we tackle this specific case first. The procedure is then used to give an approximation algorithm for the demand matching problem with a factor $3\frac{1}{2}$ approximation guarantee. This guarantee can be improved in some special cases. If the underlying graph is bipartite we obtain a factor 3 guarantee, and if the graph is a tree we obtain a factor 2 guarantee.

Consider solving the linear program on a tree. We show how to obtain an integral demand matching whose profit is at least half of the optimal fractional demand matching $\mathbf{x}$. To do this we show that T contains two disjoint integral demand matchings whose combined profit is at least that of $\mathbf{x}$. We may assume that any edge e with $x_e = 0$ has been discarded. Next, recall that the set of fractional edges, $F(\mathbf{x})$, induce a forest in T. Call the trees in this forest $T_1, \ldots, T_l$. Let $\mathbf{f}$ the vector obtained by setting all components not in F equal to 0; also let $\mathbf{h}$ be associated with the tight components and so $\mathbf{x} = \mathbf{f} + \mathbf{h}$. Notice that on its own, $\mathbf{f}$ is a feasible, but non-optimal, fractional demand matching. In particular, between any two leaves in a tree T_i there is an augmenting path P with respect to the demand matching $\mathbf{f}$. We may augment along such a path in either direction, and one of these results in an improved demand matching $\mathbf{f}'$. Moreover, this can be carried out so that either some edge becomes tight (in which case be add the edge to $\mathbf{h}$ to obtain a new set of tight edges $\mathbf{h}'$) or some edge becomes zero. Observe that $\mathbf{f}' + \mathbf{h}'$ is no longer a feasible fractional demand matching. In particular, the node constraints at the leaf nodes in T_i may now be violated. This, however, will not cause a problem as we are trying to obtain two disjoint demand matchings from this union. Note that node constraints at internal nodes in the path continue to hold.

We repeatedly augment along fractional paths between leaves in the forest induced by the remaining fractional edges (i.e. induced by $\mathbf{f}'$) until the set of fractional edges induce a matching in T. Note that the process terminates since at each step we either discard an edge or make an edge tight, thus reducing the number of fractional edges. We call this process of modification of our linear programming solution the *augmenting paths procedure*. Since the value of our (non-feasible) solution improved at each step, the solution at the end of the augmenting paths procedure has profit at least that of the optimal fractional demand matching $\mathbf{x}$. Now, divide the support for the final solution into fractional edges F^* (that form a matching) and tight edges H^*. Next, we partition the edges into two types. We say that an edge e is *bronze* with respect to v if $e \in \delta(v)$ and either i) e is in F^* or ii) v is not incident to an edge in F^*, and e was the final edge incident to v to become tight. Otherwise we say that e is a *copper* edge with respect to v. We call an edge *bronze* if it is bronze with respect to either of its endpoints, and call it *copper* otherwise.

Lemma 2. *The set of copper edges with respect to a node v are collectively feasible with regards to the associated node constraint.*

Proof. If there is no bronze edge with respect to v then all of the incident edges were tight in the initial linear program solution. Hence they are all collectively feasible at v. So let b be a bronze edge with respect to v. Let the copper edges with respect to v, ordered according to the order in which they became tight, be $c_1, c_2, \ldots, c_k$. Consider the edge c_i, and suppose it became tight whilst augmenting the path P. Note that the edge b must have been fractional at this time. Thus v was not a leaf node in some fractional tree. Therefore, v was an internal node on the path P and the associated node constraint remained satisfied after the augmentation. Thus, $c_1, c_2, \ldots, c_i$ were collectively feasible with respect to the node constraint at v. $\square$

The following theorem now shows that we have a factor 2 approximation algorithm for the case in which the underlying graph is a tree. (This is substantially improved in Sect. 6.)

Theorem 3. *A tree contains two disjoint demand matchings M_1 and M_2 whose combined profit is at least that of the optimal fractional demand matching.*

Proof. Apply the above procedure to an optimal fractional demand matching. The edges in the tree can now be partitioned into two feasible demand matchings. This is achieved by 2-coloring the edges of the tree so that the bronze edge with respect to a node v receives a different color from any of the copper edges with respect to that node. This can be done greedily, starting from an arbitrary root node in the tree, and then working out towards the leaves. Note that by Lemma 2 it follows that the demand matchings induced by each of the two color classes are feasible. The combined profit of these matchings is at least that of the optimal fractional demand matching and, therefore, at least that of the optimal demand matching. $\square$

Theorem 4. *There is a factor 3-approximation algorithm for the demand matching problem on bipartite graphs.*

Proof. Solve the linear programming relaxation. If the tight edges in the linear program provide at least one third of the profit of the optimal fractional solution (i.e., the value of the linear program) then we are done. Otherwise the fractional edges provide two-thirds of the total profit. Since the graph is bipartite, by Lemma 1, the fractional edges form a tree. Thus we can find an integral demand matching in the tree with at least half the value of the optimal fractional solution on that tree. This integral demand matching has, therefore, value at least one third that of the LP. $\square$

Theorem 5. *There is a factor $3\frac{1}{2}$-approximation algorithm for the demand matching problem on non-bipartite graphs.*

Proof. We again begin by partitioning the edge set induced by $\mathbf{x}$ into fractional edges and tight edges. We would like fractional edges to induce two demand matchings as before. However we may not be able to do this since the fractional edges may, by Lemma 1, induce components that contain a single odd cycle. Instead we obtain a total of four demand matchings by the following method. Take any odd cycle C induced by the fractional edges. Let e be the edge in C for which $d_e - x_e$ is minimized. Let e be incident to e_1 and e_2 on this cycle. Observe that if $d_e - x_e \leq \min(x_{e_1}, x_{e_2})$, then the set of tight edges still form a feasible demand matching even with the addition of e. Thus, we add e to the set of tight edges. Otherwise, suppose $d_e - x_e > x_{e_1}$, say. Then, since $d_e - x_e \leq d_{e_1} - x_{e_1}$, we have that $x_{e_1} \leq \frac{1}{2}d_{e_1}$. In this case, we place e_1 in a special set S.

By this method we remove one edge from each odd cycle, either by adding an edge to the set S or to the set of tight edges. Note that since the odd cycles are vertex disjoint, the set S induces a matching and hence induces a demand matching. The set of tight edges (plus any edges added during this process) also form a demand matching. Finally the remaining set of fractional edges induce a forest, from which we may obtain two demand matchings. Thus, we have partitioned the support of $\mathbf{x}$ into four demand matchings. Moreover, the edges in S have twice the profit that they contributed to the profit of $\mathbf{x}$. It follows that one of these demand matchings induces a profit of at least $\frac{2}{7}$ that of the linear programming solution. $\square$

4.2   Extending the Edge-Coloring

It will be useful in the following sections if we extend our edge coloring of the tree edges to include all the edges in the linear program solution (except for those edges that get discarded during the augmenting paths procedure). In particular, we will generate a 2-coloring of the edges for bipartite graphs, and a 3-coloring for non-bipartite graphs, Take the optimal fractional demand matching $\mathbf{x}$. So $\mathbf{x}$ induces a set of tight edges and a fractional subgraph. Each fractional component is a tree together (possibly) with an extra edge that induces an odd cycle. All of the tight edges induced by $\mathbf{x}$ are also termed copper with respect to both of their endpoints. The fractional edges are colored bronze, copper and red by the following augmenting paths procedure. We again augment along leaf-to-leaf fractional paths, with the additional possibility that augments may be from a leaf to itself, via the odd cycle. Note that non-simple augmentations use increments of size $\pm 2\epsilon$ along the stem of the augmenting path, whereas along the odd cycle the increments are $\pm\epsilon$. As before, we choose the direction of the swaps so as to improve the profit of the fractional demand matching. In addition, we do not alter feasibility at a node unless it was the leaf of the augment.

Note that at some point in the process a fractional component may contain no leaves i.e. it is an odd cycle. If some edge e on the cycle has the property that $x_e \leq \frac{1}{2} d_e$ then we color e *red* and remove it from the component. We then continue the augmentation procedure and color the rest of the edges copper or bronze. If no edge has this property then it can be shown that there is an edge e with the property that $d_e - x_e \leq \min(x_{e_1}, x_{e_2})$, where e is incident to e_1 and e_2 on the cycle. We color e copper with respect to both its endpoints, remove it from the component and continue the augmentation procedure. The procedure terminates when the set of fractional edges forms a matching together and all the edges are colored. It is easy to check that Lemma 2 still holds, even with respect to the enlarged for the set of copper edges generated by this extended coloring. This observation will be needed in the subsequent sections.

4.3   To Stable Sets and a Randomized Algorithm

In this section we show that the case of bipartite graphs is qualitatively different from the case of non-bipartite graphs with respect to the linear programming relaxation. In particular, we show that the integrality gap of the linear program is strictly less than 3 for bipartite graphs. This we achieve via the use of a randomized algorithm. Before presenting the randomized algorithm, we first show how the edge coloring induced by the linear program allows us to recast our problem as a stable set problem. Assume that we are in the bipartite case; hence, all the edges are colored bronze or copper (for the non-bipartite case, remove the red edges). Our stable set problem will be on a "line graph" G' whose definition is based on the copper and bronze edge interactions. The nodes of G' are the bronze and copper edges. For simplicity, we also call the nodes in G' bronze and copper. Two nodes induce an edge if their corresponding edges can not be included in the same demand matching i.e., if the edges share an endpoint v and one of the edges is bronze with respect to v. Note that, the set of bronze edges also induce a forest in G'. The copper edges induce nodes of degree at most 2 in G', since a copper edge uv can not be placed in a demand matching with the edges that are bronze with respect to u and v. It follows from Lemma 2 that a stable set in G' corresponds to a demand matching in G.

We now present a randomized algorithm for the general demand matching problem in bipartite graphs. The algorithm gives an improved approximation guarantee of 2.8125 for bipartite graphs. We work on the induced stable set problem in the associated graph G'. The randomized algorithm first selects a stable set from amongst the bronze nodes. It then greedily adds any copper node with neither endpoint in the stable set. The algorithm works as follows. Take the forest induced by the bronze nodes. For each tree in the forest pick an arbitrary root node r. We now give the nodes in the tree a $0 - 1$ labeling. First give r the label 1 with probability $\frac{1}{5}$, otherwise label it 0. We consider the other bronze nodes in the tree in increasing order of their distance to the root. Take a node v with parent u. If u has label 1, then give v the label 0. Otherwise, if u has label 0, then give v the label 1 with probability $\frac{1}{4}$ and the label 0 with probability $\frac{3}{4}$. Observe that, at the end of this labeling process, each bronze node has a probability $\frac{1}{5}$ of being labeled 1. Now the label 1 corresponds to throwing away the node. The nodes with label 0 form a collection of trees. Since trees are bipartite we obtain two stable sets from each tree. We chose at random one of these stable set to be in our final stable set. Finally the copper nodes are added into the stable set if none of their endpoints is already chosen. We now analyze the performance of this algorithm.

Theorem 6. *A randomized algorithm provides a factor* $2\frac{13}{16}$-*approximation guarantee for the demand matching problem in bipartite graphs.*

Proof. To show this we calculate the probability that a given node is chosen as part of the stable set (i.e., the probability that an edge is in the demand matching). First, note that each bronze node is not removed with probability $\frac{4}{5}$. Thus, since its bipartition class is chosen with probability $\frac{1}{2}$, it is in the stable set with probability $\frac{2}{5}$. We now consider the copper nodes. We may assume that the copper node is incident to two bronze nodes in the same tree of the bronze forest. The case in which the bronze nodes are in different trees is more beneficial. Note that each copper node induces a unique odd cycle with nodes in the bronze tree. Since our original graph was bipartite, this cycle in G' must contain at least four bronze nodes. The worst case analysis occurs when the cycle contains exactly four bronze nodes. Take a copper node with bronze neighbors u and v. We have two situations to deal with. Firstly, without loss of generality u is an ancestor of v. Secondly u and v share a common ancestor. The first case and all the possible node labelings are shown in Fig. 4. Associated with each labeling is a pair of probabilities (p, q), where p is the probability that the labeling occurs and q is the probability that (given this labeling) the copper node c can be placed in the resultant stable set. It follows that the overall probability that the copper node c is placed in the stable set is $\frac{11}{40}$.

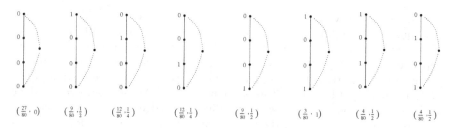

Fig. 4.

Similarly the second case is shown in Fig. 5. This also gives the same overall probability of $\frac{11}{40}$. This implies that the randomized algorithm only gives a performance guarantee of $3\frac{7}{11}$. However a simple modification allows for a better

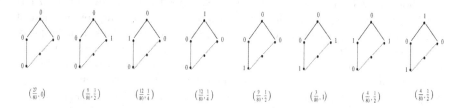

Fig. 5.

guarantee. Observe that the set of copper nodes forms a stable set. Therefore, run the randomized algorithm with probability $\frac{8}{9}$ to obtain a stable set and, with probability $\frac{1}{9}$, take the set of copper nodes to be our stable set. It now follows that each node is in the stable set with probability at least $\frac{16}{45}$. Thus we obtain an approximation guarantee of $2\frac{13}{16} = 2.8125$.   □

We remark that slightly better approximation guarantees do arise when more complex randomization schemes are applied. The randomized method may also be applied to non-bipartite graphs.

Theorem 7. *A randomized algorithm provides a factor $3\frac{5}{16}$-approximation guarantee for the demand matching problem in non-bipartite graphs.*

Proof. Recall that, after the generalized augmentation procedure is completed, the red edges have the property that $x_e \leq \frac{1}{2}d_e$. Let the value of the fractional solution after augmentation be v^*. If the demand matching induced by the red edges has value less than $\frac{16}{53}v^*$, then the value of the copper and bronze edges is at least $\frac{45}{53}v^*$. Applying the randomized algorithm to the stable set problem induced by the copper and bronze edges then produces a solution whose expected value is $\frac{16}{53}v^*$.   □

5   The Cardinality Problem

Recall that in the case of unit-profits per edge our goal is to find a maximum cardinality demand matching. For the maximum cardinality problem we show that approximation guarantees of $2\frac{1}{2}$ and 3 are obtainable for bipartite and non-bipartite graphs, respectively. We begin with the bipartite case. From the results of the previous section, it is sufficient to find, in polynomial time, a stable set in G' that contains at least two-fifths of the nodes of G'. We show this can be done in a greedy fashion. In what follows we divide the edges of G' into two classes: *tree* edges are those edges in the forest formed by the bronze nodes; the edges incident to copper nodes are termed *non-tree* edges.

Theorem 8. *In bipartite graphs, a greedy algorithm finds a demand matching whose profit is at least $\frac{2}{5}$ that of the optimal fractional demand matching.*

Proof. We build up a stable set iteratively by showing that we we can find a stable set S in G' with the following properties. Firstly, $|S| \leq 3$. Secondly, the neighbor set $\Gamma(S)$ of S has size at most $\frac{3}{2}|S|$. Thirdly, the removal of $S \cup \Gamma(S)$ from G' induces a graph of the same structure as G' i.e., the composition of a forest of bronze nodes with copper nodes of degree two. This final property allows us to repeatedly find such subsets. The theorem then follows. Naturally, we may assume that there are no isolated nodes at any stage, as we could add them directly to our stable set. One of the following five cases arises.

i) Suppose there is a bronze node v incident to two (or more) copper nodes x and y. Then select x and y to be in the stable set. Recurse on the graph obtained by removing x and y and their neighbor(s). Note that the induced graph has 5 fewer nodes. This is shown in Fig. 6i).

Suppose there is a leaf node v, with neighbor u, in the bronze tree. Then our other four cases are

ii) Neither v nor u is not incident to a copper node. Select v to be in the stable set. Recurse on the graph $G' - \{u, v\}$. See Fig. 6ii).

iii) v is not incident to a copper node but u is. Call this copper node y. Then select v and y to be in the stable set. Recurse on the graph $G' - \{u, v, y, z\}$, where z is the other neighbor of y. See Fig. 6iii).

iv) Both v and u are incident to copper nodes. Call these copper nodes x and y, respectively. Select v and w to be in the stable set. Recurse on the graph $G' - \{u, v, x, y, z\}$, where z is the other neighbor of y. Thus we select two nodes from five as required. See Fig. 6iv).

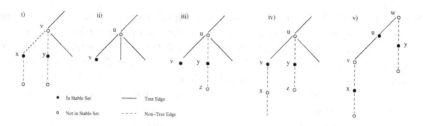

Fig. 6.

v) v is incident to a copper node x but u is not incident to a copper node. If none of i) to iv) occur for any leaf node, then we may assume that u has degree two in the tree (with neighbor w). This follows from the observation that the number of leaves in a tree is greater than the number of nodes with degree at least three. Select u and x (and the copper neighbor y of w, if it has one) to be in the stable set and recurse on the graph obtained by removing u and x (and y) as well as their neighbors. See Fig. 6v). $\qquad\square$

Applying the techniques presented earlier we obtain the following result for non-bipartite graphs.

Theorem 9. *A greedy algorithm provides a factor 3-approximation guarantee for the cardinality problem in non-bipartite graphs.* $\qquad\square$

6 A FPTAS for the Demand Matching Problem on a Tree

Recall that the knapsack problem can be formulated as a demand matching problem on a star. In addition, there exists a fully polynomial time approximation scheme (FPTAS) for the knapsack problem. In this section we describe an FPTAS for the demand matching problem when the underlying graph is a simple tree. As noted in [4], such an FPTAS is not possible for a 2-node instance with multiple edges between the nodes. Our algorithm, based on a dynamic programming approach, turns out to be an exact algorithm in the unit-profits case.

Theorem 10. *Dynamic programming gives a FPTAS for the the demand matching problem on a tree.* $\qquad\square$

Proof. Let $\mathcal{I} = (T = (V, E), d, b, p)$ be such an instance. First, choose an arbitrary root node $v_n \in V$ and create an arborescence T by orienting all edges away from v_n. Let $v_1, v_2, \ldots, v_n$ be an acyclic ordering of the nodes and, for each i, let T_i denote the subtree of T rooted at node v_i. We let h_i denote the *height* of the tree T_i.

For each $i < n$, let e_i denote the unique arc of T in $\delta^-(v_i)$. Also, from now on let $b_i := b(v_i)$. For each i, let $\mathcal{I}_i$ denote the instance obtained by restricting to T_i.

Similarly, let $\mathcal{I}_i^-$ denote the instance obtained by restricting to T_i and reducing the capacity b_i by d_{e_i}. Let α_i and β_i denote the optima for the instances $\mathcal{I}_i$ and $\mathcal{I}_i^-$, respectively. The idea is to build up a solution for the whole problem by making choices between the optimal solutions to $\mathcal{I}_i$ and $\mathcal{I}_i^-$ at each node. For motivation, notice that if M is a feasible demand matching for $\mathcal{I}_i^-$, then $M \cup e_i$ is feasible for $\mathcal{I}$. Thus, the choice between the solutions β_i and α_i corresponds to the choice of whether or not we include the edge e_i.

Consider a non-sink node v_i with out-neighbors v_{i_j}, $j = 1, 2, \ldots, r$. Let

$$S_i = \{i_j : \alpha_{i_j} < \beta_{i_j} + p_{e_{i_j}}\} \ .$$

We now define a "local" knapsack problem at v_i. The items consist of the indices $i_j \in S_i$ and each has demand $d_{e_{i_j}}$ and profit value $\beta_{i_j} + p_{e_{i_j}} - \alpha_{i_j}$. The knapsack itself has a variable capacity t. Let $\text{OPT}(t)$ denote the optimum value for this knapsack problem, and let $\Gamma(t)$ denote the set of indices chosen in some optimal solution. We claim that

$$\alpha_i = \text{OPT}(b_i) + \sum_{j=1}^{r} \alpha_{i_j} \tag{2}$$

$$\beta_i = \text{OPT}(b_i - d_{e_i}) + \sum_{j=1}^{r} \alpha_{i_j} \ .$$

Suppose that for some $j \notin S_i$, e_{i_j} is in a feasible demand matching M for $\mathcal{I}_i$. Then, since $\alpha_{i_j} \geq \beta_{i_j} + p_{e_{i_j}}$, we may discard the edge e_{i_j} and obtain a feasible demand matching of greater or equal value. It follows that $\alpha_i \leq \text{OPT}(b_i) + \sum_j \alpha_{i_j}$. A similar argument, with respect to $\mathcal{I}_i^-$, shows that $\beta_i \leq \text{OPT}(b_i - d_{e_i}) + \sum_j \alpha_{i_j}$. It is clear that the reverse inequalities also hold.

We now describe an algorithm. Take any non-sink node in T and some $\epsilon > 0$. Let $\Gamma'(t)$ denote a feasible solution of value $\text{OPT}'(t)$ to the "local" knapsack problem with $\text{OPT}(t) \leq (1 + \epsilon) \text{OPT}'(t)$. Such a subset can be constructed by using a FPTAS for the knapsack problem [10]. For each v_i we recursively define two subsets

$$\mathcal{A}_i = \bigcup_{j \in \Gamma'(b_i)} (\mathcal{B}_j \cup e_j) \cup \bigcup_{j \notin \Gamma'(b_i)} \mathcal{A}_j$$

$$\mathcal{B}_i = \bigcup_{j \in \Gamma'(b_i - d_{e_j})} (\mathcal{B}_j \cup e_j) \cup \bigcup_{j \notin \Gamma'(b_i - d_{e_j})} \mathcal{A}_j$$

In particular, if v_i is a leaf, then $\mathcal{A}_i = \mathcal{B}_i = \emptyset$.

Note that if we are building up only approximate estimates α_i' and β_i' to the true values α_i and β_i, then the sets S_i may become distorted as we move up the tree. In particular, we are concerned when we "lose" indices j, i.e., those j which become invisible because $\alpha_j < \beta_j + p_{e_j}$ but $\alpha_j' \geq \beta_j' + p_{e_j}$. We call such an index *lost*, and denote by L_i the set of lost indices for the local problem induced by v_i. We can no longer potentially gain from such lost indices when we solve the local knapsack problems. We show, nevertheless, that this loss cannot be too great. Indeed we show by induction on h_i, that

(I) $\alpha_i \leq (1+\epsilon)^{h_i} \alpha_i'$

(II) $\beta_i \leq (1+\epsilon)^{h_i} \beta_i'$

(III) $\text{OPT}(t) \leq (1+\epsilon)\text{OPT}'(t) + \sum_{j \in L_i}(1+\epsilon)^{h_i}(\alpha_j' - \alpha_j)$ for $t = b_i, b_i - d_{e_i}$.

If $h_i = 0$ the result is trivial. Suppose that claims (I), (II) and (III) hold for all $i < k$. We first prove (III). Since we apply a PTAS to obtain the set $\Gamma'(t)$, we certainly have that

$$\text{OPT}(t) \leq (1+\epsilon)\,\text{OPT}'(t) + \sum_{j \in L_k} \beta_j + p_{e_j} - \alpha_j \ .$$

However, for any lost item $j \in L_k$, we have by induction that

$$
\begin{aligned}
\beta_j + p_{e_j} - \alpha_j &= (\beta_j' + p_{e_j} - \alpha_j') + (\beta_j - \beta_j' + \alpha_j' - \alpha_j) \\
&\leq (\beta_j - \beta_j' + \alpha_j' - \alpha_j) \\
&\leq ((1+\epsilon)^{h_k - 1} - 1)\beta_j' + \alpha_j' - \alpha_j \\
&\leq ((1+\epsilon)^{h_k - 1} - 1)\alpha_j' + \alpha_j' - \alpha_j \\
&\leq (1+\epsilon)^{h_k - 1} \alpha_j' - \alpha_j \ .
\end{aligned}
$$

Thus (III) holds for $i = k$. Note next that by (III) and (2),

$$
\begin{aligned}
\alpha_i &\leq (1+\epsilon)\,\text{OPT}'(b_k) + \sum_{j \in L_k}(1+\epsilon)^{h_k - 1}\alpha_j' - \alpha_j + \sum_j \alpha_j \\
&= (1+\epsilon)\,\text{OPT}'(b_k) + \sum_{j \in L_k}(1+\epsilon)^{h_k - 1}\alpha_j' + \sum_{j \notin L_k} \alpha_j \\
&\leq (1+\epsilon)^{h_k - 1}\left(\text{OPT}'(b_k) + \sum_j \alpha_j'\right) \\
&= (1+\epsilon)^{h_k - 1}\alpha_i' \ .
\end{aligned}
$$

Thus we obtain (I). The argument for (II) is similar. Thus, $\mathcal{A}_n$ is a feasible demand matching with profit at least $\frac{1}{(1+\epsilon)^n}$ times that of the optimum demand matching for $\mathcal{I}$.

Completing the picture, we then generate FPTAS as follows. Now apply the dynamic programming procedure outlined above with $\epsilon = \frac{\log(1+\epsilon^*)}{n}$. We obtain a solution of value at least $\frac{1}{(1+\epsilon)^n}\text{OPT} \geq \frac{1}{1+\epsilon^*}\text{OPT}$. The total amount of work required is evidently bounded by a polynomial in n times the maximum time spent on one of the individual knapsack problems arising at a node. These, in turn, take time polynomial in the input size and $\frac{1}{\epsilon}$ since we invoke an FPTAS for each knapsack. For small ϵ^*, we have $\epsilon \geq \frac{\log(1+\epsilon^*)}{n} \geq \frac{\epsilon^*}{2n}$ and hence the overall running time of the algorithm is polynomial in n and $\frac{1}{\epsilon^*}$. $\square$

Corollary 1. *There is an exact algorithm for the unit-profits demand matching problem on a tree.* $\square$

Proof. Consider the node v_i with out-neighbors u_j, $j = 1, \ldots, r$. Note that in the unit-profits case, we may determine $\text{OPT}(t)$ exactly. Simply consider the edges $e_1, \ldots, e_j$ by increasing magnitude of demand, and take the largest feasible subset of the form $\{e_j\}_{j=1}^r$. Then dynamic programming gives an optimal solution for the whole instance. $\square$

Acknowledgments

This work was performed while the second author was visiting Bell Labs in the Summer 2001. The work grew, in a convoluted way, out of discussions in 1998 with Vincenzo Liberatore on maximum multicommodity flow in a path. We are thankful to Chandra Chekuri for very helpful comments including pointing out connections to [12]. The authors also thank Anupam Gupta and Santosh Vempala for their constructive remarks. The authors are especially grateful to Joseph Cheriyan for generously sharing his time and insights on the contents of the paper.

References

1. R. Ahuja, T. Magnanti and J. Orlin, *Network Flows: Theory, Algorithms and Applications*, Prentice Hall, 1993.
2. P. Berman and M. Karpinksi, "On some tighter inapproximability results", *Proc. 26th Int. Coll. on Automota, Languages, and Programming*, pp200-209, 1999.
3. A. Chakrabarti, C. Chekuri, A. Gupta and A. Kumar, "Approximation algorithms for the unsplittable flow problem", manuscript, September 2001.
4. C. Chekuri and S. Khanna, "A PTAS for the Multiple Knapsack Problem", *Proc. 11th SODA*, 2000.
5. C. Chekuri, Personal Communication, November 2000.
6. Y. Dinitz, N. Garg and M. Goemans, "On the Single-Source Unsplittable Flow Problem", *Combinatorica*, **19**, pp17-41, 1999.
7. J. Edmonds, "Maximum matching and a polyhedron with 0, 1-vertices", *Journal of Research of the Natuional Bureau of Standards (B)*, **69**, pp125–30, 1965.
8. N. Garg, V.V. Vazirani and M. Yannakakis, "Primal-Dual Approximation Algorithms for Integral Flow and Multicut in Trees with Applications to Matching and Set Cover", *Algorithmica* **18**, pp3–20, 1997.
9. A.J. Hoffman and J.B. Kruskal, "Integral boundary points of convex polyhedra", in H.W. Kuhn and A.W. Tucker eds., *Linear Inequalities and Related Systems*, Princeton University Press, pp223–246, 1956.
10. O.H. Ibarra and C.E. Kim, "Fast approximation for the knapsack and sum of subset problems", *Journal of the ACM*, **22**, pp463-468, 1975.
11. C.H. Papadimitriou, *Computational Complexity*, Addison Wesley, 1994.
12. D. Shmoys and É. Tardos, "An approximation algorithm for the generalized assignment problem", *Mathematical Programming A*, **62**, pp461–474, 1993.
13. L. Trevisan, "Non-approximability Results for Optimization Problems on Bounded Degree Instances", *Proc. 33rd STOC*, 2001.

The Single-Sink Buy-at-Bulk LP Has Constant Integrality Gap

Kunal Talwar*

University of California, Berkeley CA 94720, USA
kunal@cs.berkeley.edu

Abstract. The buy-at-bulk network design problem is to design a minimum cost network to satisfy some flow demands, by installing cables from an available set of cables with different costs per unit length and capacities, where the cable costs obey economies of scale. For the single-sink buy-at-bulk problem, [10] gave the first constant factor approximation. In this paper, we use techniques of [9] to get an improved constant factor approximation. This also shows that the integrality gap of a natural linear programming formulation is a constant, thus answering the open question in [9].

1 Introduction

Consider the problem of designing a network to allow routing of a given set of demands. The solution involves laying cables along some edges so that we can route the required demands in the resulting capacitated network. The cables installed can come from a set of available cables, where each cable type has a fixed cost per unit length and some capacity. Typically, we have economy of scale so that the cost-capacity ratio is smaller for cables with higher capacity. The goal is to find the minimum cost feasible network. This problem has been referred to as the *buy-at-bulk network design* problem. The *single-sink buy-at-bulk* problem is the special case where all flows need to be routed to a single sink. This problem includes as a special case the steiner tree problem and is therefore **NP**-hard.

This and similar network design problems have received attention in [13], [16], [5], [1], [8], [11], etc. Salman et al [17] gave a constant factor approximation for the single-sink single cable type case, and the Euclidean single sink case. Awerbuch and Azar [2] gave an $O(\log^2 n)$-approximation for the general problem, using the tree embedding techniques of Bartal [3]. An improvement to $O(\log n \log \log n)$ follows from the improved embeddings of [4,6]. On the other hand, [14] looked at a problem similar to the single cable type case, and gave a constant factor approximation using the spanner constructions of [7].

For the single sink buy-at-bulk problem, Meyerson et al [15] gave an $O(\log n)$-approximation algorithm. More recently, Garg et al [9] gave an $O(k)$ approximation, where k is the number of cable types, and Guha et al [10] gave the first $O(1)$-approximation.

---

* Research partially supported by NSF via grants CCR-0105533 and CCR-9820897.

W.J. Cook and A.S. Schulz (Eds.): IPCO 2002, LNCS 2337, pp. 475–486, 2002.
© Springer-Verlag Berlin Heidelberg 2002

In this paper, we modify the techniques of [9] and give another constant factor approximation algorithm for the single sink buy-at-bulk problem, improving on the approximation factor of [10] by one order of magnitude. Since we use a linear program rounding approach, this also shows that a natural LP formulation for the problem has a constant integrality gap, thus answering the main open question in [9].

The rest of the paper is structured as follows. In Sect. 2, we give some preliminary definitions. We give the algorithm in Sect. 3 and the analysis in Sect. 4. We conclude with some open problems in Sect. 5.

2   Preliminaries

2.1   Definitions

Single sink Buy-at-bulk problem:
Input: A graph $G = (V, E)$, a length function $l : E \to \mathbb{R}_0$, a distinguished node t (the sink), a demand function $dem : V \to \mathbb{R}_0$ which specifies the amount of flow each vertex wants to send to the sink. There are k types of cables, with costs per unit length $\sigma_0 \leq \sigma_1 \leq \ldots \sigma_{k-1}$ and capacities $u_0 \leq u_1 \leq \ldots u_{k-1}$.

Goal: Install zero or more cables on each edge and route flow dem_v from v to t in the resulting capacitated network, minimizing the total cost of the installation.

We now define another problem similar to the buy-at-bulk problem with a slightly different cost function.

Generalized deep-discount problem:
Input: A graph $G = (V, E)$, a length function $l : E \to \mathbb{R}_0$, a distinguished node t (the sink), a demand function $dem : V \to \mathbb{R}_0$ which specifies the amount of flow each vertex wants to send to the sink. There are k kinds of *discount types* with fixed costs $\sigma_0 \leq \sigma_1 \leq \ldots \sigma_{k-1}$ and incremental costs $r_0 \geq r_1 \geq \ldots r_{k-1}$ where the cost per unit length of routing flow f on discount type q is $(\sigma_q + f r_q)$.

Goal: Choose a subset of edges and a discount type for each chosen edge and route all demands to the sink (using chosen edges) such that the total cost is minimized.

2.2   Approximate Equivalence
of Buy-at-Bulk and Generalized Deep-Discount

Given a buy-at-bulk problem, with cables having (cost,capacity) tuples (σ_i, u_i), the corresponding generalized deep-discount problem has discount types with (fixed cost, incremental cost) pairs $(\sigma_i, \frac{\sigma_i}{u_i})$. Similarly a reverse reduction can be defined. Note that the cost function using a particular cable type in buy-at-bulk is a step function (See Fig. 1) and the cost function for the corresponding deep-discount problem is a straight line that is always within a factor 2 of the buy-at-bulk cost function. Hence this reduction preserves costs up to a factor of two. Hence a ρ-approximation to one problem gives a 2ρ-approximation to the other.

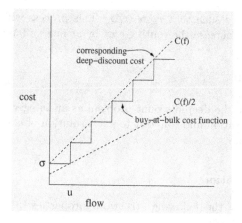

Fig. 1. Approximate equivalence of single sink buy-at-bulk and deep-discount cost functions

We shall show how to get a constant factor approximation to the generalized deep-discount problem. In the rest of the discussion, we shall use the terms cable type and discount type interchangeably. We shall use the terms *thick* and *thin* cable types to mean high and low capacity cable types respectively. We refer to the first term in the cost function (contribution from the fixed cost σ_i's) as the *building cost*, and the second term (contribuition from the incremental cost $f r_i$'s) as the *routing cost* of a solution.

2.3   Pruning of Cable Types

We first prune the set of cables so that each cable has a substantially higher fixed cost than the previous one, and a substantially lower incremental cost. This can be done with a constant factor increase in the cost of a solution. We use the following theorem from [10].

Theorem 1. *We can prune the set of cables such that*

- *The fixed costs scale, i.e.* $\sigma_{q+1} \geq \frac{1}{\epsilon}\sigma_q$
- *The incremental costs scale, i.e.* $r_{q+1} \leq \epsilon r_q$
- *Cost of OPT increases by at most* $\frac{1}{\epsilon}$

Proof. We briefly outline the proof here. First let us prune the set of cables to ensure the scaling of the fixed costs. We start with the thickest cable and for any cable type such that $\sigma_i > \epsilon\sigma_{i+1}$, we delete cable type i. In any solution, we replace cable type i by cable type $i + 1$. We continue in this manner until all cable types satisfy fixed cost scaling. In the process, we replaced some cables by thicker cable types. This increases the building cost by at most a factor of $\frac{1}{\epsilon}$ and only decreases the routing cost. Similarly, we can prune the set of cables to ensure that the incremental costs scale, starting with the thinnest cable, and deleting any cable i such that $r_i > \epsilon r_{i-1}$. This process will only decrease the building cost and increase the routing cost by at most a factor of $\frac{1}{\epsilon}$.   □

3 Algorithm

We first formalize the deep-discount problem as an integer linear program. We solve the LP relaxation and round it to get a solution of cost within a constant of the LP optimum.

3.1 LP Formulation

It is easy to show the following structural properties of an optimum of the generalized deep discount problem:

Theorem 2 ([9]). *There exists an optimum solution to the generalized deep-discount problem with the following properties.*

- **Acyclicity** *The edges installed form a tree.*
- **Monotonicity** *The discount types used by any particular flow are non-decreasing along any path from a source to the sink.*

Using the above theorem, we can now write an ILP as follows[1]

$$\text{minimize} \sum_{e \in E} \sum_{i=0}^{k-1} \sigma_i z_e^i l_e + \sum_{v_j \in V} \sum_{e \in E} \sum_{i=0}^{k-1} dem_j f_{e;i}^j r_i l_e$$

subject to:

$$\sum_{e \in Out(v_j)} \sum_{i=0}^{k-1} f_{e;i}^j \geq 1 \qquad\qquad \forall v_j \in V \qquad\qquad (1)$$

$$\sum_{e \in In(v)} \sum_{i=0}^{k-1} f_{e;i}^j = \sum_{e \in Out(v)} \sum_{i=0}^{k-1} f_{e;i}^j \quad \forall v \in V \setminus \{v_j, t\}, 1 \leq j \leq m \quad (2)$$

$$\sum_{e \in In(v)} \sum_{i=q}^{k-1} f_{e;i}^j \leq \sum_{e \in Out(v)} \sum_{i=q}^{k-1} f_{e;i}^j \quad 1 \leq q \leq k, \forall v \in V \setminus \{v_j, t\} \quad (3)$$

$$f_{e;i}^j \leq z_e^i \qquad\qquad \forall e \in E, 1 \leq i \leq k \qquad (4)$$

$$z_e^i, f_{e;i}^j \qquad \text{non negative integers .} \qquad\qquad (5)$$

---

[1] This is the same LP as the one in [9], with the constraint $\sum_{i=0}^{k-1} z_e^i \geq 1$ removed, since that applies only to the deep discount problem with $\sigma_0 = 0$. It must be noted that their algorithm can be modified so that their results hold for the generalized deep discount problem as well. However, here we further modify their algorithm and give an improved analysis that shows a constant integrality gap.

In an integer solution, the variable $f_{e;i}^j$ is 1 if the flow from source v_j uses cable type i on edge e; z_e^i is 1 if cable type i is installed on edge e. Thus the first term in the objective function is the total building cost of the cables installed and the second term is the total routing cost. The constraints (1) make sure that enough flow leaves the sources. The constraints (2) are the flow conservation constraints. The constraints (3) impose the monotonicity property in theorem 2. The constraints (4) ensure that we only use pipes that are installed. We relax the integrality constraints (5) and solve the relaxation. The fractional optimum guides our algorithm and we compare the cost of our solution against its cost.

3.2   Algorithm

The algorithm is on the same lines as [9]. We build the solution in a top down manner. Our first tree is just the sink. At each step, we augment the current tree, and install progressively thinner cables in each stage. We identify the set of vertices to be connected at each stage based on the LP fractional optimum.

At the top level the algorithm is the following. We start with $T_k = \{t\}$. We derive T_i from T_{i+1} as follows:

Identify a set of nodes to be connected: This is done as follows. Using the LP solution, we estimate the distance that flow from v_j travels on cables of type thinner than i. More formally, define $L_j^q = \sum_e f_{e;q}^j l_e$. Let $R_j^i = \sum_{q=0}^{i-1} L_j^q$. Intuitively, after traveling a distance R_j^i, flow from v_j uses discount type i or higher in the fractional optimum. For a constant γ to be determined later, we construct balls $B(v_j, \gamma R_j^i)$ around the sources, and select a set of non-overlapping balls, starting with the smallest ball. Our set of vertices to be added in this round, S_i is precisely the set of the vertices corresponding to the selected balls. For each source that we do not select, there must be one close by that we select(since we selected the balls in increasing order of radii). More precisely, for each v_l not selected, we define $\text{buddy}_i(v_l)$ to be the vertex in S_i corresponding to the ball that overlapped with $B(v_l, R_l^i)$. In our routing cost analysis, we will use the fact that $d(v_l, \text{buddy}_i(v_l)) \leq 2\gamma R_l^i$.

Build a tree connecting selected nodes: We shrink the current tree T_{i+1} to a single node t_{i+1}. We also shrink the selected balls $B(v_j, \gamma R_j^i)$ and build an approximately optimal steiner tree on these shrunk nodes. We then convert this tree to a *Light approximate shortest path tree*(LAST)[12], which increases the cost of the tree by a factor at most α but guarantees that for every v in the tree, $d_T(v, t_{i+1}) \leq \beta d(v, t_{i+1})$, where d is the shortest distance in the graph. Khuller et al [12] show how to construct (α, β)-LASTs for any $\alpha \geq 1$, $\beta \geq \frac{\alpha+1}{\alpha-1}$. This ensures that we do not pay too much for routing the flow. We install cables of type i on this LAST. This augmentation to T_{i+1} gives us T_i. We note here that in the final iteration, R_j^0 would be zero for all nodes[2] and hence we are guaranteed to connect all sources eventually.

---

[2] Note that this also implies that for the single cable case, our algorithm reduces to the single cable type algorithm in [17].

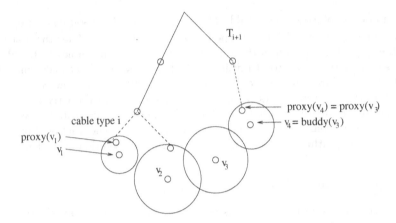

Fig. 2. Definition of proxy

Find nearby *proxies*: For the purpose of analysis, for each demand node v_l, we find a nearby node that is in the LAST. First we look at the v_j's in S_i. For each such node, there must be some node v in the ball $B(v_j, \gamma R^i_j)$ such that v is connected to T_i, since this ball was connected in the steiner tree step. We call this node the proxy $\text{proxy}_i(v_j)$ of v_j. Thus we know that for any node $v_j \in S_i$, $d(v_j, \text{proxy}_i(v_j)) \le \gamma R^i_j$. Now for every other source v_l, we define $\text{proxy}_i(v_l) = \text{proxy}_i(\text{buddy}_i(v_l))$ (see Fig. 2). Thus

$$d(v_l, \text{proxy}_i(v_l)) \le d(v_l, \text{buddy}_i(v_l)) + d(\text{buddy}_i(v_l), \text{proxy}_i(\text{buddy}_i(v_l)))$$
$$\le 2\gamma R^i_l + \gamma R^i_{\text{buddy}_i(v_l)}$$
$$\le 2\gamma R^i_l + \gamma R^i_l$$
$$= 3\gamma R^i_l \ ,$$

where the inequality in step 3 follows from the fact that we chose the balls in increasing order of radii. We shall use this fact in the routing cost analysis.

A more complete description of the algorithm is given in Fig.3.2. Here α, γ and ϵ are constants to be determined later.

4   Analysis

We seperately analyze the building and the routing cost components of our solution.

4.1   Building Cost

We show that the building cost of our solution is $O(OPT^b)$, where OPT^b is the building cost of the optimal fractional solution. The following lemma formalizes

---
Algorithm generalized-deep-discount$(\alpha, \gamma, \epsilon)$
---
1. Prune the set of available cables so that both the fixed and the incremental costs scale by ϵ
2. Write the integer linear program; Let $\mathbf{z}^*, \mathbf{f}^*$ be the optima of the fractional relaxation
3. $T_k \leftarrow \{t\}$
4. **for** $i = k-1, \ldots, 0$ **do**
5. $S_i = \phi$
6.   For every source v_j, define $R_j^i = \sum_{q=0}^{i-1} \sum_e f_{e;q}^{j*} l_e$
7.   Define L to be the set of balls $B(v_j, \gamma R_j^i)$
8.   **if** $B(v_j, \gamma R_j^i) \cap T_{i+1} \neq \phi$ **then**
9. $\text{proxy}_i(v_j) \leftarrow$ arbitrary vertex in $B(v_j, \gamma R_j^i) \cap T_{i+1}$
10.   Remove ball $B(v_j, \gamma R_j^i)$ from L
11.   Arrange the balls in L in increasing order of radii
12.   **for** every ball $B(v_j, \gamma R_j^i)$ in L **do**
13. $S_i \leftarrow S_i \cup \{v_j\}$
14.   **for** every ball $B(v_l, \gamma R_l^i)$ in L that intersects $B(v_j, \gamma R_j^i)$ **do**
15.   delete $B(v_l, \gamma R_l^i)$ from L
16. $\text{buddy}_i(v_l) \leftarrow v_j$
17.   Shrink each ball centered at a node in S_i
18.   Shrink T_{i+1}
19.   Build a 2-approximate steiner tree on these shrunk nodes
20.   Convert it to an (α, β)-LAST(for $\beta = \frac{\alpha+1}{\alpha-1}$)
21.   Unshrink T_{i+1} and install cable type i on the LAST edges to get T_i
22.   **for** every $v_j \in S_i$ **do**
23.   Unshrink ball $B(v_j, \gamma R_j^i)$
24. $\text{proxy}_i(v_j) \leftarrow$ arbitrary node in $B(v_j, \gamma R_j^i) \cap T_i$
25.   **for** every source $v_l \notin S_i$ **do**
26. $\text{proxy}_i(v_l) \leftarrow \text{proxy}_i(\text{buddy}_i(v_l))$

Fig. 3. Algorithm for the deep discount problem

the intuition that flow from j must use a type i or higher cable, after traveling a distance $\gamma \sum_{q=0}^{i-1} L_j^q$, and hence such cables must be built.

Lemma 3. *Let $S \subset V$ be a set of vertices such that $t \notin S$ and $B(v_j, \gamma R_j^i) \subset S$. Then for any feasible solution $\mathbf{f}, \mathbf{z}$,*

$$\sum_{q=i}^{k-1} \sum_{e \in Out(S)} z_e^q \geq 1 - \frac{1}{\gamma} .$$

Proof. Let $f_{q;S}^j = \sum_{e \in Out(S)} f_{q;e}^j$. Then constraint (1) and (2) in the LP together imply that $\sum_{q=0}^{k-1} f_{q;S}^j \geq 1$ for any S such that $t \notin S$. Now the flow $f_{q;S}^j$ while traveling from v_j to the boundary of S uses cable type q or thinner (constraint (3)) and travels a distance $\geq \gamma R_j^i$. Hence it contributes at least $f_{q;S}^j \cdot \gamma R_j^i$ to the quantity $\sum_{p=0}^q L_j^p$. Thus $\sum_{q=0}^{i-1} L_j^q \geq \sum_{q=0}^{i-1} f_{q;S}^j \cdot \gamma R_j^i$. However, $\sum_{q=0}^{i-1} L_j^q = R_j^i$ by definition. It follows that $\sum_{q=0}^{i-1} f_{q;S}^j \le \frac{1}{\gamma}$, which implies that $\sum_{q=i}^{k-1} f_{q;S}^j \ge 1 - \frac{1}{\gamma}$. This together with constraint (4) in the LP completes the proof. $\qquad \square$

Now let OPT_q^b be the building cost the LP optimum pays for cables of type q. We show that the total building cost is within a constant of $OPT^b = \sum_q OPT_q^b$.

The previous lemma tells us that $\frac{\gamma}{\gamma-1} \sum_{q=i}^{k-1} z_e^q$ is a feasible solution to the steiner tree linear program on the shrunk nodes in iteration i. Thus the steiner tree we build has cost at most twice as much. The cost of the ith steiner tree is therefore bounded by

$$2\frac{\gamma}{\gamma-1} \sigma_i \sum_{q=i}^{k-1} \sum_e z_e^q l_e = 2\frac{\gamma}{\gamma-1} \sum_{q=i}^{k-1} \frac{\sigma_i OPT_q^b}{\sigma_q}$$

$$\le 2\frac{\gamma}{\gamma-1} \sum_{q=i}^{k-1} \epsilon^{q-i} OPT_q^b \ .$$

where the last inequality follows from the scaling property. The total cost of the $LAST_i$ will then be at most α times the above quantity. Summing over all iterations, we get the following upper bound on the building cost of the solution

$$\sum_{i=0}^{i-1} \text{buildingcost}(LAST_i) \le \sum_{i=0}^{k-1} 2\frac{\alpha\gamma}{\gamma-1} \sum_{q=i}^{k-1} \epsilon^{q-i} OPT_q^b$$

$$= 2\frac{\alpha\gamma}{\gamma-1} \sum_{q=0}^{i-1} OPT_q^b \sum_{i=0}^{q} \epsilon^{q-i}$$

$$\le 2\frac{\alpha\gamma}{\gamma-1} \sum_{q=0}^{i-1} OPT_q^b \frac{1}{1-\epsilon}$$

$$= 2\frac{\alpha\gamma}{\gamma-1} \frac{1}{1-\epsilon} OPT^b \ .$$

Since α, γ and ϵ are constants, it follows that:

Theorem 4. *The total building cost of the solution is $O(OPT^b)$.*

4.2   Routing Cost

We now show that the total routing cost is within a constant factor of the cost of the LP optimum. For each source v_j, let C_j^q be the amount that the LP optimum spends on routing one unit of flow on cable type q. Let $C_j = \sum_q C_j^q$ be the total per unit routing cost of flow from v_j. We show that our per unit routing cost for flow from v_j is within a constant of C_j.

Lemma 5. *For any source vertex v_j, the cost of routing a unit amount of flow is $O(C_j)$*

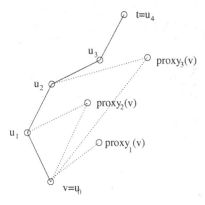

Fig. 4. Routing cost analysis

Proof. Consider the path from v_j to t, as made up of sub-paths (u_i, u_{i+1}) where $u_0 = v_j$, $u_k = t$ and path (u_i, u_{i+1}) uses cable type i (see Fig. 4). Thus u_i is the first node in the path from v_j to t that lies in T_i. Then the total routing cost per unit flow is $\sum_{q=0}^{k-1} r_q d_T(u_i, u_{i+1})$ where d_T denotes the distance function in the tree T_0 that we output. Note that $d_T(u_i, u_{i+1})$ must be bounded by $\beta d(u_i, T_{i+1})$, since we build a LAST at each stage. Moreover, we know that $\text{proxy}_{i+1}(v_j)$ lies in T_{i+1}. Hence we can conclude that $d_T(u_i, u_{i+1}) \leq \beta d(u_i, \text{proxy}_{i+1}(v_j))$. Also the way we selected the balls ensures that $d(v_j, \text{proxy}_{i+1}(v_j)) \leq 3\gamma R_j^i$. Moreover, recall that $R_j^i = \sum_{q=0}^{i-1} L_j^i$. We now bound the distance $d_T(u_i, u_{i+1})$ in terms of L_j^i's.

Claim. For u_i's defined as above,

$$d_T(u_i, u_{i+1}) \leq 3\gamma\beta \sum_{q=0}^{i} (1 + \beta)^{i-q} L_j^q \ .$$

Proof. By induction on i. Base case (i=0) follows from the above discussion. For the induction step:

$$
\begin{aligned}
d_T(u_i, u_{i+1}) &\leq \beta d(u_i, \text{proxy}_{i+1}(v_j)) && \text{(by LAST construction)} \\
&\leq \beta\big[d(u_i, u_{i-1}) + \ldots + d(u_1, u_0) + d(u_0, \text{proxy}_{i+1}(v_j))\big] \\
& && \text{(by triangle inequality)} \\
&\leq \beta \sum_{q=0}^{i-1} d(u_q, u_{q+1}) + \beta d(u_0, \text{proxy}_{i+1}(v_j)) \\
&\leq \beta \sum_{q=0}^{i-1} 3\gamma\beta \sum_{s=0}^{q} (1 + \beta)^{q-s} L_j^s + \beta \cdot 3\gamma \sum_{q=0}^{i} L_j^q \\
& && \text{(by I.H. and def. of proxy)}
\end{aligned}
$$

$$= 3\gamma\beta[\beta\sum_{q=0}^{i-1}\sum_{s=0}^{q}(1+\beta)^{q-s}L_j^s + \sum_{s=0}^{i}L_j^s]$$

$$= 3\gamma\beta[\sum_{s=0}^{i-1}L_j^s(\beta\sum_{q=s}^{i-1}(1+\beta)^{q-s}+1) + L_j^i] \qquad \text{(rearranging)}$$

$$= 3\gamma\beta\sum_{s=0}^{i}(1+\beta)^{i-s}L_j^s \ .$$

Hence the induction holds $\qquad\qquad\qquad\qquad\qquad\qquad\qquad$ □

Thus the per unit cost of routing on cable type i is

$$r_i d_T(u_i, u_{i+1}) \leq r_i \cdot 3\gamma\beta\sum_{s=0}^{i}(1+\beta)^{i-s}L_j^s$$

$$= 3\gamma\beta\sum_{s=0}^{i}(1+\beta)^{i-s}(\frac{r_i}{r_s})C_j^s$$

$$\leq 3\gamma\beta\sum_{s=0}^{i}(\epsilon(1+\beta))^{i-s}C_j^s \ .$$

Now the total routing cost is given by

$$\sum_{i=0}^{k-1}r_i d_T(u_i, u_{i+1}) \leq 3\gamma\beta\sum_{i=0}^{k-1}\sum_{s=0}^{i}(\epsilon(1+\beta))^{i-s}C_j^s$$

$$= 3\gamma\beta\sum_{s=0}^{k-1}[C_j^s\sum_{i=s}^{k-1}(\epsilon(1+\beta))^{i-s}]$$

$$= 3\gamma\beta\sum_{s=0}^{k-1}[C_j^s\sum_{t=0}^{k-1-s}(\epsilon(1+\beta))^t]$$

$$\leq 3\gamma\beta\frac{1}{1-\epsilon(1+\beta)}\sum_{s=0}^{k-1}C_j^s$$

$$= \frac{3\gamma\beta}{1-\epsilon(1+\beta)}C_j \ .$$

Hence the lemma holds. $\qquad\qquad\qquad\qquad\qquad\qquad\qquad$ □

Thus it follows that

Theorem 6. *The total routing cost of our solution is* $O(OPT^r)$.

Therefore the algorithm outputs a constant factor approximation to the optimum, and furthermore the integrality gap of the LP is constant.

5   Conclusion

Plugging values $(\alpha, \gamma, \epsilon)$ to be $(5, 2, \frac{1}{4})$, we get an approximation factor of 216 for the original buy-at-bulk problem. This improves upon the algorithm of [10] where the constant is roughly 2000. An open problem is to improve the approximation factor to a more reasonable constant.

Moreover, without the single sink restriction(i.e. for the general buy-at-bulk problem), the best approximation factor known is $O(\log n \log \log n)$ ($O(\log n)$ for the single cable type case). It would be desirable to get improved approximation algorithms/hardness results for this problem.

Acknowledgements

The author would like to thank Satish Rao for several helpful discussions and suggestions, and the anonymous referees for their helpful comments. Thanks are also due to Jittat Fakcharoenphol for his careful reading of the paper.

References

1. M. Andrews, L. Zhang. "The access network design problem", *Proceedings of the 39th Annual IEEE Symposium on Foundations of Computer Science*, 42-49, 1998.
2. B. Awerbuch, Y. Azar. "Buy-at-bulk network design", *Proceedings of the 38th Annual IEEE Symposium on Foundations of Computer Science*, 542-547, 1997.
3. Y. Bartal. "Probabilistic approximation of metric spaces and its algorithmic applications", *Proceedings of the 37th Annual IEEE Symposium on Foundations of Computer Science*, 184-193, 1996.
4. Y. Bartal. "On approximating arbitrary metrics by tree metrics", *Proceedings of the 30th Annual ACM Symposium on Theory of Computing*, 161-168, 1998.
5. A. Balakrishnan, T. Magnanti, P. Mirchandani. "Network Design", in *Anottated bibliographies in combinatorial optimization, M. Dell'Amico, F. Maffioli and S. Martello(eds.), John Wiley and Sons, New York*, 311-334, 1997.
6. M. Charikar, C. Chekuri, A. Goel, S. Guha. "Approximating a finite metric by a small number of tree metrics, *Proceedings of the 39th Symposium on Foundations of Computer Science*, 379-388, 1998.
7. B. Chandra, G. Das, G. Narasimhan, J. Soares. "New sparseness results for graph spanners", *Proc. of 8th Symposium on Computational Geometry*, 192-201, 1992.
8. R. Epstein. "Linear programming and capacitated network loading", Ph.D. Thesis, MIT, 1998.
9. N. Garg, R. Khandekar, G. Konjevod, R. Ravi, F.S. Salman, A. Sinha. "On the integrality gap of a natural formulation of the single-sink buy-at-bulk network design problem", *Proceedings of the Eighth International Conference on Integer Programming and Combinatorial Optimization, Lecture Notes in Computer Science 2081*, Springer, 170-184, 2001.
10. S. Guha, A. Meyerson, K. Munagala. "A constant factor approximation for the single sink edge installation problem", *Proceedings of the 33rd Annual ACM Symposium on Theory of Computing*, 383-388, 2001.

11. R. Hassin, R. Ravi, F.S. Salman. "Approximation algorithms for a capacitated network design problem", *Approximation algorithms for combinatorial optimization, Third International workshop, APPROX 2000, Proceedings, Lecture Notes in Computer Science 1913*, Springer, 167-176, 2000.
12. S. Khuller, B. Raghavachari, N. Young. "Balancing minimum spanning and shortest path trees", *Algorithmica*, 14(4):305-321, 1994.
13. P. Mirchandani. Ph.D. thesis, MIT, 1989.
14. Y. Mansour, D. Peleg. "An approximation algorithm for minimum-cost network design", The Weizman Institute of Science Tech. Report CS94-22.
15. A. Meyerson, K. Munagala, S. Plotkin. "Cost-distance: Two metric network design", *Proceedings of the 41st Annual IEEE Symposium on Foundations of Computer Science*, 624-630, 2000.
16. T. Magnanti, P. Mirchandani, R. Vachani, "Modeling and solving the two-facility capacitated network loading problem", *Operations Research* 43, 142-157, 1995.
17. F.S. Salman, J. Cheriyan, R. Ravi, S. Subramanian."Buy-at-bulk network design: Approximating the single sink edge installation problem", *Proceedings of the 8th Annual ACM-SIAM Symposium on Discrete Algorithms*, 619-628, 1997.

Author Index

Lecture Notes in Computer Science

For information about Vols. 1–2257
please contact your bookseller or Springer-Verlag

Vol. 2293: J. Renz, Qualitative Spatial Reasoning with Topological Information. XVI, 207 pages. 2002. (Subseries LNAI).

Vol. 2294: A. Cortesi (Ed.), Verification, Model Checking, and Abstract Interpretation. Proceedings, 2002. VIII, 331 pages. 2002.

Vol. 2295: W. Kuich, G. Rozenberg, A. Salomaa (Eds.), Developments in Language Theory. Proceedings, 2001. IX, 389 pages. 2002.

Vol. 2296: B. Dunin-Kęplicz, E. Nawarecki (Eds.), From Theory to Practice in Multi-Agent Systems. Proceedings, 2001. IX, 341 pages. 2002. (Subseries LNAI).

Vol. 2297: R. Backhouse, R. Crole, J. Gibbons (Eds.), Algebraic and Coalgebraic Methods in the Mathematics of Program Construction. Proceedings, 2000. XIV, 387 pages. 2002.

Vol. 2298: I. Wachsmuth, T. Sowa (Eds.), Gesture and Language in Human-Computer Interaction. Proceedings, 2001. XI, 323 pages. 2002. (Subseries LNAI).

Vol. 2299: H. Schmeck, T. Ungerer, L. Wolf (Eds.), Trends in Network and Pervasive Computing – ARCS 2002. Proceedings, 2002. XIV, 287 pages. 2002.

Vol. 2300: W. Brauer, H. Ehrig, J. Karhumäki, A. Salomaa (Eds.), Formal and Natural Computing. XXXVI, 431 pages. 2002.

Vol. 2301: A. Braquelaire, J.-O. Lachaud, A. Vialard (Eds.), Discrete Geometry for Computer Imagery. Proceedings, 2002. XI, 439 pages. 2002.

Vol. 2302: C. Schulte, Programming Constraint Services. XII, 176 pages. 2002. (Subseries LNAI).

Vol. 2303: M. Nielsen, U. Engberg (Eds.), Foundations of Software Science and Computation Structures. Proceedings, 2002. XIII, 435 pages. 2002.

Vol. 2304: R.N. Horspool (Ed.), Compiler Construction. Proceedings, 2002. XI, 343 pages. 2002.

Vol. 2305: D. Le Métayer (Ed.), Programming Languages and Systems. Proceedings, 2002. XII, 331 pages. 2002.

Vol. 2306: R.-D. Kutsche, H. Weber (Eds.), Fundamental Approaches to Software Engineering. Proceedings, 2002. XIII, 341 pages. 2002.

Vol. 2307: C. Zhang, S. Zhang, Association Rule Mining. XII, 238 pages. 2002. (Subseries LNAI).

Vol. 2308: I.P. Vlahavas, C.D. Spyropoulos (Eds.), Methods and Applications of Artificial Intelligence. Proceedings, 2002. XIV, 514 pages. 2002. (Subseries LNAI).

Vol. 2309: A. Armando (Ed.), Frontiers of Combining Systems. Proceedings, 2002. VIII, 255 pages. 2002. (Subseries LNAI).

Vol. 2310: P. Collet, C. Fonlupt, J.-K. Hao, E. Lutton, M. Schoenauer (Eds.), Artificial Evolution. Proceedings, 2001. XI, 375 pages. 2002.

Vol. 2311: D. Bustard, W. Liu, R. Sterritt (Eds.), Soft-Ware 2002: Computing in an Imperfect World. Proceedings, 2002. XI, 359 pages. 2002.

Vol. 2312: T. Arts, M. Mohnen (Eds.), Implementation of Functional Languages. Proceedings, 2001. VII, 187 pages. 2002.

Vol. 2313: C.A. Coello Coello, A. de Albornoz, L.E. Sucar, O.Cairó Battistutti (Eds.), MICAI 2002: Advances in Artificial Intelligence. Proceedings, 2002. XIII, 548 pages. 2002. (Subseries LNAI).

Vol. 2314: S.-K. Chang, Z. Chen, S.-Y. Lee (Eds.), Recent Advances in Visual Information Systems. Proceedings, 2002. XI, 323 pages. 2002.

Vol. 2315: F. Arhab, C. Talcott (Eds.), Coordination Models and Languages. Proceedings, 2002. XI, 406 pages. 2002.

Vol. 2316: J. Domingo-Ferrer (Ed.), Inference Control in Statistical Databases. VIII, 231 pages. 2002.

Vol. 2317: M. Hegarty, B. Meyer, N. Hari Narayanan (Eds.), Diagrammatic Representation and Inference. Proceedings, 2002. XIV, 362 pages. 2002. (Subseries LNAI).

Vol. 2318: D. Bošnački, S. Leue (Eds.), Model Checking Software. Proceedings, 2002. X, 259 pages. 2002.

Vol. 2319: C. Gacek (Ed.), Software Reuse: Methods, Techniques, and Tools. Proceedings, 2002. XI, 353 pages. 2002.

Vol. 2320: T. Sander (Ed.), Security and Privacy in Digital Rights Management. Proceedings, 2001. X, 245 pages. 2002.

Vol. 2322: V. Mařík, O. Stěpánková, H. Krautwurmová, M. Luck (Eds.), Multi-Agent Systems and Applications II. Proceedings, 2001. XII, 377 pages. 2002. (Subseries LNAI).

Vol. 2323: À. Frohner (Ed.), Object-Oriented Technology. Proceedings, 2001. IX, 225 pages. 2002.

Vol. 2324: T. Field, P.G. Harrison, J. Bradley, U. Harder (Eds.), Computer Performance Evaluation. Proceedings, 2002. XI, 349 pages. 2002.

Vol 2326: D. Grigoras, A. Nicolau, B. Toursel, B. Folliot (Eds.), Advanced Environments, Tools, and Applications for Cluster Computing. Proceedings, 2001. XIII, 321 pages. 2002.

Vol. 2327: H.P. Zima, K. Joe, M. Sato, Y. Seo, M. Shimasaki (Eds.), High Performance Computing. Proceedings, 2002. XV, 564 pages. 2002.

Vol. 2329: P.M.A. Sloot, C.J.K. Tan, J.J. Dongarra, A.G. Hoekstra (Eds.), Computational Science – ICCS 2002. Proceedings, Part I. XLI, 1095 pages. 2002.

Vol. 2330: P.M.A. Sloot, C.J.K. Tan, J.J. Dongarra, A.G. Hoekstra (Eds.), Computational Science – ICCS 2002. Proceedings, Part II. XLI, 1115 pages. 2002.

Vol. 2331: P.M.A. Sloot, C.J.K. Tan, J.J. Dongarra, A.G. Hoekstra (Eds.), Computational Science – ICCS 2002. Proceedings, Part III. XLI, 1227 pages. 2002.

Vol. 2332: L. Knudsen (Ed.), Advances in Cryptology – EUROCRYPT 2002. Proceedings, 2002. XII, 547 pages. 2002.

Vol. 2334: G. Carle, M. Zitterbart (Eds.), Protocols for High Speed Networks. Proceedings, 2002. X, 267 pages. 2002.

Vol. 2335: M. Butler, L. Petre, K. Sere (Eds.), Integrated Formal Methods. Proceedings, 2002. X, 401 pages. 2002.

Vol. 2336: M.-S. Chen, P.S. Yu, B. Liu (Eds.), Advances in Knowledge Discovery and Data Mining. Proceedings, 2002. XIII, 568 pages. 2002. (Subseries LNAI).

Vol. 2337: W.J. Cook, A.S. Schulz (Eds.), Integer Programming and Combinatorial Optimization. Proceedings, 2002. XI, 487 pages. 2002.